Ceramics in the Circular Economy for a Sustainable World

Ceramics in the Circular Economy for a Sustainable World

Guest Editors

Francesco Baino
Pardeep Gianchandani
Enrico Fabrizio
Bartolomeo Megna
Manuela Ceraulo

MDPI

Basel • Beijing • Wuhan • Barcelona • Belgrade • Novi Sad • Cluj • Manchester

Guest Editors
Francesco Baino
Department of Applied Science and Technology
Politecnico di Torino
Turin
Italy

Pardeep Gianchandani
Department of Applied Science and Technology
Politecnico di Torino
Turin
Italy

Enrico Fabrizio
Dipartimento Energia (DENERG)
Politecnico di Torino
Torino
Italy

Bartolomeo Megna
Dipartimento di Ingegneria
Università di Palermo
Palermo
Italy

Manuela Ceraulo
Department of Engineering
University of Palermo
Palermo
Italy

Editorial Office
MDPI AG
Grosspeteranlage 5
4052 Basel, Switzerland

This is a reprint of the Special Issue, published open access by the journal *Ceramics* (ISSN 2571-6131), freely accessible at: https://www.mdpi.com/journal/ceramics/special_issues/P2VK1V185K.

For citation purposes, cite each article independently as indicated on the article page online and as indicated below:

Lastname, A.A.; Lastname, B.B. Article Title. *Journal Name* **Year**, *Volume Number*, Page Range.

ISBN 978-3-7258-7188-9 (Hbk)
ISBN 978-3-7258-7189-6 (PDF)
https://doi.org/10.3390/books978-3-7258-7189-6

Contents

About the Editors

Francesco Baino

Francesco Baino is an Associate Professor of Materials Science and Technology in the Department of Applied Science and Technology (DISAT), Politecnico di Torino (Italy), where he teaches materials science, materials degradation, and porous materials courses for BS, MS, and PD students. He has been a member of the American Ceramic Society since 2018. He is currently an Associate Editor for the journals *International Journal of Applied Ceramic Technologies* (Wiley) and *Frontiers in Bioengineering* (Frontiers Media Group). His current research interests include biomaterials with a special focus on bioceramics and bioactive glasses, ceramics for sustainable applications in the circular economy, as well as processing and testing of advanced ceramics. He has published more than 260 peer-reviewed scientific journal articles, 40 book chapters, 100 abstracts or contributions in congress proceedings, and is an author of 3 patents. His H-index is 64, with the total number of citations above 13,000 (Scopus database). According to a scientometric study performed by Stanford University, since 2021, he has been listed among the 2% top-cited scientists in the world.

Pardeep Gianchandani

Pardeep Gianchandani is an Assistant Professor in the Department of Textile Engineering at Mehran University of Engineering and Technology (MUET), Pakistan, and has also held an academic appointment in the same role at Politecnico di Torino, Italy. He earned his Ph.D. in Materials Science from Politecnico di Torino, where his research focused on advanced joining technologies for ceramics and ceramic matrix composites (CMCs) for aerospace and high-temperature energy applications. His expertise lies in advanced ceramic materials, additive manufacturing, sustainable materials engineering, and functional surface modification. His work integrates high-performance materials development with resource-efficient and circular economy principles. In addition to his research activities, he teaches materials science, yarn engineering, textile manufacturing, and process control at undergraduate and postgraduate levels. He is an Associate Fellow of AdvanceHE (UK), reflecting his commitment to excellence in higher education teaching and learning.

Enrico Fabrizio

Enrico Fabrizio is a Full Professor of Environmental Technical Physics in the Department of Energy at Politecnico di Torino, Italy. His work focuses on performance simulation and optimization of zero-energy/zero-emission buildings, energy systems for buildings, and environmental control of buildings. His research is conducted through numerical developments and experimental monitoring to promote energy-sustainable building design in cities and the surrounding area. Since 2023, he has been Coordinator of the Technical Commission CT 272 "Automation and control systems for energy management and comfort in buildings" of the Italian Thermotechnical Committee. He is Deputy Editor-in-Chief of the journal *Sustainable Cities and Society*.

Bartolomeo Megna

Bartolomeo Megna is an Associate Professor in Materials Science and Technology at the Department of Engineering of the University of Palermo, devoted to research on the formulation and characterization of innovative lime mortars, geopolymers, and materials for the conservation of cultural heritage. His research on Cultural Heritage is also focused on characterization of constitutive materials by means of several analytical techniques, Raman and FTIR spectroscopy, XRD, Optical and Scanning Electron Microscopy, and Thermogravimetric Analysis, to better understand the relationship between microscopic and macroscopic characteristics of materials. These skills allowed him to effectively collaborate with researchers in other fields, like catalysis, photocatalysis, and surface treatments of metals. Coherently with research activity, he teaches courses on materials science and technology for students in the Conservation and Restoration of Cultural Heritage and Buildings Engineering.

Manuela Ceraulo

Manuela Ceraulo is an Assistant Professor in Materials Science and Technology at the Department of Engineering of the University of Palermo. Her research activity focuses on polymeric, biodegradable, composite and biocomposite materials, geopolymers, and nanocomposites, with the goal of developing and characterizing new materials for diverse applications, paying particular attention to environmental sustainability. Materials are processed mainly through extrusion and mixing, then characterized from rheological, mechanical, and thermal perspectives, including permeability analysis, fiber spinning, and film processing to assess application potential, and studies on mechanical strength and conductivity for geopolymers specifically designed for thermal energy storage.

ceramics

MDPI

Editorial

Ceramics in the Circular Economy for a Sustainable World

Pardeep Kumar Gianchandani [1,2], Enrico Fabrizio [3], Bartolomeo Megna [4], Manuela Ceraulo [4] and Francesco Baino [1,*]

1 Institute of Materials Physics and Engineering, Department of Applied Science and Technology (DISAT), Politecnico di Torino, Corso Duca degli Abruzzi 24, 10129 Turin, Italy; pkg_05@hotmail.com
2 Department of Textile Engineering, Mehran University of Engineering & Technology, Jamshoro 76062, Sindh, Pakistan
3 Department of Energy (DENERG), Politecnico di Torino, Corso Duca degli Abruzzi 24, 10129 Torino, Italy; enrico.fabrizio@polito.it
4 Department of Engineering, University of Palermo, Viale delle Scienze, 90128 Palermo, Italy; bartolomeo.megna@unipa.it (B.M.); manuela.ceraulo@unipa.it (M.C.)
* Correspondence: francesco.baino@polito.it

1. Introduction to the Field of This Special Issue

The transition toward a circular economy is one of the most pressing challenges and opportunities of our time, requiring fundamental shifts in how we produce, consume, and manage materials [1]. Within this context, the ceramics industry, a sector traditionally associated with high energy consumption and significant raw material extraction, is undergoing a remarkable transformation [2,3]. This Special Issue, "Ceramics in the Circular Economy for a Sustainable World", showcases cutting-edge research that demonstrates how ceramic materials can become key enablers of sustainability through waste valorisation, resource efficiency, and innovative processing technologies.

The global ceramics industry faces mounting pressure to reduce its environmental footprint while meeting growing demand for construction materials, refractories, and functional ceramics [4,5]. Traditional ceramic production relies heavily on virgin raw materials such as clays, feldspars, and quartz, alongside energy-intensive firing processes that contribute substantially to CO_2 emissions [6,7]. Simultaneously, modern societies generate vast quantities of industrial, agricultural, mining, and construction waste that pose environmental and economic burdens. The convergence of these challenges has catalysed a new research paradigm: transforming diverse waste streams into valuable ceramic feedstocks and products [8,9].

Recent advances in this field reveal several converging research directions. First, there is a systematic effort to replace virgin clays and minerals with industrial and biological residues in traditional ceramic formulations. Second, researchers are developing alternative consolidation routes, e.g., alkali-activated materials and geopolymers, that reduce or eliminate high-temperature firing requirements, thus saving energy and fuels. Third, the field is moving toward engineered porous and lightweight ceramics that serve multiple functions in insulation, filtration, and environmental remediation [10,11]. Fourth, life cycle assessment (LCA) methodologies are increasingly being applied to quantify the environmental benefits of waste-derived ceramics, although standardisation still remains a challenge [12].

The scope of waste materials being valorised in ceramic applications is remarkably diverse. Industrial wastes include foundry sand, ceramic shell waste, glass cullet, metallurgical slags, and bottom/fly ash from waste incineration. Mining residues encompass bauxite tailings, granite powder, ornamental stone fines, and phosphate mine waste rocks [13].

Agricultural and biogenic waste ranges from rice husk ash and coffee grounds to sunflower shell ash and banana fibres. Construction and demolition waste [14,15] provides another significant feedstock stream, including crushed concrete, brick fragments, and gypsum waste [16,17].

This Special Issue brings together 23 original research articles that exemplify the breadth and depth of current innovations in circular ceramic production. The contributions span multiple continents and address diverse waste streams, ceramic applications, and sustainability dimensions. Collectively, these papers advance both fundamental understanding and practical implementation pathways for ceramics in the circular economy. They demonstrate that waste valorisation in ceramics is not merely about disposal (landfilling) reduction but it represents a strategic opportunity to create high-performance materials with reduced environmental impact, lower production costs, and enhanced functionality.

2. Overview of the Contributions Collected in This Special Issue

This Special Issue collects 4 comprehensive reviews and 19 research articles. We would like to present here a small selection of the highly significant papers included in this Special Issue, which well summarise its key topics and scientific areas.

Six papers focus on the utilisation of glass cullet/waste glass as well as the vitrification of bottom ash/fly ash to obtain functional products, such as porous glasses for thermal insultation in buildings or glass-ceramics. Glass in an exceptionally versatile material but its recycling or reuse pose several challenges, predominantly associated with the heterogeneity in the chemical composition of various glass types and the glass particle size distribution, both things that dramatically affect the efficiency and feasibility of recycling procedures. Tameni and Bernardo [18] comprehensively discuss the emerging trends in the recycling and upcycling of waste glass through the utilisation of glass in the production of geopolymers and alkali-activated materials, thereby efficiently converting waste into high-value products.

Six other papers focus on the use of waste ceramics or recycled materials in the production of cement, concrete, mortar, and masonry, which are all key elements in construction. Santos Nascimento et al. [14] investigates the use of gypsum waste from civil construction as a partial substitute for cement in soil–cement formulations in an effort to produce eco-friendly bricks, which aligns with the paradigms of the circular economy. Strategies like this one, which aim at reducing cement usage, are key from the viewpoint of sustainability as cement production greatly contributes to greenhouse emissions (e.g., primarily as a consequence of carbon dioxide derived from the thermal decomposition of limestone).

The challenges associated with the use of ceramics in the circular economy are important not only on the Earth but also... in space! In a highly interesting paper, Ellery [19] discusses the strategies that can be adopted on for full industrialisation of the Moon, based on a circular lunar industrial ecology, and examines the contribution of ceramics, with a special focus on additive manufacturing technologies applied using these inorganic materials. Popular and relatively affordable 3D printing methods, such as direct ink writing, are less suitable on the Moon compared to the Earth as they typically require polymers, which are scarce on the Moon, for the production of printable inks. This suggests that full industrialisation of the Moon cannot be completed until the problem of ceramic manufacturing is resolved, most likely in conjunction with polymer synthesis from potential carbon sources.

Sustainable approaches applied to ceramics are also highly appealing with regard to improving the quality of life and wellbeing of humans. Torre et al. [20] propose the use of polyphenols from grape pomace to impart antibacterial and anti-inflammatory properties to calcium phosphate bioceramics to be implanted in bone defects, potentially accelerating tissue healing and regeneration.

3. Invitation to the Readership

The continuous industrial growth and demand for ever more products that come as consequences of growing populations worldwide have led to the over-exploitation of natural resources and are associated with the generation of huge amounts of waste, pollution, and carbon dioxide introduced into the atmosphere. Traditional economic models are linear—i.e., resources are consumed, goods are fabricated, and waste is produced and eventually disposed of—resulting in an obvious impact on the environment and on human health. In an attempt to avoid these criticalities, a new paradigm has emerged based on the circular economy, which implies a change in mindset and the implementation of fundamental changes throughout the value chain, including new approaches to product design and manufacturing technologies, new business models, novel strategies to preserve natural resources and turn waste into new (secondary) resources, as well as new behaviours, common practices, and education. Ceramics can indeed play a vital role in such a change in paradigm, and this Special Issue demonstrates that once more. As Editors, we hope that the valuable contributions gathered together in this Special Issue can stimulate the discussion among various stakeholders involved in the ceramics community and in sustainable development, including researchers from academia, industry, and government. This exchange of ideas, methods, and results will be key to coping with the needs of a world that has finite resources but (almost) infinite desires.

Acknowledgments: This Special Issue is guest-edited by scientists working within the framework of a project funded under the National Recovery and Resilience Plan (NRRP), Mission 4 Component 2 Investment 1.3—Call for tender No. 1561 of 10 November 2 022 of Ministero dell'Università e della Ricerca (MUR), Italy; the project was funded by the European Union—NextGenerationEU. Project details: project code PE0000021, Concession Decree No. 1561 of 10 November 2022 adopted by Ministero dell'Università e della Ricerca (MUR), CUP E13C22001890001, Project title "Network 4 Energy Sustainable Transition—NEST".

Conflicts of Interest: The authors declare no conflicts of interest.

References

1. Davies, M.L. Introduction to the circular economy themed collection. *RSC Sustain.* **2025**, *3*, 1036–1038. [CrossRef]
2. da Silva, S.B.G.; Barros, M.V.; Radicchi, J.Â.Z.; Puglieri, F.N.; Piekarski, C.M. Opportunities and challenges to increase circularity in the product's use phase. *Sustain. Futures* **2024**, *8*, 100297. [CrossRef]
3. Vázquez Calvo, V.L.; Korzeb, Z.; Alonso-Fariñas, B.; Morillo Aguado, J. Circular economy transition under the revised Industrial Emissions Directive: A review of challenges and enablers. *J. Clean. Prod.* **2025**, *533*, 146779. [CrossRef]
4. Furszyfer Del Rio, D.D.; Sovacool, B.K.; Foley, A.M.; Griffiths, S.; Bazilian, M.; Kim, J.; Rooney, D. Decarbonizing the ceramics industry: A systematic and critical review of policy options, developments and sociotechnical systems. *Renew. Sustain. Energy Rev.* **2022**, *157*, 112081. [CrossRef]
5. Nasare, R.; Shah, V.; Joshi, T.; Dave, U.; Ghaffar, S.H. Exploring sustainable utilization of ceramic waste in heat-resistant concrete: A comprehensive review. *Int. J. Appl. Ceram. Technol.* **2025**, *22*, e15021. [CrossRef]
6. Almeida, M.; Simões, F.; Dias, F.; Amado, A. Ceramic industry contribution to a circular economy. In Proceedings of the Congress of Innovation on Sustainable Construction CINCOS, Lisbon, Portugal, 3–4 November 2016.
7. Muthaiyan, U.M.; Lankapalli, C.; Thangavel, B.; Arumugam, K. Integrating ceramic wastes into concrete: Sustainable disposal and resource optimization strategies. *Environ. Prot. Eng.* **2025**, *51*, 53–72. [CrossRef]
8. Peng, X.; Jiang, Y.; Chen, Z.; Osman, A.I.; Farghali, M.; Rooney, D.W.; Yap, P.-S. Recycling municipal, agricultural and industrial waste into energy, fertilizers, food and construction materials, and economic feasibility: A review. *Environ. Chem. Lett.* **2023**, *21*, 765–801. [CrossRef]
9. Srivastava, A.; Pandey, S.; Shahwal, R.; Sur, A. Recycling of waste into useful materials and their energy applications. In *Microbial Niche Nexus Sustaining Environmental Biological Wastewater and Water-Energy-Environment Nexus*; Springer: Berlin/Heidelberg, Germany, 2025; pp. 251–296.

10. Aguiar, M.C.d.; Gadioli, M.C.B.; Sant'Ana, M.A.K.; Almeida, K.M.d.; Vidal, F.W.H.; Vieira, C.M.F. Red Ceramics Produced with Primary Processing Fine Waste of Ornamental Stones According to the Circular Economy Model. *Sustainability* **2022**, *14*, 12887. [CrossRef]
11. Terrones-Saeta, J.M.; Domínguez, V.; Ramos, D.; Romero, E.; Asensio-Lozano, J. Analysis of the Mechanical Properties of Structural Ceramics Made from Aggregate Washing Sludge and Manganese Mining Waste. *Materials* **2024**, *17*, 4427. [CrossRef] [PubMed]
12. Ünsal, Z.; Ekinci, M.Ö.; İlcan, H.; Şahin, O.; Selçuk, S.; Şahmaran, M. Life cycle assessment of geopolymer materials utilizing construction and demolition waste. *Environ. Res.* **2025**, *285*, 122359. [CrossRef] [PubMed]
13. Almendro-Candel, M.B.; Jordán Vidal, M.M. Glasses, Frits and Glass-Ceramics: Processes and Uses in the Context of Circular Economy and Waste Vitrification. *Coatings* **2024**, *14*, 346. [CrossRef]
14. Nascimento, E.S.S.; Oliveira, H.A.d.; Santos, C.P.d.; Gomes, M.d.A.; Valerio, M.E.G.; Macedo, Z.S. Sustainable Use of Gypsum Waste for Applications in Soil–Cement Bricks: Mechanical, Environmental, and Durability Performance. *Ceramics* **2025**, *8*, 83. [CrossRef]
15. Murali, G.; Hassas, N.; Abdelgader, H.S. Ceramic waste as a sustainable cementitious resource: Pathways to cleaner and high-performance concrete. *Clean. Mater.* **2025**, *18*, 100352. [CrossRef]
16. Marín Cortés, S. Towards a More Sustainable and Circular Ceramic Tile Industry Through the Valorization of Construction and Demolition Waste. Ph.D. Thesis, Autonomous University of Madrid, Madrid, Spain, 2025.
17. Allaoui, D.; Majdoubi, H.; Haddaji, Y.; Nadi, M.; Mansouri, S.; Oumam, M.; Tamraoui, Y.; Alami, J.; Hannache, H.; Manoun, B. Valorization of ceramic sanitary waste into Resilient phosphoric acid-based geopolymers for sustainable construction: Thermal, mechanical, and microstructural properties. *Ceram. Int.* **2024**, *50*, 8875–8889. [CrossRef]
18. Tameni, G.; Bernardo, E. Alkali Activation of Glass for Sustainable Upcycling: An Overview. *Ceramics* **2025**, *8*, 108. [CrossRef]
19. Ellery, A. Ceramics—The Forgotten but Essential Ingredients for a Circular Economy on the Moon. *Ceramics* **2025**, *8*, 107. [CrossRef]
20. Torre, E.; Morra, M.; Cassinelli, C.; Iviglia, G. Anti-Bacterial and Anti-Inflammatory Effects of a Ceramic Bone Filler Containing Polyphenols from Grape Pomace. *Ceramics* **2024**, *7*, 975–988. [CrossRef]

ceramics

MDPI

Article

Analysis of the Environmental Compatibility of the Use of Porcelain Stoneware Tiles Manufactured with Waste Incineration Bottom Ash

Luigi Acampora [1], Giulia Costa [1,*], Iason Verginelli [1], Francesco Lombardi [1], Claudia Mensi [2] and Simone Malvezzi [3]

[1] Laboratory of Environmental Engineering, Department of Civil Engineering and Computer Science Engineering, University of Rome "Tor Vergata", Via del Politecnico 1, 00133 Rome, Italy; acampora@ing.uniroma2.it (L.A.); verginelli@ing.uniroma2.it (I.V.); lombardi@ing.uniroma2.it (F.L.)
[2] A2A Ambiente S.p.A., Cascina Maggiore, 20084 Lacchiarella, Italy; claudia.mensi@a2a.it
[3] A2A Ambiente S.p.A., Via Lamarmora 230, 25124 Brescia, Italy; simone.malvezzi@a2a.it
* Correspondence: costa@ing.uniroma2.it; Tel.: +39-067-259-7007

Abstract

In line with circular economy principles and the reduction of primary material exploitation, waste-to-energy (WtE) by-products such as bottom ash (BA) are increasingly being used as raw materials in cement and ceramics manufacturing. However, it is critical to verify that the final product presents not only adequate technical properties but also that it does not pose negative impacts to the environment and human health during its use. This study investigates the environmental compatibility of the use of ceramic porcelain stoneware tiles manufactured with BA as partial replacement of traditional raw materials, with a particular focus on the leaching behavior of the tiles during their use, and also after crushing to simulate their characteristics at their end of life. To evaluate the latter aspect, compliance leaching tests were performed on crushed samples and compared with Italian End-of-Waste (EoW) thresholds for the use of construction and demolition waste as recycled aggregates. Whereas, to assess the environmental compatibility of the tiles during the utilization phase, a methodology based on the application of monolithic leaching tests to intact tiles, and the evaluation of the results through multi-scenario human health risk assessment and the analysis of the main mechanisms governing leaching at different stages, was employed. The results of the study indicate that the analyzed BA-based tiles showed no significant increase in the release of potential contaminants compared to traditional formulations and fully complied with End-of-Waste criteria. The results of the monolith tests used as input for site-specific risk assessment, simulating worst-case scenarios involving the potential contamination of the groundwater, indicated negligible risks to human health for both types of tiles, even considering very conservative assumptions. As for differences in the release mechanisms, tiles containing BA exhibited a shift toward depletion-controlled leaching and some differences in early element release compared to the ones with a traditional formulation.

Keywords: bottom ash; ceramics; leaching; circular economy; risk assessment

1. Introduction

The increasing global production of municipal solid waste has led to a growing interest in waste-to-energy (WtE) technologies such as incineration, which significantly reduces the waste volume while recovering energy. However, this process generates various by-products, the most significant in terms of weight being bottom ash (BA) [1].

Bottom ash accounts for roughly 20% of the total weight of waste incinerated in WtE plants [2] and makes up around 80% of the solid residues produced by such facilities [3]. Until recently, in Italy and other European countries, this by-product was primarily disposed of in landfills; however, over the past two decades, significant efforts have been made to find more sustainable and valuable uses for this residue [4]. The mineral fraction of BA is widely used as a replacement for sand or aggregates in concrete and asphalt mixtures [5,6], as well as in unbound applications such as road sub-base layers [4,7,8]. Beyond these uses, BA has also been explored as a raw material in cement production [9] and, more recently, in ceramic manufacturing for products like tiles, bricks, refractories, glass, and sanitary ware [10].

Ceramic materials present an excellent opportunity for the integration of alternative and more sustainable raw materials like bottom ash from incineration of municipal solid waste. This is mainly due to the high-temperature manufacturing process, which allows the stabilization and inertization of potential inorganic contaminants present in the BA. Additionally, ceramics manufacturing can incorporate industrial by-products without requiring significant modifications to processing methods. As a result, the use of BA in ceramics formulations represents a promising opportunity for sustainable material innovation and the practical implementation of circular economy strategies. However, it is fundamental to assess the effect of BA addition on product performance, durability, and environmental compatibility in terms of potential negative impacts on the environment and human health.

The preprocessing stage is essential for preparing bottom ash before its incorporation into ceramic formulations. This typically involves processes such as sieving to select particles of the optimal size, improving homogeneity, and removing ferrous and non-ferrous metals. Additionally, particle size reduction through milling or grinding is generally performed to enhance the reactivity and compatibility of the BA with other ceramic raw materials. These steps are largely consistent with traditional ceramic processing techniques [11], which commonly include washing, sieving, ball milling, drying, and compaction/pelletizing. Through these treatments, BA, as well as the other raw materials, can be more uniformly and effectively integrated into the ceramic matrix, ensuring better control over the final properties of the product.

After preprocessing, thermal treatment is essential for manufacturing the final ceramic product. This process, generally named firing, can be classified into three main categories: melting, sintering, or vitrification [1,12–14]. Melting and vitrification involve the complete fusion of the material at very high temperatures (typically between 1000 and 1500 °C) [1,11], which makes them less suitable for traditional ceramic processes. In contrast, sintering, the most widely used thermal treatment process for ceramics manufacturing, is performed at lower temperatures, typically between 700 and 1200 °C [1,10], involving the partial fusion of the material and resulting in improved density, mechanical strength, and durability, while preserving the structural integrity of the final product. It is widely applied in the production of various types of ceramic materials, including those that incorporate bottom ash as an additive. Sintering allows the BA to bond with the other raw materials without fully melting, enhancing the final product's performance [11].

Several studies have analyzed key technical parameters to evaluate the feasibility of incorporating bottom ash in the formulation of ceramic materials, focusing on aspects such as sintering behavior, mechanical properties, and environmental behavior, particularly in relation to the release of potential contaminants, such as metals, metalloids, and salts, and the durability of the product [10,15].

Studies on the sintering behavior of bottom ash have demonstrated that the temperature employed significantly influences the product's density, mineralogical transformations, and metal encapsulation. A study identified 1080 °C as the temperature at which the maximum density is achieved [16]. Similar findings were reported by Cheeseman et al. [17], who examined BA that was milled, pelletized, and sintered at temperatures between 900 °C and 1080 °C, confirming the improved densification and reduction in porosity of the treatment product. Other research highlighted that the use of pre-vitrified BA reduces the sintering temperatures by approximately 20 °C compared to non-vitrified BA, facilitating crystallization and improving material stability [18,19]. Regarding the mineralogical composition, it was observed that ceramics composed of 60% by weight of BA and 40% by weight of industrial clays exhibit a different sintering process. In these materials, the liquid phase forms through transformations involving CaO, SiO_2, and Al_2O_3, rather than feldspar melting as in traditional ceramic formulations [20].

The influence of the variation in BA incorporation percentages on the properties of ceramic materials has been extensively investigated. Research on ceramic materials containing up to 20% BA has consistently shown that higher BA contents lead to increased water absorption, while reducing shrinkage, bending strength, and wear resistance [21]. Dagnew [22], who examined BA percentages ranging from 0 to 60%, identified 30% BA as the optimal content for maximizing flexural strength and water absorption, with the breaking strength reaching a plateau at this percentage. Another study indicated that incorporating up to 60% BA can still lead to a product with mechanical properties comparable to commercial products [20]. Specifically, a mixture of 60% BA and 40% clay, sintered at temperatures between 1190 and 1240 °C, demonstrated high crystallinity, low water absorption, and good mechanical resistance. Experimental work on glass–ceramic materials has also confirmed pre-vitrified BA's potential to enhance mechanical properties, including bending strength and Young's modulus, also when combined with other industrial waste and sintered at high temperatures [23].

The environmental impact of the utilization of the mineral fraction of BA for different applications has been widely studied, particularly through the analysis of the leaching behavior of the initial residues and of the obtained products, and also through life cycle assessment (LCA) studies comparing the manufacturing of the products with or without the use of BA. As for the latter, LCA has been employed to quantify the environmental footprint of BA utilization in ceramics manufacturing both as a replacement for feldspar sand in conventional materials [24] and for the production of frit for ceramic glaze [25], making use of data regarding energy consumption, raw material savings, emissions, and the management of the solid residues. The results of these studies indicate that BA incorporation can lead to a decrease in environmental impacts compared to traditional formulations, particularly related to the avoidance of the quarrying and transport of the substituted raw materials and to the avoidance of BA landfilling [24].

Leaching refers to the process by which substances are released from a material into the environment when in contact with an eluent such as water. The alkalinity and potential leaching of inorganic contaminants from BA, such as metals, metalloids, and salts, are the critical issues that are considered when assessing the potential adverse environmental effects related to BA use, since the leached pollutants could affect the quality of water resources and hence human and ecosystem health. In materials with low permeability, such as ceramics, contaminants primarily diffuse through the material, which often limits the rate of leaching. This is particularly relevant for monolithic materials, which tend to exhibit slower leaching rates due to the lower specific surface area in contact with the extracting liquid compared to granular materials. Studies have shown that sintered BA in granular form exhibits a reduced acid neutralization capacity, especially at low pH values, and its neutralization capacity decreases when there is an increase in the sintering temperature [7,8]. Multiple investigations have reported a significant reduction in the leaching of metals and metalloids (Ni, Cr, Cd, Pb) and major elements (Ca, Mg, Na, K) as a result of their encapsulation in glassy and crystalline phases during sintering [16–18,26]. In particular, one study reported an up to 90% reduction in Ni release and up to 99% decrease in the leaching of other metals from sintered BA, even for acidic environments, under which these metals typically exhibit the highest release from untreated bottom ash [16]. Only in one case the leaching of elements such as Cr, Zn, and Cd was reported to have been slightly higher from sintered products compared to untreated bottom ash [17]. Regarding vitrified bottom ash, studies have shown a significant improvement in their leaching behavior, with metal and metalloid release reduced to nearly zero and a one-order-of-magnitude decrease in the leaching of macro-elements such as Ca and Mg [18,26].

Although the literature on the environmental behavior of bottom ash employed in granular form is extensive, studies on the leaching behavior of monolithic materials obtained employing bottom ash as one of the raw materials is instead quite limited. This is especially true for ceramic materials incorporating BA, despite the growing interest in sustainable construction materials and waste valorization options. Studies on monolithic applications have demonstrated that leaching from BA-containing monolithic products is significantly reduced, particularly when BA is incorporated into concrete, asphalt, or other construction materials [27–30]. Clavier et al. [31] also observed acceptable leaching values for BA used in cement production as a kiln feed ingredient, replacing traditional kiln feed components, further supporting the potential of BA utilization for manufacturing sustainable construction materials.

In summary, the research on incorporating bottom ash into ceramic materials consistently shows that BA can be an effective alternative raw material, providing benefits such as reduced sintering temperature, enhanced mechanical properties, and reduced environmental impacts. However, further studies are necessary to assess the long-term environmental behavior of BA-containing monolithic materials, especially in ceramic products.

This study primarily aims to determine whether the partial replacement of traditional materials with bottom ash in porcelain stoneware tile formulations can be considered as an environmentally compatible solution, i.e., that does not lead to additional risks to the environment or human health. The main objective is to understand whether this substitution leads to significant variations in terms of chemical composition, leaching behavior, and compliance with regulatory requirements, when available. First, the total content of major elements and trace contaminants in tiles manufactured with BA and in those obtained with the traditional formulation was performed. This allowed to assess whether the use of BA could significantly alter the chemical composition of the material, potentially affecting its environmental compatibility. The focus then shifted to the end-of-life stage of the tiles. Here, the utilization potential of BA-containing ceramics as recycled aggregates in the construction sector at the end of their life (i.e., after demolition) was assessed. Compliance leaching tests were conducted on crushed tile samples manufactured with or without BA. The resulting concentrations were compared with the Italian End-of-Waste (EoW) criteria for the utilization of construction and demolition waste as recycled aggregates. This step was essential to verify that the BA-based tiles would meet environmental requirements not only during their use, but also when reintroduced into the material cycle after disposal. To evaluate the environmental behavior of the materials during their service life, monolith leaching tests were carried out on intact tile samples. This approach enables to perform a realistic evaluation of the materials' environmental behavior under actual use conditions. In particular, depending on the utilization scenario, the amount of eluent and pH conditions may change, affecting in turn the release behavior of the material. The release of potential contaminants has been shown to vary depending on the characteristics of the element, but also on those of the solid matrix. It is hence important to identify the mechanisms controlling the release of the different constituents of the ceramic tile and how they may be modified if BA is employed in the formulation of the ceramics. The results of these tests were then integrated into a site-specific human health risk assessment study designed to simulate worst-case scenarios and assess the potential impact on groundwater quality and human health.

By examining variations in the chemical composition and the leaching mechanisms of the products manufactured with or without BA, the study aims to provide a clear and evidence-based framework for evaluating if the integration of BA into ceramic formulations could be a viable solution also from an environmental compatibility perspective, on the basis of the release behavior of the product during its use and at its end of life. This contributes to the broader goal of promoting resource efficiency and circularity in the ceramics industry, while ensuring long-term safety and environmental protection.

2. Materials and Methods

2.1. Analyzed Samples

In this study, ceramics samples were obtained from an Italian company that promotes a new production approach inspired by circular economy principles in the ceramics sector. The company has started a test production line for porcelain stoneware tiles to be used for outdoor pavement applications, incorporating 30% by weight of waste-to-energy bottom ash, used as a partial replacement for traditional raw materials rich in silico–aluminate compounds.

Before being used, the bottom ash underwent several processing steps to ensure its suitability for ceramics production. These steps included grinding to achieve a finer particle size, as well as the removal of both ferrous and non-ferrous metals. Once processed, the pre-treated bottom ash was added in the production line where it was blended with conventional raw materials typically used in traditional porcelain stoneware production and subsequently subjected to the standard firing process.

All tests were carried out both on stoneware tiles made only with traditional materials and on those containing bottom ash from WtE plants. Three replicates of each type of tile were provided by the company and are referred to as A, B, and C in this paper. A picture and the dimensions and surface area of each type of tile are reported in Table 1. As can be observed, the tiles manufactured with BA presented a slightly lighter brown color compared to those manufactured with the traditional formulation.

The following terminology is used throughout the paper:

- Sample SP: refers to porcelain stoneware tiles made with the traditional formulation;
- Sample SP-BA: refers to porcelain stoneware tiles containing 30% by weight of WtE bottom ash.

In addition to the monolithic samples, the company also provided three crushed samples of each type of porcelain stoneware tile formulation, named D, E, and F in this paper, resulting in six crushed samples overall. These were specifically employed to carry out the standardized batch compliance tests on the crushed samples to analyze the leaching behavior of recycled aggregates that may be obtained at the end of life of the ceramic product.

Table 1. Analyzed materials.

SP: Traditional Formulation				SP-BA: Containing 30% by Weight of WtE Bottom Ash			
Width [cm]	**Length [cm]**	**Height [cm]**	**Surface Area [cm^2]**	**Width [cm]**	**Length [cm]**	**Height [cm]**	**Surface Area [cm^2]**
9.9	9.9	6	433	10	10	6.3	452

2.2. Total Content

The total content of metals, metalloids, and major constituents of the two types of analyzed tiles was determined using an alkaline fusion process. This technique enables the complete decomposition of complex matrices into forms that are more easily dissolvable and analyzable. The process involves a high-temperature reaction between an inorganic flux and the sample matrix. Some of the most used fluxes include lithium metaborate ($LiBO_2$), lithium tetraborate ($Li_2B_4O_7$), boric acid (H_3BO_3), sodium peroxide (Na_2O_2), sodium carbonate (Na_2CO_3), and sodium hydroxide (NaOH) [32,33]. In order to perform the alkaline fusion process effectively, it is essential that the samples present a very fine particle size to ensure the complete and homogeneous fusion of the material. For this reason, a portion of the crushed samples provided by the company was subjected to an additional milling step aimed at improving homogeneity and reducing the particle size to the required level. All six materials (three replicates for traditional formulation samples (SP) and three replicates of samples containing bottom ash (SP-BA)) were milled to a particle size lower than 200 µm using a RETSCH RS 200 disk vibration mill (Retsch GmbH, Haan, Germany) set at 700 rpm for 10 s.

In preparation for the alkaline fusion process, approximately 0.3 g of each finely ground sample were weighed and transferred into platinum melting pots. Subsequently, 2.1 g of lithium tetraborate ($Li_2B_4O_7$) (Sigma-Aldrich, St. Louis, MO, USA) were added to each melting pot, corresponding to a flux-to-sample ratio of 7:1. The mixture was fully homogenized and initially heated in a preheated muffle furnace (LT/ME-271000/M model Linetronic Technologies SA, Mendrisio, Switzerland) at approximately 500 °C. The temperature was then increased to 1050 °C and kept for 1 h to ensure the complete fusion of the sample with lithium tetraborate. Upon completion of the fusion step, the molten material from each platinum melting pot was rapidly contacted with a 4% by weight nitric acid (HNO_3) (Sigma-Aldrich, St. Louis, MO, USA) solution in distilled water and allowed to dissolve under constant mixing for about 60 min. This step facilitated the complete dissolution of the solidified fused material into a clear solution. Each solution was then diluted to a final volume of 200 mL using the 4% nitric acid solution. The resulting solutions were filtered through 0.45 µm membrane filters in cellulose acetate to remove any remaining particulate matter before subsequent analytical determination.

2.3. Leaching Tests

2.3.1. Compliance Leaching Test

Batch compliance leaching tests are commonly performed on granular materials to evaluate their leaching behavior at native pH and for a set liquid-to-solid ratio. In this study, we applied these tests to crushed tile samples to simulate potential end-of-life scenarios, where the tiles, which could be used as recycled aggregates, may be subject to greater surface area exposure. The concentrations obtained were compared to the thresholds set by the Italian End-of-Waste (EoW) criteria for the use of construction and demolition waste as recycled aggregates (Decree No. 152/2022) [34].

The batch tests were performed applying the EN 12457-2 [35] standard compliance test at the native pH of the samples. The tested samples were provided by the company and consisted of crushed samples of both traditionally formulated tiles and tiles containing bottom ash. According to the standard, the particle size of the samples must be below 4 mm. If more than 20% of the weight of the sample exceeded this size, the coarser material was crushed further in the lab to ensure the sample met the required particle size specification. For each test, a liquid-to-solid ratio (L/S) of 10 L/kg was used. The samples were contacted with deionized water for a period of 24 h in an end-over-end Lauda Varioshake VS 20 OH tumbling device (LAUDA DR. R. WOBSER GMBH & CO. KG, Lauda-Königshofen, Germany) to allow for the leaching process to reach equilibrium conditions. These tests were performed in triplicate.

2.3.2. Monolith Leaching Test

The monolith test was conducted following the EN 15863:2015 [36] standard procedure, which is specifically designed to assess the leaching behavior of monolithic waste under dynamic conditions. The leaching procedure begins by suspending the test samples in tanks filled with deionized water. The samples were carefully placed in the leaching tanks, ensuring that there was a minimum distance of 2 cm from both the bottom and the walls of the container, with at least 2 cm of the sample submerged under the water surface. A support system was used to maintain this precise position throughout the experiment. This setup ensures that a controlled concentration gradient is established between the material's surface and the water, which serves as the driving force for mass transport, facilitating the release of inorganic elements over time.

To maintain a consistent leaching environment, the leaching fluid is periodically renewed according to a standardized schedule (Supporting Information, Table S1). The volume of deionized water required for each test was determined using the standardized liquid-to-surface ratio ($L/A = 8 \pm 2$ mL/cm^2) required by the standard, where L represents the volume of the leaching agent (deionized water) and A corresponds to the exposed surface area of the sample.

Applying this ratio, the calculated volume of deionized water used for the SP samples ($A = 433$ cm^2) was 3469 mL, while for the SP-BA samples ($A = 452$ cm^2), it was 3616 mL. At each leaching interval, the test samples were removed and transferred to new tanks containing an equal volume of fresh deionized water. The collected eluates underwent pH and electrical conductivity measurements before filtration through a 0.45 μm cellulose acetate membrane. The results can be expressed either as mg of released constituent per m^2 of the sample's surface area, or as mg of the constituent per liter of eluate for each leaching interval.

2.4. *Analysis of the Obtained Solutions*

The filtered solutions collected from the alkaline fusion procedure, compliance leaching tests, and monolith tests were analyzed to determine the concentrations of major and trace elements (Al, As, B, Ba, Be, Ca, Cd, Co, Cr, Cu, Fe, K, Mg, Mn, Mo, Na, Ni, Pb, Sb, Se, Si, V, and Zn) by Inductively Coupled Plasma Optical Emission Spectroscopy (ICP-OES) (Agilent 5800, Agilent Technologies, Santa Clara, CA, USA), and of anions (chlorides, sulfates, nitrates, and fluorides) by Ion Chromatography (IC) (Thermo Scientific Dionex ICS-1100, Thermo Fisher Scientific, Waltham, MA, USA). Specifically, for the ICP-OES analysis the following measurement conditions were employed: a radio frequency of 1.3 kW, axial viewing mode, a read time of 15 s, and the following flows: nebulizer 0.7 L/min, plasma 12 L/min, and auxiliary 1 L/min. For the IC measurements, the following conditions were used: a AS22 column with an AG22 guard column (Thermo Fisher Scientific, Waltham, MA, USA), a DRS 600-4 mm suppressor, a 4.5 mM Na_2CO_3–1.4 mM $NaHCO_3$ (Sigma-Aldrich, St. Louis, MO, USA) eluent with a flow rate of 1.2 mL/min, current of 33 mA, and injection volume of 25 μL. The limits of quantification (LOQ) of all of the analyzed constituents are reported in Table S2 of the Supporting Information section. In the figures presented in Section 3, the LOQ values are visually distinguished by a diagonal pattern or by a gray dashed line, enabling a clear comparison with the other values.

2.5. Assessement of the Results of the Monolith Leaching Tests

In this section, the methodologies employed to evaluate the results obtained from the leaching monolith tests on both types of tile samples in order to identify the main mechanisms controlling the contaminants' release and to perform the human health risk assessment are described.

2.5.1. Analysis of the Main Release Mechanisms

To assess the release mechanisms resulting from the results of the monolith leaching tests, various alternative approaches are reported in the literature. Experimental data can be interpreted by fitting a regression curve to the cumulative release over time, or by using a logarithmic plot of the cumulative flux vs. time [37]. In this study, we opted for the latter procedure, adopting the analytical framework proposed by De Groot and Van der Sloot [38] which is also incorporated into the Dutch standard for monolith leaching tests NEN 7375:2004 [39]. This methodology is widely applied for the evaluation of the release mechanisms for monolithic waste materials [29,37,40–43] and involves plotting the cumulative mass of constituent released per unit of surface area (mg/m^2) as a function of the cumulative leaching time on a double logarithmic (log–log) scale.

In this study, the analysis was conducted to assess whether incorporating bottom ash into the porcelain stoneware tiles could modify the release mechanisms of certain constituents, compared to the ones resulting for the traditional tile formulation. The inclusion of BA could in fact potentially change the solubility or mobility of specific elements, influencing how they are bound within the ceramic matrix and, in turn, modifying their release behavior.

The analysis can be carried out both considering the entire dataset or for defined time-specific intervals to capture temporal variations in the leaching behavior. The leaching intervals analyzed, which correspond to different time intervals (see Table S1 in the Supporting Information section) were the following:

- Total range: full dataset encompassing all sampling fractions (leaching interval from 1 to 8);
- Initial leaching phase: early-stage fractions (leaching interval from 1 to 3);
- Intermediate phase: mid-leaching stage (leaching interval from 3 to 6);
- Final phase: later-stage fractions (leaching interval from 5 to 8).

Linear regression analysis was applied to both the total leaching period and its segmented stages (initial, intermediate, and final) to determine the slope of the trend line. This slope provides insight into the prevailing release mechanism and how it evolves over time [38]. A slope < 0.35 during the total leaching period is generally indicative of surface wash-off, reflecting the rapid dissolution of readily soluble species from the product surface. In the segmented intervals, a slope < 0.35 has a distinct interpretation depending on the release stage: in the initial phase, it indicates surface wash-off, characterized by the rapid release of readily available constituents. Conversely, in the intermediate and final phases, the same slope reflects depletion-controlled release, driven by the gradual exhaustion of leachable constituents and by a corresponding decline in the release rate. Slopes between 0.35 and 0.65 are associated with diffusion-controlled release. This interpretation is consistent for both total and segmented analyses. Notably, Van der Sloot and Dijkstra [44] affirmed that a slope of 0.5 is characteristic of a purely diffusion-controlled mechanism, reinforcing this threshold as a meaningful indicator. Finally, slopes > 0.65 suggest a dissolution-driven mechanism, where the solubility of the product matrix is sufficiently high that the dissolution of a constituent at the surface exceeds the rate of diffusion through the internal pore structure.

In some cases, the slopes of the lines interpolating the leaching data could result in being very close to the threshold values that define the classification limits of each mechanism. Such rigid thresholds may lead to uncertainties in mechanism assignment, because even small variations in the measurements or slight experimental fluctuations can shift the classification from one mechanism to another. To better account for this, the 95% confidence interval (CI) of the regression slope was calculated. Only when the CI fell entirely within the boundaries of a single mechanism was the mechanism considered clearly defined. When the CI overlapped two mechanisms, the dominant mechanism was considered to be potentially either one of these, and when it spanned three mechanisms, the release mechanism was classified as not defined (ND). This approach allows a more objective and statistically supported interpretation of the prevailing release mechanisms, especially for borderline cases.

2.5.2. Risk Assessment Procedure

Given the nature of the contaminants characteristic of the investigated type of residues (i.e., metals, metalloids, and salts) and the utilization scenario assumed (use of the tiles for outdoor paving), in the human health risk assessment, leaching to the underlying groundwater was considered as the only potentially relevant migration pathway of the contaminants [45–47]. Other potential exposure routes, such as accidental ingestion, dermal contact, or dust inhalation, were not considered relevant in this case. Accidental ingestion is in fact unlikely to occur, as the product is in the form of tiles. Similarly, dermal contact is not considered a concern, since the tiles remain intact during normal use and do not disintegrate or release particles. This minimizes the potential for skin adsorption of any contaminants that may be bound within the tile matrix. Furthermore, the inhalation of vapors, which can be a concern for certain organic waste types, is not relevant in this case, since the contaminants of interest are non-volatile, meaning they do not pose a risk through vapor release.

Considering these factors, the risk assessment was carried out based on the following target contaminants, which include a range of metals, metalloids, and inorganic ions typically found in BA at varying concentrations: Al, As, B, Ba, Be, Cd, Co, Cr, Cu, Mn, Mo, Ni, Pb, Sb, V, Zn, chlorides, fluorides, sulfates, and nitrates. The results obtained from the monolith leaching test were used as input data for the assessment.

In particular, the procedure followed for conducting the assessment was structured into four steps:

- Definition of the exposure scenario, i.e., the conditions under which the product is used (for outdoor pavements in this case);
- Calculation of the predicted concentration of a specific contaminant in the environmental compartment (PEC) using the respective maximum concentration resulting from the monolith leaching test and standardized fate and transport models;

- Calculation of the concentration of the specific contaminant that is compatible with an acceptable risk, or Predicted No-Effect Concentration (PNEC);
- Calculation of the risk by comparing the PEC to the PNEC value for each contaminant; if the PEC/PNEC ratio is below 1, the human health risk can be considered acceptable.

Utilization Scenarios Considered for the Risk Assessment

The first step of the risk assessment procedure consists of defining the conceptual model. This involves identifying the source of contamination, the pathways through which contaminants migrate across different environmental compartments, and the potential exposure mechanisms to determine the receptors that may be affected.

Specifically, the following two different utilization scenarios were considered (see Figure 1):

- **Scenario 0**: representing a "no restrictions" scenario, where direct contact between the porcelain stoneware tiles and the groundwater is assumed. This situation is characterized by the absence of attenuation in the unsaturated zone and of dilution in the groundwater.
- **Scenario 1**: assumes the use of the porcelain stoneware tiles for the construction of the outdoor paving of a large area. As an example, we considered the size of St. Peter's square in Vatican City (320 m of length by 240 m of width). In this case, the attenuation in the unsaturated zone and dilution in the groundwater were considered.

It is important to highlight that Scenario 0, although considered as the most conservative case, represents a condition that is unlikely to occur in practice. This scenario was included to evaluate if the material could be used in free-use conditions without any restrictions.

The Point of Compliance (POC) represented in the figure indicates the location where the risk assessment was carried out by comparing the Predicted Environmental Concentration (PEC) with the Predicted No-Effect Concentration (PNEC). In this study, the POC was assumed to be in the groundwater table directly beneath the tile layer.

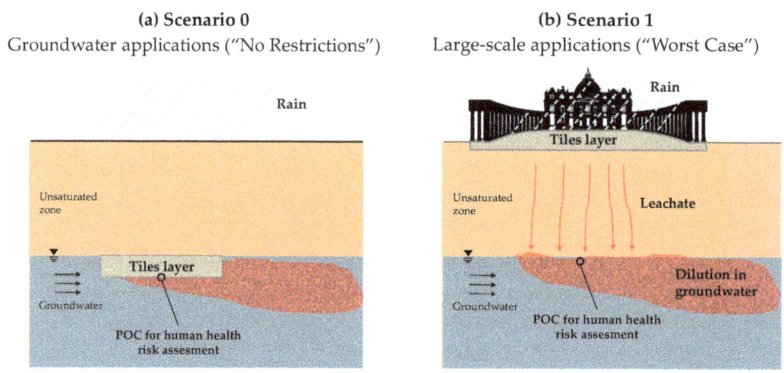

Figure 1. Scheme of the proposed scenarios for porcelain stoneware tiles utilization considering (**a**) direct application in the groundwater, or (**b**) large-scale deployment in the unsaturated zone.

Human Health Risk Calculation

The Predicted Environmental Concentration (PEC) in the environmental compartment was then determined using the maximum concentration obtained from the eight leaching intervals of the monolith tests for each target contaminant, applying a standardized fate and transport model.

Specifically, the concentration in the groundwater (PEC) for each target contaminant was calculated as follows:

$$\mathrm{PEC} = \frac{\mathrm{C_{max\ leachate}}}{\mathrm{LF}} \quad (1)$$

where $\mathrm{C_{max\ leachate}}$ (mg/L) is the maximum concentration of the target contaminant obtained from the monolith test across all leaching intervals and LF (-) is the dimensionless leaching factor. The LF considers the dilution and attenuation of the contaminant concentration during its transport from the tile layer to the groundwater. In particular, the leaching factor (LF) can be calculated according to the following:

$$\mathrm{LF} = \frac{\mathrm{SAM}}{\mathrm{LDF}} \quad (2)$$

where LDF (-) is the Leachate Dilution Factor, which represents the dilution of the contaminant concentration when it moves from the unsaturated zone to the groundwater [48]:

$$\mathrm{LDF} = 1 + \frac{\mathrm{v_{gw}} \cdot \delta_{\mathrm{gw}}}{\mathrm{I_{eff}} \cdot \mathrm{W}} \quad (3)$$

where δ_{gw} (m) is the groundwater mixing zone height, v_{gw} (m/s) the groundwater Darcy velocity, W the width of the source area longitudinal to the groundwater flow, and I_{eff} (m/s) the effective water infiltration rate. I_{eff} can be estimated by multiplying the annual precipitation rate, P (expressed as cm/y), by an empirical coefficient for water infiltration, β (-), which depends on the specific soil type considered, and by an infiltration coefficient, η (-), which accounts for the water infiltration through the paved surface.

SAM (-) is the Soil Attenuation Model, which considers the sorption of constituents from the leachate onto clean soil [49]. SAM is calculated by the following equation:

$$\mathrm{SAM} = \frac{\mathrm{d}}{\mathrm{d} + \mathrm{L_{gw}}} \quad (4)$$

where d (m) is the thickness of the tile layer and L_{gw} (m) the water table depth (see Figure S1 in Supporting Information for a graphical explanation of the parameters).

The input parameters used for the calculation of the leaching factor are reported in the Supporting Information section (Table S3). Most of these values refer to the conservative default values suggested by the Italian guidelines for the application of risk assessment to contaminated sites [50].

By applying these values to Equations (2)–(4) for the calculation of the leaching factor (LF), the overall dilution factors were obtained for the two different utilization scenarios:

- Scenario 0 ("no restrictions" scenario): LF = 1 (i.e., no attenuation nor dilution of the leachate concentration in the groundwater).
- Scenario 1 ("worst-case" scenario): LF = 484.

The leaching factors were then applied to the maximum concentrations resulting from the monolith laboratory tests (Table S4 in the Supporting Information section) to estimate the Predicted Environmental Concentration (PEC) using Equation (1).

The PEC (mg/L) was subsequently compared to the Predicted No-Effect Concentration, PNEC (mg/L), which represents the risk-based concentration limit for the specific contaminant in the groundwater. The PNEC was derived using risk-based criteria, specifically considering the water ingestion pathway for a child receptor. The assessment accounted for both toxicological and carcinogenic effects. The toxicological parameters were obtained from the Italian ISS-INAIL database. For this assessment, the Risk-net software (version 3.1.1), developed by the University of Rome "Tor Vergata" was used [51]. Note that if the risk-based limit values calculated with Risk-net resulted lower than the Italian limit values, these limits were adopted, as required by Italian regulations concerning these specific cases. The results are presented in Table S5 of the Supporting Information section.

Thus, the risk for groundwater for each contaminant, R_{GW} (-), was calculated as follows:

$$R_{GW} = \frac{\text{PEC}}{\text{PNEC}} \quad (5)$$

If R_{GW} was lower than 1, the risk was considered acceptable, as this condition indicates that the PEC for the target contaminant was below the corresponding PNEC.

3. Results

3.1. Total Content of Major Elements and of Potential Contaminants

The results obtained from the total content analysis, performed after alkaline fusion of the samples, provide a comprehensive overview of the elemental composition of the two types of samples. This method allows for the complete dissolution of the ceramic matrix, ensuring the quantification of all of the non-volatile elements, including those bound within crystalline or amorphous phases. The results expressed in grams of constituent per kilogram of product are reported in Figure 2, which shows the average concentrations obtained from three replicates, along with the corresponding standard deviations.

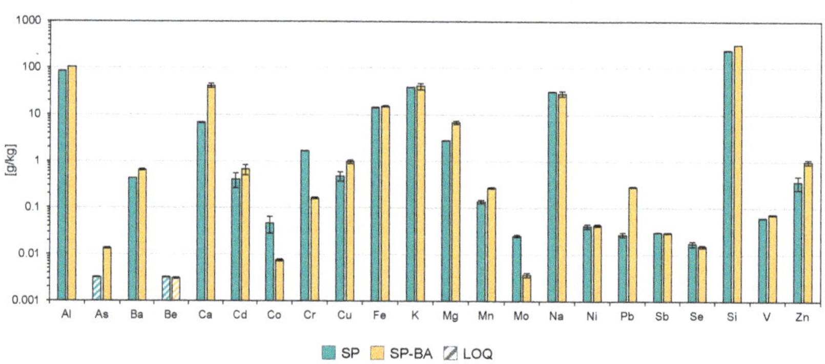

Figure 2. Total content results in g/kg. Traditional porcelain stoneware tiles are represented in green, while those containing bottom ash (SP-BA) are shown in yellow. LOQ values, indicated by bars filled with a diagonal pattern, are reported for the elements that presented a concentration below the LOQ. Error bars indicate the standard deviation of three independent replicates.

Silicon and aluminum were the most abundant elements retrieved in both samples, with silicon concentrations exceeding 300 g/kg in the SP-BA formulation and aluminum levels around 100 g/kg for both types of samples. These high levels are attributable to the use of feldspar sand and clay in the tiles' formulation, but also to the presence of the BA in samples SP-BA, since all of these materials are rich in silicates and aluminosilicates [52,53]. Potassium, sodium, iron, calcium, and magnesium were also present in significant amounts. The addition of BA appeared to slightly increase the concentration of some elements, including Si, Ca, Mg, Ba, Cu, Mn, Pb, and Zn, suggesting a contribution of these elements from the BA source. Conversely, elements such as Co, Cr, and Mo showed to be present in lower amounts in the SP-BA sample, indicating a dilution effect or lack of contribution from the BA. Trace elements like As, Be, Mo, Ni, Sb, Se, and V were found at much lower concentrations, i.e., below 0.1 g/kg, for both materials.

Overall, from the perspective of the total content, the incorporation of bottom ash did not appear to cause significant variations in elemental composition compared to the traditional formulation, except for a one-order-of-magnitude-higher content of Pb in the SP-

BA samples that, however, exhibited a one-order-of-magnitude-lower Cr and Mo content compared to the SP tiles. The concentrations of major and trace elements remained largely consistent, suggesting that the partial substitution of traditional materials with BA did not substantially alter the chemical composition of the final ceramic product.

3.2. Compliance Leaching Test

The compliance leaching test results and the comparison with the limits set by the Italian regulation for the utilization of construction and demolition materials as recycled aggregates are shown, respectively, in Figure 3 and Table 2. The results include all of the analyzed elements, while the comparison only considers the regulated elements.

As shown in Figure 3, on average, the tiles containing bottom ash exhibited a pH increase of 0.7 units, which could influence the release of certain elements. The leaching of many of the constituents, however, remained below the quantification limit (LOQ) in both cases. The addition of BA appeared to slightly increase the eluate concentration of a few elements, including B, Mg, Mn, and chlorides, suggesting that the use of the BA as raw material increased their contents or that changes in pH may have affected their release. In contrast, for other elements such as Cu, Fe, Mo, and Zn, the addition of BA seemed to have reduced their release. Overall, it does not seem that the addition of BA negatively affected the leaching behavior of the material in its granular form. In fact, the results showed a similar release behavior for the two tile formulations, indicating that BA incorporation did not significantly affect the environmental behavior of the ground tile material.

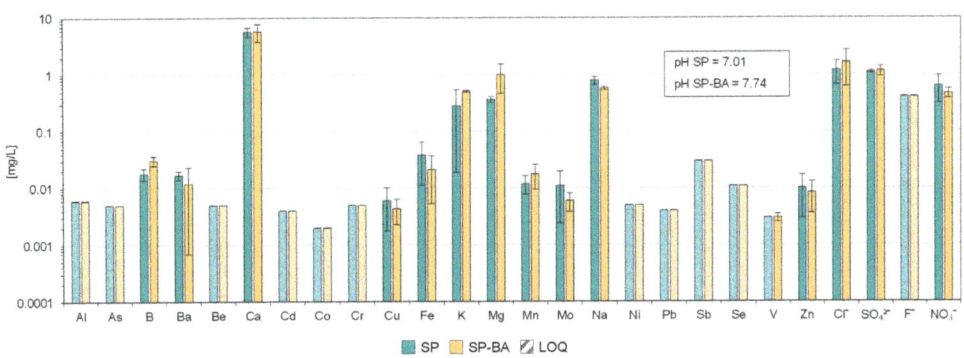

Figure 3. Compliance leaching test results obtained for the traditional porcelain stoneware tiles (in green) and for the product containing bottom ash (in yellow). LOQ values, indicated by bars filled with a diagonal pattern, are reported for the constituents that presented a concentration below the LOQ. Error bars indicate the standard deviation of three independent replicates.

The comparison of the results of the compliance leaching test with the Italian End-of-Waste (EoW) criteria for construction and demolition (C&D) waste use as recycled aggregates is shown in Table 2, which reports the results obtained for the three replicates of the crushed samples, indicated as D, E, and F. Some of the analyzed elements (Al, B, Ca, Fe, K, Mg, Mn, Na, Sb) are not included in Table 2, as the Italian End-of-Waste (EoW) Decree 152/22 does not report limit values for these constituents. Examining the concentrations resulting from each replicate, it can be noted that for all constituents, apart from Ba, chlorides, and sulfates, at least one of the replicates was below the quantification limit. In some cases, e.g., Zn, Cu, and nitrates, one of the replicates presented a higher concentration than the others; this result may be related to possible heterogeneity within the samples that in this case were only crushed and not finely milled as in the case of the total composition analysis. Nonetheless, the absence of any exceedances, even for the single replicates, demonstrates that both materials may be suitable for use as recycled aggregates at the end of their application as pavement tiles.

Table 2. Comparison of compliance leaching test results with Italian EoW concentration limits (EoW Decree 152/22) for the utilization of recycled aggregates from C&D waste. The green cell background indicates values below the limits. Values written in gray represent the LOQ (limit of quantification). Values highlighted in red indicate exceedance of the limits. D, E, and F represent the replicates of the crushed tile samples.

Constituents	Limits (EoW Decree 152/22) [mg/L]	Compliance Leaching Test Results [mg/L]					
		SP			SP-BA		
		D	E	F	D	E	F
As	0.05	0.005	0.005	0.005	0.005	0.005	0.005
Ba	1	0.021	0.015	0.015	0.028	0.004	0.004
Be	0.1	0.005	0.005	0.005	0.005	0.005	0.005
Cd	0.005	0.004	0.004	0.004	0.004	0.004	0.004
Co	0.25	0.002	0.002	0.002	0.002	0.002	0.002
Cr	0.05	0.005	0.005	0.005	0.005	0.005	0.005
Cu	0.05	0.003	0.003	0.021	0.007	0.004	0.002
Ni	0.01	0.005	0.005	0.005	0.005	0.005	0.005
Pb	0.05	0.004	0.004	0.004	0.004	0.004	0.004
Se	0.01	0.011	0.011	0.011	0.011	0.011	0.011
V	0.25	0.003	0.003	0.003	0.004	0.003	0.003
Zn	3	0.005	0.005	0.077	0.053	0.005	0.005
Chlorides	750	1.21	0.65	1.72	3.08	0.81	1.03
Sulfates	750	1.17	1.06	1.06	1.15	1.39	0.92
Fluorides	1.5	0.4	0.4	0.4	0.4	0.4	0.4
Nitrates	50	1.08	0.4	0.4	0.56	0.4	0.4

C < LIM
C > LIM
LOQ

To further assess the leaching behavior of each type of material, for each element, the released percentage was calculated as the ratio between the maximum concentration leached during the compliance leaching test and the corresponding total content in the material determined by the alkaline fusion procedure. The results are presented in Figure 4. It should be noted that this evaluation was performed only for constituents for which the total content was determined, i.e., for all constituents apart from boron, since lithium tetraborate was used in the alkaline fusion procedure, and the anions analyzed in the leachate solutions, since their total contents cannot be determined through the employed method. In cases where the maximum concentration detected in the leachate corresponded to the instrumental quantification limit (LOQ) (i.e., Al, As, Be, Cd, Co, Cr, Ni, Pb, Sb, Se, and V), this value was used as a conservative estimate for the calculation.

As shown, the majority of the elements exhibited leaching percentages well below 0.1%, indicating a very low release. A few elements displayed slightly higher percentages around 1–2%. Among these, elements such as As, Be, Sb, and Se were detected at concentrations lower than the respective LOQs in the compliance leaching test. In these cases, the release percentages may have been hence overestimated. This conservative assumption was intentionally adopted to ensure that even potentially critical elements were evaluated under worst-case conditions. Overall, these results clearly confirm the effective immobilization of the constituents within the ceramic matrix, demonstrating that the incorporation of bottom ash did not compromise the material's leaching behavior even after the material was ground.

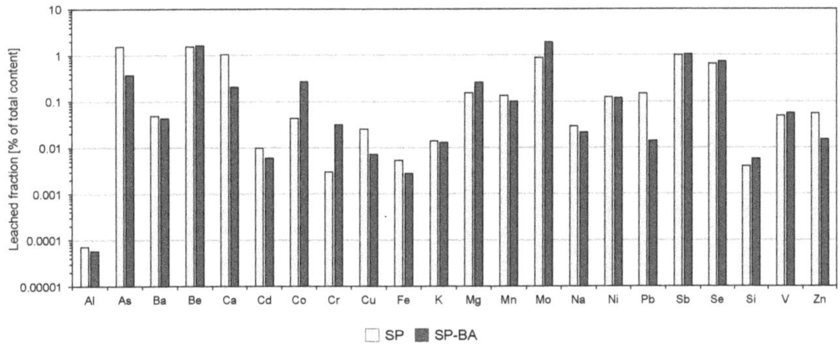

Figure 4. Leached percentage for each element, calculated as the ratio between the maximum concentration resulting from the compliance leaching test and the total content measured for SP and SP-BA samples.

Such limited release is most likely attributable to the high temperature used in porcelain stoneware tile production, which promotes the formation of a dense structure and the inertization of the bottom ash. High temperatures in fact promote the incorporation of elements, including those introduced via bottom ash, into both the crystalline and the amorphous phases of the ceramic matrix. This process significantly reduces the mobility of the elements, effectively stabilizing them within the structure of the product. The encapsulation of potentially hazardous elements limits their solubility under the leaching test conditions [26].

Even for elements present in relatively high concentrations in terms of total content (e.g., Si, Al, Ca, K, and Fe), the leached percentages were negligible. This confirms the high chemical stability of the products. Importantly, the inclusion of bottom ash in the SP-BA formulation did not lead to an increase in the release of constituents of potential environmental concern, even after the material was ground, reinforcing the conclusion that 30% by weight of bottom ash could be incorporated into the ceramic matrix without compromising its leaching performance at the end of life of the product.

3.3. Monolith Leaching Test

The leaching behavior of monolithic samples of traditional porcelain stoneware tiles (SP) and tiles containing bottom ash (SP-BA) was analyzed through three replicate tests. The results obtained for pH and conductivity in each leaching interval are presented in Table 3 as averages of the replicates; the correlated standard deviations are also shown. For both tile formulations, the pH remained relatively stable, with pH values generally falling in the slightly acidic range (values around 5–6), showing only minor fluctuations and a slight increase over time, especially for the last sampling interval.

Conductivity also fluctuated, without a clear trend in this case, indicating small variations in ion concentrations. Notably, the conductivity values remained very low, typically in the range of 2–8 μS/cm, reflecting minimal ion dissolution from the samples. No significant differences were observed between the two types of tiles, suggesting that the addition of this secondary raw material did not significantly affect pH or conductivity.

Figure 5a,b reports the average concentrations of the investigated elements released during the monolith leaching tests, along with the corresponding standard deviations (SD), based on three replicates. The constituents resulting consistently below the respective limits of quantification (LOQ) across all replicates and time points were omitted from the figure; the LOQ values are reported in Table S2 in the Supporting Information section.

Table 3. pH and conductivity of the eluates from the monolith leaching tests for each of the eight standardized time intervals. The table reports the average values obtained from the tests along with the corresponding standard deviations (SD).

Duration of the Time Interval	SP: Traditional Formulation				SP-BA: Tiles Containing WtE Bottom Ash			
	pH -		Conductivity µS/cm		pH -		Conductivity µS/cm	
	Average	SD	Average	SD	Average	SD	Average	SD
6 h ± 15 min	5.41	0.04	4.4	0.5	5.42	0.043	3.7	0.1
18 h ± 45 min	5.36	0.02	6.8	1.5	5.49	0.017	3.6	0.6
1 d and 6 h ± 1 h	5.54	0.02	3.0	0.7	5.50	0.028	4.6	1.4
1 d and 18 h ± 2 h	5.66	0.02	3.2	0.4	5.65	0.024	2.7	0.4
5 d ± 6 h	5.94	0.03	3.9	0.6	5.97	0.028	3.8	0.5
7 d ± 8 h	6.01	0.03	4.0	0.2	6.02	0.012	3.9	0.3
20 d ± 24 h	5.66	0.05	4.9	0.3	5.83	0.049	5.6	0.5
28 d ± 24 h	6.40	0.03	5.8	0.8	6.48	0.025	7.0	0.2

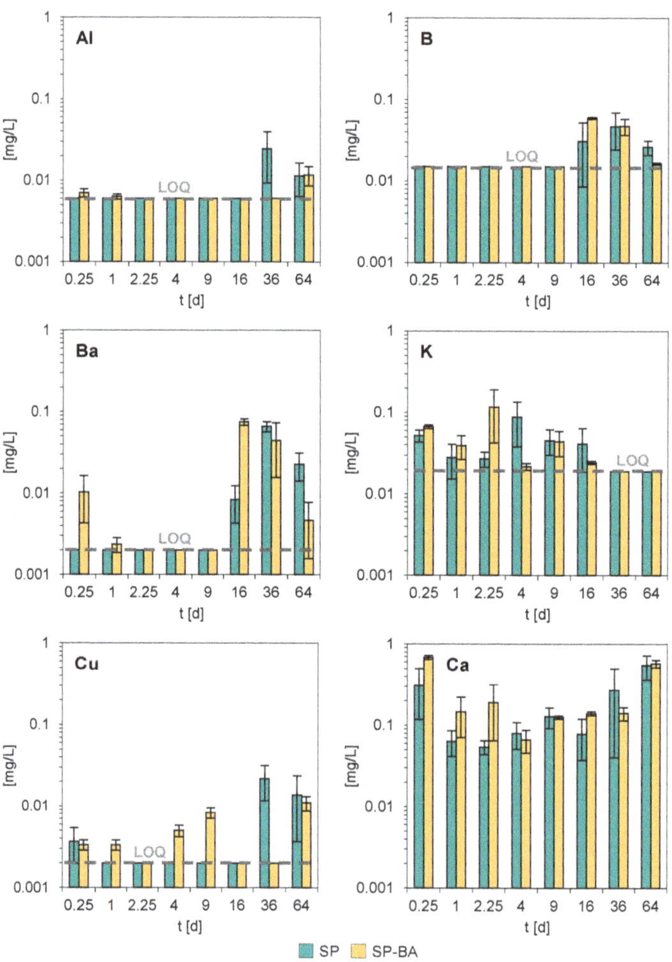

(a)

Figure 5. *Cont.*

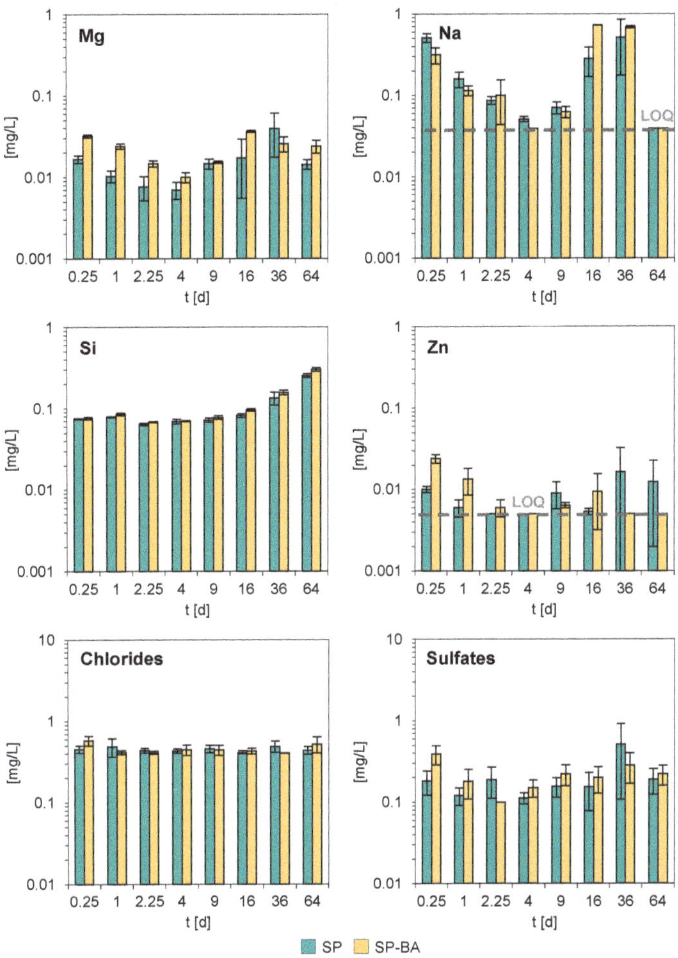

(b)

Figure 5. (**a**) and (**b**). Results of the monolith leaching test. The limit of quantification (LOQ) of each constituent, if relevant, is reported as a dashed gray line. Error bars indicate the standard deviation of three independent replicates.

Also in this case, the constituents that presented the highest concentrations in the collected leachates were major elements such as calcium, potassium, magnesium, sodium, and silicon. For most of the other analyzed elements, the concentrations in the eluates remained consistently close or below the respective LOQs throughout the entire testing period. Notably, the samples containing bottom ash (SP-BA) did not display significant differences in their dynamic leaching behavior compared to the traditional SP formulation. A slightly higher release was found in some leaching intervals for Ba, Ca, Mg, and Zn, which is consistent with the results of the total content analysis, which indicated that these elements were more abundant in the SP-BA samples. This minor increase, however, did not affect the overall environmental performance of the material. As already observed for pH and conductivity, the elemental release trends remained comparable for both formulations. This suggests that the incorporation of bottom ash did not significantly alter the leaching characteristics of the tiles. The overall stability of both formulations during all tested intervals, combined with the absence of substantial variations in the release of potentially hazardous elements, supports the conclusion that the addition of bottom ash did not negatively impact the environmental performance of the porcelain stoneware tiles. Additionally, the tests showed a good reproducibility between the three replicates, with consistent leaching trends observed throughout the experiment.

The very limited release exhibited by both types of products suggests a high stability of these elements within the ceramic matrix, also considering that some of them, e.g., Fe, Mn, and Pb, were present in significant concentrations, as shown by the results of the total content analysis (see Figure 2). The differences between the total contents and the leached concentrations from the monolith test highlight again the strong binding of these elements within the tile structure. In fact, as for the compliance leaching test (see Section 3.2), also for the monolith tests, the maximum concentrations obtained for each constituent, considering all eight leaching intervals and three replicates, were divided by the respective total contents to obtain the release percentages. Since the maximum concentrations observed in the monolith tests were slightly lower than those resulting from the compliance test, the calculated leachate percentages, shown in Figure S2 of the Supporting Information section, were even lower in this case.

3.4. Assessment of the Results of the Monolith Leaching Tests

3.4.1. Identification of the Main Release Mechanisms

Figure 6a,b reports the release mechanisms resulting for 10 constituents for both the traditional and BA-containing tiles, following the methodology described in Section 2.5.1. Constituents that consistently exhibited concentrations below the respective LOQs, or those for which detectable concentrations were only found in one or two leaching intervals, were excluded from the analysis.

The graphs reported in the previously mentioned figures focus on the three temporal phases of leaching (initial, intermediate, and final) and show the evolution over time in a log–log scale of the average cumulative release resulting from the three replicate samples. When considering the cumulative release values, a trend can be noted: tiles incorporating bottom ash tended to release higher amounts over time of several elements, such as Ba, Ca, Cu, K, Mg, Zn, and SO_4^{2-}. For other elements like chlorides and Si, the cumulative release appeared very similar between the two materials. Only Na showed a lower release from BA-containing tiles, possibly indicating a stronger retention or incorporation within less soluble phases. As for the release mechanisms, as can be seen, for all constituents, more than one type of mechanism prevailed depending on the leaching interval considered. Furthermore, differences in the main mechanisms governing release were highlighted for the tiles containing BA compared to those with the traditional formulation.

The release of constituents from a solid material typically occurs through four distinct mechanisms: surface wash-off, dissolution, diffusion, and depletion, depending on the characteristics of the constituent and of the material and on other parameters such as the liquid-to-solid ratio and pH.

Considering the 95% confidence interval (CI) analysis, which allows a more rigorous assessment of the uncertainty in mechanism assignment, clearly defined single mechanisms could be identified in only a few cases. Surface wash-off (SW), occurring exclusively in the initial phase, was clearly identified for Ca and Na in SP tiles, and for Ba and Na in BA-containing tiles, reflecting the immediate availability of these elements at the surface. Diffusion (DF) was observed as a single mechanism only for Ba in the initial phase for SP tiles. Depletion (DP), which occurs when the source of an element within the tile matrix becomes progressively exhausted, leading to a decrease in the release rate over time, was observed more frequently than other single mechanisms. In SP tiles, DP was clearly identified for Ca (intermediate), K (final), and Cl^- (final). In BA-containing tiles, it occurred for Ca (intermediate), K (intermediate and final), Zn (intermediate and final), Cl^- (final), and SO_4^{2-} (intermediate and final). Overall, the DP mechanism was more pronounced for SP-BA tiles, particularly during the final phases of leaching, highlighting a stronger tendency for progressive exhaustion of certain elements in the tile matrix.

In many cases, the release behavior could not be clearly attributed to a single mechanism. When the CI overlapped two mechanisms (e.g., DP/DF), the dominant process was considered to be potentially combined. This situation often arose when the slope of the linear regression was very close to the thresholds defining mechanism boundaries (e.g., 0.35 or 0.65), so that even minor fluctuations in the leaching data could shift the apparent mechanism.

In other cases, the CI spanned three mechanisms, and the release mechanism was classified as not defined (ND). These ND cases indicate that the observed release may have been influenced by more than one process without a clearly dominant mechanism.

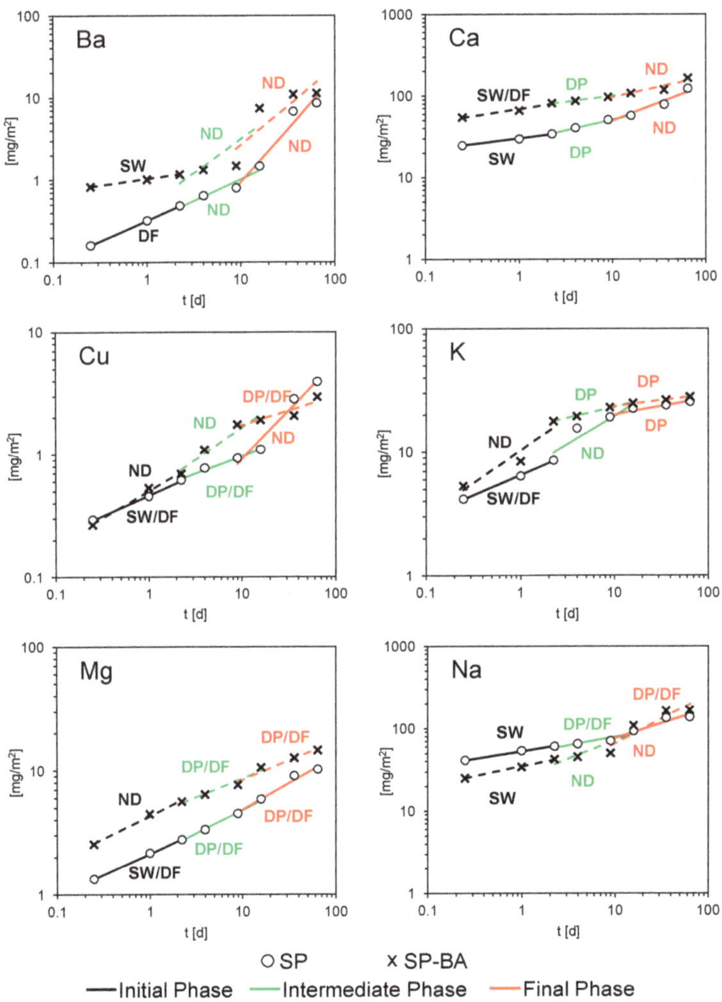

(a)

Figure 6. *Cont.*

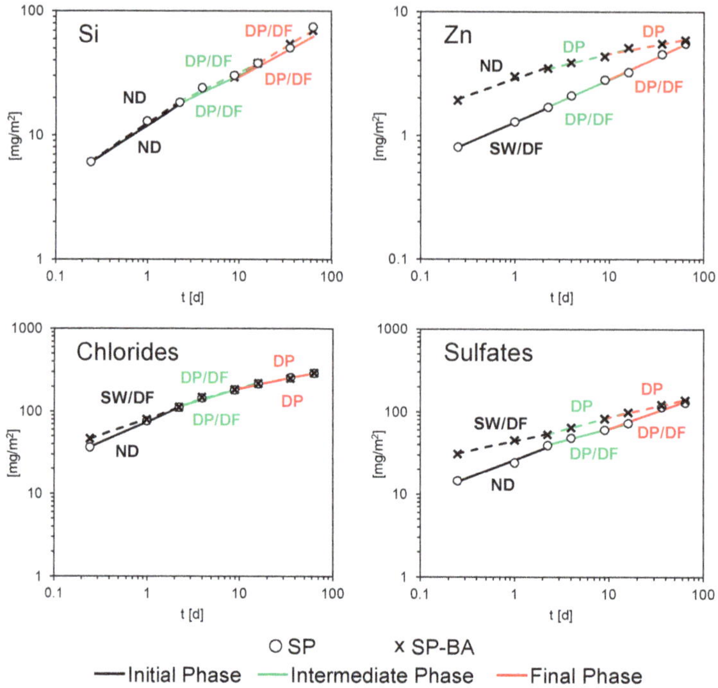

(**b**)

Figure 6. (**a**) and (**b**). Cumulative release vs. time of different constituents from porcelain stoneware tiles. Solid lines refer to the traditional formulation (SP), while dashed lines represent the formulation containing bottom ash (SP-BA). The colors indicate the release phase: black = initial, green = intermediate, red = final. The identified release mechanisms are reported as labels: SW = surface wash-off, DP = depletion, DF = diffusion, DS = dissolution, ND = not defined.

A particularly distinct behavior was observed for copper. In BA-containing tiles, copper was released more quickly during the early and intermediate stages, resulting in higher cumulative release concentrations than the SP tiles at corresponding time intervals. During these phases, the release could not be clearly attributed to a single mechanism, likely reflecting contributions from different mechanisms. In contrast, SP tiles showed a lower initial release of Cu, with surface wash-off and combined diffusion–depletion observed in the intermediate phase. However, in the final stage, the release from SP tiles increased, ultimately exceeding that of the SP-BA tiles. Overall, this pattern indicates that BA-containing tiles released more copper at the beginning of the leaching process, but over the long term, the total release became greater for traditional SP tiles.

A similar behavior, though less pronounced, was also observed for Ba, Ca, K, Na, Zn, and sulfates: SP-BA tiles showed a higher initial release and a slope that decreased over time, whereas SP tiles presented lower initial values but a higher release in the final stage. As time progressed, the release curves of the two tile types tended to converge, and in the case of copper, as already discussed, the release from traditional SP tiles eventually exceeded that of the SP-BA tiles.

In these cases, the influence on the release mechanisms of the presence of bottom ash in the tile matrix may be related to microstructural features (e.g., porosity and specific surface area), which were not investigated here and, to our knowledge, have not been reported in the literature, warranting focused attention in future research. Concerning the comparison of the results of this study with those reported in previous works, it should be considered that the approach that was employed has been widely applied in the literature to investigate the leaching behavior of various monolithic matrices incorporating waste materials. These include hazardous waste stabilized through cement-based solidification processes, such as municipal solid waste incineration fly ash, filter ashes, metal sludge, and filter cakes derived from wastewater treatment [41]. Other examples involve bottom ash stabilized with Portland cement concrete and hot mix asphalt [29], as well as concrete made with cement from cement kiln co-processing of hazardous waste [40]. In these systems, diffusion is commonly identified as the predominant release mechanism throughout all leaching phases for many metals. In some cases, it has been observed that deviations from diffusion-controlled behavior may occur in the later stages of leaching, often as a result of changes in environmental conditions. In particular, shifts in pH can significantly affect the release of certain elements, indicating a transition in the mechanisms controlling leaching [41]. In this study, the incorporation of bottom ash showed to reduce early surface availability and shift the leaching mechanism towards depletion. At the same time, the tiles containing BA showed to be characterized by a higher early cumulative release for certain elements such as Cu, due to more sustained diffusion in the early phases, contrasted by the delayed release behavior observed in the latter stages of leaching. For a comprehensive summary of all the release mechanisms, including the total cumulative release, the reader is referred to the table in the Supporting Information section (Table S6), which includes a detailed overview of the data.

3.4.2. Human Health Risk Assessment

The results obtained from the monolith leaching tests were employed to assess the potential impact on water resources associated with the use of the tiles. This evaluation was carried out through a risk assessment analysis, considering the two utilization scenarios described in section "Utilization Scenarios Considered for the Risk Assessment". The first scenario (Scenario 0) involves the use of the porcelain stoneware tiles in direct contact with the groundwater, while the second one (Scenario 1) involves using the material for paving a large outdoor area. As mentioned, the size of St. Peter's Square in Vatican City was considered. To perform the risk assessment, the maximum concentrations obtained from the eight leaching intervals from the monolith tests for each target contaminant (Supporting Information, Table S4) were used as the starting point. This approach ensures a precautionary evaluation by considering the highest concentrations of leached substances measured, which serve as the basis for estimating the potential impact on groundwater quality. By considering the leaching factor (see section "Human Health Risk Calculation") resulting for each scenario, the Predicted Environmental Concentrations (PEC) in the groundwater were determined.

In Figure 7a,b, the risks for the groundwater (R_{GW}) resulting from each constituent are reported for the two scenarios. These were calculated as the ratios between the PEC and the PNEC values for each relevant constituent. For more details on these parameters and their determination, refer to section "Human Health Risk Calculation". For the constituents whose maximum leaching concentration from the monolith test was below the respective LOQ, the risk was estimated by assuming the concentration to be equal to the LOQ value.

The results show that, in both scenarios and for both types of tiles, the calculated risks remained consistently below the acceptability threshold of 1 (i.e., PEC < PNEC). This suggests that the leached concentrations would not pose a significant concern to groundwater quality under any of the considered conditions. In addition, the SP-BA formulation showed risk levels comparable to those resulting for the traditional SP tiles. This confirms that the addition of bottom ash would not increase the environmental risk and can be considered safe from the perspective of its leaching behavior during use and also in terms of its long-term impacts on groundwater.

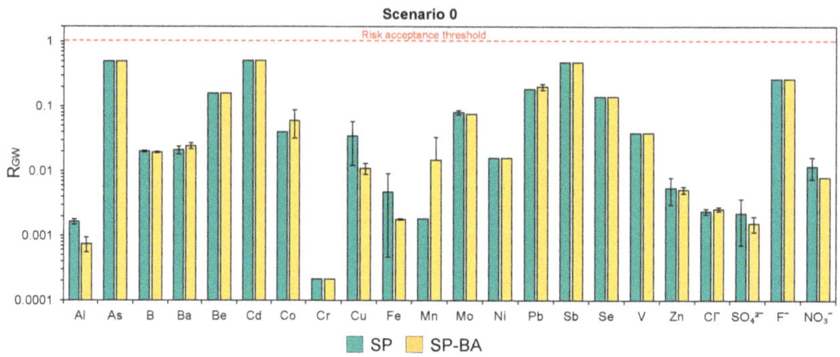

(a)

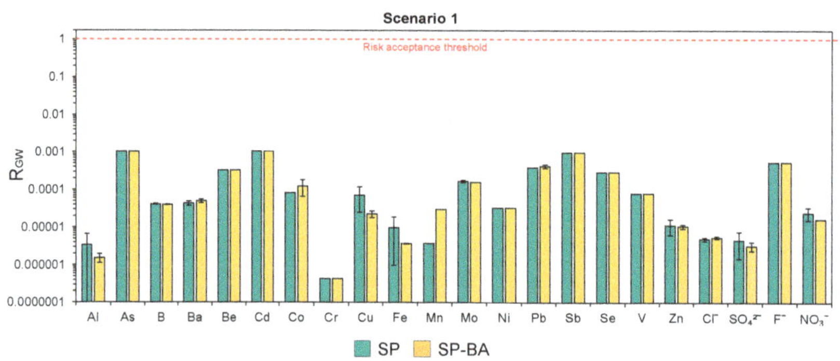

(b)

Figure 7. (**a**). Risk assessment results for Scenario 0, porcelain stoneware tiles in direct contact with the groundwater. (**b**). Risk assessment results for Scenario 1, porcelain stoneware tiles used for the outdoor pavement of an area the size of St. Peter's Square in Vatican City. Error bars indicate the standard deviation.

Notably, even in Scenario 0 (an extreme, unrealistic case for which higher risks are expected due to direct contact with the groundwater and the absence of natural attenuation processes), the risk for all elements remained below the acceptable limit. In some cases, the difference was of several orders of magnitude (e.g., Al, Cr, Ni, Zn, chlorides, and sulfates). This result is particularly relevant, as the absence of a significant risk even under these unrealistic assumptions demonstrates a strong safety margin, supporting the feasibility of the use of the BA-based tiles without restrictions. Hence, in this case, the product could be considered as a viable End-of-Waste product suitable for free use from an environmental perspective.

Additionally, a statistical analysis was conducted to globally compare the risk levels obtained for the SP and SP-BA samples, considering all of the involved analytes. This comparison was made by determining the 5th, 25th, 50th, 75th, and 95th percentiles of all the calculated risks for the water resource. The results are illustrated in Figure 8. To carry out this analysis, a box plot was created, offering an immediate visual summary of the dispersion of the values in the dataset.

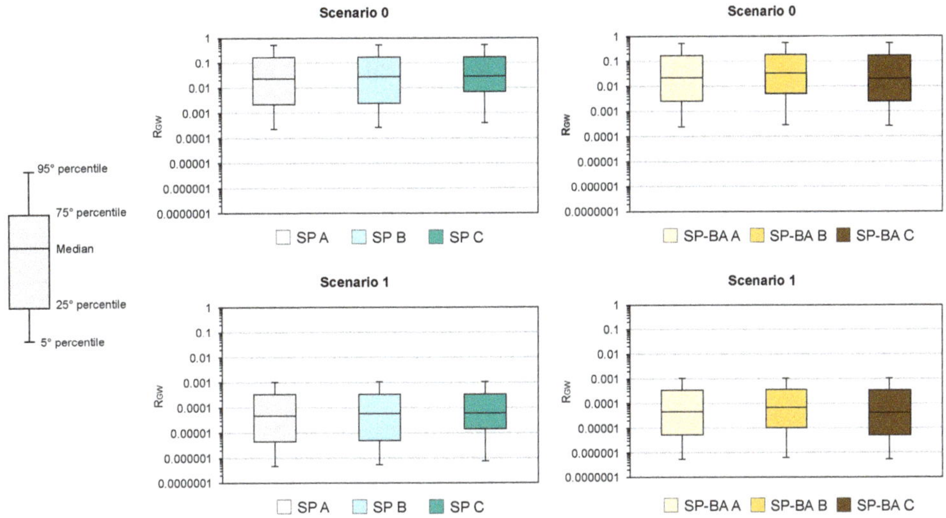

Figure 8. Statistical analysis of the R_{GW} results for samples SP and SP-BA in Scenario 0 and Scenario 1.

The statistical analysis showed that the calculated risks were very similar among the three replicates in both scenarios. Specifically, referring to Scenario 0, it can be seen that 50% of the data showed a risk lower by one to two orders of magnitude than the acceptance threshold, both for the traditional samples (SP) and for the samples containing incineration residues (SP-BA). The results suggest that there would be no significant differences in terms of risks posed by both types of porcelain stoneware tiles, indicating that the tested materials were comparable in terms of their environmental impact and safety for use. For Scenario 1, 50% of the data resulted within four to five orders of magnitude below the limit for both materials. Furthermore, the 95th percentile consistently yielded values at least three orders of magnitude below the risk threshold of 1.

The overall distribution of the risk values indicates a low level of risk, with only a very small fraction of the data approaching the threshold for both types of samples. This suggests that the addition of BA in the tile samples did not significantly affect the leaching behavior of the material and hence did not lead to an increased human health risk related to its use in outdoor pavements.

4. Conclusions

This study aimed to evaluate the environmental viability of incorporating bottom ash (BA) as a raw material for porcelain stoneware tile manufacturing. The primary objective was to assess whether the addition of BA would negatively impact the chemical properties or the environmental behavior of the tiles. By comparing the properties of BA-based tiles to the traditional porcelain formulation, we sought to determine whether this waste material could serve as a sustainable alternative material in tile production while maintaining compliance with environmental standards. Furthermore, this work aimed to fill a gap in the existing literature by conducting monolith leaching tests on ceramic materials containing BA, an area that has been scarcely explored to date.

In the initial phase, we focused on characterizing both materials through elemental composition analysis to determine whether the addition of 30% BA by weight in the tile formulation would significantly alter their chemical properties of the product. This analysis demonstrated that the incorporation of BA did not result in substantial changes in the elemental content of the tiles.

Subsequently, a standardized batch compliance leaching test was performed on crushed tile samples to assess the material's compliance with the Italian End-of-Waste (EoW) criteria for construction and demolition waste (Legislative Decree 27 September 2022). This test was performed to assess if at the end-of-life stage of the tiles they could be used as recycled aggregates in construction from an environmental compliance point of view. The results showed that the BA-containing ceramics met the EoW thresholds, indicating their suitability for use also in granular form.

Following these assessments, more detailed evaluations were carried out to assess potential risks during the use phase of the product. In particular, the monolith leaching test on intact tiles was conducted to evaluate their behavior under direct exposure conditions in scenarios such as the use for outdoor paving. The results were then integrated into a human health risk assessment methodology to quantify the potential impacts resulting under assumed exposure scenarios. The comparison between the two tile types also allowed us to identify differences in release mechanisms. Although a single mechanism was not always clearly identifiable, BA-containing tiles showed a reduced early surface availability, leading to a shift toward depletion-controlled leaching. These tiles also presented a higher early release of some elements, likely due to increased diffusion at the beginning of the leaching process. Despite these variations in release behavior, the contaminant concentrations measured in the leachate led to no significant differences in risk between the two types of tile formulations, which, under all the conditions assumed, remained below the risk acceptance threshold of 1. These findings provide robust evidence supporting the environmental compatibility and safety of the free use of the analyzed BA-based tiles.

In conclusion, the results of this study demonstrate that the integration of bottom ash from Waste-to-Energy processes into the porcelain stoneware tiles examined in this study represents an environmentally viable solution. The addition of BA showed not to negatively impact the chemical stability nor the environmental performance of the tiles. The results highlight the potential of BA to be employed as a secondary raw material in ceramics manufacturing, contributing to circular economy strategies and reducing the need for waste disposal.

Supplementary Materials: The following supporting information can be downloaded at: https://www.mdpi.com/article/10.3390/ceramics8030116/s1. Figure S1: Reference framework for the SAM and LDF calculation; Figure S2: Leached percentage of each element, calculated as the ratio between the maximum leached concentration from the monolith test and the total content for SP and SP-BA samples; Table S1: Leaching solution renewal schedule; Table S2: Limits of quantification (LOQ) for the analyzed elements; Table S3: Input values for Scenario 1. For Scenario 0, dilution and attenuation factors do not apply, as no reduction in concentration is assumed (LF = 1); Table S4: The maximum concentration obtained from the monolith test for the target contaminants. The LOQ values are shown in grey; Table S5: Limit concentrations considered for the groundwater (PNEC values); Table S6: Release mechanisms resulting from the results of the monolith test for traditional tiles (SP) and tiles containing bottom ash (SP-BA).

Author Contributions: Conceptualization, G.C., I.V., F.L., C.M. and S.M.; methodology, G.C. and I.V.; software, I.V.; formal analysis, L.A.; investigation, L.A.; data curation, L.A.; writing—original draft preparation, L.A. and G.C.; writing—review and editing, I.V.; supervision, F.L., C.M. and S.M.; resources, C.M. and S.M.; project administration, G.C. and F.L.; funding acquisition, G.C. and F.L. All authors have read and agreed to the published version of the manuscript.

Funding: This research was partially funded by A2A Ambiente S.p.A.

Data Availability Statement: The original contributions presented in this study are included in the article and Supplementary Material. Further inquiries can be directed to the corresponding author.

Acknowledgments: The authors gratefully acknowledge A2A Ambiente S.p.A. and the Italian Ministry of University and Research (MUR) for funding the industrial PhD scholarship of Luigi Acampora through the PON program (Ministerial Decree No. 1061/2021, dated 10 August 2021), supported by the FSE REACT-EU initiative. The authors wish also to thank Alessia Falsetti for her assistance in supervising all chemical analyses and Sara D'Onofrio who participated in the experimental activities as part of her Master's degree thesis.

Conflicts of Interest: Claudia Mensi and Simone Malvezzi were employed by A2A Ambiente S.p.A. A2A Ambiente S.p.A. provided funding assistant for this research. The remaining authors declare that the research was conducted in the absence of any commercial or financial relationships that could be construed as a potential conflict of interest.

References

1. Sabbas, T.; Polettini, A.; Pomi, R.; Astrup, T.; Hjelmar, O.; Mostbauer, P.; Cappai, G.; Magel, G.; Salhofer, S.; Speiser, C.; et al. Management of Municipal Solid Waste Incineration Residues. *Waste Manag.* **2003**, *23*, 61–88. [CrossRef] [PubMed]
2. CEWEP Bottom Ash Fact Sheet. 2019. Available online: https://www.cewep.eu/wp-content/uploads/2017/09/FINAL-Bottom-Ash-factsheet.pdf (accessed on 16 August 2025).
3. Chimenos, J.M.; Segarra, M.; Fernández, M.A.; Espiell, F. Characterization of the Bottom Ash in Municipal Solid Waste Incinerator. *J. Hazard. Mater.* **1999**, *64*, 211–222. [CrossRef]
4. Blasenbauer, D.; Huber, F.; Lederer, J.; Quina, M.J.; Blanc-Biscarat, D.; Bogush, A.; Bontempi, E.; Blondeau, J.; Chimenos, J.M.; Dahlbo, H.; et al. Legal Situation and Current Practice of Waste Incineration Bottom Ash Utilisation in Europe. *Waste Manag.* **2020**, *102*, 868–883. [CrossRef] [PubMed]
5. Cho, B.H.; Nam, B.H.; An, J.; Youn, H. Municipal Solid Waste Incineration (MSWI) Ashes as Construction Materials—A Review. *Materials* **2020**, *13*, 3143. [CrossRef]
6. Verbinnen, B.; Billen, P.; Van Caneghem, J.; Vandecasteele, C. Recycling of MSWI Bottom Ash: A Review of Chemical Barriers, Engineering Applications and Treatment Technologies. *Waste Biomass Valor* **2017**, *8*, 1453–1466. [CrossRef]
7. Hjelmar, O.; Holm, J.; Crillesen, K. Utilisation of MSWI Bottom Ash as Sub-Base in Road Construction: First Results from a Large-Scale Test Site. *J. Hazard. Mater.* **2007**, *139*, 471–480. [CrossRef]
8. Lynn, C.J.; Ghataora, G.S.; Dhir Obe, R.K. Municipal Incinerated Bottom Ash (MIBA) Characteristics and Potential for Use in Road Pavements. *Int. J. Pavement Res. Technol.* **2017**, *10*, 185–201. [CrossRef]
9. Yin, K.; Ahamed, A.; Lisak, G. Environmental Perspectives of Recycling Various Combustion Ashes in Cement Production—A Review. *Waste Manag.* **2018**, *78*, 401–416. [CrossRef]
10. Silva, R.V.; De Brito, J.; Lynn, C.J.; Dhir, R.K. Use of Municipal Solid Waste Incineration Bottom Ashes in Alkali-Activated Materials, Ceramics and Granular Applications: A Review. *Waste Manag.* **2017**, *68*, 207–220. [CrossRef]
11. Lindberg, D.; Molin, C.; Hupa, M. Thermal Treatment of Solid Residues from WtE Units: A Review. *Waste Manag.* **2015**, *37*, 82–94. [CrossRef]
12. Chandler, A.J.; Eighmy, T.T.; Hjelmar, O.; Kosson, D.S.; Sawell, S.E.; Vehlow, J.; Van der Sloot, H.A.; Hartlén, J. *Municipal Solid Waste Incinerator Residues*; Elsevier: Amsterdam, The Netherlands, 1997; Volume 67, ISBN 0-08-053718-9.
13. Lam, C.H.K.; Ip, A.W.M.; Barford, J.P.; McKay, G. Use of Incineration MSW Ash: A Review. *Sustainability* **2010**, *2*, 1943–1968. [CrossRef]
14. Dou, X.; Ren, F.; Nguyen, M.Q.; Ahamed, A.; Yin, K.; Chan, W.P.; Chang, V.W.-C. Review of MSWI Bottom Ash Utilization from Perspectives of Collective Characterization, Treatment and Existing Application. *Renew. Sustain. Energy Rev.* **2017**, *79*, 24–38. [CrossRef]
15. Yeo, R.J.; Sng, A.; Wang, C.; Tao, L.; Zhu, Q.; Bu, J. Strategies for Heavy Metals Immobilization in Municipal Solid Waste Incineration Bottom Ash: A Critical Review. *Rev. Environ. Sci. Biotechnol.* **2024**, *23*, 503–568. [CrossRef]
16. Bethanis, S.; Cheeseman, C.R.; Sollars, C.J. Effect of Sintering Temperature on the Properties and Leaching of Incinerator Bottom Ash. *Waste Manag. Res.* **2004**, *22*, 255–264. [CrossRef] [PubMed]
17. Cheeseman, C.R.; Makinde, A.; Bethanis, S. Properties of Lightweight Aggregate Produced by Rapid Sintering of Incinerator Bottom Ash. *Resour. Conserv. Recycl.* **2005**, *43*, 147–162. [CrossRef]
18. Rambaldi, E.; Esposito, L.; Andreola, F.; Barbieri, L.; Lancellotti, I.; Vassura, I. The Recycling of MSWI Bottom Ash in Silicate Based Ceramic. *Ceram. Int.* **2010**, *36*, 2469–2476. [CrossRef]
19. Conte, S.; Buonamico, D.; Magni, T.; Arletti, R.; Dondi, M.; Guarini, G.; Zanelli, C. Recycling of Bottom Ash from Biomass Combustion in Porcelain Stoneware Tiles: Effects on Technological Properties, Phase Evolution and Microstructure. *J. Eur. Ceram. Soc.* **2022**, *42*, 5153–5163. [CrossRef]

20. Schabbach, L.M.; Andreola, F.; Barbieri, L.; Lancellotti, I.; Karamanova, E.; Ranguelov, B.; Karamanov, A. Post-Treated Incinerator Bottom Ash as Alternative Raw Material for Ceramic Manufacturing. *J. Eur. Ceram. Soc.* **2012**, *32*, 2843–2852. [CrossRef]
21. Lin, D.-F.; Wang, W.-J.; Chen, C.-W.; Lin, K.-L. Applying Mixture of Municipal Incinerator Bottom Ash and Sewage Sludge Ash for Ceramic Tile Manufacturing. *Materials* **2021**, *14*, 3863. [CrossRef]
22. Dagnew, E.H. Alternative Resource of Incineration Bottom Ash for Ceramic Tile Production. *Int. J. Ceram. Eng. Sci.* **2022**, *4*, 281–285. [CrossRef]
23. Appendino, P.; Ferraris, M.; Matekovits, I.; Salvo, M. Production of Glass–Ceramic Bodies from the Bottom Ashes of Municipal Solid Waste Incinerators. *J. Eur. Ceram. Soc.* **2004**, *24*, 803–810. [CrossRef]
24. Costa, G.; Acampora, L.; Spagnuolo, R.; Piacenti, L.; Lombardi, F. Evaluating by LCA the Environmental Impact Savings Related to the Use of Waste Incineration Bottom Ash in Porcelain Tiles Manufacturing: The Role of Metals Recovery. In Proceedings of the WasteLCA_3: Life Cycle Sustainability Assessment for Waste Management and Resource Optimization, Calabria, Italy, 5–10 June 2022.
25. Barberio, G.; Buttol, P.; Masoni, P.; Scalbi, S.; Andreola, F.; Barbieri, L.; Lancellotti, I. Use of Incinerator Bottom Ash for Frit Production. *J. Ind. Ecol.* **2010**, *14*, 200–216. [CrossRef]
26. Cheeseman, C.R.; Monteiro Da Rocha, S.; Sollars, C.; Bethanis, S.; Boccaccini, A.R. Ceramic Processing of Incinerator Bottom Ash. *Waste Manag.* **2003**, *23*, 907–916. [CrossRef] [PubMed]
27. Eighmy, T.T.; Crimi, D.; Whitehead, I.E.; Zhang, X.; Gress, D.L. The Influence of Monolith Physical Properties and Integrity on Diffusional Leaching Behavior of Asphaltic Pavements Constructed with MSW Combustion Bottom Ash. In *Studies in Environmental Science*; Elsevier: Amsterdam, The Netherlands, 1997; Volume 71, pp. 125–147, ISBN 978-0-444-82771-5.
28. Gonzalez, M.L.; Blanc, D. Technical and Environmental Assessment of MSWI Bottom-Ash as Secondary Raw Material in Road Construction. In Proceedings of the CIGOS 2013-Construction et Développement durable, Lyon, France, 4–5 April 2013.
29. Roessler, J.G.; Townsend, T.G.; Ferraro, C.C. Use of Leaching Tests to Quantify Trace Element Release from Waste to Energy Bottom Ash Amended Pavements. *J. Hazard. Mater.* **2015**, *300*, 830–837. [CrossRef]
30. Taoufiq, L. Concrete Made with Treated Bottom Ash: Mechanical and Environmental Study. *J. Mater. Cycles Waste Manag.* **2021**, *23*, 1102–1113. [CrossRef]
31. Clavier, K.A.; Watts, B.; Liu, Y.; Ferraro, C.C.; Townsend, T.G. Risk and Performance Assessment of Cement Made Using Municipal Solid Waste Incinerator Bottom Ash as a Cement Kiln Feed. *Resour. Conserv. Recycl.* **2019**, *146*, 270–279. [CrossRef]
32. Lindahl, P.; Olszewski, G.; Eriksson, M. Total Dissolution of Environmental Samples for the Determination of Uranium and Thorium by ICP-MS and Alpha-Particle Spectrometry. *Appl. Radiat. Isot.* **2022**, *181*, 110103. [CrossRef]
33. Awaji, S.; Nakamura, K.; Nozaki, T.; Kato, Y. A Simple Method for Precise Determination of 23 Trace Elements in Granitic Rocks by ICP-MS after Lithium Tetraborate Fusion. *Resour. Geol.* **2006**, *56*, 471–478. [CrossRef]
34. Decree No. 152/2022, Italian End-of-Waste (EoW) Criteria for the Use of Construction and Demolition Waste as Recycled Aggregates. Available online: https://www.gazzettaufficiale.it/eli/id/2022/10/20/22G00163/SG (accessed on 17 August 2025).
35. *EN 12457-2:2002*; Characterisation of Waste-Leaching-Compliance Test for Leaching of Granular Waste Materials and Sludges-Part 2: One Stage Batch Test at a Liquid to Solid Ratio of 10 l/Kg for Materials with Particle Size Below 4 Mm (Without or with Size Reduction). CEN: Brussels, Belgium, 2002. Available online: https://standards.cencenelec.eu/dyn/www/f?p=CEN:110:0::::FSP_PROJECT,FSP_ORG_ID:14487,2046877&cs=1E17B76E9588A8F99D58DF6F2C1B14C6A (accessed on 17 August 2025).
36. *EN 15863:2015*; Characterization of Waste-Leaching Behaviour Test for Basic Characterization-Dynamic Monolithic Leaching Test with Periodic Leachant Renewal, Under Fixed Conditions. CEN: Brussels, Belgium, 2015. Available online: https://standards.cencenelec.eu/dyn/www/f?p=CEN:110:0::::FSP_PROJECT,FSP_ORG_ID:39449,2046877&cs=17A7E761EC879812AAE6ED700EA7A1A86 (accessed on 17 August 2025).
37. Torras, J.; Buj, I.; Rovira, M.; De Pablo, J. Semi-Dynamic Leaching Tests of Nickel Containing Wastes Stabilized/Solidified with Magnesium Potassium Phosphate Cements. *J. Hazard. Mater.* **2011**, *186*, 1954–1960. [CrossRef]
38. de Groot, G.; van der Sloot, H. Determination of Leaching Characteristics of Waste Materials Leading to Environmental Product Certification. In *Stabilization and Solidification of Hazardous, Radioactive, and Mixed Wastes: 2nd Volume*; Gilliam, T., Wiles, C., Eds.; ASTM International: West Conshohocken, PA, USA, 1992; ISBN 978-0-8031-1443-2.
39. *NEN 7375:2004*; Leaching Characteristics of Moulded or Monolithic Building and Waste Materials—Determination of Leaching of Inorganic Components with the Diffusion Test—Solid Sample. Nederlands Normalisatie-Instituut (NEN): Delft, The Netherlands, 2004.
40. Yang, Y.; Yang, Y.; Wang, Q.; Huang, Q. Release of Heavy Metals from Concrete Made with Cement from Cement Kiln Co-Processing of Hazardous Wastes in Pavement Scenarios. *Environ. Eng. Sci.* **2010**, *28*, 35–42. [CrossRef]
41. Van Der Sloot, H.A.; Van Zomeren, A.; Meeussen, J.C.L.; Seignette, P.; Bleijerveld, R. Test Method Selection, Validation against Field Data, and Predictive Modelling for Impact Evaluation of Stabilised Waste Disposal. *J. Hazard. Mater.* **2007**, *141*, 354–369. [CrossRef] [PubMed]

42. Štulović, M.; Ivšić-Bajčeta, D.; Ristić, M.; Kamberović, Ž.; Korać, M.; Anđić, Z. Leaching Properties of Secondary Lead Slag Stabilized/Solidified with Cement and Selected Additives. *Environ. Protect. Eng.* **2013**, *39*, 149–163. [CrossRef]
43. Iacobescu, R.I.; Cappuyns, V.; Geens, T.; Kriskova, L.; Onisei, S.; Jones, P.T.; Pontikes, Y. The Influence of Curing Conditions on the Mechanical Properties and Leaching of Inorganic Polymers Made of Fayalitic Slag. *Front. Chem. Sci. Eng.* **2017**, *11*, 317–327. [CrossRef]
44. Van der Sloot, H.A.; Dijkstra, J.J. *Development of Horizontally Standardized Leaching Tests for Construction Materials: A Material Based or Release Based Approach? Identical Leaching Mechanisms for Different Materials*; ECN: Petten, The Netherlands, 2004.
45. Birgisdóttir, H.; Pihl, K.A.; Bhander, G.; Hauschild, M.Z.; Christensen, T.H. Environmental Assessment of Roads Constructed with and without Bottom Ash from Municipal Solid Waste Incineration. *Transp. Res. Part D Transp. Environ.* **2006**, *11*, 358–368. [CrossRef]
46. Olsson, S.; Kärrman, E.; Gustafsson, J.P. Environmental Systems Analysis of the Use of Bottom Ash from Incineration of Municipal Waste for Road Construction. *Resour. Conserv. Recycl.* **2006**, *48*, 26–40. [CrossRef]
47. Shih, H.; Ma, H. Assessing the Health Risk of Reuse of Bottom Ash in Road Paving. *Chemosphere* **2011**, *82*, 1556–1562. [CrossRef] [PubMed]
48. *E2081-00*; ASTM Standard Guide for Risk-Based Corrective Action. American Society for Testing and Materials: West Conshohocken, PA, USA, 2000.
49. Connor, J.A.; Bowers, R.L.; Paquette, S.M.; Newell, C.J. *Soil Attenuation Model for Derivation of Risk-Based Soil Remediation Standards*; Groundwater Services Inc.: Houston, TX, USA, 1997.
50. ISPRA. Criteri Metodologici per L'applicazione Dell'analisi Assoluta di Rischio ai Siti Contaminati (Revisione 2, Marzo 2008). Available online: https://www.isprambiente.gov.it/files/temi/siti-contaminati-02marzo08.pdf (accessed on 19 April 2025).
51. Verginelli, I. Risk-Net v 3.2-User Guide. Available online: https://www.reconnet.net/Docs/Risk-net%20ver%203_2/User%20guide%20Risk-net%203.2%20Pro.pdf (accessed on 19 April 2025).
52. Grim, R.E. Clay Mineralogy: The Clay Mineral Composition of Soils and Clays Is Providing an Understanding of Their Properties. *Science* **1962**, *135*, 890–898. [CrossRef] [PubMed]
53. EUROFEL–European Association of Feldspar Producers (Member of IMA-Europe), What Is Feldspar? Available online: https://ima-europe.eu/wp-content/uploads/2020/06/Feldspar_An-WEB-2011.pdf (accessed on 16 April 2025).

ceramics

MDPI

Article

Life Cycle Assessment of Industrial Glass Cullet Recycling Process Based on Alkaline Activation

Elena Battiston [1], Francesco Carollo [2], Giulia Tameni [2], Enrico Bernardo [2] and Anna Mazzi [1,*]

[1] SAM.lab (Sustainability Assessment and Management Laboratory), Department of Industrial Engineering, University of Padova, 35131 Padova, Italy; elena.battiston@phd.unipd.it

[2] CERAM Glass (Advanced Ceramic and Glass Manufacturing), Department of Industrial Engineering, University of Padova, 35131 Padova, Italy; francesco.carollo@unipd.it (F.C.); giulia.tameni@phd.unipd.it (G.T.); enrico.bernardo@unipd.it (E.B.)

* Correspondence: anna.mazzi@unipd.it; Tel.: +39-049-8271611

Abstract

To mitigate the issue of accumulating glass waste, an advanced process has been developed for the production of glass foams via alkaline activation, employing industrial glass cullet as the primary raw material. This method contributes to circular economy strategies by enabling high-value upcycling of secondary raw materials. The aim of the study is to conduct an environmental assessment of this recycling process using the Life Cycle Assessment (LCA). The analysis is performed with SimaPro software, adopting the ReCiPe impact assessment method, which allows for the quantification of 18 impact categories. Four distinct foaming processes were compared to determine the most environmentally preferable option and a sensitivity analysis was conducted to assess how variations in energy sources influence the environmental performance. The findings indicate that the scenario involving hardening at 40 °C for seven days results in the highest environmental burdens. Specifically, in the Human Carcinogenic Toxicity category, the normalized impacts for this process are approximately an order of magnitude greater. Electricity consumption is identified as the primary contributor to the overall impact. The sensitivity analysis underscores that utilizing photovoltaic panels reduces impacts. Future developments will focus on expanding the system boundaries to provide a more comprehensive understanding and supporting informed decision-making.

Keywords: life cycle assessment; industrial glass waste; sustainability; alkaline activation; glass foams

1. Introduction

Recycling glass from segregated urban waste is highly beneficial, as it leads to substantial reductions in raw material usage and energy consumption during the melting process (with pre-formed glass facilitating the reaction of mineral materials) [1]. However, "ideal" recycling, where glass cullet is fully reused in the production of the same glass products (referred to as "closed-loop recycling"), is not practically achievable with discarded fractions landfilled up to an amount of 200 million tons/year worldwide [2,3].

A central issue in glass recycling lies in maintaining the quality of the final recycled product. The effectiveness of recycled glass depends largely on the purity of the cullet [4,5]. Crushed glass from municipal waste collection often requires an expensive and complex sorting process to separate pieces by color and remove impurities such as metal, plastic, or ceramics before it can be used as a viable alternative to raw minerals. Glass fractions

 https://doi.org/10.3390/ceramics8030109

containing these impurities are typically diverted to landfills [6,7]. In some cases, even glass with minimal contamination may be discarded, especially when original glass products are no longer being manufactured, as seen with glass from dismantled cathode ray tubes (such as those from outdated TV and computer screens) [8–10].

Further complexity arises in recycling specialized glass compositions such as boro-alumino-silicate or borosilicate glass used in pharmaceutical containers. These products are typically formed through precise thermo-mechanical processes and are not easily reintegrated into standard glass production streams [11]. For decades, the pharmaceutical industry has relied on borosilicate glass vials due to their exceptional chemical stability and, in principle, infinite recyclability [12–14]. However, these glasses are synthesized from high-purity raw materials and shaped into preforms (e.g., tubes, rods) in specialized facilities [15], making closed-loop recycling practically unfeasible. These challenges have driven interest in open-loop recycling, where discarded glass is repurposed into different products rather than remelted into the original ones [8]. While not preserving the same material function, open-loop recycling offers important environmental advantages such as reducing landfill volumes and conserving natural resources and can lead to highly sustainable alternatives, especially when revenues from the new products outweigh transformation costs. When this happens, open-loop recycling can be accurately described as upcycling [16].

Glass foams (or cellular glasses) are valued for their unique combination of properties, including high strength-to-density ratios, large surface areas, high permeability, low specific heat, and excellent thermal and acoustic insulation [9,17]. They also boast superior chemical and thermal resistance compared to polymeric foams, making them more suitable for thermal insulation (glass foams are non-flammable and flame-resistant) [18]. The value of these materials can be further enhanced by optimizing the manufacturing process. While the energy-intensive melting process can be avoided by utilizing viscous flow sintering of glass powders at significantly lower temperatures, the cost and environmental impact of foaming agents remain controversial [18,19].

A promising approach to improving glass foams involves separating the foaming and sintering steps, according to gel casting technology [20]. This method can be applied to solutions derived from sol-gel processes, as well as suspensions of glass powders [21]. In this process, air bubbles are incorporated into the mixture through intensive mechanical stirring, assisted by surfactants, to form a homogeneous cellular structure. This structure is stabilized first by the progressive hardening (gelation) of the slurry and then by the sintering process. Unlike traditional glass foams, gel-cast foams can retain the open-celled structure created during foaming, which is crucial for high-value applications like sorbents for organic dyes, recognized as key polluting agents of wastewater [13,22]. Recent studies have focused on adapting gel casting technology to produce foams from recycled glass [8]. Instead of using costly organic polymerization methods (which involve adding monomers, cross-linkers, and catalysts), gelation of glass slurries can be achieved through simple alkali activation. Rincon et al. demonstrated that soda-lime glass slurries in alkaline media (such as KOH aqueous solutions) exhibit distinct pseudoplastic behavior due to the formation of calcium silicate hydrate (C-S-H) gel on the surface of the glass particles [6]. At low shear rates, this gel helps bond the particles together, trapping air bubbles introduced through high shear-rate stirring at low temperatures. Sintering can then take place at 700–800 °C, much lower than the temperatures used for traditional soda-lime glass foams produced with foaming agents [23].

The 'inorganic gel casting' process has also been successfully applied to other types of glass, including the production of glass-ceramic foams [23]. Specifically, alkali activation of glass powders may be followed by sinter-crystallization, which combines viscous flow sintering with crystallization. The precipitation of crystals increases the apparent viscosity of the glass during sintering, helping to "freeze" the open-celled structure formed at lower temperatures.

Foams based on gel casting actually employ some organics. They are not vital for gelation but, in the form of surfactants, support the incorporation of air bubble during intensive mechanical stirring of alkali-activated glass suspensions. These organic compounds may be avoided with an improved hardening mechanism. As reported by Tameni et al. [12], activated glass powders may undergo gelation even independently from the formation of binding compounds (such as C-S-H gels). Adjacent glass particles may 'weld' with each other by condensation reactions involving Si-OH, Al-OH and B-OH groups at the surface. The establishment of strong chemical bonds (e.g., Si-O-Si), bridging adjacent particles, results in insoluble porous monoliths (e.g., resisting to the attack of boiling water), even in the absence of any viscous flow sintering treatment after drying. Porous monoliths, yielded by organic-free glass suspensions, may turn into foams through the inclusion of compounds (e.g., perborates) releasing gas during drying (in a condition of 'direct foaming'). Surfactants may help in homogenizing the pore distribution but in much lower quantity than in the case of foams from gel casting.

The study of the alkaline activation (AA) process to produce foams from recycled cullet glass cannot ignore the environmental impacts assessment. Before developing this process, even on an industrial scale, it is important to evaluate the performance not only from a technological but also an environmental point of view [24]. The Life Cycle Assessment (LCA) methodology was developed to measure the impacts of a good and/or service [25]. Starting from the input-output analysis, the quantification of impacts is obtained for each phase of a product's life cycle and for each phase of a process. Bjørn et al. summarized the strengths of LCA into four main characteristics: it adopts a life cycle perspective, it covers a wide range of environmental issues, it is a quantitative analysis, and it is based on science [26]. LCA is widely used as an ex-post evaluation method across all sectors, providing a standardized approach to assess the environmental impacts of products, processes, and services after their development and implementation [27,28].

The methodology is also applied for scaling up an emerging technology and for comparing its performance and potential with that of the more established incumbent technology [29–32]. Many studies present the advantages and opportunities that derive from the application of the LCA analysis already in the design phase [33]. First of all, the possibility of optimizing processes and supporting the strategic decision-making process [34–36]. After that, making resources and consumption of the process more efficient and reducing costs. Furthermore, the LCA study can be applied to meet regulatory and certification requirements [37–39]. In the construction sector, several products have already been discussed in the literature, accompanied by an LCA study [40–42]. Comparative LCA studies on alkali-activated glass foam processes are currently lacking in the literature [43]. Indeed, most existing works focus on a single type of process and do not offer a systematic comparison between alternative foaming methods. This deficiency limits understanding of relative environmental performances across different process configurations and highlights the novelty and importance of the present study in filling this gap with robust comparative data.

Driven by these motivations, this study aims to quantify the environmental impacts of the four innovative alkaline activation processes studied at the University of Padova. The objective is to understand which is the most environmentally preferable and how environmental performance can be improved before upscaling. The main research questions may be summarized as follows:

- What are the environmental impacts of the four alkaline activation processes?
- How much do the environmental impacts change when using a different energy source?

The analysis of the environmental impact associated with the alkaline activation processes is conducted through the LCA methodology. The Section 2 describes this approach and presents the objective and scope of the study as well as the data collection. The Results paragraph contains the environmental impact assessment, and further insights are presented through sensitivity analysis. The comparison between the results obtained for four processing variants demonstrates that it is possible to determine the environmental impacts and intervene to reduce them before implementing the industrial-scale process. In the Discussion, ample space is dedicated to the limitations and future opportunities of the research.

2. Materials and Methods

The LCA methodology is employed to quantify the environmental impacts associated with the glass foam production process currently under investigation at the laboratory scale. The analysis is carried out in accordance with general standards ISO 14040 and ISO 14044, as well as EN 15804:2012, which is specific to the construction sector [25,44,45]. The methodology consists of four phases: Goal and Scope Definition, Life Cycle Inventory (LCI), Life Cycle Impact Assessment (LCIA), and Interpretation of Results. Conducting an LCA study is an iterative process, where the initial scope is often refined or revised as new information emerges during inventory, impact assessment, and interpretation phases [46].

2.1. Goal and Scope Definition

The aim of the LCA study is to measure the environmental impacts associated with four alkaline activation processes of industrial waste glass cullet for the production of glass foam. The study is aimed at both researchers exploring processes to improve the performance of glass foams and industry companies interested in optimizing the efficiency of their production processes. The selected functional unit is 1 kg of glass foam produced through these processes. The intended function is the use of the product as an insulating material in the construction sector. Using the same functional unit, defining the same function and adopting the same system boundaries is essential to enable a meaningful comparison of the results. Also, using the same data sources and the same impact assessment method makes the studies comparable. The evaluation must be fair to ensure the reliability of the results and the reproducibility of the analysis. The assumptions in Table 1 ensure that the differences arising from the study result from actual impact differences and not from methodological or data discrepancies. The data related to material and energy consumption were obtained through direct measurements and/or technical datasheets of the machinery and equipment actually used in the process. Regarding the type of energy, it reflects the electricity mix fed into the Italian market, as provided by official national sources and consistent with the Ecoinvent database.

Table 1. Assumptions related to the production of 1 kg of glass foam.

Elements	Process	Assumptions
Functional unit	A, B, C, D	1 kg of glass foam produced
Function	A, B, C, D	Insulating material in construction sector
System boundaries	A, B, C, D	Gate-to-gate approach
Data sources	A, B, C, D	Primary data, Technical datasheet or Ecoinvent
Impact assessment method	A, B, C, D	ReCiPe 2016 v1.1 midpoint Egalitarian

2.1.1. Description of the Alkali Activation Process

The primary raw material employed in all the foams' formulations is boro-alumino-silicate glass, referred as BASG. Boro-alumino-silicate glass exhibits a specific chemical composition which is detailed in Table 2 [47]. The BASG utilized in the present study was kindly sourced by Stevanato Group (Piombino Dese, Padova, Italy) in the form of vials. These glass vials were subjected to milling (Pulverisette 6, Fritsch GmbH, Idar-Oberstein, Germany) and subsequently sieved to achieve a particle size less than 75 μm (IRIS FTL-0300, Filtra Vibraction, Badalona, Spain). The alkali activation was performed by suspending BASG powder in a blended solution of sodium hydroxide (NaOH, Sigma-Aldrich, Gillingham, UK) and potassium hydroxide (KOH, Sigma-Aldrich, Gillingham, UK), specifically at 2.5 mol/L solution (50:50 mol%). This methodology is consistent with previous studies in this field, ensuring effective activation of the glass substrate [13,48].

Table 2. Chemical composition of boro-alumino-silicate glass.

	SiO_2 [wt%]	B_2O_3 [wt%]	Al_2O_3 [wt%]	CaO [wt%]	Na_2O [wt%]	K_2O [wt%]	BaO [wt%]
BASG	72	12	7	1	6	2	<0.1

The liquid-to-solid ratio for the preparation of samples is fixed at 0.5. The precursors are mixed for 3 h at a rotational speed of 500 rpm using a mechanical stirrer (Argolab AM20-D, Modena, Italy). This mixing is designed to ensure homogeneity in the resulting slurry, thereby facilitating optimal reactions during the subsequent processing stage.

The preparation of alkali-activated foams is distinguished by the methodologies employed in the foaming process as well as the associated curing strategies. These curing strategies may vary in terms of temperature, duration, and environment conditions, all of which are critical determinants of the stability and performance of the resulting foams.

In the present study, four distinct industrial glass cullet recycling process based on alkaline activation are considered:

A. Following the protocol outlined by Rincon et al. [23], after the AA process, the resulting slurry is transferred into closed polystyrene molds. The samples are subsequently dried in an oven at 75 °C for 2 h using a Bio-tech oven (Biomedica Elettronica, Padova, Italy) to facilitate preliminary gelation. Afterward, Triton X-100 (polyoxyethylene octyl phenyl ether—$C_{14}H_{22}O(C_2H_4O)_n$, n = 9–10, Sigma-Aldrich, Gillingham, UK) was incorporated to the suspension at a concentration of 4 wt%. Aeration was achieved through vigorous mechanical mixing at 2000 rpm for 10 min. The resultant foams were removed from the molds after an additional drying phase at 75 °C for 24 h, after which they underwent thermal firing at 700 °C for 1 h (BSF-laboratory chamber furnace, Elite Thermal Systems Limited, Leicestershire, UK), with a controlled heating rate of 10 °C/min.
B. Direct foaming experiments are carried out by introducing sodium perborate ($NaBO_3{\cdot}H_2O$, Sigma Aldrich, Schnelldorf, Germany) and sodium dodecyl sulphate (SDS, $CH_3(CH_2)_{11}OSO_3Na$, Sigma Aldrich, Schnelldorf, Germany) into a mixture at a concentration of 1 wt% relative to the mass of the glass powder. Homogenization was achieved via mechanical mixing for 10 min at 2000 rpm. The drying phase was extended to 24 h at 40 °C, followed by thermal treatment at 600 °C for 1 h [13].
C. The experimental conditions and methodologies are comparable to those outlined in process B. The type and quantity of foaming agent employed remained consistent. Notably, no additional sintering process was applied during this phase, as the hardening process occurs just upon drying at nearly room temperature (40 °C), for seven days [12], in a condition of 'cold consolidation'.
D. To mitigate the time-consuming nature of the hardening phase, the integration of microwave technology is explored. After the direct foaming expressed as in process B, the specimens are left for 1 day and later subjected to microwave heating (microwave

oven MJ3965BPS, LG, Samsung MS23F300EEK, Samsung Electronics Italia S.p.A, Milano, Italy) for 5 min at 400 W [49].

2.1.2. System Boundaries of Processes

In line with the study's objective, the system boundaries of the processes include only the glass foam production phases (Figure 1). This "gate-to-gate" approach is consistent for comparative studies with a focus on a specific stage of the life cycle [50–52]. The use phase and the end-of-life (EoL) of the product are excluded. The underlying assumption is that all the glass foams produced exhibit the same chemical and physical performance. It is assumed that the use phase and disposal occur in the same manner for all the cases, resulting in equivalent environmental impacts. Therefore, when comparing the four processes, these stages would not introduce significant differences in the results, making it reasonable to exclude them from the system boundaries.

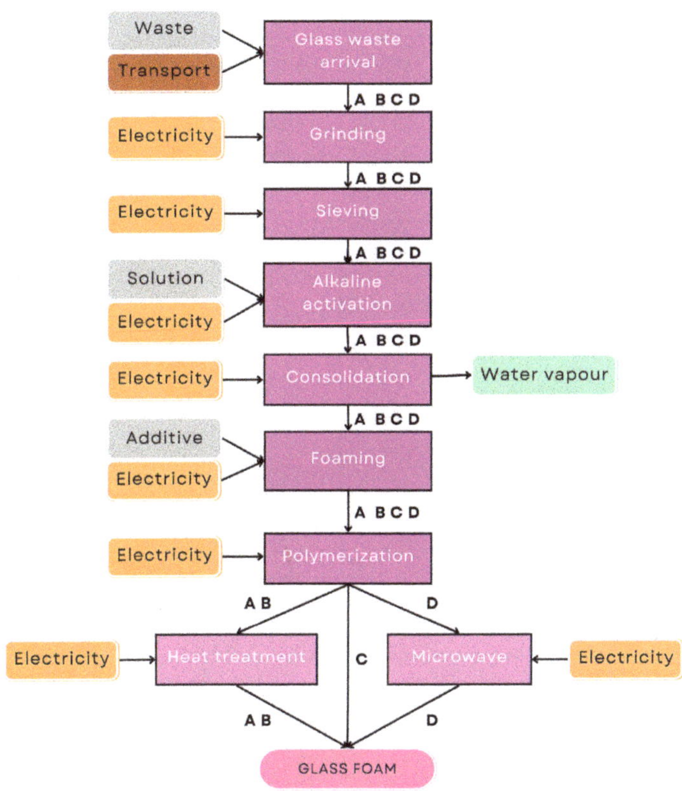

Figure 1. System boundaries of the processes A, B, C, and D.

For each phase of the process under consideration, it is essential to identify the resource flows entering and exiting the process. Inputs may include materials (such as raw materials, semi-finished goods, and auxiliary materials) or natural resources (air, water, energy). Outputs, on the other hand, typically include emissions (to air, water, or soil), waste, and by-products. In this specific case, the main resources used are electricity for machinery and transports, and materials (glass waste, alkaline solution and additives), while the output consists of water vapor generated by the heating processes. It is assumed that water vapor is released only during the consolidation phase of the process.

2.2. Life Cycle Inventory

During the inventory phase, an input-output analysis is conducted for each phase of the alkaline activation processes. The quantification of inputs and outputs must relate to the functional unit and must always be balanced. Primary data is preferred and other information can be sought from the literature or from databases available in the software, which are constantly updated. In this case, the data is collected with the support of the technicians who are involved in the process implementation and laboratory tests. The characterization of the materials used and their quantities is ensured through direct measurements of the process. Missing information is supplemented with data from technical sheets of the equipment and data available in the Ecoinvent 3.10 database. Energy is modelled based on the Italian electricity mix of 2024 (Electricity, low voltage (IT) market for, APOS, System). The energy consumption value for the processes was calculated based on the rated power of the machine, as specified in its technical datasheet, and the actual operating time required to complete the process. Transportation is assumed to be carried out using a heavy commercial vehicle (Transport, Freight, lorry > 32 metric ton, EURO5, Cut-off, System). Table 3 presents the input collected for all four processes. The model is simplified by applying a cut-off criterion of 1% of the product mass. Materials below this threshold are excluded from the system, as their contribution is considered negligible or not relevant. It is assumed that no material loss occurs between phases and that from 1 kg of industrial glass waste, 1 kg of glass foam is produced. According to experimental data, the alkaline solution and additives are in the same quantity as the steam released during the process. While additional materials such as activators (NaOH, KOH, surfactants) are added, their mass is compensated by the evaporation of water during the process, resulting in an approximately equal final mass of glass foam. The only material leaving the process is therefore assumed to be water vapor.

Table 3. Input of the processes related to the production of 1 kg of glass foam.

Step	Process	Input
1-arrival of waste	A, B, C, D	Boro-alumino-silicate glass; 1 kg
	A, B, C, D	Road transport; 30 km
2-grinding	A, B, C, D	Italian energy mix; 0.75 kWh
3-sieving	A, B, C, D	Italian energy mix; 0.10 kWh
4-alkaline activation	A, B, C, D	Italian energy mix; 0.15 kWh
	A, B, C, D	NaOH and KOH; 0.5 kg
5-consolidation	A, B, C, D	Italian energy mix, 1.60 kWh
6-foaming	A, B, C, D	Italian energy mix, 0.008 kWh
	A,	$C_{14}H_{22}O(C_2H_4O)n$, n = 9–10; 0.04 kg
	B, C, D	$NaBO_3 \cdot H_2O$ and $CH_3(CH_2)_{11}OSO_3Na$; 0.01 kg
7-polymerization	A, B, D	Italian energy mix; 19.2 kWh
	C	Italian energy mix; 134.4 kWh
8-heat treatment	A, B	Italian energy mix; 1.50 kWh
9-microwave	D	Italian energy mix; 0.037 kWh

2.3. Life Cycle Impact Assessment and Interpretation

The environmental impact assessment is conducted using the Simapro software with the available version (SimaPro 10.1.0.4). The ReCiPe 2016 v1.1 midpoint Egalitarian perspective allows a multicriteria assessment. This method is one of the most widely applied in the literature [53,54]. The Egalitarian approach in SimaPro is rigorous and precautionary as it considers a longer time horizon. For this reason, it is often used in contexts that go beyond GWP analysis alone, such as for multi-criteria environmental assessments, as in this case. The mid-point visualization provides greater scientific robustness and lower uncertainty than endpoint models. In fact, the mid-point approach provides an evaluation for 18 impact categories: global warming (GWP), stratospheric ozone depletion (SOD), ionizing radiation (IR), ozone formation—human health (OFHH), fine particulate matter (FPMF), ozone formation—terrestrial ecosystem (OFTE), terrestrial acidification (TA), freshwater eutrophication (FE), marine eutrophication (ME), terrestrial ecotoxicity (TET), freshwater ecotoxicity (FET), marine ecotoxicity (MET), human toxicity—carcinogenic (HCT), human toxicity—non-carcinogenic (HNCT), land use (LU), mineral resource scarcity (MRS), fossil resource scarcity (FRS), and water consumption (WC). Choosing the Egalitarian (E) approach over the Hierarchist (H) and Individualist (I) approach means achieving results that emphasize long-term environmental responsibility.

3. Results and Discussions

The results are presented as the environmental profile of each process, a comparison of the environmental profiles of the four processes, and a sensitivity analysis. The latter is carried out using a different energy mix as the source for electrical equipment.

3.1. Life Cycle Inventory Results

The first result of the LCA conducted in SimaPro provides a tree or network representation that allows for the identification of input flows of the analyzed process. Figure 2 shows the tree diagram of Process A. In Figure 2, the acronyms a, b, c, d define the processes to which the phases are attributed. Eight subprocesses are identified, but grinding and sieving are not shown because only the main contributions in terms of quantity (above 0.5%) are highlighted. Each step contributes a different percentage to the total value, and the polymerization step is the dominant contributor, representing 83.04%. The diagram clearly shows that energy-intensive operations, particularly polymerization, are the primary hotspots in the production phase.

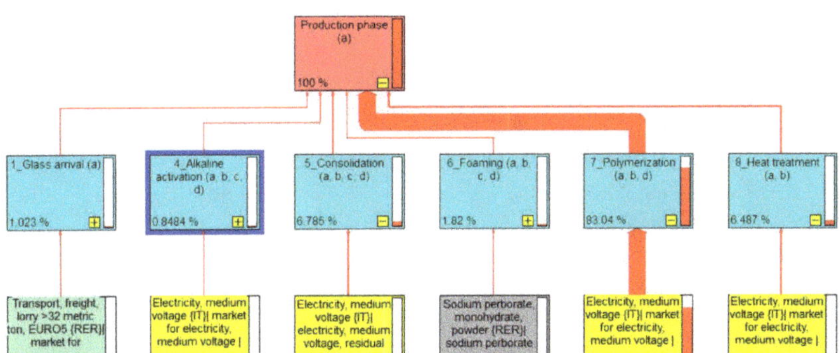

Figure 2. Tree diagram representing the input flows of the process A.

3.2. Life Cycle Impact Assessment Results

The results of the environmental impact assessment of the four alkaline activation processes are reported in Figure 3: Figure 3a represents the results in absolute terms while Figure 3b represents the results in normalized form. Looking at the characterization, process C has the highest impacts for all 18 impact categories. The impact assessment shows that the impacts are due to the use of energy rather than to the type of input and output materials. The processes with the highest energy consumption are also those with the highest impacts. The results show that the process in which hardening was performed into an oven for seven days (process C) is clearly the worst. Its impacts are significant in all the categories analyzed in the ReCipe:2016 method. The other three cases have a similar environmental profile, and the process that involves the use of the microwave (process D) is confirmed as the best.

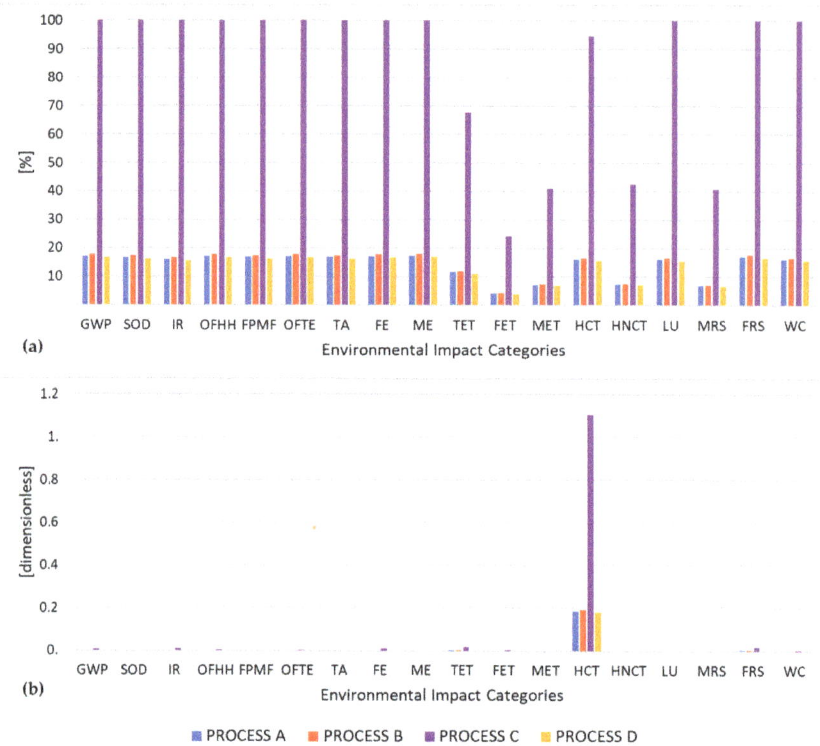

Figure 3. (**a**) Graphical results of the LCIA of four alkaline activation processes. (**b**) Graphical results of the LCIA of four alkaline activation processes in normalized form.

Since the units of measurement for impact categories vary widely (e.g., kg CO_2, kg SO_2 eq, m^3 of water, etc.), normalization converts all results into dimensionless values, making them comparable to one another. This representation helps identify the most significant impact categories for a given system. Normalized results highlight that the Human Carcinogenic Eco-toxicity category is the most relevant. This fact depends on the large use of energy which mostly affects this category.

Other recent investigations have shown that it is possible to reduce the environmental impact of the process by eliminating the sintering phase. In this regard, cold consolidation upon weak alkali activation of glass powders allows foams with competitive mechanical properties to be obtained at nearly room temperature [13]. Drying treatment of unfired glass foams can be greatly reduced by microwave heating, saving energy and leading to more chemically stable products [49].

3.3. Sensitivity Analysis

A sensitivity analysis is relevant when scientists are studying the performance of processes, in particular before up-scaling. It is used to evaluate how impacts change if a variable is changed. The choice to change the energy source is justified by the fact that this is the component that most influences the results. The starting hypothesis is that solar energy can be used to operate all the equipment through the use of already-installed photovoltaic (PV) panels. Therefore, the electricity from the Italian energy mix (Electricity, low voltage [IT], market for, APOS, System) is replaced with renewable energy from photovoltaic panels (Electricity, low voltage [IT], electricity production, photovoltaic, 3 kWp, slanted-roof installation, multi-Si panels, mounted, Cut-off, System). Furthermore, it is a plausible change in the case of building an industrial-scale plant.

Table 4 shows the comparison between the environmental performances of the four processes, initially using the Italian energy mix (Mix) and then energy from photovoltaic panels (PV). The use of energy from photovoltaics is preferable for most impact categories. The only impact categories that are exceptions are MRS, HNCT, MET, FET, and TET. The first measures the effects related to the availability of mineral resources, while the others measure the toxic impact of chemicals. This result is justified by the use of silicon-based photovoltaic panels. Therefore, these categories are more influenced by the chemical substance and material flows within the system. Additionally, they are particularly linked to long-term impacts that are difficult to reverse. The result is thus explained by the use of polycrystalline silicon photovoltaic panels. Even with the use of solar energy, case C proves to be worse for all the impact categories compared to case A, B, and D.

Table 4. Results of the sensitivity analysis of the processes.

	Process A		Process B		Process C		Process D	
	Mix	PV	Mix	PV	Mix	PV	Mix	PV
GWP	17.2719	3.5211	17.824	3.5843	100	20.3544	16.767	3.3686
SOD	16.776	2.6563	17.332	2.6887	100	15.3279	16.2686	2.5261
IR	16.044	2.3608	16.6579	2.4366	100	13.8092	15.5858	2.2903
OFHH	17.2646	6.178	17.7544	6.2728	100	35.6697	16.6965	5.8947
FPMF	16.8444	9.4965	17.3918	9.7759	100	56.5377	16.3292	9.1744
OFTE	17.263	6.027	17.7529	6.1165	100	34.784	16.695	5.7477
TA	16.8444	7.6399	17.4114	7.8709	100	45.5077	16.349	7.3868
FE	17.1324	14.9438	17.7158	15.4624	100	89.0582	16.6574	14.5157
ME	17.4473	11.9418	18.0169	12.183	100	68.4856	16.9623	11.5959
TET	11.7015	16.837	11.8814	17.218	67.768	100	11.1626	16.1533
FET	4.1011	16.4889	4.2329	17.0897	24.4053	100	3.9735	16.0233
MET	7.1044	16.6557	7.2897	17.2074	41.0687	100	6.8552	16.1425
HCT	16.11	16.8113	16.5735	17.3087	94.7238	100	15.5683	16.245
HNCT	7.397	16.6655	7.5917	17.2169	42.6434	100	7.1409	16.1521
LU	16.3363	5.1039	16.7882	5.113	100	29.0885	15.7179	4.8046
MRS	6.9983	16.6622	7.2039	17.2343	40.9521	100	6.7698	16.1697
FRS	17.26	2.7176	17.8134	2.7526	100	15.5826	16.7562	2.5875
WC	16.2135	6.2732	16.828	6.497	100	37.3995	15.7582	6.0996

4. Conclusions

This study examines the environmental profile of industrial glass cullet recycling using the alkaline activation process. Through an LCA, the environmental performance of different process was assessed, with the aim of identifying critical phases and opportunities for impact reduction. The findings highlight that the most environmentally burdensome scenario is the one in which hardening was carried out in an oven for seven days. This result is consistent with expectations, as processes with higher energy demands generally correspond to greater environmental impacts. In particular, categories such as climate change, fossil fuel depletion, and particulate matter formation are directly influenced by the intensity of energy consumption during production phases. A significant reduction in overall environmental impacts can be observed when renewable energy sources are used in place of conventional ones. However, this positive effect is not uniform across all impact categories. In fact, impact categories related to mineral resource depletion and human/ecotoxicological effects tend to worsen in scenarios involving the use of polycrystalline silicon photovoltaic panels. Although renewables reduce emissions during use, their production phase involves intensive extraction and processing of raw materials, which affects the overall balance in those specific categories. This observation underlines a common issue in LCA called burden-shifting. While improvements in one area (e.g., lower greenhouse gas emissions) may be achieved, they can sometimes lead to unintended trade-offs in others, especially when alternative technologies introduce new forms of resource consumption or pollution.

The main limitation of the study relates to the availability of primary data, particularly regarding energy use during certain processing steps. While the technical datasheets provide a useful baseline, they are associated with a degree of uncertainty, which should be addressed in future research.

Author Contributions: Conceptualization, E.B. (Enrico Bernardo) and A.M.; methodology, E.B. (Elena Battiston) and A.M.; software, E.B. (Elena Battiston) and A.M.; validation, A.M., formal analysis, A.M.; investigation, F.C. and G.T.; resources, F.C., G.T., and E.B. (Enrico Bernardo); data curation, E.B. (Elena Battiston), F.C., and G.T.; writing—original draft preparation, E.B. (Elena Battiston), F.C., and G.T.; writing—review and editing, E.B. (Enrico Bernardo) and A.M.; visualization, E.B. (Elena Battiston), F.C., and G.T.; supervision, E.B. (Enrico Bernardo) and A.M.; project administration, E.B. (Enrico Bernardo); funding acquisition, E.B. (Enrico Bernardo). All authors have read and agreed to the published version of the manuscript.

Funding: This research was funded by the Italian Ministry of University and Research (MUR) in the framework of the GLASS_Trea.S.U.Res (GLASS-based TREAtments for Sustainable Upcycling of inorganic RESidues) project (PRIN 2022 PNRR project #P2022S4TK2).

Institutional Review Board Statement: Not applicable.

Informed Consent Statement: Not applicable.

Data Availability Statement: Data will be made available on request.

Acknowledgments: The authors thank Stevanato Group (Piombino Dese, Italy) for supplying BASG glass.

Conflicts of Interest: The authors declare no conflicts of interest. Stevanato Group (Piombino Dese, Italy) provided BASG glass. This company and funders had no role in the design of the study; in the collection, analyses, or interpretation of data; in the writing of the manuscript; or in the decision to publish the results.

Abbreviations

The following abbreviations are used in this manuscript:

AA	Alkaline Activation
BASG	Boro-alumino-silicate glass
CE	Circular Economy
E	Egalitarian
EoL	End of Life
FPMF	Fine Particulate Matter Formation
FRS	Fossil Resource Scarcity
FET	Freshwater Eco-Toxicity
FE	Freshwater Eutrophication
GWP	Global warming potential
H	Hierarchist
HCT	Human Carcinogenic toxicity

HNCT	Human Non-Carcinogenic Toxicity
I	Individualist
IR	Ionizing Radiation
LU	Land Use
LCA	Life Cycle Assessment
LCI	Life Cycle Inventory
LCIA	Life Cycle Impact Assessment
MET	Marine Eco-Toxicity
ME	Marine Eutrophication
MRS	Mineral Resource Scarcity
OFHH	Ozone Formation, Human Health
OFTE	Ozone Formation, Terrestrial Ecosystems
PV	Photovoltaic
SOD	Stratospheric Ozone Depletion
TA	Terrestrial Acidification
TE	Terrestrial Eco-Toxicity
WC	Water Consumption

References

1. Curran, M.A. Life Cycle Impact Assessment. In *Life Cycle Assessment Handbook: A Guide for Environmentally Sustainable Products*; Curran, M.A., Ed.; Wiley: Hoboken, NJ, USA, 2012; Chapter 16. [CrossRef]
2. Adekomaya, O.; Majozi, T. Mitigating environmental impact of waste glass materials: Review of the existing reclamation options and future outlook. *Environ. Sci. Pollut. Res.* **2021**, *28*, 10488–10502. [CrossRef] [PubMed]
3. Fu, C.; Liang, J.; Yang, G.; Dagestani, A.A.; Liu, W.; Luo, X.; Zeng, B.; Wu, H.; Huang, M.; Lin, L.; et al. Recycling of waste glass as raw materials for the preparation of self-cleaning, light-weight and high-strength porous ceramics. *J. Clean. Prod.* **2021**, *317*, 128395. [CrossRef]
4. Bonifazi, G.; Serranti, S. Imaging spectroscopy based strategies for ceramic glass contaminants removal in glass recycling. *Waste Manag.* **2006**, *26*, 627–639. [CrossRef]
5. Farcomeni, A.; Serranti, S.; Bonifazi, G. Non-parametric analysis of infrared spectra for recognition of glass and glass ceramic fragments in recycling plants. *Waste Manag.* **2008**, *28*, 557–564. [CrossRef]
6. Rincón, A.; Giacomello, G.; Pasetto, M.; Bernardo, E. Novel 'inorganic gel casting' process for the manufacturing of glass foams. *J. Eur. Ceram. Soc.* **2017**, *37*, 2227–2234. [CrossRef]
7. Delbari, S.A.; Hof, L.A. Glass waste circular economy—Advancing to high-value glass sheets recovery using industry 4.0 and 5.0 technologies. *J. Clean. Prod.* **2024**, *462*, 142629. [CrossRef]
8. Rincón, A.; Marangoni, M.; Cetin, S.; Bernardo, E. Recycling of inorganic waste in monolithic and cellular glass-based materials for structural and functional applications. *J. Chem. Technol. Biotechnol.* **2016**, *91*, 1946–1961. [CrossRef]
9. Singh, N.; Li, J.; Zeng, X. Solutions and challenges in recycling waste cathode-ray tubes. *J. Clean. Prod.* **2016**, *133*, 188–200. [CrossRef]
10. Baek, C.R.; Kim, H.D.; Jang, Y.-C. Exploring glass recycling: Trends, technologies, and future trajectories. *Environ. Eng. Res.* **2025**, *30*, 240241. [CrossRef]
11. Guadagnino, E.; Guglielmi, M.; Nicoletti, F. Glass: The best material for pharmaceutical packaging. *Int. J. Appl. Glass Sci.* **2022**, *13*, 281–291. [CrossRef]
12. Tameni, G.; Lago, D.; Kaňková, H.; Buňová, L.; Kraxner, J.; Galusek, D.; Dawson, D.M.; Ashbrook, S.E.; Bernardo, E. Alkaline attack of boro-alumino-silicate glass: New insights of the molecular mechanism of cold consolidation and new applications. *Open Ceram.* **2025**, *21*, 100726. [CrossRef]
13. Mehta, A.; Karbouche, K.; Kraxner, J.; Elsayed, H.; Galusek, D.; Bernardo, E. Upcycling of Pharmaceutical Glass into Highly Porous Ceramics: From Foams to Membranes. *Materials* **2022**, *15*, 3784. [CrossRef] [PubMed]
14. Bristogianni, T.; Oikonomopoulou, F. Glass up-casting: A review on the current challenges in glass recycling and a novel approach for recycling 'as-is' glass waste into volumetric glass components. *Glass Struct. Eng.* **2023**, *8*, 255–302. [CrossRef]
15. Iacocca, R.G.; Allgeier, M. Corrosive attack of glass by a pharmaceutical compound. *J. Mater. Sci.* **2007**, *42*, 801–811. [CrossRef]
16. Singh, P.J.; Sung, K.; Cooper, T.; West, K.; Mont, O. Challenges and opportunities for scaling up upcycling businesses—The case of textile and wood upcycling businesses in the UK. *Resour. Conserv. Recycl.* **2019**, *150*, 104439. [CrossRef]
17. Shi, X.; Liao, Q.; Liu, L.; Deng, F.; Chen, F.; Wang, F.; Zhu, H.; Zhang, L.; Liu, C. Utilizing multi-solid waste to prepare and characterize foam glass ceramics. *Ceram. Int.* **2023**, *49*, 35534–35543. [CrossRef]

18. Zhang, J.; Liu, B.; Zhang, S. A review of glass ceramic foams prepared from solid wastes: Processing, heavy-metal solidification and volatilization, applications. *Sci. Total Environ.* **2021**, *781*, 146727. [CrossRef]
19. Meylan, G.; Ami, H.; Spoerri, A. Transitions of municipal solid waste management. Part II: Hybrid life cycle assessment of Swiss glass-packaging disposal. *Resour. Conserv. Recycl.* **2014**, *86*, 16–27. [CrossRef]
20. Sepúlveda, P.; Jones, J.R.; Hench, L.L. In vitro dissolution of melt-derived 45S5 and sol-gel derived 58S bioactive glasses. *J. Biomed. Mater. Res.* **2002**, *61*, 301–311. [CrossRef]
21. Wu, Z.Y.; Hill, R.G.; Yue, S.; Nightingale, D.; Lee, P.D.; Jones, J.R. Melt-derived bioactive glass scaffolds produced by a gel-cast foaming technique. *Acta Biomater.* **2011**, *7*, 1807–1816. [CrossRef]
22. Bruzzoniti, M.C.; Appendini, M.; Rivoira, L.; Onida, B.; Del Bubba, M.; Jana, P.; Soraru, G.D. Polymer-derived ceramic aerogels as sorbent materials for the removal of organic dyes from aqueous solutions. *J. Am. Ceram. Soc.* **2018**, *101*, 821–830. [CrossRef]
23. Romero, A.R.; Tamburini, S.; Taveri, G.; Toušek, J.; Dlouhy, I.; Bernardo, E. Extension of the "Inorganic Gel Casting" Process to the Manufacturing of Boro-Alumino-Silicate Glass Foams. *Materials* **2018**, *11*, 2545. [CrossRef]
24. Riondet, L.; Rio, M.; Perrot-Bernardet, V.; Zwolinski, P. Emerging technologies upscaling: A framework for matching LCA practices with upscaling archetypes. *Sustain. Prod. Consum.* **2024**, *50*, 347–363. [CrossRef]
25. *ISO 14044*; Environmental Management—Life Cycle Assessment—Requirements and Guidelines. International Organization for Standardization: Geneva, Switzerland, 2006.
26. Bjørn, A.; Owsianiak, M.; Molin, C.; Laurent, A. Main characteristics of LCA. In *Life Cycle Assessment*; Hauschild, M., Rosenbaum, R., Olsen, S., Eds.; Springer: Cham, Switzerland, 2018. [CrossRef]
27. Hetherington, A.C.; Borrion, A.L.; Griffiths, O.G.; Panoutsou, N.L.; Webb, J.S.; Kendrick, A.M. Use of LCA as a development tool within early research: Challenges and issues across different sectors. *Int. J. Life Cycle Assess.* **2014**, *19*, 130–143. [CrossRef]
28. Liu, M.; Zhu, G.; Tian, Y. The historical evolution and research trends of life cycle assessment. *Green Carbon* **2024**, *2*, 425–437. [CrossRef]
29. Cucurachi, S.; van der Giesen, C.; Guinée, J. Ex-ante LCA of emerging technologies. *Procedia CIRP* **2018**, *69*, 463–468. [CrossRef]
30. Tan, L.; Mandley, S.J.; Peijnenburg, W.; der Loop, S.L.W.-V.; Giesen, D.; Legradi, J.B.; Shen, L. Combining ex-ante LCA and EHS screening to assist green design: A case study of cellulose nanocrystal foam. *J. Clean. Prod.* **2018**, *178*, 494–506. [CrossRef]
31. Georgiades, M.; Kilcan, C.Ö.; Giels, M.; Stewart, D.I.; Zhan, A.; Muñoz, J.F.; Vandepitte, N.; Dewulf, S. Ex-ante life cycle assessment of bauxite residue vitrification technology. *Int. J. Life Cycle Assess.* **2025**, *30*, 1896–1911. [CrossRef]
32. Bergerson, A.; Brandt, A.; Cresko, J.; Carbajales-Dale, M.; MacLean, H.L.; Matthews, H.S.; McCoy, S.; McManus, M.; Miller, S.A.; Morrow, W.R.; et al. Life cycle assessment of emerging technologies: Evaluation techniques at different stages of market and technical maturity. *J. Ind. Ecol.* **2020**, *24*, 11–25. [CrossRef]
33. Moni, S.M.; Mahmud, R.; High, K.; Carbajales-Dale, M. Life cycle assessment of emerging technologies: A review. *J. Ind. Ecol.* **2020**, *24*, 52–63. [CrossRef]
34. Subal, L.; Braunschweig, A.; Hellweg, S. The relevance of life cycle assessment to decision-making in companies and public authorities. *J. Clean. Prod.* **2024**, *435*, 140520. [CrossRef]
35. Barbhuiya, S.; Das, B.B. Life Cycle Assessment of construction materials: Methodologies, applications and future directions for sustainable decision-making. *Case Stud. Constr. Mater.* **2023**, *19*, e02326. [CrossRef]
36. Pryshlakivsky, J.; Searcy, C. Life Cycle Assessment as a decision-making tool: Practitioner and managerial considerations. *J. Clean. Prod.* **2021**, *309*, 127344. [CrossRef]
37. European Commission. Recommendation on the Use of Common Methods to Measure and Communicate the Life Cycle Environmental Performance of Products and Organisations. (2013/179/EU). Available online: https://eur-lex.europa.eu/legal-content/EN/TXT/?uri=CELEX%3A32013H0179 (accessed on 30 July 2025).
38. Directive 2009/125/EC of the European Parliament and of the Council of 21 October 2009 Establishing a Framework for the Setting of Eco-Design Requirements for Energy-Related Products. Available online: https://eur-lex.europa.eu/legal-content/EN/TXT/?uri=CELEX:32009L0125 (accessed on 30 July 2025).
39. *ISO 14025:2006*; Environmental Labels and Declarations—Type III Environmental Declarations—Principles and Procedures. International Organization for Standardization: Geneva, Switzerland, 2006.
40. Farahzadi, L.; Tellnes, L.G.F.; Shafei, B.; Kioumarsi, M. Life-cycle environmental assessment of ultra-high-performance concrete with sustainable materials and fiber substitutions. *Clean. Eng. Technol.* **2024**, *23*, 100846. [CrossRef]
41. Mazzi, A.; Sciarrone, M.; Bernardo, E. Environmental performance of glass foam as insulation material from waste glass with the alkali activation process. *Heliyon* **2023**, *9*, 19001. [CrossRef]
42. Rodrigues, C.; König, J.; Freire, F. Prospective life cycle assessment of a novel building system with improved foam glass incorporating high recycled content. *Sustain. Prod. Consum.* **2023**, *36*, 161–170. [CrossRef]
43. Nodehi, M.; Aguayo, F.; Madey, N.; Zhou, L. A comparative review of polymer, bacterial-based, and alkali-activated (also geopolymer) binders: Production, mechanical, durability, and environmental impacts life cycle assessment (LCA). *Constr. Build. Mater.* **2024**, *422*, 135816. [CrossRef]

44. *ISO 14040*; Environmental Management—Life Cycle Assessment—Principles and Framework, 2nd ed. International Organization for Standardization: Geneva, Switzerland, 2006.
45. *EN 15804:2012+A2:2019*; Sustainability of Construction Works—Environmental Product Declarations—Core Rules for the Product Category of Construction Products. European Committee for Standardization (CEN): Brussels, Belgium, 2019.
46. European Commission. Joint Research Centre. *Institute for Environment and Sustainability, International Reference Life Cycle Data System (ILCD) Handbook: General Guide for Life Cycle Assessment: Detailed Guidance. LU: Publications Office, 2010.* Available online: https://data.europa.eu/doi/10.2788/38479 (accessed on 15 May 2025).
47. Yue, Y.; Tuheen, M.I.; Du, J. Borosilicate glasses. In *Encyclopedia of Materials: Technical Ceramics and Glasses*; Pomeroy, M., Ed.; Elsevier: Amsterdam, The Netherlands, 2021; pp. 519–539. [CrossRef]
48. Tameni, G.; Cammelli, F.; Elsayed, H.; Stangherlin, F.; Bernardo, E. Upcycling of Boro-Alumino-Silicate Pharmaceutical Glass in Sustainable Construction Materials. *Detritus* **2022**, *20*, 17–21. [CrossRef]
49. Tameni, G.; Carollo, F.; Cavazzini, A.M.; Forzan, M.; Bernardo, E. Microwave assisted cold consolidation of alkali activated suspension of glass waste powders. *Mater. Lett.* **2025**, *389*, 138354. [CrossRef]
50. Zhang, S.; Xu, C.; Xie, R.; Yu, H.; Sun, M.; Li, F. Environmental assessment of fabric wet processing from gate-to-gate perspective: Comparative study of weaving and materials. *Sci. Total Environ.* **2023**, *857*, 159495. [CrossRef]
51. Pražanová, A.; Fridrich, M.; Weinzettel, J.; Knap, V. Gate-to-gate life cycle assessment of lithium-ion battery recycling pre-treatment. *Clean. Environ. Syst.* **2025**, *16*, 100263. [CrossRef]
52. Karadirek, I.E.; Erkaya, O.; Ciggin, A.S. Comparative life cycle assessment of sewage sludge drying by solar and thermal drying technologies. *Waste Manag.* **2025**, *201*, 114826. [CrossRef] [PubMed]
53. Huijbregts, M.A.J.; Steinmann, Z.J.N.; Elshout, P.M.F.; Stam, G.; Verones, F.; Vieira, M.; Zijp, M.; Hollander, A.; van Zelm, R. ReCiPe2016: A harmonised life cycle impact assessment method at midpoint and endpoint level. *Int. J. Life Cycle Assess.* **2017**, *22*, 138–147. [CrossRef]
54. Rybaczewska-Błażejowska, M.; Jezierski, D. Comparison of ReCiPe 2016, ILCD 2011, CML-IA baseline and IMPACT 2002+ LCIA methods: A case study based on the electricity consumption mix in Europe. *Int. J. Life Cycle Assess.* **2024**, *29*, 1799–1817. [CrossRef]

ceramics

MDPI

Article

Transforming Rice Husk Ash into Road Safety: A Sustainable Approach to Glass Microsphere Production

Ingrid Machado Teixeira, Juliano Pase Neto, Acsiel Budny, Luis Enrique Gomez Armas, Chiara Valsecchi and Jacson Weber de Menezes *

Graduate Program in Engineering, Federal University of Pampa, Av. Tiarajú, 810, Alegrete 97546-550, RS, Brazil; ingridmachado.aluno@unipampa.edu.br (I.M.T.); julianopase.aluno@unipampa.edu.br (J.P.N.); acsielbudny.aluno@unipampa.edu.br (A.B.); luisarmas@unipampa.edu.br (L.E.G.A.); chiaravalsecchi@unipampa.edu.br (C.V.)

* Correspondence: jacsonmenezes@unipampa.edu.br

Abstract

Glass microspheres are essential components in horizontal road markings due to their retroreflective properties, enhancing visibility and safety under low-light conditions. Traditionally produced from soda-lime glass made with high-purity silica from sand, their manufacturing raises environmental concerns amid growing global sand scarcity. This study explores the viability of rice husk ash (RHA)—a high-silica byproduct of rice processing—as a sustainable raw material for microsphere fabrication. A glass composition containing 70 wt% SiO_2 was formulated using RHA and melted at 1500 °C. Microspheres were produced through flame spheroidization and characterized following the Brazilian standard NBR 16184:2021 for Type IB beads. The RHA-derived microspheres exhibited high sphericity, appropriate size distribution (63–300 µm), density of 2.42 g/cm^3, and the required acid resistance. UV-Vis analysis confirmed their optical transparency, and the refractive index was measured as 1.55 ± 0.03. Retroreflectivity tests under standardized conditions revealed performance comparable to commercial counterparts. These results demonstrate the technical feasibility of replacing conventional silica with RHA in glass microsphere production, aligning with circular economy principles and promoting sustainable infrastructure. Given Brazil's significant rice production and corresponding RHA availability, this approach offers both environmental and socio-economic benefits for road safety and material innovation.

Keywords: rice husk ash; glass microspheres; sustainable road marking; circular economy in materials

1. Introduction

Horizontal road markings play a critical role in modern transportation systems, serving as visual guides that regulate traffic flow, delineate lanes, and enhance road safety, especially under low-visibility conditions [1,2]. However, the lack of adequate road markings or their premature deterioration is a persistent issue worldwide, contributing to a significant number of road traffic accidents. According to estimates from the World Health Organization (WHO) [3], over 1.1 million people die annually in road traffic crashes, and many of these incidents are exacerbated by insufficient or ineffective road signage and demarcation.

To enhance the visibility and durability of road markings, glass microspheres are incorporated into paints and thermoplastic materials. These microspheres function as retroreflective elements, reflecting the incident light from vehicle headlights directly back to drivers, thereby improving nighttime and low-light visibility driving [4–6]. The optical performance of these microspheres depends largely on their physical and chemical properties, including morphology, refractive index, and composition [6–9].

Soda-lime glass is the predominant material used in their production due to its favorable optical properties and ease of processing [10–12]. However, traditional glass manufacturing relies heavily on high-purity silica sourced from natural sand deposits, a resource under increasing scarcity. Recent studies predict that by 2060, the world may face a significant shortage of suitable sand for use in glass production and construction, due to rapid urbanization and unsustainable extraction practices [13–15]. This scenario underscores the urgent need to identify alternative and renewable sources of silica.

Rice husk, a byproduct of the rice milling industry, represents approximately 20% of the total weight of rice grains [16,17]. When properly treated and burned under controlled conditions, it produces rice husk ashes (RHAs), a residue that can contain up to 99.6 wt% of amorphous silica (SiO_2), which is comparable to high-purity quartz sand (95–99.9 wt% SiO_2) used in glass production [16–19]. Unlike sand, the silica in RHA is mostly amorphous, offering higher reactivity and lower melting temperatures, which are advantageous in glassmaking [18]. On the other hand, it was already demonstrated that RHA industrially calcinated without any further purification treatments was employed successfully in the production of conventional glass [20,21]. Furthermore, RHA is a renewable and low-cost by-product, presenting a sustainable alternative to sand mining and consequently offers itself as a promising raw material for glass synthesis, aligning with the principles of waste valorization and circular economy [18]. Brazil, as one of the largest producers globally—with a harvest of over 10 million metric tons annually—generates a substantial quantity of rice husk and, consequently, RHA [22,23]. Utilizing this abundant biomass residue for high-value applications such as glass microsphere production could place Brazil as a strategic player in the transition towards more sustainable infrastructure materials. Moreover, this approach supports the objectives of the United Nations 2030 Agenda for Sustainable Development, particularly Goals 9 (Industry, Innovation, and Infrastructure), 11 (Sustainable Cities and Communities), and 12 (Responsible Consumption and Production) [24].

In the Brazilian context, the classification and technical standards for glass microspheres are defined by NBR 16184:2021 [11], which includes the specification of Type IB microspheres. This nomenclature is specific to Brazilian regulatory frameworks and may differ from international classifications. Type IB microspheres are typically applied as "premix" materials, incorporated into road marking compounds prior to application, and are required to meet stringent criteria regarding size distribution, morphology, density, chemical resistance, and retroreflective performance. Unlike drop-on microspheres, which provide immediate retroreflectivity, Type IB microspheres become optically active only after partial wear of the paint layer, gradually emerging to maintain reflectivity as the surface coating degrades. This delayed-action mechanism ensures sustained visibility of horizontal road markings throughout their lifecycle, contributing to improved long-term road safety.

In this study, we propose the first study, to the best of our knowledge, to report the synthesis and characterization of Type IB glass microspheres produced in accordance with a technical standard, using glass derived from rice husk ash, an agricultural residue. The goal is to evaluate the technical viability of this novel and alternative silica source by comparing the physical, chemical, and optical properties of the resulting microspheres with those of commercially available products and the standard NBR 16184:2021. Thus, this work contributes to both road safety innovation and the sustainable management of agricultural residues, offering a circular economy solution with positive environmental and socio-economic implications for Brazil and beyond.

2. Materials and Methods

2.1. Glass Fabrication

For the production of glass samples, rice husk ash (RHA) was sourced from a local supplier. The glass batch composition, in weight percentages (wt%), was defined as follows: (70.00) SiO_2:(13.00) Na_2O:(7.00) CaO:(10.00) B_2O_3. Sodium and calcium carbonate, together with Borax (sodium tetraborate decahydrate), were used as raw materials. To achieve the target composition, the oxide content of the RHA—previously characterized by X-ray Fluorescence (XRF) Spectroscopy (Malvern Panalytical, Epsilon 1, São Paulo, Brazil) and CHN analysis—was also taken into account. The raw materials were weighed individually and homogenized using a mortar and pestle. The homogenized mixture was transferred to a platinum–gold crucible and subjected to a melting procedure in a muffle furnace using three heating ramps:

(1) Heating at 10 °C/min to 900 °C, with a 1 h dwell time;
(2) Heating at 7 °C/min to 1200 °C, with a 1 h dwell time;
(3) Heating at 5 °C/min to 1500 °C, held for 3 h.

The dwell times were chosen in order to fully remove bubbles from carbonates and allow for a more homogenized matrix. After melting, the molten glass was poured onto a stainless-steel plate preheated to 480 °C and immediately transferred to an annealing furnace at the same temperature for 1 h to relieve internal stresses. A total of 200 g of RHA-derived glass was produced.

2.2. Glass Characterization

The fabricated glass samples were optically characterized using two distinct techniques. First, their transmittance properties were evaluated by UV-Visible spectrophotometry in the wavelength range of 300 to 750 nm. Prior to this analysis, the glass surfaces were sanded and polished using a metallographic polisher (Fortel, PLF model, São Paulo, Brazil) to ensure suitable optical quality. The refractive index was subsequently determined using the Brewster's angle method [25], which identifies the incidence angle at which reflectance of p-polarized light reaches a minimum. This measurement was carried out using a custom-built experimental setup consisting of a 650 nm TM-polarized laser and a goniometer (Thorlabs, XRNR1, São Carlos, Brazil), also used as a sample holder, with a 0.1° angular resolution.

2.3. Microsphere Production

In order to produce the Type IB microspheres, the glass was crushed with a mortar and pestle, and the resulting fragments were sieved using mesh sizes of 63 µm, 150 µm, 212 µm, and 300 µm to obtain the desired particle size distribution. Fragments retained in the coarser sieves were re-ground and re-sieved until they fell within the specified size range. Particles smaller than 63 µm were stored for the production of new glass. Different manufacturing methods can be implemented for microsphere production [12]: in this work, a lab-made horizontal flame spheroidization method was applied. The apparatus consisted of a horizontal stainless-steel pipe (1485 mm length, 84 mm inner diameter) connected to a vertical stainless-steel tube (1000 mm height, 162 mm inner diameter) via a 90° elbow. Crushed glass fragments were fed through a funnel positioned above the flame. The particles were heated by two atmospheric burners made of galvanized steel, positioned in parallel to increase the particle residence time within the flame. The burners were supplied by a 13 kg LPG cylinder, operating at an outlet pressure of 1 atm. Air intake regulation was achieved through a vertical slit located at the rear of each burner, equipped with an adjustable screw mechanism. This configuration allowed for fine control of the air-to-gas ratio by modifying the slit opening, enabling efficient aspiration of ambient air by the LPG flow and ensuring proper gas mixing. As a result, the flame characteristics could be effectively adjusted to optimize the heating conditions. Upon exposure to the flame (approx. 1000 °C), the fragments softened and spheroidized due to surface tension while flying, and they were collected at the bottom of the vertical tube in an aluminum tray. Figure 1 presents a photograph and a schematic diagram illustrating the details of the manufacturing process employed. The used method proved suitable for the fabrication of microspheres with diameters up to 300 μm.

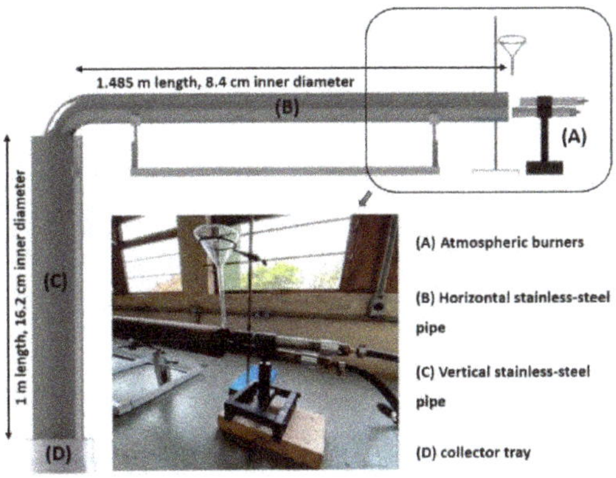

Figure 1. Schematic of the microsphere production apparatus.

2.4. Microsphere Characterization

2.4.1. Chemical Composition and Density

The chemical composition of the microspheres produced from RHA-based glass was analyzed and compared with commercial Type IB (Premix) microspheres, using the Brazilian standard NBR 16184:2021 as reference [11]. Elemental composition was analyzed by XRF spectroscopy. Three replicates of each sample type were analyzed, and the average values were reported. The primary goal of this analysis was to quantify analysis mainly aimed at determining the silica content and the potential toxic elements in the glass composition. The density of the microspheres was measured via the liquid displacement method using isopropyl alcohol as the immersion medium. For each measurement, 5.0 g of microspheres were submerged in 20.0 mL of alcohol, and density was calculated as the ratio of mass to the displaced alcohol volume.

2.4.2. Morphology and Particle Size Distribution

Morphological features and particle size distribution of the microspheres were assessed using Scanning Electron Microscopy (SEM). To evaluate the particle size distribution as well as the morphology of the elements generated, the concept of statistical sample size (n) was applied [26]. Sample size provides the number of observations that must be evaluated in a given population. Here, the population corresponds to the total quantity of microspheres, which can be considered infinite, while the sample size represents the minimum number of microspheres that should be evaluated for statistical significance. To determine a statistically significant sample sizes for the microsphere morphological analysis, a confidence level of 95%, population proportion $\pi = 0.5$, and a 5% margin of error were considered, resulting in a required sample size of at least 385 elements. Consequently, 400 elements were analyzed for each sample type in triplicate, totaling 1200 particles per sample. Microspheres were classified as spherical, ovoid, twin-shaped, or shards. This classification enabled the quantification of production quality and the identification of defects or irregularities associated with the manufacturing process.

2.4.3. Retroreflectivity Measurements

Retroreflectivity measurements were conducted using an Easylux Classic horizontal retroreflectometer (Easylux GmbH, Bocholt, Germany), compliant with national road marking standards [27–29]. Retroreflectivity data were collected for the 15 m geometry, where the vehicle is set to be 15 m away from the road sign observation point. Two test conditions were employed:

(1) Black Dry Surface: Medium-density fiberboard (MDF) panels (34 cm × 10 cm) were painted with matte black acrylic paint to represent a low-reflectivity background; in this case, the microspheres were not anchored in the black paint surface, allowing the determination of the intrinsic retroreflectivity.
(2) White Painted Surface: Water-based white acrylic road marking paint was applied to MDF panels using a polyester foam roller. A uniform thickness (~0.6 mm) was ensured using a metal depth gauge. A fixed quantity of 1.7 g of microspheres (equivalent to 50 g/m^2) was manually deposited onto both test surfaces: (1) matte black (dry) and (2) white painted. To ensure consistent coverage, the application was performed inside a closed chamber, which allowed any loose microspheres to be collected and uniformly reapplied. For the black dry surface, where the microspheres remained unanchored, ten retroreflectivity measurements were taken per sample (RHA-based and commercial microspheres), moving the panel among measurements to account for different configurations for the bead positions. In contrast, for the white painted surface, where the microspheres adhered to the substrate, three retroreflectivity measurements were conducted for each sample. The average and standard deviation of the retroreflectivity measurements values were reported as the final result.

2.4.4. Hydrochloric Acid and Calcium Chloride Corrosion Resistance

The Brazilian standard NBR 16184:2021 recommends a visual inspection of the microsphere surface after hydrochloric acid and calcium chloride exposure, specifically to assess whether the surface has become dull or cloudy, which may indicate surface degradation. Acid resistance was evaluated by immersing 10.0 g of microspheres in hydrochloric acid (pH 5.0–5.3) for 90 h. Calcium chloride ($CaCl_2$) resistance was assessed by submerging the microsphere in a $CaCl_2$ solution (1 N) for 3 h, following the standard [11]. In both cases, after immersion, the microspheres were rinsed with distilled water and air-dried. Optical microscopy at (100× and 200× magnification) was used to examine changes in surface shine brightness and integrity. These results were qualitatively compared with untreated control samples to assess chemical durability. In order to quantify any surface degradation, ten retroreflectivity measurements were made for the samples before and after treatment. Measurements were also carried out using the 15 m geometry and the unpainted matte black surface to assess the intrinsic retroreflection of the microspheres. The microsphere density was again chosen to be 1.7 g per test area (50 g/m^2). The granulometry range was chosen to be between 150 and 180 μm to guarantee less dispersion between the retroreflectivity measurements.

3. Results

3.1. Rice Husk Ash Composition and Bulk Glass Measurements

Table 1 presents the chemical composition of the rice husk ash used for glass production. Initially, the carbon content was quantified through CHN elemental analysis. The oxide composition of the ash was determined by XRF, with the total percentage normalized to 100%, considering the carbon content. As shown, the ash contains a high percentage of silicon dioxide, similar to results presented in other works [30], which enabled its use as a silica source in glass fabrication. To achieve the target glass composition, the silica content was adjusted by increasing the mass of the ashes relative to pure silica.

Table 1. RHA composition in weight percentage.

RHA Composition (wt%)						
SiO_2	Al_2O_3	MgO	CaO	P_2O_5	C *	Others < 0.1
88.3	1.4	3.1	0.3	0.5	6.3	0.1

* quantified by CHN.

Once the glass samples were fabricated, they were characterized using UV-Vis spectrophotometry, and the refractive index was determined based on the Brewster angle method. Figure 2 presents the UV-Vis spectrum of a sample with 3 mm thickness, highlighting the visible region of the electromagnetic spectrum (400–700 nm), with the inset of a typical sample obtained under the selected fabrication conditions. As observed, from 400 nm to 700 nm, the glass exhibits very high and nearly constant transmittance, which characterizes it as colorless. The measured refractive index, accounting for the goniometer uncertainty, was 1.55 ± 0.03, a typical value for silica-based glasses [25].

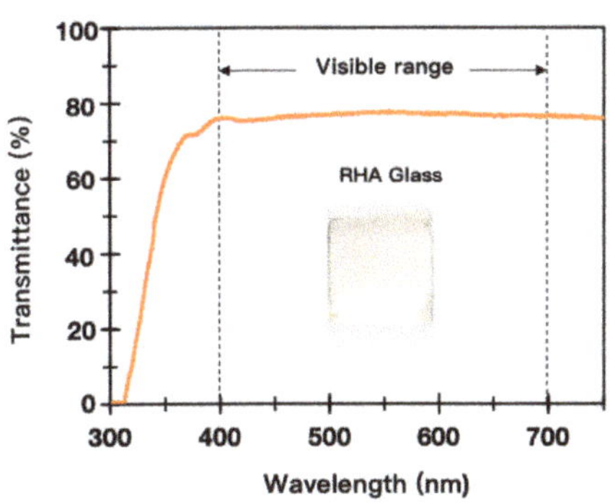

Figure 2. UV-Vis spectrum for a 3 mm RHA glass sample.

3.2. Microspheres Measurements

Table 2 presents the XRF results for both rice husk ash (RHA-based) and commercial microspheres. Overall, it is observed that both samples exhibit a high percentage of silicon dioxide and do not contain toxic oxides such as As_2O_3 (arsenic oxide), PbO (lead oxide), or Sb_2O_3 (antimony oxide).

Table 2. XRF results for RHA-based and commercial microspheres.

Microspheres Composition (wt%)					
	SiO_2	**Na_2O**	**CaO**	**Toxic Oxides (As_2O_3, PbO, Sb_2O_3)**	**Others**
From RHA	72.8	2.8	20.5	ND	3.9
Commercial	72.7	4.7	21. 0	ND	1.6

ND = Not detected.

Figure 3 shows scanning electron microscopy images of the microspheres. Figure 3a displays a representative image of commercial microspheres, where different geometrical outcomes resulting from the production process can be identified, such as spherical microspheres (predominant), shards, ovoids, and twin-shaped particles. Figure 3b, in turn, presents a representative image of RHA-based microspheres. In this image, the particles exhibit a predominantly spherical morphology, with diameters ranging from approximately 50 to 180 μm.

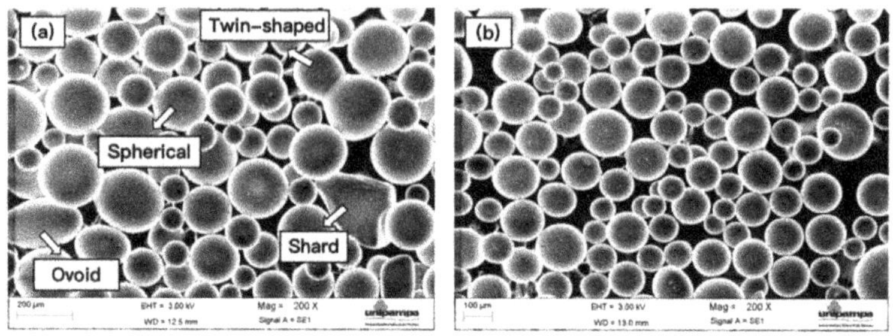

Figure 3. SEM images of (**a**) commercial microspheres and found shapes, and (**b**) RHA microspheres.

The morphology and size distribution of glass microspheres are critical parameters for their use in horizontal road markings. Morphological analysis determines whether the particles exhibit a spherical geometry, as deviations from sphericity can diminish the retroreflective efficiency [31]. Meanwhile, particle size distribution enables the categorization of microspheres within specific dimensional ranges, directly influencing their suitability for different application methods [11]. Table 3 presents the morphological analysis results for both RHA-derived and commercial microspheres. In both cases, the evaluation was performed in triplicate, totaling 1200 particles. As observed, particles with a perfect spherical shape predominate in both samples when compared to other morphological types. Notably, the high proportion of spherical RHA microspheres indicates that the chosen glass composition is well suited for producing glass beads intended for road marking applications.

Table 3. Morphology of RHA-derived and commercial microspheres.

Sample	Total Counts	Spheres (%)	Ovoids/Twinned (%)	Shards (%)
RHA Glass	1200	91 ± 3	7 ± 2	1.5 ± 0.2
Commercial Glass	1200	86 ± 3	11 ± 3	2.1 ± 0.4

During the morphological analysis that resulted in Table 3, it was also possible to assess the diameter of the produced microspheres. In this case, only particles with a spherical shape were considered. Figure 4 displays the percentage distribution of microspheres retained by different sieves as a function of the sieve aperture. The dashed black lines indicate the size limits defined by the Brazilian standard for Premix-type (Type IB) microspheres. As shown, the particle size distributions of both the commercial and RHA-derived samples fall within the specified range, confirming their compliance with national regulations and classifying the manufactured microspheres as suitable for Premix-type applications.

Regarding mass density, the results indicate that both types of glass exhibit values consistent with typical soda-lime glasses, with 2.42 ± 0.01 g/cm^3 for the rice husk ash (RHA) microspheres and 2.43 ± 0.01 g/cm^3 for the commercial microspheres [32].

After validating the production process with respect to the physical characteristics and chemical composition, the retroreflectivity performance of both RHA and commercial microspheres was evaluated on matte black (unpainted) and white (painted) surfaces. The measurements were carried out using the 15 m geometry, which is the standard configuration adopted by most road authorities in Brazil [29]. Typically, Type IB glass beads are premixed into the road marking paint at a concentration of 250 g/L [33]. Here, a lower density of 1.7 g per test area (equivalent to 85 g/L) was selected for both samples. This reduced amount was intentionally chosen to prevent bead overlap, which could affect the consistency of the measurements. The primary objective of these measurements was not to achieve the highest possible retroreflectivity [34], but rather to compare the retroreflective performance of the RHA-based microspheres with that of their commercial counterparts of the same type. For the unpainted configuration test case, a matte black board served as the reference surface, whereas for the painted configuration case, the reference consisted of the white paint film, both without microspheres. Figure 5 presents the retroreflectivity results obtained for both samples, considering the particle size distribution described in Figure 4. As expected, both sample types showed substantial increases in retroreflectivity compared to their respective controls. More importantly, the results revealed that RHA-based microspheres exhibited retroreflectivity values comparable to those of commercial glass beads, supporting their potential implementation in road marking applications.

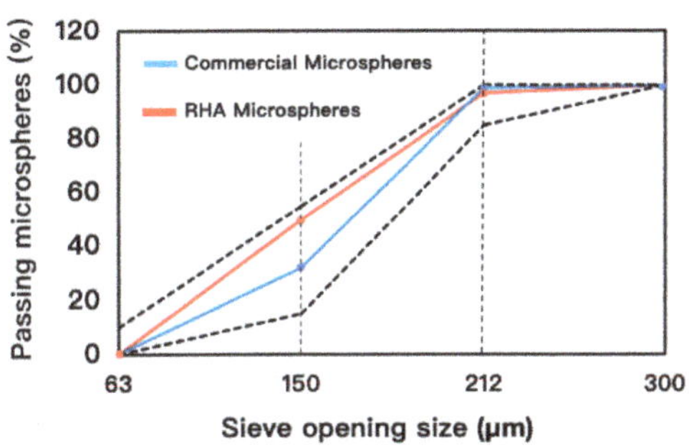

Figure 4. Size distribution of glass microspheres for RHA-derived and commercial microspheres.

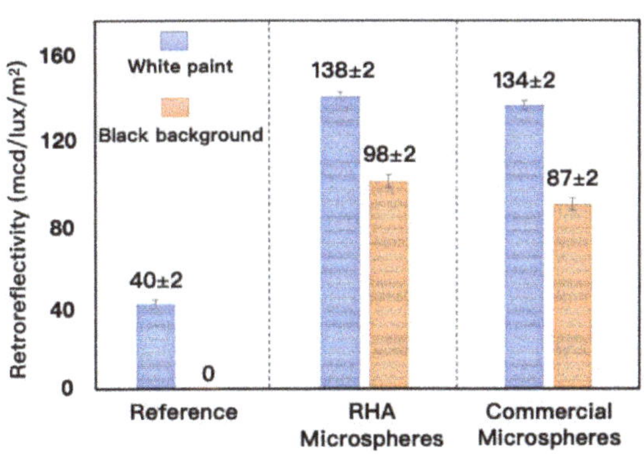

Figure 5. Retroreflectivity measurements to the RHA and commercial microspheres.

Moreover, the evaluation of glass microspheres under chemical attack from hydrochloric acid and calcium chloride resistance test is essential to assess the microspheres durability against chemical agents commonly found in the environment, such as acid rain or cleaning products used in traffic areas [11,12]. These agents can degrade the microsphere surface over time, potentially altering key properties such as surface integrity, adhesion to paint, and their retroreflective performance, reducing road marking visibility and thereby increasing the risk of traffic accidents. Figure 6a,b and 6c,d shows representative images of the RHA microspheres before and after immersion in hydrochloric acid and calcium chloride, respectively. No visible alterations in the surface appearance (shine or hue) were observed for either sample, indicating resistance to hydrochloric acid and $CaCl_2$-induced surface degradation. To complement this analysis, retroreflectivity measurements were also carried out before and after each chemical exposure to verify whether the optical performance was compromised. The results, as reported in Figure 6e, also showed no significant change, confirming the chemical stability and performance reliability of the RHA glass microspheres under both hydrochloric acid and $CaCl_2$ exposure.

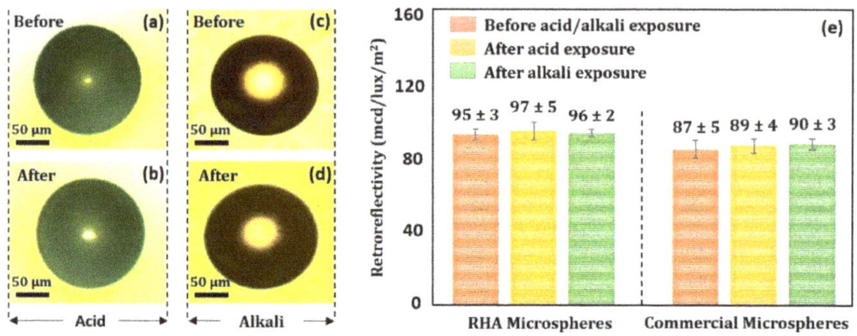

Figure 6. Microspheres optical images: (**a**) before and (**b**) after hydrochloric acid immersion, (**c**) before and (**d**) after calcium chloride immersion; (**e**) corresponding retrorreflectivity measurements.

4. Discussion

The physical, chemical, and optical properties of the microspheres produced using rice husk ash as a raw material were evaluated with reference to the NBR 16184:2021 standard, as well as in comparison with commercial microspheres of the same type. Tables 4 and 5 summarize the requirements of the Brazilian standard NBR 16184:2021 for Type IB glass beads, along with the results obtained for both RHA-based and commercial microspheres presented in this work, respectively.

Table 4. Requirements of the Brazilian standard NBR 16184:2021 for Type IB glass beads.

Type IB Microspheres (Premix)			
SiO_2 content		$\geq$65%	
Toxic elements (As, Pb, Sb)		Maximum 200 mg/kg	
Color		Colorless	
Refractive index		$\geq$1.5	
Granulometry (φ)		63 µm $\leq \varphi \leq$ 250 µm	
Morphology	Ovoid/twin Maximum 20%	Shards Maximum 3%	Spherical $\geq$77%
Density (g/cm^3)		$2.4 \leq d \leq 2.6$	
Resistance to hydrochloric acid		Surface without fogging	

As can be observed, regarding the silica content and the presence of toxic elements, the XRF results revealed SiO_2 contents above 72%, which exceeds the minimum threshold established by the standard. Additionally, no toxic elements such as As, Pb, or Sb were detected in either sample. Regarding color, UV-Vis spectrophotometry demonstrated high transmittance throughout the visible spectrum, indicating that the produced glass can be classified as colorless. Moreover, visual inspection of the RHA and commercial microspheres revealed a similar appearance in terms of coloration. The refractive index of the produced glass was measured as 1.55 $\pm$ 0.03 at a wavelength of 650 nm (the upper boundary of the visible spectrum). Although this measurement was performed at a single wavelength, physical models for homogeneous materials such as glass indicate that the refractive index typically increases for shorter wavelengths [25,35]. Therefore, it can be stated that the refractive index across the entire visible range is at least 1.55 ± 0.03, which meets the requirements of NBR 16184:2021. Although it has not been possible to measure the refractive index of the commercial microspheres, considering that the base glass of these microspheres is soda-lime, its bulk refractive index is <1.52 in the visible spectrum [25]. The higher refractive index observed in the glass derived from rice husk ash (RHA), when compared to commercial soda-lime glass, is likely attributed to the presence of boron in its composition: borossilicate glass can present higher refractive indexes [36]. This can also explain the slightly higher retroreflectivity values measured for the RHA-based microspheres compared to the commercial ones, as reported by other authors [37].

Table 5. Compiled results obtained for RHA-based and commercial microspheres.

Sample Properties	RHA Microspheres	Commercial Microspheres
SiO_2 Content (%)	72.8	72.7
Toxic Elements (As, Pb, Sb)	ND	ND
Color	Colorless	Colorless
Refractive Index	1.55 ± 0.03	–
Granulometry (φ)	Categorized as Type IB	Categorized as Type IB
Morphology		
— Ovoid/Twin (%)	7 ± 2	11 ± 3
— Shards (%)	1.5 ± 0.2	2.1 ± 0.4
— Spherical (%)	91 ± 3	86 ± 3
Density (g/cm^3)	2.42 ± 0.01	2.43 ± 0.01
Resistance to Hydrochloric Acid	Surface without fogging	Surface without fogging

ND = Not detected.

In terms of particle size, both the RHA and commercial microspheres fall within the 63–300 μm range, classifying them as Type IB microspheres according to the standard. This makes them suitable for incorporation into road marking paint prior to application. Moreover, the RHA-based glass formulation employed in this study resulted in a high percentage of spherical particles when compared to other outcomes (shards, ovoids, twin-shaped), fulfilling the morphology criteria established by NBR 16184:2021, and demonstrating the thermochemical feasibility to use this agricultural waste as a source material. Furthermore, both the measured density and resistance to chemical attack (hydrochloric acid and calcium chloride) were within the acceptable limits set by the standard. These results collectively demonstrate that the microspheres produced from rice husk ash exhibit comparable properties to those of commercial glass beads and comply fully with all specifications outlined in the Brazilian regulation. Although the comparison in this work is based on a Brazilian standard, minor formulation adjustments could be made to ensure compliance with international standards as well. Overall, the RHA presents itself as a valuable alternative for glass retroreflective microspheres production.

5. Conclusions

This study demonstrates the technical feasibility of producing Type IB glass microspheres for road marking applications using soda-lime glass derived from rice husk ash (RHA). All physical, chemical, morphological, and optical characterizations confirmed that the microspheres fabricated from this renewable silica source meet the specifications outlined in the Brazilian standard NBR 16184:2021. Key parameters—such as the silica content (72.8 wt%), the absence of toxic element, refractive index (1.55 ± 0.03), density (2.42 ± 0.01 g/cm^3), particle size distribution (63 μm–250 μm), spherical morphology (91 ± 3%), chemical resistance (no fogging), and retroreflectivity (138 ± 2 mcd/lux/m^2)—

were found to be comparable, equivalent, or superior to those of commercial products and in accordance with the standard. Beyond meeting technical criteria, the process offers a sustainable and scalable route for converting agricultural waste into high-value materials for infrastructure. The successful transformation of RHA into glass microspheres supports the principles of circular economy and waste valorization, while reducing reliance on finite natural resources such as high-purity sand. Given the abundance of rice production in countries like Brazil, this approach presents an environmentally responsible and economically viable alternative for large-scale implementation in the road marking industry. Overall, this work highlights a promising pathway for enhancing road safety through sustainable materials innovation, contributing to the broader goals of responsible production, sustainable cities, and infrastructure resilience.

Author Contributions: Conceptualization, J.W.d.M. and C.V.; methodology, J.W.d.M.; validation, J.W.d.M. and C.V.; formal analysis, L.E.G.A.; investigation, I.M.T., J.P.N. and A.B.; resources, L.E.G.A., C.V. and J.W.d.M.; data curation, I.M.T., J.P.N. and A.B.; writing—original draft preparation, I.M.T., J.P.N. and L.E.G.A.; writing—review & editing, C.V. and J.W.d.M.; visualization, J.P.N. and L.E.G.A.; supervision, J.W.d.M.; funding acquisition, C.V. and J.W.d.M. All authors have read and agreed to the published version of the manuscript.

Funding: The authors would like to thank FAPERGS-Research Support Foundation of the State of Rio Grande do Sul, Brazil (Grant n. 22/2551-0000841-0) and National Council for Scientific and Technological Development: 407045/2021-0.

Institutional Review Board Statement: Not applicable.

Informed Consent Statement: Not applicable.

Data Availability Statement: The original contributions presented in this study are included in the article. Further inquiries can be directed to the corresponding author(s).

Conflicts of Interest: The authors declare no conflict of interest.

Abbreviations

The following abbreviations are used in this manuscript:

WHO	World Health Organization
RHA	Cinza da Casca de Arroz
NBR	Brazilian Standard
Uv-Vis	Ultraviolet-Visible (spectrophotometry)
XRF	X-ray Fluorescence
CHN	Carbon, Hydrogen, Nitrogen (elemental analysis)
SEM	Scanning Electron Microscopy
LPG	Liquefied Petroleum Gas
MDF	Medium-Density Fiberboard
TM	Transverse Magnetic (polarization mode of laser light)
UN	United Nations

References

1. Burghardt, T.E.; Mosböck, H.; Pashkevich, A.; Fiolic, M. Horizontal Road Markings for Human and Machine Vision. *Transp. Res. Procedia* **2020**, *48*, 3622–3633. [CrossRef]
2. Babic, D.; Fiolic, M.; Babic, D.; Gates, T. Road markings and their impact on driver behaviour and road safety: A systematic review of current findings. *J. Adv. Transp.* **2020**, *2020*, 7843743. [CrossRef]
3. World Health Organization. Road Traffic Injuries. Available online: https://www.who.int/health-topics/road-safety (accessed on 15 February 2025).
4. Plainis, S.; Murray, I.; Pallikaris, I. Road traffic casualties: Understanding the night-time death toll. *Injury Prev.* **2006**, *12*, 125–138. [CrossRef] [PubMed]

5. Burghardt, T.E.; Popp, R.; Helmreich, B.; Reiter, T.; Bohm, G.; Pitterle, G.; Artmann, M. Visibility of various road markings for machine vision. *Case Stud. Constr. Mater.* **2021**, *15*, e00579. [CrossRef]
6. Qian, H.; Lu, X.; Kou, P.; He, R. Influence of construction technology on the retroreflective performance of two-component traffic marking. *Vibroeng. Procedia* **2022**, *47*, 35–41. [CrossRef]
7. Mazzoni, L.N.; Vasconcelos, K.; Albarracín, O.; Bernucci, L.; Linhares, G. Field Data Analysis of Pavement Marking Retroreflectivity and Its Relationship with Paint and Glass Bead Characteristics. *Appl. Sci.* **2024**, *14*, 4205. [CrossRef]
8. Salles, L.S.; Pereira, D.S.; Teixeira, D.L.K.; Specht, L.P. Road markings retroreflectivity experimental assessment: Observations on rainfall, dirt, retroreflectometer geometry and minimum requirements. In Proceedings of the Transportation Research Board 95th Annual Meeting, Washington, DC, USA, 10–14 January 2016.
9. Lee, S.; Koh, E.; Jeon, S.; Kim, R.E. Pavement marking construction quality inspection and night visibility estimation using computer vision. *Case Stud. Constr. Mater.* **2024**, *20*, e02953. [CrossRef]
10. Lee, M.Y.; Tan, J.; Heng, J.Y.; Cheeseman, C. A Comparative Study of Production of Glass Microspheres by using Thermal Process. *IOP Conf. Ser. Mater. Sci. Eng.* **2017**, *205*, 012022. [CrossRef]
11. *NBR 16184*; Road Horizontal Signaling—Glass Beads and Microbeads—Requirements and Test Methods. ABNT—Brazilian Association of Technical Standards: Rio de Janeiro, Brazil, 2021.
12. Mahmoud, M.; Kraxner, J.; Elsayed, H.; Bernardo, E.; Galusek, D. Fabrication and environmental applications of glass microspheres: A review. *Ceram. Int.* **2023**, *49*, 39745–39759. [CrossRef]
13. Zhong, X.; Deetman, S.; Tukker, A.; Behrens, P. Increasing material efficiencies of buildings to address the global sand crisis. *Nat. Sustain.* **2022**, *5*, 389–392. [CrossRef]
14. Cao, Z.; Masanet, E. Material efficiency to tackle the sand crisis. *Nat. Sustain.* **2022**, *5*, 370–371. [CrossRef]
15. United Nations Environment Programme (UNEP). *Sand and Sustainability: 10 Strategic Recommendations to Avert a Crisis*; UNEP: Nairobi, Kenya, 2022. Available online: https://www.unep.org/resources/report/sand-and-sustainability-10-strategic-recommendations-avert-crisis (accessed on 18 December 2024).
16. Zou, Y.; Yang, T. Rice Husk, Rice Husk Ash and Their Applications. In *Rice Bran and Rice Bran Oil*; Elsevier: Amsterdam, The Netherlands, 2019; pp. 207–246. [CrossRef]
17. Sulaiman. Rice Husk, Rice Husk Ash, and Their Applications. In *Rice Husk Biomass: Processing, Properties and Applications*; Jawaid, M., Parmar, B., Eds.; Springer: Singapore, 2025; pp. 43–54. [CrossRef]
18. Nzereogu, P.U.; Omah, A.D.; Ezema, F.I.; Iwuoha, E.I.; Nwanya, A.C. Silica Extraction from Rice Husk: Comprehensive Review and Applications. *Hybrid Adv.* **2023**, *4*, 100111. [CrossRef]
19. Xia, M.; Yang, X.; Hou, Z. Preparation of High-Purity Quartz Sand by Vein Quartz Purification and Characteristics: A Case Study of Pakistan Vein Quartz. *Minerals* **2024**, *14*, 727. [CrossRef]
20. Marçal, J.M.M.; Chim, M.R.; Teixeira, I.M.; Armas, L.E.; Buchner, S.; Vasconcellos, M.; Menezes, J.W.; Valsecchi, C. Characterization of rice husk ash glass produced at low temperature as a sustainable alternative for the glass industry. *Int. J. Environ. Sci. Technol.* **2025**. [CrossRef]
21. Kaewkhao, J.; Limsuwan, P. Utilization of rice husk fly ash in the color glass production. *Procedia Eng.* **2012**, *32*, 670–675. [CrossRef]
22. United States Department of Agriculture (USDA). Brazil Rice Area, Yield and Production. Available online: https://ipad.fas.usda.gov/countrysummary/Default.aspx?crop=Rice&id=BR (accessed on 2 March 2025).
23. Silva, L.A.; Santos, I.F.S.; Machado, G.O.; Tiago Filho, G.L.; Barros, R.M. Rice Husk Energy Production in Brazil: An Economic and Energy Extensive Analysis. *J. Clean. Prod.* **2021**, *290*, 125188. [CrossRef]
24. UN General Assembly. Transforming Our World: The 2030 Agenda for Sustainable Development. 2015. Available online: https://sdgs.un.org/2030agenda (accessed on 15 December 2024).
25. Synowicki, R.A.; Johs, B.D.; Martin, A.C. Optical properties of soda-lime float glass from spectroscopic ellipsometry. *Thin Solid Films* **2011**, *519*, 2907–2913. [CrossRef]
26. Verma, J.P.; Verma, P. *Determining Sample Size and Power in Research Studies*; Springer: Singapore, 2020.
27. *EN 1436:2018*; Road Marking Materials—Road Marking Performance for Road Users and Test Methods. European Committee for Standardization: Brussels, Belgium, 2018.
28. *E1710-18*; Standard Test Method for Measurement of Retroreflective Pavement Marking Materials with CEN Prescribed Geometry Using a Portable Retroreflectometer. ASTM—American Society for Testing and Materials: West Conshohocken, PA, USA, 2018.
29. *NBR 14723*; Road Horizontal Signaling—Measurement of Retroreflectivity Using Handheld Equipment with 15 m Geometry—Test Method. ABNT—Brazilian Association of Technical Standards: Rio de Janeiro, Brazil, 2020.
30. Endale, S.A.; Taffese, W.Z.; Vo, D.; Yehualaw, M.D. Rice Husk Ash in Concrete. *Sustainability* **2023**, *15*, 137. [CrossRef]
31. Smadi, O.; Hawkins, N.; Bektas, B.A.; Carlson, P.; Pike, A.; Davies, C. Recommended Laboratory Test for Predicting the Initial Retroreflectivity of Pavement Markings from Glass Bead Quality. *Transp. Res. Rec.* **2014**, *2440*, 94–102. [CrossRef]

32. Shelby, J.E. Compositions and Properties of Commercial Glasses. In *Introduction to Glass Science and Technology*; Royal Society of Chemistry: Cambridge, UK, 2005; pp. 262–274. [CrossRef]
33. *NBR 13699*; Road Horizontal Signaling—Water-Based Acrylic Resin Paint. ABNT—Brazilian Association of Technical Standards: Rio de Janeiro, Brazil, 2021.
34. Zhang, G.; Hummer, J.E.; Rasdorf, W. Impact of bead density on paint pavement marking retroreflectivity. *J. Transp. Eng.* **2010**, *136*, 773–781. [CrossRef]
35. Smith, D.Y.; Shiles, E.; Inokuti, M. Refraction and dispersion in optical glass. *Nucl. Instrum. Methods Phys. Res. B* **2004**, *218*, 170–175. [CrossRef]
36. Pavliukevich, Y.G.; Papko, L.F.; Trusova, E.E. Optical Glass with Increased Refractive Index Based on the $BaO–B_2O_3–SiO_2$ System. *Chem. Chem. Eng.* **2024**, *2024*, 1. [CrossRef]
37. Wenzel, K.M.; Burghardt, T.E.; Pashkevich, A.; Buckermann, W.A. Glass Beads for Road Markings: Surface Damage and Retroreflection Decay Study. *Appl. Sci.* **2022**, *12*, 2258. [CrossRef]

ceramics

MDPI

Article

Sustainable Use of Gypsum Waste for Applications in Soil–Cement Bricks: Mechanical, Environmental, and Durability Performance

Elvia Soraya Santos Nascimento [1], Herbet Alves de Oliveira [2], Cochiran Pereira dos Santos [3], Maria de Andrade Gomes [3], Mário Ernesto Giroldo Valerio [1] and Zélia Soares Macedo [1,*]

[1] Postgraduate Program in Materials Science, Federal University of Sergipe, São Cristóvão 49107-230, SE, Brazil; elviasoraya@academico.ufs.br (E.S.S.N.); megvalerio@academico.ufs.br (M.E.G.V.)
[2] Department of Civil Engineering, Federal Institute of Sergipe, Aracaju 49055-260, SE, Brazil; herbet.oliveira@ifs.edu.br
[3] Physics Department, Federal University of Sergipe, São Cristóvão 49107-230, SE, Brazil; cochiran@academico.ufs.br (C.P.d.S.); mg.lpcmufs@gmail.com (M.d.A.G.)
* Correspondence: zelia.macedo@academico.ufs.br

Abstract

This study investigates the use of gypsum waste from civil construction as a partial substitute for cement in soil–cement formulations, aiming to produce eco-friendly bricks aligned with circular economy principles. Formulations were prepared using a 1:8 cement–soil ratio, with gypsum replacing cement in proportions ranging from 5% to 40%. The raw materials were characterized in terms of chemical composition, crystalline phases, plasticity, and thermal behavior. Specimens, molded by uniaxial pressing into cylindrical bodies and cured for either 7 or 28 days, were evaluated for compressive strength, water absorption, durability, and microstructure. Water absorption remained below 20% in all samples, with an average value of 16.20%. Compressive strength after 7 days exhibited a slight reduction with increasing gypsum content, ranging from 16.36 MPa (standard formulation) to 13.74 MPa (40% gypsum), all meeting the quality standards. After 28 days of curing, the formulation containing 10% gypsum achieved the highest compressive strength (26.7 MPa), surpassing the reference sample (25.2 MPa). Mass loss during wetting–drying cycles remained within acceptable limits for formulations incorporating up to 20% gypsum. Notably, samples with 5% and 10% gypsum demonstrated superior mechanical performance, while the 20% formulation showed performance comparable to the standard formulation. These findings indicate that replacing up to 20% of cement with gypsum waste is a technically and environmentally viable approach, supporting sustainable development, circular economy, and reduction of construction-related environmental impacts.

Keywords: soil–cement bricks; gypsum waste; circular economy; sustainable construction

1. Introduction

In recent years, sustainable practices have gained prominence across various sectors, due to the urgent need to preserve natural resources and mitigate environmental impacts. The concept of sustainable development, as defined by the United Nations (UN) World Commission on Environment and Development in 1983, emphasizes the importance of meeting present needs without compromising the ability of future generations to meet their own. To achieve this goal, it is essential to manage key elements in production chains by reducing the consumption of natural resources, minimizing energy use, decreasing waste generation, and promoting reuse and recycling. Solid waste management plays a critical role in advancing the UN's Sustainable Development Goals (SDGs), particularly SDG 11 (Sustainable Cities and Communities), SDG 12 (Responsible Consumption and Production), and SDG 13 (Climate Action), by enabling resource recovery and fostering circular economy strategies [1,2].

The construction sector is one of the most impactful industries in terms of resource consumption and waste generation. In response, the industry has increasingly adopted innovative practices to embed sustainability into its production chain [3]. A notable advancement is the use of eco-friendly materials, such as soil–cement bricks, which offer a cost-effective and environmentally conscious alternative to conventional masonry materials. These bricks can be produced on site using locally sourced materials, thereby reducing transportation demands and associated energy consumption [4]. Further contributions to the sustainability of soil–cement bricks include the incorporation of recycled aggregates and industrial by-products [5], as well as the partial substitution of traditional binders such as Portland cement with alternative materials [6,7]. However, challenges remain in optimizing production processes to meet the required standards for compressive strength and durability [5,8]. Recent studies have explored the addition of different waste materials to soil–cement bricks, such as coffee husks, paper and pulp industry waste, sugarcane bagasse ash, and leather powder, with varied outcomes regarding their technical viability [9–11].

One underexplored waste material in this context is gypsum, which is a major byproduct of civil construction. The global gypsum production ranges from 30 to 150 million tons annually [12], with Brazil accounting for approximately 1.3 million tons [13]. It is estimated that up to 47% of gypsum used in coatings, drywall cutting, and demolition ends up discarded [14]. Recycling gypsum is challenging due to sulfate contamination and landfill emissions [15], yet it remains essential to reduce resource depletion and environmental risks [16].

Existing studies have primarily focused on reusing gypsum waste in plasters and mortars, typically as a partial replacement for either conventional gypsum or Portland cement [17,18]. However, its application in soil–cement bricks remains unexplored. Moreover, there is a notable gap in the literature regarding the effects of gypsum waste on the mechanical properties, microstructure, and long-term durability of construction materials. In this context, the present study aims to determine safe and effective incorporation levels of gypsum waste in soil–cement bricks by evaluating their physical, mechanical, and microstructural properties across varying proportions of recycled gypsum.

2. Materials and Methods

2.1. Raw Materials

The materials used in this study were: natural soil, Portland cement (CP V-ARI), and gypsum waste from construction activities. The soil was collected from northeastern Brazil and has a friable texture, light reddish color, and is commonly used in local construction. After collection, it was dried in an oven at 60 $\pm$ 5 °C. The particle size distribution was determined according to ASTM D6913/D6913M [19], with approximately 40% of the material passing through a 0.075 mm sieve. The specific gravity was measured using the pycnometer method, following DNER-ME-093/94 [20]. Atterberg limits, determined according to ASTM D4318 [21], yielded a liquid limit (LL) of 32% and a plasticity index (PI) of 8.0. The plasticity index (PI) of the soil was 8.0, and the liquid limit (LL) was 32%, classifying it as a medium-plasticity material, considered suitable for soil–cement brick production [22,23]. According to ABCP (1985) and NBR 10833 [24], suitable soils for soil–cement brick production must have LL < 45% and PI < 18%. The soil used in this study satisfies these requirements. Based on these parameters, it is classified as a low- to medium-plasticity inorganic clay (CL) under the Unified Soil Classification System (USCS), and as A-2-6 in the Transportation Research Board (TRB) system—granular soils with clayey fine fractions that influence plastic behavior.

The cement used was CP V-ARI, a high early strength Portland cement suitable for precast and fast-setting applications. It was characterized for initial and final setting times (ASTM C191 [25]), specific gravity (ASTM C188 [26]), specific surface area using the Blaine air permeability method (NBR 16372 [27]), and compressive strength (ASTM C109/C109M [27]).

The gypsum waste was collected from companies manufacturing and installing prefabricated gypsum boards. It consisted mainly of calcium sulfate hemihydrate ($CaSO_4{\cdot}0.5\ H_2O$), typically used in interior coatings. During construction, this material is hydrated to form calcium sulfate dihydrate ($CaSO_4{\cdot}2H_2O$). The collected waste was crushed to approximately 1 cm, oven-dried at 60 ± 5 °C to remove residual moisture, and then ground using a ball mill. The resulting powder was sieved through a No. 100 ASTM mesh (150 μm) and stored in sealed plastic containers to prevent moisture uptake and carbonation. No chemical treatments were applied.

The decision to use this material was based on its local abundance and chemical similarity to commercial gypsum. Its reuse aligns with sustainability principles by reducing landfill disposal and promoting circular economy practices.

2.2. Specimen Preparation

Specimens weighing 200 g were uniaxially pressed into cylindrical shapes with approximate dimensions of 6 cm in diameter and 4 cm in height, using a moisture content of 10%, as reported in the literature [28–32]. The initial formulations contained 90% soil and 10% cement. However, specimens prepared with this mixture did not meet the minimum compressive strength required by the standard (2 MPa). To improve performance, part of the soil was replaced with fine aggregate (sand), and the compaction pressure was varied to identify optimal conditions. Figure 1 presents the compressive strength of specimens with different sand-to-soil ratios, pressed at 5 MPa, 10 MPa, and 15 MPa. The results indicate that the formulation containing 10% sand, 80% soil, and 10% cement, compacted at 10 MPa, yielded the highest compressive strength among the combinations tested.

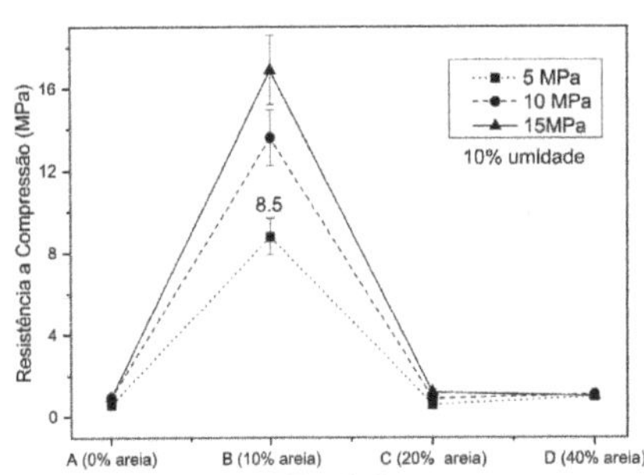

Figure 1. Compressive strength of formulations containing 10% cement and varying sand content (0% to 40% by total mass), used as a partial replacement for soil. The optimal performance was achieved with 10% sand, indicating a balance between workability and particle packing.

After molding, the specimens were air-cured under natural environmental conditions, with room temperature of approximately 25 °C and relative humidity ranging from 70% to 82%. No artificial humidification system was used. The samples were stored in a sheltered area, protected from direct sunlight and wind, following NBR 8492 [33]. This curing method reflects common practices in the construction industry and allows a realistic evaluation of performance under local conditions.

Five formulations were prepared, incorporating varying proportions of gypsum residue as a partial replacement for cement. The compositions are shown in Table 1. The reference formulation without gypsum was labeled SF (standard formulation), and the gypsum-containing mixes were designated G5, G10, G20, and G40, corresponding to 0.5%, 1%, 2%, and 4% gypsum, respectively.

Table 1. Formulations used in the experiments (% by mass).

Compositions	Soil	Cement	Sand	Gypsum
SF	80	10	10	-
G5	80	9.5	10	0.5
G10	80	9	10	1
G20	80	8	10	2
G40	80	6	10	4

2.3. Mineralogical and Chemical Characterization

X-ray diffraction (XRD) patterns of the cement, gypsum, and soil samples were obtained using a Rigaku DMAX 100 diffractometer (Tokyo, Japan) with CuKα radiation (λ = 1.5418 Å). Scans were performed in continuous mode over a 2θ range of 5° to 70° at a scan rate of 1°/min. The crystalline phases were identified by comparing experimental patterns with the ICSD crystallographic database.

Chemical composition analysis was performed using energy-dispersive X-ray fluorescence (EDXRF) with a Shimadzu EDX-7000 spectrometer (Kyoto, Japan) under vacuum, using a 10 mm collimator and Mylar capsule both supplied by the equipment manufacturer. Operating conditions included voltages of 50 kV (Al-U) and 15 kV (Na-Sc), with a 100 s acquisition time and ~30% dead time.

2.4. Thermal Analysis

Thermogravimetric analysis (TGA) of the soil was conducted using a TA Instruments SDT Q600 (New Castle, DE, USA) under a nitrogen atmosphere. Approximately 10 mg of sample was heated in a platinum crucible at a rate of 10 °C/min from room temperature to 1000 °C.

2.5. Scanning Electron Microscopy (SEM) Procedures

The morphology of the soil–cement composites after 28 days of curing was analyzed using a JEOL JSM-6510LV scanning electron microscope (SEM) (Akishima, Japan) at CM-Nano/UFS (Proposal #080/2023). For these analyses, the samples were dried at 60 ± 5 °C to remove residual moisture and manually fractured to expose the internal microstructure. The fractured surfaces were mounted on metallic stubs using carbon tape and coated with a thin layer of gold using a sputter coater to minimize charging effects and ensure adequate image quality. This preparation method enabled clear microstructural visualization and reliable elemental analysis under high-vacuum conditions. The SEM operated at 15 kV in secondary electron imaging (SEI) mode.

2.6. Mechanical and Physical Testing

2.6.1. Compressive Strength Test

The compressive strength of the cylindrical specimens was evaluated after 7 and 28 days of curing using a Contenco HD-200T hydraulic press (São José da Lapa, Brazil) with a maximum load capacity of 2000 kN. The loading rate was kept between 1.0 MPa/s and 1.5 MPa/s, in line with the guidelines of NBR 5739 [34]. Tests were conducted on five samples per formulation. The results are presented as mean values with standard deviations. Compressive strength tests were performed at 7 and 28 days of curing for all formulations. Intermediate curing ages (e.g., 14 days) were not included in this study, as the focus was on early-age and long-term mechanical performance.

2.6.2. Water Absorption and Apparent Density

Water absorption and apparent density were measured after 28 days of curing, in accordance with NBR 8492 [35] and using three specimens per formulation. The dry mass (M) was first recorded, followed by immersion in distilled water for 24 h to determine the wet mass (MW). Water absorption was calculated as the difference between wet and dry mass (NBR 10836 [36]). Apparent density (Db) was calculated as the ratio between dry mass and volume (Db = M/V), where the volume was determined from direct measurements of height and diameter, with 0.02 mm precision. Water absorption tests were performed on five specimens per formulation, and the results are presented as mean values with standard deviations.

2.6.3. Durability Test Under Wet–Dry Cycles

Durability was assessed by simulating weather exposure through alternating wetting and drying cycles (NBR 13554 [37]). Three specimens from each formulation were immersed in water for 6 h and then oven-dried for 18 h at 70 °C. This procedure was repeated for six cycles, and the mass loss was recorded before and after each cycle (NBR 8492 [34]). Durability-related tests were conducted on five specimens for each composition, and the results were expressed as mean values and standard deviations.

2.7. *Statistical Analysis*

The normality of the data was verified, and a one-way analysis of variance (ANOVA) was conducted to assess the influence of different gypsum substitution levels (0%, 5%, 10%, 20%, and 40%) on compressive strength and water absorption at 7 and 28 days of curing. Statistical analysis was conducted using Paleontological Statistics (PAST) software version 4.13 [38], with a significance level (p-value) of ≤ 0.05.

The null hypothesis (H_0) assumed that there were no statistically significant differences among the treatment group means, while the alternative hypothesis (H_1) posited that at least one group mean differed significantly. When ANOVA results indicated significance ($p < 0.05$), Tukey's post-hoc test was applied to identify specific group pairs with statistically significant differences.

These statistical procedures were essential to support the interpretation of the experimental results and ensure the reliability of the conclusions drawn from the data.

3. Results

3.1. *Physical and Chemical Characterization of Raw Materials*

Table 2 presents the specific density, bulk density, and specific surface area of the raw materials, which were obtained from triplicate measurements and are reported as mean values with standard deviations. The soil showed a specific density of 2.71 g/cm^3, which suggests the presence of dense minerals such as quartz or iron oxides. Materials with high specific density generally exhibit lower porosity, a factor that can positively influence mechanical strength [39]. The values obtained for cement and sand fall within the typical range for these materials. Regarding specific surface area, cement and soil showed similar values, while the gypsum residue exhibited a significantly higher value. This indicates that the gypsum particles are smaller and more uniformly distributed, which may influence hydration kinetics, mechanical strength, and durability in cementitious systems [40].

Table 2. Specific density, bulk density, and specific surface area of raw materials used in soil–cement–gypsum formulations.

	Specific Density (g/cm^3)	Bulk Density (g/cm^3)	Specific Surface Area (cm^2/g)
Soil	2.71 ± 0.02	1.15 ± 0.01	4243 ± 100
Gypsum	2.25 ± 0.02	1.10 ± 0.02	5500 ± 150
Cement	3.10 ± 0.02	1.40 ± 0.02	4400 ± 120
Sand	2.63 ± 0.01	1.65 ± 0.01	-

Table 3 provides the chemical composition of the materials, obtained via EDXRF. The soil is predominantly composed of silica (SiO_2—39.39%), with relevant amounts of calcium oxide (CaO—22.97%), iron oxide (Fe_2O_3—18.45%), and alumina (Al_2O_3—9.40%). The organic matter content, estimated by loss-on-ignition (LOI), was 0.64%. These values are consistent with compositions containing clay minerals, quartz, feldspars, and limestone [41,42]. A SiO_2/Al_2O_3 ratio of 4.33 suggests a high proportion of free quartz and accessory minerals. The presence of alkaline oxides (Na_2O + K_2O) indicates feldspars or mica [40]. Silica contributes to drying and reduces plasticity, while alumina improves thermal stability and cohesion [41]. Cement was mainly composed of calcium oxide (CaO—80.95%), with smaller fractions of silica (SiO_2—8.46%) and iron oxide (Fe_2O_3—5.95%). Sand consisted mostly of silica (SiO_2—89.79%), and gypsum was primarily composed of calcium oxide (CaO—68.01%) and sulfur trioxide (SO_3—29.97%).

Table 3. Chemical composition of the raw materials determined by energy-dispersive X-ray fluorescence (EDXRF). Oxide content values are expressed in weight percentage, with an uncertainty of ±0.02 wt%.

Component	Soil	Cement	Sand	Gypsum
CaO	22.97	80.95	1.54	68.01
SiO_2	39.39	8.46	89.79	0.63
Fe_2O_3	18.45	5.95	1.721	-
SO_3	0.11	2.06	0.32	29.97
K_2O	6.95	1.60	0.66	0.30
Al_2O_3	9.40	-	1.82	-
TiO_2	2.09	0.37	3.56	-
Loss on ignition	0.64	0.61	0.59	1.09

3.2. Mineralogical and Phase Analysis

Figure 2 presents the XRD patterns of soil, cement, and gypsum. Cement exhibited characteristic peaks of alite (C_3S) and larnite (C_2S), confirming its expected mineralogy [43]. The gypsum sample was predominantly composed of calcium sulfate hemihydrate ($CaSO_4 \cdot 0.5\ H_2O$—80.8%), with 19.2% calcium sulfate dihydrate and minor peaks of anhydrous calcium sulfate [44,45]. The soil showed quartz as the dominant mineral (30.8–61.7%), followed by illite and calcite, consistent with literature findings [46,47].

Figure 3 displays the X-ray diffraction patterns of the soil–cement formulations with varying contents of recycled gypsum. The presence of crystalline phases such as quartz (Q), calcite (C), gypsum (G), ettringite (E), illite (I), paragonite (P), and inesite (In) was identified. Ettringite formation becomes more pronounced in the samples with higher gypsum content, indicating the influence of calcium sulfate addition on the hydration process. The persistence of gypsum peaks in the formulations containing recycled gypsum also confirms the presence of unreacted sulfate, particularly in mixtures with higher replacement levels. These results support the role of gypsum in modifying the mineralogical composition of the composite and suggest potential effects on its microstructure and mechanical performance, as discussed in the following sections.

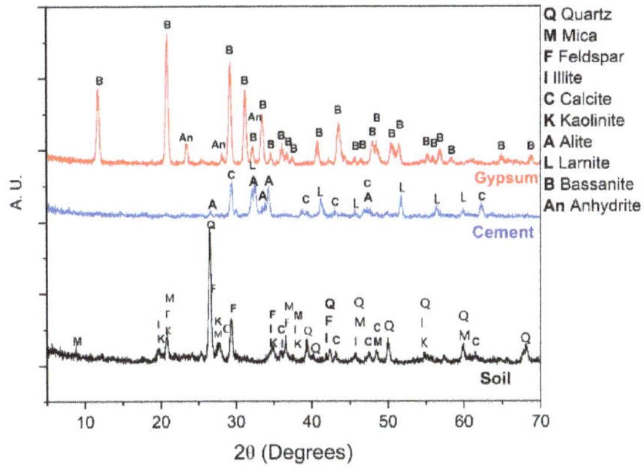

Figure 2. X-ray diffraction (XRD) patterns of soil, cement, and gypsum, highlighting the dominant crystalline phases and their compatibility for hydration reactions.

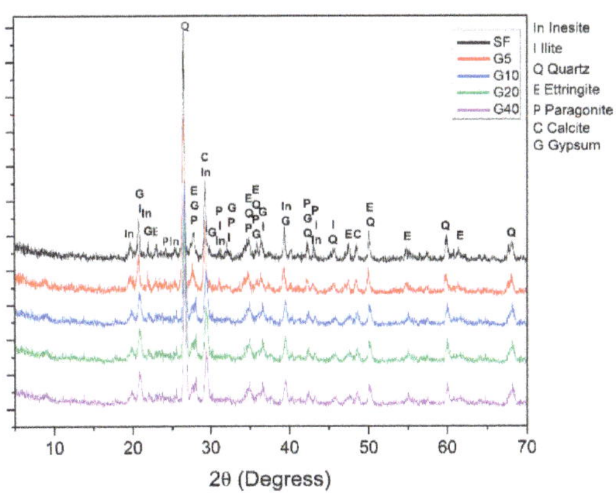

Figure 3. X-ray diffraction patterns of the soil–cement formulations with varying recycled gypsum content. The diffractograms reveal the presence of crystalline phases such as ettringite, gypsum, quartz, calcite, illite, paragonite, and inesite. The increased intensity of ettringite peaks in the samples with gypsum highlights the influence of sulfate addition on phase formation and potential effects on the material's performance.

The crystalline phases observed in Figure 3 were quantitatively estimated using Match! Software version 4.2 Build 324, and summarized in Table 4. The presence of crystalline phases such as quartz, paragonite, inesite, ettringite, calcite, gypsum, and illite was confirmed. The absence of portlandite ($Ca(OH)_2$) may be due to its consumption in pozzolanic reactions involving silica, alumina, and iron oxide, forming phases like C-S-H, C-A-S-H, and C-A-F-H [48,49]. Portlandite may also be present in amorphous form or have been consumed by carbonation, forming calcite [50,51].

Table 4. Proportions of the phases identified by XRD using the Match! software.

Mineral Phase	Concentration Range (%)
Quartz	24.7–38.0
Paragonite	18.4–22.8
Inesite	10.1–15.8
Ettringite	8.0–10.0
Calcite	12.5–16.2
Gypsum	0.9–2.5
Ilite	16.9–20.1

3.3. Particle Size Distribution and Plasticity

Figure 2 shows the particle size distribution of the raw materials. Proper grading is essential to optimize compaction, reduce porosity, and improve strength. According to ABCP (1985) and NBR 10833 [24], soils suitable for producing soil–cement bricks must meet the following criteria: 100% of the soil must pass through a No. 4 sieve (4.8 mm), and 10% to 50% of the sample must pass through a No. 200 sieve (0.075 mm). As shown in Figure 4, all materials passed 100% through the No. 4 sieve. For the No. 200 sieve, cement and soil passed approximately 40%, while gypsum passed about 10%. Therefore, all materials meet the granulometric classification requirements.

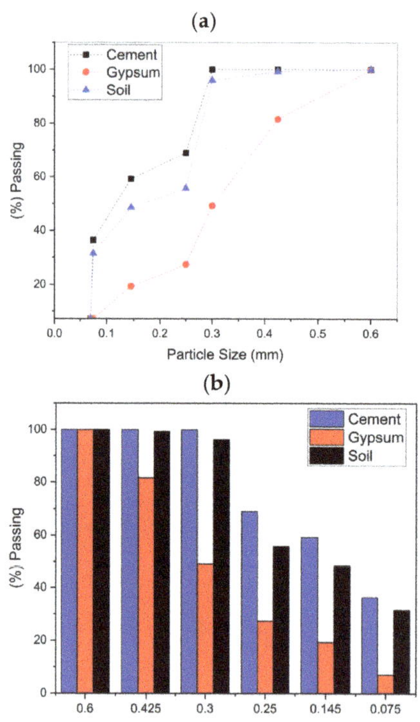

Figure 4. Particle size distribution: (**a**) for the soil–cement brick components according to ASTM D6913/D6913M, with dashed lines provided as a guide to the eye; (**b**) histogram showing the proportion of fine and coarse fractions contributing to compaction and mechanical behavior.

The plasticity characteristics of the soil play an important role in its suitability for stabilization. Soils with higher plasticity index (PI) generally exhibit reduced compressive strength due to less efficient formation of cementitious compounds, such as C-S-H. In this study, the soil met the PI and LL limits of ABCP (1985) and NBR 10833, as shown in Section 2.1. Its classification under USCS (CL) and TRB (A-2-6) reflects moderate cohesion and favorable compaction behavior.

3.4. Thermal Analysis of Gypsum and Soil

Figures 5 and 6 show the TG/DTG curves of gypsum and soil, respectively. For gypsum, a major mass loss occurs at approximately 139 °C, associated with the dehydration of calcium sulfate dihydrate to form hemihydrate ($CaSO_4 \cdot 0.5H_2O$). Between 90 °C and 140 °C, continuous transformations occur, eventually leading to the formation of anhydrite above 170 °C. This process is influenced by environmental conditions such as humidity and the presence of acids or salts, which may accelerate or delay the transformation [52,53].

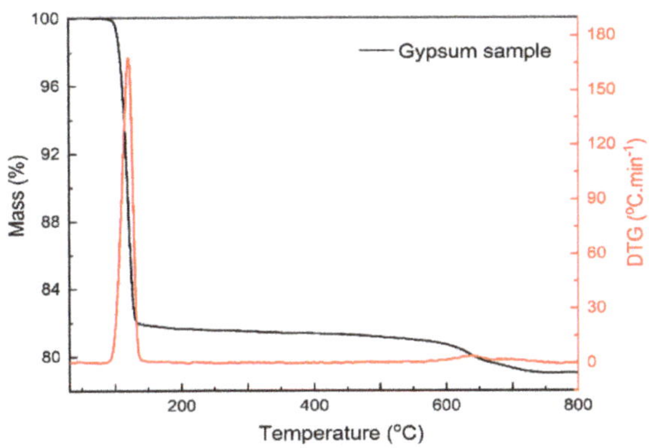

Figure 5. TG/DTG curves of gypsum. Mass loss events in the range of 100–200 °C are associated with dehydration of gypsum phases. A secondary peak near 560 °C suggests decomposition of minor components.

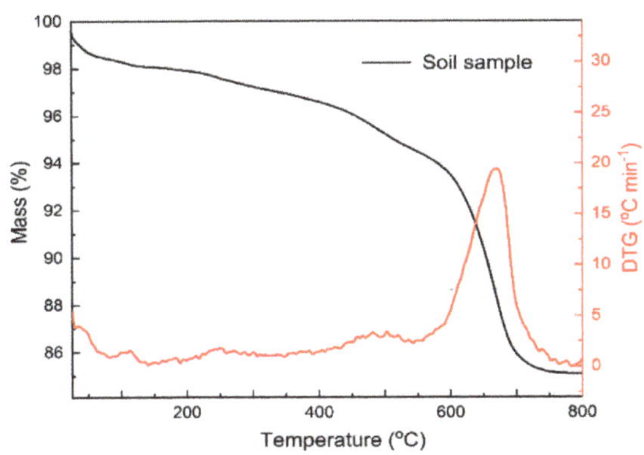

Figure 6. TG/DTG analysis of soil showing thermal events attributed to moisture loss and dehydroxylation of clay minerals, relevant for understanding reactivity and processing behavior.

Above 180 °C, the conversion of anhydrite III to the more stable anhydrite II begins. A secondary mass loss observed near 560 °C may be related to the thermal decomposition of residual impurities in the recycled gypsum, such as contaminant oxides or traces of organic matter. These components may not have been detected in the XRD analysis but are consistent with the EDS and chemical composition data in Table 2. Some authors also associate these events with late-stage dehydration or minor carbonate decomposition [54–57].

For the soil sample, mass loss below 120 °C is attributed to the evaporation of free water. In the range of 120 °C to 430 °C, the loss corresponds to the release of water bound to exchangeable cations and hydrated mineral phases. Between 430 °C and 600 °C, the decomposition of aluminum silicates occurs, while from 600 °C to 800 °C, the thermal degradation of calcium silicates becomes evident [58].

Although thermal behavior is not directly linked to curing at ambient temperatures, these analyses help to elucidate the stability and reactivity of the phases involved, especially under potential exposure to high temperatures during processing or long-term use. In addition, the distinct dehydration behavior of gypsum confirms its potential for early reactivity, which is consistent with the accelerated ettringite formation observed in microstructural and mechanical analyses (Sections 3.5 and 3.6).

3.5. Mechanical Performance

3.5.1. Compressive Strength

Figure 7 shows the compressive strength of the specimens at 7 and 28 days, confirming the continued hydration and densification of the matrix. Although no data were collected at 14 days, the increasing trend indicates effective curing progression across formulations. All formulations exceeded the 2 MPa minimum required. The G10 sample (10% gypsum) showed the highest strength (26.5 MPa), surpassing the standard formulation (25.2 MPa). Statistical analysis confirmed significant differences ($F = 199.02$, $p < 0.05$). This strength increase is attributed to several synergistic mechanisms. The addition of gypsum introduces sulfate ions (SO_4^{2-}), which accelerate early hydration of aluminate phases, leading to ettringite formation. Ettringite refines the pore structure and improves matrix connectivity, enhancing strength [37,59,60]. Gypsum also acts as a fine filler, improving particle packing and reducing porosity, contributing to a denser, more homogeneous microstructure. However, this benefit is limited to moderate gypsum contents. Higher replacement levels (e.g., 20% or 40%) increase porosity and create weak zones that reduce strength. G40, for instance, showed a strength of 20 MPa.

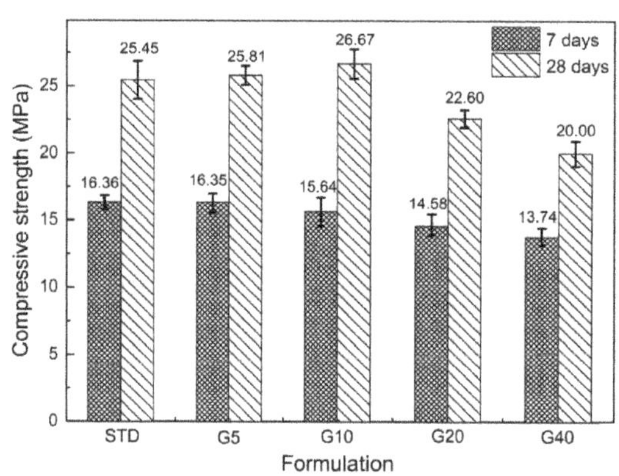

Figure 7. Compressive strength at 7 and 28 days for each formulation. Bars represent standard deviation from five specimens. No intermediate age was tested. All formulations exceeded the 2 MPa minimum required, but the 10% gypsum formulation showed improved strength, while 40% gypsum reduced mechanical performance.

Additionally, in systems with high gypsum content, the lack of free CaO—consumed during earlier hydration reactions—may prevent further formation of C-S-H and ettringite. The excess sulfate may behave as an inert phase, further compromising matrix development. Nonetheless, all mixtures demonstrated acceptable compressive strength.

Formulations replacing cement with fly ash and chemically activated slag mixtures showed compressive strength of 25–33 MPa at 28 days, but they required a more intensive processing. This refers mainly to the ultrafine grinding of the materials, thermal curing of the fly ash, preparation of alkaline activating solutions, and controlled use of superplasticizer additives, which demand greater energy input, precise processing steps, and specialized equipment [61,62]. Formulations with ceramic powder or calcined brick waste typically achieve strengths between 15 and 30 MPa, also showing improvements in durability and microstructural refinement [6,63]. Additionally, non-traditional residues such as shredded masks and rubber fibers have shown benefits in terms of toughness and tensile strength in soil–cement composites, although compressive strength usually remains below 22 MPa [64,65]. The high strength obtained for G10 formulation, together with the simplified processing and lower environmental impact, shows that the use of recycled gypsum in soil–cement bricks can technically compete with established residues, offering advantages in sustainability and ease of use.

3.5.2. Water Absorption

Figure 8 presents the water absorption results for specimens cured for 7 days. Values increased slightly with gypsum content, ranging from 15.71% (SF) to 16.54% (G40). Despite this increase, all values remained below 20%, which is the maximum allowed by NBR 10834 [66], for samples cured for 7 days. Statistical analysis (ANOVA) confirmed a significant difference between the means (F = 332.80, $p < 0.05$). However, the residue-containing samples present water absorption that complies with the standard, although slightly higher than the result obtained for the SF specimen. The water absorption of soil–cement bricks is directly related to porosity and pore connectivity [67]. The high silica (SiO_2—39.39%) content in the soil and the formation of C-S-H (calcium silicate hydrate) contribute to the densification of the material's microstructure, reducing pore connectivity and, consequently, water absorption. Additionally, the chemical reaction of cement with aluminosilicates in the soil leads to the formation of cementitious products such as C-S-H and C-A-S-H (calcium aluminosilicate hydrate), which bind the particles and minimize pore interconnectivity [68,69].

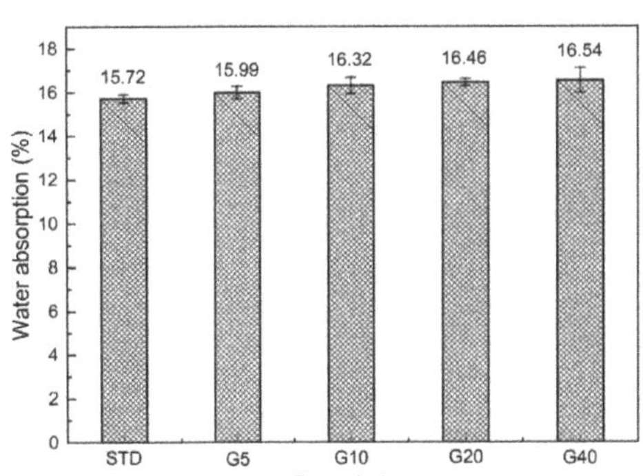

Figure 8. Water absorption of specimens after 7 days of curing. Results indicate increasing porosity with higher gypsum content, though all remain within acceptable standards.

With the partial replacement of cement by gypsum, there is a reduction in the formation of dense hydration products, which may explain the increased porosity and water absorption in gypsum-based formulations. This is consistent with findings in the literature, which report higher pore connectivity in composites with increased gypsum content due to limited formation of compact hydrates [70]. Recycled gypsum composites also tend to exhibit greater water uptake and reduced water resistance, especially when unreacted particles or agglomerates are present in the microstructure [71].

Physical aspects also contribute to this behaviour. Gypsum particles are generally angular and less spherical than cement grains, which reduces particle packing efficiency and generates additional capillary pores within the matrix [39]. Moreover, the mismatch in shrinkage between the gypsum and the surrounding matrix can lead to microcracking, thereby contributing to an increase in interconnected porosity [72]. These chemical, microstructural, and physical mechanisms together explain the statistically significant increase in water absorption observed with higher gypsum content, while still remaining within acceptable limits.

3.5.3. Durability

Figure 9 shows the durability results of the specimens cured for 28 days and evaluated by mass loss after six wetting and drying cycles. Samples SF, G5, G10, and G20 remained below the 10% mass loss limit defined by NBR 13553 [73] for A-2-6 soils. In contrast, G40 exhibited a loss of 23.86%, indicating compromised durability. This behavior is associated with excessive ettringite formation in high-sulfate systems. While moderate ettringite content improves early strength by filling pores, excess leads to expansion and cracking. Studies show that when ettringite forms in small capillary pores, it can cause swelling and structural degradation [74,75]. Conversely, the presence of SiO_2 and Al_2O_3 in the soil promotes formation of stable C-S-H and C-A-S-H phases, less susceptible to leaching. These phases enhance microstructural stability and reduce water ingress. Thus, in formulations with lower gypsum content, pozzolanic reactions dominate, leading to durable structures.

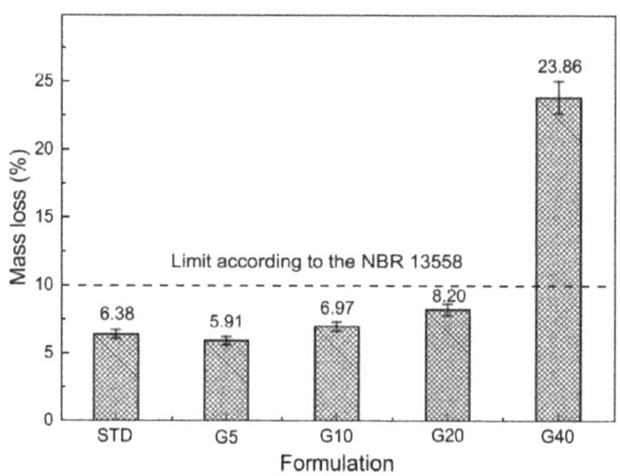

Figure 9. Durability of bricks evaluated by mass loss after six wetting–drying cycles. Only the 40% gypsum formulation exceeded the acceptable threshold.

The balance between ettringite formation and matrix densification is critical. Up to 20% gypsum maintained integrity, but 40% substitution disrupted the structure. The mechanical and absorption data corroborate these findings, highlighting the role of sulfate levels in durability performance.

3.6. Microstructural Analysis

The microstructural evaluation by scanning electron microscopy (SEM) provided key insights into the internal morphology and hydration products of the soil–cement composites. Figure 10 displays representative SEM images of samples SF, G10, and G40. The relevant features are highlighted in the Figure: orange rectangles indicate C-S-H-rich regions with dense morphology, red circles mark visible pores within the matrix, yellow circles show acicular ettringite crystals, burgundy rectangles denote large unreacted gypsum particles in the G40 formulation, and the yellow rectangle highlights more elongated ettringite crystals with characteristic acicular morphology. These features are consistent with the morphological differences described in the subsequent paragraphs and help illustrate the relationship between gypsum content, microstructure, and material performance.

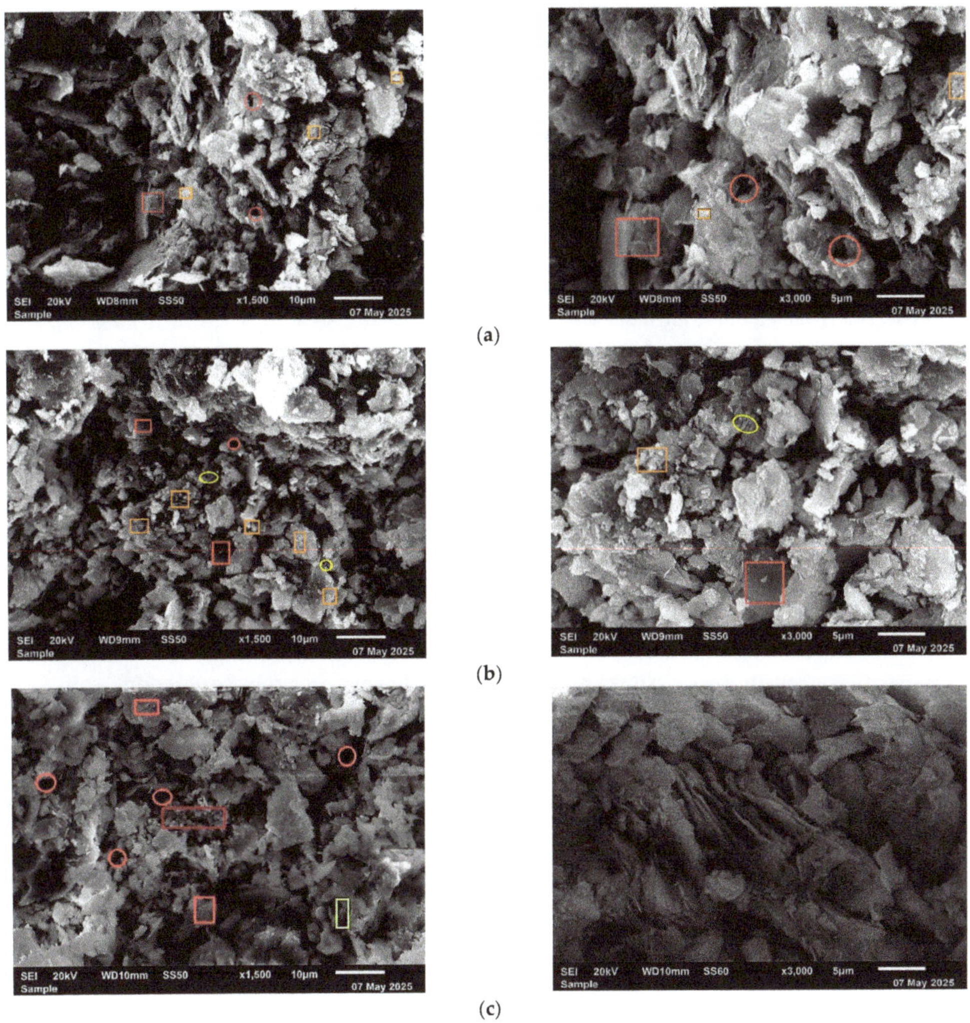

Figure 10. Scanning electron microscopy (SEM) images of (**a**) standard formulation (SF), (**b**) sample with 10 wt% gypsum (G10), and (**c**) sample with 40 wt% gypsum (G40), taken at 1500× and 3000× magnification. Scale bars represent 10 μm and 5 μm, respectively. Highlighted features include: orange rectangles indicating dense calcium silicate hydrate (C–S–H) regions, red circles marking visible pores, yellow circles denoting acicular ettringite crystals, burgundy rectangles identifying unreacted gypsum agglomerates in G40, and a yellow rectangle showing elongated ettringite with typical needle-like morphology. The SF sample presents a compact matrix with low porosity; G10 shows a denser structure with well-dispersed ettringite; G40 exhibits high porosity, microcracking, and unreacted gypsum, reflecting reduced microstructural integrity.

The SF sample exhibited a relatively compact matrix with a moderate presence of C-S-H (calcium silicate hydrate) phases and limited porosity. The absence of significant ettringite formation is consistent with the lower sulfate availability in this formulation. This microstructural pattern correlates with the intermediate mechanical strength and low water absorption observed in this sample.

In contrast, the G10 sample displayed a denser and more cohesive matrix with enhanced particle bonding. Needle-like ettringite crystals were observed dispersed throughout the structure, filling microvoids and contributing to a refined pore network. These features support the higher compressive strength and reduced water absorption of G10. The coexistence of ettringite and C-S-H appears to enhance matrix integrity and is indicative of an optimal balance between early hydration reactions and long-term stability. Similar findings have been reported in the literature [76,77], emphasizing the synergistic effect of moderate gypsum incorporation in promoting early ettringite formation without compromising the long-term stability of the cementitious matrix.

The G40 sample, however, exhibited microstructural degradation. SEM analysis revealed large unreacted gypsum agglomerates, significant porosity, and microcracks—particularly in regions of ettringite accumulation. These features are associated with internal stress from excessive ettringite formation and lower matrix cohesion. The porous structure explains the reduced mechanical strength and increased water absorption seen in this formulation. Similar mechanisms have been reported in high-sulfate cementitious systems, where overproduction of ettringite contributes to expansion, cracking, and durability loss [78,79].

The EDS analysis (Figure 11) further supported the SEM findings. For G10, the EDS spectra showed clear peaks for calcium (Ca), silicon (Si), and aluminum (Al), consistent with the formation of C-S-H and C-A-S-H phases—essential for mechanical performance and durability. Conversely, the EDS spectrum for G40 revealed a more heterogeneous elemental distribution, with visible concentrations of unreacted gypsum and lower levels of hydration products. The irregular distribution of elements such as iron (Fe), aluminum (Al), silicon (Si), magnesium (Mg), and sodium (Na) suggests the presence of residual soil minerals and incomplete hydration, particularly in G40.

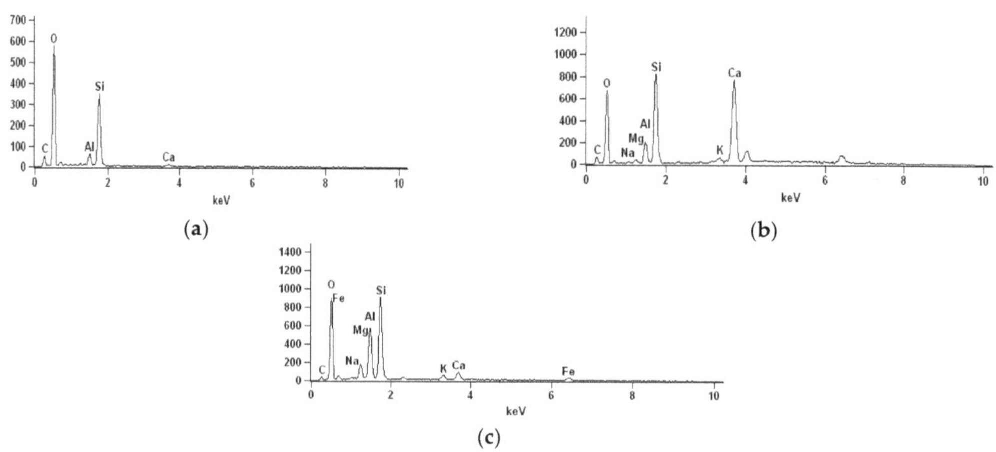

Figure 11. EDS spectra of the fracture surface of soil–cement specimens cured for 28 days: (**a**) SF; (**b**) G10; (**c**) G40. Spectra confirm the presence of calcium, silicon, aluminum, and sulfur, supporting the formation of C-S-H and ettringite, with variations correlating to gypsum content.

Furthermore, the presence of free quartz, as evidenced by the high SiO_2/Al_2O_3 molar ratio (ranging from 8.66 to 10.53, based on XRD), may act as an inert phase that creates discontinuity zones within the matrix, promoting pore formation and increasing water absorption. These effects are especially apparent in G40 and contribute to its inferior performance.

The absence of portlandite ($Ca(OH)_2$) in the XRD patterns further supports the hypothesis that this phase was consumed in pozzolanic reactions or transformed into calcite ($CaCO_3$) via carbonation. This is corroborated by the presence of calcite peaks in the XRD patterns and carbonate bands in the FTIR spectra (Section 3.7). It is also possible that portlandite exists in an amorphous or poorly crystalline state, escaping detection by XRD.

A methodological limitation should be noted regarding the EDS analysis. Since the spectra were acquired from fracture surfaces, the data may be affected by surface roughness and uneven morphology, which can limit spatial resolution and compositional accuracy. While the results provide useful qualitative insights, future studies should consider using polished cross-sections to improve the reliability of hydration product identification.

3.7. FTIR Analysis

FTIR analysis was performed to investigate the functional groups and chemical bonds present in the composites and to complement the microstructural findings. Figure 12 shows the FTIR spectra for the standard formulation (SF) and for samples with 10% (G10) and 40% (G40) gypsum replacement. In the 3600–3000 cm^{-1} region, SF displayed a narrow band at ~3640 cm^{-1} corresponding to the O–H stretching of portlandite ($Ca(OH)_2$), which was less intense or absent in G10 and G40, suggesting consumption through pozzolanic or carbonation reactions. A broad band near 3450 cm^{-1}, attributed to O–H stretching in C–S–H and adsorbed water, was more prominent in G10 and G40, indicating a higher degree of hydration and water retention [46,80]. The band near 1650 cm^{-1}, more visible in G10 and G40, is related to H–O–H bending vibrations of retained molecular water. Between 1550 and 1250 cm^{-1}, all samples exhibited carbonate (C–O) bands, with higher intensity in G10 and G40, consistent with partial carbonation and supported by calcite peaks in XRD. These observations confirm the transformation of portlandite into calcite and the presence of carbonation products. In the region between 1150 and 900 cm^{-1}, sulfate groups (SO_4^{2-}) and Si–O–Si vibrations overlapped. The sulfate bands were stronger in gypsum-containing samples, particularly G10, suggesting ettringite formation. Slightly lower intensity in G40 may reflect reduced hydration efficiency or overlapping effects of unreacted gypsum. Finally, the 900–700 cm^{-1} region showed Al–O and Si–O bending modes from clay minerals (e.g., illite) and aluminosilicate hydrates (C–A–S–H) [45,81]. These bands were present in all spectra, but weaker in G40, suggesting lower formation of hydration products and a greater proportion of unreacted or inert phases, as also observed in SEM and XRD analyses

Overall, the FTIR spectra reinforce the findings from XRD and SEM/EDS. Moderate gypsum content (G10) favors the formation of cementitious phases and carbonation products associated with improved performance, while excessive gypsum (G40) appears to reduce hydration efficiency and promote porosity.

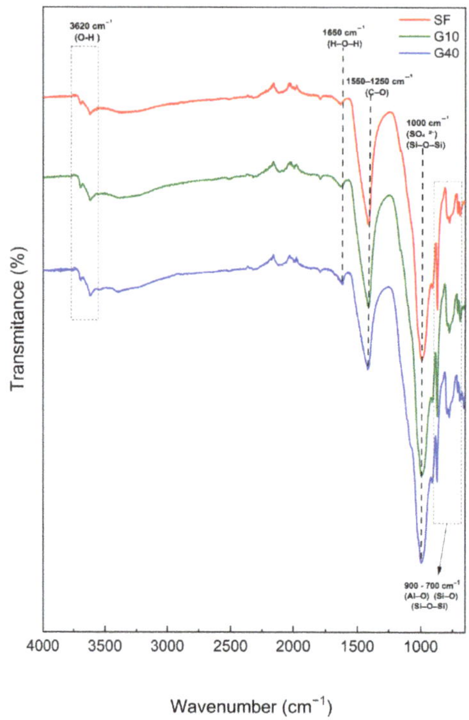

Figure 12. FTIR spectra of soil–cement composites with 0% (SF), 10% (G10), and 40% (G40) gypsum. Key bands include O–H, H–O–H, carbonate (C–O), sulfate (SO_4^{2-}), and silicate (Si–O–Si) vibrations, indicating hydration and carbonation processes.

3.8. Environmental Implications and Limitations Related to Gypsum Impurities

The use of recycled gypsum waste as a partial substitute for Portland cement in soil–cement bricks represents an environmentally favorable strategy. It contributes to reducing CO_2 emissions, energy consumption, and landfill disposal associated with construction and demolition waste. The production of 1 kg of Portland cement emits approximately 0.9 kg of CO_2 [82], primarily due to the calcination of limestone and the high energy demands of clinker production [83,84]. With 10% cement replacement and an average of 0.3 kg of cement per brick, the use of recycled gypsum in 75 bricks would avoid the emission of approximately 20.25 kg of CO_2, highlighting its environmental benefit at scale. The processing of recycled gypsum involves only low-energy steps, drying, grinding, and sieving, making it a more sustainable alternative compared to cement production. In line with successful examples from other sectors, such as mining and metallurgy, that demonstrate that circular economy models can be enhanced by integrating waste materials into cement-based systems through institutional synergy and localized circular flows [85], the reuse of recycled gypsum supports circular economy goals by promoting waste valorization and resource efficiency.

However, an important methodological limitation of this study must be acknowledged: the chemical composition of the recycled gypsum waste was not fully characterized in terms of potential impurities. These impurities may include organic matter, soluble salts such as chlorides or sulfates, and heavy metals, which can significantly influence hydration kinetics, the formation of cementitious products, and the long-term performance and environmental safety of the bricks. While the gypsum used in this work was visually selected and thermally treated at 60 ± 5 °C to remove moisture, no chemical treatments or specific analyses were conducted to quantify impurity levels. The presence of these impurities may delay the setting time or cause premature stiffening depending on salt concentration, reduce mechanical strength by interfering with hydration reactions, compromise durability due to enhanced leaching or efflorescence, and pose environmental risks if contaminants migrate from the bricks during use or disposal. Given these concerns, future studies should include comprehensive chemical characterization of recycled gypsum from different sources, using techniques such as XRF, TGA, and leaching tests, to evaluate impurity content and assess its impact on strength, durability, and environmental compliance. Additionally, comparing untreated and purified gypsum could provide further insight into the necessity and feasibility of pre-treatment.

Despite this limitation, the formulations with 10% gypsum substitution showed adequate performance in mechanical, physical, and durability tests, suggesting that the impurity level in the material used was not critical. Further studies should be done to assess the long-term leaching and environmental impact of potential impurities in recycled gypsum.

4. Conclusions

This study demonstrated that replacing up to 20% of Portland cement with recycled gypsum waste in soil–cement bricks is technically feasible and environmentally advantageous. Replacing up to 20% of Portland cement with recycled gypsum waste in soil–cement bricks proved technically feasible and compliant with Brazilian standards, with the 10% gypsum formulation (G10) achieving the highest compressive strength after 28 days of curing. This result was supported by microstructural and mineralogical analyses, which indicated the formation of denser cementitious matrices rich in C-S-H and C-A-S-H phases.

Although water absorption increased slightly with higher gypsum contents, all values remained within acceptable limits. Durability was also maintained in formulations with up to 20% gypsum, while the 40% formulation (G40) showed signs of performance loss due to excessive ettringite formation and carbonation-related degradation.

Beyond technical validation, this approach contributes to sustainability goals by reducing cement consumption, diverting gypsum waste from landfills, and lowering carbon emissions. However, proportions above 20% should be avoided to preserve the mechanical and durability performance of the bricks.

Author Contributions: Conceptualization, E.S.S.N., H.A.d.O., Z.S.M. and C.P.d.S.; Methodology, E.S.S.N., H.A.d.O., M.d.A.G. and Z.S.M.; Software, E.S.S.N. and H.A.d.O.; Validation, E.S.S.N., C.P.d.S. and H.A.d.O.; Formal analysis, E.S.S.N.; Investigation, E.S.S.N., H.A.d.O. and M.d.A.G.; Resources, M.E.G.V., Z.S.M. and H.A.d.O.; Data curation, E.S.S.N., M.E.G.V. and Z.S.M.; Writing—original draft preparation, E.S.S.N.; Writing—review and editing, H.A.d.O., Z.S.M., C.P.d.S. and M.d.A.G.; Visualization, E.S.S.N., C.P.d.S., M.d.A.G., M.E.G.V. and Z.S.M.; Supervision, Z.S.M.; Project administration, H.A.d.O. and Z.S.M.; Funding acquisition, Z.S.M. All authors have read and agreed to the published version of the manuscript.

Funding: This research was funded by CNPq grant number 409048/2021-6.

Institutional Review Board Statement: Not applicable.

Informed Consent Statement: Not applicable.

Data Availability Statement: The data is available at: https://data.mendeley.com/datasets/75mm7y7h99/1, DOI: 10.17632/75mm7y7h99.1 Accessed on 1 April 2025.

Acknowledgments: The authors thank CAPES and CNPq for their financial support, and the Soil and Materials Laboratories (IFS), CMNano (UFS), and LMDCEM for providing infrastructure and technical assistance essential for this study.

Conflicts of Interest: The authors declare no conflict of interest.

References

1. Mariyam, S.; Cochrane, L.; Al-Ansari, T.; McKay, G. A framework to support localized solid waste management decision making: Evidence from Qatar. *Environ. Dev.* **2024**, *50*, 100986. [CrossRef]
2. Beccarello, M.; Di Foggia, G. Sustainable Development Goals data-driven local policy: Focus on SDG 11 and SDG 12. *Adm. Sci.* **2022**, *12*, 167. [CrossRef]
3. Vilela, A.P.; Eugênio, T.M.C.; de Oliveira, F.F.; Mendes, J.F.; Ribeiro, A.G.C.; Vaz, L.E.V.d.S.B.; Mendes, R.F. Technological properties of soil-cement bricks produced with iron ore mining waste. *Constr. Build. Mater.* **2020**, *262*, 120883. [CrossRef]
4. da Silva, M.V.V.; Garcia, R.L.M.; de Siqueira, R.A.; Júnior, G.A.d.S. Técnicas de estabilização e reforço sustentável do solo. *Rev. FT* **2024**, *29*, 29–30. [CrossRef]
5. Xiong, W.; Chen, Y.; Xu, J.; Zhang, Z.; Liang, C. Reuse of engineering waste soil and recycled fine aggregate to manufacture eco-friendly unfired clay bricks: Experimental assessment, data-driven modeling and environmental friendliness evaluation. *Case Stud. Constr. Mater.* **2023**, *19*, e02608. [CrossRef]
6. Hashim, A.A.; Anaee, R.A.; Nasr, M.S. Improving the mechanical, corrosion resistance, microstructural and environmental performance of recycled aggregate concrete using ceramic waste powder as an alternative to cement. *Ceramics* **2025**, *8*, 11. [CrossRef]
7. Rocha Silva, B.; Junior, H.S.M.; Filho, A.J.d.S.; Santos, A.M.S.; de Melo, F.M.C.; Almeira, V.G.d.O.; de Oliveira, H.A. Soil cement brick production process: Literature review. *MOJ Civ. Eng.* **2023**, *7*, 19–26. [CrossRef]
8. Ângelo, F.A.; Simões, G.F. Tijolos ecoeficientes de barro cru com resíduos sólidos e efluente industrial utilizando tecnologias não convencionais. *Ambiente Construído* **2023**, *23*, 131–148. [CrossRef]
9. de Castro, E.D.; Villella, L.S.; Mendes, L.M.; Mendes, R.F.; Ribeiro, A.G.C.; Júnior, J.B.G.; Rabelo, G.F. Analysis of the coffee peel application over the soil-cement bricks properties. *Coffee Sci.* **2019**, *14*, 12. [CrossRef]
10. Azevedo, A.R.G.; Marvila, T.; Fernandes, W.J.; Alexandre, J.; Xavier, G.; Zanelato, E.; Cerqueira, N.; Pedroti, L.; Mendes, B. Assessing the potential of sludge generated by the pulp and paper industry in assembling locking blocks. *J. Build. Eng.* **2019**, *23*, 334–340. [CrossRef]
11. Araújo, V.D.; Martins, V.A.; Cavalheri, P.S.; Honório, L.; Lima, P.D.M.; Filho, F.J.C.M. Incorporation of tannery waste and sugarcane bagasse ash in soil–cement bricks. *Rev. AIDIS De Ing. Y Cienc. Ambient.* **2022**, *15*, 154. [CrossRef]
12. Ahmed, A.; Ugai, K.; Kamei, T. Laboratory and field evaluations of recycled gypsum as a stabilizer agent in embankment construction. *Soils Found.* **2011**, *51*, 975–990. [CrossRef]
13. Azevedo, A.C.; Delgado, J.M.P.Q.; Neves, T.H.C.; Costa e Silva, A.J. Adhesion of gypsum plaster coatings: Experimental evaluation. In *Building Pathology and Rehabilitation*; Springer: Berlin/Heidelberg, Germany, 2021; pp. 41–66. [CrossRef]
14. Camarini, G.; Pimentel, L.L.; De Sá, N.H.R. Assessment of the material loss in walls renderings with β-hemihydrate paste. *Appl. Mech. Mater.* **2011**, *71–78*, 1242–1245. [CrossRef]
15. Jiménez-Rivero, A.; García-Navarro, J. Exploring factors influencing post-consumer gypsum recycling and landfilling in the European Union. *Resour. Conserv. Recycl.* **2017**, *116*, 116–123. [CrossRef]
16. Antunes, M.L.P.; de Sá, A.B.; Oliveira, P.S.; Rangel, E.C. Utilization of gypsum from construction and demolition waste in Portland cement mortar. *Cerâmics* **2019**, *65* (Suppl. S1), 1–6. [CrossRef]
17. Torres-Ortega, R.; Torres-Sanchez, D.L.; López-Lara, T. Mechanical properties of hydraulic concretes with partial replacement of Portland cement by pozzolans obtained from agro-industrial residues: A review. *Heliyon* **2024**, *11*, e41004. [CrossRef] [PubMed]
18. Puerto, J.D.; Uribe, S.; Ayala, L.; Padilla, A.; Rodriguez, A. Reliability of reusing gypsum flat board grinded waste as a conventional plaster replacement for buildings. *Sustainability* **2024**, *16*, 7889. [CrossRef]
19. *ASTM D6913*; Standard Test Methods for Particle-Size Distribution (Gradation) of Soils Using Sieve Analysis. ASTM International: West Conshohocken, PA, USA, 2017.
20. *ASTM D4318-18*; Standard Test Methods for Liquid Limit, Plastic Limit, and Plasticity Index of Soils. ASTM International: West Conshohocken, PA, USA, 2018.
21. *DNER-ME 093/94*; Solos—Determinação da Densidade Real. DNER: Rio de Janeiro, Brazil, 1994.
22. Kang, H.; Kang, S.-H.; Moon, J. A comparative investigation of hydration reaction of Portland cement with the use of alkanolamine as a chemical admixture or grinding agent under laboratory condition. *J. Build. Eng.* **2024**, *88*, 109214. [CrossRef]
23. Axthammer, D.; Lange, T.; Dengler, J.; Gädt, T. Early hydration and viscoelastic properties of tricalcium aluminate pastes influenced by soluble sodium salts. *Cem. Concr. Res.* **2025**, *190*, 107788. [CrossRef]

24. *NBR 10833*; Fabricação de Tijolo e Bloco de Solo-Cimento com Utilização de Prensa Manual ou Hidráulica—Procedimento. ABNT: Rio de Janeiro, Brazil, 2012.
25. *ASTM C191*; Standard Test Methods for Time of Setting of Hydraulic Cement by Vicat Needle. ASTM International: West Conshohocken, PA, USA, 2023.
26. *ASTM C188*; Standard Test Method for Density of Hydraulic Cement. ASTM International: West Conshohocken, PA, USA, 2022.
27. *NBR 16372*; 2015—Cimento Portland e Outros Materiais em pó—Determinação da Finura pelo Método de Permeabilidade ao ar (Método de Blaine). ABNT: Rio de Janeiro, Brazil, 2015.
28. *ASTM C109/C109M*; Standard Test Method for Compressive Strength of Hydraulic Cement Mortars. ASTM International: West Conshohocken, PA, USA, 2023.
29. Santa Rosa, J.M.S.; Silva, R.G.P.; Lafayette, K.P.V. Sustainable Alternative for the Production of Soil Cement Bricks. *J. Manag. Sustain.* **2023**, *13*, 45. [CrossRef]
30. Ashour, T.; Korjenic, A.; Korjenic, S. Equilibrium Moisture Content of Earth Bricks Biocomposites Stabilized with Cement and Gypsum. *Cem. Concr. Compos.* **2015**, *59*, 18–25. [CrossRef]
31. Metzker, S.L.O.; Sabino, T.P.F.; Mendes, J.F.; Ribeiro, A.G.C.; Mendes, R.F. Soil-cement bricks development using polymeric waste. *Environ. Sci. Pollut. Res.* **2022**, *29*, 21034–21048. [CrossRef] [PubMed]
32. Kongkajun, N.; Laitila, E.A.; Ineure, P.; Prakaypan, W.; Cherdhirunkorn, B.; Chakartnarodom, P. Soil-Cement Bricks Produced from Local Clay Brick Waste and Soft Sludge from Fiber Cement Production. *Case Stud. Constr. Mater.* **2020**, *13*, e00448. [CrossRef]
33. Leonel, R.F.; Folgueras, M.V.; Dalla Valentina, L.V.O.; Prim, S.R.; Prates, G.A.; Caraschi, J.C. Characterization of Soil-Cement Bricks with Incorporation of Used Foundry Sand. *Cerâmica* **2017**, *63*, 329–335. [CrossRef]
34. *NBR 8492*; Tijolo de Solo-Cimento—Análise Dimensional, Determinação da Resistência à Compressão e da Absorção de Água—Método de Ensaio. ABNT: Rio de Janeiro, Brazil, 2012.
35. *NBR 5739*; Cimento Portland—Determinação da Resistência à Compressão. ABNT: Rio de Janeiro, Brazil, 2007.
36. *NBR 10836*; Bloco de Solo-Cimento Sem Função Estrutural—Análise Dimensional, Determinação da Resistência à Compressão e da Absorção de Água—Método de Ensaio. ABNT: Rio de Janeiro, Brazil, 2012.
37. *NBR 13554*; 2012—Solo-Cimento—Ensaio de Durabilidade por Molhagem e Secagem—Método de Ensaio. ABNT: Rio de Janeiro, Brazil, 2012.
38. Hammer, Ø. Harper, D.A.T.; Ryan, P.D. PAST: Paleontological Statistics software package for education and data analysis. *Palaeontol Electron.* **2001**, *4*, 1–9. Available online: http://palaeo-electronica.org/2001_1/past/issue1_01.htm (accessed on 27 June 2025).
39. Chernyshova, I.V.; Lesovik, V.S.; Fediuk, R.S.; Amran, M.; Vatin, N.I. Enhancement of Fresh Properties and Performances of the Eco-Friendly Gypsum-Cement Composite (EGCC). *Constr. Build. Mater.* **2020**, *260*, 120462. [CrossRef]
40. Alves, M.L.X.F.N.; Barros, S.V.A.; de Oliveira, F.N. Physical and Chemical Characterization of Soil for Use in Ecological Bricks. In *Uniting Knowledge: Integrated Scientific Research for Global Development*; Seven Publicações: São José dos Pinhais, Brazil, 2024; Volume 2, pp. 532–539. [CrossRef]
41. Qi, J.; Ma, W.; Song, C. Influence of Freeze—Thaw on Engineering Properties of a Silty Soil. *Cold Reg. Sci. Technol.* **2008**, *53*, 397–404. [CrossRef]
42. Melnick, V.; Pianaro, S.A.; Cava, S.; Tebcherani, S.M. Application of Oil Shale Mining By-Products as Raw Materials in the Determination of the Vitrification Curve of Red Porcelain Stoneware Tiles by Dilatometric Method. *Appl. Clay Sci.* **2010**, *50*, 311–314. [CrossRef]
43. Peffers, K.; Tuunanen, T.; Rothenberger, M.A.; Chatterjee, S. A Design Science Research Methodology for Information Systems Research. In *Design Science Research: A Method for Science and Engineering*; Peffers, K., Rothenberger, M.A., Kuechler, B., Eds.; CRC Press: Boca Raton, FL, USA, 2023; pp. 11–28. [CrossRef]
44. Harrison, T.; Jones, M.R.; Lawrence, D. The Production of Low Energy Cements. In *Lea's Chemistry of Cement and Concrete*, 5th ed.; Hewlett, P.C., Liska, M., Eds.; Butterworth-Heinemann: Oxford, UK, 2019; pp. 341–361. [CrossRef]
45. Reigl, S.; Van Driessche, A.E.S.; Wagner, E.; Montes-Hernandez, G.; Mehringer, J.; Koltzenburg, S.; Kunz, W.; Kellermeier, M. Toward More Sustainable Hydraulic Binders: Controlling Calcium Sulfate Phase Selection via Specific Additives. *ACS Sustain. Chem. Eng.* **2023**, *11*, 8450–8461. [CrossRef]
46. Stawski, T.M.; Besselink, R.; Chatzipanagis, K.; Hövelmann, J.; Benning, L.G.; Van Driessche, A.E.S. Nucleation Pathway of Calcium Sulfate Hemihydrate (Bassanite) from Solution: Implications for Calcium Sulfates on Mars. *J. Phys. Chem. C* **2020**, *124*, 8411–8422. [CrossRef]
47. Gitirana, G.F.N.; Bandeira, A.P.N.; Dourado, K.C.A.; Souza Neto, J.B. *Solos Não Saturados no Contexto Geotécnico*, 2nd ed.; Editora Cubo: São Carlos, Brazil, 2023. [CrossRef]
48. Macioski, G.; Soto, N.T.A.; Medeiros, M.H.F.d.; Hoppe Filho, J.; Araújo, M.S.d.; Cerri, J.A. Portlandite consumption by red ceramic waste due to alkali activation reaction. *Ambiente Construído* **2021**, *21*, 7–21. [CrossRef]

49. Kleib, J.; Amar, M.; Aouad, G.; Bourbon, X.; Benzerzour, M.; Abriak, N.-E. The use of Callovo-Oxfordian Argillite as a raw material for Portland cement clinker production. *Buildings* **2022**, *12*, 1421. [CrossRef]
50. Bošnjak, J.; Ožbolt, J.; Hahn, R. Permeability Measurement on High Strength Concrete without and with Polypropylene Fibers at Elevated Temperatures Using a New Test Setup. *Cem. Concr. Res.* **2013**, *53*, 104–111. [CrossRef]
51. Park, S.-S.; Suh, E.-H.; Chae, K.-H.; Jang, S.-K.; Kim, J.-H. A Study on Soil Cementation and Calcite Precipitation with Clay as a Medium. *J. Korean Geotech. Soc.* **2015**, *31*, 17–27. [CrossRef]
52. Krause, F.; Renner, B.; Coppens, F.; Dewanckele, J.; Schwotzer, M. Reactivity of gypsum-based materials subjected to thermal load: Investigation of reaction mechanisms. *Materials* **2020**, *13*, 1427. [CrossRef]
53. Azimi, G.; Papangelakis, V.G. Mechanism and kinetics of gypsum–anhydrite transformation in aqueous electrolyte solutions. *Hydrometallurgy* **2011**, *108*, 122–129. [CrossRef]
54. Voigt, W.; Freyer, D. Solubility of anhydrite and gypsum at temperatures below 100 °C and the gypsum-anhydrite transition temperature in aqueous solutions: A re-assessment. *Front. Nucl. Eng.* **2023**, *2*, 1208582. [CrossRef]
55. Al Disi, Z.A.; Sadooni, F.; Al-Kuwari, H.A.-S.; Bontognali, T.R. Microbially influenced formation of anhydrite at low temperature. *Sci. Total Environ.* **2023**, *902*, 165820. [CrossRef]
56. Engbrecht, D.C.; Hirschfeld, D.A. Thermal analysis of calcium sulfate dihydrate sources used to manufacture gypsum wallboard. *Thermochim. Acta* **2016**, *639*, 173–185. [CrossRef]
57. Lira, M.C.d.A.; Neiva, L.S. Investigative study of the potentialities of noble applications for the Brazilian calcium sulfate α-hemihydrate. *Pesqui. E Ensino Em Ciências Exatas E Da Nat.* **2020**, *4*, e1516. [CrossRef]
58. Vipulanandan, C.; Mohammed, A. XRD and TGA, swelling and compacted properties of polymer treated sulfate contaminated CL soil. *J. Test. Eval.* **2016**, *44*, 2270–2284. [CrossRef]
59. Cai, H.; Chen, C.; Wei, S.; Zhu, S. Strength development of cemented soil cured in water-air conditions at varied temperatures: Experimental investigation and model characterization. *J. Mater. Civ. Eng.* **2023**, *35*, 3. [CrossRef]
60. Zak, P.; Ashour, T.; Korjenic, A.; Korjenic, S.; Wu, W. The influence of natural reinforcement fibers, gypsum and cement on compressive strength of earth bricks materials. *Constr. Build. Mater.* **2016**, *106*, 179–188. [CrossRef]
61. Zhu, Y.; Jia, Y.; Xie, X.S.; Xu, J. Study on the Macroscopic Properties and Microstructure of High Fly Ash Content Alkali-Activated Fly Ash Slag Concrete Cured at Room Temperature. *Materials* **2025**, *18*, 547. [CrossRef] [PubMed]
62. Rohman, M.N.; Simanihuruk, B.; Dewita, H. Karakteristik Beton Mutu FC 30 Dengan Menggunakan Limbah Peleburan Baja (Slag) Dan Pembakaran Batubara (Fly Ash). *Pros. Semin. Nas. Rekayasa Dan Teknol.* **2025**, *5*, 61–65. [CrossRef]
63. Kirgiz, M.S.; Khatib, J.; Galdino, A.G.S.; Kenaï, S.; Ashteyat, A.M.; Campilho, R.D.S.G.; González-Lezcano, R.A.; Bamigboye, G.; Nagaprasad, N.; Ramaswamy, K. Waste Marble Sludge and Calcined Clay Brick Powders in Conventional Cement Farina Production for Cleaner Built Environment. *Dent. Sci. Rep.* **2025**, *15*, 3467. [CrossRef]
64. Li, Y.; Yan, X.; Wan, M.P.; Zhou, J.; Liu, J. Research on Recycling and Utilization of Shredded Waste Mask Fibers to Prepare Sustainable Engineered Cementitious Composites. *Buildings* **2025**, *15*, 402. [CrossRef]
65. Jafari, M.M.; Bagheripour, M.H.; Yaghoubi, E. Prediction of Strength Properties of Reinforced and Stabilized Sandy Soil as a Building Foundation Material. *Buildings* **2025**, *15*, 382. [CrossRef]
66. *NBR 10834*; 2012—Bloco de Solo-Cimento Sem Função Estrutural—Requisitos. ABNT: Rio de Janeiro, Brazil, 2012.
67. Dulal, P.; Maharjan, S.; Timalsina, M.P.; Maharjan, Y.; Giri, A.; Tamang, A. Engineering properties of cement-stabilized compressed earth bricks. *J. Build. Eng.* **2023**, *77*, 107453. [CrossRef]
68. Thennarasan Latha, A.; Murugesan, B.; Thomas, B.S. Compressed stabilized earth block incorporating municipal solid waste incinerator bottom ash as a partial replacement for fine aggregates. *Buildings* **2023**, *13*, 1114. [CrossRef]
69. Corrêa-Silva, M.; Cristelo, N.; Rouainia, M.; Araújo, N.; Miranda, T. Constitutive behaviour of a clay stabilised with alkali-activated cement based on blast furnace slag. *Sustainability* **2022**, *14*, 13736. [CrossRef]
70. Kamran, F.; Basavarajappa, M.; Bala, N.; Hashemian, L. Laboratory Evaluation of Stabilized Base Course Using Asphalt Emulsion and Asphaltenes Derived from Alberta Oil Sands. *Constr. Build. Mater.* **2021**, *283*, 122735. [CrossRef]
71. Dong, W.; Li, W.; Tao, Z.; Wang, K. Piezoresistive Properties of Cement-Based Sensors: Review and Perspective. *Constr. Build. Mater.* **2019**, *203*, 146–163. [CrossRef]
72. Chen, H.; Zhou, S.; Zhang, J.; Liu, X.; Wu, H. Effect of fly ash and gypsum on drying shrinkage and mechanical properties of one-part alkali-activated slag mortar. *Struct. Concr.* **2025**, *26*, 1275–1288. [CrossRef]
73. *NBR 13553*; Materiais Para Emprego em Parede Monolítica de Solo-Cimento Sem Função estrutural—Requisitos. ABNT: Rio de Janeiro, Brasil, 2012.
74. Höglund, L.O. Some notes of ettringite formation in cementitious materials; Influence of hydration and thermodynamic constraints for durability. *Cem. Concr. Res.* **1992**, *22*, 217–228. [CrossRef]
75. Gu, Y.; Martin, R.-P.; Metalssi, O.O.; Fen-Chong, T.; Dangla, P. Pore size analyses of cement paste exposed to external sulfate attack and delayed ettringite formation. *Cem. Concr. Res.* **2019**, *123*, 105766. [CrossRef]

76. Wang, L.; Wang, Q.; Cai, J.; Zhou, F.; Cheng, Z.; Zhang, J. Experimental Study on the Mechanical Properties and Microstructural Mechanisms of Coal Gangue-Based Cementitious Materials Synergistically Activated by Desulfurization Gypsum and Lime. *Polymers* **2025**, *17*, 932. [CrossRef] [PubMed]
77. Shi, C.; Zhang, Z.; Yang, J.; Zhang, J. An Eco-Friendly Phosphogypsum-Based Cementitious Material. *Front. Phys.* **2022**, *10*, 892037. [CrossRef]
78. Fu, Y.; Beaudoin, J.J. Mechanisms of Delayed Ettringite Formation in Portland Cement Systems. *ACI Mater. J.* **1996**, *93*, 327–333. [CrossRef]
79. Li, L.; Darquennes, A.; Hannawi, K.; Che, C. Effects of Sulfates on the Hydration of Portland Cement—A Review. *Constr. Build. Mater.* **2021**, *270*, 121134. [CrossRef]
80. Taman, M.; Abd Elaty, M.A.; Behiry, R.N. Codes Applicability of Estimating the FRC Compressive Strength by the Core-Drilling Method. *Constr. Build. Mater.* **2022**, *330*, 127227. [CrossRef]
81. Wild, S.; Khatib, J.M. Portlandite consumption in metakaolin cement pastes and mortars. *Cem. Concr. Res.* **1997**, *27*, 137–146. [CrossRef]
82. Habert, G.; d'Espinose de Lacaillerie, J.B.; Roussel, N. An environmental evaluation of geopolymer based concrete production: Reviewing current research trends. *J. Clean. Prod.* **2010**, *19*, 1229–1238. [CrossRef]
83. Scrivener, K.L.; John, V.M.; Gartner, E.M. Eco-efficient cements: Potential economically viable solutions for a low-CO_2 cement-based materials industry. *Cem. Concr. Res.* **2018**, *114*, 2–26. [CrossRef]
84. Habert, G. Environmental Impact of Portland Cement Production. In *Eco-Efficient Concrete*; Woodhead Publishing Series in Civil and Structural Engineering; Woodhead Publishing: Cambridge, UK, 2013; pp. 3–25. [CrossRef]
85. Born, K. Exploring the Spatial Dynamics of Circular Economy Transitions: Insights and Lessons from Chile's Mining Territories. *Prog. Econ. Geogr.* **2025**, *3*, 100041. [CrossRef]

ceramics

MDPI

Article

Sunflower Shells Biomass Fly Ash as Alternative Alkali Activator for One-Part Cement Based on Ladle Slag

Aleksandar Nikolov *, Vladislav Kostov, Nadia Petrova, Liliya Tsvetanova, Stanislav V. Vassilev and Rositsa Titorenkova

Institute of Mineralogy and Crystallography, Bulgarian Academy of Sciences, Acad. G. Bonchev Str., bl. 107, 1113 Sofia, Bulgaria; vkytin@imc.bas.bg (V.K.); lilicvetanova79@abv.bg (L.T.); rositsatitorenkova@imc.bas.bg (R.T.)

* Correspondence: drsashko@imc.bas.bg

Abstract

This study explores the synergistic potential of ladle slag (LS) and sunflower shell fly ash (SSFA) in alkali-activated binder systems, focusing on their chemical and mineralogical characteristics and the influence of SSFA addition on the mechanical performance of LS-based pastes. X-ray fluorescence and XRD analysis revealed that LS is rich in CaO and latent hydraulic phases such as γ-belite and mayenite, while SSFA is dominated by K_2O, SO_3, and KCl/K_2SO_4 phases, reflecting its biomass origin. Infrared spectroscopy and thermal analysis confirmed the presence of carbonate, hydroxide, and hydrate phases, with SSFA exhibiting more complex thermal behavior due to volatile-rich composition. When used alone, LS produced weak binders; however, a 10 wt% SSFA addition tripled compressive strength to nearly 30 MPa, indicating a significant activation effect. Further increases in SSFA content led to strength reduction, likely due to increased porosity and excess salts. Microstructural analysis showed that SSFA promotes the formation of AFm phases such as Friedel's salt and hydrocalumite, altering hydration pathways and enhancing early strength through chemical activation and carbonation processes. The findings highlight the potential of combining LS and SSFA as a sustainable binder system, offering a waste-derived alternative for low-carbon construction materials.

Keywords: ladle slag; sunflower shell; biomass fly ash; alkali activation; one part; waste

1. Introduction

The iron and steel industry is a pillar of the worldwide economy and an energy-intensive consumer. In 2021, the global crude steel production reached 1958.4 million metric tons (Mt), of which the European Union accounted for 152.8 Mt [1]. The production of iron and steel is accompanied by large amounts of metallurgical slag generated during smelting [2]. According to different steelmaking methods, steel slag can be subdivided into the following: conversion slag or basic oxygen furnace slag resulting from the refinement of crude iron; electric arc furnace slag from recycled scrap and/or crude iron; argon oxygen decarburization slag derived from the decarburization and desulfurization of stainless steel; ladle furnace slag from the refinement of steel in a ladle furnace; and other secondary metallurgical slags [3]. Each slag type possesses specific properties and potential applications. Certain slags are well-utilized, such as ground granulated blast furnace slag as a supplementary cementitious material [4] or aggregated slags utilized in road base construction and asphalt mixtures [5,6]. Other low-reactive slags such as ladle slag hardly

https://doi.org/10.3390/ceramics8030079

find application outside the steel industry and consequently are stockpiled outside [7]. The ladle slag annual output has exceeded 20 Mt [8].

The ladle metallurgy furnace is part of a secondary steelmaking process of high-alloyed steel and an additional refining step after the basic oxygen furnace or electric arc furnace process [9]. The production of steel generally includes primary steelmaking and secondary treatment procedures. After the primary treatment process, the generated molten steel is cast into a ladle furnace. The ladle furnace heats the steel using graphite electrodes [10]. The ladle slag commonly consist of dicalcium silicate (C_2S), mayenite ($C_{12}Al_7$), periclase (MgO), bredigite ($C_{12}MgS_4$), free lime (CaO), etc. [11]. Since the ladle slag is produced from a slow cooling process, the dicalcium silicate is transformed from β-C_2S to γ-C_2S. This transformation is accompanied by volume expansion and self-powdering, which means that ladle slag is unsuitable to be used as an aggregate [12]. Moreover, the γ-C_2S is considered as scarcely reactive with water at normal conditions [13]. Still, ladle slag implies a certain potential as construction material, despite rather weak hydration and along with other difficulties (inconsistent chemical composition, potential leaching problems, volume stability due to MgO content) in being used as binding material. However, it was found that the binding properties of ladle slag, in particular γ-C_2S, can be greatly enhanced through alkaline activation [14–16].

Alkaline activation is based on the reaction between solid aluminosilicate precursors under alkaline conditions [17]. The alkaline activation of slags has been widely discussed and promoted as a component of the toolkit of 'sustainable cementing binder systems' [18]. However, several authors used different ladle slags to obtain alkali-activated materials. The most common activators used are a mixture of sodium silicate and sodium hydroxide [7,19,20]; sodium silicate and potassium hydroxide [21]; and potassium silicate [22]. The alkali-activated ladle slag showed high compressive strength (up to about 65 MPa) and durability [23]. However, some authors reported rather ineffective activations of ladle slag with NaOH, Na_2SO_4, and/or sodium silicate [24,25]. However, despite the relatively respectable results using alkali silicates activators, the latter are expensive and have significant environmental impact [26]. This triggers the need for extensive research into alternative alkaline activators based on waste or by-products [27]. The effective replacement of commercial sodium silicates was achieved by the hydrothermal dissolution of waste glass [28], rice husk ash [29], silica fume [30], sugarcane straw ash [31], etc. Direct usage of alkaline waste solution as an activator by utilizing aluminum anodizing etching solution [32] and Bayer liquor [33] is also possible. Another approach is using dry activators and to design a one-part alkali-activated cement [34]. In this method the activator solution is replaced by solid materials consisting of alkali or alkali–earth components in a reactive form. Among the dry activators derived from waste products, different types of biomass ash are suitable as activators due to their high alkaline character [35]. For example, several biomass ashes were used as dry activators, such as maize stalk and cob ashes [36], olive stone ash [37], almond-shell ash [38], wood ash [39], and coffee husk ash [40]. Additionally, one-part cement formulations are considerably more user-friendly, requiring only the addition of water similar to conventional Portland cement. To the best of our knowledge, the use of sunflower shell fly ash as an alternative activator in alkali-activated binders has not been previously reported.

The present study aims to explore the potential of sunflower shell fly ash (SSFA) as a dry alkali activator for one-part alkali-activated cement based on ladle slag. Given the high alkalinity and reactive silica content of biomass ashes, SSFA is investigated as an alternative to conventional liquid activators, which are often costly and environmentally damaging. This study evaluates the physicochemical interactions between ladle slag and SSFA, assessing the phase evolution, microstructure, and mechanical properties of the resulting cementitious material. The findings contribute to the development of cost-effective one-part alkali-activated binders, promoting the valorization of industrial and agricultural waste in sustainable construction applications.

2. Materials and Methods

The main raw material for the preparation of alkali-activated material was ladle slag (LS). The ladle slag is a by-product of secondary refinement steel, produced in the industrial complex of Stomana Industries, S.A, Bulgaria, and provided by Aeiforos Bulgaria S.A, Pernik, Bulgaria. The LS was temporarily stored in stockpiles at a steel plant yard. A representative sample of 10 kg ladle slag was collected, dried to constant mass at 80 °C, and grinded via steel ball milling for 1 h. The role of the dry activator in the present study was taken by biomass fly ash from sunflower shells (SSFA) generated from the combustion of sunflower shell cake in a boiler producing water steam. The SSFA was collected from a bag cyclone filter at the Biodiesel Plant of the company Astra Bioplant EOOD company in the town of Slivo Pole, Ruse region, Bulgaria. The SSFA was dried to a constant mass at 80 °C and milled for 1 h in a ceramic ball mill. The SSFA exhibited a bright white color, which is associated with minimal residual carbon (see Supplementary Materials—Figure S1). The alkali-activated mixtures were obtained by using tap water.

The chemical composition of the precursors was determined by XRF using pressed pellets, analyzed on Rigaku Supermini 200WD apparatus, Rigaku Corporation, Osaka, Japan. The powder X-ray diffractograms were obtained on Empyrean (Malvern Panalytical, Almelo, The Netherlands) diffractometer using CuKα radiation at 40 kV and 30 mA. The infrared spectra were measured using the Zn-Se ATR accessory and Tensor 37 (Bruker, Berlin, Germany) spectrometer, averaging over 128 scans with ± 2 cm^{-1} spectral resolution. The thermal analyses were carried out on the DSC-TG analyser SETSYS2400 (SETARAM, Caluire, France) under the following conditions: temperature range from 20 to 1000 °C, in a static air atmosphere, with a heating rate of 10 °C min^{-1}, and 10–15 mg sample mass.

Compressive strength was measured on 3 cubic specimens with one side area of 10 cm^2. The density of each series was calculated using dry mass and measuring the dimensions of three specimens per series by digital caliper with a resolution of 0.01 mm.

3. Results

3.1. Characterization of the Precursors

The chemical compositions of the two raw materials (LS and SSFA) used in the presented study are shown in Table 1. The dominant oxides in the LS sample were CaO, SiO_2, and Al_2O_3. Additionally, the magnesium oxide (MgO) content was measured at 6.5 wt%. A certain amount of sulfur was also detected, calculated as sulfate (SO_3).

Table 1. Chemical composition of ladle slag (LS) and biomass fly as from sunflower shells (SSFA), determined by XRF, in wt%.

	Na_2O	MgO	Al_2O_3	SiO_2	SO_3	Cl	K_2O	CaO	TiO_2	MnO	Fe_2O_3	Others
Ladle slag	-	6.50	15.50	15.60	2.45	-	0.10	55.50	0.70	0.80	2.69	0.16
SSFA	0.03	2.33	0.04	0.22	14.02	4.78	61.34	15.43	0.04	0.03	0.12	-

On the other hand, the SSFA exhibits a distinct chemical profile, being exceptionally rich in potassium (K_2O), which accounts for over 61 wt%. This is in line with the composition of sunflower shells ash reported by Lokare at al. [41] and Vassilev et al. [42], with K_2O content of = 45,1%, and 49,8%, respectively. They also report a significant amount of P_2O_5 content (7–10%), a fraction of which is discarded in the fly ash. Some authors report high content of SiO_2 for SSA, which is probably due to contamination with sand and dust during the harvesting and processing of the sunflowers [43,44]. Notable quantities of alkaline–earth oxides, such as CaO (15.43 wt%) and MgO (2.33 wt%), were identified in SSFA. Sulfur and chlorine were also present in significant amounts, with SO_3 reaching 14.02 wt% and Cl reaching 4.78 wt%. Elements such as S, Cl, and K are enriched in fly ashes due to their significant volatile nature during biomass combustion and often are transported by flue gases and the subsequent condensation in fly ashes [45].

The high calcium content in LS, combined with the alkali and alkali–earth nature of SSFA, suggests their potential for synergistic interactions in cementitious or geopolymer applications.

The mineral composition of the precursors is presented in Figure 1. The LS exhibited a well-crystalline structure, indicative of slow cooling during its formation. The dominant crystalline phase in LS was γ-belite (Ca_2SiO_4), a calcium disilicate phase that forms under slow cooling conditions and contributes to the material's latent hydraulic properties [46]. The second major phase identified was mayenite ($Ca_{12}Al_{14}O_{33}$), a calcium aluminate phase known for its high reactivity and rapid setting [47]. Additionally, periclase (MgO) was detected, often regarded for its role in volume stability on hydration [48]. Gehlenite ($Ca_2Al_2SiO_7$), which is active phase in hydration, was also presented to enhance the reactivity of the slag in cementitious systems. Minor amounts of strätlingite ($Ca_2Al_2SiO_7{\cdot}8H_2O$), katoite ($Ca_3Al_2(SiO_4)(OH)_8$), and brucite ($Mg(OH)_2$) were also detected, probably as a product of natural hydration due to outside pile storage of the LS at the steel producing plant. Calcite was also detected, probably as a product of natural carbonation.

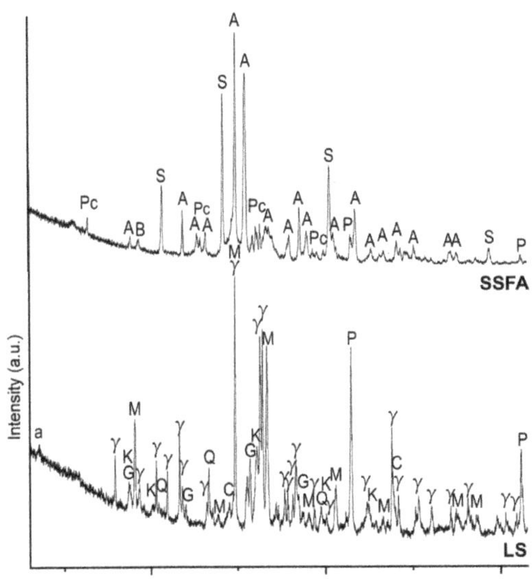

Figure 1. Powder XRD of the raw materials: biomass fly ash from sunflower shells (SSFA) and ladle slag (LS). Legend: A—arcanite, a—strätlingite, B—brucite, C—calcite, G—gehlenite, K—katoite; Pc—potassium carbonate hydrate ($K_2CO_3.1.5H_2O$), P—periclase, Q—quartz, M—mayenite, S—sylvite, γ—γ-belite (Ca_2SiO_4).

In contrast, the mineral composition of biomass fly ash from sunflower shells (SSFA) was dominated by arcanite (K_2SO_4) and sylvite (KCl). The high potassium content in SSFA promotes natural carbonation processes, resulting in the formation of potassium carbonate hydrate ($K_2CO_3{\cdot}1.5H_2O$). The strong hygroscopic nature of sylvite accelerates carbonation, further facilitating the transformation to potassium-bearing carbonate phases. Despite the dominance of crystalline potassium salts, the chemical and mineral composition results suggest that a certain amount of the potassium remains in an amorphous phase, as inferred from the high K_2O content determined by XRF. A minor amount of brucite was also detected.

The infrared spectra of raw materials LS and SSFA are presented in Figure 2. The two starting materials have a complex phase composition, whereby the peaks in the spectra overlap and make interpretation difficult. However, some conclusions can be drawn and contribute to the description of the source materials.

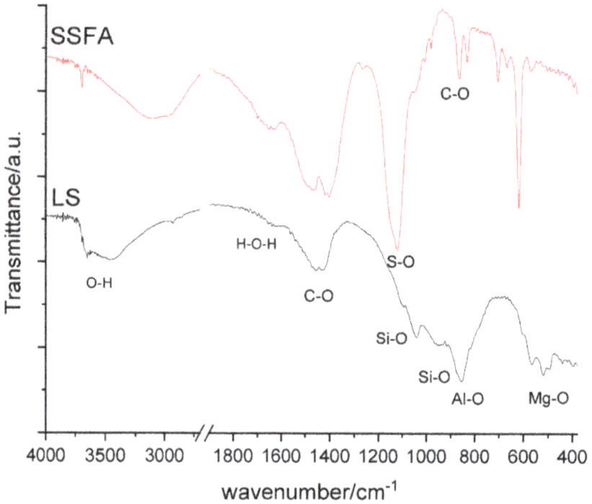

Figure 2. Infrared spectra of the raw materials: biomass fly ash from sunflower shells (SSFA), and ladle slag (LS).

The most intensive absorption peak of the LS is centered at 855 cm^{-1}, which is in the range of the Al-O stretching vibrations of AlO_4 tetrahedra in mayenite [49]. The asymmetric Al-O stretching of other aluminosilicates presented in the ladle slag (LS) also contribute in the range of 800-870 cm^{-1}. Silicate minerals already detected by powder diffraction data are γ-belite—(γ-Ca_2SiO_4), gehlenite ($CaAl(AlSiO_7)$), katoitite ($Ca_3Al_2SiO_4$ OH_8) and strätlingite ($Ca_2Al_2SiO_7{\cdot}8H_2O$). In general, minerals containing silicate groups exhibit characteristic infrared absorption bands due to stronger Si–O stretching and less intensive bending vibrations. These bands typically appear in the range of 800–1200 cm^{-1} for stretching vibrations and 400–600 cm^{-1} for bending vibrations. Bands due to Si–O stretching are visible near 960 and 1040 cm^{-1}. At such frequencies one would expect the intense absorption bands of belite, gehlenite, and other silicate phases present in the sample [50,51]. Strätlingite is a hydrated calcium aluminum silicate hydrate (also called gehlenite hydrate) that forms in hydrated cement systems and is important in cement chemistry. The infrared spectrum shows peaks related to water molecules at 3450 and 1640 cm^{-1}, which could be attributed to this phase. Strong peaks in the range of the carbonate stretching region at 1458 and 1429 cm^{-1} indicate the presence of carbonate minerals. Calcite is confirmed by the infrared absorption at 2926w, 2860w, 1429s, and a shoulder at 870 cm^{-1} [52]. Brucite ($MgOH_2$) is detected by the strong and sharp peak at 3660 cm^{-1} due to O-H stretching vibrations [53].

The most intensive sharp peaks in the infrared spectrum of SSFA (Figure 2) at 1119 cm^{-1} and 620 cm^{-1} are indicative of potassium sulfate arcanite (K_2SO_4) [54]. The strong absorption at 1400-1470 cm^{-1} and at 830-870 cm^{-1} correspond to asymmetric stretching (ν_3) and out-of-plane bending (ν_2) modes of the carbonate ion, respectively. These carbonate peaks and characteristic absorption bands of water at 3100–2967 cm^{-1} and at 1630 cm^{-1} confirm the presence of potassium carbonate sesquihydrate [55]. The strong and sharp peak at 3690 cm^{-1} is characteristic of hydroxyl group stretching in brucite.

The thermal behaviors of both precursors LS and SSFA are shown in Figure 3. The LS exhibits insignificant mass losses on the TG curve and endothermic effects on the DSC curve within the temperature range of 25–400 °C, indicating the presence of a minor amount of hydrate phases due to weathering of the ladle slag. Minor endo-effects around 160 °C and 210 °C were related to strätlingite [56], while katoite decomposed at about 310 °C [57]. The distinct endothermic peak at 389 °C is characteristic for brucite dehydration [58]. The endothermal event at about 700 °C is due to the presence of a carbonate mineral (calcite), the amount of which, calculated from the TG curve, is below 0.5%. The major mineral phases identified by PXRD such as γ-belite, mayenite, gehlenite, and periclase are thermally stable in this temperature region without expressions on the registered curves. In contrast to LS, SSFA shows more complicated thermal behavior. The only identified hydrate phase by PXRD was $KCO_3 \cdot 1.5HO$, which during heating from 25 to 300 °C dissociates to anhydrous form. The decomposition endo-effect of K_2CO_3 is around 900 °C, accompanied by mass loss. It is known that the suitable sintering temperature for this compound should be below the decomposition temperature [59]. The endothermal peak at about 380 °C is related to brucite dehydration. The sharp endo-effect at 660 °C and 725 °C are probably attributed to the two-step melting of KCl [60]. The melting point of pure KCl is 776 °C, but when this potassium salt is presented together with other salts, this temperature is considerably lower due to possible eutectics [61]. The KCl decomposition starts immediately after its melting. In the temperature range of 350–600 °C, SSFA showed no significant mass loss or exothermic peaks, indicating a very low content of unburned carbon.

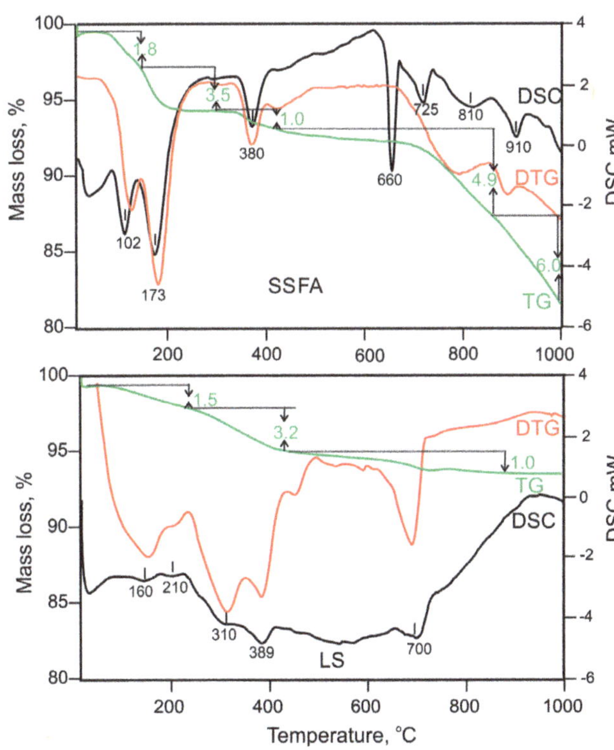

Figure 3. DSC-TG(DTG) curves of both precursors—LS and SSFA.

3.2. Influence of Sunflower Shells Fly Ash Addition to Ladle Slag

The influence of different amounts of SSFA addition (from 0 up to 30%) to ladle slag was investigated. The composition design of the prepared series is presented in Table 2. The ladle slag and SSFA were placed in a laboratory planetary mixer to obtain a homogenous dry mixture. The water-to-solid ratios of all series were fixed to 0.35. The fresh alkali-activated paste was stirred for 2 min and poured in steel molds wrapped in polyethylene foil. The samples were demolded after one day and stored under laboratory conditions. The properties of the obtained alkali-activated materials were examined on the 28th day.

Table 2. Composition design and physical properties of the prepared pastes.

Series	LS	SSFA	Water to Solid	Compressive Strength, MPa	Density, g/cm^3
LBFA0	100	0	0.35	9.5 ± 0.5	1.591 ± 0.006
LBFA10	90	10	0.35	29.6 ± 0.2	1.569 ± 0.001
LBFA20	80	20	0.35	22.4 ± 0.3	1.560 ± 0.002
LBFA30	70	30	0.35	11.6 ± 0.2	1.548 ± 0.004

3.2.1. Physical Properties

The hydration of ladle slag (LS) alone resulted in the formation of a relatively weak material, achieving a compressive strength of approximately 10 MPa (Table 2). This indicates that LS, when used independently, exhibits limited self-cementing properties. However, the incorporation of 10 wt% SSFA significantly enhanced the compressive strength, increasing it by approximately threefold to 29.6 MPa. This remarkable improvement suggests that the presence of SSFA contributes to enhanced reaction kinetics and possibly the formation of additional binding phases. Further increasing the SSFA content to 20 wt% led to a decrease in strength (22.4 MPa), although the material still exhibited a notable improvement compared to the reference LS paste. A more pronounced reduction in mechanical performance was observed at 30 wt% SSFA, where the compressive strength dropped to 11.6 MPa, approaching the strength of the pure LS system. This decline may be attributed to an excessive sulfates and chlorides, or increased porosity associated with higher SSFA content.

In addition to mechanical properties, density measurements revealed that the alkali-activated pastes had relatively low bulk density, which further decreased with the progressive addition of SSFA. The reference paste (LBFA0) exhibited a density of 1.591 g/cm^3, and the density decreased to 1.569 g/cm^3, 1.560 g/cm^3, and 1.548 g/cm^3 for LBFA10, LBFA20, and LBFA30, respectively. This trend suggests that the incorporation of SSFA introduces a degree of porosity or reduces the packing density of the solid matrix, likely due to the physical and chemical nature of SSFA and reactions with LS.

The results demonstrate that a moderate substitution of 10 wt% SSFA optimally enhances strength, while excessive replacement leads to a decline in mechanical performance.

3.2.2. Microstructural Characterization

The primary reactive phases within the ladle slag were identified as mayenite ($Ca_{12}Al_{14}O_{33}$) and gehlenite ($Ca_2Al_2SiO_7$). Upon hydration of the ladle slag, strätlingite [$(Ca_2Al_2SiO_7)\cdot 8H_2O$] was formed as a dominant phase (LBFA0), followed by the relative increase in katoite phase (Figure 4). Hemicarboaluminate (Hc) [$3CaO\cdot Al_2O_3\cdot Ca(OH)(CO_3)_{0.5}\cdot xH_2O$] and monocarboaluminate (Mc) [$3CaO\cdot Al_2O_3\cdot CaCO_3\cdot xH_2O$] were also formed as products of natural carbonation.

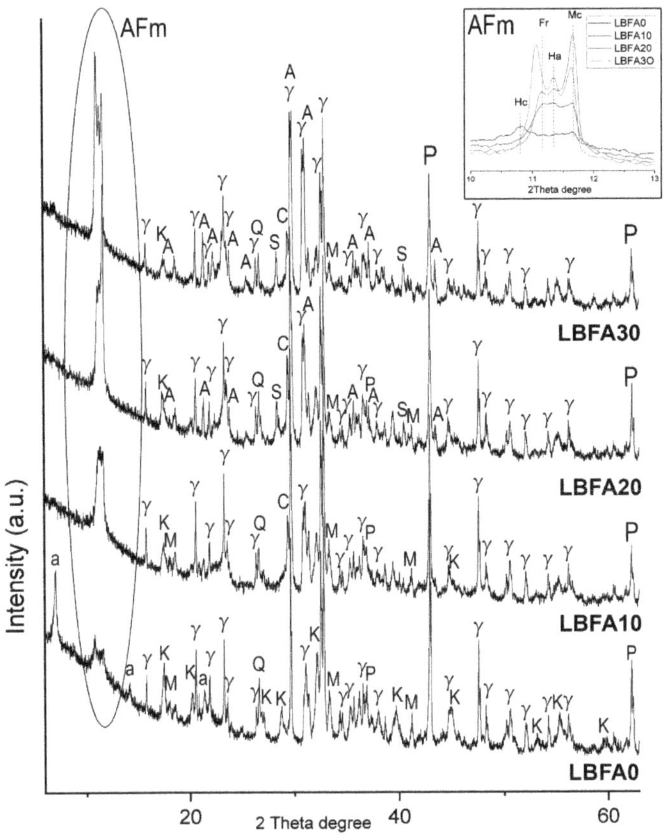

Figure 4. Powder XRD of the alkali-activated material based on ladle slag and biomass ash from sunflower shells. Legend: A—arcanite (PDF 00-044-1414), a—strätlingite (PDF 00-029-0285), AFm—family of hydrated calcium aluminate phases, C—calcite (PDF 01-076-2713), Fr—Friedel's salt(PDF 00-028-0207), Ha—hydrocalumite (PDF 00-031-0245), Hc—hemicarboaluminate (PDF 00-041-0221), Mc—monocarboaluminate (PDF 00-014-0083), K—katoite (PDF 01-074-3033), M—mayenite (PDF 00-048-1882), P—periclase (PDF 00-087-0651), Q—quartz (PDF 01-070-2517), S—sylvite (PDF 00-000-0587), γ—γ-belite (dicalcium silicate) (PDF 01-080-4279).

The incorporation of SSFA altered the formation of strätlingite and katoite. Instead, the presence of chloride ions in SSFA promoted the crystallization of AFm phases—a family of hydrated calcium aluminate phases representing a layered double hydroxide with a representative formula $[Ca_2(Al,Fe)(OH)_6]\cdot X\cdot nH_2O$, where X equals an exchangeable single charged anion (such as chloride) or half of a double charged one (sulfate, carbonate). The AFm phases exhibit variable compositions and significantly influence cement performance, including strength, durability, and chemical stability [62]. The main phases detected in the LS activated by SSFA were hydrocalumite $[3CaO\cdot Al_2O_3\cdot Ca(OH)_2\cdot 10H_2O]$, Friedel's salt $[3CaO\cdot Al_2O_3\cdot CaCl_2\cdot 10H_2O]$, and Mc. The amount of Mc increased, suggesting enhanced carbonation due to SSFA addition. Presence of sylvite and arcanite in the system increased the hygroscopicity of the material, which accelerated the natural carbonation [63]. Ettringite and monosulfoaluminate phases were not detected. The formation of hydrotalcite phase is also possible but its main peak overlaps with monocarboaluminate [64]. The presented AFm phases (Friedel's salt, hydrocalumite, Hc, Mc) are referred as stable at normal conditions and contribute to the mechanical strength of the obtained material [65,66].

The γ-belite ($2CaO{\cdot}SiO_2$) and periclase (MgO) remained relatively inert even after the addition of SSFA, indicating their lack of hydration under the given conditions. Periclase (MgO) is often negatively associated with the volume stability of cementitious systems due to its potential late-stage hydration. The potassium-rich environment did not significantly promote the periclase hydration. The stability of periclase in this system can be attributed to its "dead-burned" nature. Dead-burned periclase refers to MgO that has undergone high-temperature sintering—typically around 1500 °C—within the molten mass of the kiln, resulting in a highly crystalline, dense, and refractory phase with significantly reduced reactivity [67]. Dead-burned MgO exhibits minimal hydration in conventional cementitious systems due to its low solubility and slow reaction kinetics [68]. Periclase and γ-belite stayed relatively inert even at SSFA addition.

The FTIR measurements were conducted to investigate the effect of varying SSFA additions (0–30%) to ladle slag. Infrared spectra of the alkali-activated material based on LS and SSFA are shown in Figure 5. The main phases already detected in LBFA0 upon hydration and carbonation of LS are strätlingite ($Ca_2Al_2SiO_7{\cdot}8H_2O$), katoite ($Ca_3Al_2SiO_4OH_8$), hemicarboaluminate [$3CaO{\cdot}Al_2O_3{\cdot}Ca(OH)(CO_3)_{0.5}{\cdot}xH_2O$], and monocarboaluminate [$3CaO{\cdot}Al_2O_3{\cdot}CaCO_3{\cdot}xH_2O$]. In a mixture containing these phases, the IR spectrum displays overlapping bands in the O-H stretching range at around 3400 cm^{-1} due to hydroxyl groups and water molecules, carbonate stretching, and bending near 1420 and 870 cm^{-1}, as well as peaks in the range 1100–500 cm^{-1} due to Si-O and Al-O stretching and bending vibrations. The addition of SSFA leads to a clear trend: the appearance of peaks at 1117 and 620 cm^{-1} in LBFA10, which intensify progressively up to LBFA30, corresponding to the vibrations of the sulfate group. This is a main spectral indication of the increasing amount of arcanite (K_2SO_4) with the increasing amount of added SSFA, since the mineral remains stable.

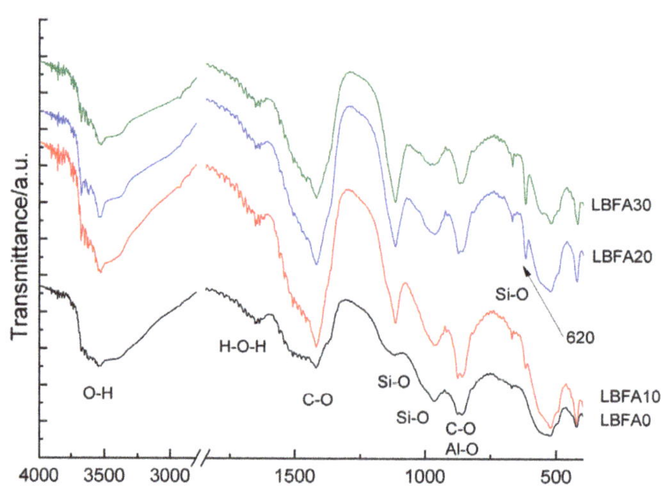

Figure 5. Infrared spectra of the alkali-activated material (LBFA) based on ladle slag (LS) and biomass ash from sunflower shells (SSFA).

The thermal behavior of alkali-activated material based on ladle slag and biomass ash from sunflower shells is presented in Figure 6. As was mentioned, the hydration of LS alone and especially in combination with SSFA promotes the crystallization of AFm phases—a family of hydrated calcium aluminate phases and hemi-, monocarboaluminates. This provokes the manifestation of a number of processes during heating. The data concerning the mass losses and the related endothermal events could be, in general, separated into three temperature regions: (I) from 25 to 200 °C; (II) from 200 to 500 °C; and (III) from 500 to 1000 °C. This grouping depends on the specific thermal behavior of the series compounds under study and therefore it is not exactly fixed. In the first temperature region, a dehydration process occurs. In the second one, along with the completion of the dehydration, a dehydroxylation process also proceeds. While in the third region, mainly a decarbonation of the carbonate-containing phases and likely desulfation takes place. Such observations have also been noted by other authors [69,70]. In the first temperature region, the dehydration processes are manifested by a well-defined endothermal effect at about 150–160 °C, which is related to strätlingite dehydration the and AFm phases [56]. The small shoulder at 200 °C observed in series LBFA0 is characteristic for strätlingite [71]. In the second region, the endothermal events are more complicated due to the existence of dehydroxylation. The endo-effects around 317 °C are related to katoite dehydration, which is most pronounced at series LBFA0 [57]. All of the series showed endo-effects at about 387 °C, which is related to brucite dehydraton. The endothermal effect at about 286 °C, most pronounced in series LBFA20, is related to Friedel salt dehydration. The Friedel salts recrystallize to calcium chloroaluminate at about 670 °C [72]. It becomes clear that the presence of hydrated calcium aluminate minerals is quite diverse, especially since a hydrated component is also present in the carbonate or semi-carbonate phases. In the third temperature region, decarbonation and desulfation mainly occur, displayed by an endothermal effect at around 700 °C and beyond. For both series, LBFA20 and LBFA30, where arcanite was identified, an endothermal peak at 583–584 °C has been observed due to β to α phase transformation [73].

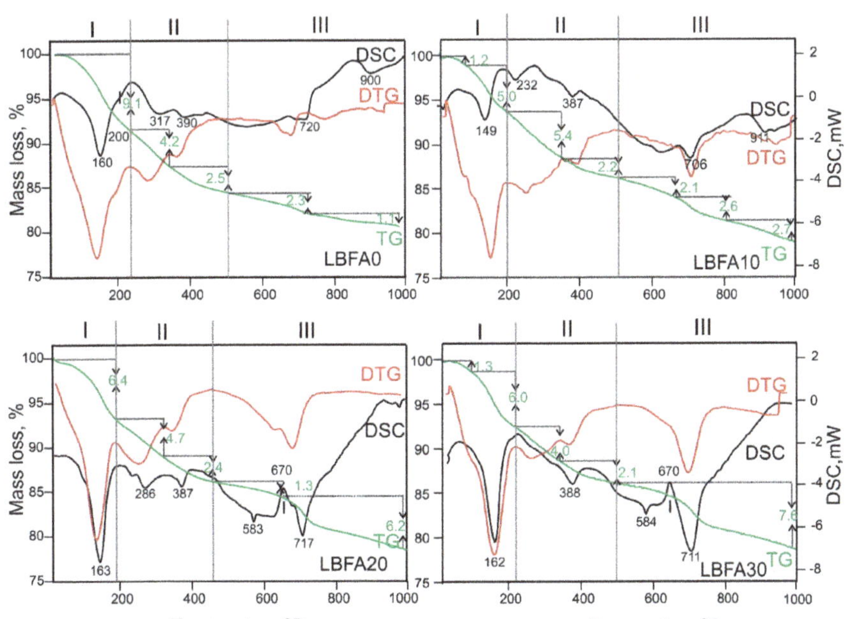

Figure 6. DSC-TG(DTG) curves of LBFA0, LBFA10, LBFA20, and LBFA30.

The presented thermal data for the LBFA series in Table 3 show a decreasing amount of hydrate components in the direction from LBFA0 to LBFA30. However, in the same direction, the carbonate component increases and, in general, there is an increase in total mass loss with the increasing presence of biomass fly ash into the ladle slag.

Table 3. Mass loss during dehydration–dehydroxylation and decarbonation processes.

Series	(I) Dehydration–(II) Dehydroxylation, Mass %, (Process-Related New Phases)	(III) Decarbonization, Mass % (Process-Related New Phase)	Total Loss, Mass %
LBFA0	15.8 (strätlingite, katoite)	3.4 (hemi-, monocarboaluminate, calcite)	19.2
LBFA10	13.8 (AFm)	7.4 (hemi-, monocarboaluminate, calcite)	21.2
LBFA20	13.5 (AFm)	7.5 (hemi-, monocarboaluminate, calcite)	21.0
LBFA30	13.4 (AFm)	7.6 (hemi-, monocarboaluminate, calcite)	21.0

3.3. Leaching Experiments

The leaching behavior of sulfates, chlorides, and alkali is a crucial aspect in evaluating the environmental impact and durability of blended cementitious materials. In the ladle slag–biomass fly ash (LBFA) system, the release of sulfate, chloride, and potassium ions into a pore solution is governed by phase formation and stability, solubility equilibria, and external exposure conditions. To evaluate the leaching behavior ladle slag–biomass fly ash (LBFA), hardened material samples of LBFA20 were ground to powder and washed several times with distilled water. The washed powder was dried and chemical compositions were determined by XRF (Table 4). The results revealed a significant reduction in sulfur, chlorine, and potassium concentrations, indicating that these elements were not strongly bonded into stable crystalline phases and were instead present in soluble or weakly bound amorphous phases. Only a partial amount of chlorine (~26%) and sulfate (~12%) stayed after the washing test. Interestingly, potassium bearing phases were not detected at any stage of the analysis, suggesting that potassium was primarily presented in an amorphous phase or in soluble ionic form. X-ray diffraction analysis of the washed sample confirmed that AFm phases in LBFA20 remained stable (Figure 7). The remaining sulfur and chlorine are likely incorporated into the AFm phases, which play a crucial role in binding chlorides and sulfates, thereby enhancing the material's ability to prevent steel corrosion [74,75]. However, most of the soluble components in the SSFA do not participate in the formation of any or enough stable phases, thus we can state that the full potential of SSFA was not achieved at the current addition levels.

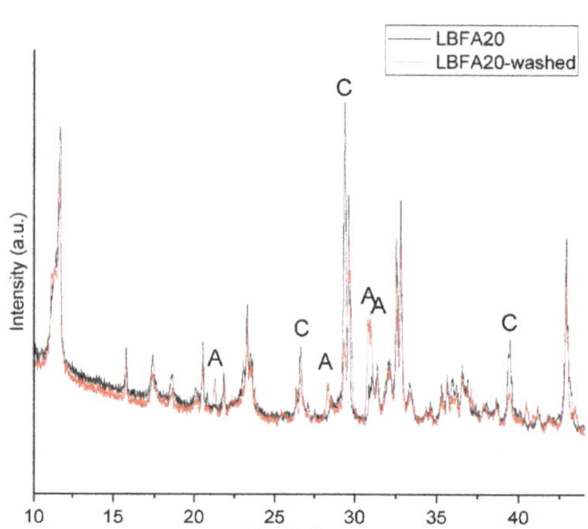

Figure 7. Powder XRD diffraction of series LBFA20 before and after washing experiments. Legend showing the difference phases: A—Arcanite, C—calcite.

Table 4. Leaching of sulfates and chlorides after washing with water.

	MgO	Al_2O_3	SiO_2	SO_3	Cl	K_2O	CaO	TiO_2	MnO	Fe_2O_3	Others
LBFA20	5.67	10.9	9.28	5.91	0.983	13.4	48.6	0.654	0.868	2.84	0.895
LBFA20-washed	6.97	11.0	11.8	0.692	0.254	1.93	60.3	0.912	1.22	3.78	1.142

4. Conclusions

This study explored the potential of sunflower shell biomass fly ash (SSFA) as a dry alkali activator for one-part alkali-activated ladle slag binder. The chemical and mineralogical analysis of the precursors revealed that ladle slag (LS) is primarily composed of γ-belite, mayenite, and periclase, while SSFA is rich in potassium-bearing phases such as arcanite and sylvite.

The results showed that addition of SSFA to ladle slag mixtures improved the early hydration processes and contributed to higher compressive strength compared to normal hydration of the ladle slag. The optimal addition of SSFA was 10%, at which the compressive strength reached 30 MPa, indicating the effectiveness of SSFA as an alternative activator.

The high alkalinity of SSFA facilitated the partial dissolution of the LS phases, leading to the formation of cementitious reaction products. The presence of chorine and sulfur contribute to formation of more AFm phases such as hydrocalumite, Friedel's salt, hemicarboaluminate, and monocarboaluminate, which contributed to higher mechanical strength.

However, most of the soluble components in the SSFA do not participate in the formation of any or enough stable phases, thus we can state that the full potential of SSFA as alkali activator was not achieved at higher proportions of SSFA addition.

By valorizing both industrial and agricultural waste materials, this study supports the development of environmentally friendly cement alternatives, contributing to circular economy principles and sustainable construction practices. Future research should focus on evaluating the long-term stability, durability, and potential leaching of heavy metals to further assess the material's suitability for sustainable construction applications

Supplementary Materials: The following supporting information can be downloaded at: https://www.mdpi.com/article/10.3390/ceramics8030079/s1, Figure S1. Images of raw materials: LS, SSFA.

Author Contributions: Conceptualization, A.N.; methodology, A.N., V.K., N.P., L.T., S.V.V. and R.T.; software, A.N., V.K., N.P., L.T. and R.T.; validation, A.N., V.K., N.P., L.T., S.V.V. and R.T.; formal analysis, A.N., V.K., N.P., L.T. and R.T.; investigation, A.N., V.K., N.P., L.T., S.V.V. and R.T.; resources, A.N.; data curation, A.N., V.K., N.P., L.T., S.V.V. and R.T.; writing—original draft preparation, A.N., V.K., N.P., L.T. and R.T.; writing—review and editing, A.N., S.V.V. and R.T.; visualization, A.N., V.K., N.P., L.T. and R.T.; supervision, A.N. and R.T.; project administration, R.T.; funding acquisition, R.T. All authors have read and agreed to the published version of the manuscript.

Funding: This work has been carried out in the framework of the National Science Program "Critical and strategic raw materials for a green transition and sustainable development", approved by the Resolution of the Council of Ministers No. 508/18 July 2024 and funded by the Ministry of Education and Science (MES) of Bulgaria.

Institutional Review Board Statement: Not applicable.

Informed Consent Statement: Not applicable.

Data Availability Statement: The original contributions presented in this study are included in the article/Supplementary Material. Further inquiries can be directed to the corresponding author(s).

Acknowledgments: The team acknowledge Stomana Industry, S.A, Bulgaria, Aeiforos Bulgaria S.A., and Astra Bioplant EOOD, Bulgaria, companies for provision of the raw materials.

Conflicts of Interest: The authors declare no conflicts of interest. Stomana Industry, S.A, Bulgaria, Aeiforos Bulgaria S.A., and Astra Bioplant EOOD, Bulgaria, companies provided the raw materials. These companies had no role in design of the study; in the collection, analyses or interpretation of data; in the writing of the manuscript; or in the decision to publish the results.

Abbreviations

The following abbreviations are used in this manuscript:

AFm	A family of hydrated calcium aluminate phases
SSFA	Sunflower shells fly ash
LS	Ladle slag

References

1. World Steel Association. *Steel Statistical Yearbook 2022: A Cross-Section of Steel Industry Statistics 2012–2021*; World Steel Association: Brussels, Belgium, 2022.
2. Chen, J.; Xing, Y.; Wang, Y.; Zhang, W.; Guo, Z.; Su, W. Application of iron and steel slags in mitigating greenhouse gas emissions: A review. *Sci. Total Environ.* **2022**, *844*, 157041. [CrossRef] [PubMed]
3. Baalamurugan, J.; Kumar, V.G.; Chandrasekaran, S.; Balasundar, S.; Venkatraman, B.; Padmapriya, R.; Raja, V.B. Recycling of steel slag aggregates for the development of high density concrete: Alternative & environment-friendly radiation shielding composite. *Compos. Part B Eng.* **2021**, *216*, 108885. [CrossRef]
4. Raut, S.R.; Saklecha, P.; Kedar, R. Review on ground granulated blast-furnace slag as a supplementary cementitious material. *Int. J. Comput. Appl.* **2015**, *975*, 8887.
5. Aiban, S.A. Utilization of steel slag aggregate for road bases. *J. Test. Eval.* **2006**, *34*, 65–75. [CrossRef]
6. Kumar, H.; Varma, S. A review on utilization of steel slag in hot mix asphalt. *Int. J. Pavement Res. Technol.* **2021**, *14*, 232–242. [CrossRef]
7. Pinheiro, C.; Rios, S.; da Fonseca, A.V.; Fernández-Jiménez, A.; Cristelo, N. Application of the response surface method to optimize alkali activated cements based on low-reactivity ladle furnace slag. *Constr. Build. Mater.* **2020**, *264*, 120271. [CrossRef]
8. Wu, L.; Li, H.; Mei, H.; Rao, L.; Wang, H.; Lv, N. Generation, utilization, and environmental impact of ladle furnace slag: A minor review. *Sci. Total Environ.* **2023**, *895*, 165070. [CrossRef]
9. Brand, A.S.; Fanijo, E.O. A review of the influence of steel furnace slag type on the properties of cementitious composites. *Appl. Sci.* **2020**, *10*, 8210. [CrossRef]
10. Najm, O.; El-Hassan, H.; El-Dieb, A. Ladle slag characteristics and use in mortar and concrete: A comprehensive review. *J. Clean. Prod.* **2021**, *288*, 125584. [CrossRef]
11. Shi, C. Steel slag—Its production, processing, characteristics, and cementitious properties. *J. Mater. Civ. Eng.* **2004**, *16*, 230–236. [CrossRef]
12. Serjun, V.Z.; Mirti, B.; Mladenovič, A. Evaluation of ladle slag as a potential material for building and civil engineering. *Mater. Technol.* **2013**, *47*, 543–550.
13. Taylor, H. *Cement Chemistry*; Thomas Telford Publishing: London, UK, 1997.
14. Yan, Z.; Jiang, Y.; Yin, K.; Wang, L.; Pan, T. Enhancement of Hydration Activity and Microstructure Analysis of γ-C_2S. *Materials* **2023**, *16*, 6762. [CrossRef] [PubMed]
15. Wang, Q.; Yang, J.; Yan, P. Influence of initial alkalinity on the hydration of steel slag. *Sci. China Technol. Sci.* **2012**, *55*, 3378–3387. [CrossRef]
16. Shi, C. Characteristics and cementitious properties of ladle slag fines from steel production. *Cem. Concr. Res.* **2002**, *32*, 459–462. [CrossRef]
17. Krivenko, P.V. Alkaline cements. In Proceedings of the 1st International Conference on Alkaline Cements and Concretes, Kiev, Ukraine, 11–14 October 1994; pp. 11–129.
18. Provis, J.L. Alkali-activated materials. *Cem. Concr. Res.* **2018**, *114*, 40–48. [CrossRef]
19. Wang, W.-C.; Wang, H.-Y.; Tsai, H.-C. Study on engineering properties of alkali-activated ladle furnace slag geopolymer. *Constr. Build. Mater.* **2016**, *123*, 800–805. [CrossRef]
20. Murri, A.N.; Rickard, W.; Bignozzi, M.; Van Riessen, A. High temperature behaviour of ambient cured alkali-activated materials based on ladle slag. *Cem. Concr. Res.* **2013**, *43*, 51–61. [CrossRef]
21. Adesanya, E.; Ohenoja, K.; Kinnunen, P.; Illikainen, M. Alkali activation of ladle slag from steel-making process. *J. Sustain. Metall.* **2017**, *3*, 300–310. [CrossRef]
22. Češnovar, M.; Traven, K.; Horvat, B.; Ducman, V. The potential of ladle slag and electric arc furnace slag use in synthesizing alkali activated materials; the influence of curing on mechanical properties. *Materials* **2019**, *12*, 1173. [CrossRef]

23. Adesanya, E.; Ohenoja, K.; Kinnunen, P.; Illikainen, M. Properties and durability of alkali-activated ladle slag. *Mater. Struct.* **2017**, *50*, 255. [CrossRef]
24. Xu, B.; Yi, Y. Use of ladle furnace slag containing heavy metals as a binding material in civil engineering. *Sci. Total Environ.* **2020**, *705*, 135854. [CrossRef] [PubMed]
25. Lancellotti, I.; Ponzoni, C.; Bignozzi, M.C.; Barbieri, L.; Leonelli, C. Incinerator bottom ash and ladle slag for geopolymers preparation. *Waste Biomass Valor.* **2014**, *5*, 393–401. [CrossRef]
26. Passuello, A.; Rodríguez, E.D.; Hirt, E.; Longhi, M.; Bernal, S.A.; Provis, J.L.; Kirchheim, A.P. Evaluation of the potential improvement in the environmental footprint of geopolymers using waste-derived activators. *J. Clean. Prod.* **2017**, *166*, 680–689. [CrossRef]
27. Adesanya, E.; Perumal, P.; Luukkonen, T.; Yliniemi, J.; Ohenoja, K.; Kinnunen, P.; Illikainen, M. Opportunities to improve sustainability of alkali-activated materials: A review of side-stream based activators. *J. Clean. Prod.* **2021**, *286*, 125558. [CrossRef]
28. Torres-Carrasco, M.; Puertas, F. Waste glass as a precursor in alkaline activation: Chemical process and hydration products. *Constr. Build. Mater.* **2017**, *139*, 342–354. [CrossRef]
29. Tchakouté, H.K.; Rüscher, C.H.; Kong, S.; Kamseu, E.; Leonelli, C. Geopolymer binders from metakaolin using sodium waterglass from waste glass and rice husk ash as alternative activators: A comparative study. *Constr. Build. Mater.* **2016**, *114*, 276–289. [CrossRef]
30. Billong, N.; Oti, J.; Kinuthia, J. Using silica fume based activator in sustainable geopolymer binder for building application. *Constr. Build. Mater.* **2021**, *275*, 122177. [CrossRef]
31. Moraes, J.; Font, A.; Soriano, L.; Akasaki, J.; Tashima, M.; Monzó, J.; Borrachero, M.V.; Payá, J. New use of sugar cane straw ash in alkali-activated materials: A silica source for the preparation of the alkaline activator. *Constr. Build. Mater.* **2018**, *171*, 611–621. [CrossRef]
32. Nikolov, A.; Nugteren, H.; Rostovsky, I. Optimization of geopolymers based on natural zeolite clinoptilolite by calcination and use of aluminate activators. *Constr. Build. Mater.* **2020**, *243*, 118257. [CrossRef]
33. Jamieson, E.; van Riessen, A.; McLellan, B.; Penna, B.; Kealley, C.; Nikraz, H. *Introducing Bayer Liquor-Derived Geopolymers*; Elsevier: Amsterdam, The Netherlands, 2017; pp. 159–189.
34. Luukkonen, T.; Abdollahnejad, Z.; Yliniemi, J.; Kinnunen, P.; Illikainen, M. One-part alkali-activated materials: A review. *Cem. Concr. Res.* **2018**, *103*, 21–34. [CrossRef]
35. Vassilev, S.V.; Baxter, D.; Andersen, L.K.; Vassileva, C.G. An overview of the chemical composition of biomass. *Fuel* **2010**, *89*, 913–933. [CrossRef]
36. Peys, A.; Rahier, H.; Pontikes, Y. Potassium-rich biomass ashes as activators in metakaolin-based inorganic polymers. *Appl. Clay Sci.* **2016**, *119*, 401–409. [CrossRef]
37. Font, A.; Soriano, L.; Tashima, M.M.; Monzó, J.; Borrachero, M.V.; Payá, J. One-part eco-cellular concrete for the precast industry: Functional features and life cycle assessment. *J. Clean. Prod.* **2020**, *269*, 122203. [CrossRef]
38. Soriano, L.; Font, A.; Tashima, M.M.; Monzó, J.; Borrachero, M.V.; Bonifácio, T.; Payá, J. Almond-shell biomass ash (ABA): A greener alternative to the use of commercial alkaline reagents in alkali-activated cement. *Constr. Build. Mater.* **2021**, *290*, 123251. [CrossRef]
39. Hassan, H.S.; Abdel-Gawwad, H.; Vásquez-García, S.; Israde-Alcántara, I.; Flores-Ramirez, N.; Rico, J.; Mohammed, M.S. Cleaner production of one-part white geopolymer cement using pre-treated wood biomass ash and diatomite. *J. Clean. Prod.* **2019**, *209*, 1420–1428. [CrossRef]
40. Lima, F.S.; Gomes, T.C.F.; de Moraes, J.C.B. Novel one-part alkali-activated binder produced with coffee husk ash. *Mater. Lett.* **2022**, *313*, 131733. [CrossRef]
41. Lokare, S.S.; Dunaway, J.D.; Moulton, D.; Rogers, D.; Tree, D.R.; Baxter, L.L. Investigation of ash deposition rates for a suite of biomass fuels and fuel blends. *Energy Fuels* **2006**, *20*, 1008–1014. [CrossRef]
42. Vassilev, S.V.; Vassileva, C.G.; Baxter, D. Trace element concentrations and associations in some biomass ashes. *Fuel* **2014**, *129*, 292–313. [CrossRef]
43. Amin, M.; Zeyad, A.M.; Agwa, I.S.; Heniegal, A.M. Effect of peanut and sunflower shell ash on properties of sustainable high-strength concrete. *J. Build. Eng.* **2024**, *89*, 109208. [CrossRef]
44. Shahbazpanahi, S.; Faraj, R.H. Feasibility study on the use of shell sunflower ash and shell pumpkin ash as supplementary cementitious materials in concrete. *J. Build. Eng.* **2020**, *30*, 101271. [CrossRef]
45. Nikolov, A.; Kostov-Kytin, V.; Tarassov, M.; Tsvetanova, L.; Jordanov, N.B.; Karamanova, E.; Rostovsky, I. Characterization of cement kiln dust from Bulgarian cement plants. *J. Chem. Technol. Metall.* **2025**, *60*, 455–463. [CrossRef]
46. Lyu, H.; Hao, L.; Zhang, S.; Poon, C.S. High-performance belite rich eco-cement synthesized from solid wastes: Raw feed design, sintering temperature optimization, and property analysis. *Resour. Conserv. Recycl.* **2023**, *199*, 107211. [CrossRef]
47. He, Z.; Li, Y. The influence of mayenite employed as a functional component on hydration properties of ordinary Portland cement. *Materials* **2018**, *11*, 1958. [CrossRef] [PubMed]
48. Zheng, X.; Cordonnier, B.; Zhu, W.; Renard, F.; Jamtveit, B. Effects of confinement on reaction-induced fracturing during hydration of periclase. *Geochem. Geophys. Geosyst.* **2018**, *19*, 2661–2672. [CrossRef]

49. Tolkacheva, A.; Shkerin, S.; Plaksin, S.; Vovkotrub, E.; Bulanin, K.; Kochedykov, V.; Ordinartsev, D.; Gyrdasova, O.; Molchanova, N. Synthesis of dense ceramics of single-phase mayenite ($Ca_{12}Al_{14}O_{32}$)O. *Russ. J. Appl. Chem.* **2011**, *84*, 907–911. [CrossRef]
50. Horgnies, M.; Chen, J.; Bouillon, C. Overview about the use of Fourier transform infrared spectroscopy to study cementitious materials. *WIT Trans. Eng. Sci.* **2013**, *77*, 251–262.
51. Fernández Carrasco, L.; Torrens Martín, D.; Morales, L.; Martínez Ramírez, S. *Infrared Spectroscopy—Materials Science, Engineering and Technology*; InTech: Rijeka, Croatia, 2012; pp. 357–372. [CrossRef]
52. Gunasekaran, S.; Anbalagan, G.; Pandi, S. Raman and infrared spectra of carbonates of calcite structure. *J. Raman Spectrosc.* **2006**, *37*, 892–899. [CrossRef]
53. Braterman, P.S.; Cygan, R.T. Vibrational spectroscopy of brucite: A molecular simulation investigation. *Am. Mineral.* **2006**, *91*, 1188–1196. [CrossRef]
54. Chukanov, N.V. The Application of IR Spectroscopy to the Investigation of Minerals. In *Infrared Spectra of Mineral Species: Extended Library*; Springer: Dordrecht, The Netherlands, 2014; pp. 1–19.
55. Schutte, C.; Buijs, K. The infra-red spectra of K_2CO_3 and its hydrates. *Spectrochim. Acta* **1961**, *17*, 921–926. [CrossRef]
56. Okoronkwo, M.U.; Glasser, F.P. Stability of strätlingite in the CASH system. *Mater. Struct.* **2016**, *49*, 4305–4318. [CrossRef]
57. Litwinek, E.; Madej, D. Structure, microstructure and thermal stability characterizations of C_3AH_6 synthesized from different precursors through hydration. *J. Therm. Anal. Calorim.* **2020**, *139*, 1693–1706. [CrossRef]
58. Chen, Y.; Zhou, T.; Fang, H.; Li, S.; Yao, Y.; He, Y. A novel preparation of nano-sized hexagonal $Mg(OH)_2$. *Procedia Eng.* **2015**, *102*, 388–394. [CrossRef]
59. Sabzevari, M.; Sajjadi, S.A.; Moloodi, A. Physical and mechanical properties of porous copper nanocomposite produced by powder metallurgy. *Adv. Powder Technol.* **2016**, *27*, 105–111. [CrossRef]
60. Vassilev, S.V.; Baxter, D.; Vassileva, C.G. An overview of the behaviour of biomass during combustion: Part I. Phase-mineral transformations of organic and inorganic matter. *Fuel* **2013**, *112*, 391–449. [CrossRef]
61. Broström, M.; Enestam, S.; Backman, R.; Mäkelä, K. Condensation in the KCl–NaCl system. *Fuel Process. Technol.* **2013**, *105*, 142–148. [CrossRef]
62. Matschei, T.; Lothenbach, B.; Glasser, F. The AFm phase in Portland cement. *Cem. Concr. Res.* **2007**, *37*, 118–130. [CrossRef]
63. Rai, R.K.; Serra-Maia, R.; Shi, Y.; Psarras, P.; Vojvodic, A.; Stach, E.A. Enhanced mineral carbonation on surface functionalized MgO as a Proxy for mine tailings. *Environ. Sci. Nano* **2025**, *12*, 2630–2646. [CrossRef]
64. Khan, M.; Kayali, O.; Troitzsch, U. Chloride binding capacity of hydrotalcite and the competition with carbonates in ground granulated blast furnace slag concrete. *Mater. Struct.* **2016**, *49*, 4609–4619. [CrossRef]
65. Damidot, D.; Glasser, F. Thermodynamic investigation of the CaO—Al_2O_3—$CaSO_4$—$CaCO_3$-H_2O closed system at 25 °C and the influence of Na_2O. *Adv. Cem. Res.* **1995**, *7*, 129–134. [CrossRef]
66. Damidot, D.; Glasser, F. Investigation of the CaO-Al_2O_3-SiO_2-H_2O system at 25 °C by thermodynamic calculations. *Cem. Concr. Res.* **1995**, *25*, 22–28. [CrossRef]
67. Landy, R.A. Magnesia refractories. *Mech. Eng.* **2004**, *178*, 109.
68. Thomas, J.J.; Musso, S.; Prestini, I. Kinetics and activation energy of magnesium oxide hydration. *J. Am. Ceram. Soc.* **2014**, *97*, 275–282. [CrossRef]
69. Madej, D. Hydration, carbonation and thermal stability of hydrates in $Ca_{7-x}Sr_xZrAl_6O_{18}$ cement. *J. Therm. Anal. Calorim.* **2018**, *131*, 2411–2420. [CrossRef]
70. Petkova, V.; Stoyanov, V.; Pelovski, Y. TG–DTG–DTA in studying white self-compacting cement mortars. *J. Therm. Anal. Calorim.* **2012**, *109*, 797–806. [CrossRef]
71. Gartner, E.; Walenta, G.; Morin, V.; Termkhajornkit, P.; Baco, I.; Casabonne, J. Hydration of a belite-calciumsulfoaluminate-ferrite cement: Aether™. In Proceedings of the 13th International Congress on the Chemistry of Cement, Madrid, Spain, 3–8 July 2011.
72. Birnin-Yauri, U.; Glasser, F. Friedel's salt, $Ca_2Al(OH)_6(Cl, OH)\cdot 2H_2O$: Its solid solutions and their role in chloride binding. *Cem. Concr. Res.* **1998**, *28*, 1713–1723. [CrossRef]
73. Wang, Z.; Yang, W.; Liu, H.; Jin, H.; Chen, H.; Su, K.; Tu, Y.; Wang, W. Thermochemical behavior of three sulfates ($CaSO_4$, K_2SO_4 and Na_2SO_4) blended with cement raw materials (CaO-SiO_2-Al_2O_3-Fe_2O_3) at high temperature. *J. Anal. Appl. Pyrolysis* **2019**, *142*, 104617. [CrossRef]
74. Lv, Z.; Tan, H.; Liu, X.; Chen, P.; Wang, Y.; Liang, W.; Hong, J. Chloride binding of AFm in the presence of Na^+, Ca^{2+} and Ba^{2+}. *Constr. Build. Mater.* **2023**, *364*, 129804. [CrossRef]
75. Jun, Y.; Yoon, S.; Oh, J.E. A comparison study for chloride-binding capacity between alkali-activated fly ash and slag in the use of seawater. *Appl. Sci.* **2017**, *7*, 971. [CrossRef]

ceramics

MDPI

Article

First Test and Characterizations on Urban Glass Waste with Waste-Derived Carbon Fiber Treated to Realize Foam Glass for Possible Construction Applications

Zakim Hussain [1,*], Seyed Mostafa Nouri [1], Matteo Sambucci [1,2] and Marco Valente [1,2]

[1] Department of Chemical Engineering, Materials, Environment, Sapienza University of Rome, 00184 Rome, Italy; seyedmostafa.nouri@uniroma1.it (S.M.N.); matteo.sambucci@uniroma1.it (M.S.); marco.valente@uniroma1.it (M.V.)
[2] INSTM Reference Laboratory for Engineering of Surface Treatments, UdR Rome, Sapienza University of Rome, 00184 Rome, Italy
* Correspondence: zakim.hussain@uniroma1.it

Abstract: Urban glass waste is a significant by-product of residential areas, while scrap carbon fiber is a prevalent industrial by-product. This study explores an innovative approach to valorize these materials by producing foam glass (FG) for versatile applications, particularly in construction. A key challenge in FG production is enhancing its properties to meet increasingly stringent application-specific standards. The properties of FG are intrinsically linked to its porous structure, which depends on factors such as the foaming process. The oxidation of carbon fibers at high temperatures can induce a foaming effect, creating a porous matrix in the glass. This research investigates the effect of powdered recycled carbon fiber (PRCF)—an alternative method for recovering waste carbon fiber as a foaming agent for FG. PRCF was added at concentrations of 0.5%, 1%, and 1.5% by mass relative to powdered waste glass. Increasing PRCF content enhanced foaming and improved porosity, with total porosity rising from 47.18% at 0.5% PRCF to 65.54% at 1.5% PRCF, accompanied by a 50% reduction in compressive strength and a 68% decrease in thermal conductivity. The results demonstrate the feasibility of large-scale FG production with enhanced properties, achieved without substantial additional investment and by recovering two waste materials. This process supports sustainable development by promoting waste valorization and advancing circular economy principles.

Keywords: urban glass; recycled carbon fiber; foam glass; waste valorization

1. Introduction

Foam glass is a porous material characterized by numerous closed or interconnected pores, typically ranging in size from nanometers to micrometers. FG exhibits remarkable properties, including low density, high strength, and excellent insulating ability [1–4], heat resistance, chemical resistance, low water absorption, and a practically unlimited service life. These superior properties, compared to those of organic and fibrous insulation materials, have attracted growing interest and expanded its range of applications [5,6]. This material can be produced with a high degree of purity, and its pore sizes can be controlled and adjusted to suit specific applications.

Foam glass is made by combining glass powder with a foaming agent (graphite (C), silicon carbide (SiC), sodium carbonate (Na_2CO_3), and calcium carbonate ($CaCO_3$)), which release gas at high temperatures, forming a lightweight, porous material with good insulation and structural properties [7]. Common glass powder includes waste glasses and industrial byproducts. FG quality depends on key process parameters such as sintering temperature, heating rate, glass powder particle size, and foaming agent type, all of which affect pore formation, mechanical strength, and thermal performance. The optimal control of these factors is essential for producing high-quality foam glass [8–10]. The effectiveness of a foaming agent in FG production depends on its gas-releasing behavior and interaction with the glass matrix [11,12]. Both the type of foaming agent and the particle size distribution of glass powder critically influence foam morphology, thermal insulation, and mechanical properties [12]. Finer glass powder enhances pore uniformity, strength, and insulation, while coarser glass powder leads to irregular pores and reduced structural integrity [13].

FG is available in various forms, including powders, granules, and larger-shaped pieces, making it highly versatile for diverse applications [14]. In the construction industry, FG is used as a lightweight fill material, a water retention medium for landscaping, a subgrade improvement material, and an aggregate for lightweight concrete [15–18]. FG's closed-porous structure makes it ideal for thermal insulation and water-resistant applications, while its open-porous structure provides sound insulation and water absorption capabilities, making it useful in metro stations and gardens [19].

Over the last two decades, researchers have increasingly focused on using waste glass to produce FG. This approach aligns with the principles of the circular economy, transforming waste into valuable resources and reducing environmental impact [20–22]. But the relationship between the foaming agent, glass powder particle size, and desired foam glass properties highlights the importance of controlling pore size distribution to optimize performance across applications. However, the precise functional links between key factors such as pore size distribution, heating rate, density, and sintering temperature, and FG properties like compressive strength and porosity remain insufficiently explored. This gap has hindered the efficient design and production of foam glass. Therefore, a systematic methodology is urgently needed to optimize FG preparation by accounting for these critical variables. However, enhancing FG's properties remains a key research challenge to meet evolving application standards.

This study explores the feasibility of reusing urban glass waste by incorporating powdered recycled carbon fiber as a foaming agent for FG production. Experimental investigations focused on the effect of PRCF concentration on the foaming and density of foam glass for enhancing FG's performance for mechanical and structural applications.

2. Materials and Methods

2.1. Raw Materials

2.1.1. Powdered Waste Glass

Urban waste glass bottles were collected from public waste bins in Rome and washed with water to remove plastic labels. The bottles were then crushed into small pieces using a hammer. Subsequently, 200 g of these fragments were ground into a fine powder using a Retsch PM 100 planetary ball (Retsch GmbH, Haan, Germany) mill at 250 rpm for 30 min. The milling process involved six large steel balls (19.05 mm in diameter, 29 g per ball) and thirty-five smaller steel balls (6.35 mm in diameter, 1.25 g per ball). The procedure for preparing powdered waste glass is illustrated in Figure 1.

The chemical composition of the glass powder (Figure 2), analyzed using Energy-Dispersive X-ray Spectroscopy (Octane Elect EDS system, Edax, Mahwah, NJ, USA), confirmed the predominant presence of silicon (Si) and oxygen (O), the main components of silica glass. The presence of gold (Au) indicates that the sample was coated with a thin gold layer, a common practice to prevent charging effects during SEM/EDS analysis by making the sample conductive.

Figure 1. Procedure for the production of powered waste glass from urban waste glass.

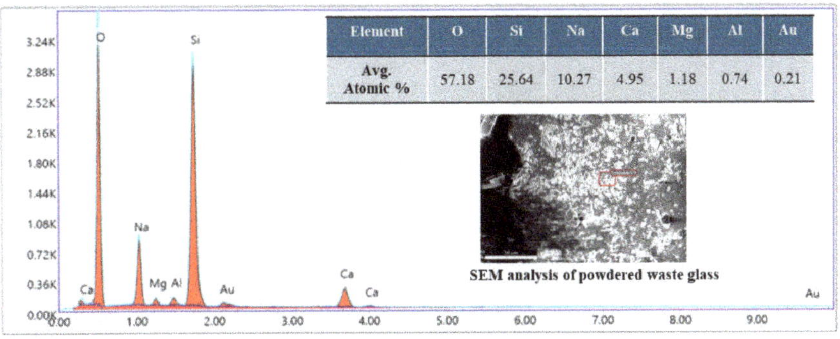

Figure 2. Powdered glass compositional analysis using EDS along selected spot via SEM.

Secondary elements, including sodium (Na), magnesium (Mg), aluminum (Al), and calcium (Ca), were also detected, possibly due to modifications, contamination, or the material being soda-lime-based silica glass. Calcium (Ca) appears at multiple positions, suggesting the presence of calcium-containing impurities or additives such as CaO or $CaCO_3$.

2.1.2. Powdered Recycled Carbon Fiber

Figure 3 presents the flowchart to produce powdered recycled carbon fiber, implemented in this study. This material is obtained by milling secondary waste carbon fiber, a by-product of woven/non-woven fabric production [23]. The waste, appearing as fluffy carbon agglomerates of different dimensions, was supplied by Carbon Task Srl (Biella, Italy).

During fabric production, 30–50 kg of secondary waste carbon fiber is generated per ton of processed recycled carbon fiber (RCF). The morphology of powdered recycled carbon fiber was examined using a scanning electron microscope (SEM) (Tescan MIRA 3 FEG-SEM, Brno, Czech Republic), revealing small fiber particles with non-uniform shapes after ball milling. This material was previously used by the authors as a filler for producing 3D printable thermoplastic composites [24]. Its density is reported as 1.917 g/cm^3, and the average fiber length is 30 ± 12 µm.

Figure 3. PRCF production flow chart.

2.2. FG Sample Preparation

To prepare the dry mixture, 15 g of powdered waste glass was manually mixed in a becker with powdered recycled carbon fiber at weight concentrations of 0.5%, 1%, and 1.5%. Subsequently, 30 drops of tap water were added and stirred thoroughly to achieve homogeneity. The wet mixture was placed into molds and compressed using a hand press to form samples. The samples were dried at 100 °C temperature in the oven for 24 h to remove the moisture. After that samples underwent thermal treatment in a Lindberg oven (Lindberg/MPH, Riverside, MI, USA) at 950 °C for 20 min. Four samples were prepared for each powdered recycled carbon fiber concentration. Note that the drying stage was performed solely to measure the density of the dried state of the material. FG production can proceed without this step, with wet samples directly placed in the 950 °C oven. Figure 4 illustrates the preparation process of foam glass samples.

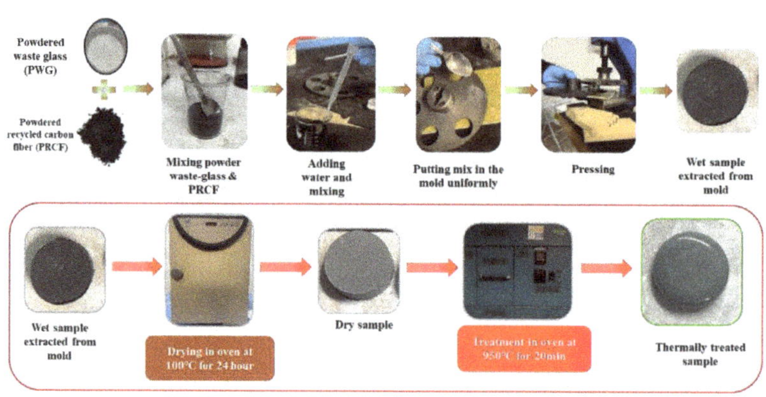

Figure 4. The schematic diagram for sample preparation.

2.3. Testing Techniques

2.3.1. Setup for Optical Microscopic Analysis of Pores

To obtain detailed morphological information of pores in foamed glass, the samples were first polished first, and then the microscopic observations were undertaken using a Nikon Eclipse L150 optical microscope (Nikon Corporation, Tokyo, Japan) at 5× magnifications, respectively. In the above technique, FG samples were placed under the lens and analyzed to identify sizes of pores at 20 different points. The microphotographs were acquired using software Lucia v3.

2.3.2. Density and Porosity Measurements

Bulk density, true density, open porosity, closed porosity, and total porosity were determined based on the method described by described by previous researchers [25], with some modifications. For bulk density measurements, a modified approach was employed. In contrast to previous researchers, who used regular cubic samples and a geometric method, samples in this study were coated with paraffin wax to prevent water infiltration into open pores. The coating was applied by brushing melted paraffin onto the sample surfaces. Density was measured using a commercial density determination kit (Mettler Toledo ME54, Columbus, OH, USA). The bulk density of the uncoated samples was calculated by subtracting the mass and volume contributions of the paraffin coating.

True density was assessed using six smaller samples, each approximately half the size of the original samples, with two samples allocated per powdered recycled carbon fiber percentage. These samples were distinct from those used for bulk density measurements. Measurements were averaged, and the mean value was reported as the true density.

Open porosity was determined by extending the soaking time in bi-distilled water to 18 h, ensuring complete water penetration into the open pores. This deviates from the original method by Zhai et al. [25], which utilized a 2 h soaking duration. The reported bulk density and porosity values represent the averages obtained from three samples, each nearly half the size of the originally prepared specimens for each formulation.

2.3.3. Setup for Thermal Conductivity Analysis

A C-Therm TCi thermal analyzer (C-Therm Technologies, Fredericton, NB, Canada) was used to assess the thermal conductivity (Figure S1) of the FG samples according to [26]. Measurements were made at room temperature of FG samples. The thermal conductivity values for each glass-PRCF formulation were obtained by considering two test samples on which three scans were performed.

2.3.4. Setup for Compression Puncture Test

The compression puncture tests (Figure S2) were carried out for each sample with the universal mechanical machine Zwick/Roell Z010 (Zwick/Roell, Ulm, Germany) equipped with a load cell of 10 kN. The diameter of the compression stainless-steel puncture tester was 10 mm, and a rigid PVC support was used to prevent contact between the puncture and the metal plate, thereby avoiding any damage to the system. The trend of each percentage of PRCF sample during the compression test was recorded with a speed of 1 mm/min. The results of the compression test were expressed in terms of maximum compression load. Four specimens were tested for each formulation, and the average values, along with their standard deviations, were calculated.

2.3.5. Microstructure Analysis via SEM

The FG sample was cut into very small pieces measuring 2×3 mm^3 and gold-coated using an Edwards S150B sputter coater (Edwards Ltd., Burgess Hill, UK) to improve electrical conductivity before placing it in the SEM. The porous structure and undecomposed PRCF particles in the FG sample were investigated.

3. Results and Discussion

The amount of foaming agent, such as carbon, has a significant impact on the foaming effects of the final products. Based on previous works, we altered the PRCF amount from

0.5 wt.% to 1.5 wt.% under a sintering temperature of 950 °C to investigate the amount of PRCF that yields an optimal combination of mechanical and physical properties [27–29].

The images of polished and fractured surfaces of FG samples (Figure 5) reveal a correlation between the PRCF ratio and the foaming of samples. The foaming effect is created due to the oxidation of carbon atoms during the heating process, forming a porous structure in the softened glass matrix. When the ratio of carbon increases, more CO_2 is produced, leading to an increase in foaming inside the glass matrix. This phenomenon underscores the impact of the PRCF ratio on foaming formation [30].

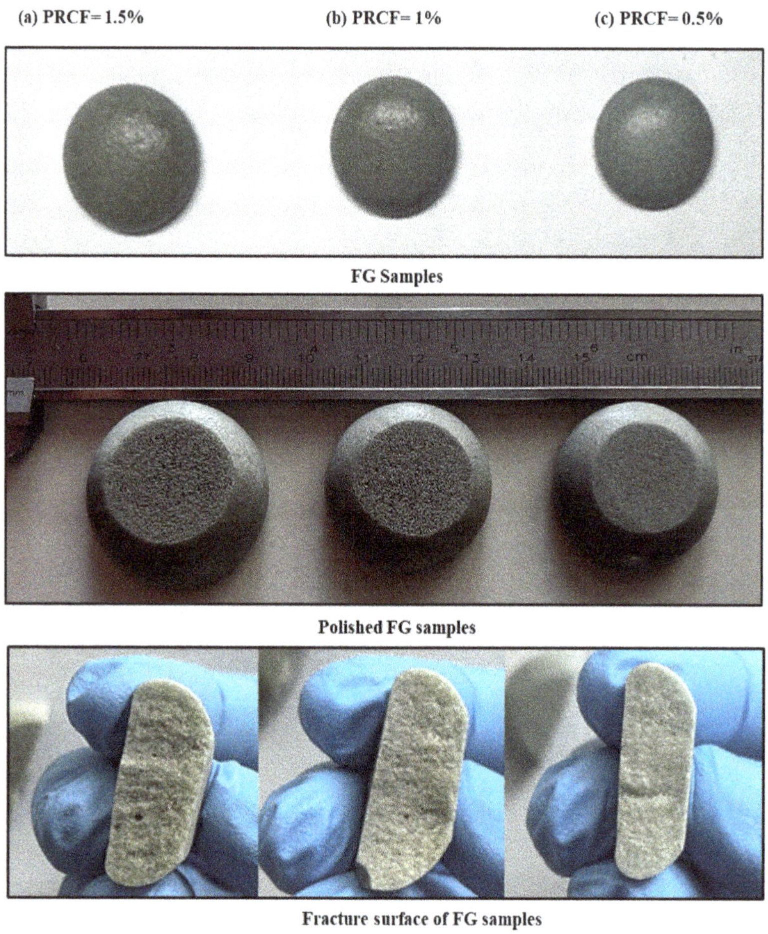

Figure 5. FG samples with different PRFG percentages.

3.1. *Optical Microscopic Analysis of Pores*

The optical microscopic analysis of polished FG samples demonstrated the morphologies of pores, and it can also be seen in Figure 6 that there are still undecomposed PRCF particles in the red selected areas. The size of pores also enlarged with respect to increasing the percentage of PRCF particles. Figure 6 also demonstrated that more pores are distributed narrowly with a small diameter (μm) range in the FG sample, having a lower PRCF percentage as compared to samples having higher PRCF ratio.

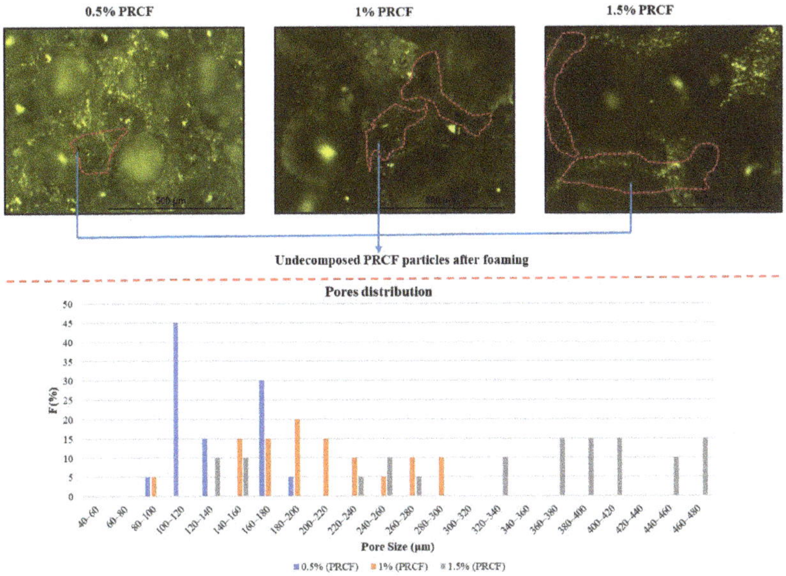

Figure 6. Microscopical analysis of polished surfaces of FG samples and pores distribution.

3.2. Thermal Conductivity Analysis

The bar chart shown by Figure 7 illustrates the relationship between the %PRCF content and the thermal conductivity (k) of FG.

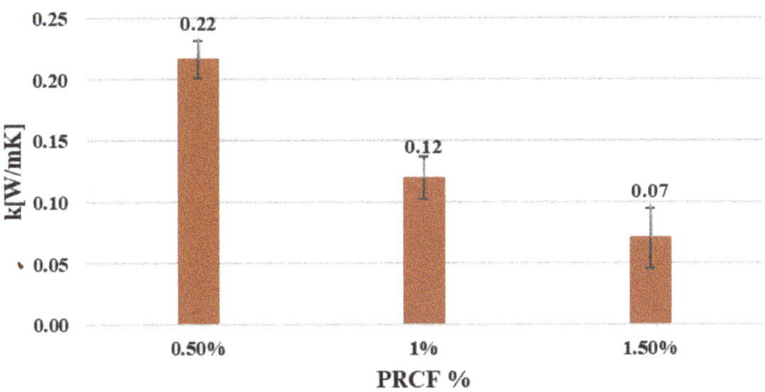

Figure 7. Thermal conductivity analysis of FG samples.

The data shows a clear decreasing trend in thermal conductivity as the PRCF percentage increases. At 0.5% PRCF, the thermal conductivity was 0.22 W/mK, which dropped to 0.12 W/mK for 1% PRCF and further decreased to 0.07 W/mK for 1.5% PRCF. It indicated that PRCF acts as a thermal insulator in foamed glass, likely due to increased porosity, changes in microstructure, or reduced heat conduction pathways [31]. This makes foamed glass with higher PRCF content potentially more suitable for applications requiring low thermal conductivity, such as thermal insulation in construction and industrial applications.

3.3. SEM Analysis

Figure 8 shows the SEM surface morphology of foam glass products, where the porous structure is created as a function of the carbon amount. When the PRCF decomposed under

950 °C, a nearly small- and large-sized almost spherical pore structure is achieved, but there are still undecomposed PRCF particles with irregular geometry which is increased with respect to a higher PRCF ratio of 1 wt.% and 1.5 wt.% at the same temperature. The physical and mechanical properties of the FG samples have been characterized in previous research. It is shown that both compressive strength and bulk density are minimal when the PRCF percentage is increasing; this should be attributable to the highest total porosity of the pore structure, which indicates a porous structure that leads to the lowest strength and density [25].

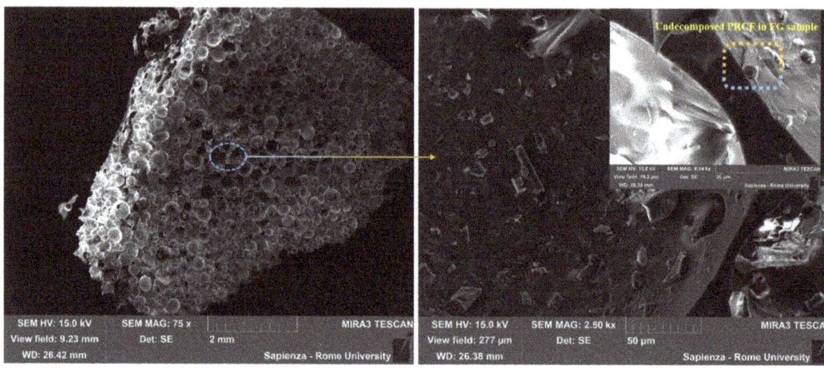

Figure 8. SEM of porous FG sample having undecomposed PRCF.

3.4. Compression Puncture Test

The relationship between the maximum compression load (F_{max} in N) and the percentage of PRCF in FG samples is illustrated in Figure 9. The compression load decreases as the PRCF percentage increases in foamed glass. The highest compressive strength (1944.13 N) is observed at 0.5% PRCF, while the lowest strength (960.60 N) is recorded at 1.5% PRCF.

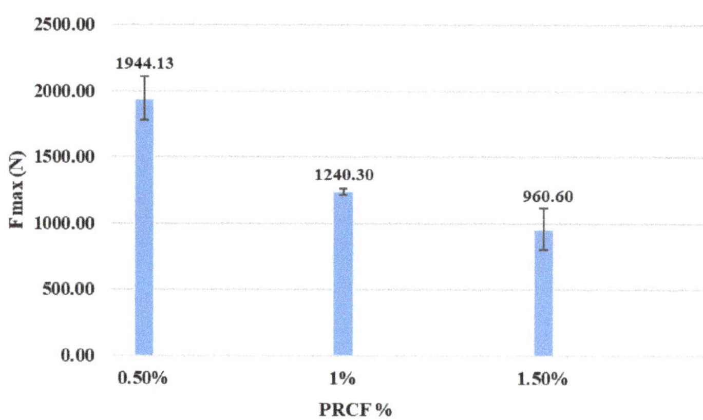

Figure 9. Compression puncture test of FG samples.

This suggests that while higher PRCF content improves insulation, it may compromise structural integrity. The decomposition of carbon in high-temperature conditions may not actually strengthen the silica network. Instead, it could potentially weaken it or alter its structure. The release of gases could cause expansion or even structural collapse, leading to a lower density [32]. Further research is needed to explore ways to enhance mechanical properties while retaining the insulation benefits of PRCF-infused foamed glass.

3.5. Density and Porosity

Figure 10 illustrates the true and bulk density values of foam glass samples. The bulk density exhibited a notable decrease as the percentage of PRCF increased.

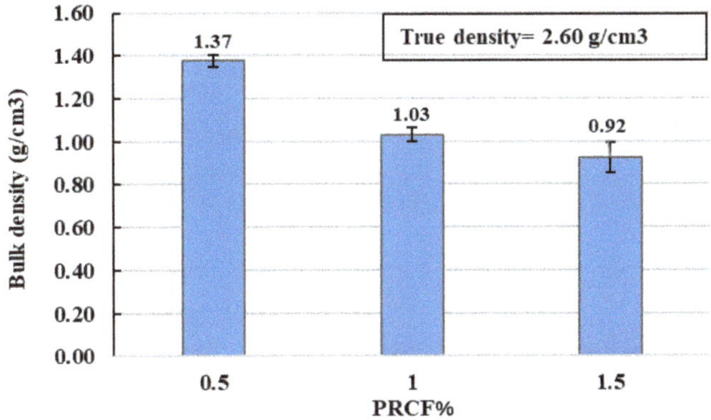

Figure 10. True and bulk densities of foam glass.

Specifically, it dropped from 1.37 g/cm^3 at 0.5% PRCF to 0.92 g/cm^3 at 1.5% PRCF, reflecting a 33% reduction. This trend indicates that increasing PRCF content enhances pore formation within the glass matrix, resulting in a more porous structure. The decrease in bulk density aligns with the expected effects of foaming agents, which reduce the overall mass per unit volume of the material, thereby contributing to its lightweight properties.

However, increasing the PRCF content also affects the viscosity of molten glass. At higher concentrations, PRCF reinforces the silica network, increasing its internal resistance to flow because of a higher quantity of undecomposed PRCF particles. The SEM investigation in the above section clarified this aspect in Figure 11. This phenomenon could potentially limit further foaming if the viscosity becomes too high, restricting the bubble formation [33–35].

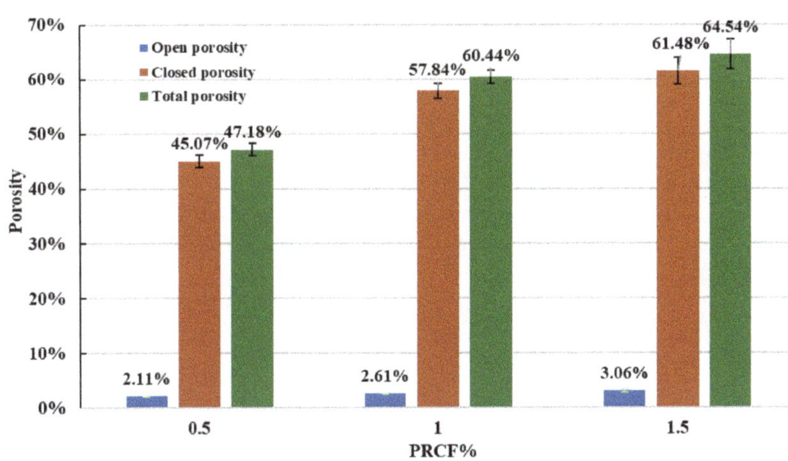

Figure 11. Open, closed, and total porosity of foam glass samples.

Figure 11 showcases the variations in open, closed, and total porosities of the foam glass samples. The porosity characteristics exhibited significant changes with increasing

PRCF content. Total porosity increased from 47.18% at 0.5% PRCF to 65.54% at 1.5% PRCF, with closed porosity being the dominant component. Specifically, closed porosity rose from 45.07% to 61.48% over this range, while open porosity showed a modest increase from 2.11% to 3.06%. The prevalence of closed porosity is beneficial for thermal insulation applications, as it minimizes heat transfer through convective pathways [36]. Additionally, the marginal increase in open porosity suggests that most pores remain isolated rather than interconnected, which is advantageous for maintaining structural integrity and reducing moisture penetration [37].

The relationship between PRCF content and porosity can be attributed to the role of carbon fibers as nucleation sites for gas bubbles during the sintering process. As PRCF decomposes at high temperatures, it releases gases that become trapped within the glass matrix, forming closed pores [38,39]. The increase in pore size observed with higher PRCF percentages supports this mechanism, suggesting that more extensive gas release leads to larger, more numerous pores. This controlled pore formation is crucial for achieving the desired balance between thermal insulation performance and mechanical stability in foam glass materials [36,37]. The significant enhancement in total porosity, coupled with the low bulk density, underscores the potential of PRCF as an effective foaming agent in the production of lightweight, thermally efficient foam glasses [38].

The results validate the effectiveness of PRCF as a foaming agent for FG production, demonstrating its potential for lightweight and porous structural applications. Additionally, the successful use of urban glass waste and secondary carbon fiber waste supports sustainability efforts by reducing landfill waste and providing a cost-effective solution for producing environmentally friendly FG materials. The findings also underscore the importance of optimizing FG processing parameters to achieve desirable physical and mechanical characteristics. The developed methodology offers a sustainable approach to manufacturing FG, promoting waste recycling, and advancing circular economic principles.

4. Conclusions

This study demonstrated the feasibility of producing foam glass using powdered urban waste glass and powdered recycled carbon fibers as a foaming agent. Bubble growth and the viscous flow of the vitreous melt occurred throughout the entire foam formation process. The main factors influencing the foam structure were sintering temperature, sintering duration, and the weight percentage of the foaming agent relative to the glass powder. From the experimental and characteristics analysis, the following key findings were obtained.

- The results revealed a direct correlation between PRCF concentration and the reduction in the density of FG samples. Density decreased from 1.37 g/cm^3 at 0.5% PRCF to 0.92 g/cm^3 at 1.5% PRCF, representing a 33% reduction. This trend indicates that increasing PRCF content promotes pore formation within the glass matrix, leading to a more porous structure.
- Increasing the PRCF percentage resulted in a greater degree of foaming, driven by CO_2 generation from PRCF oxidation. This resulted in a significant reduction in compressive strength by 50% and thermal conductivity by 68%, as PRCF content increased from 0.5% to 1.5%.
- The findings also confirmed that PRCF effectively induces porosity in the glass matrix. Total porosity rising from 47.18% at 0.5% PRCF to 65.54% at 1.5% PRCF, which is supporting its potential application in lightweight construction materials.

Moreover, the successful use of urban glass waste and secondary carbon fiber waste aligns with sustainability objectives by promoting waste valorization and the principles of a circular economy. Future studies will explore the effects of different curing temperatures, mechanical properties, thermal conductivity, and microstructural characteristics to further optimize FG production and enhance its applicability in the construction and insulation industries.

Supplementary Materials: The following supporting information can be downloaded at: https://www.mdpi.com/article/10.3390/ceramics8020073/s1. Figure S1. Experimental setup for thermal conductivity, and Figure S2. Setup for the compression puncture test.

Author Contributions: Conceptualization, M.S. and M.V.; methodology, Z.H., S.M.N. and M.S.; validation, Z.H., S.M.N. and M.S.; investigation, Z.H. and S.M.N.; resources, M.V.; writing—original draft preparation Z.H. and S.M.N.; writing—review and editing, S.M.N. and M.S.; supervision, M.S. and M.V.; project administration, M.S. and M.V. All authors have read and agreed to the published version of the manuscript.

Funding: This research received no external funding.

Institutional Review Board Statement: Not applicable.

Informed Consent Statement: Not applicable.

Data Availability Statement: The original contributions presented in this study are included in the article. Further inquiries can be directed to the corresponding author.

Acknowledgments: The authors would like to thank Christian Scopinich (Carbon Task Srl) for providing PRCF implemented in the research activity. A heartfelt acknowledgement also goes to Jacopo Tirillò and Fabrizio Sarasini for their availability and support in thermal conductivity measurements.

Conflicts of Interest: The authors declare no conflict of interest.

Abbreviations

The following abbreviations are used in this manuscript:

FG	Foam glass
PRCF	Powdered recycled carbon fiber
RCF	Recycled carbon fiber
SEM	Scanning electron microscope

References

1. Guo, Y.; Zhang, Y.; Huang, H.; Meng, K.; Hu, K.; Hu, P.; Wang, X.; Zhang, Z.; Meng, X. Novel glass ceramic foams materials based on red mud. *Ceram. Int.* **2014**, *40*, 6677–6683. [CrossRef]
2. Liu, Y.; Xie, J.; Hao, P.; Shi, Y.; Xu, Y.; Ding, X. Study on factors affecting properties of foam glass made from waste glass. *J. Renew. Mater.* **2020**, *9*, 237–253. [CrossRef]
3. König, J.; Lopez-Gil, A.; Cimavilla-Roman, P.; Rodriguez-Perez, M.A.; Petersen, R.R.; Østergaard, M.B.; Iversen, N.; Yue, Y.; Spreitzer, M. Synthesis and properties of open- and closed-porous foamed glass with a low density. *Constr. Build. Mater.* **2020**, *247*, 1+18574. [CrossRef]
4. Caniato, M.; Kyaw Oo D'Amore, G.; Kaspar, J.; Gasparella, A. Sound absorption performance of sustainable foam materials: Application of analytical and numerical tools for the optimization of forecasting models. *Appl. Acoust.* **2020**, *161*, 107166. [CrossRef]
5. Ivanov, K.S. Preparation and Properties of Foam Glass-ceramic from Diatomite. *J. Wuhan Univ. Technol. Mater. Sci. Ed.* **2018**, *33*, 273–277. [CrossRef]
6. Zhu, M.; Ji, R.; Li, Z.; Wang, H.; Liu, L.L.; Zhang, Z. Preparation of glass ceramic foams for thermal insulation applications from coal fly ash and waste glass. *Constr. Build. Mater.* **2016**, *112*, 398–405. [CrossRef]
7. Scheffler, M.; Colombo, P. *Cellular Ceramics: Structure, Manufacturing, Properties and Applications*; Wiley-VCH: Weinheim, Germany; John Wiley: Hoboken, NJ, USA, 2005.
8. Kaz'mina, O.V.; Vereshchagin, V.I.; Abiyaka, A.N. Raw Materials Prospects for Use of Finely Disperse Quartz Sands in Production of Foam-Glass Crystalline Materials. *Glass Ceram.* **2008**, *65*, 319–321. [CrossRef]
9. Vancea, C.; Lazău, I. Glass foam from window panes and bottle glass wastes. *Cent. Eur. J. Chem.* **2014**, *12*, 804–811. [CrossRef]

10. Zhang, J.; Liu, B.; Zhang, S. A review of glass ceramic foams prepared from solid wastes: Processing, heavy-metal solidification and volatilization, applications. *Sci. Total Environ.* **2021**, *781*, 146727. [CrossRef]
11. Naser, M.Z.; Thai, S.; Thai, H.T. Evaluating structural response of concrete-filled steel tubular columns through machine learning. *J. Build. Eng.* **2021**, *34*, 101888. [CrossRef]
12. Chen, N.; Zhang, S.; Pan, X.; Zhou, S.; Zhao, M. Foaming mechanism and optimal process conditions of foamed glass based on thermal analysis. *J. Porous Mater.* **2020**, *27*, 621–626. [CrossRef]
13. Davraz, M.; Koru, M.; Akdağ, A.E.; Kılınçarslan Delikanlı, Y.E.; Çabuk, M. An investigation of foaming additives and usage rates in the production of ultra-light foam glass. *J. Therm. Anal. Calorim.* **2022**, *147*, 3567–3576. [CrossRef]
14. Zhu, B.; Zhang, Z.; Zhang, W.; Wu, Y.; Zhang, J.; Imran, Z.; Zhang, D. Synthesis and Applications of Porous Glass. *J. Shanghai Jiaotong Univ. Sci.* **2019**, *24*, 681–698. [CrossRef]
15. Harrison, E.; Berenjian, A.; Seifan, M. Recycling of waste glass as aggregate in cement-based materials. *Environ. Sci. Ecotechnol.* **2020**, *4*, 100064. [CrossRef]
16. Khan, M.N.N.; Saha, A.K.; Sarker, P.K. Reuse of waste glass as a supplementary binder and aggregate for sustainable cement-based construction materials: A review. *J. Build. Eng.* **2020**, *28*, 101052. [CrossRef]
17. Nistratov, A.V.; Klimenko, N.N.; Pustynnikov, I.V.; Vu, L.K. Thermal Regeneration and Reuse of Carbon and Glass Fibers from Waste Composites. *Emerg. Sci. J.* **2022**, *6*, 967–984. [CrossRef]
18. Huang, D.; Sun, P.; Gao, P.; Liu, G.; Wang, Y.; Chen, X. Study on the effect and mechanism of alkali–silica reaction expansion in glass concrete. *Sustainability* **2021**, *13*, 10618. [CrossRef]
19. König, J.; Nemanič, V.; Žumer, M.; Petersen, R.R.; Østergaard, M.B.; Yue, Y.; Suvorov, D. Evaluation of the contributions to the effective thermal conductivity of an open-porous-type foamed glass. *Constr. Build. Mater.* **2019**, *214*, 337–343. [CrossRef]
20. Mohajerani, A.; Vajna, J.; Cheung, T.H.H.; Kurmus, H.; Arulrajah, A.; Horpibulsuk, S. Practical recycling applications of crushed waste glass in construction materials: A review. *Constr. Build. Mater.* **2017**, *156*, 443–467. [CrossRef]
21. Heriyanto Pahlevani, F.; Sahajwalla, V. Synthesis of calcium silicate from selective thermal transformation of waste glass and waste shell. *J. Clean. Prod.* **2016**, *172*, 3019–3027. [CrossRef]
22. Popov, M.; Zakrevskaya, L.; Vaganov, V.; Hempel, S.; Mechtcherine, V. Performance of lightweight concrete based on granulated foamglass. *IOP Conf. Ser. Mater. Sci. Eng.* **2015**, *96*, 012017. [CrossRef]
23. Sambucci, M.; Nouri, S.M.; Tayebi, S.T.; Valente, M. Synergic Effect of Recycled Carbon Fibers and Microfibrillated Cellulose Gel for Enhancing the Mechanical Properties of Cement-Based Materials. *Gels* **2023**, *9*, 981. [CrossRef] [PubMed]
24. Valente, M.; Sambucci, M.; Rossitti, I.; Abruzzese, S.; Sergi, C.; Sarasini, F.; Tirillò, J. Carbon-Fiber-Recycling Strategies: A Secondary Waste Stream Used for PA6,6 Thermoplastic Composite Applications. *Materials* **2023**, *16*, 5436. [CrossRef] [PubMed]
25. Zhai, C.; Zhong, Y.; Liu, J.; Zhang, J.; Zhu, Y.; Wang, M.; Yeo, J. Customizing the properties of borosilicate foam glasses via additions under low sintering temperatures with insights from molecular dynamics simulations. *J. Non-Cryst. Solids* **2022**, *576*, 121273. [CrossRef]
26. Adjei, E.A.; Amos-Abanyie, S.; Omer, S. Characterization of Thermo-Physical Properties of Cement-Based Blocks of Varied Sand Types Using Cost-Effective Enhancement Approach. *Open J. Energy Effic.* **2020**, *9*, 14–30. [CrossRef]
27. Zhai, C.; Zhong, Y.; Li, Z.; Wang, X.; Zhao, L.; Pan, L.; Zhang, J. Preparation and characterization of mechanical properties of foam glass for artificial floating island carrier. *Adv. Mech. Eng.* **2015**, *7*, 1687814015589660. [CrossRef]
28. Zhai, C.; Li, Z.; Zhu, Y.; Zhang, J.; Wang, X.; Zhao, L.; Pan, L.; Wang, P. Effects of Sb_2O_3 on the mechanical properties of the borosilicate foam glasses sintered at low temperature. *Adv. Mater. Sci. Eng.* **2014**, *2014*, 703194. [CrossRef]
29. König, J.; Petersen, R.R.; Yue, Y.; Suvorov, D. Gas-releasing reactions in foam-glass formation using carbon and Mn_xO_y as the foaming agents. *Ceram. Int.* **2017**, *43*, 4638–4646. [CrossRef]
30. Kim, H.D.; Baek, C.R.; Jang, Y.C. Advancing glass recycling and environmental applications with porous glass: A mini-review. *J. Mater. Cycles Waste Manag.* **2024**, *26*, 2620–2633. [CrossRef]
31. Qin, Z.; Li, G.; Tian, Y.; Ma, Y.; Shen, P. Numerical simulation of thermal conductivity of foam glass based on the steady-state method. *Materials* **2018**, *12*, 54. [CrossRef]
32. Osfouri, M.; Simon, A. Study on the thermal conductivity and density of foam glass. *Pollack Period.* **2023**, *18*, 126–131. [CrossRef]
33. Rani, M.; Choudhary, P.; Krishnan, V.; Zafar, S. A review on recycling and reuse methods for carbon fiber/glass fiber composites waste from wind turbine blades. *Compos. Part B Eng.* **2021**, *215*, 108768. [CrossRef]
34. Lu, G.; Chen, Y.; Yan, Q.; Zhan, H.; Mao, P.; Liu, D. Investigation of microstructure and properties in short carbon fiber reinforced silica-based ceramic cores via atmosphere sintering. *J. Eur. Ceram. Soc.* **2021**, *41*, 7339–7347. [CrossRef]
35. Yuan, M.; Li, Z.; Teng, Z. Progress and prospects of recycling technology for carbon fiber reinforced polymer. *Front. Mater.* **2024**, *11*, 1484544. [CrossRef]
36. Liu, H.; Zhao, X. Thermal Conductivity Analysis of High Porosity Structures with Open and Closed Pores. *Int. J. Heat Mass Transf.* **2022**, *183*, 122089. [CrossRef]

37. Lin, H.; Shen, Q.; Ma, M.; Ji, R.; Guo, H.; Qi, H.; Xing, W.; Tang, H. 3D Printing of Porous Ceramics for Enhanced Thermal Insulation Properties. *Adv. Sci.* **2024**, *12*, 2412554. [CrossRef]
38. Lv, D.; Li, X.; Wang, L.; Du, J.; Zhang, J. Effect of carbon as foaming agent on pore structure of foam glass. *Adv. Mater. Res.* **2010**, *105*, 765–768. [CrossRef]
39. Østergaard, M.B.; Cai, B.; Petersen, R.R.; König, J.; Lee, P.D.; Yue, Y. Impact of pore structure on the thermal conductivity of glass foams. *Mater. Lett.* **2019**, *250*, 72–74. [CrossRef]

ceramics

MDPI

Article

Synergistic Effects of Furfurylated Natural Fibers and Nanoclays on the Properties of Fiber–Cement Composites

Thamires Alves da Silveira [1], Felipe Vahl Ribeiro [2], Cristian Conceição Gomes [1], Arthur Behenck Aramburu [3], Sandro Campos Amico [3], André Luiz Missio [1] and Rafael de Avila Delucis [1,*]

[1] Post-Graduate Program in Materials Science and Engineering, Federal University of Pelotas, Pelotas 96010-610, Brazil; thaamiresasilveira@gmail.com (T.A.d.S.); cristianconceicao8@gmail.com (C.C.G.); andreluizmissio@gmail.com (A.L.M.)
[2] Undergraduate Course in Wood Industry Engineering, Federal University of Pelotas, Pelotas 96010-610, Brazil; felipe.vs.ribeiro@gmail.com
[3] Post-Graduate Program in Mining, Metallurgical and Materials Engineering, Federal University of Rio Grande do Sul, Porto Alegre 90650-001, Brazil; arthuraramburu@gmail.com (A.B.A.); amico@ufrgs.br (S.C.A.)
* Correspondence: rafael.delucis@ufpel.edu.br

Abstract: Fiber–cement composites have been increasingly studied for sustainable construction applications, but durability issues—particularly fiber degradation in alkaline environments—remain a challenge. This study aimed to evaluate the individual and combined effects of furfurylated sisal fibers and nanoclay additions on the physical and mechanical performance of autoclaved fiber–cement composites, seeking to enhance fiber durability and matrix compatibility. All the composites were formulated with CPV-ARI cement and partially replaced with agricultural limestone to reduce the environmental impact and production costs. Sisal fibers (2 wt.%) were chemically modified using furfuryl alcohol, and nanoclays—both hydrophilic and surface-functionalized—were incorporated at 1% and 5% of cement weight. The composites were characterized for physical properties (density, water absorption, and apparent porosity) and mechanical performance (flexural and compressive strength, toughness, and modulus). Furfurylation significantly improved fiber–matrix interaction, leading to higher flexural strength and up to 100% gain in toughness. Nanoclay additions reduced porosity and increased stiffness, particularly at 5%, though excessive content showed diminishing returns. The combination of furfurylated fibers and functionalized nanoclay provided the best results in maintaining a compact microstructure, reducing water absorption, and improving mechanical resilience. Optical microscopy confirmed improved fiber dispersion and interfacial bonding in composites containing furfurylated fibers and functionalized nanoclay. These findings highlight the effectiveness of integrating surface-treated natural fibers with pozzolanic additives to enhance the performance and longevity of fiber–cement composites.

Keywords: fiber–cement composites; nanoclay; pozzolanic reaction; furfuryl alcohol; mechanical properties

1. Introduction

The construction industry is widely recognized as a major contributor to environmental degradation, from raw material extraction to waste generation. According to Munaro et al. [1], construction accounts for nearly one-third of global energy consumption. Among construction materials, cementitious products represent a significant environmental burden, with Portland cement production alone responsible for approximately 7% of global CO_2 emissions [2]. Reducing greenhouse gas emissions, limiting the use of non-renewable resources, and minimizing the generation of non-biodegradable waste in the cement industry are critical steps toward achieving sustainability in this economically and socially vital sector. Cement-based materials, due to their inorganic and non-metallic nature, are conventionally classified as ceramic materials within the field of materials science. This classification reinforces their relevance in discussions on ceramic materials for sustainable development.

One approach to making cement-based materials more sustainable involves reinforcing them with natural fibers. The use of plant fibers, such as jute and sisal, has historical roots but became prominent in the 1940s with the production of non-structural precast components [3]. Jaber et al. [4] highlight that natural fibers help address several inherent weaknesses in cementitious matrices, including brittleness, poor crack resistance, limited flexural deformability, and low flexural strength relative to compressive strength. Numerous studies report improvements in mechanical performance—such as increased flexural and compressive strength, enhanced ductility, reduced cracking, and improved toughness—following the incorporation of natural fibers [5,6]. These fibers may also reduce the density and thermal conductivity of the composite as their content increases [7].

However, several limitations still hinder the performance of fiber-reinforced cement composites. These include increased setting time, reduced workability, and most notably, fiber degradation in the highly alkaline environment (pH > 12) produced by cement hydration products like $Ca(OH)_2$ [3]. Alkaline degradation primarily affects lignin and hemicellulose, exposing holocellulose components (cellulose and hemicellulose) to further deterioration. Over time, even the cellulose breaks down, leaving the fiber ineffective as reinforcement. This degradation accelerates at elevated temperatures (above 75 °C) [8]. As highlighted by Pickering et al. [9], fiber degradation also compromises durability, increases permeability, and weakens the fiber–matrix bond. These effects reduce mechanical strength, alter fiber morphology (shorter and thinner fibers), and disrupt the chemical composition, resulting in reduced adherence between fibers and the cement matrix [10]. To mitigate these issues, pozzolanic additions and surface treatments are commonly used.

Pozzolans help lower matrix alkalinity by reacting with the alkaline ions released during cement hydration, such as NaOH and KOH, forming non-expansive compounds that densify the matrix and reduce its aggressiveness toward fibers [11]. This reduction in alkalinity is particularly important for preserving the integrity of lignocellulosic fibers, which are highly susceptible to degradation in high-pH environments. By consuming calcium hydroxide ($Ca(OH)_2$), pozzolanic materials contribute to the additional formation of calcium silicate hydrate (C–S–H) gel, thus enhancing the mechanical strength and dimensional stability of the composite [11,12]. Furthermore, pozzolans reduce porosity and refine the pore structure, which limits water ingress and mitigates the leaching of alkaline components over time [13].

Among the available pozzolans, nanoclays are particularly promising due to their high silica and alumina content—approximately 60% of total composition—and their ability to react with calcium hydroxide to form additional C–S–H gel and calcium aluminate hydrates. This reaction improves both early and long-term strength and durability. Nanoclays also enhance microstructural refinement through their filler effect and hydration acceleration, making them highly effective for improving cement-based materials. Recent studies have supported these findings. Hakamy et al. [12,14] demonstrated that the addition of 1 wt.% of organomodified nanoclay (Cloisite 30B—Sigma-Aldrich, St. Louis, MO, USA) to cement composites reinforced with hemp fabric significantly improved flexural strength, fracture toughness, and matrix density while reducing porosity. The nanoclay not only acted as a filler but also facilitated a more effective pozzolanic reaction, enhancing the bond between the fiber and the matrix. Wei and Meyer [15] explored the combined effect of metakaolin and nanoclay in sisal fiber–cement composites. Their findings showed that a partial replacement of cement with 3 wt.% nanoclay and 27 wt.% metakaolin reduced the CH content by more than 64%, suppressed ettringite formation, and increased the interfacial bond strength and pull-out energy of the fibers by 131% and 196%, respectively.

In parallel, chemical fiber treatments are used to limit $Ca(OH)_2$ deposition on fiber surfaces. The treatment of natural fibers before their incorporation into fiber–cement composites is a well-established strategy to enhance their compatibility with the cementitious matrix and improve composite durability. Methods like acetylation, mercerization, and silanization are commonly reported in the literature [16]. For instance, Ajouguim et al. [17] demonstrated that chemically treated alfa fibers (e.g., alkaline and hydrothermal treatments) increased the flexural strength of cement mortars by up to 44% and improved fiber–matrix adhesion, confirmed through shear stress measurements. Similarly, Biskri et al. [18] investigated the incorporation of chemically treated sisal fibers using urea and ethylenediaminetetraacetic acid and observed a 17–23.4% increase in flexural strength along with denser interfacial microstructures and higher cement hydrate development. Fonseca et al. [19] studied the hybrid reinforcement of fiber–cement composites using jute fibers and cellulose nanofibrils, reporting synergistic improvements in mechanical strength, porosity reduction (up to 75%), and enhanced post-weathering performance, particularly for compositions combining both reinforcement scales.

One environmentally friendly treatment option is furfurylation, a process in which furfuryl alcohol (FAL) reacts with a catalyst within the fiber structure, leading to the formation of poly(furfuryl alcohol). This treatment, widely used for wood modification, enhances fiber dimensional stability, hydrophobicity, and chemical resistance—key properties for protecting fibers in alkaline cementitious environments [20]. In a recent study by Aramburu et al. [20], microfibrillated cellulose treated with furfuryl alcohol was incorporated into cementitious pastes, resulting in significant improvements in fresh and hardened properties. The furfurylated MFC enhanced early-age compressive strength by up to 149%, mitigated setting time delays, and increased durability under wet–dry cycles.

Based on that, the combination of pozzolanic additions and fiber surface treatments in fiber–cement composites represents a promising approach, with several encouraging results already reported in the literature, to enhance both mechanical performance and long-term durability. While pozzolanic materials contribute to matrix densification, a reduction in free calcium hydroxide, and decreased alkalinity, fiber treatments improve dimensional stability, reduce water absorption, and strengthen fiber–matrix interfacial bonding. This dual approach may lead to composites with higher flexural strength, improved toughness, and reduced porosity. Recent studies support these findings. Mejia-Ballesteros et al. [21] observed that thermally treated eucalyptus fibers combined with a pozzolanic matrix containing activated coal waste led to increased modulus of rupture and durability after aging cycles. Hakamy et al. [22] demonstrated that alkali-treated hemp fabrics reinforced with calcined nanoclay yielded substantial gains in fracture toughness and flexural strength due to improved matrix compactness and fiber adhesion. Thepruttana et al. [23] further showed that coating sisal fibers with natural rubber latex and expanded perlite, a pozzolanic additive, significantly enhanced flexural performance and reduced porosity, highlighting the synergy between fiber modification and matrix refinement.

Therefore, despite its potential, furfurylation has been scarcely investigated in the context of natural fiber reinforcement for fiber–cement systems. Additionally, there is limited literature exploring the combined use of fiber surface treatments and pozzolanic additions as a strategy to prevent fiber mineralization and enhance the mechanical performance of fiber cements. This study deals with novel fiber–cement composites reinforced with chemically treated sisal fibers using furfurylation and further modified by incorporating two types of nanoclays—one hydrophilic and the other surface-functionalized. Physical properties such as density, water absorption, and porosity, as well as mechanical performance including flexural and compressive strength, toughness, and modulus, are systematically assessed. Fracture toughness is also measured using notched specimens. Through this approach, the study investigates the potential synergistic interaction between fiber treatment and pozzolanic addition to improve fiber durability and overall composite performance under autoclave curing conditions.

2. Materials and Methods

2.1. Materials

The cement used in this study was a high-early-strength Portland cement (CPV-ARI) from the Cauê brand (InterCement Brasil S.A, Cezarina, Goiás, Brazil), chosen due to its purity and absence of mineral additives that could interfere with the pozzolanic reactions or fiber interaction. Agricultural limestone from Dagoberto Barcellos (Dagoberto Barcellos S.A, Caçapava do Sul, Rio Grande do Sul, Brazil) was used as a partial replacement for the cement to reduce the environmental impact and material cost, a practice already common in industrial fiber–cement production.

Two nanoclays were used: (i) a hydrophilic bentonite nanoclay (Nanomer® PGV), consisting mainly of unmodified montmorillonite, with a maximum particle size of 25 μm, and (ii) a surface-functionalized nanoclay (Nanomer® I.31PS), based on montmorillonite modified with octadecylamine (15–35 wt.%) and aminopropyltriethoxysilane (0.5–5 wt.%), with particles up to 20 μm. Both were acquired from Sigma-Aldrich (Laboratory and Industrial Chemical—St. Louis, MO, USA) and incorporated into the cement matrix using the same dry mixing method.

The sisal fibers were commercially obtained from Compel, a supplier located in Arujá, São Paulo. They were delivered as twisted yarns and manually cut to 2 cm in length before use. Their selection was based on their low cost, native availability in Brazil, and good reinforcement potential. The fibers were characterized morphologically using an Opton TNB-04T-PL optical microscope (Opton—São Paulo, Brazil) and analyzed using ImageJ software (Version 1.54d) for dimensional measurements of 30 samples. The fibers were also chemically characterized using standard wet-chemistry methods for quantifying moisture, ash, extractives, acid-insoluble lignin, and holocellulose content (by difference). Fourier Transform Infrared Spectroscopy (FTIR) analysis was conducted on dried fiber powder compacted into KBr pellets, with 32 scans per sample in the 4000–1000 cm^{-1} range, at 4 cm^{-1} resolution. Thermal stability was assessed via thermogravimetric analysis (TGA) under inert nitrogen atmosphere from 50 °C to 250 °C at a heating rate of 10 °C/min, using a TA Instruments Q50 device (TA Instrutments—New Castle, DE, USA). Particle size distribution was determined using a CILAS 1180 laser (CILAS—Aubagne, France) diffraction analyzer, with isopropyl alcohol as the dispersant and continuous stirring at 200 rpm.

2.2. Simulation of Alkaline Fiber Degradation

To simulate the degradation process of sisal fibers caused by the alkaline environment of the cementitious matrix, an induced degradation reaction was carried out using a sodium hydroxide (NaOH) solution (98% purity; Sigma-Aldrich). NaOH was added to distilled water until it reached a pH of 12 (approximately 1.7 wt.% NaOH relative to water), which is representative of the alkalinity typically found in cementitious systems. The following material mixtures, including fibers, were immersed in the NaOH solution: (i) fibers only; (ii) fibers + nanoclay; and (iii) fibers + nanoclay + acrylic acid (AF) + maleic anhydride

(MA). The MA (98% purity, Sigma-Aldrich) was used as a catalyst for the AF (98% purity, Sigma-Aldrich). After 60 days of exposure to the induced alkaline degradation, the samples were analyzed by FTIR, using the same equipment and parameters previously described.

2.3. Fabrication of Fiber–Cement Composites

The fiber–cement formulations were developed based on the mass proportions established by Tonoli et al. [24]. A fiber content of 2 wt.% was adopted, as supported by prior analyses and recommendations from the literature [3,5,24]. All fiber–cement pastes were prepared with a fixed water-to-cement (w/c) ratio of 0.40, which was determined using a standard consistency test in accordance with ASTM C187 [25]. These parameters ensured consistency across all the formulations and were selected to align with established practices in fiber–cement composite research. To prevent excessive water absorption during mixing, all fibers were pre-moistened. The amount of water used for moistening was measured and subtracted from the total water added during paste preparation to ensure consistent water/cement ratios. Table 1 presents detailed information on the nomenclature and mass proportions applied for all the fiber–cement composites from this study.

Table 1. Formulation of the fiber–cement composites.

Nomenclature	Cement (wt.%)	Limestone (wt.%)	Sisal (wt.%)	Water (wt.%)	Nanoclay (wt.%)	FAL (wt.%)	MA (wt.%)
CON	62.91	10.43	1.5	25.16	0.0	0.0	0.0
NC1	62.52	10.37	1.49	25.01	0.63	0.0	0.0
NC5	60.99	10.11	1.45	24.4	3.05	0.0	0.0
FAL50	62.42	10.35	1.49	24.97	0.0	0.74	0.04
NC1FAL50	62.03	10.28	1.48	24.81	0.62	0.74	0.04
NC1FAL50P	62.03	10.28	1.48	24.81	0.62	0.74	0.04
NC5FAL50	60.53	10.04	1.44	24.21	3.03	0.72	0.04
FNC1FAL50	62.03	10.28	1.48	24.81	0.62	0.74	0.04

Where FAL and MA are furfuryl alcohol and maleic anhydride, respectively; CON = control (no fibers or additives); NC1 = 1 wt.% nanoclay; FAL50 = furfurylated fibers (50 wt.%); NC1FAL50 = combination of 1 wt.% nanoclay and furfurylated fibers (50 wt.%); NC1FAL50P = fibers pre-treated and with 1 wt.% nanoclay and furfuryl alcohol (50 wt.%); NC5FAL50 = combination of 5 wt.% nanoclay and furfuryl alcohol (50 wt.%); FNC1FAL50 = combination of 1 wt.% surface-functionalized nanoclay and furfuryl alcohol (50 wt.%).

The materials were first manually homogenized and then mixed using a planetary mortar mixer (Electromechanical planetary mortar mixer—Fortminas—Sao Paulo, Brazil): 3 min at 140 rpm, followed by 1 min at 220 rpm. In mixtures containing furfuryl alcohol (FAL), the fibers were blended with FAL, maleic anhydride (MA), and water before being added to the cement paste. In one specific group (NC1FAL50P), the fibers were pre-treated with FAL and MA, followed by oven curing at 80 °C for 24 h. The color change to reddish brown indicated successful in situ polymerization of the FAL into poly(furfuryl alcohol), which increases fiber hydrophobicity and resistance to alkaline degradation. These treated fibers were then incorporated into the cement paste as in the other groups.

All pastes were cast into prismatic molds (40 × 40 × 160 mm) in two layers and compacted on a vibrating table. In specimens used for fracture toughness testing, a cylindrical metal rod (8 mm diameter) was gently pressed into the center of the fresh paste to form a consistent notch, which remained after hardening. After molding, the specimens were demolded following 5 days of air curing and then subjected to hydrothermal curing in an autoclave at 0.15 MPa and 100 °C for 8 h, as described elsewhere [6]. This also ensured the polymerization of any remaining FAL within the fiber structure.

2.4. Characterization of Fiber–Cement Composites

The final setting time was determined using a Vicat apparatus (ASTM C191 [26]). The initial setting time could not be reliably measured due to the interference of fibers with the needle. Workability was assessed using the flow table method in accordance with ASTM C230/C230M [27]. A truncated cone mold with a base diameter of 10 cm, top diameter of 7 cm, and height of 6 cm was placed at the center of the flow table and filled with the fresh paste in a single lift without compaction. The mold was then lifted vertically, and the table was dropped 25 times within 15 s. The spread diameter was measured in three directions, and the average was taken as the flow value for each mixture.

Bulk density was determined by measuring the oven-dry mass of the specimen and its geometric volume. To assess apparent porosity and water absorption, the specimens were immersed in water for 24 h. Their mass was recorded while submerged, saturated surface-dry, and oven-dry. These values allowed the estimation of the open pore volume and the amount of water absorbed. The calculations followed the ASTM C948 [28] procedure, which relates these masses to determine the accessible voids and water-holding capacity of the composite, as also described by Gutiérrez et al. [29].

Three-point bending and compression tests were conducted to evaluate the mechanical behavior of the composites, following the procedures defined by ASTM C348 [30]. Flexural tests were performed using a universal testing machine (EMIC DL 30000—Instron Brasil, São José dos Pinhais, Paraná, Brazil) with a span of 100 mm and a loading rate of 50 N/s. Rectangular prismatic specimens were positioned on two lower supports, and a central load was applied until failure. During the test, the equipment continuously recorded the applied force (F) and the vertical displacement (δ) at the upper face of the specimen. These raw data were used to calculate the flexural stress (σ_x), strain (ε_x), and flexural toughness. The flexural stress at a given load was calculated using Equation (1). The corresponding flexural strain was determined using Equation (2), which relates the specimen's height, span length, and measured displacement. The flexural strength (σ_f), defined as the maximum stress sustained by the specimen before failure, was obtained using the maximum applied load (F_{max}) in Equation (3). Finally, the flexural toughness was computed as the area under the stress–strain curve using OriginPro 2022 software (version 9.9.0.225):

$$\sigma_x = \frac{3FL}{2bh^2} \quad (1)$$

$$\varepsilon_x = \frac{6h\delta}{L^2} \quad (2)$$

$$\sigma_f = \frac{3F_{max}L}{2bh^2} \quad (3)$$

where σ_x = flexural stress at a given load (MPa); ε_x = flexural strain (dimensionless); σ_f = flexural strength (MPa); F = applied load (N); F_{max} = maximum applied load (N); L = span length (mm); b = specimen width (mm); h = specimen height (mm); δ = vertical displacement at the midpoint (mm).

After the flexural test, one side of each fractured specimen was subjected to a uniaxial compression test, also following ASTM C348 [30]. The compression test was carried out using the same universal testing machine, with prismatic platens (40 mm × 40 mm) and a loading rate of 500 N/s. During the test, load and displacement were recorded and converted into stress and strain. The compressive strength (σ_c) was determined by Equation (4). The compressive modulus of elasticity (E_c) was calculated as the slope of the linear portion of the stress–strain curve, defined between 0.5 MPa and 30% of σ_c, using Equation (5):

$$\sigma_c = \frac{F_{max}}{A} \quad (4)$$

$$E_c = \frac{\Delta\sigma}{\Delta\varepsilon} \quad (5)$$

where σ_c = compressive strength (MPa); E_c = compressive modulus of elasticity (MPa); F_{max} = maximum applied load (N); A = cross-sectional area of the specimen under compression (mm^2); $\Delta\sigma$ = variation in stress within the linear region (MPa); $\Delta\varepsilon$ = corresponding variation in strain (dimensionless).

Fracture toughness was measured using notched specimens in flexural configuration, as per ISO 12135 [31]. The test span was set to 120 mm. The critical load required to initiate crack propagation was used in conjunction with specimen geometry to calculate fracture toughness. Although originally developed for metallic materials, this standard has also been applied to cementitious composites with appropriate adaptations, as seen in previous studies [7,32]. All calculations were performed using OriginPro 2022 (version 9.9.0.225), which was also used to apply correction factors and determine notch depth and geometry parameters. Following specimen fracture, selected samples were subjected to morphological evaluation using optical microscopy (OM). The analysis was conducted with a Dino-Lite PRO series microscope (AnMO Eletronics Corporation—Hsinchu City, Taiwan).

2.5. Characteristics of the Raw Materials

Figure 1 shows the morphological, chemical, and structural characterization of the sisal fiber used as reinforcement in this study. In Figure 1a, the optical micrograph reveals a rough and irregular surface, which can improve mechanical interlocking with the cementitious matrix. This topography is characteristic of natural lignocellulosic fibers and contributes to better anchoring within the matrix. In Figure 1b, the FTIR spectrum highlights the presence of functional groups typical of cellulose, hemicellulose, and lignin. The broad absorption band around 3340 cm^{-1} corresponds to O–H stretching from hydroxyl groups, confirming the hydrophilic nature of the fiber [33]. The bands near 2900 cm^{-1} refer to C–H stretching in methyl and methylene groups, while the peak around 1735 cm^{-1} is associated with C=O stretching, mainly from hemicellulose [34]. Bands in the region between 1030 and 1160 cm^{-1} are related to C–O stretching and C–O–C linkages in polysaccharides, confirming the presence of holocellulose [34]. The presence of these functional groups corroborates the fiber's hydrophilic nature, which affects the water demand of cementitious mixtures [35]. These chemical groups are directly related to the high water-absorption capacity observed for the fiber–cement composites, as discussed later.

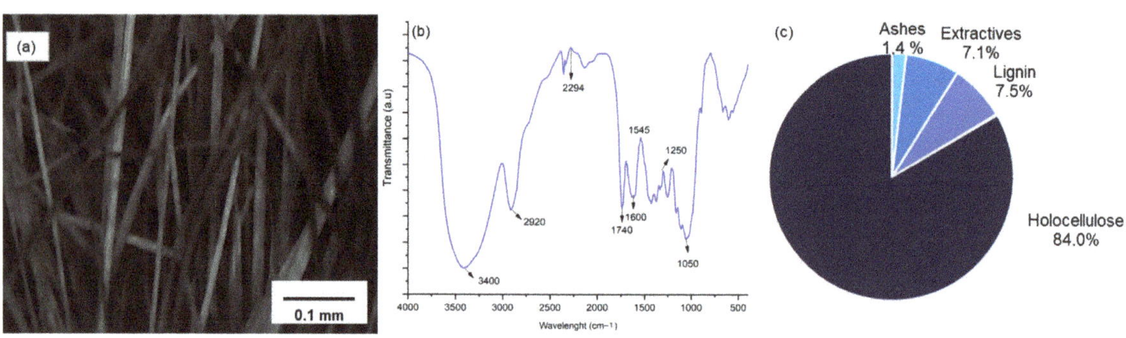

Figure 1. Optical micrograph (**a**), FTIR spectrum (**b**), and chemical composition (**c**) of sisal fiber.

Figure 1c presents the semi-quantitative chemical composition of the sisal fiber, showing a high holocellulose content (71.03%), moderate lignin (13.01%), and low ash content (1.34%). These values are in agreement with those reported in the literature for sisal fibers and justify their selection in the present work due to their balance between strength, availability, and sustainability. To better understand the role of each component used in the formulation of the composites, Table 2 summarizes the main physical and chemical characteristics of the raw materials, which were reported by each respective supplier.

Table 2. Summary of raw material characteristics.

	Density (g/cm^3)	Main Compounds	Size	Function in Composite
Cement	~3.15	Ca, Si, Al and Fe	D50 ≈ 13 μm	Main binder and source of C–S–H
Limestone	~2.70	$CaCO_3$ with high Ca content	D50 ≈ 25 μm	Filler that reduces cement use and overall cost
Hydrophilic nanoclay	≤2.60	Montmorillonite	≤25 μm	Pozzolanic additive
Functionalized nanoclay	≤2.60	Montmorillonite + OA/APTES	≤20 μm	Pozzolanic additive
Untreated sisal fiber	~1.45	Cellulose, hemicelluloses, and lignin	2 cm length and 100–300 μm diameter	Reinforcement

However, the presence of the hemicellulose and hydroxyl groups increases the susceptibility of the fibers to alkaline degradation, especially under autoclave curing conditions. This explains the need for chemical treatments such as furfurylation, which are intended to reduce hydroxyl group activity and improve durability within the cementitious matrix.

Figure 2 presents the thermal and particle size characterization of the cement and agricultural limestone used in the formulations. In Figure 2a, the thermogravimetric analysis (TGA) curves of the cement CPV-ARI display mass loss stages corresponding to the thermal decomposition of each component. The cement sample exhibits three main decomposition regions: the first below 150 °C, associated with the evaporation of free water and the dehydration of C–S–H gels; the second between 400 and 500 °C, attributed to the decomposition of calcium hydroxide ($Ca(OH)_2$); and the third between 600 and 750 °C due to the decarbonation of calcium carbonate ($CaCO_3$). These thermal events are typical of Portland cement and confirm its purity and hydration potential. The limestone, in turn, shows a predominant weight loss starting around 650 °C, indicating the decarbonation of $CaCO_3$ as its main thermal event. These observations are consistent with findings by Tawfik and Abd-El-Razik [36]. This confirms its high calcium content and its potential as a partial replacement for cement in the mixtures.

Figure 2b shows the particle size distribution obtained by laser diffraction. The cement exhibited a finer distribution with a median particle diameter (D50) of approximately 13 μm, while the agricultural limestone showed a broader distribution and larger particles, with a D50 of approximately 25 μm. The smaller particle size of the cement contributes to a higher surface area for hydration reactions, which is essential for early-age strength development. On the other hand, the coarser limestone particles provide a filler effect, improving the packing density of the paste and promoting matrix densification. The difference in granulometry between the two materials justifies the observed changes in rheology and physical properties of the composites, especially regarding flowability and porosity, as discussed in later sections. Moreover, the particle size contrast plays a significant role in the mechanical synergy between the binder components. When compared to the nanoclays used, both with particle sizes ≤ 25 μm and ≤20 μm, it is clear that the nanoclays fall within or below the size range of cement and limestone. Their ultrafine nature and high specific surface area allow them not only to fill voids between larger particles but also to participate in pozzolanic reactions, especially in the case of the functionalized nanoclay.

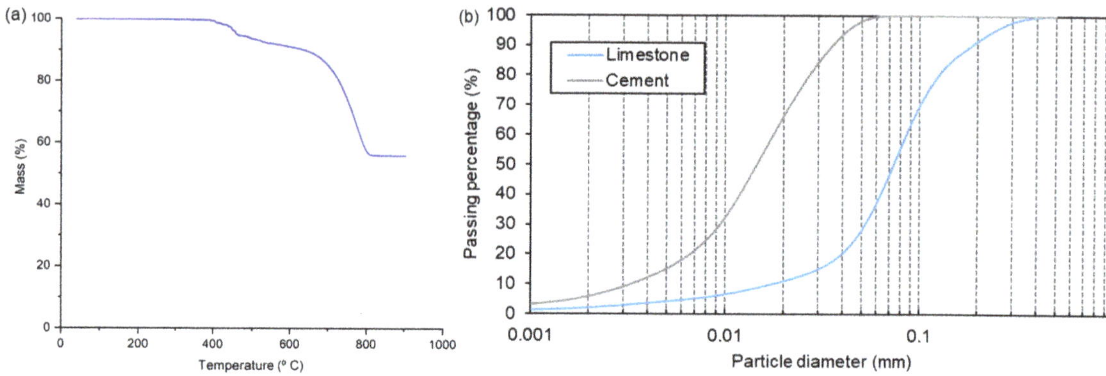

Figure 2. Thermogravimetric analysis of limestone (**a**), and particle size distribution (**b**) of cement and limestone.

3. Results and Discussion

3.1. Physicochemical Evaluation of the Treated Sisal Fibers

The resistance of sisal fibers to alkaline hydrolysis was evaluated for untreated fibers, fibers treated with FAL, and fibers exposed to nanoclay, preserving the proportions used in the composites. The FTIR results are shown in Figure 3. For comparative purposes, the spectra of the untreated fiber and pure nanoclay were also included.

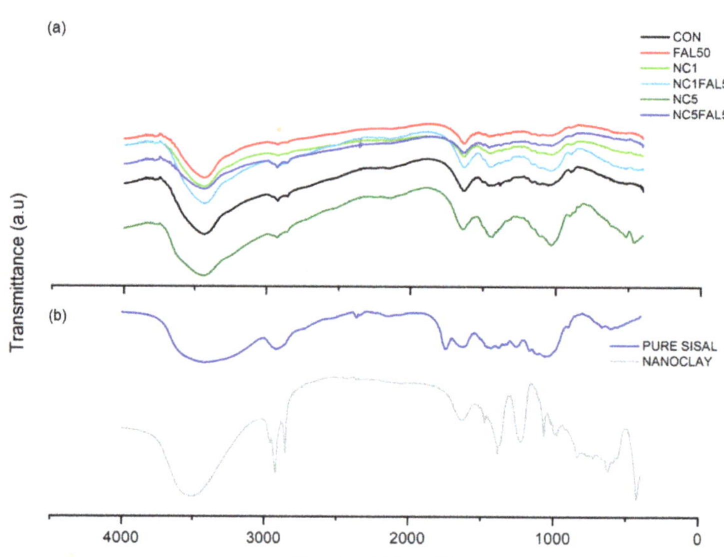

Figure 3. Comparative FTIR spectra showing chemical changes in sisal fiber and nanoclay after alkaline treatment: (**a**) alkali-treated samples; (**b**) untreated references.

In lignocellulosic materials subjected to alkaline hydrolysis, structural alterations often manifest through variations in the intensity and shape of the C–H deformation bands between 1300 and 1500 cm^{-1}. This degradation process can promote the transformation of aldehyde and ketone groups into carboxylic acids (–COOH), giving rise to a new C=O stretching band around 1600 cm^{-1}. Simultaneously, shifts in the C–O (1000–1300 cm^{-1}) and C=C (1500–1600 cm^{-1}) regions further signal chemical modifications in the fiber's structure [37].

The introduction of aliphatic groups from FAL results in increased intensity of the C–H stretching bands (2800–3000 cm^{-1}). A distinctive absorption band also emerges near 1600 cm^{-1}, attributed to ketone groups from FAL and maleic anhydride. Alterations in the C=C stretching band in this region further confirm the fiber's chemical transformation. The polymerization of FAL is supported by the appearance of a pronounced carbonyl peak near 1600 cm^{-1} [38].

Absorption bands around 2900 cm^{-1} are associated with methyl (C–H) stretching, while those near 1350 cm^{-1} reflect the bending vibrations of the C–H bonds in natural fibers. The literature reports that these bands typically intensify after alkaline treatment due to lignin and hemicellulose removal [39]. In the present study, however, such intensification was not observed, suggesting that both the furfurylation treatment and nanoclay incorporation were effective in mitigating chemical degradation and preserving the fiber structure. Additionally, bands attributed to the aromatic vibrations of lignin and hemicellulose—around 1500 and 1450 cm^{-1}—were observed with reduced intensity [39]. The broad absorption between 3000 and 3700 cm^{-1}, characteristic of hydrogen bonding, confirmed the widespread presence of hydroxyl groups in all compositions.

The presence of polymerized FAL is further evidenced by the formation of diketone structures, as indicated by the distinct carbonyl absorption near 1700 cm^{-1}, consistent with the findings of Pranger et al. [38]. An intensified peak at 1600 cm^{-1} also supports the presence of 2,5-disubstituted furan rings, reinforcing the successful incorporation of FAL into the fiber matrix [40].

Alkali-treated fibers exhibited spectra with similar peak positions but significantly reduced intensities, especially around 3000 cm^{-1}, 1500 cm^{-1}, and 1625 cm^{-1}. The near disappearance of these peaks suggests substantial degradation of aromatic compounds such as lignin, hemicellulose, and pectins [37]. Compared to the untreated fiber, the reduction in O–H and C–H stretching intensities implies the cleavage of hydroxyl bonds. The broad band from 4000 to 3000 cm^{-1} further confirms the presence of hydroxyl groups in carbohydrate structures [41]. The peak observed at 1800 cm^{-1} corresponds to C–O stretching in acetyl groups from hemicellulose. Interestingly, its intensity remained unchanged after NaOH treatment, indicating that the alkali treatment did not significantly reduce the hemicellulose content in the sisal fibers. Peaks between 1000 and 1050 cm^{-1} are related to hydroxyl group stretching in cellulose and β-glycosidic bonds—features that persisted even after chemical modification, suggesting that such treatments did not compromise these components [42].

For nanoclay, prominent peaks at 3500 and 3200 cm^{-1}—associated with hydroxyl groups and hydrogen bonding—were detected in the composite spectra. This suggests that nanoclay interacted with the cementitious matrix, possibly forming hydrogen bonds with water or other components, which could enhance adhesion and improve the composite's mechanical performance [41]. Altogether, the chemical treatments appear to have protected the fiber structure effectively. However, when comparing all spectra, signs of degradation due to the alkaline solution were evident, and no clear protective effect could be attributed to either the nanoclay or furfurylation under the specific test conditions.

3.2. Characteristics of the Fiber–Cement Composites

Figure 4 presents the flow diameter of the various fiber–cement pastes measured on the flow table, allowing the evaluation of their workability under a constant water-to-cement ratio. The reference CPV-ARI cement paste (CON) shows a flow diameter of

264.67 mm, while the incorporation of 2 wt.% untreated sisal fiber (Control) led to an increase of approximately 9% in flow diameter. This unexpected gain in workability may be related to the slight deflocculation of particles due to fiber interaction, as also observed by Aramburu et al. [20].

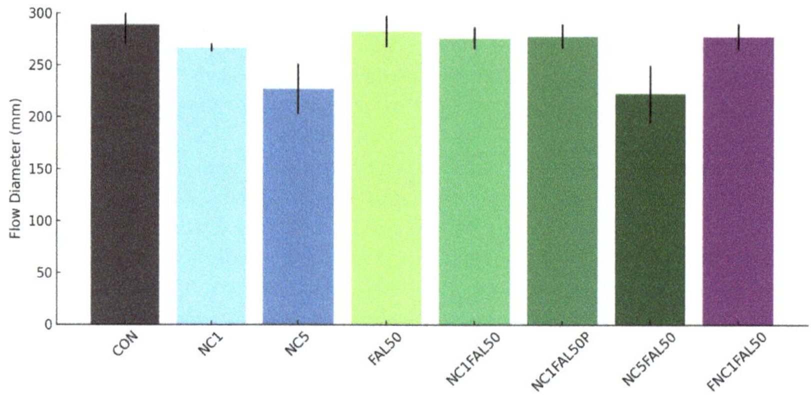

Figure 4. Flow diameter of the fiber–cement pastes.

The addition of 50% furfurylated fiber (FAL50) resulted in the highest spread, with a flow diameter of 282.50 mm, representing a 6.7% increase compared to the control paste. This suggests that furfurylation improved fiber dispersion and compatibility with the matrix, likely due to the reduction in polar hydroxyl groups and increased hydrophobicity. However, when 1 wt.% of nanoclay was added to the FAL50 system (NC1FAL50), the flow diameter slightly decreased to 266.50 mm, indicating minimal interference of nanoclay in this context. This aligns with Hisseine et al. [43], who reported reduced fluidity in cement pastes containing nano-additives. The PGV nanoclay likely absorbs free water, reducing workability.

In contrast, pastes with 5 wt.% nanoclay (NC5 and NC5FA50) showed reductions of 21% and 23%, respectively, compared to the control paste, evidencing a marked loss of fluidity. This behavior is associated with the high specific surface area of the nanoclay and its strong water-absorption capacity, which increases the internal friction and reduces paste mobility—a phenomenon also reported by Kunchariyakun et al. [44].

Interestingly, the paste with functionalized nanoclay (FNC1FAL50) also showed a reduced flow diameter compared to the FAL50 and NC1FAL50 groups, though the decrease was less pronounced than in NC5. This may be due to better particle dispersion promoted by the organosilane groups in the modified clay, which mitigates some of the water demand and agglomeration effects.

The control paste with fibers delayed the setting, a behavior corroborated by Onuaguluchi et al. [45], who reported that fibers slow hydration initially but can accelerate it later. Jiao et al. [46] explained that hydroxyl and carboxyl groups in fibers form complexes with Ca^{2+}, decreasing the number of active hydration sites and delaying early C–S–H formation.

Figure 5 shows the physical properties of the composites after autoclave curing, including water absorption, apparent porosity, and bulk density. In general, the incorporation of untreated sisal fibers increased both water absorption and porosity compared to the cement-only reference (CON). Specifically, the control paste containing 2 wt.% untreated fiber showed a water absorption increase of 11.2% and porosity increase of 10.3%, which is consistent with the intrinsic hydrophilicity and irregular geometry of the fiber, which induces void formation during mixing and curing.

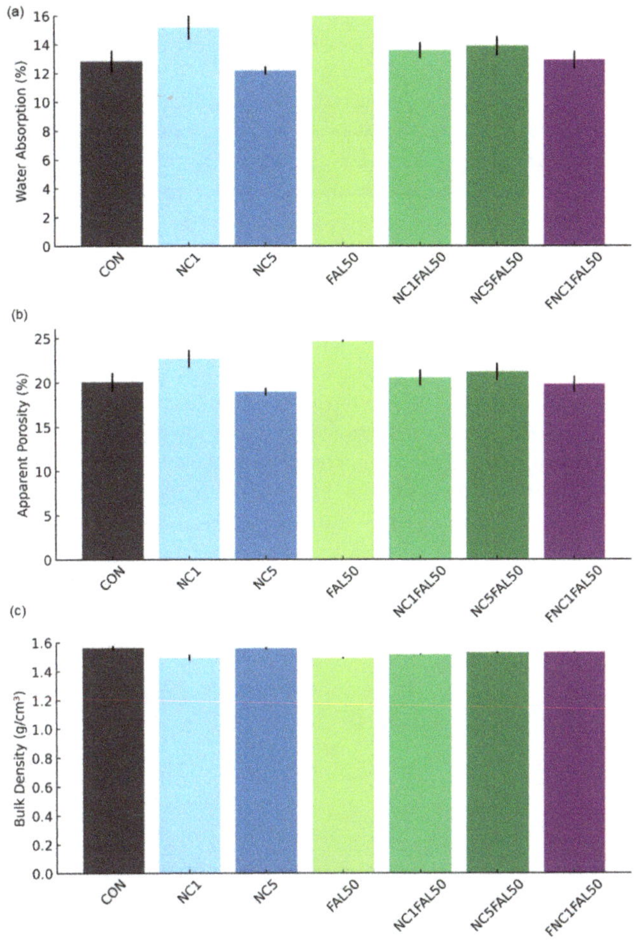

Figure 5. Water absorption (**a**), apparent porosity (**b**), and apparent bulk density (**c**) of the fiber–cement composites.

The incorporation of 1 wt.% nanoclay (NC1) further increased water absorption and porosity, reaching values 13% and 23% higher than for the control, respectively. This behavior may be attributed to partial agglomeration and inefficient dispersion of the nanoclay, which limited its ability to fill the micropores effectively and instead contributed to capillary porosity. However, the group with 5 wt.% nanoclay (NC5) exhibited reduced porosity and water absorption compared to NC1, indicating a saturation effect in which the increased nanoclay content led to better pore filling and matrix densification. This phenomenon aligns with the findings of Hisseine et al. [43], who reported that an optimal threshold exists for nanoclay addition, beyond which its pozzolanic activity becomes more effective.

The FAL50 group showed 25% and 33% increases in absorption and porosity, respectively, consistent with findings in fiber–cements with untreated natural fibers [47,48]. This result suggests that the furfurylation process, although beneficial for chemical stability, may have increased the surface roughness or micro-cracking of the fiber during curing, which contributed to higher moisture uptake. Nonetheless, when nanoclay was added to the furfurylated systems (NC1FAL50 and NC5FAL50), these values decreased and approached those of the control group. The NC5FAL50 group, in particular, exhibited a 22% reduction in porosity compared to FAL50, evidencing the synergistic sealing effect of nanoclay on the porous matrix.

Interestingly, the FNC1FAL50 composite—which combines furfurylated fibers and functionalized nanoclay—showed one of the lowest water-absorption values among all the modified groups, 7.8% lower than those of the control. This suggests that the functionalized nanoclay not only reacted pozzolanically but also promoted superior particle dispersion and fiber encapsulation, contributing to a more impermeable and compact matrix.

As for bulk density, most modified composites showed slightly lower values compared to the control, particularly the groups with isolated fiber or nanoclay addition. The exceptions were NC5 and NC5FAL50, which maintained densities similar to the control group due to better packing and fewer voids. The FNC1FAL50 group also demonstrated a relatively high density among the modified composites, confirming that functionalization improved the filler efficiency of the nanoclay and its integration into the matrix.

Figure 6 presents the stress–strain curves obtained from the three-point bending tests for the different fiber–cement composites. All curves display the expected behavior for fiber-reinforced cementitious materials: an initial linear elastic region, a peak stress point, and a post-peak softening region associated with progressive fiber pull-out and rupture. The control composite (CON) showed a modest post-peak response, indicative of limited ductility and energy absorption, despite the presence of untreated sisal fibers. This matches the mechanisms described by Claramunt et al. [49], where fiber bridging delays crack propagation.

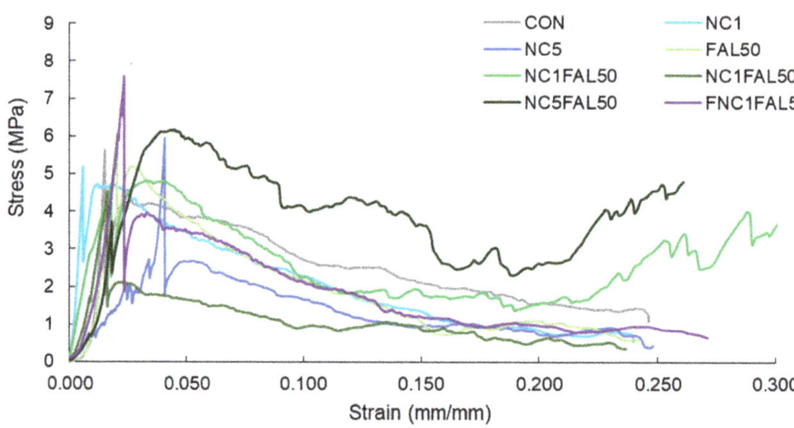

Figure 6. Representative stress–strain curves from the flexural tests of the fiber–cement composites.

The addition of 1 wt.% nanoclay (NC1) did not significantly alter the shape of the stress–strain curve but resulted in a slightly higher stress level in the linear region, suggesting improved stiffness. In contrast, the NC5 composite exhibited a more brittle response, with a sharper drop in stress after the peak. This behavior is consistent with the saturation mechanism discussed earlier, in which excess nanoclay may lead to agglomeration and ineffective stress transfer across the matrix.

The incorporation of furfurylated fibers (FAL50) resulted in a broader and more pronounced post-peak region, with higher strain values compared to CON, indicating an enhancement in toughness and ductility. This effect became even more significant in the NC1FAL50 composite, which showed the most extended plastic region among all the samples. The curve of NC1FAL50 suggests that the chemical treatment improved fiber–matrix interfacial bonding, enabling better stress redistribution and delaying catastrophic failure.

On the other hand, the pre-treatment group (NC1FAL50P) displayed a much steeper drop in stress after the peak, with reduced strain capacity, reflecting that furfurylation conducted before the composite mixing stage may have negatively affected fiber dispersion or flexibility. This result highlights the importance of the timing and method in applying chemical treatments to natural fibers.

The NC5FAL50 group also exhibited an extended post-peak response, though slightly less than that of NC1FAL50. This composite still benefited from the combined effects of pozzolanic activity and fiber modification, indicating synergy between the two strategies. Most notably, the FNC1FAL50 composite revealed one of the most favorable profiles, combining moderate peak stress with a prolonged post-peak response, which is characteristic of materials capable of sustaining loads beyond initial cracking.

These findings are consistent with studies by Claramunt and Ardanuy [5,49], which reported similar ductile behavior in composites with properly treated fibers, and by Lima et al. [50], who observed extended strain capacity due to gradual fiber pull-out. Overall, the results from Figure 6 demonstrate that the combination of furfurylation and nanoclay incorporation—particularly when functionalized—effectively enhances the ductile performance of fiber–cement composites under flexural loading.

Figure 7 presents the flexural performance of the composites through three key parameters: (a) flexural strength, (b) maximum deflection, and (c) flexural toughness. Together, these data allow a comprehensive evaluation of the influence of nanoclay content and fiber chemical treatment on mechanical behavior under bending loads.

In Figure 7a, the flexural strength results show that the composites NC5, NC5FAL50, and NC1FAL50 exhibited significant improvements compared to the control (CON). NC5 achieved an increase of 19.3%, NC5FAL50 increased by 12.84%, and NC1FAL50 by 15.72%. These gains are attributed to the pozzolanic and filler effects of nanoclays, especially when used at 5%, as they enhance matrix densification and improve interfacial adhesion. The superior performance of NC1FAL50, despite having only 1% nanoclay, reinforces the role of the furfurylation treatment in strengthening fiber–matrix bonding. Conversely, the NC1FAL50P composite—where the fiber was furfurylated prior to incorporation—did not show the same gains, indicating that premature treatment may have affected fiber integrity or interaction with the cement matrix.

In Figure 7b, the maximum deflection results highlight the ductility of the composites. The NC1FAL50 sample presented the highest deflection, approximately 17% greater than that of the control, demonstrating enhanced deformability prior to rupture. This behavior reflects the synergistic effect of nanoclay and in situ furfurylation on fiber preservation and stress redistribution. In contrast, the NC5 and NC1FAL50P groups exhibited the lowest deflections, both showing reductions around 20% compared to the control. These results align with the hypothesis of nanoclay saturation and premature stiffening caused by excess additive or pre-treatment, respectively. The FNC1FAL50 sample displayed a high deflection value, similar to that of NC1FAL50, confirming the efficacy of combining functionalized nanoclay with furfurylated fibers.

In Figure 7c, the toughness values—calculated from the area under the stress–strain curve—show the clearest distinction between groups. The NC5FAL50 composite demonstrated a remarkable increase in toughness, nearly doubling the control value (approximately +100%), indicating an exceptional capacity for energy absorption prior to failure. This result underscores the synergistic role of nanoclay and fiber modification when both are properly dosed. NC1FAL50 showed a moderate gain, while FAL50 alone and NC5 showed toughness losses, confirming that neither furfurylation nor nanoclay addition alone was sufficient to significantly enhance this property. The FNC1FAL50 sample again exhibited a robust response, with toughness values around 45% higher than those of the control, validating the strategy of combining pozzolanic and chemical interventions.

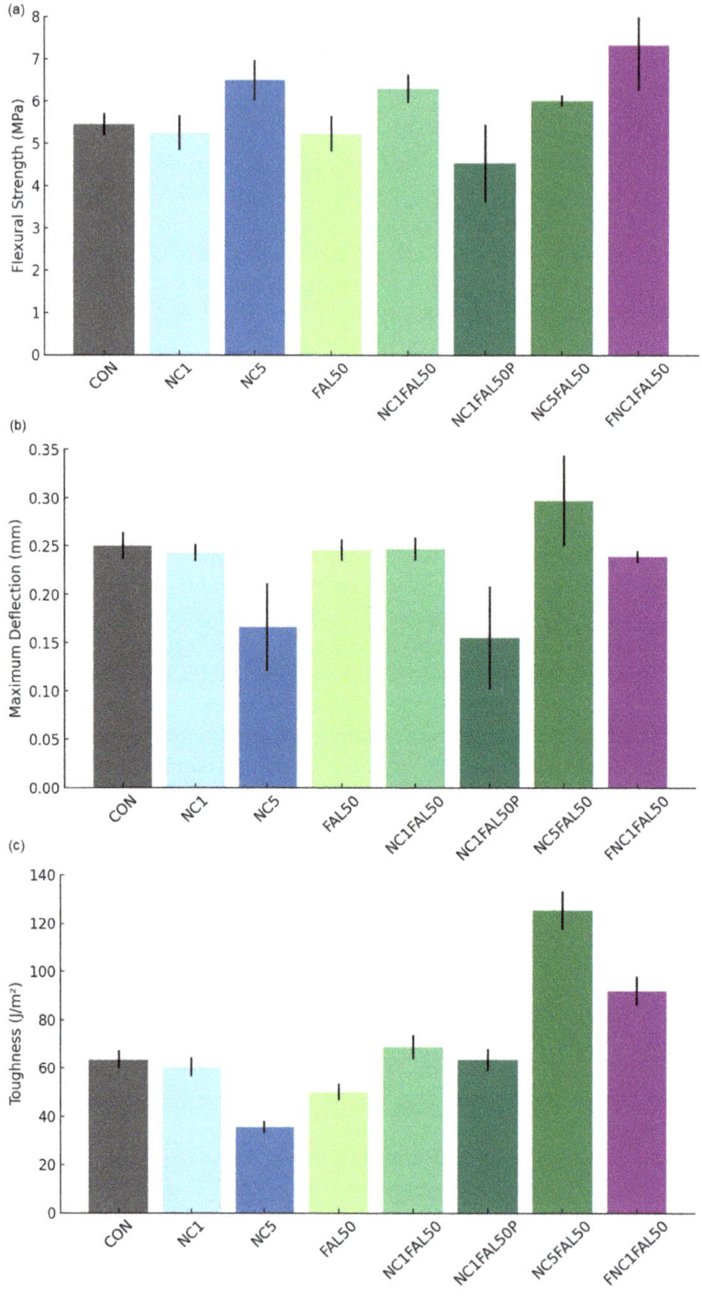

Figure 7. Flexural strength (**a**), maximum deflection (**b**), and flexural toughness (**c**) of the fiber–cement composites.

Figure 8 shows the stress–deflection curves obtained from the compressive strength tests of the fiber–cement composites. All the curves follow the expected pattern for cementitious materials reinforced with natural fibers, displaying an initial linear elastic region followed by a peak stress, a softening phase, and, finally, failure. In fact, this shape is common in fiber–cements with short fibers, as described by Pereira et al. [51]. However, distinct differences in curve shapes and slopes allow for the qualitative assessment of stiffness and ductility across the experimental groups.

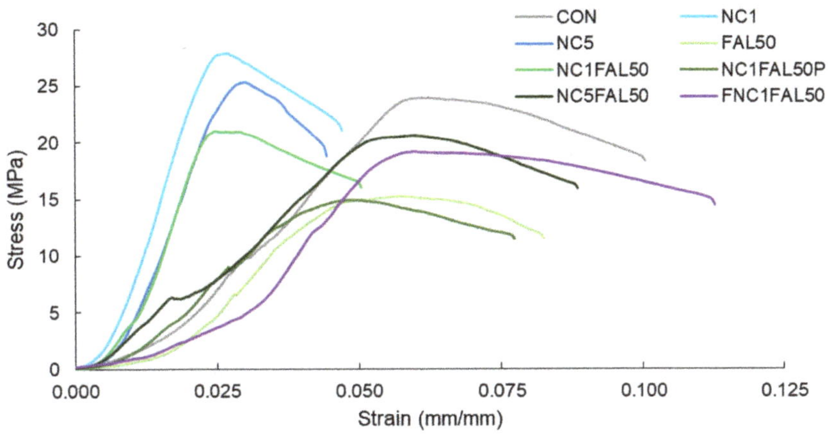

Figure 8. Representative stress–deflection curves from the compressive tests of the fiber–cement composites.

The control (CON) composite exhibited a typical sharp peak followed by a steep stress drop, indicating brittle behavior and low energy absorption after the yield point. In contrast, composites containing furfurylated fibers, particularly NC1FAL50 and FNC1FAL50, showed broader post-peak regions and more gradual stress reductions, suggesting enhanced plastic deformation and increased resistance to crack propagation. These groups also exhibited a more extended plateau, indicative of better stress redistribution provided by the treated fibers.

The NC1 composite presented the highest initial slope among all the curves, reflecting its superior compressive modulus and stiffness, likely a result of the optimal pozzolanic activity of 1% nanoclay. Conversely, NC5 showed a less pronounced slope and a more abrupt failure, likely associated with excess nanoclay saturation, agglomeration, and possible interference with proper cement hydration.

Notably, the NC1FAL50P composite showed a relatively brittle curve, with reduced ductility and abrupt fracture after the stress peak. This behavior indicates that the pretreatment of the fiber—although chemically protective—may have compromised its structural integrity or interaction with the cement matrix. FAL50 and NC5FAL50 also showed moderately ductile behavior, although less pronounced than NC1FAL50 and FNC1FAL50, reinforcing the role of functionalized nanoclay in preserving fiber efficiency during compressive loading.

Overall, the stress–strain curves from the compressive testing suggest that the best combination of stiffness and deformability was achieved by composites containing 1% nanoclay and furfurylated fibers added in situ, with the FNC1FAL50 and NC1FAL50 samples showing the most favorable mechanical response. These results are consistent with prior findings in the literature on natural fiber–cement systems and confirm the beneficial synergistic effect of combining surface treatment with nanostructured pozzolanic additives.

Figure 9a displays the compressive strength results of the fiber–cement composites. The control and NC1 groups achieved the highest average values, around 23 MPa, indicating that the addition of 1% nanoclay contributed positively to compressive strength. This matches findings by Melo Filho et al. [52] and Krishna et al. [53], who observed strength reductions with poorly dispersed or degraded fibers. In contrast, the FAL50 and NC5 groups experienced reductions of approximately 13% and 17%, respectively, compared to the control. This reduction can be associated with either the partial substitution of cement by furfurylated fiber content (in FAL50) or the excess nanoclay in NC5, which may have caused particle agglomeration and impaired cement hydration due to a saturation effect. These phenomena lead to poor dispersion and inefficient pozzolanic reactions, corroborating the literature that warns of mechanical impairments in systems with excessive nanostructured pozzolanic materials.

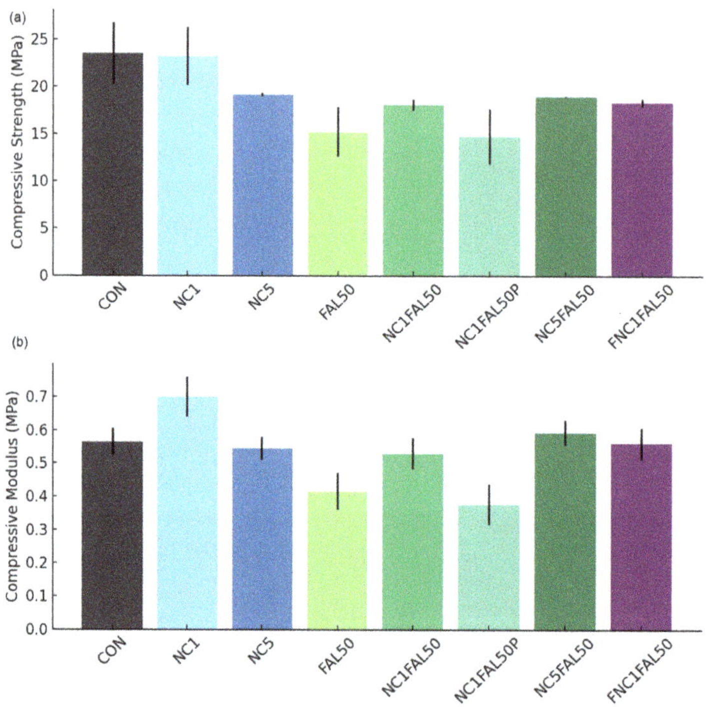

Figure 9. Compressive strength (**a**) and compressive modulus (**b**) of the fiber–cement composites.

Interestingly, the NC1FAL50 composite maintained compressive strength values statistically similar to those of the control, suggesting that the negative effect of fiber addition was mitigated by the synergistic interaction between the 1% nanoclay and the pozzolanic filler effect. However, the pre-treated NC1FAL50P group exhibited a strength drop of approximately 21%, likely due to damage caused to the fiber structure during furfurylation prior to mixing. The NC5FAL50 group also showed a decline, reinforcing the idea that 5% nanoclay oversaturates the matrix, limiting hydration and weakening the mechanical integrity of the composite. The FNC1FAL50 sample, incorporating functionalized nanoclay and in situ-treated fibers, presented a balanced result close to that of the control, indicating that this combination may prevent significant degradation of compressive performance.

Several studies corroborate the present compressive strength range. For instance, Kunchariyakun et al. [44] achieved similar strength values using autoclaved composites reinforced with untreated natural fibers. Similarly, Choi and Choi [54] reported compressive strengths around 25 MPa in cement matrices incorporating 2% natural fiber content. Pereira et al. [51] observed a slight strength reduction (~3 MPa) with short sisal fibers (25 mm), yet the overall structural integrity was preserved across a broader deformation range. These comparisons validate the consistency of the current results and the viability of these composites for structural applications.

Figure 9b shows the compressive modulus (modulus of elasticity in compression) of the same composites. The NC1 sample stood out with the highest modulus, showing a 174% increase compared to the control, confirming the efficiency of 1% nanoclay in enhancing rigidity through pore refinement and better packing. The NC1FAL50 composite followed with the second-highest modulus, suggesting that the nanoclay remained effective even in the presence of furfurylated fibers. Although the FAL50 sample displayed higher deformability, its stiffness was lower than that of the control, likely due to the reduction in the cement fraction and the lower modulus of the fiber material.

In contrast, the NC5FAL50 and NC1FAL50P composites showed the lowest stiffness among all the groups, suggesting that either excess nanoclay or aggressive fiber pre-treatment impaired the structural integrity of the matrix. This decrease in modulus reinforces the concept of an optimal nanoclay dosage threshold, beyond which the expected filler and pozzolanic effects are no longer beneficial. Overall, the data demonstrate that the judicious combination of nanoclay and fiber treatments can significantly affect compressive behavior, especially stiffness, which is critical for load-bearing applications.

Figure 10 illustrates the fracture toughness results of the fiber–cement composites, which reflect the material's ability to resist crack propagation under stress. Among all groups, the FAL50 composite exhibited the highest toughness value, surpassing the control by approximately 23%. This indicates that the presence of furfurylated fibers—despite partially replacing cement—was effective in increasing energy absorption during crack initiation and propagation. The enhanced performance may be attributed to the improved fiber–matrix interface achieved through furfurylation, which likely limited the interfacial debonding and allowed greater fiber bridging during fracture.

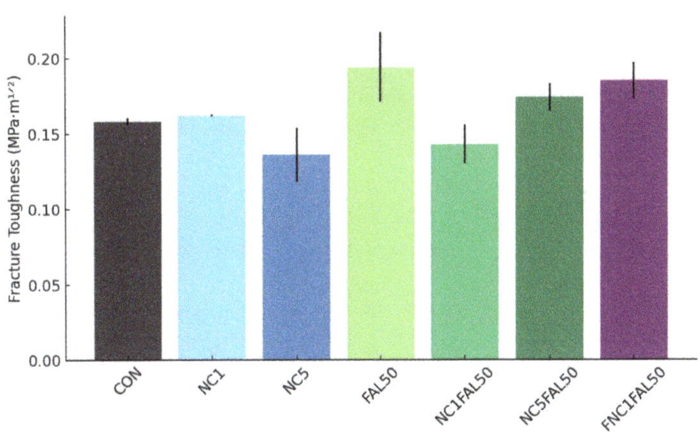

Figure 10. Fracture toughness of the fiber–cement composites.

On the other hand, composites containing only nanoclay (NC1 and NC5) did not exhibit significant improvements in fracture toughness compared to the control, suggesting that nanoclay alone, whether at 1% or 5%, did not meaningfully contribute to energy dissipation under fracture. This result highlights that while nanoclay may refine the microstructure and enhance stiffness, it does not directly enhance fracture energy unless paired with mechanisms like fiber bridging.

Notably, the NC1FAL50 and NC1FAL50P composites showed lower toughness values than FAL50, and even lower than the control in the case of NC1FAL50P, indicating that the expected synergistic benefits of combining nanoclay and furfurylated fiber were not fully realized in this property. In the pre-treated group (NC1FAL50P), the decreased toughness may be due to over-hardening or embrittlement of the fibers caused by the ex situ furfurylation process, which could have compromised their ability to deform and arrest cracks. Similarly, the NC5FAL50 composite exhibited a modest toughness gain, suggesting that excessive nanoclay may hinder effective fiber–matrix interaction by saturating the system or causing particle agglomeration, thus limiting fiber activation during crack propagation.

Interestingly, the FNC1FAL50 composite, which incorporated functionalized nanoclay and in situ-treated fibers, demonstrated a toughness level close to that of the control. This suggests that while the treatment and filler may have preserved the fiber structure and microstructural integrity, the chemical modifications might have reduced the number of active sites for energy absorption mechanisms (e.g., fiber pull-out or rupture). Therefore, although this formulation proved promising in other mechanical aspects, its contribution to fracture resistance was neutral.

A possible explanation, supported by Melo Filho et al. [52], is that excessive pozzolanic material can reduce matrix deformability, limiting fiber movement during fracture and negating toughness gains. Despite these observations, all values fall within ranges reported by other studies involving natural fiber composites. For example, Choi and Choi [54] also noted increased toughness through fiber reinforcement, though they stressed the need for careful control of the interface chemistry. Therefore, the fracture behavior observed here aligns well with the literature findings and further supports the potential of fiber treatments like furfurylation when properly integrated into the composite system.

Figure 11 illustrates the microstructural differences among selected fiber–cement composites, emphasizing the effects of fiber treatment and nanoclay incorporation on fiber dispersion, interfacial adhesion, and matrix densification. In the control composite (CON), the image reveals poorly distributed fibers with visible interfacial gaps and voids, suggesting limited fiber–matrix interaction and inadequate encapsulation. These morphological features are consistent with the mechanical results, where the control group displayed lower toughness and limited post-peak deformation capacity. For the NC1FAL50 composite, the micrograph shows a more homogeneous fiber dispersion and improved fiber alignment, along with fewer interfacial voids. This reflects the synergistic effect of 1 wt.% nanoclay and furfurylated fiber treatment, which promoted better fiber wetting and interfacial bonding. These observations align with the mechanical findings, where this group exhibited enhanced flexural strength, ductility, and a moderate increase in toughness. The FNC1FAL50 composite stands out, with a compact microstructure and fibers well integrated within the cement matrix and minimal voids or microcracks observed. The improved interaction is attributed to the surface-functionalized nanoclay, which enhances compatibility and dispersion, as well as the in situ furfurylation, which preserves fiber integrity. These microstructural characteristics corroborate the superior mechanical performance and reduced porosity noted for this formulation, reaffirming the benefits of combining targeted chemical modifications with pozzolanic nanomaterials.

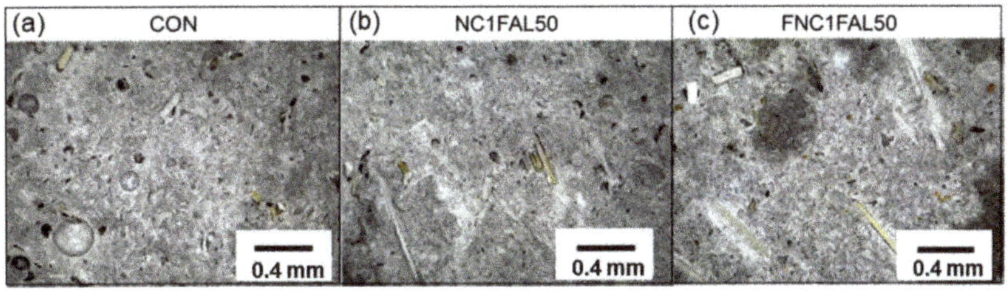

Figure 11. Representative optical micrographs of fractured fiber–cement composites: (**a**) control (CON), (**b**) composite with 1 wt.% nanoclay and furfurylated fibers (NC1FAL50), and (**c**) composite with functionalized nanoclay and furfurylated fibers (FNC1FAL50).

4. Conclusions

This study aimed to improve the performance and durability of autoclaved fiber–cement composites by combining furfurylated sisal fibers and pozzolanic nanoclays. The incorporation of 2 wt.% furfurylated sisal fibers increased flexural strength by 15.7%, maximum deflection by 17%, and toughness by 45% compared to the untreated fiber reference. These improvements were attributed to enhanced interfacial bonding and the fibers' greater resistance to alkaline degradation.

Among the pozzolanic additives, the functionalized nanoclay modified with octadecylamine and aminopropyltriethoxysilane (OA/APTES) yielded superior performance relative to unmodified bentonite. When used at 1 wt.% in combination with furfurylated fibers (FNC1FAL50), it resulted in one of the lowest water-absorption values (7.8% below the control), reduced porosity by 22%, and improved flexural strength and compressive modulus without compromising workability or density. Morphological analysis by optical microscopy revealed that the combination of furfurylated fibers and functionalized nanoclay led to a more compact matrix and improved fiber–matrix integration, corroborating the mechanical and physical property results. These outcomes confirm the synergistic effect between the fiber treatment and the enhanced pozzolanic and filler properties of the nanoclay.

Furthermore, the partial replacement of Portland cement by agricultural limestone—at a ratio of approximately 14%—reduced the environmental impact of the formulations without significantly affecting physical or mechanical performance. The findings support the feasibility of combining sustainable additives and fiber treatments to overcome the limitations typically associated with natural fiber–cement systems. Nevertheless, the study has some limitations, such as the absence of microstructural analysis via SEM and the restriction of curing and aging conditions to a single autoclave regime. Future investigations should focus on the following:

(i) optimization of the furfurylation process for other lignocellulosic fibers;
(ii) evaluation of long-term durability;
(iii) exploration of additional pozzolanic nanomaterials to enhance interfacial compatibility;
(iv) evaluation of workability retention over time.

Author Contributions: Conceptualization, R.d.A.D. and A.L.M.; methodology, T.A.d.S., F.V.R. and C.C.G.; validation, T.A.d.S. and F.V.R.; formal analysis, F.V.R. and C.C.G.; investigation, T.A.d.S. and F.V.R.; resources, A.B.A. and R.d.A.D.; writing—original draft preparation, T.A.d.S., F.V.R., C.C.G. and R.d.A.D.; writing—review and editing, A.L.M., S.C.A. and R.d.A.D.; visualization, F.V.R., A.B.A. and S.C.A.; supervision, A.L.M. and R.d.A.D.; project administration, R.d.A.D.; funding acquisition, A.L.M. and R.d.A.D. All authors have read and agreed to the published version of the manuscript.

Funding: This work was supported by Rio Grande do Sul State Research Foundation (FAPERGS), Coordination for the Improvement of Higher Education–CAPES (code 001), and the National Council for Scientific and Technological Development–CNPq.

Institutional Review Board Statement: Not applicable.

Informed Consent Statement: Not applicable.

Data Availability Statement: The data presented in this study are available on request from the corresponding author.

Conflicts of Interest: The authors declare no conflicts of interest.

References

1. Munaro, M.R.; Tavares, S.F.; Bragança, L. Towards Circular and More Sustainable Buildings: A Systematic Literature Review on the Circular Economy in the Built Environment. *J. Clean. Prod.* **2020**, *260*, 121–134. [CrossRef]
2. Chong, B.W.; Othman, R.; Ramadhansyah, P.J.; Doh, S.I.; Li, X. Properties of Concrete with Eggshell Powder: A Review. *Phys. Chem. Earth* **2020**, *120*, 102951. [CrossRef]
3. Claramunt, J.; Ardanuy, M.; García-hortal, J.A.; Filho, R.D. The Hornification of Vegetable Fibers to Improve the Durability of Cement Mortar Composites. *Cem. Concr. Compos.* **2011**, *33*, 586–595. [CrossRef]
4. Jaber, H.A.; Mahdi, R.S.; Hassan, A.K. Influence of Eggshell Powder on the Portland Cement Mortar Properties. *Mater. Today Proc.* **2020**, *20*, 391–396. [CrossRef]
5. Ardanuy, M.; Claramunt, J.; Toledo Filho, R.D. Cellulosic Fiber Reinforced Cement-Based Composites: A Review of Recent Research. *Constr. Build. Mater.* **2015**, *79*, 115–128. [CrossRef]
6. Insaurriaga, G.L.; Gomes, C.C.; Ribeiro, F.V.; Calegaro, G.L.; Silveira, T.A.; Cruz, L.F.; Cruz, J.A.; Amico, S.C.; Delucis, R.A. Effect of Hybridization of Carbon Fibers on Mechanical Properties of Cellulose Fiber–Cement Composites: A Response Surface Methodology Study. *C* **2024**, *10*, 41. [CrossRef]
7. Hamdaoui, O.; Limam, O.; Ibos, L.; Mazioud, A. Thermal and Mechanical Properties of Hardened Cement Paste Reinforced with Posidonia-Oceanica Natural Fibers. *Constr. Build. Mater.* **2021**, *269*, 121–339. [CrossRef]
8. Santana, H.A.; Amorim, N.S.; Ribeiro, D.V.; Cilla, M.S.; Dias, C.M.R. Vegetable Fibers Behavior in Geopolymers and Alkali-Activated Cement Based Matrices: A Review. *J. Build. Eng.* **2021**, *44*, 103291. [CrossRef]
9. Pickering, K.L.; Efendy, M.G.A.; Le, T.M. A Review of Recent Developments in Natural Fibre Composites and Their Mechanical Performance. *Compos. Part A Appl. Sci. Manuf.* **2016**, *83*, 98–112. [CrossRef]
10. da Silva, E.J.; Marques, M.L.; Velasco, F.G.; Fornari, C., Jr.; Luzardo, F.M.; Tashima, M.M. A New Treatment for Coconut Fibers to Improve the Properties of Cement-Based Composites—Combined Effect of Natural Latex/Pozzolanic Materials. *Sustain. Mater. Technol.* **2017**, *12*, 44–51. [CrossRef]
11. Wei, J.; Meyer, C. Degradation Mechanisms of Natural Fiber in the Matrix of Cement Composites. *Cem. Concr. Res.* **2015**, *73*, 1–16. [CrossRef]
12. Hakamy, A.; Shaikh, F.U.A.; Low, I.M. Characteristics of Hemp Fabric Reinforced Nanoclay-Cement Nanocomposites. *Cem. Concr. Compos.* **2014**, *50*, 27–35. [CrossRef]
13. von Ameln Luzzardi, F.J.; Bosel Wally, G.; Costa Magalhães, F.; Behenck Aramburu, A.; Alves da Silveira, T.; de Avila Delucis, R. Lightweight Concrete Containing Rigid Polyurethane Foam and Three Different Ashes: Assessing Mechanical and Physical Properties. *Aust. J. Civ. Eng.* **2025**, 1–12. [CrossRef]
14. Hakamy, A.; Shaikh, F.; Low, I.M. Microstructures and Mechanical Properties of Hemp Fabric Reinforced Organoclay-Cement Nanocomposites. *Constr. Build. Mater.* **2013**, *49*, 298–307. [CrossRef]
15. Wei, J.; Meyer, C. Sisal Fiber-Reinforced Cement Composite with Portland Cement Substitution by a Combination of Metakaolin and Nanoclay. *J. Mater. Sci.* **2014**, *49*, 7604–7619. [CrossRef]
16. Delucis, R.d.A.; Almeida, J.H.S., Jr. Influence of Chemical Treatments on the Thermal Properties of Natural Fiber-Reinforced Hybrid Composites (NFRHC). In *Natural Fiber-Reinforced Composites*; Wiley: Hoboken, NJ, USA, 2022; pp. 291–307.
17. Ajouguim, S.; Page, J.; Djelal, C.; Saâdi, L. Effect of Treated Alfa Fibers on Hydration Kinetics, Mechanical, and Adhesion Properties of Fiber Cement Composite. *J. Build. Eng.* **2023**, *71*, 106558. [CrossRef]
18. Biskri, Y.; Babouri, L.; Boukhelf, F.; Charradi, K.; Annaba, K.; El Mendili, Y. On the Physical-Mechanical Behavior of Fiber Cement Composite: Effect of Chemical Treatment of Sisal Fibers. *J. Build. Eng.* **2025**, *101*, 111978. [CrossRef]
19. Fonseca, C.S.; Scatolino, M.V.; Silva, L.E.; Martins, M.A.; Guimarães Júnior, M.; Tonoli, G.H.D. Valorization of Jute Biomass: Performance of Fiber–Cement Composites Extruded with Hybrid Reinforcement (Fibers and Nanofibrils). *Waste Biomass Valoriz.* **2021**, *12*, 5743–5761. [CrossRef]

20. Aramburu, A.B.; Rossetto, H.L.; Magalhães, W.L.E.; Trindade, G.H.; de Avila Delucis, R. Cement Matrix Reinforced with Microfibrillated Cellulose Treated with Furfuryl Alcohol. *Cellulose* **2023**, *30*, 10297–10312. [CrossRef]
21. Mejia-Ballesteros, J.E.; Rodier, L.; Filomeno, R.; Savastano, H.; Fiorelli, J.; Rojas, M.F. Influence of the Fiber Treatment and Matrix Modification on the Durability of Eucalyptus Fiber Reinforced Composites. *Cem. Concr. Compos.* **2021**, *124*, 104280. [CrossRef]
22. Hakamy, A.; Shaikh, F.U.A.; Low, I.M. Thermal and Mechanical Properties of NaOH Treated Hemp Fabric and Calcined Nanoclay-Reinforced Cement Nanocomposites. *Mater. Des.* **2015**, *80*, 70–81. [CrossRef]
23. Thepruttana, S.; Patthanavarit, J.; Hankoy, M.; Kitiwan, M.; Keawprak, N.; Tunthawiroon, P. Enhancement of Flexural Strength in Fiber–Cement Composites through Modification of Sisal Fiber with Natural Rubber Latex and Expanded Perlite. *Buildings* **2024**, *14*, 1067. [CrossRef]
24. Tonoli, G.H.D.; Rodrigues Filho, U.P.; Savastano, H.; Bras, J.; Belgacem, M.N.; Rocco Lahr, F.A. Cellulose Modified Fibres in Cement Based Composites. *Compos. Part A Appl. Sci. Manuf.* **2009**, *40*, 2046–2053. [CrossRef]
25. *ASTM C187*; Standard Test Method for Amount of Water Required for Normal Consistency of Hydraulic Cement Paste. ASTM International: Washington, DC, USA, 2016; pp. 1–3. [CrossRef]
26. *ASTM C191*; Standart Test Methods for Time of Setting of Hydraulic Cement by Vicat Needle. ASTM International: Washington, DC, USA, 2021. [CrossRef]
27. *ASTM C230*; Standard Specification for Flow Table for Use in Tests of Hydraulic Cement. ASTM International: Washington, DC, USA, 2020. [CrossRef]
28. *ASTM C948*; Standart Test Method for Dry and Wet Bulk Density, Water Absorption, and Apparent Porosity of Thin Sections of Glass-Fiber Reinforced Concrete. ASTM International: Washington, DC, USA, 2023. [CrossRef]
29. Gutiérrez, M.A.U.; da Silveira, T.A.; Ribeiro, F.V.; Aramburu, A.B.; Gomes, C.C.; Gonçalves, M.R.F.; Delucis, R.A. Thermally Treated Granite Slurry Waste as a Mitigating Agent for Sisal Fiber Mineralization in Fiber-cement Composites. *Int. J. Appl. Ceram. Technol.* **2025**, e15157. [CrossRef]
30. *ASTM C348*; Standard Test Method for Flexural Strength of Hydraulic-Cement Mortars. ASTM International: Washington, DC, USA, 2014; pp. 1–6. [CrossRef]
31. *ISO 12135:2021*; Metallic Materials—Unified Method of Test for the Determination of Quasistatic Fracture Toughness. International Organization for Standardization: Geneva, Switzerland, 2021; p. 100.
32. Ghafari, S.; Nejad Moghadas, F. R-Curve Behavior and Crack Propagation Properties of Asphalt Concrete at Low Temperatures. *J. Civ. Eng. Manag.* **2015**, *21*, 559–570. [CrossRef]
33. Luo, R.; Liu, H.; Duan, P.; Ye, J.; Chen, W.; Zhang, Z.; Wang, A. Sulfate Resistance and Microstructure of Metakaolin Geopolymer Reinforced by Cellulose Nanofiber and Wollastonite. *J. Build. Eng.* **2023**, *64*, 105580. [CrossRef]
34. Coldebella, R.; Gentil, M.; Berger, C.; Costa, H.W.D.; Pedrazzi, C.; Labidi, J.; Delucis, R.A.; Missio, A.L. Nanofibrillated Cellulose-Based Aerogels Functionalized with Tajuva (Maclura Tinctoria) Heartwood Extract. *Polymers* **2021**, *13*, 908. [CrossRef]
35. Elfaleh, I.; Abbassi, F.; Habibi, M.; Ahmad, F.; Guedri, M.; Nasri, M.; Garnier, C. A Comprehensive Review of Natural Fibers and Their Composites: An Eco-Friendly Alternative to Conventional Materials. *Results Eng.* **2023**, *19*, 101–271. [CrossRef]
36. Tawfik, A.; Abd-El-Razik, R.A. Design, Fabrication, and Characterization of Distinguished Lightweight and Sustainable Building Materials. *J. Build. Eng.* **2022**, *46*, 103–712. [CrossRef]
37. Motaung, T.E.; Mngomezulu, M.E.; Hato, M.J. Effects of Alkali Treatment on the Poly(Furfuryl) Alcohol—Flax Fibre Composites. *J. Thermoplast. Compos. Mater.* **2016**, 48–60. [CrossRef]
38. Pranger, L.A.; Nunnery, G.A.; Tannenbaum, R. Mechanism of the Nanoparticle-Catalyzed Polymerization of Furfuryl Alcohol and the Thermal and Mechanical Properties of the Resulting Nanocomposites. *Compos. Part B* **2012**, *43*, 1139–1146. [CrossRef]
39. Motaung, T.E.; Linganiso, L.Z.; Kumar, R.; Anandjiwala, R.D. Agave and Sisal Fibre-Reinforced Polyfurfuryl Alcohol Composites. *J. Thermoplast. Compos. Mater.* **2017**, *30*, 1323–1343. [CrossRef]
40. Jia, Y.; Fiedler, B. Influence of Furfuryl Alcohol Fiber Pre-Treatment on the Moisture Absorption and Mechanical Properties of Flax Fiber Composites. *Fibers* **2018**, *6*, 59. [CrossRef]
41. Saw, S.; Sarkhel, G.; Arup, C. Surface Modification of Coir Fibre Involving Oxidation of Lignins Followed by Reaction with Furfuryl Alcohol: Characterization and Stability. *Appl. Surf. Sci.* **2011**, *257*, 3763–3769. [CrossRef]
42. Bekele, A.E.; Lemu, H.G.; Jiru, M.G. Experimental Study of Physical, Chemical and Mechanical Properties of Enset and Sisal Fibers. *Polym. Test.* **2022**, *106*, 107453. [CrossRef]
43. Hisseine, O.A.; Omran, A.F.; Tagnit-Hamou, A. Influence of Cellulose Filaments on Cement Paste and Concrete. *J. Mater. Civ. Eng.* **2018**, *30*, 04018109. [CrossRef]
44. Kunchariyakun, K.; Sinyoung, S.; Kajitvichyanukul, P. Comparative Microstructures and Mechanical Properties of Mortar Incorporating Wood Fiber Waste from Various Curing Conditions. *Case Stud. Constr. Mater.* **2022**, *16*, e00855. [CrossRef]
45. Onuaguluchi, O.; Panesar, D.K.; Sain, M. Properties of Nanofibre Reinforced Cement Composites. *Constr. Build. Mater.* **2014**, *63*, 119–124. [CrossRef]

46. Jiao, L.; Su, M.; Chen, L.; Wang, Y.; Zhu, H.; Dai, H. Natural Cellulose Nanofibers As Sustainable Enhancers in Construction Cement. *PLoS ONE* **2016**, *11*, e0168422. [CrossRef]
47. Morton, J.H.; Cooke, T.; Akers, S.A.S. Performance of Slash Pine Fibers in Fiber Cement Products. *Constr. Build. Mater.* **2010**, *24*, 165–170. [CrossRef]
48. Hasan, K.M.F.; Horváth, P.G.; Alpár, T. Lignocellulosic Fiber Cement Compatibility: A State of the Art Review Lignocellulosic Fiber Cement Compatibility: A State of the Art Review. *J. Nat. Fibers* **2022**, *19*, 5409–5434. [CrossRef]
49. Claramunt, J.; Fernández-carrasco, L.J.; Ventura, H.; Ardanuy, M. Natural Fiber Nonwoven Reinforced Cement Composites as Sustainable Materials for Building Envelopes. *Constr. Build. Mater.* **2016**, *115*, 230–239. [CrossRef]
50. Lima, P.R.L.; Santos, H.M.; Camilloto, G.P.; Cruz, R.S. Effect of Surface Biopolymeric Treatment on Sisal Fiber Properties and Fiber-Cement Bond. *J. Eng. Fiber. Fabr.* **2017**, *12*, 59–71. [CrossRef]
51. Pereira, M.V.; Fujiyama, R.; Darwish, F.; Alves, G.T. On the Strengthening of Cement Mortar by Natural Fibers. *Mater. Res.* **2015**, *18*, 177–183. [CrossRef]
52. Filho, J.D.A.M.; Silva, F.D.A.; Toledo Filho, R.D. Degradation Kinetics and Aging Mechanisms on Sisal Fiber Cement Composite Systems. *Cem. Concr. Compos.* **2013**, *40*, 30–39. [CrossRef]
53. Krishna, N.K.; Prasanth, M.; Gowtham, R.; Karthic, S.; Mini, K.M. ScienceDirect Enhancement of Properties of Concrete Using Natural Fibers. *Mater. Today Proc.* **2018**, *5*, 23816–23823. [CrossRef]
54. Choi, H.; Choi, Y.C. Setting Characteristics of Natural Cellulose Fiber Reinforced Cement Composite. *Constr. Build. Mater.* **2021**, *271*, 121910. [CrossRef]

ceramics

MDPI

Article

Eco-Fired Bricks from Phosphate Mine Waste Rocks: The Effects of Marble Waste Powder on the Physical and Microstructural Properties

Ayoub Bayoussef [1,2], Samira Moukannaa [3,*], Mohamed Loutou [2,4], Yassine Taha [3], Mostafa Benzaazoua [3] and Rachid Hakkou [2,3,*]

1 Higher School of Technology of Fkih Ben Salah, Sultane Moulay Sliman University of Beni-Mellal, Hay Tighnari, National Road to Casablanca N°11, P.O. Box 336, Beni-Mellal 23000, Morocco; a.bayoussef@usms.ac.ma

2 Materials and Environmental Chemistry Laboratory (IMED-Lab), Faculty of Science and Technology Gueliz, Cadi Ayyad University, Avenue A. Elkhattabi, BP549, Marrakech 40000, Morocco

3 Geology and Sustainable Mining Institute (GSMI), Mohammad VI Polytechnic University, Lot 660. Hay Moulay Rachid, Ben Guerir 43150, Morocco

4 Laboratory of Molecular Chemistry, Materials and Environment (LCM2E), Department of Chemistry, Faculty Multidisciplinary Nador, University Mohamed Premier, B.P. 300, Selouane, Nador 62700, Morocco

* Correspondence: samira.moukannaa@um6p.ma (S.M.); r.hakkou@uca.ac.ma (R.H.)

Abstract: Nowadays, the global brick industry utilizes billions of cubic meters of clay soil annually, resulting in the massive consumption of non-renewable resources. This study explores the viability of utilizing red marl from phosphate mining waste rocks for fired brick production. Ecofriendly fired bricks produced from 100% side streams (red marly clays (RM) and marble waste powder (MWP)) were prepared, pressed, dried at 105 °C, and then fired at 1100 °C for 1 h. The effects of marble waste powder addition (up to 30 wt%) on the physical, mechanical, mineralogical, and microstructural properties of the fired bricks were explored. The main results show that fired bricks with high compressive strength of a maximum of 39 MPa could be prepared with a mixture of red marl and 10 wt% of marble waste powder. The thermal conductivity was decreased by marble waste addition (from 0 to 30%) and was reduced from 0.93 W/m.k to 0.53 W/m.k; however, the compressive strength was also decreased to reach a minimum of 17 MPa. The firing shrinkage and density were also reduced with 30% marble waste by 41% and 18%, respectively. Therefore, red marly clays and marble waste could be promising raw materials for eco-fired brick production.

Keywords: red marl; mine waste rock valorization; marble waste; fired bricks; thermal conductivity; density

1. Introduction

The circular economy concept has become essential in addressing the growing challenges associated with waste and by-product generation, disposal, and management in various industries, including the mining sector. Traditional operating methods focusing on resource extraction, use, and disposal generally lead to environmental and economic concerns [1]. In contrast, the circular economy approach prioritizes resource recovery, reuse, and recycling to minimize waste and maximize resource utilization [2,3]. Adopting a circular economy can effectively tackle the issues related to solid waste management and help to preserve natural resources, especially for the mining industry, where large quantities of by-products and waste are generated [4,5].

https://doi.org/10.3390/ceramics8020048

Phosphate mining is one of the industries known for its continuous growth due to the increasing demand for phosphate ore owing to its crucial utilization in medicine and agriculture. During phosphate open-pit mining, large volumes of non-exploitable interlayers and cover layers are blasted and disposed of as waste rocks in stockpiles. Around 780 to 3120 Mt of waste rocks are estimated to be generated yearly and disposed of in large stockpiles at mining sites [1]. These waste rocks, comprising topsoil, flint, limestones, marl, and clays, are chemically stable; however, they can induce several environmental impacts associated with landscape degradation and reductions in arable land [6,7]. It is worth noting that the chemical and mineralogical composition of these waste rocks resembles that of several raw materials, such as clays, limestone, and aggregates used in the construction sector. This implies that the reuse of phosphate waste rocks as secondary available resources for construction material production could help to reduce the disposal of these by-products owing to the huge and increasing demand of the construction industry for resources, thus preserving natural resources.

Among these types of waste, clay-rich by-products from phosphate mining represent promising candidates as raw materials in construction. Numerous studies have been conducted in this regard, attempting to reuse clay-rich phosphate by-products for construction material manufacturing. Studies have demonstrated the suitability of producing ceramics and bricks using phosphate mining clay by-products, while similar studies have emphasized the necessity of fly ash incorporation into the clay by-product brick mix to improve the mechanical properties of the fired bricks [8,9]. Clay by-products from phosphate mining have demonstrated their viability as raw materials for different construction material types, e.g., as co-binders for ecological binder production, such as alkali-activated materials [10,11]; as supplementary cementitious materials to substitute clinkers [12]; and as binders for lightweight aggregates [13]. However, their chemical and mineralogical composition makes clay by-products from phosphate interlayers seem more suitable for fired brick or ceramic manufacturing. Previous studies in this regard have demonstrated the viability of this by-product for fired brick production, yet the firing shrinkage rate was high, varying from 16 to 21% [14]. Marble waste has been proven to increase the porosity, reduce the density, and decrease the shrinkage of fired bricks [15,16].

Marble waste is produced in large quantities during marble block cutting and polishing and is often disposed of near residential areas [17,18]. This side stream is mainly composed of calcium, magnesium, silicon, and aluminum oxides and is frequently generated as a fine powder, which can represent a source of several environmental issues, such as air pollution, reducing the fertility of natural soil, and surface water contamination [19,20]. Hence, its incorporation into fired bricks from phosphate waste rock might be an interesting research pathway to produce 100% side-stream-based fired bricks, while improving the firing shrinkage and density and conserving their outstanding mechanical properties.

The present study investigates the potential of manufacturing ecological fired bricks by combining red marly clay (RM) from phosphate by-products with marble waste powder (MWP). Although prior studies have established the individual potential of phosphate mine waste and marble waste powder for brick production, the synergistic combination of these two waste materials remains unexplored. Our innovative formulation was designed to achieve dual objectives: (1) enhancing the thermal insulation properties through the controlled porosity generated by marble powder's carbonate decomposition and (2) enabling effective waste valorization by capitalizing on the complementary characteristics of both industrial by-products. This approach simultaneously addresses two critical environmental challenges while advancing the development of sustainable construction materials with an improved insulating capacity, aligning with the circular economy principles. Marble waste was incorporated at different percentages (up to 30 wt%) (Table 1) in the brick mix.

The effects of marble addition on the phase formation, microstructure, density, shrinkage, thermal conductivity, and compressive strength of the fired bricks were studied.

Table 1. Experimental design.

Sample ID	Red Marl (wt %)	Marble Waste Powder (wt %)	Water Content (wt%)
MP 0	100	0	10
MP 10	90	10	10
MP 20	80	20	10
MP 30	70	30	10

2. Materials and Methods

The red marl (Figure 1) was sourced from phosphate mines in the Guantour Basin, Morocco, while marble powder (Figure 1) was obtained from a local industrial area in Marrakech, Morocco. Upon receipt, the materials underwent crushing and grinding to reduce the particle size, followed by sieving through a 200 μm sieve to retain the finer fraction.

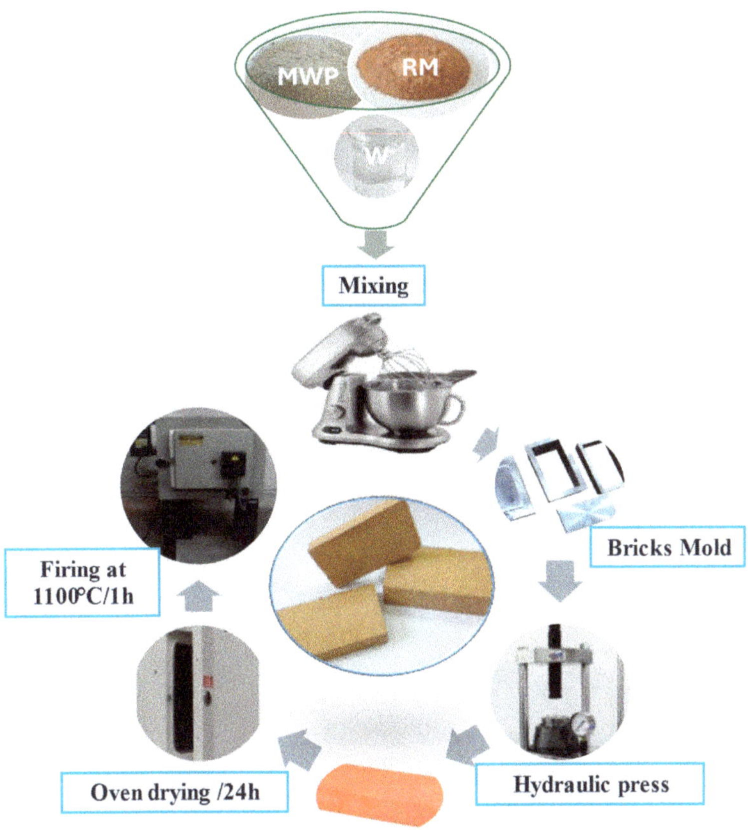

Figure 1. Fired brick production.

Both the raw marl and marble powder were characterized in terms of their mineralogical and chemical compositions.

Three different formulations were prepared using variable proportions of red marl and marble powder (10%, 20%, and 30% by weight), in addition to a reference brick composed of marl only (Table 1). Each formulation was thoroughly mixed for 10 min to ensure the even distribution of the powders. Next, water was gradually added (10% by weight of the dry mass) to the mixture to achieve the optimum moisture content for further processing. The moistened mixture was then pressed using a laboratory hydraulic press (Figure 1).

After pressing, the wet products were air-dried at ambient temperature for 48 h, followed by overnight oven drying at 105 °C. The final firing step involved introducing the dried products into a Nabertherm furnace and heating them to 1100 °C over 1 h, with a heating rate of 2 °C/min.

Characterization Techniques

The raw materials and final products were characterized using complementary analytical techniques to assess their chemical, mineralogical, structural, and thermal properties. X-ray fluorescence (XRF) spectroscopy with an Epsilon 4 instrument (Almelo, the Netherlands) and inductively coupled plasma optical emission spectroscopy (ICP-OES) using a device from PerkinElmer (Waltham, MA, USA) were employed to quantify major and trace elements, respectively, providing the basis for understanding their chemical compositions. Mineralogical phases were identified using X-ray diffraction (XRD) with a RIGAKU diffractometer (Tokyo, Japan) ($\lambda K\alpha = 1.54056$ Å), operating at 30 kV and 10 mA, with diffractograms recorded over a 2θ range of 5° to 70° (step size: 0.02°). These results complemented the chemical results, offering insights into the crystalline phases present. Thermogravimetric analysis (TGA) was conducted using the LABSYS Evo equipment (Lyon, France), heating the samples from room temperature to 1100 °C at 5 °C/min to evaluate their thermal stability and decomposition behavior, which correlated with the material's composition and structural characteristics. Finally, scanning electron microscopy (SEM) with a TESCAN VEGA3 instrument (Brno, Czech Republic), coupled with energy-dispersive X-ray spectroscopy (EDS), was employed to examine the microstructures and elemental distributions of the fired samples, linking the morphological observations to chemical and mineralogical data. These integrated techniques provided a comprehensive understanding of the samples' properties. In addition, the shaped samples were subjected to compressive strength measurements using an Instron 3369 (Norwood, MA, USA), at a speed of 0.1 mm/min. The results of the mechanical strength tests are reported as the mean values of three to five repeated tests. Water absorption was determined by weighing the samples before and after immersion in water for 24 h, according to standard European procedures, while density measurements were carried out using Archimedes' principle.

3. Results and Discussion

3.1. Characterization of Raw Materials

The XRD diagrams and mineralogical compositions of the marl and marble powder are illustrated in Figure 2. It reveals that the main phases constituting the marl material were dolomite (PDF#01-084-1208), quartz (PDF#00-046-1045), and palygorskite (PDF#96-900-5567), with the minor presence of illite (PDF#84e1220), fluoroapatite (PDF#00-015-0876), and hematite (PDF#00-033-0664). Marble powder contained calcite, albite, quartz, and a minor amount of dolomite [21]. The identified crystalline phases were in agreement with the microstructural analysis results (Figure 3) and the chemical composition (Tables 2 and 3).

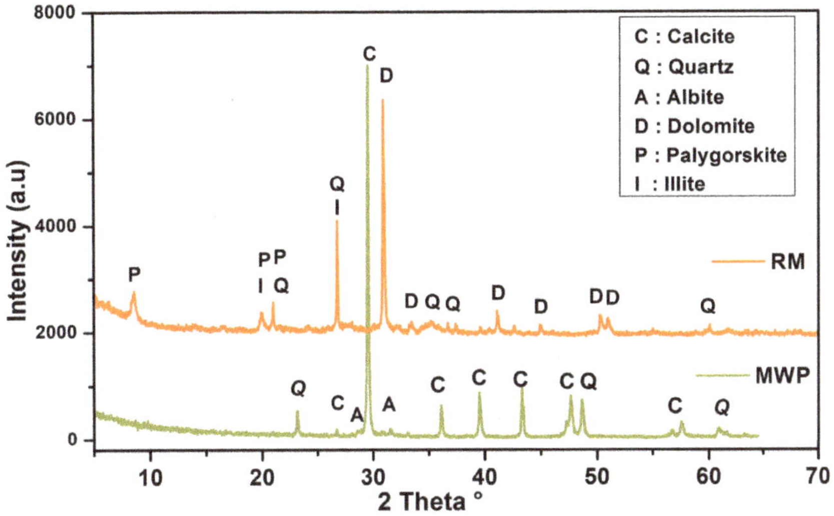

Figure 2. XRD of RM and MWP.

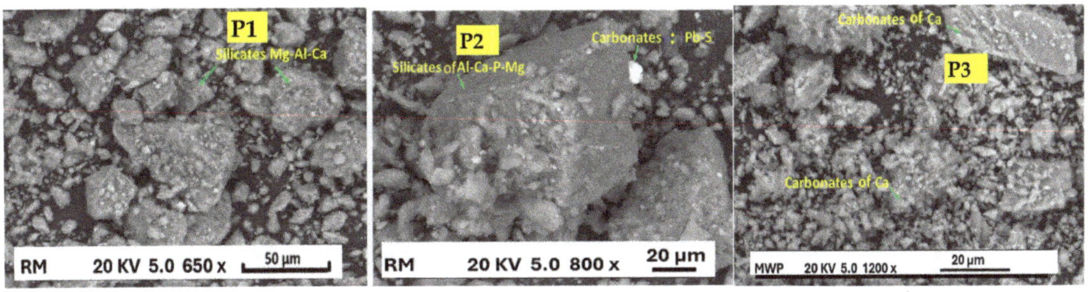

(**a**) (**b**)

Figure 3. SEM analysis of raw materials of (**a**) RM and (**b**) MWP.

Table 2. Chemical analysis of RM and MWP.

Element (%)	SiO_2	Al_2O_3	Fe_2O_3	CaO	MgO	K_2O	TiO_2	P_2O_5	S	Na_2O
RM	38.16	8.57	3.54	14.97	11.64	1.78	0.45	1.96	0.32	0.2
MWP	1.38	0.76	0.81	54.49	1.46	0.49	0.02	<0.01	0.57	0.16

Table 3. Chemical composition in terms of trace elements of RM and MWP.

Element (g/t)	As	Ba	Cd	Co	Cr	Cu	Ni	Pb	Zn
RM	24	133	<2	<8	718	114	172	<26	2270
MWP	90	21	<2	<8	111	81	101	<26	1009

The chemical analysis of RM and MWP through X-ray fluorescence spectroscopy (XRF) (Table 2) and inductively coupled plasma (ICP) analysis (Table 3) offers comprehensive insights into the elemental compositions of the raw materials employed in brick production. Notably, RM exhibits significant proportions of SiO_2, Al_2O_3, and Fe_2O_3, while MWP is characterized by the prominent presence of CaO. The high concentration of calcium oxide in MWP suggests a potential fluxing effect during the firing process, influencing the sintering behavior, while the average presence of both CaO and MgO in RM will give rise to a complex interaction during heat treatment. Both act as fluxes, reducing the resulting melting temperature and improving the thermal and mechanical properties of the final product [22,23]. Furthermore, the ICP analysis reveals variations in trace elements like copper (Cu), nickel (Ni), and zinc (Zn) (Table 3). These variations may impact the final properties of the bricks, influencing factors such as the color, strength, and thermal conductivity. The distinct concentrations of copper and zinc in RM compared to MWP may contribute to divergent material behavior during firing, potentially affecting the mechanical and thermal performance of the resulting bricks [24,25]. Understanding these chemical subtleties is crucial in refining the manufacturing process and optimizing the performance of eco-friendly fired bricks.

The SEM examinations of the raw materials are displayed in Figure 3a,b. Both micrographs showed variability in the particle size and surface features. This heterogeneity is common in raw marls and affects the packing density and initial porosity during the sintering process. The irregular and granular morphology observed is characteristic of raw marls, reflecting minimal processing. Natural marls often exhibit aggregates of fine particles, which might have formed through compaction or natural weathering processes. The granular appearance and lack of a smooth particle surface might indicate the presence of non-marl materials or impurities, such as quartz, calcite, or organic matter. This statement corroborates the XRD-related results. The presence of silicates (Mg-Al-Ca), as also shown in Figure 3 and Table 4, indicates that the marl contains a mix of mineral phases, which play a significant role in its plasticity and fluxing during the sintering process.

Table 4. EDS results of the raw materials.

	Oxygen	Magnesium	Aluminium	Silicon	Calcium	Phosphorus	Carbon	Sum
P1	35.49	2.17	4.96	14.19	3.26	0	0	60.06
P2	62.67	1.79	1.13	3.35	23.84	4.31	0	97.10
P3	59.4	0	0	0	35.4	0	14.87	110.24

3.2. Thermal Analysis (TG-DTG)

A thermogravimetric analysis (TGA) was carried out to examine the weight loss of the marl and marble powder during high-temperature firing, as shown in Figure 4a. For the red marl, TG showed two main weight losses. The first (−5%) occurred between 30 °C and 180 °C due to the removal of adsorbed water, confirmed by a DTG endothermic peak between 100 °C and 200 °C (Figure 4b) [26]. A small endothermic peak at 300 °C and another small and large peak at 400 °C were attributed to the release of bound water in palygorskite [27] and to organic matter decomposition and the dehydroxylation of marl minerals, respectively [28]. The second major TG weight loss (−15%), observed between 558 °C and 864 °C, corresponded to the dihydroxylation of Mg-OH and Ca-OH groups; this phenomenon corresponds to the decomposition of dolomite and dihydroxylation of palygorskite, as explained by previous studies [29–31]. This was evidenced by two intense, overlapping DTG peaks between 600 °C and 700 °C [32]; see Figure 4b.

For the marble powder (Figure 4a), TG revealed minor weight loss (−5%) between 100 °C and 200 °C, linked to surface water evaporation [33], and significant weight loss (−45%) between 600 °C and 900 °C, attributed to $CaCO_3$ decomposition into CaO and CO_2 [34]. The DTG analysis confirmed these processes, showing a small endothermic peak at 600 °C for initial carbonate destabilization and an intense peak between 700 °C and 900 °C for the main decomposition of calcium carbonate [35]. These thermal behaviors highlight the structural and compositional differences between the two materials, with red marl undergoing more complex transformations due to its multi-mineral composition, while the marble powder exhibits a dominant decomposition event related to its high carbonate content.

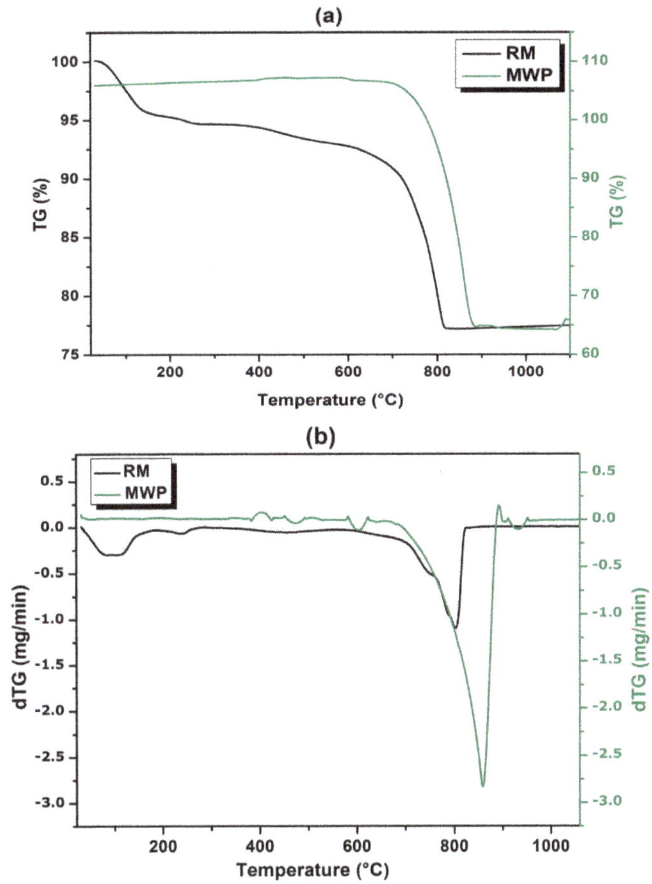

Figure 4. Thermal analysis: (**a**) TGA and (**b**) DTG of RM and MWP.

3.3. Physical Properties of Fired Bricks

The physical properties of building materials, such as the mechanical strength, density, water absorption, and thermal conductivity, are critical indicators of their quality and suitability for construction applications. In this study, cylindrical brick samples were tested, and the results were compared with those of a reference sample to assess the impact of marble powder substitution. The test results demonstrate the variations in the compressive strength, water absorption, density, and shrinkage with increasing substitution rates of marl with marble powder (Table 5 and Figure 5). The incorporation of marble powder significantly influenced these properties. The compressive strength decreased from 68.57 MPa in the reference sample to 17 MPa in bricks with 30% marble powder, while water absorption increased from 5% to 15.8%. These trends are primarily attributed to increased porosity, which will be further explored and discussed in the context of the microstructural analysis in the following section [36]. The XRD patterns and SEM micrographs show diminished amorphous halo signatures with higher marble powder content, potentially indicating the formation of fewer vitreous phases during firing. As these glassy phases normally enhance the structural integrity, their apparent reduction correlates with the observed mechanical property trends. The slight reduction in the firing shrinkage of the bricks with higher marble powder content also supports the observed increase in porosity. This phenomenon is likely due to the decomposition of carbonates during firing, which leaves voids in the brick structure, as well as the reduced content of clay minerals in the mixture. Since clay minerals are primarily responsible for shrinkage during firing, their smaller proportion in marble-rich mixes further contributed to the diminished shrinkage. Similarly, the density of the bricks exhibited a slight decline with marble powder incorporation, reflecting the less compact microstructure caused by carbonate decomposition and subsequent pore formation. This effect was further amplified by the greater calcite content in the initial mix, which led to higher mass loss during firing (due to CO_2 release) and ultimately contributed to the lighter bricks. Normally, the density would be expected to show a more marked variation with MWP addition, similarly to the compressive strength and/or water absorption. This is probably due to the antagonistic effect of sintering with the formation of porosity, leading to a more or less compacted structure.

Table 5. Physical and mechanical properties of brick samples.

MWP Content (%)	Water Absorption (%)	Density (g/cm^3)	Shrinkage (%)	Compressive Strength (MPa)	LOI (%)
10.00	8.91	1.77	3.16	39.74	22.88
20.00	13.19	1.75	2.46	29.40	25.26
30.00	15.80	1.71	1.76	17.09	27.67

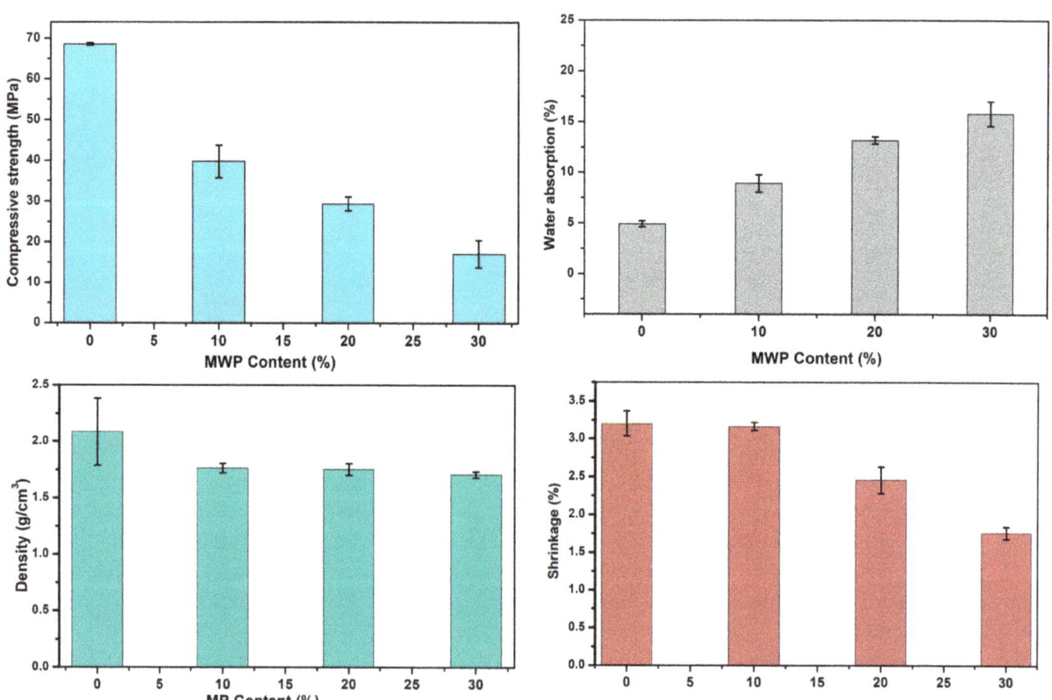

Figure 5. Physical properties of fired bricks.

Figure 6 illustrates the relationships among various sintered brick properties, revealing a strong correlation between the compressive strength and water absorption ($R^2 = 0.93$) and between the loss on ignition (LOI) and firing shrinkage ($R^2 = 0.8$). These findings align with the observed trends: lower compressive strength corresponds to higher water absorption, while reduced loss on ignition, potentially influenced by other factors, is linked to decreased shrinkage due to the bricks' microstructural characteristics [37]. It is important to note that these correlations, including factors like the pore distribution, crack propagation, and phase transitions during firing, are supported by the data but may also involve additional influences.

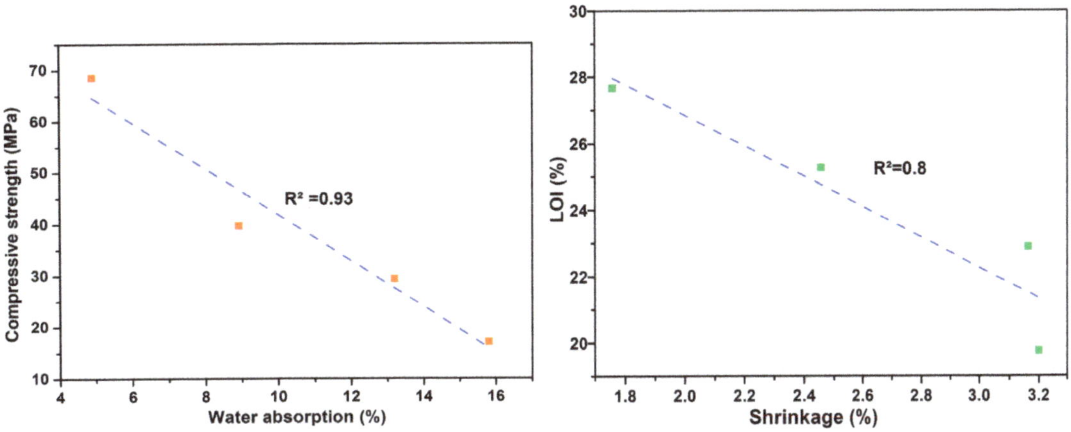

Figure 6. Correlations between physical properties.

Moreover, the integration of carbonate-rich marble powder resulted in reduced thermal conductivity compared to the reference products, as depicted in Figure 7. This decline in thermal conductivity can be attributed to the increased porosity of the bricks, which inherently influences their thermal properties [38,39]. Porosity introduces voids in the ceramic matrix, which are typically filled with air or another gas [40]. Air has significantly lower thermal conductivity (~0.024 W/m·K) compared to a solid ceramic material. Air pockets act as thermal insulators, disrupting the continuous solid pathways for heat transfer and scattering heat carriers [41–43]. Consequently, the thermal insulation properties were significantly enhanced, showcasing a remarkable improvement of up to 42%. This finding underscores the potential of marble powder incorporation to enhance the thermal performance of fired bricks, thereby contributing to improved energy efficiency and overall sustainability in construction applications.

3.4. Microstructural Analysis of Fired Bricks

The mineralogical characterization results of the manufactured bricks are displayed in Figure 8. It reveals significant mineralogical transformations during firing. Neoformed phases such as diopside (PDF#00-041-1370), akermanite (COD#96-900-6116), gehlenite (PDF#01-089-5917), enstatite (COD#96-900-5543), and rondorfite were identified, indicating reactions between the original minerals and the decomposition products of carbonates, among other components, during sintering.

$$CaO + MgO + 2SiO_2 \rightarrow CaMgSi_2O_6 \text{ (Diopside)} \quad (1)$$

$$2CaO + MgO + 2SiO_2 \rightarrow Ca_2MgSi_2O_7 \text{ (Akermanite)} \quad (2)$$

$$MgO + SiO_2 \rightarrow MgSiO_3 \text{ (Enstatite)} \quad (3)$$

$$8CaO + MgO + 4SiO_2 + 2Cl_2 \rightarrow Ca_8Mg(SiO_4)4Cl_2 \text{ (Rondorfite)} \quad (4)$$

$$2CaO + Al_2O_3 + SiO_2 \rightarrow Ca_2Al_2SiO_7 \text{ (Gehlenite)} \quad (5)$$

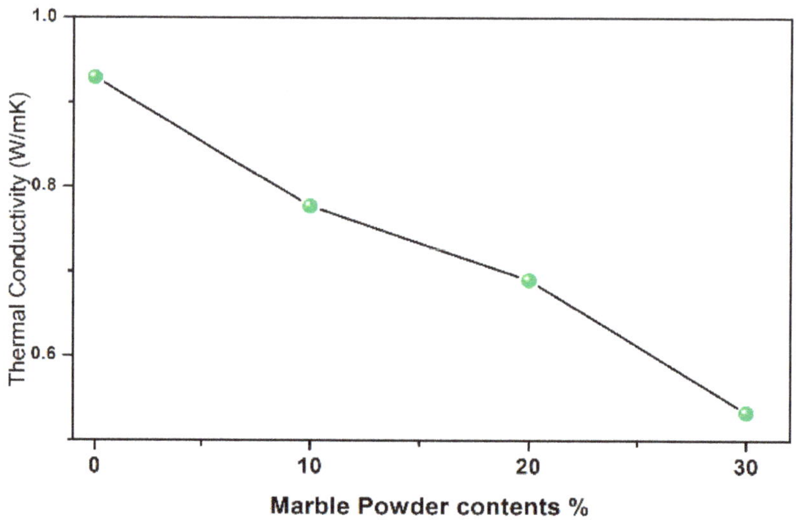

Figure 7. Thermal conductivity of fired bricks.

The formation of gehlenite and akermanite can be attributed to reactions between calcite ($CaCO_3$ from marble powder) and silica-rich phases like quartz or illite in the marl [32]. Similarly, the decomposition of dolomite and calcite at high temperatures liberates CaO and CO_2 [44], with the CaO reacting with aluminosilicates to form gehlenite and diopside. Palygorskite undergoes dehydroxylation, releasing MgO, which contributes to enstatite and diopside formation, while illite decomposition primarily yields K_2O [45,46]. Rondorfite formation is likely a result of interactions involving fluorapatite, providing fluorine for stabilization, and calcium and magnesium from calcite and dolomite. The presence of hematite may also play a catalytic role in phase formation [47,48].

These mineralogical changes are directly linked to the physical properties of the bricks, highlighting the interplay between phase development and structural integrity during sintering. The newly formed phases, such as diopside and akermanite, contribute significantly to improving the thermal and mechanical stability of the bricks by providing a robust crystalline framework. These phases are particularly valuable in enhancing their durability, making them suitable for structural applications.

However, the observed reduction in vitreous phases, coupled with the formation of micropores and cracks, as observed in the SEM images (Figures 9 and 10), introduces structural vulnerabilities. This phenomenon correlates with decreased compressive strength and increased water absorption as the marble powder content increases. The decomposition of carbonates plays a pivotal role, leaving voids that elevate the porosity and weaken the brick's matrix. This increased porosity accounts for the decline in density and firing shrinkage, which are critical parameters in evaluating the quality and performance of sintered bricks.

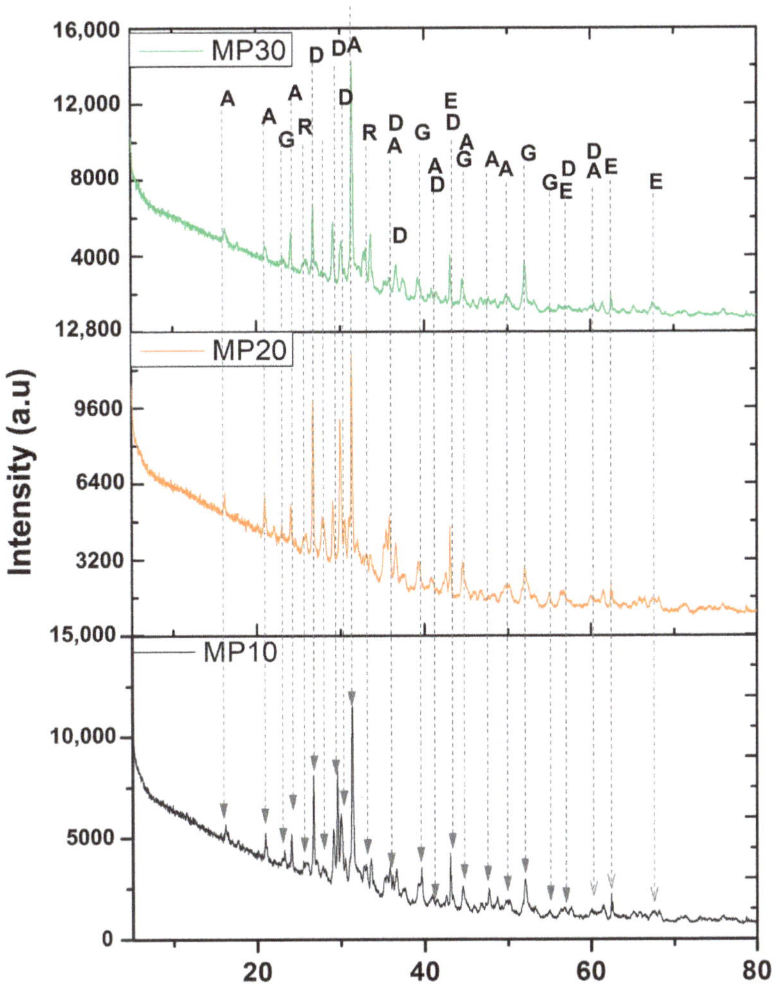

Figure 8. X-ray crystallography spectra of fired bricks.

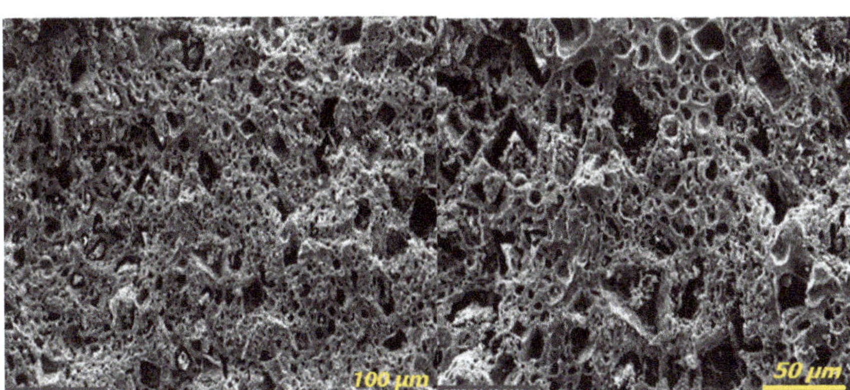

Figure 9. SEM of reference bricks produced with RM.

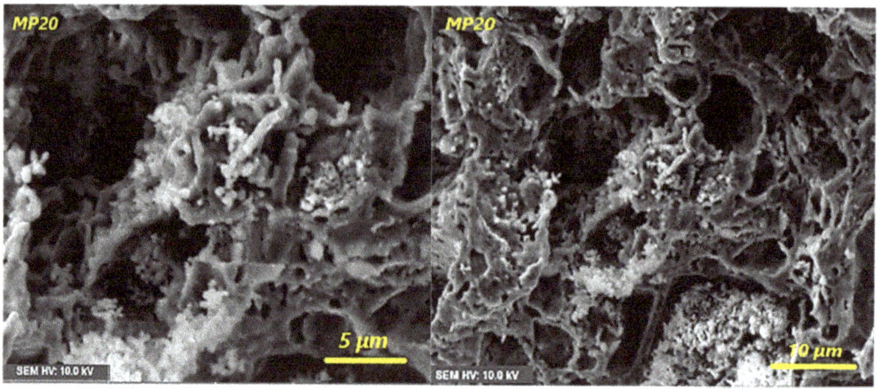

Figure 10. SEM of bricks containing 20% of MWP.

Therefore, the mineralogical evolution during firing exerts a dual influence: while newly formed stable phases enhance the thermal and mechanical properties, the structural weaknesses induced by porosity counteract these benefits. Achieving an optimal balance between these opposing effects is essential in tailoring the sintering process to produce bricks with desired properties. These findings emphasize the need for a detailed understanding of the mineral interactions and decomposition mechanisms to effectively valorize marl-rich mine by-products and maximize their potential in sustainable brick production.

A scanning electron microscope (SEM) was employed to investigate the internal microstructural characteristics of the bricks under study. The resulting micrographs revealed prominent microstructural features and exhibited heterogeneous porosity within the samples. Additionally, the micrographs (refer to Figures 9 and 10) depicted the presence of crystalline particles with varying textures and sizes, further highlighting the complexity of the brick microstructure. The observed heterogeneous porosity can be attributed to the decomposition and subsequent loss of carbonates within the samples. This decomposition leads to the formation of voids and irregularities within the brick matrix, which compromise the overall structural integrity and result in decreased compressive strength and density. These findings are corroborated by the SEM micrographs, which illustrate the presence of a highly irregular and porous microstructure.

The diverse textures and sizes of the crystalline particles observed in the micrographs suggest variations in their composition and distribution across the bricks. These variations can stem from two primary factors: the inherently heterogeneous nature of the raw mine by-products used (marl lithologies and marble waste powders) and the firing conditions during sintering. Notably, these differences in crystalline structure and porosity influence how stress is distributed across the brick matrix and contribute to the material's performance under various loading and environmental conditions.

Moreover, the microstructural features observed, such as the development of micropores and cracks and the presence of non-uniform carbonate decomposition, further explain the changes in the bricks' physical properties. These porosities act as stress concentrators, exacerbating mechanical weaknesses, leading to reduced compressive strength, and increasing the water absorption rate [40,49]. Understanding these microstructural characteristics can provide insights for the tailoring of the sintering process or the modification of the raw material composition to optimize these sustainable bricks' mechanical and physical properties [50,51].

These observations underscore the importance of monitoring mineralogical transformations and their microstructural effects during sintering to better control the performance and functionality of mine by-products valorized into construction materials.

Overall, the SEM analysis provides valuable insights into the intricate microstructures of the bricks, shedding light on the factors influencing their mechanical behavior and highlighting potential avenues for optimization in future manufacturing processes.

4. Conclusions

In conclusion, this study demonstrates the successful development of environmentally friendly bricks with enhanced mechanical properties and thermal insulation by utilizing waste-derived raw materials from the mining industry. The comprehensive characterization and experimental results confirm the feasibility and effectiveness of integrating red marl from phosphate by-products and marble waste powder into the brick production process. This innovative approach addresses pressing environmental challenges by valorizing mining waste, reducing the reliance on natural resources traditionally used in brick manufacturing and minimizing waste disposal issues.

The findings highlight that the incorporation of marble waste powder and its associated mineralogical transformations during firing significantly influence the physical and mechanical properties of the resulting bricks. The produced bricks exhibited competitive compressive strength and thermal insulation performance, demonstrating their potential to meet contemporary construction standards while serving as a sustainable alternative to conventional brick materials. This approach not only reduces the dependency on natural marl but also offers both ecological and economic advantages by mitigating the environmental impacts of mining waste.

This study emphasizes the critical role of interdisciplinary research in addressing sustainability challenges and advancing innovative solutions for the construction industry. The findings pave the way for the development of advanced, eco-friendly building materials that align with global sustainability goals and efforts to reduce environmental footprints.

While the lab-scale bricks showed no bloating defects, industrial-scale production with thicker formats may experience CO_2 entrapment during calcite decomposition, potentially causing internal pressures and structural defects [52]. This risk rises with both the brick thickness and marble powder (MWP) content. Future research should focus on (1) formulation optimization, particularly in controlling the marble powder (MWP) particle size distribution to mitigate rapid gas release through more uniform carbonate decomposition; (2) optimizing raw material formulations, particularly in determining the optimal marble powder (MWP) dosage and firing regime to balance porosity development and bloating risks; (3) scaling up production processes for industrial applications—our planned tests will systematically evaluate the maximum MWP incorporation levels before defect formation, temperature thresholds for controlled decarbonation, and firing cycle modifications to facilitate gradual CO_2 release [53,54]; and (4) assessing long-term durability under diverse environmental conditions. These efforts will ensure the safe, widespread adoption of this sustainable building material, supporting greener construction practices.

Author Contributions: Conceptualization, A.B. and R.H.; methodology, A.B. and R.H.; software, A.B. and Y.T.; validation, R.H.; formal analysis, A.B. and S.M.; investigation, A.B., S.M., M.L. and Y.T.; resources, R.H.; data curation, A.B. and M.L.; writing—original draft preparation, A.B., S.M. and M.L.; writing—review and editing, A.B., S.M., M.L., Y.T., M.B. and R.H.; visualization, A.B., M.L. and M.B.; supervision, R.H.; project administration, R.H.; funding acquisition, R.H. All authors have read and agreed to the published version of the manuscript.

Funding: This research was funded by the OCP Foundation through the APRD Research Program, grant number Eco-MAT 21–24.

Informed Consent Statement: Not applicable.

Data Availability Statement: The original contributions presented in this study are included in the article. Further inquiries can be directed to the corresponding author(s).

Acknowledgments: The authors would like to thank the Moroccan Ministry of Higher Education, Scientific Research and Innovation and the OCP Foundation for funding this work through the APRD Research Program under the Eco-MAT 21–24 grant.

Conflicts of Interest: The authors declare no conflicts of interest.

Abbreviations

The following abbreviations are used in this manuscript:

RM	Red Marl
MWP	Marble Waste Powder

References

1. Beniddar, H.; El Machi, A.; El Abbassi, F.-E.; Taha, Y.; Benzaazoua, M.; Hakkou, R. Sustainable utilization of phosphate mine waste rocks as sand substitutes in cement mortar production. *Constr. Build Mater.* **2024**, *438*, 136949. [CrossRef]
2. Hu, S.; Wang, J.; Xia, X. Study on development of circular economy in Hefei. In Proceedings of the 3rd International Conference on Environmental Technology and Knowledge Transfer, Hefei, China, 13–14 May 2010.
3. Yuan, H.; Wang, J. A system dynamics model for determining the waste disposal charging fee in construction. *Eur. J. Oper. Res.* **2014**, *237*, 988–996. [CrossRef]
4. Olejarczyk, M.; Rykowska, I.; Urbaniak, W. Management of Solid Waste Containing Fluoride—A Review. *Materials* **2022**, *15*, 3461. [CrossRef] [PubMed]
5. Pandey, V.; Panda, S.K.; Singh, V.K. Preparation and characterization of high-strength insulating porous bricks by reusing coal mine overburden waste, red mud and rice husk. *J. Clean. Prod.* **2024**, *469*, 143134. [CrossRef]
6. Moukannaa, S.; Bagheri, A.; Benzaazoua, M.; Sanjayan, J.G.; Pownceby, M.I.; Hakkou, R. Elaboration of alkali activated materials using a non-calcined red clay from phosphate mines amended with fly ash or slag: A structural study. *Mater. Chem. Phys.* **2020**, *256*, 123678. [CrossRef]
7. Oubaha, S.; El Machi, A.; Mabroum, S.; Taha, Y.; Benzaazoua, M.; Hakkou, R. Recycling of phosphogypsum and clay by-products from phosphate mines for sustainable alkali-activated construction materials. *Constr. Build. Mater.* **2024**, *411*, 134262. [CrossRef]
8. Mouih, K.; Hakkou, R.; Taha, Y.; Benzaazoua, M. Performances of compressed stabilized bricks using phosphate waste rock for sustainable construction. *Constr. Build. Mater.* **2023**, *388*, 131577. [CrossRef]
9. Bayoussef, A.; Oubani, M.; Loutou, M.; Taha, Y.; Benzaazoua, M.; Manoun, B.; Hakkou, R. Manufacturing of high-performance ceramics using clays by-product from phosphate mines. *Mater. Today Proc.* **2021**, *37*, 3994–4000. [CrossRef]
10. Gavali, H.R.; Ralegaonkar, R.V. Design development of sustainable alkali-activated bricks. *J. Build. Eng.* **2020**, *30*, 101302. [CrossRef]
11. Boughriet, A.; Allahdin, O.; Poumaye, N.; Doyemet, G.; Tricot, G.; Revel, B.; Ouddane, B.; Wartel, M. Alkali-Activated Brick Aggregates as Industrial Valorized Wastes: Synthesis and Properties. *Ceramics* **2023**, *6*, 108. [CrossRef]
12. Bahhou, A.; Taha, Y.; El Khessaimi, Y.; Idrissi, H.; Hakkou, R.; Amalik, J.; Benzaazoua, M. Use of phosphate mine by-products as supplementary cementitious materials. *Mater. Today Proc.* **2021**, *37*, 3781–3788. [CrossRef]
13. Loutou, M.; Hajjaji, M.; Mansori, M.; Favotto, C.; Hakkou, R. Phosphate sludge: Thermal transformation and use as lightweight aggregate material. *J. Environ. Manag.* **2013**, *130*, 354–360. [CrossRef]
14. Loutou, M.; Taha, Y.; Benzaazoua, M.; Daafi, Y.; Hakkou, R. Valorization of clay by-product from moroccan phosphate mines for the production of fired bricks. *J. Clean Prod.* **2019**, *229*, 169–179. [CrossRef]
15. Ahmad, S.; Shah, M.U.H.; Ullah, A.; Shah, S.N.; Rehan, M.S.; Khan, I.A.; Ahmad, M.I. Sustainable Use of Marble Waste in Industrial Production of Fired Clay Bricks and Its Employment for Treatment of Flue Gases. *ACS Omega* **2021**, *6*, 22559–22569. [CrossRef]
16. Galetakis, M.; Soultana, A. A review on the utilisation of quarry and ornamental stone industry fine by-products in the construction sector. *Constr. Build Mater.* **2016**, *102*, 769–781. [CrossRef]

17. Tunc, E.T. Recycling of marble waste: A review based on strength of concrete containing marble waste. *J. Environ. Manag.* **2019**, *231*, 86–97. [CrossRef] [PubMed]
18. Khan, M.; Khan, B.; Shahzada, K.; Khan, S.; Wahab, N.; Ahmad, M.I. Conversion of Waste Marble Powder into a Binding Material. *Civ. Eng. J.* **2020**, *6*, 431–445. [CrossRef]
19. Das, D.; Gołąbiewska, A.; Rout, P.K. Geopolymer bricks: The next generation of construction materials for sustainable environment. *Constr. Build Mater.* **2024**, *445*, 137876. [CrossRef]
20. Wang, Y.; Xiao, J.; Zhang, J.; Duan, Z. Mechanical Behavior of Concrete Prepared with Waste Marble Powder. *Sustainability* **2022**, *14*, 4170. [CrossRef]
21. Vardhan, K.; Goyal, S.; Siddique, R.; Singh, M. Mechanical properties and microstructural analysis of cement mortar incorporating marble powder as partial replacement of cement. *Constr. Build. Mater.* **2015**, *96*, 615–621. [CrossRef]
22. Rehman, M.U.; Ahmad, M.; Rashid, K. Influence of fluxing oxides from waste on the production and physico-mechanical properties of fired clay brick: A review. *J. Build. Eng.* **2020**, *27*, 100965. [CrossRef]
23. Eliche-Quesada, D.; Leite-Costa, J. Use of bottom ash from olive pomace combustion in the production of eco-friendly fired clay bricks. *Waste Manag.* **2016**, *48*, 323–333. [CrossRef] [PubMed]
24. Kazmi, S.M.S.; Munir, M.J.; Patnaikuni, I.; Wu, Y.F.; Fawad, U. Thermal performance enhancement of eco-friendly bricks incorporating agro-wastes. *Energy Build* **2018**, *158*, 1117–1129. [CrossRef]
25. Dai, Z.; Wu, Y.; Hu, L.; Zhang, W.; Mao, L. Evaluating physical-mechanical properties and long periods environmental risk of fired clay bricks incorporated with electroplating sludge. *Constr. Build. Mater.* **2019**, *227*, 116716. [CrossRef]
26. Hofmann, E.; Németh, T.; Bidló, A. Thermal analysis of soils formed on limestone in the Bükk Mountains, North Hungary. *J. Agrokémia És Talajt* **2018**, *67*, 5–22. [CrossRef]
27. Wang, S.; Gainey, L.; Wang, X.; Mackinnon, I.D.R.; Xi, Y. Influence of palygorskite on in-situ thermal behaviours of clay mixtures and properties of fired bricks. *Appl. Clay Sci.* **2022**, *216*, 106384. [CrossRef]
28. Collard, F.-X.; Blin, J. A review on pyrolysis of biomass constituents: Mechanisms and composition of the products obtained from the conversion of cellulose, hemicelluloses and lignin. *J. Renew. Sustain. Energy Rev.* **2014**, *38*, 594–608. [CrossRef]
29. Carazo, E.; Borrego-Sánchez, A.; García-Villén, F.; Sánchez-Espejo, R.; Viseras, C.; Cerezo, P.; Aguzzi, C. Adsorption and characterization of palygorskite-isoniazid nanohybrids. *Appl. Clay Sci.* **2018**, *160*, 180–185. [CrossRef]
30. Kuang, W.; Facey, G.A.; Detellier, C. Dehydration and rehydration of palygorskite and the influence of water on the nanopores. *Clays Clay Miner.* **2004**, *52*, 635–642. [CrossRef]
31. Dong, W.; Lu, Y.; Wang, W.; Zong, L.; Zhu, Y.; Kang, Y.; Wang, A. A new route to fabricate high-efficient porous silicate adsorbents by simultaneous inorganic-organic functionalization of low-grade palygorskite clay for removal of Congo red. *Microporous Mesoporous Mater.* **2019**, *277*, 267–276. [CrossRef]
32. Harrati, A.; Arkame, Y.; Manni, A.; Aqdim, S.; Zmemla, R.; Chari, A.; El Bouari, A.; Hassani, I.-E.E.A.E.; Sdiri, A.; Hassani, F.O.; et al. Akermanite-based ceramics from Moroccan dolomite and perlite: Characterization and in vitro bioactivity assessment. *Open Ceram.* **2022**, *10*, 100276. [CrossRef]
33. Pramono, E.; Pratiwi, W.; Wahyuningrum, D.; Radiman, C.L. Thermal properties of Bentonite Modified with 3-aminopropyltrimethoxysilane. *IOP Conf. Ser. Mater. Sci. Eng.* **2018**, *333*, 12084. [CrossRef]
34. Qian, H.; Kai, W.; Hongde, X. A novel perspective of dolomite decomposition: Elementary reactions analysis by thermogravimetric mass spectrometry. *J. Thermochim. Acta* **2019**, *676*, 47–51. [CrossRef]
35. Scott, S.A.; Dennis, J.S.; Davidson, J.F.; Hayhurst, A.N. Thermogravimetric measurements of the kinetics of pyrolysis of dried sewage sludge. *J. Fuel* **2006**, *85*, 1248–1253. [CrossRef]
36. Zhuginissov, M.T.; Nurlybayev, R.E.; Orynbekov, Y.S.; Zhumadilova, Z.O.; Khamza, Y.Y.; Bulenbayev, M.Z. The Influence of the Burning Environment on the Properties of Ceramic Products Based on Fusible Raw Materials. *Ceramics* **2023**, *6*, 872–885. [CrossRef]
37. Ozturk, S.; Sutcu, M.; Erdogmus, E.; Gencel, O. Influence of tea waste concentration in the physical, mechanical and thermal properties of brick clay mixtures. *Constr. Build. Mater.* **2019**, *217*, 592–599. [CrossRef]
38. Singh, D.; Kumar, R.; Nighot, N.S.; Rajput, A.; Prajapati, A.; Singh, B.K.; Kirgiz, M.S.; Srinivasaraonaik, B.; Mishra, R.K.; Khan, S.; et al. A comprehensive review on valorisation of octal by-product as supplementary admixtures in the production of fired and unfired bricks. *Constr. Build. Mater.* **2023**, *408*, 133641. [CrossRef]
39. Munir, M.J.; Abbas, S.; Nehdi, M.L.; Kazmi, S.M.S.; Khitab, A. Development of Eco-Friendly Fired Clay Bricks Incorporating Recycled Marble Powder. *J. Mater. Civ. Eng.* **2018**, *30*, 04018069. [CrossRef]
40. Kadir, A.A.; Mohajerani, A. Effect of heating rate on gas emissions and properties of fired clay bricks and fired clay bricks incorporated with cigarette butts. *Appl. Clay Sci.* **2015**, *104*, 269–276. [CrossRef]
41. Wang, C.; Wang, J.; Li, Q.; Xu, S.; Yang, J. A review on recent development of foam Ceramics prepared by particle-stabilized foaming technique. *Adv. Colloid Interface Sci.* **2024**, *330*, 103198. [CrossRef]

42. Evcin, A.; Ersoy, B.; Çiftçi, H. Utilization of Marble and Boron Waste in Brick Products. *Int. J. Comput. Exp. Sci. Eng.* **2019**, *5*, 19–22. [CrossRef]
43. Makrygiannis, I.; Karalis, K. Optimizing Building Thermal Insulation: The Impact of Brick Geometry and Thermal Coefficient on Energy Efficiency and Comfort. *Ceramics* **2023**, *6*, 1449–1466. [CrossRef]
44. Gunasekaran, S.; Anbalagan, G. Thermal decomposition of natural dolomite. *Bull. Mater. Sci.* **2007**, *30*, 339–344. [CrossRef]
45. Bozkaya, Ö.; Yalçın, H. Mineral Chemistry of Low-Temperature Phyllosilicates in Early Paleozoic Metaclastic Rocks, Eastern Tauride Belt, Türkiye. *Minerals* **2022**, *12*, 1088. [CrossRef]
46. Meunier, A.; Velde, B. *Illite: Origins, Evolution and Metamorphism*; Springer: Berlin/Heidelberg, Germany, 2010.
47. Li, M.; Su, P.; Guo, Y.; Zhang, W.; Mao, L. Effects of SiO_2, Al_2O_3 and Fe_2O_3 on leachability of Zn, Cu and Cr in ceramics incorporated with electroplating sludge. *J. Environ. Chem. Eng.* **2017**, *5*, 3143–3150. [CrossRef]
48. Souza, A.E.; Teixeira, S.R.; Santos, G.T.A.; Costa, F.B.; Longo, E. Reuse of sugarcane bagasse ash (SCBA) to produce ceramic materials. *J. Environ. Manag.* **2011**, *92*, 2774–2780. [CrossRef]
49. Aouba, L.; Bories, C.; Coutand, M.; Perrin, B.; Lemercier, H. Properties of fired clay bricks with incorporated biomasses: Cases of Olive Stone Flour and Wheat Straw residues. *Constr. Build. Mater.* **2016**, *102*, 7–13. [CrossRef]
50. Menezes, R.R.; Ferreira, H.S.; Neves, G.A.; de L. Lira, H.; Ferreira, H.C. Use of granite sawing wastes in the production of ceramic bricks and tiles. *J. Eur. Ceram. Soc.* **2005**, *25*, 1149–1158. [CrossRef]
51. Subari, F.E. Utilization of Marble and Limestone Wastes for Making the Eksposed Bricks Based Synthetic Wollastonite. *J. Teknol. Miner. Dan Batubara* **2016**, *12*, 171–178.
52. Cultrone, G.; Rodriguez-Navarro, C.; Sebastian, E.; Cazalla, O.; De La Torre, M.J. Carbonate and silicate phase reactions during ceramic firing. *J. Eur. J. Mineral.* **2001**, *13*, 621–634. [CrossRef]
53. Karaman, S.; Ersahin, S.; Gunal, H. Firing temperature and firing time influence on mechanical and physical properties of clay bricks. *J. Sci. Ind. Res.* **2006**, *65*, 153–159.
54. Tsega, E. Effects of Firing Time and Temperature on Physical Properties of Fired Clay Bricks. *Am. J. Civ. Eng.* **2017**, *5*, 21. [CrossRef]

ceramics

MDPI

Article

Physical and Mechanical Properties of Fired Bricks from Amazon Bauxite Tailings with Granite Powder

Igor A. R. Barreto * and Marcondes L. da Costa *

Program for Post-Graduation in Geology and Geochemistry, Institute of Geosciences, Federal University of Pará Belém, Belém 66075-110, PA, Brazil
* Correspondence: igorrochaq@gmail.com (I.A.R.B.); marcondeslc@gmail.com (M.L.d.C.)

Abstract: In the Amazon region, bauxite processing generates significant quantities of clay mineral-rich tailings, which pose a major challenge for bauxite mining operations. This study explores the use of bauxite tailings to produce fired bricks and evaluates their properties. Using a Box–Behnken experimental design, nine specimens were prepared with varying granite content (0%, 5%, 10%, 20%, and 30%) and fired at three different temperatures: 800 °C, 900 °C, and 1000 °C. The bauxite tailings contain gibbsite, kaolinite, Al-goethite, and hematite, while the granite powder comprises quartz, potassium feldspar, sodium plagioclase, muscovite, and occasionally kaolinite. Linear shrinkage values remained within recommended limits, below 8%. Apparent porosity (AP) results ranged from 60.2% to 72%, with maximum water absorption reaching 23.6%. The compressive strength of bricks without granite addition was 11.9 MPa at 900 °C, with the highest value recorded at 14.9 MPa at 800 °C when granite was added. These findings indicate that bauxite tailings, when supplemented with pulverized granite, exhibit promising potential for fired brick production.

Keywords: mining tailings; ceramics; kaolinite; circular economy

1. Introduction

The red ceramics sector is still an important source of essential materials for the construction industry and other sectors [1–6]. Red ceramics include blocks, roof tiles, solid bricks (fired bricks), drainage pipes, slabs, green walls, hollow elements, and expanded clay [7]. The term red ceramic is applied to these materials because of the red coloration that appears during firing and remains afterward, conferred by the newly formed minerals such as hematite (Fe_2O_3) and maghemite [7]. The raw materials used to produce red ceramics generally come from natural sources (rocks and soils), industrial waste, and even synthetic materials.

Clay is fundamental in the raw material used to produce red ceramics. It consists mainly of illite and kaolinite, with restricted montmorillonite, among others, which are responsible for the property of "plasticity" when it is mixed with water, which facilitates its handling in the shaping of different types of ceramic pieces and their various applications [1,3]. Certain clay minerals contribute to the plasticity of a ceramic piece and play other roles. For instance, illite, which contains K_2O in its composition, acts as a flux during the firing process [3].

A material used in the production of ceramics is unlikely to consist solely of clay minerals, even when it originates from natural sources. Such materials typically contain a variety of other constituents, including minerals like muscovite, goethite, hematite, quartz, and anatase, as well as organic matter and even sulfides such as pyrite. These components significantly influence the properties of the ceramic material, either positively or negatively. For instance, minerals such as quartz and anatase impart a non-plastic character to the ceramic body, whereas goethite, hematite, and muscovite can act as fluxing agents during the firing process, enhancing the mechanical strength of the final product [3]. The presence of organic matter is not desirable, as it can cause pores to form inside the ceramic piece, reducing its strength.

The material used to produce red ceramics does not always contain all the essential minerals to achieve the properties needed to produce high-quality ceramics, meaning that two or more raw materials must be mixed in pre-defined proportions to obtain a product with the right properties. Similar technological products are only suitable for specific applications if their raw materials possess distinct characteristics, often achieved through the use of additives. For instance, materials like Markforged 17-4 PH and BASF Ultrafuse 316L enable the production of stainless steel, while fibers enhance composite materials. Additionally, Artificial Intelligence can optimize material compositions and processing parameters to achieve target properties. In the production of biodegradable or recyclable polymers, biomass serves as a sustainable raw material. Numerous applications require specific additives to ensure the materials meet performance standards and achieve desirable properties [8–13].

In Paragominas, Pará, Brazil, the company Mineração Paragominas S.A. extracts and processes bauxite from regional deposits as a raw material for the production of metallic aluminum in Barcarena, Pará. During the bauxite beneficiation process—which includes crushing, grinding, and washing—a significant amount of clay-rich material is generated and mixed with water. For every ton of bauxite ore processed, approximately 0.35 tons of this clay residue are produced. Until recently, this by-product was disposed of in tailings basins.

The tailing consists mainly of kaolinite (the predominant mineral), gibbsite, hematite, goethite, quartz, and anatase. This mineralogical composition, which is dominated by clay, combined with the fact that it is available in large volumes in the tailing's ponds and previous laboratory research, demonstrates the very attractive potential of this waste for applications in civil construction, such as bricks, cement, and mortar, among others [14]. Despite these aspects, this material currently has no applications and could be considered a potential problematic waste for bauxite mining.

Because of the points listed above, this study aims to assess the technical feasibility of producing "burnt brick" from these clay residues from bauxite mining in Paragominas, with and without the addition of pulverized granite. The choice of granite powder is intended to demonstrate the potential of fine waste resulting from the production of gravel from these rocks in quarries in the region.

2. Materials and Methods

2.1. Raw Materials

The raw material used in the development of this study consisted of a representative sample of bauxite tailing (ALB-1) from the Mineração Paragominas S.A company of Paragominas Pará, PA, Brazil. Additionally, pulverized granite (GP-1) from a quarry in the region was used as an additive.

2.2. The Characterization of the Samples

2.2.1. Mineralogical Composition

The mineralogical phases of the sample were determined using X-ray diffraction (XRD) in accordance with the powder method. A BRUKER diffractometer, model D2 PHASER, equipped with a θ/θ goniometer, a radius of 141.1 nm, and a copper anode emitting at a characteristic wavelength of 1.54 Å (8.047 keV, Cu-Kα1) was utilized, operating at a maximum power of 300 W (30 kV × 10 mA). A linear Lynxeye detector with a 5° 2θ aperture and 192 channels was employed for data collection. The analyses were conducted at the Laboratory of Mineralogy, Geochemistry, and Applications (LAMIGA-UFPA) within the Institute of Geosciences at the Federal University of Pará (UFPA).

2.2.2. Chemical Composition with X-Ray Fluorescence Spectrometry

The major chemical elements were determined using X-ray fluorescence spectrometry (XRF) through the preparation of a fused pellet. The pellet was prepared with a sample-to-flux ratio (lithium tetraborate) established according to the XRF79C_10 method. Analyses were conducted at the Chemical Analysis Laboratory of SGS GEOSOL, located in Vespasiano, Minas Gerais, Brazil.

The loss on ignition (LOI) was determined using the gravimetric method, involving the calcination of previously dried samples at 1000 °C.

2.2.3. Thermogravimetric Analysis (TGA) and Differential Scanning Calorimetry (DSC)

Thermogravimetric and differential scanning calorimetry (TGA/DSC) analyses were conducted using a NETZSCH model STA 449 F5 Jupiter, which features a simultaneous thermal analyzer equipped with a vertical cylindrical oven. The analyses were performed under a nitrogen flow rate of 50 mL/s, a heating rate of 10 °C/min, and a temperature range from 30 °C to 1100 °C. This procedure was carried out at the Laboratory of Mineralogy, Geochemistry, and Applications (LAMIGA-UFPA) within the Institute of Geosciences at the Federal University of Pará (UFPA).

2.3. Formulation and Characterization of Fired Brick

The bauxite tailing (ALB-1) and granite (GP-1) samples were initially dried at 110 °C for 24 h. Subsequently, the granite sample underwent a preparation process that included crushing, grinding, and sieving through an 80-mesh sieve to produce a uniform powder. The ALB-1 sample, however, did not undergo particle size reduction. For the formulation of test specimens, both samples were combined in various weight ratios (wt%), ranging from 0 to 30% granite powder and 70 to 100% ALB-1, to achieve a final mass of 30 g (Table 1). This blend of ALB-1 and GP-1 was then transferred to a porcelain dish, where 10% water (g/g) was added and the mixture was homogenized using a pestle for 5 min.

Table 1. Experimental matrix and real and coded values of the independent variables.

Experiment	Temperature (°C)	Percentage of Granite Powder (wt/wt)
C1	1000	20
C2	1000	5
C3	900	0
C4	800	5
C5	800	20
C6	900	30
C7	900	10
C8	900	10
C9	900	10

The prepared mass was then used to form test specimens (CPs) in steel molds shaped as parallelepipeds with dimensions of 50.0 mm × 20.0 mm × 10.0 mm. Specimen compaction was performed via single-action uniaxial pressing at 30 kgf/cm^2 using a hydraulic press

(Karl Kolb, model PW-40). Following pressing, the specimens were dried at 105 °C for 24 h and subsequently fired at varying final temperatures (800, 900, and 1000 °C) (Figure 1).

Figure 1. Specimens conformed according to Doehlert Design. Legend: C4 and C5: fired at 800 °C; C2 and C1: fired at 1000 °C; C3, C6, C7, C8 and C9: fired at 900 °C.

The selection of granite percentage and firing temperatures was guided by a Doehlert experimental matrix with two variables, including three replicates at the central point. The temperature parameters set were 800, 900, and 1000 °C, with granite powder proportions evaluated at 0, 5, 10, 20, and 30 wt%. The experimental design and conditions are summarized in Table 1, resulting in a total of nine experimental trials.

2.3.1. Water Absorption

This parameter was evaluated following an adaptation of the ASTM C20-00 standard [15], based on the relationship between the dry and wet weights of the specimens. The ceramic specimens were immersed in water for at least 24 h to obtain the wet weight. The wet weight was measured after the surface water was removed from the specimens.

2.3.2. Apparent Porosity

The apparent porosity of the specimens was determined by the ASTM C20-00 standard [15].

2.3.3. Compressive Strength

The compressive strength of the specimens was evaluated using a universal testing machine, model 300/15 kN, equipped with Servo-Plus Evolution control from MSTEST. The testing was conducted at a speed of 1 MPa/s, with an initial load of 1 N and a maximum load set to 30% of the specimen's capacity. The tests were carried out in the Concrete Laboratory of the School of Architecture at the Federal University of Pará (UFPA).

3. Results

3.1. Raw Materials

3.1.1. Mineralogical Composition

The XRD patterns of granite powder (GP-1) and the tailing sample (ALB-1) are shown in Figure 2. The granite powder is mainly composed of quartz, potassium feldspar, plagioclase, muscovite and kaolinite. Sample ALB-1 consists of kaolinite, gibbsite, Al-goethite, and hematite.

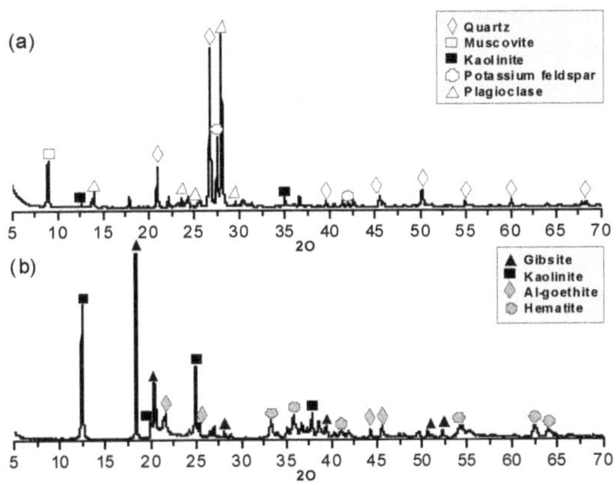

Figure 2. X-ray diffraction mineralogy patterns of GP-1 (**a**) and ALB-1 (**b**).

3.1.2. Chemical Composition

The chemical composition of the two samples is given in Table 2. Sample ALB-1 consists of Al_2O_3 (41.12%), SiO_2 (23%), and Fe_2O_3 (15.35%), while sample GP-1 consists of 74.4% SiO_2, 14.42% Al_2O_3, 4.45% K_2O and 3.86% Na_2O.

Table 2. Chemical composition of raw materials (wt%).

Sample	SiO_2	Al_2O_3	Fe_2O_3	TiO_2	K_2O	Na_2O	LOI	Other	Total
ALB-1	23	41.12	15.35	2.15	0.02	0.01	19.18	0.37	101.2
GRT-1	74.4	14.42	1.05	0.11	4.45	3.86	0.87	1.09	100.25

In sample ALB-1, the percentage of Al_2O_3 from the gibbsite and kaolinite domain, partial Al-goethite, SiO_2 from kaolinite, and Fe_2O_3 from goethite and hematite is reported. The percentage of SiO_2 in sample GP-1 is associated with silicates such as quartz, potassium feldspar, plagioclase, muscovite, and kaolinite.

The concentration of Al_2O_3 is associated with potassium feldspar, plagioclase, muscovite, and kaolinite. The percentage of K_2O is related to potassium feldspar, and the percentage of Na_2O is associated with plagioclase, probably albite.

3.1.3. Thermal Behavior

The thermal behavior of the samples can be seen in Figure 3. Sample ALB-1 shows three endothermic events in connection with mass loss (TG curve) and one exothermic event. In sample ALB-1, the first endothermic event at 293 °C is characteristic of the dehydroxylation of the mineral gibbsite; the event at 358 °C refers to the dehydroxylation of goethite, forming the mineral hematite; the third endothermic event at 510 °C is due to the loss of structural water from the mineral kaolinite [4]; and the exothermic event at 980 °C is characteristic of the formation of the mullite phase. Sample GP-1 showed no significant mass loss (≈0.4%) and the DSC curve shows only one endothermic event at 576 °C. The endothermic event at 573 °C observed in sample GP-1 is due to the structural transformation of alpha quartz (α-SiO_2) into beta quartz (β-SiO_2) [16].

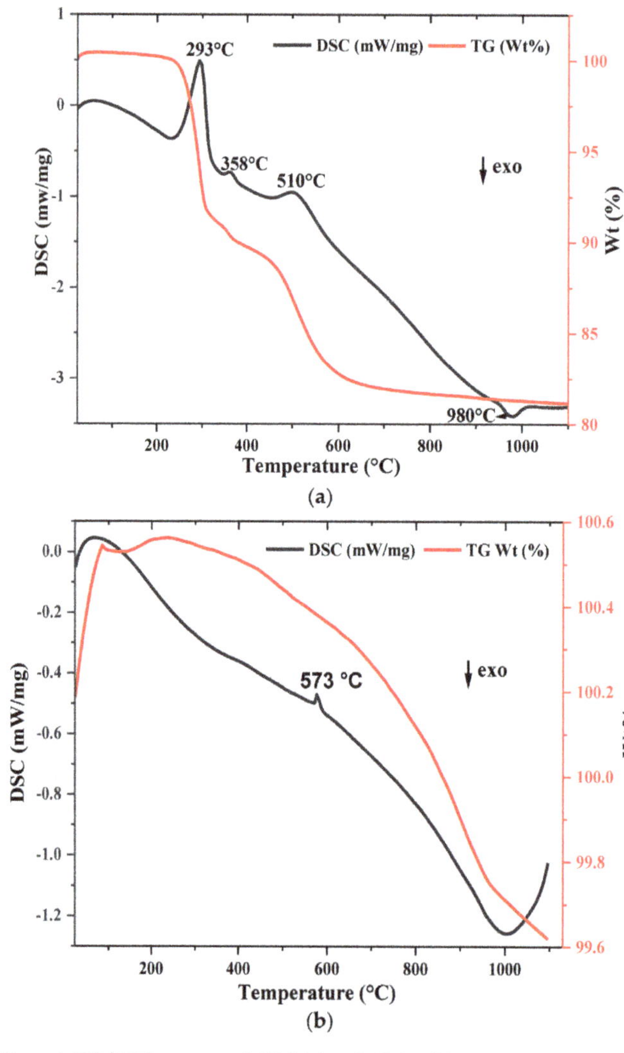

Figure 3. TG/DSC patterns of GP-1 (**a**) and ALB-1 (**b**). **Exo:** exothermic.

3.2. Fired Bricks (Specimens)

3.2.1. Mineralogical Composition

The composition of the specimens is shown in Figure 4. Quartz, plagioclase, and potassium feldspar were identified in all diffractograms for all compositions with granite components (C3 is the only one that does not contain granite), even if they were more "broadened" compared to the fresh sample (plagioclase and feldspar). Another mineral that was "preserved" during firing was hematite, which was present in all compositions.

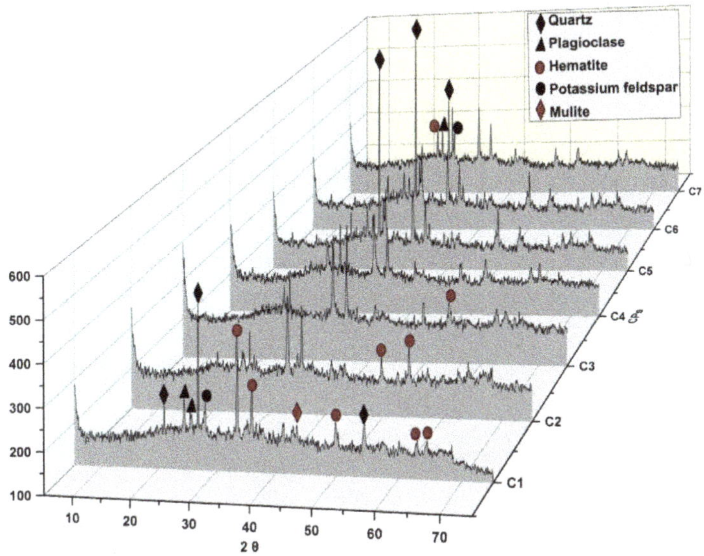

Figure 4. X-ray diffraction patterns of specimens (C1, C2, C3, C4, C5, C6 and C7).

The kaolinite, muscovite, gibbsite, and Al-goethite minerals present in the raw samples collapsed during firing. Most of the kaolinite may have given rise to amorphous materials (metakaolinite) or been transformed into mullite [3], as shown by the small peaks in the samples fired at 900 and 1000 °C.

3.2.2. Linear Firing Shrinkage (RLq)

The linear shrinkage of the test specimens (CP) is shown in Table 3. The values of the linear shrinkage are in the range of 1.19–3.71%. All the values are below the maximum allowed (8.0%) for good-quality bricks [17]. Linear shrinkage can be used as an indication of crack formation during the firing process of fired bricks [17,18].

Table 3. Linear retraction of the specimen (%).

Experiment	Dry Length	Length After Calcination	Linear Shrinkage (%)
C1	51.82	50.55	2.51
C2	51.81	50.24	3.13
C3	51.93	50.07	3.71
C4	51.79	50.96	1.63
C5	51.79	51.18	1.19
C6	51.93	51.23	1.37
C7	51.78	50.68	2.15 ± 0.25

3.2.3. Water Absorption and Apparent Porosity

Figure 5 illustrates the results of water absorption and apparent porosity. Water absorption is a fundamental parameter that indicates the potential of ceramic materials [19,20]. The highest apparent porosity value (72%) was obtained in experiment 3 (0% granite and 900 °C temperature) and the lowest values (60.7 and 60.2%) were observed in experiments 5 (20% granite and 900 °C temperature) and 6 (20% granite and 900 °C temperature), respectively. Regarding the water absorption results, the highest value was observed in experiment 3 (23.6%) and the lowest in experiment 6 (19.7%).

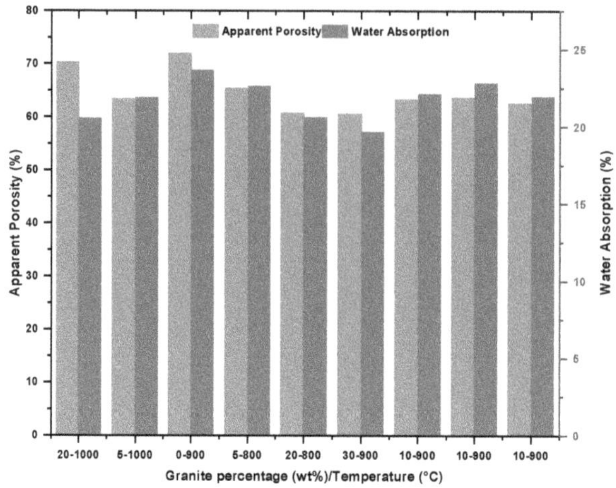

Figure 5. Apparent porosity and water absorption of specimens.

3.2.4. Compressive Strength

Figure 6 shows the results of the compressive strength of the specimen's compositions tested. The lowest results were obtained at 1000 °C (4.6 and 5.3 MPa) with a granite percentage of 20 and 6%, respectively. The best results were obtained at 900 °C without the addition of granite (11.9 MPa) and at 800 °C with the addition of 6% (14.9 MPa) and 20% granite (10.1 MPa). The standard deviation of the experiments was evaluated by the replicates of compositions 7 (6.43 MPa), 8 (7.14 MPa), and 9 (6.13 MPa) with an average of 6.55 ± 0.49 MPa. All the compressive strength results are above the values established by the ABNT and ASTM C55-17 standards [15,21–23].

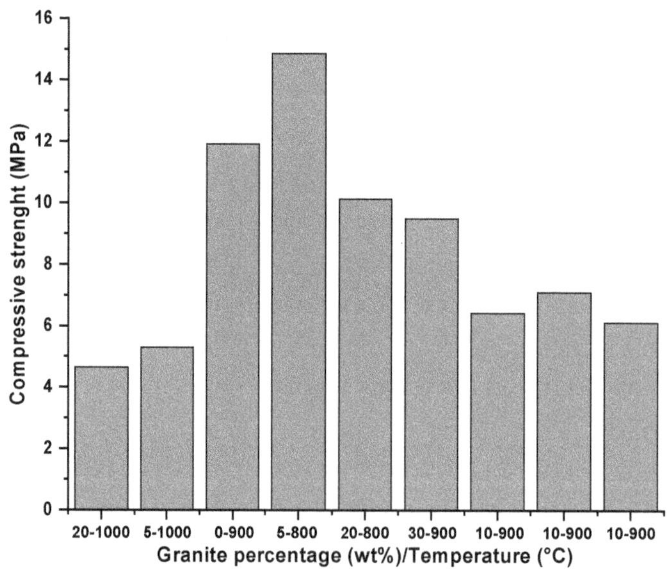

Figure 6. Compressive strength of specimens.

4. Discussion

4.1. Raw Materials

4.1.1. Mineralogical Composition and Chemical Composition

While kaolinite is one of the most important minerals used to produce ceramic materials, as it imparts plasticity to the ceramic body and contributes to resistance after firing [24,25], iron oxide minerals (Al-goethite and hematite) impart a reddish color; during firing, Al-goethite transforms into hematite. The contribution of kaolinite for ceramic is also described by [7]. The minerals in GP-1, mainly potassium feldspar, plagioclase, and muscovite, will mix to reduce porosity during the heating of the ceramic masses due to the presence of alkali metals, which will undergo a melting process [7,17].

The chemical composition of the samples, which is determined by the minerals present, has a direct effect on the technological properties of the ceramic product [1]. The presence of Al_2O_3, SiO_2, K_2O, and Fe_2O_3 in appropriate proportions is essential for establishing good properties in ceramic pieces [7].

4.1.2. Thermal Behavior

The events observed in sample ALB-1, which are associated with the loss of water forming the minerals gibbsite, goethite, and kaolinite, contribute to the formation of cracks during the heating of CPs produced only with this sample. On the other hand, no volatiles are observed in the granite sample. Due to this behavior, adding percentages of GP-1 to CPs with ALB-1 helps to reduce cracking during heating.

4.2. Specimens

4.2.1. Mineralogical Composition

Muscovite may have initiated its fusion process and contributed to the formation of a liquid phase during firing, thus enhancing the strength of the ceramic pieces. This behavior is essential to ensure suitable technological properties for ceramics. Other authors have already highlighted this characteristic [7].

The mineral gibbsite collapses and forms non-crystalline Al_2O_3, resulting in the refractory nature of the ceramic pieces. Al-goethite transforms into hematite, further accentuating the reddish color of the ceramic pieces.

4.2.2. Linear Firing Shrinkage (RLq)

The highest linear shrinkage value was obtained in CP-3, which was expected due to the predominance of kaolinite in the ALB-1 sample. The lowest linear shrinkage values were found in CP-5 (1.19%) and CP-6 (1.37%), which contain 30% and 20% granite, respectively. The higher value for CP-6 is due to the higher firing temperature (900 °C) than CP-5 (800 °C). These low linear shrinkage values are a consequence of the chemical/mineralogical composition of the granite, which contributes to reducing the plasticity of the CPs. This behavior was also identified by [7], who attributed the high linear shrinkage to the presence of kaolinite, while the reduced linear shrinkage was influenced by the presence of muscovite and other minerals that contribute to the formation of a liquid phase.

4.2.3. Water Absorption and Apparent Porosity

The results show a direct correlation between apparent porosity and water absorption, with the higher percentage of porosity directly influencing greater water absorption in the compositions studied. Although the incorporation of granite contributes to the formation of cracks and consequently to greater water absorption [15], its incorporation into the compositions studied reduced water absorption compared to the ceramic mass with only bauxite washing clay (experiment 3). The results are close to the maximum values allowed by the standards (ABNT 15270-3, 2005a; ABNT 15310, 2005b) for ceramic bricks ($\leq$20%) and tiles ($\leq$22%) [21,23].

The increase in temperature contributed to the formation of greater porosity in the specimens fired at 900 and 1000 °C [2]. On the other hand, when the same temperature is analyzed with different percentages of granite, it is evident that the addition of granite favored a reduction in porosity and consequently a reduction in water absorption. All points mentioned are influenced by the predominance of kaolinite in the ALB-1 sample and the addition of the liquid phase from GP-1, as similarly defined by [7].

4.2.4. Compressive Strength

The temperature and percentage of granite directly influenced the gain and loss of resistance. At 800 °C, the sintering process was most effective in increasing resistance compared to 900 and 1000 °C. This behavior suggests that the release of volatile species during heating may have contributed to the formation of cracks, reducing the strength of the specimens at higher temperatures (900 and 1000 °C). A similar behavior was reported by [7], where higher temperatures decreased the strength of the specimens. Concerning the percentage of granite, it is clear that the addition of a low percentage was more favorable for gaining strength, but this gain was not as pronounced when compared to the strength value obtained with the ABT-1 sample alone. These characteristics suggest that the addition of a low percentage of granite contributes to reducing the plasticity of the ABT-1 sample and with the possible formation of a liquid phase after 900 °C, which contribute to a gain in strength, but that the percentage of granite above 10% may have reduced the plasticity to levels that jeopardize the strength of the ceramic mass [16,18,26].

5. Conclusions

The ALB sample (bauxite tailing) exhibits a composition similar to that of raw materials commonly used in the production of fired bricks, primarily due to the presence of minerals such as kaolinite and hematite. Additional minerals found in the sample, including gibbsite and Al-goethite, also contribute to the final properties of the bricks.

The technological properties of the bricks were directly influenced by the studied compositions. In terms of linear shrinkage, the addition of 20% and 30% granite resulted in the lowest shrinkage values, measuring 1.19% and 1.37%, respectively. Regarding apparent porosity, the incorporation of 20% granite at 900 °C yielded the lowest porosity values (60.7% and 60.2%), while the highest porosity was observed in specimens without granite. As for water absorption, all compositions complied with the requirements defined by relevant standards. In terms of compressive strength, granite additions ranging from 6% to 20% yielded satisfactory results, with compressive strengths of 14.9 MPa and 10.1 MPa, respectively. Although the incorporation of granite improved the ceramic properties, even the specimens without granite exhibited favorable characteristics, particularly in compressive strength (11.9 MPa). However, their water absorption exceeded the limits established by the ABNT standard.

The addition of granite powder partially altered the properties of the fired bricks. The incorporation of 6% granite powder proved to be the most beneficial, improving both water absorption and compressive strength. Thus, considering the overall characteristics of the ALB sample, its use in fired brick production—enhanced with granite powder as an additive—is technically feasible. This approach may contribute to reducing porosity and increasing mechanical strength in the final product.

The application of fired bricks presents a promising alternative for the reuse of bauxite tailings due to several advantages: fired brick manufacturing is a low-cost process, does not require environmentally harmful additives, and enables the complete utilization of raw materials without generating additional waste post production.

Nonetheless, further investigation into the use of alternative additives is recommended to better understand their influence on red ceramics formulated with bauxite tailings, especially considering that variations in granite content and firing temperatures were not explored in this study.

Author Contributions: I.A.R.B.: Conceptualization, Methodology, Validation, Investigation, Formal analysis, Data curation, Writing—original draft, Writing—review & editing, Visualization; M.L.d.C.: Conceptualization, Methodology, Investigation, Writing—review & editing, Visualization, Funding acquisition, Project administration. All authors have read and agreed to the published version of the manuscript.

Funding: This research was supported by Hydro Company and the Dean of Research and Graduate Studies (PROPESP) of the Universidade Federal do Pará (UFPA).

Data Availability Statement: Data is contained within the article.

Acknowledgments: We thank the Hydro Company for providing the raw materials used in this study and financial support with the fieldwork.

Conflicts of Interest: The authors declare no conflict of interest. The funders had no role in the design of the study; in the collection, analyses, or interpretation of data; in the writing of the manuscript; or in the decision to publish the results.

References

1. Bai, M.; Xiao, J.; Gao, Q.; Shen, J. Utilization of Construction Spoil and Recycled Powder in Fired Bricks. *Case Stud. Constr. Mater.* **2023**, *18*, e02024. [CrossRef]
2. Sarani, N.A.; Kadir, A.A.; Din, M.F.M.; Amiza Hashim, A.; Hassan, M.I.H.; Hamid, N.J.A.; Hashar, N.N.H.; Hissham, N.F.N.; Johan, S.F.S.M. Physical-Mechanical Properties and Thermogravimetric Analysis of Fired Clay Brick Incorporating Palm Kernel Shell for Alternative Raw Materials. *Constr. Build. Mater.* **2023**, *376*, 131032. [CrossRef]
3. Wang, S.; Gainey, L.; Mackinnon, I.D.R.; Allen, C.; Gu, Y.; Xi, Y. Thermal Behaviors of Clay Minerals as Key Components and Additives for Fired Brick Properties: A Review. *J. Build. Eng.* **2023**, *66*, 105802. [CrossRef]
4. Adazabra, A.N.; Viruthagiri, G.; Yirijor, J. Combined Effects of Biomass Bottom Ashes and Spent Charcoal on Characteristics of Fired Clay Bricks. *Constr. Build. Mater.* **2023**, *399*, 132570. [CrossRef]
5. Makrygiannis, I.; Karalis, K. Optimizing Building Thermal Insulation: The Impact of Brick Geometry and Thermal Coefficient on Energy Efficiency and Comfort. *Ceramics* **2023**, *6*, 1449–1466. [CrossRef]
6. Makrygiannis, I.; Tsetsekou, A.; Papastratis, O.; Karalis, K. Assessing the Effects of Refuse-Derived Fuel (RDF) Incorporation on the Extrusion and Drying Behavior of Brick Mixtures. *Ceramics* **2023**, *6*, 2367–2385. [CrossRef]
7. Gadioli, M.C.B.; de Aguiar, M.C.; Vidal, F.W.H.; Sant'Ana, M.A.K.; de Almeida, K.M.; Giori, A.J.N. Incorporation of Ornamental Stone Waste in the Manufacturing of Red Ceramics. *Materials* **2022**, *15*, 5635. [CrossRef] [PubMed]
8. Kedziora, S.; Decker, T.; Museyibov, E.; Morbach, J.; Hohmann, S.; Huwer, A.; Wahl, M. Strength Properties of 316L and 17-4 PH Stainless Steel Produced with Additive Manufacturing. *Materials* **2022**, *15*, 6278. [CrossRef]
9. Gonçalves, R.M.; Martinho, A.; Oliveira, J.P. Recycling of Reinforced Glass Fibers Waste: Current Status. *Materials* **2022**, *15*, 1596. [CrossRef]
10. Badini, S.; Regondi, S.; Pugliese, R. Unleashing the Power of Artificial Intelligence in Materials Design. *Materials* **2023**, *16*, 5927. [CrossRef]
11. Freeland, B.; McCarthy, E.; Balakrishnan, R.; Fahy, S.; Boland, A.; Rochfort, K.D.; Dabros, M.; Marti, R.; Kelleher, S.M.; Gaughran, J. A Review of Polylactic Acid as a Replacement Material for Single-Use Laboratory Components. *Materials* **2022**, *15*, 2989. [CrossRef] [PubMed]
12. Bellucci, D.; Cannillo, V. Low-Temperature Sintering of a New Bioactive Glass Enriched with Magnesium Oxide and Strontium Oxide. *Materials* **2022**, *15*, 6263. [CrossRef]
13. Balti, S.; Boudenne, A.; Yahya, K.; Hamdi, N. Advancing Reinforcement of Sustainable Gypsum Composites: High-Performance Design by Reusing Waste Materials. *Mater. Today Sustain.* **2024**, *27*, 100946. [CrossRef]
14. Negrão, L.B.A.; da Costa, M.L.; Pöllmann, H. Waste Clay from Bauxite Beneficiation to Produce Calcium Sulphoaluminate Eco-Cements. *Constr. Build. Mater.* **2022**, *340*, 127703. [CrossRef]

15. Dong, Y.; Guo, W.; Jiang, C.; Shao, Y.; Zhang, L.; Wang, D.; Lu, X.; Huang, S.; Cheng, X. Using CaO as a Modifier Agent to Optimize the Pore Structure of Foamed Ceramics from Granite Scrap. *Ceram. Int.* **2023**, *49*, 13443–13451. [CrossRef]
16. Adazabra, A.N.; Viruthagiri, G.; Atingabono, J. Developing Fired Clay Bricks by Incorporating Scrap Incinerated Waste and River Dredged Sediment. *Process Saf. Environ. Prot.* **2023**, *179*, 108–123. [CrossRef]
17. Adazabra, A.N.; Viruthagiri, G. Effect of Variegated Biosolids Incorporation on the Technological Properties of Fired Clay Bricks Using Taguchi Method. *Case Stud. Constr. Mater.* **2023**, *19*, e02314. [CrossRef]
18. Wahab, R.A.A.; Mohammad, M.; Mazlan, M.; Yaki, A.N.A.; Bahari, N.S.S.; Fadzli, S.N.A.M.; Zahanis, Z.H.B.; Zaid, M.H.M. Study on the Physical and Mechanical Properties of Low Energy Consumption Fired Industrial Waste Clay Bricks from Eggshells and Rice Husks. *Mater. Today Proc.* **2023**, *75*, 79–83. [CrossRef]
19. Mohammad, M.; Abdul Wahab, R.A.; Mazlan, M.; Nisa Syuhaidah Mohamad Fazil, N.; Suraya Hanim Ibrahim, N.; Najiha Muhamad Nizam, U.; Humaidi Abu Hanifah, A.; Hafiz Mohd Zaid, M. Physical and Mechanical Properties of Fired Industrial Waste-Clay Bricks from Clam Shells and Soda Lime Silica Glass. In *Materials Today: Proceedings*; Elsevier Ltd.: Amsterdam, The Netherlands, 2022; Volume 75, pp. 151–155.
20. ABNT—Associação Brasileira de Normas Técnicas. *Alvenaria Estrutural—Blocos Cerâmicos Parte 3: Métodos de Ensaio*; Associação Brasileira de Normas Técnicas: Rio de Janeiro, RJ, Brazil, 2017; Volume 34.
21. ABNT—Associação Brasileira de Normas Técnicas. *NBR 15270-1: Componentes Cerâmicos. Parte 1: Blocos Cerâmicos Para Alvenaria de Vedação—Terminologia e Requisitos*; Associação Brasileira de Normas Técnicas: Rio de Janeiro, RJ, Brazil, 2005; Volume 15.
22. ABNT—Associação Brasileira de Normas Técnicas. *NBR 15270-2: Componentes Cerâmicos Parte 2: Blocos Cerâmicos Para Alvenaria Estrutural—Terminologia e Requisitos*; Associação Brasileira de Normas Técnicas: Rio de Janeiro, RJ, Brazil, 2017; Volume 15.
23. *ASTM C20-00*; Standard Test Methods for Apparent Porosity, Water Absorption, Apparent Specific Gravity, and Bulk Density of Burned Refractory Brick and Shapes by Boiling Water. American Society for Testing and Materials: West Conshohocken, PA, USA, 2015; pp. 1–3. [CrossRef]
24. Adazabra, A.N.; Viruthagiri, G.; Foli, B.Y. Evaluating the Technological Properties of Fired Clay Bricks Incorporated with Palm Kernel Shell. *J. Build. Eng.* **2023**, *72*, 106673. [CrossRef]
25. Zhao, J.; Wang, W.; Luan, Z.; He, M. First-Principles Analysis on Phase Transition, Atomic, Electronic, and Mechanical Properties of Kaolinite under Pressures. *Phys. B Condens. Matter.* **2024**, *674*, 415554. [CrossRef]
26. Pacheco, G.R.C.; Gonçalves, G.E.; Lins, V. de F.C. Design of Magnesia–Spinel Bricks for Improved Coating Adherence in Cement Rotary Kilns. *Ceramics* **2021**, *4*, 652–666. [CrossRef]

ceramics

MDPI

Article

Enhancing the Thermal Insulation Properties of Clay Materials Using Coffee Grounds and Expanded Perlite Waste: A Sustainable Approach to Masonry Applications

Ioannis Makrygiannis [1,2,*], Konstantinos Karalis [3] and Ploutarchos Tzampoglou [4]

1 SABO Clay Laboratory & Research Department, SABO S.A., Vassiliko–Chalkida, 34002 Evia, Greece
2 School of Mining Engineering and Metallurgy, National Technical University of Athens, Zografou Campus, 15780 Athens, Greece
3 Institute of Geological Sciences, University of Bern, CH-3012 Bern, Switzerland; konstantinos.karalis@geo.unibe.ch
4 Department of Civil & Environmental Engineering, University of Cyprus, Nicosia 1678, Cyprus; tzampogloup@gmail.com
* Correspondence: ymakrigiannis@sabo.gr

Abstract: The development of energy-efficient and sustainable construction materials is essential for reducing environmental impact and enhancing building performance. This study investigates the incorporation of coffee grounds and expanded perlite waste—two underutilized industrial byproducts—into clay-based ceramics to improve thermal insulation while maintaining mechanical integrity. Unlike previous studies that explore these additives separately or in impractically high dosages, this research focuses on their combined effect at low, industrially viable ratios to ensure large-scale feasibility. Four clay mixtures were analyzed: a reference clay (TZ), clay with coffee grounds (TZCF), clay with expanded perlite waste (TZPW), and clay with both additives (TZCFPW). Laboratory testing and computational fluid dynamics (CFD) simulations were employed to assess the physical, mechanical, and thermal properties of these formulations. The results indicated that coffee grounds increased plasticity, while expanded perlite waste reduced it, requiring adjustments in processing parameters. Both additives contributed to lower shrinkage and drying sensitivity, improving dimensional stability during production. Although mechanical strength declined due to increased porosity—most notably in the TZPW mixture—the fired bending strength remained within acceptable limits for masonry applications. The most significant finding was the substantial improvement in thermal performance, with all the modified formulations exhibiting reduced thermal conductivity and enhanced insulation. The best performance was observed in the TZPW mixture, which demonstrated the lowest thermal conductivity, highest thermal resistance, and optimal U-values in masonry wall testing, confirming its potential for energy-efficient construction. CFD simulations further validated these enhancements, providing detailed insights into heat transfer mechanisms. These findings demonstrate the feasibility of repurposing industrial waste materials to create scalable, eco-friendly building products. Future research should refine formulation ratios to optimize the balance between strength and insulation, ensuring widespread adoption in sustainable construction.

Keywords: coffee waste; expanded perlite waste; thermal properties; clay bricks; environment

1. Introduction

The demand for energy-efficient building materials has intensified in response to growing environmental concerns, rising energy costs, and the urgent need for sustainable waste management solutions. The construction industry, as a major contributor to resource consumption and CO_2 emissions, must adopt innovative approaches to mitigate its environmental impact [1]. Among the various strategies for sustainable construction, integrating industrial byproducts into building materials has gained significant attention, as it simultaneously reduces waste accumulation and enhances the functional properties of masonry products [2].

The brick and tile industry, with its large-scale production and high energy demands, presents an ideal opportunity for incorporating alternative materials into clay formulations. Modern brick factories typically produce at least 700 tons of fired bricks per day, requiring substantial amounts of raw materials. Therefore, identifying viable, locally available waste additives that can be integrated into large-scale production is crucial for achieving sustainability without compromising efficiency [3]. One key factor in this pursuit is reducing the thermal conductivity of bricks to enhance energy performance in buildings. By modifying the ceramic microstructure through lightweight, pore-forming additives, it is possible to improve insulation while maintaining mechanical integrity [4].

Coffee grounds and expanded perlite waste are two promising materials for this purpose. Coffee grounds, widely available in Mediterranean and coffee-consuming regions, are often discarded as waste despite their natural porosity and potential for enhancing insulation [5]. While prior studies have explored their use in brick production, the high replacement levels examined in some cases are impractical for industrial applications due to the sheer volume required to sustain daily production demands. Additionally, expanded perlite waste—an underutilized byproduct generated during the thermal expansion of perlite—is a fine, lightweight material with excellent insulating properties. Unlike standard expanded perlite, which is costly, expanded perlite waste is an industrial residue that is typically discarded and can be sourced at little to no cost. Its successful integration into ceramic bodies would provide a sustainable solution by transforming an otherwise unusable byproduct into a valuable resource for construction [6].

Previous research has demonstrated the effectiveness of lightweight organic additives in modifying ceramic properties. Studies incorporating olive stone residue and paper mill sludge have shown improved insulation and optimized porosity in clay bricks [7]. However, limited research has been conducted on the combined use of coffee grounds and expanded perlite waste, particularly in low percentages suitable for industrial applications. The novelty of this study lies in the strategic blending of these two waste materials at controlled ratios to achieve optimal thermal insulation without severely compromising mechanical strength. Unlike prior studies, which often focus on high additive content, this research seeks to develop a practical formulation that aligns with large-scale brick production requirements while maintaining environmental benefits [8].

To evaluate these effects, four distinct clay mixtures were developed and tested: TZ (reference clay), TZCF (clay with coffee grounds), TZPW (clay with perlite waste), and TZCFPW (clay with both additives). The study systematically examines key properties, including the plasticity, shrinkage, drying sensitivity, bending strength, water absorption, thermal conductivity, and density of fired bricks. Unlike previous works that primarily focus on laboratory-scale properties, this research also evaluates the real-world application by assessing the thermal resistance (R-value) and thermal transmittance (U-value) of masonry walls constructed with these materials [9]. Furthermore, computational fluid dynamics (CFD) simulations were conducted to provide a more detailed analysis of heat transfer and energy efficiency within the masonry walls, offering deeper insights into their performance in practical construction scenarios.

By integrating coffee grounds and expanded perlite waste into clay formulations, this study presents a novel and practical approach to improving the energy efficiency of masonry materials. The findings contribute to the advancement of eco-friendly construction by balancing mechanical strength with enhanced thermal insulation, offering a viable solution for reducing energy consumption in buildings [10]. This research not only minimizes reliance on virgin raw materials but also introduces a scalable and industry-compatible method for sustainable brick production, aligning with global efforts toward energy-efficient and environmentally responsible construction practices.

2. Materials

2.1. Brick Products

Vertically perforated bricks are a type of clay brick characterized by one or more perforations running vertically through their center, extending from the top to the bottom of the brick. Unlike hollow bricks with horizontal perforations, these bricks feature circular or rectangular openings that run in a vertical orientation. The vertical perforations enhance the brick's surface area, facilitating improved air circulation and providing better thermal insulation [11]. This design leverages natural convection, where hot air rises and cold air sinks, to create air currents within the walls. As a result, vertically perforated bricks help maintain indoor temperatures, keeping interiors cooler in summer and warmer in winter. These bricks come in various sizes and shapes, including standard rectangular forms, corner pieces, and interlocking designs, making them suitable for a range of applications, such as load-bearing and non-load-bearing walls or for decorative purposes. For this study, SABO S.A. provided a design for a new vertical clay block introduced to the Greek market, as can be seen in Figure 1. The brick features dimensions of 240 mm in height, 300 mm in width, and 250 mm in length, and its performance was analyzed as part of the investigation.

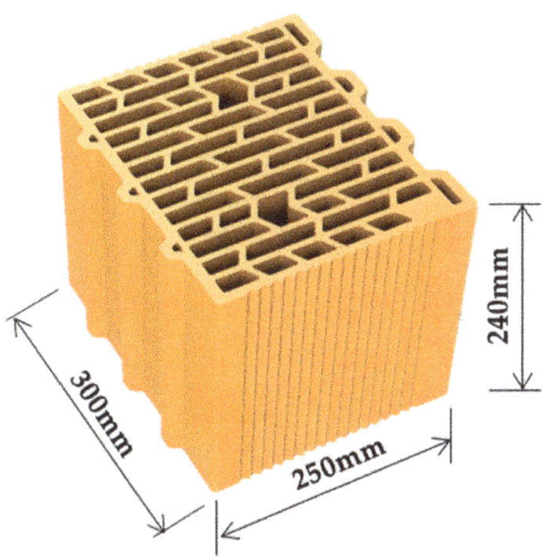

Figure 1. Vertically perforated brick in the Greek market.

2.2. Construction Mixture Components

a. Characteristics of the Clay Material

The study utilized a type of clay commonly used in industrial brick production in the Evia region, referred to as TZ clay, as seen in Figure 2a. This clay is classified as inorganic with moderate plasticity according to ISO 14688-2:2017, as its liquid limit falls within the range of 35–50% [12]. The key physical properties of TZ clay include its plastic limit, liquid limit, and plasticity index. The clay's moderate plasticity makes it suitable for applications in brick and tile manufacturing [13].

(**a**) (**b**) (**c**)

Figure 2. Materials used for the testing procedure: (**a**) TZ clay material after grinding; (**b**) coffee grounds residue; (**c**) expanded perlite residue.

The chemical composition of TZ clay was determined using atomic absorption spectrometry (AAS) in compliance with ISO 26845:2016 [14]. The analysis revealed a silica-rich composition, with SiO_2 as the predominant oxide at 56.45%, followed by notable levels of Al_2O_3 (16.72%), CaO (5.65%), and Fe_2O_3 (5.08%). Minor oxides such as MgO, K_2O, and Na_2O were also present, along with a loss on ignition (LOI) of 9.34%. This composition underscores the clay's suitability for thermal processing in ceramic applications [15].

Particle size analysis was performed in accordance with ASTM D422-63 (2007), showing a wide range of particle sizes, from 2 mm to 2 μm [16]. The results reveal that the clay consists of 44.28% clay-sized particles (<2 μm), 40.37% silt (20–2 μm), 9.35% fine sand (63–20 μm), and 6% coarse sand (>63 μm). This distribution indicates a high proportion of fine particles, which contributes to the clay's workability and plasticity.

The clay was transported from the manufacturing site to the laboratory with an average humidity content of 8.15%, ensuring it was in a semi-moist state suitable for further processing. The average density of the clay was measured at 1781 kg/m^3, following the ASTM D698-12 standard [17]. This density value reflects the compact nature of the material and its potential for forming dense ceramic products after firing [18].

The combined data on physical properties, chemical composition, and particle size distribution provide a comprehensive characterization of TZ clay. These properties highlight its suitability for the industrial-scale production of bricks and tiles, particularly in achieving the desired plasticity and durability during the forming and firing processes. The consistency of the material's composition and properties further ensures its reliability in manufacturing applications [19].

b. Characteristics of the Solid Waste Additive

i. Coffee grounds

Used coffee grounds, a byproduct of the coffee brewing process, present a promising sustainable additive for ceramic masses. Rich in organic compounds, this residue can act as a pore-forming agent during the firing process, as the organic material burns out, leaving behind a porous structure. Incorporating used coffee grounds into ceramic formulations not only enhances the material's thermal insulation properties but also promotes waste valorization by repurposing an abundant organic residue [20]. This approach aligns with sustainable material development practices, reducing reliance on virgin raw materials while contributing to improved energy efficiency in ceramic products.

The coffee grounds residue used in this study had a particle size of less than 0.25 mm and an initial humidity of 61% and can be seen in Figure 2b. With a calorific value of 4300 kcal/kg at a water content of 10%, this organic residue demonstrates significant energy potential. Its fine particle size and high organic content make it suitable as a pore-forming additive in ceramic masses, where it burns out during firing to create a porous structure, enhancing the thermal insulation properties of the final product.

ii. Expanded perlite residue

Expanded perlite residue, commonly recognized as the fine white dust generated during the thermal expansion of perlite, is an underutilized byproduct of the perlite manufacturing process. This residue consists of lightweight, highly porous particles with excellent thermal and insulating properties, similar to those of expanded perlite but in a finer form [21]. Traditionally considered waste, it poses disposal challenges while representing a potential resource for enhancing the thermal insulation of composite materials. By incorporating expanded perlite residue into clay-based mixtures, it is possible to develop sustainable, energy-efficient building materials that leverage the insulating properties of perlite while reducing the environmental footprint associated with industrial waste.

The expanded perlite residue utilized in the testing procedure exhibited a grain size of less than 0.25 mm and a loose bulk density of 60 kg/m^3, as seen in Figure 2c. These fine particles are derived from the thermal expansion process of raw perlite and possess distinct physical and chemical characteristics. The chemical composition of the perlite is 76.0% SiO_2, 13.2% Al_2O_3, 1.0% CaO, 1.2% Fe_2O_3, 0.5% MgO, 2.5% K_2O, and 0.5% Na_2O, and it has a loss on ignition (LOI) of 4.0%. This composition highlights the silica-rich nature of perlite, which contributes to its favorable thermal properties and lightweight structure.

The physical properties of perlite further enhance its suitability for thermal applications. Expanded perlite appears as white or gray odorless particles, with a bulk density of 575 kg/m^3 and a specific gravity of 2.2 g/cm^3. It exhibits a maximum moisture content of 1%, a neutral pH of 7.5, and a softening point ranging from 870 °C to 1000 °C. Additionally, the material contains 49.22% active silica, which influences its performance in insulation and composite applications. These properties underline the versatility of expanded perlite residue as a functional additive for improving the thermal efficiency of composite materials.

Detailed microphotographs of the materials used in the tests are presented in Figure 3. These visual representations offer a deeper insight into the ceramic mass, highlighting its microscopic characteristics and further enriching the study's analysis.

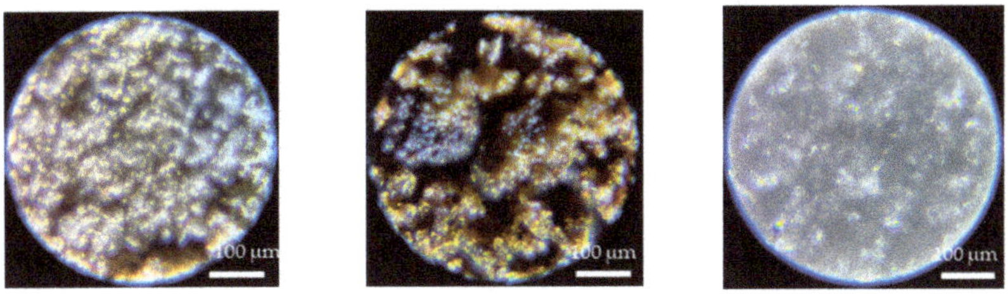

(a) (b) (c)

Figure 3. (**a**) Surface of the TZ material; (**b**) surface of the coffee grounds residue; (**c**) surface of the expanded perlite residue.

3. Methods

The clay material (referred to as TZ) was procured and processed in the laboratory [22]. Initially, the raw clay was ground and subjected to a semi-wet preparation process. The prepared clay was then extruded into solid specimens, which underwent a drying process using an eight-step program over a 24 h period. After drying, the specimens were fired at a peak temperature of 900 °C utilizing a controlled 10-step firing program. These initial firing results were used as the baseline for comparison.

The eight-step heating program follows a controlled temperature increase from 40 °C to 80 °C in sequential stages. The process starts at 40 °C and gradually rises through intermediate steps at 44 °C, 50 °C, 55 °C, 60 °C, and 70 °C before reaching 75 °C and finally 80°C. Each step allows for sufficient heat distribution and structural adjustments, minimizing thermal stress and ensuring a stable transformation of the material.

The 10-step heating program begins at 20 °C and progressively increases to 900 °C before undergoing a controlled cooling phase. The temperature rises through key stages at 150 °C, 300 °C, 480 °C, and 700 °C and finally reaches 900 °C, ensuring optimal material processing. A stabilization period occurs at 900 °C before the cooling phase starts, gradually lowering the temperature through 650 °C and 400 °C before returning to 20 °C. This controlled approach ensures uniform heating, reduces thermal shock, and enhances the material's final properties.

Three modified clay mixtures were then constructed, each incorporating specific additives, as seen in Table 1.

Table 1. Mixtures constructed during the study.

Materials	TZ	TZCF	TZPW	TZCFPW
Clay material	100%	97%	97%	97%
Expanded perlite waste	-	3%	-	1.5%
Coffee grounds waste	-	-	3%	1.5%

The same preparation, drying, and firing procedures were applied to these mixtures as the baseline samples to maintain consistency. All testing and analysis were performed according to the industrial standards for brick and tile production.

3.1. Bulk Density and Thermal Conductivity Coefficient Measurement

The bulk density of the fired samples was determined using the Archimedes method, in accordance with the ASTM C373-14a standard [23]. The thermal conductivity (λ10, dry, mat) of the fired clay samples was then calculated in accordance with EN1745:2012 [24]. The thermal conductivity values were used to calculate the thermal transmittance (U-value) and thermal coefficient of the brick samples. These thermal properties were applied in a model to simulate the heat transfer rate through a building wall. This simulation aimed to evaluate the energy efficiency and thermal comfort implications of using these modified bricks in construction. The calculated U-values and thermal coefficients were also used to assess the thermal performance of a building wall, providing insight into the potential application of these materials and their compliance with thermal performance standards.

The study aimed to explore the feasibility of incorporating coffee grounds and expanded perlite waste into brick and tile production. By assessing both the physical and thermal properties of the materials, the research sought to determine their suitability for improving energy efficiency and sustainability in construction.

3.2. Preparation

In the initial phase of preparing each composite formulation, the moisture content of the clay and additives was determined. Moisture content represents the percentage of water in a material, expressed as a percentage of the wet material's weight. To determine the moisture content, the sample was placed in a laboratory vessel and dried in an electric dryer (Ceramic Instruments, Fiorano Modenese, Italy, model SCN/400/DG) at 105 °C for

24 h. Once a constant weight was achieved, the sample was reweighed, and the weight of the laboratory vessel was subtracted from the total.

Before mixing, the materials underwent pre-crushing using a jaw hammer (Ceramic Instruments, Fiorano Modenese, Italy, model A92) with a jaw opening of 2 mm. The crushed materials were then passed through a laboratory roller mill (Verdes, Barcelona, Spain, 080 model) with an adjustable separation of 1.2 mm. Paper sludge was combined with the required extruding water and mixed for 24 h. The clay material was weighed according to the mixing ratio, considering the moisture content, and homogenized in a kneading mixer with the necessary preparation water. The amount of water added was adjusted until the desired plasticity index was achieved (between 0.7 and 0.85 Pfefferkorn), as determined by Pfefferkorn's test.

The Pfefferkorn plasticity method involves evaluating the deformation of a sample by the fall of a calibrated plate on the test specimen [25]. The deformation is measured in millimeters, with one scale for deformation and another for calculating the test specimen's deformation using Pfefferkorn's theory. For this study, the Ceramic Instruments, Fiorano Modenese, Italy, 01CI4540 Pfefferkorn plasticity tester was used, and the calculation method followed the standard procedure. The amount of water added during the pre-mixture process depended on the absorptivity of the clay and the extrusion process required for the final product.

3.3. Extrusion

The homogenized mixture was extruded through a vacuum extruder to form rectangular samples with standard dimensions. The extrusion was carried out using the HANDLE, Mühlacker, Germany, KHS-Type: PZVM8b model extruder. The wet mixture was placed in the feeding chamber, where it passed through a pre-extruder mixer and an air vacuum chamber before reaching the extruder's output. The pressure was continuously monitored with a gauge. The extruder had interchangeable molds for producing the desired shape and size. The vacuum pressure was consistently maintained at 0.8 kp/cm^2 during the extrusion process. The final dimensions of the extruded samples were 120 mm $\times$ 20 mm $\times$ 20 mm (L $\times$ W $\times$ H). A total of thirty samples were produced, with fifteen samples per mixture [26].

3.4. Drying

The extruded samples were marked and subjected to a controlled drying process in an electric oven (Ceramic Instruments, Fiorano Modenese, Italy, model SCN/400/DG). The drying procedure consisted of three distinct phases: the humidity phase, the shrinkage critical point phase, and the final drying phase, each requiring specific attention to avoid issues such as cracks or deformation [27].

Humidity phase: High ambient humidity was maintained in the dryer to keep the surface pores of the bricks open. This phase was the most critical, as premature surface cracking or fragility could occur if not properly managed.

Shrinkage critical point phase: During this phase, the drying shrinkage should be complete before the temperature rises. The temperature increase was gradual to prevent cracking [28].

Final drying phase: The goal was to reduce the remaining moisture as much as possible, and all the drying regulations were adjusted to suit the mixture and its behavior.

The drying process aimed to maintain open surface pores in the beginning to allow the moisture to evaporate from the interior of the bricks. This was essential to prevent cracks or structural fragility in the second phase, where the temperature increases and the humidity decreases [29].

The drying sensitivity of the formulated mixtures was evaluated using Bigot's curve method. This assessment involved subjecting test samples to controlled conditions within a laboratory dryer, maintaining a constant temperature of 25 °C and a stable humidity level of 75%. Bigot's curve, a graphical representation, depicted the variation in water content in relation to linear shrinkage over a 24 h period.

According to Bigot's curve method, the drying process occurs in two distinct stages. The first stage is marked by a consistently high drying rate, as indicated by a strong linear correlation. This rapid moisture loss continues until a critical point is reached. In the subsequent stage, the drying rate progressively declines, and shrinkage slows down as the process nears completion. During the experiments, the humidity of the drying air was carefully tracked. A humidity sensor (TMI Orion SA, Castelnau-le-Lez, France, CeriDry) was used to measure the relative humidity inside the dryer at five-minute intervals, following the detailed measurement protocols established by Makrygiannis and Karalis. The drying sensitivity level was assessed using the CSB index, which serves as an indicator of drying sensitivity based on Bigot's methodology. The categorization of the sensitivity levels is provided in Table 2.

Table 2. Classification of drying sensitivity according to Bigot's CSB index.

Classification of CSB	
<1.0	Insensitive
1.0–1.5	Medium sensitive
1.5–2.0	Sensitive
>2.0	Highly sensitive

3.5. Firing

The firing process was carried out using an electric gradient kiln (Nabertherm GmbH, Lilienthal, Germany, model GR1300/13), which was computer-controlled and followed a ten-step time-duration program. The firing process was designed to ensure the appropriate preheating, maximum temperature firing, and cooling of the samples. The samples were heated to a peak temperature of 900 °C, a typical firing temperature for building bricks and tiles, and held for three hours. The preheating and cooling phases were carefully managed, particularly near 573 °C, to avoid issues from quartz inversion.

The temperature increased at a rate of 0.7 to 1.16 °C/min, depending on the firing zone and the programmed procedure. The total firing cycle, from cold to cold, lasted 24 h. Throughout the firing process, the mixtures showed no issues, and all the samples were fully dried before being loaded into the kiln. The samples were dried at 105 °C for 24 h in a dryer before firing to ensure complete moisture removal.

3.6. Thermal Performance of the Fired Clay Bricks

The thermal behavior of clay bricks can be analyzed using two different approaches: computational fluid dynamics (CFD) simulations or a simplified method based on standards such as EN ISO 10211:2007 [30], and EN ISO 6946 [31]. In CFD modeling, the governing partial differential equations are transformed into algebraic equations through discretization, allowing them to be solved using direct or iterative numerical solvers [32].

For fluid motion, the conservation of mass is represented by the following continuity equation:

$$\frac{\vartheta}{\vartheta t}\rho + \nabla\cdot(\rho u) = 0 \quad (1)$$

while the momentum balance equation, accounting for fluid density variations, is given by the following:

$$\frac{\vartheta}{\vartheta t}(\rho u) + \nabla \cdot (\rho u u) = -\nabla P + \nabla \cdot (\mu \nabla u) + \rho g \quad (2)$$

where u is the velocity vector field (m s^{-1}), ρ is the fluid density (kg m^{-3}), ∇ is the nabla operator, representing spatial differential operations such as gradient (∇u) and divergence (∇·A), P is the pressure (Pa), and μ is the dynamic viscosity (Pa s).

The temperature distribution in the system is governed by the following energy conservation equation:

$$\rho C_p \frac{\vartheta T}{\vartheta t} + (\rho C_p u \cdot \nabla) T = \nabla \cdot (k \nabla \mathrm{T}) \quad (3)$$

where T is the temperature (K), Cp is the heat capacity (J kg^{-1} K^{-1}), and k is the thermal conductivity (W/m·K).

Additionally, radiative heat transfer is expressed through the following radiative heat flux equation:

$$Q = \sigma \left(T_{max}^4 - T_{min}^4 \right) \quad (4)$$

where T is the temperature (K) and σ is the Stefan–Boltzmann constant.

The second approach, based on standardized methods, simplifies the analysis by determining the effective thermal properties of air cavities. This is particularly complex when dealing with non-rectangular cavities. To address this challenge, irregularly shaped cavities are first transformed into equivalent rectangular ones, after which their thermal conductivity is calculated using the convective heat transfer coefficient (ha) and the radiative heat transfer coefficient (h).

The thermal properties of clay and air, considered as homogeneous continua, are listed in Table 3. Figure 4 illustrates the tested ceramic product and the respective domains for all four tested mixtures. The ceramic product is thermally insulated at both the top and bottom.

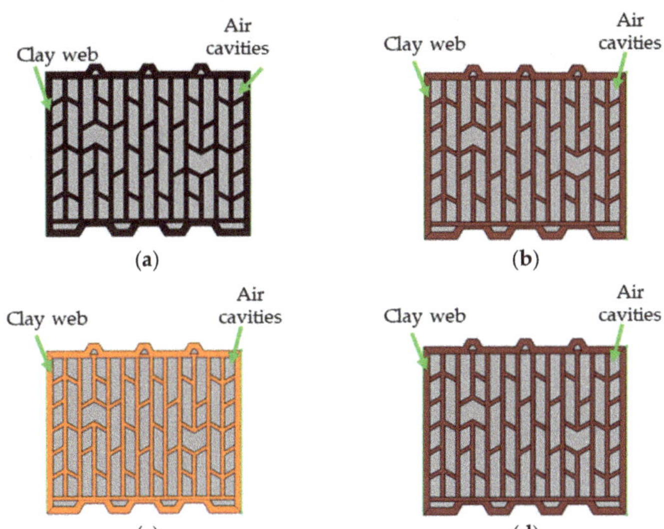

Figure 4. Domains of the vertically perforated bricks: (**a**) TZ clay, (**b**) TZCF mixture, (**c**) TZCFPW mixture, (**d**) TZPW mixture.

Table 3. Properties of the materials used in the computations.

Properties	Porous Clay	Air
Density TZ (kg/m^3)	1720	1.23
Density TZCF (kg/m^3)	1641	1.23
Density TZPW (kg/m^3)	1612	1.23
Density TZCFPW (kg/m^3)	1625	1.23
Heat Capacity (J/(kg·K))	1000	1008
Thermal Conductivity TZ (W/(m·K))	0.52	Geometry and Temperature Dependent
Thermal Conductivity TZCF (W/(m·K))	0.49	Geometry and Temperature Dependent
Thermal Conductivity TZPW (W/(m·K))	0.47	Geometry and Temperature Dependent
Thermal Conductivity TZCFPW (W/(m·K))	0.48	Geometry and Temperature Dependent

3.7. Masonry Build Performance

The thermal performance of a masonry wall is determined by calculating its overall heat transfer coefficient (U), which depends on the thermal resistance of the external and internal surfaces as well as the plaster and brick layers. The values of the external (R_{se}) and internal (R_{si}) surface resistance vary based on wall construction and environmental conditions, while the thickness (t) and thermal conductivity (λ) of the plaster and brick layers are essential for determining U. Once calculated, the U-value quantifies the rate of heat loss through the wall, which can then be used to estimate heat loss by considering the wall's area and the temperature difference between the indoor and outdoor environments. This analysis helps assess a building's energy efficiency and identify potential areas for improvement. Enhancing thermal performance can be achieved by optimizing the thickness and material composition of the plaster and brick layers as well as reducing thermal resistance at the wall's internal and external surfaces, as seen in Table 4, ultimately improving the overall energy efficiency.

Table 4. Data and materials used in the computations.

Layer	Thickness (mm)	Thermal Conductivity (W/m·K)	Thermal Resistance (m^2·K/W)
R_{si}			0.13
Plaster	20	0.87	0.023
Brick	Depends on the dimensions and the calculations through CFD of the bricks		
Plaster	20	0.87	0.023
R_{se}			0.04

R_{si}: refers to the thermal resistance of the interior surface of the wall, which includes all the materials on the inner side of the insulation layer. This includes items such as drywall, plaster, and paint. R_{se}: refers to the thermal resistance of the exterior surface of the wall, which includes all the materials on the outer side of the insulation layer. This includes items such as cladding, air cavities, and weather-resistant barriers. Plaster: A typical building material used for coating walls and ceilings.

4. Results

The results for the baseline clay (TZ) and modified clay mixtures (TZCF, TZPW, and TZCFPW) are summarized in Table 5 and analyzed based on key physical and thermal properties.

Table 5. Mixture proportions and gathered results.

		TZ	TZCF	TZPW	TZCFPW
Plasticity by Pfefferkorn		0.79	0.82	0.70	0.73
Added initial water	Wt.-%	17.44	21.28	22.06	21.94
Linear dry shrinkage	%	5.07	4.27	4.98	4.79
Drying sensitivity (Bigot's curve)	CSB	1.32	1.11	1.04	1.08
Bending strength/dry	Kg/cm^2	117.40	93.78	59.92	70.34
Peak firing temperature	°C	900	900	900	900
Bending strength/fired	Kg/cm^2	156.00	147.14	137.16	140.91
Water absorption	%	9.45	10.97	11.57	11.09
Body density of fired material	Kg/m^3	1720	1641	1612	1625
Thermal coductivity coefficient	W/m·K	0.52	0.49	0.47	0.48

4.1. Water Content and Plasticity

The plasticity index, measured using the Pfefferkorn method, showed variations across the mixtures. The baseline clay (TZ) exhibited a plasticity index of 0.79, while the addition of coffee grounds (TZCF) improved the plasticity to 0.82. Conversely, the addition of expanded perlite waste (TZPW) and the combination of both additives (TZCFPW) resulted in reduced plasticity values of 0.70 and 0.73, respectively. This indicates that coffee grounds contribute positively to the plasticity of the clay mixture, while expanded perlite reduces it [33].

The initial water added during preparation increased with the addition of additives. The baseline clay (TZ) required 17.44 wt.% water, whereas TZCF, TZPW, and TZCFPW required 21.28 wt.%, 22.06 wt.%, and 21.94 wt.%, respectively. This trend reflects the increased absorptivity of the additives, particularly expanded perlite. Both plasticity and water content can be seen in Figure 5 for all the constructed mixtures [34].

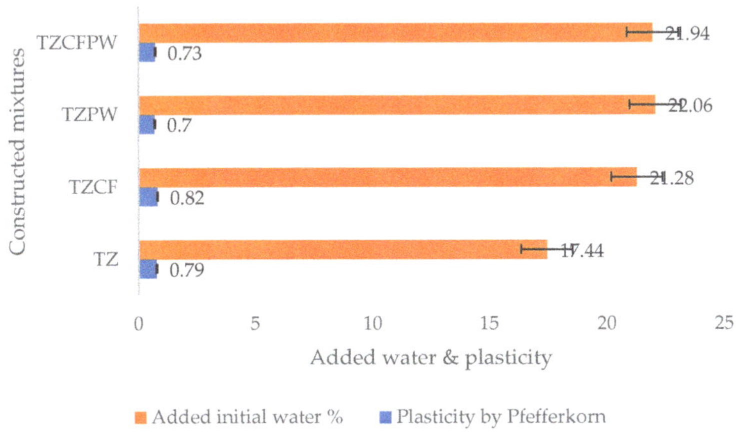

Figure 5. Plasticity by Pfefferkorn and necessary extrusion water for all mixtures.

4.2. Drying Sensitivity and Linear Shrinkage

Linear drying shrinkage was highest for the baseline clay (5.07%) and decreased in the modified mixtures. TZCF showed the lowest shrinkage (4.27%), followed by TZCFPW (4.79%) and TZPW (4.98%). The lower shrinkage values indicate that the additives help mitigate deformation during drying, particularly coffee grounds [35].

Drying sensitivity, represented by the Bigot's curve shrinkage coefficient (CSB), was lowest for TZPW (1.04) and TZCFPW (1.08) compared to the baseline clay (1.32) and TZCF (1.11). This suggests that the additives, especially expanded perlite waste, reduce the risk of cracking during drying by stabilizing the mixture's structure [36]. The results for both drying sensitivity and shrinkage can be seen in Figure 6.

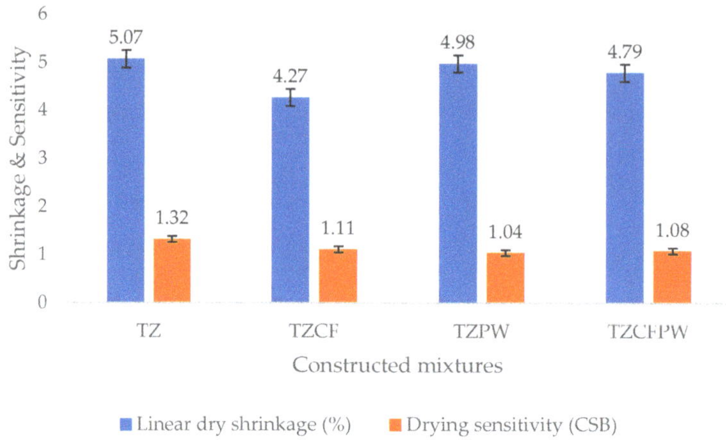

Figure 6. Linear drying shrinkage and drying sensitivity according to Bigot's curve for all mixtures.

4.3. Bending Strength

The dry bending strength of the baseline clay was the highest (117.40 kg/cm^2), while the modified mixtures showed reduced dry bending strength, with values of 93.78 kg/cm^2 (TZCF), 59.92 kg/cm^2 (TZPW), and 70.34 kg/cm^2 (TZCFPW). After firing, the bending strength of all the mixtures improved, with the baseline clay again exhibiting the highest value (156 kg/cm^2), as can be seen in Figure 7 [37]. The fired bending strengths of the modified mixtures were slightly lower, at 147.14 kg/cm^2 (TZCF), 137.16 kg/cm^2 (TZPW), and 140.91 kg/cm^2 (TZCFPW).

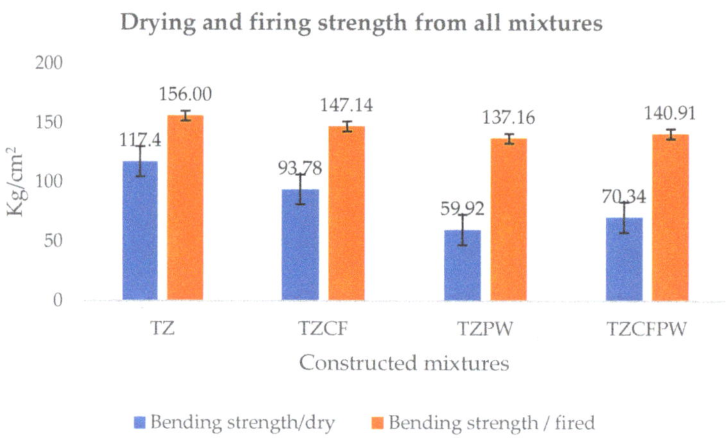

Figure 7. Drying and firing strength for all mixtures.

This decrease in strength can be attributed to the reduced density and increased porosity caused by the additives, which introduce voids during firing [38].

4.4. Density and Water Absorption

The body density of the fired material decreased across the modified mixtures, with the baseline clay having the highest density (1720 kg/m^3). TZCF, TZPW, and TZCFPW exhibited densities of 1641 kg/m^3, 1612 kg/m^3, and 1625 kg/m^3, respectively. The reduction in density reflects the introduction of lightweight additives, which increase porosity.

Similarly, water absorption increased in the modified mixtures, with TZPW showing the highest absorption (11.57%) compared to TZ (9.45%). TZCF and TZCFPW displayed intermediate values of 10.97% and 11.09%, respectively. These results can be seen in Figure 8 and align with the increased porosity observed in the modified mixtures.

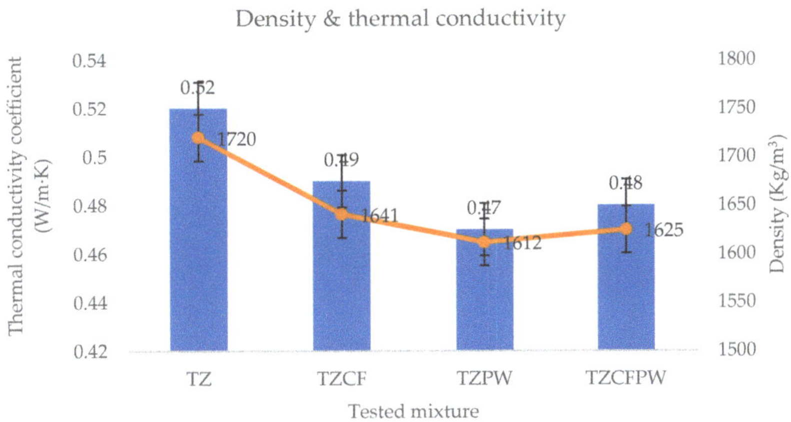

Figure 8. Density and thermal conductivity coefficient for all mixtures.

4.5. Thermal Performance of the Bricks

The results as measured from the CFD software and seen in Figures 9 and 10 indicate that modifying clay with additives such as coffee grounds and expanded perlite waste significantly enhances its thermal insulation properties. The baseline clay exhibited the highest thermal conductivity at 0.52 W/m·K, meaning it allowed more heat to pass through than the modified mixtures. In contrast, the addition of alternative materials reduced the thermal conductivity, with the lowest value recorded for the TZPW mixture at 0.47 W/m·K. This decrease suggests that the modified materials introduce greater porosity into the clay matrix, reducing its ability to conduct heat and improving its insulating performance.

Among the modified mixtures, TZCF, which contains coffee grounds, showed a thermal conductivity of 0.49 W/m·K, demonstrating a moderate improvement over the baseline clay. The TZCFPW mixture, which incorporates both coffee grounds and expanded perlite waste, exhibited an even lower thermal conductivity of 0.48 W/m·K. However, the best insulation performance was observed in the TZPW mixture, which contains only expanded perlite waste, achieving the lowest thermal conductivity of 0.47 W/m·K. This suggests that expanded perlite waste plays a crucial role in enhancing insulation by increasing the number of air pockets within the material.

The results that can be seen in Table 6 and Figure 11 also highlight how these modifications influence the overall thermal behavior of the bricks and walls. The brick thermal resistance (R_{brick}) increased from 1.8119 m^2·K/W in the baseline clay to 1.8971 m^2·K/W in TZPW, indicating a greater ability to resist heat flow. Correspondingly, the overall heat transfer coefficient of the brick (U_{brick}) decreased, further confirming the improved insulation capacity. Lower U-values suggest that less heat is transferred through the material, which is beneficial for maintaining stable indoor temperatures and reducing energy consumption for heating and cooling [39].

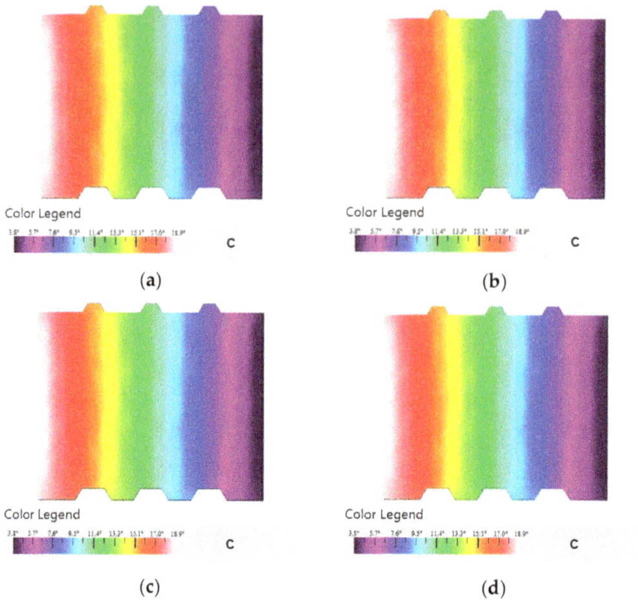

Figure 9. Vertically perforated bricks: (**a**) temperature distribution in degrees Celsius constructed from TZ clay, (**b**) temperature distribution in degrees Celsius constructed from TZCF mixture, (**c**) temperature distribution in degrees Celsius constructed from TZCF mixture, (**d**) temperature distribution in degrees Celsius constructed from TZCFPW mixture.

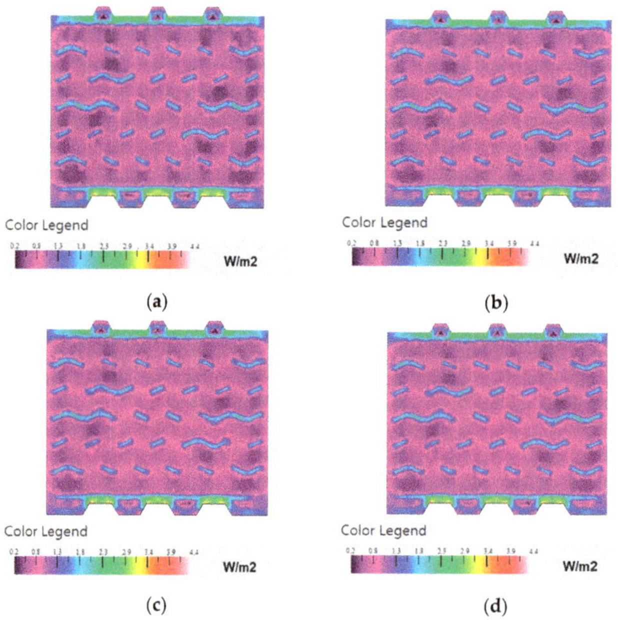

Figure 10. Vertically perforated bricks: (**a**) isothermal contour in degrees Celsius constructed from TZ clay, (**b**) isothermal contour in degrees Celsius constructed from TZCF mixture, (**c**) isothermal contour in degrees Celsius constructed from TZCFPW mixture, (**d**) isothermal contour in degrees Celsius constructed from TZPW mixture.

Table 6. Thermal performance of the constructed mixtures in the tested bricks.

Mixtures	U_{brick} $W/m^2{\cdot}K$	R_{brick} $m^2{\cdot}K/W$	λ_{brick} $W/m{\cdot}K$
TZ	0.5519	1.8119	0.182
TZCF	0.5372	1.8617	0.177
TZPW	0.5271	1.8971	0.174
TZCFPW	0.5322	1.8791	0.176

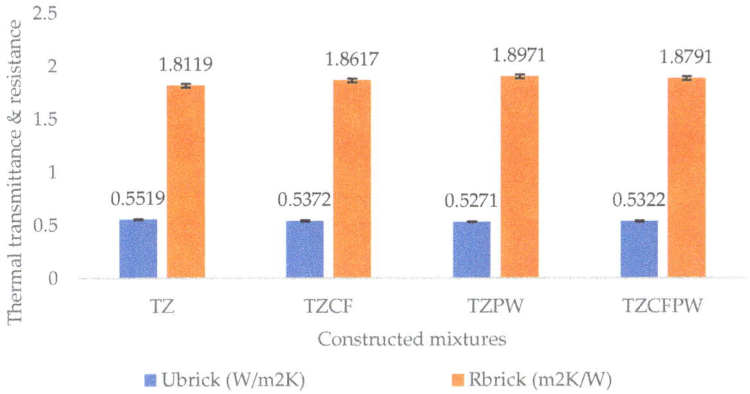

Figure 11. Thermal transmittance (U_{brick}) and thermal resistance (R_{brick}) of the tested brick for the constructed mixtures.

4.6. Masonry Built Performance

The thermal performance of a masonry wall is determined using the thermal coefficients of the materials involved, such as the tested clay mixtures and plaster, along with the thermal resistance of the interior and exterior air films (R_{si} and R_{se}) [40]. The overall thermal resistance of the wall is calculated by summing the thermal resistance of each material layer, including the mortar joints and the constructed mixtures, as seen in Figure 12.

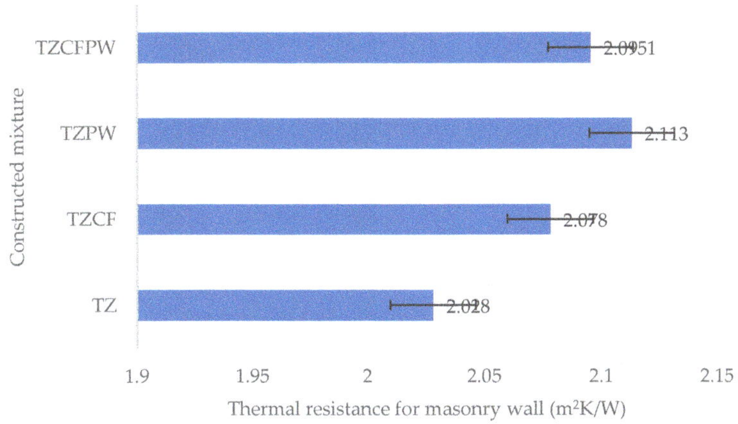

Figure 12. Thermal resistance (R_{wall}) of a wall constructed with bricks from the tested mixtures.

From this, the U-value of the wall is derived by dividing its thermal conductivity by its total thermal resistance. The U-value is crucial in assessing heat loss through the wall, which helps evaluate the building's energy efficiency [41]. The results of these calculations are shown above.

The overall thermal transmittance of the wall (U_{wall}), as can be seen in Tables 7–10, decreased from 0.5039 W/m^2·K in the baseline clay wall to 0.4847 W/m^2·K in the TZPW-modified wall. This indicates that using TZPW bricks can reduce heat loss through building envelopes, enhancing the energy efficiency of structures. The improved thermal resistance of the modified bricks suggests that incorporating waste materials like coffee grounds and expanded perlite waste can lead to sustainable and energy-efficient construction materials [42].

Table 7. Masonry built performance for vertically perforated bricks constructed from TZ clay.

Layer	Thickness (mm)	Thermal Conductivity (W/m·K)	Thermal Resistance (m^2·K/W)
R_{si}			0.13
Plaster	20	0.87	0.023
Vertically perforated brick	300	0.182	1.8119
Plaster	20	0.87	0.023
R_{se}			0.04
Total resistance:		**2.028**	m^2·K/W
U-Value		**0.493**	W/m^2·K

Table 8. Masonry built performance for vertically perforated bricks constructed from the TZCF mixture.

Layer	Thickness (mm)	Thermal Conductivity (W/m·K)	Thermal Resistance (m^2·K/W)
R_{si}			0.13
Plaster	20	0.87	0.023
Vertically perforated brick	300	0.177	1.8617
Plaster	20	0.87	0.023
R_{se}			0.04
Total resistance:		**2.078**	m^2·K/W
U-Value		**0.481**	W/m^2·K

Table 9. Masonry built performance for vertically perforated bricks constructed from the TZPW mixture.

Layer	Thickness (mm)	Thermal Conductivity (W/m·K)	Thermal Resistance (m^2·K/W)
R_{si}			0.13
Plaster	20	0.87	0.023
Vertically perforated brick	300	0.174	1.8971
Plaster	20	0.87	0.023
R_{se}			0.04
Total resistance:		**2.113**	m^2·K/W
U-Value		**0.473**	W/m^2·K

Table 10. Masonry built performance for vertically perforated bricks constructed from the TZCFPW mixture.

Layer	Thickness (mm)	Thermal Conductivity (W/m·K)	Thermal Resistance (m^2·K/W)
R_{si}			0.13
Plaster	20	0.87	0.023
Vertically perforated brick	300	0.176	1.8791
Plaster	20	0.87	0.023
R_{se}			0.04
Total resistance:		**2.0951**	m^2·K/W
U-Value		**0.477**	W/m^2·K

Overall, these findings demonstrate that adding coffee grounds and expanded perlite waste to clay bricks is an effective strategy for enhancing their thermal performance [43]. The introduction of porosity through these waste materials reduces thermal conductivity, increases thermal resistance, and lowers the heat transfer coefficient. These modifications not only improve the insulation properties of masonry walls but also contribute to sustainability by utilizing industrial byproducts and reducing waste. Future research could explore further optimization of these mixtures to balance insulation performance with mechanical strength and durability.

5. Discussion

The results of this study highlight the impact of incorporating coffee grounds and expanded perlite waste into clay mixtures on the physical, mechanical, and thermal properties of masonry materials. These modifications introduced notable variations in plasticity, shrinkage, strength, density, and insulation performance, demonstrating both the benefits and trade-offs associated with these waste-derived additives [44]. While the addition of these materials led to a reduction in mechanical strength, the significant improvement in thermal efficiency makes these modified mixtures highly suitable for energy-efficient masonry applications.

The plasticity of the mixtures varied depending on the type of additive used. Coffee grounds increased plasticity (0.82), attributed to their organic composition and ability to retain water, whereas expanded perlite waste reduced plasticity (0.70) due to its lightweight and highly porous nature, which limits its binding ability with clay particles. The increased initial water demand in all the modified mixtures further supports the idea that these additives enhance the absorptivity of the clay, necessitating additional water during preparation. However, this increased absorptivity did not lead to excessive shrinkage. Instead, both additives contributed to reduced linear drying shrinkage, with the most significant reduction observed in the TZCF mixture (4.27%). The lower shrinkage values indicate that these waste materials help stabilize the clay matrix, reducing deformation risks during the drying process.

In terms of mechanical performance, the introduction of coffee grounds and expanded perlite waste resulted in a decline in both dry and fired bending strengths. The baseline clay (TZ) exhibited the highest strength, while TZPW (expanded perlite waste) had the lowest. The dry bending strength decreased from 117.40 kg/cm^2 in TZ to 59.92 kg/cm^2 in TZPW, while the fired bending strength was highest in TZ (156.00 kg/cm^2) and lowest in TZPW (137.16 kg/cm^2). This reduction can be attributed to the increased porosity introduced by the additives, which weakens the structural integrity of the bricks. The lower strength of TZPW compared to the other mixtures is primarily due to its higher porosity, as the addition of expanded perlite waste increases the number of air voids within the ceramic matrix. These air pockets act as stress concentrators, reducing the overall density and making the material more susceptible to fracture under load. Despite this reduction in strength, the fired bending strength values remained within acceptable limits for non-load-bearing masonry applications, demonstrating that these materials remain viable for non-structural and thermally insulating uses [45].

One of the key advantages of incorporating coffee grounds and expanded perlite waste was the significant enhancement of thermal insulation properties. The decrease in body density across the modified mixtures directly correlated with improved thermal performance, as lower-density materials exhibit reduced thermal conductivity. The thermal conductivity coefficient decreased from 0.52 W/m·K in TZ to 0.47 W/m·K in TZPW, reinforcing the effectiveness of expanded perlite waste in enhancing insulation. This finding aligns with previous studies on lightweight ceramic materials, where increased porosity effectively reduced heat transfer. The improved thermal resistance of the modified bricks further supports their potential use in energy-efficient building designs, as they can contribute to reducing heating and cooling demands.

The analysis of masonry wall performance confirmed the practical benefits of these modifications. The U-values of the walls constructed with modified bricks were consistently lower than those of the baseline clay, indicating better insulation and reduced heat transfer. The TZPW wall exhibited the lowest U-value, highlighting the superior thermal performance of expanded perlite waste in improving energy efficiency. This enhanced insulation capability makes these walls particularly suitable for sustainable construction, especially in regions with extreme temperature variations, where reducing heat loss and maintaining indoor thermal comfort is essential.

From an economic and technical perspective, the utilization of waste materials such as coffee grounds and expanded perlite waste presents a cost-effective and sustainable alternative for the brick industry. Industrial-scale applications of these findings could be particularly relevant in brick manufacturing plants with capacities of 500 to 700 tons/day of fired bricks, where optimizing material formulations for energy-efficient products could lead to significant energy savings and a reduction in raw material consumption. Future research should focus on optimizing the balance between mechanical and thermal properties, potentially by exploring different ratios of waste additives or combining them with other reinforcements to enhance structural performance while maintaining superior insulation properties. Additionally, further investigation into large-scale production feasibility, durability assessments, and long-term environmental impact analyses will be crucial in ensuring the widespread adoption of these sustainable clay-based materials in the construction industry.

The analysis of masonry wall performance confirmed the practical benefits of these modifications. The U-values of the walls constructed with modified bricks were consistently lower than those of the baseline clay, indicating better insulation and reduced heat transfer. The TZPW wall exhibited the lowest U-value, reinforcing the superior thermal performance of expanded perlite waste in improving energy efficiency.

The computational fluid dynamics (CFD) simulations further validated these experimental findings by providing a detailed analysis of heat transfer and airflow within the masonry walls. The simulations demonstrated that the increased porosity resulting from the addition of coffee grounds and expanded perlite waste led to a more pronounced reduction in thermal conductivity. The air pockets formed within the ceramic matrix acted as insulating barriers, reducing the rate of heat transfer through the walls. Additionally, the CFD analysis highlighted a more uniform temperature distribution across the TZPW wall, minimizing the thermal bridging effects that are commonly observed in traditional clay bricks.

Moreover, the CFD results revealed that the modified bricks contributed to a slower heat accumulation and dissipation process, which is particularly beneficial in regions experiencing extreme temperature fluctuations. The ability to retain heat during colder periods and resist rapid heat penetration in warmer conditions enhances the overall energy efficiency of buildings. This effect is particularly evident in the TZPW walls, where the lower thermal diffusivity contributes to improved indoor thermal stability.

In summary, the integration of coffee grounds and expanded perlite waste into clay formulations offers a sustainable and effective method for enhancing the thermal insulation of masonry materials. While there is a trade-off in mechanical strength, the significant improvement in thermal performance makes these materials well suited to energy-efficient building applications. Future research should focus on optimizing the balance between mechanical and thermal properties, potentially by exploring different ratios of waste additives or combining them with other reinforcements to enhance structural performance while maintaining superior insulation properties. Furthermore, advanced CFD simulations with varying environmental conditions and real-world case studies could provide deeper insights into the long-term performance of these materials in diverse climatic settings.

6. Conclusions

This study investigated the effects of incorporating coffee grounds and expanded perlite waste into clay mixtures on the physical, mechanical, and thermal properties of masonry materials. The results demonstrated that while these additives influenced plasticity, water absorption, shrinkage, and strength, their most significant impact was on thermal insulation performance. The addition of coffee grounds increased the plasticity of the clay, whereas expanded perlite waste reduced it. Both additives contributed to lower linear drying shrinkage and drying sensitivity, indicating their stabilizing effect on the clay matrix, making the production process more efficient and reducing the risk of defects.

The mechanical strength of the modified mixtures was lower than that of the baseline clay, with expanded perlite waste contributing to the greatest reduction due to increased porosity. However, despite this decline, the fired bending strength values remained within acceptable ranges for non-load-bearing masonry applications, with TZCF, TZPW, and TZCFPW achieving bending strengths of approximately 147 kg/cm^2, 137 kg/cm^2, and 141 kg/cm^2, respectively. The reduction in density and increase in water absorption further confirmed the introduction of porosity, which directly influenced the thermal properties of the materials.

Thermal performance analysis revealed that incorporating waste additives significantly improved insulation capabilities. The thermal conductivity of the modified mixtures decreased, with the TZPW mixture exhibiting the lowest value at 0.174 W/m·K, representing a 4.4% improvement over the reference clay. The masonry wall performance analysis further confirmed these benefits, showing reduced U-values and improved thermal resistance in walls constructed from the modified bricks. The TZPW mixture provided the highest energy efficiency, making it the most effective option for thermal insulation, particularly in regions with extreme temperature variations.

From an economic and technical standpoint, the integration of these waste materials presents a viable and sustainable solution for the brick industry. Industrial-scale brick factories, with production capacities of 500 to 700 tons of fired bricks per day, require significant raw material inputs—approximately 850 tons of clay and additives daily. While the large-scale use of coffee grounds alone may be impractical due to supply limitations, combining them with expanded perlite waste offers a more feasible approach. Expanded perlite waste, an industrial byproduct currently underutilized, is available at little to no cost and could provide a cost-effective means of reducing raw material expenses while improving insulation properties. Future research should focus on industrial-scale validation, optimizing the balance between mechanical strength and thermal efficiency for large-scale implementation. Further work should also explore the long-term durability and environmental impact of these modified bricks, ensuring their viability for widespread adoption in sustainable construction.

Author Contributions: Conceptualization, I.M. and K.K.; methodology, I.M.; validation, I.M., P.T. and K.K.; formal analysis, I.M. and K.K.; investigation, P.T.; resources, P.T. and K.K.; data curation, I.M. and K.K.; writing—original draft preparation, I.M.; writing—review and editing, P.T., I.M. and K.K.; supervision, I.M. All authors have read and agreed to the published version of the manuscript.

Funding: This research received no external funding.

Institutional Review Board Statement: Not applicable.

Informed Consent Statement: Not applicable.

Data Availability Statement: The original contributions presented in this study are included in the article. Further inquiries can be directed to the corresponding author.

Conflicts of Interest: Author Ioannis Makrygiannis is employed by the company SABO S.A. The remaining authors declare that the research was conducted in the absence of any commercial or financial relationships that could be construed as a potential conflict of interest.

References

1. Makrygiannis, I.; Karalis, K. Optimizing Building Thermal Insulation: The Impact of Brick Geometry and Thermal Coefficient on Energy Efficiency and Comfort. *Ceramics* **2023**, *6*, 1449–1466. [CrossRef]
2. Rzepa, K.; Wons, W.; Reben, M. Building ceramics with improved thermal insulation parameters. *E3S Web Conf.* **2016**, *10*, 00082. [CrossRef]
3. Hossain, S.S.; Roy, P.K. Sustainable ceramics derived from solid wastes: A review. *J. Asian Ceram. Soc.* **2020**, *8*, 984–1009. [CrossRef]
4. Makrygiannis, I.; Tsetsekou, A. Effect of Expanded Perlite in the Brick Mixture on the Physicochemical Thermal Properties of the Final Products. *J. Compos. Sci.* **2022**, *6*, 211. [CrossRef]
5. Eliche-Quesada, D.; Pérez-Villarejo, L.; Iglesias-Godino, F.J.; Martínez-García, C.; Corpas-Iglesias, F.A. Incorporation of coffee grounds into clay brick production. *Adv. Appl. Ceram.* **2011**, *110*, 225–232. [CrossRef]
6. Demir, I. Effect of organic residues addition on the technological properties of clay bricks. *Waste Manag.* **2008**, *28*, 622–627. [CrossRef]
7. Makrygiannis, I.; Tsetsekou, A. Efficient Recovery of Solid Waste Units as Substitutes for Raw Materials in Clay Bricks. *Recycling* **2022**, *7*, 75. [CrossRef]
8. Torosyan, V.F.; Torosyan, E.S.; Yakutova, V.A.; Antyufeev, V.K. The development of a modified composition of ceramic mass for the production of bricks. *IOP Conf. Ser. Mater. Sci. Eng.* **2016**, *124*, 012145. [CrossRef]
9. Vijayan, D.S.; Mohan, A.; Revathy, J.; Parthiban, D.; Varatharajan, R. Evaluation of the impact of thermal performance on various building bricks and blocks: A review. *Environ. Technol. Innov.* **2021**, *23*, 101577. [CrossRef]
10. Makrygiannis, I.; Tsetsekou, A.; Papastratis, O.; Karalis, K. Assessing the Effects of Refuse-Derived Fuel (RDF) Incorporation on the Extrusion and Drying Behavior of Brick Mixtures. *Ceramics* **2023**, *6*, 2367–2385. [CrossRef]
11. Lourenço, P.B.; Vasconcelos, G.; Medeiros, P.; Gouveia, J. Vertically perforated clay brick masonry for loadbearing and non-loadbearing masonry walls. *Constr. Build. Mater.* **2010**, *24*, 2317–2330.
12. *ISO 14688-2:2017*; Geotechnical Investigation and Testing–Identification and Classification of Soil–Part 2: Principles for a Classification. International Organization for Standardization: Geneva, Switzerland, 2017.
13. Kanaoka, S.; Shibasaki, Y.; Maeda, T.; Watamura, S.; Jinno, Y. Evaluation of Plasticity of Ceramic Substrate by Pfefferkorn Method and Atterberg Method. *J-STAGE* **1981**, *21*, 37–46. [CrossRef]
14. *ISO 26845:2016*; Determination of Thermal Conductivity of Porous Materials–Standard Test Method. International Organization for Standardization: Geneva, Switzerland, 2016.
15. Lourenço, P.B.; Fernandes, F.M.; Castro, F. Handmade Clay Bricks: Chemical, Physical and Mechanical Properties. *Int. J. Archit. Herit.* **2010**, *4*, 38–58. [CrossRef]

16. *ASTM D422-63*; Standard Test Method for Particle-Size Analysis of Soils. ASTM International: West Conshohocken, PE, USA, 2007.
17. *ASTM D698-12*; Standard Test Methods for Laboratory Compaction Characteristics of Soil Using Standard Effort. ASTM International: West Conshohocken, PE, USA, 2012.
18. Hansen, W.; Kung, J.H. Pore structure and frost durability of clay bricks. *Mater. Struct.* **1988**, *21*, 443–447. [CrossRef]
19. Nigay, P.M.; Sani, R.; Cutard, T.; Nzihou, A. Modeling of the thermal and mechanical properties of clay ceramics incorporating organic additives. *Mater. Sci. Eng. A* **2017**, *708*, 375–382. [CrossRef]
20. Ahmad, S.; Iqbal, Y.; Muhammad, R. Effects of coal and wheat husk additives on the physical, thermal and mechanical properties of clay bricks. *Boletín Soc. Española Cerámica Vidrio* **2017**, *56*, 131–138. [CrossRef]
21. Hamza, A.; Kocserha, I. The effect of expanded perlite on fired clay bricks. *J. Phys. Conf. Ser.* **2020**, *1527*, 012032. [CrossRef]
22. Fiala, J.; Mikolas, M.; Krejsova, K. Full Brick, History and Future. *IOP Conf. Ser. Earth Environ. Sci.* **2019**, *221*, 012139. [CrossRef]
23. *ASTM C373-14a*; Standard Test Method for Water Absorption, Bulk Density, Apparent Porosity, and Apparent Specific Gravity of Fired Whiteware Products. ASTM International: West Conshohocken, PE, USA, 2014.
24. *EN 1745:2012*; Masonry and Masonry Products—Methods for Determining Thermal Properties. European Committee for Standardization (CEN): Brussels, Belgium, 2012.
25. Andrade, F.A.; Al-Qureshi, H.A.; Hotza, D. Measuring the plasticity of clays: A review. *Appl. Clay Sci.* **2011**, *51*, 1–7. [CrossRef]
26. Händle, F. (Ed.) *Extrusion in Ceramics, Engineering Materials and Processes*; Springer: Berlin/Heidelberg, Germany, 2007. [CrossRef]
27. de Souza Tavares, F.V.; de Melo Cavalcante, A.M.; de Figueiredo, M.J.; Vilela, A.F.; Carmo de Lima, A.R.; Azerêdo, L.P.M.; de Lima, A.G.B. On the Modeling of Drying Process in an Industrial Tunnel Dryer. *Defect Diffus. Forum* **2020**, *400*, 51–56. [CrossRef]
28. Mancuhan, E.; Özen, S.; Sayan, P.; Sargut, S.T. Experimental investigation of green brick shrinkage behavior with Bigot's curves. *Dry. Technol.* **2016**, *34*, 1535–1545. [CrossRef]
29. Tavares, F.; Neto, S.; Barbosa, E.; Lima, A.; Silva, C. Drying of Ceramic Hollow Bricks in an Industrial Tunnel Dryer: A Finite Volume Analysis. *Int. J. Multiphysics* **2014**, *8*, 297–312. [CrossRef]
30. *EN ISO 10211:2007*; Thermal Bridges in Building Construction—Heat Flows and Surface Temperatures—Detailed Calculations. International Organization for Standardization (ISO): Geneva, Switzerland, 2007.
31. *EN ISO 6946*; Building Components and Building Elements—Thermal Resistance and Thermal Transmittance—Calculation Methods. International Organization for Standardization (ISO): Geneva, Switzerland, 2017.
32. Karalis, K.T.; Karkalos, N.; Cheimarios, N.; Antipas, G.S.E.; Xenidis, A.; Boudouvis, A.G. A CFD analysis of slag properties, electrode shape and immersion depth effects on electric submerged arc furnace heating in ferronickel processing. *Appl. Math. Model.* **2016**, *40*, 9052–9066. [CrossRef]
33. Kushnierova, L.; Ghiassi, B.; Khine, M. Investigation of the Role of Clay Types and Porous Aggregate Effects on the Hygric Properties of Earth-Based Plasters. In Proceedings of the 18th International Brick and Block Masonry Conference, Birmingham, UK, 21–24 July 2024; Milani, G., Ghiassi, B., Eds.; Springer Nature: Cham, Switzerland, 2025; pp. 902–913.
34. Touil, M.; Lachheb, A.; Saadani, R.; Sanbi, M.; Talidi, A.; Rahmoune, M. Experimental investigation on the combined effect of the water mixing ratio and the addition of spent coffee grounds on plaster's thermo-mechanical properties. *Therm. Sci. Eng. Prog.* **2022**, *36*, 101488. [CrossRef]
35. Zhao, N.; Liu, Z.; Yu, T.; Yan, F. Spent coffee grounds: Present and future of environmentally friendly applications on industries-A review. *Trends Food Sci. Technol.* **2024**, *143*, 104312. [CrossRef]
36. Gualtieri, A.F.; Ricchi, A.; Lassinantti Gualtieri, M.; Maretti, S.; Tamburini, M. Kinetic study of the drying process of clay bricks. *J. Therm. Anal. Calorim.* **2016**, *123*, 153–167. [CrossRef]
37. Tsega, E. Effects of Firing Time and Temperature on Physical Properties of Fired Clay Bricks. *AJCE* **2017**, *5*, 21. [CrossRef]
38. Cobo-Ceacero, C.J.; Moreno-Maroto, J.M.; Guerrero-Martínez, M.; Uceda-Rodríguez, M.; López, A.B.; García, C.M.; Cotes-Palomino, T. Effect of the addition of organic wastes (cork powder, nut shell, coffee grounds and paper sludge) in clays to obtain expanded lightweight aggregates. *Boletín Soc. Española Cerámica Vidrio* **2023**, *62*, 88–105. [CrossRef]
39. Ismaiel, M.; Chen, Y.; Cruz-Noguez, C.; Hagel, M. Thermal resistance of masonry walls: A literature review on influence factors, evaluation, and improvement. *J. Build. Phys.* **2022**, *45*, 528–567. [CrossRef]
40. Uygunoğlu, T.; Özgüven, S.; Çalış, M. Effect of plaster thickness on performance of external thermal insulation cladding systems (ETICS) in buildings. *Constr. Build. Mater.* **2016**, *122*, 496–504. [CrossRef]
41. Karalis, K.; Karalis, N.; Karkalos, N.; Antipas, G.S.E.; Xenidis, A. Computational fluid dynamics analysis of a three-dimensional electric submerged arc furnaceoperation. *Preprint* **2020**. [CrossRef]
42. Dondi, M.; Mazzanti, F.; Principi, P.; Raimondo, M.; Zanarini, G. Thermal Conductivity of Clay Bricks. *J. Mater. Civ. Eng.* **2004**, *16*, 8–14. [CrossRef]
43. Wakili, K.G.; Tanner, C. U-value of a dried wall made of perforated porous clay bricks: Hot box measurement versus numerical analysis. *Energy Build.* **2003**, *35*, 675–680. [CrossRef]

44. Zanelli, C.; Marrocchino, E.; Guarini, G.; Toffano, A.; Vaccaro, C.; Dondi, M. Recycling Construction and Demolition Residues in Clay Bricks. *Appl. Sci.* **2021**, *11*, 8918. [CrossRef]
45. Coletti, C.; Cultrone, G.; Maritan, L.; Mazzoli, C. How to face the new industrial challenge of compatible, sustainable brick production: Study of various types of commercially available bricks. *Appl. Clay Sci.* **2016**, *124–125*, 219–226. [CrossRef]

ceramics

MDPI

Article

Nanoflower-Shaped ZnO Nanoparticles Reduced with Blueberry Waste and Their Evaluation of Malachite Green Dye Degradation

Iván Balderas-León [1], Jorge Manuel Silva-Jara [1,*], Miguel Ángel López-Álvarez [2], Pedro Ortega-Gudiño [3], Arturo Mendoza-Galván [4], Omar Fabela-Sánchez [5], Rocío Ivette López-Roa [1] and María Esther Macías-Rodríguez [1,*]

1 Departamento de Farmacobiología, Centro Universitario de Ciencias Exactas e Ingenierías, Universidad de Guadalajara, Blvd. Marcelino García Barragán 1421, Guadalajara 44430, Jalisco, Mexico; ivan.balderas@academicos.udg.mx (I.B.-L.)
2 Departamento de Ingeniería Mecánica, Centro Universitario de Ciencias Exactas e Ingenierías, Universidad de Guadalajara, Blvd. Marcelino García Barragán 1421, Guadalajara 44430, Jalisco, Mexico; miguel.lalvarez@academicos.udg.mx
3 Departamento de Ingeniería Química, Centro Universitario de Ciencias Exactas e Ingenierías, Universidad de Guadalajara, Blvd. Marcelino García Barragán 1421, Guadalajara 44430, Jalisco, Mexico; pedro.ogudino@academicos.udg.mx
4 Cinvestav-Unidad Querétaro, Libramiento Norponiente 2000, Queretaro 76230, Queretaro, Mexico
5 Departamento de Química Macromolecular y Nanomateriales, Centro de Investigación en Química Aplicada, Enrique Reyna H. 140, San José de los Cerritos, Saltillo 25294, Coahuila, Mexico; omar.fabela@ciqa.edu.mx
* Correspondence: jorge.silva@academicos.udg.mx (J.M.S.-J.); mesther.macias@academicos.udg.mx (M.E.M.-R.)

Abstract: This work focuses on the green synthesis of zinc oxide nanoparticles (ZnO NPs) to be used as a photocatalyst, preparing a blueberry (*Vaccinium corymbosum*) extract as a reducing agent, zinc chloride as a precursor, and NaOH as precipitating agent (1, 7 and 14 M). Characterization techniques included X-ray diffraction (XRD), transmission electron microscopy (TEM), ultraviolet-visible (UV-Vis), Fourier-transform infrared (FTIR), X-ray photoelectron (XPS), and Raman spectroscopy. The XRD analysis confirmed the hexagonal wurtzite crystalline structure of ZnO NPs, while the SEM and TEM revealed a flower-like morphology. Moreover, the ZnO NPs exhibited a band gap energy in the range of 3.17 to 3.27 eV, depending on the concentration of NaOH. Regarding the photocatalytic activity, the synthesized ZnO NPs showed a photocatalytic efficiency greater than 90%, degrading malachite green (MG) dye in 60 min under sunlight. The proposed photocatalytic mechanism includes reactive oxygen species (ROS) generation (hydroxyl radicals ($OH^\bullet$) and superoxide radicals (O^{2-})). The XRD and Raman analysis also revealed that the ZnO NPs maintained structural integrity after repeated photocatalytic cycles, emphasizing their stability and suitability for practical applications.

Keywords: ZnO nanoparticles; photocatalysis; green synthesis; nanoflower morphology; malachite green dye

1. Introduction

The green synthesis of metallic nanoparticles (NPs), particularly zinc oxide nanoparticles (ZnO NPs), offers a simple, eco-friendly approach to producing nanomaterials for various applications, especially in photocatalysis. The increasing demand for sustainable and cost-effective methods has prompted the use of natural materials, which exhibit minimal hazardous environmental effects and low toxicity for human health [1,2]. Utilizing non-toxic resources like blueberry (*Vaccinium corymbosum*) extracts reduces environmental impact and capitalizes on byproducts from local markets. This practice diminishes food waste, encourages sustainable practices, and aligns with food sovereignty using local agricultural byproducts. Another significant advantage of ZnO NPs is their relatively low production cost and abundant raw materials compared to other nanomaterials. For example, titanium dioxide NPs (TiO_2 NPs), TiO_2 NPs-modified derivatives, and rare earth element NPs have notably higher production costs. While copper oxide NPs (CuO NPs) produced through green synthesis are moderately priced, the cost-effectiveness of ZnO NPs makes them more accessible for various industrial applications [1]. Additionally, they are biocompatible, making them suitable for different applications in multiple fields, including medicine, solar cells, textiles, cosmetics, and wastewater treatment, where their efficiency and safety are highly valued [3,4].

One of the critical environmental challenges that ZnO NPs can tackle is wastewater contamination from harmful dyes used in industries like textiles and printing, which pose significant risks to the ecosystem and human health. The pigments and dyes in wastewater inhibit the photosynthetic process of aquatic plants by blocking direct sunlight. Organic dyes employed in the textiles and food industries are a significant source of environmental pollution, containing highly toxic and carcinogenic substances that threaten humans and natural ecosystems. Malachite green (MG) is a crystalline powder highly soluble in water. It is commonly used as a food coloring additive in the textile, paper, and cosmetic industries. However, it has mutagenic and genotoxic effects that can harm aquatic life and biological systems even at low concentrations. The US Food and Drug Administration (FDA) has banned its use due to these risks, yet production continues, leading to environmental damage. Addressing the impact of MG is urgent, with efforts focused on mitigating its toxic effects. This discrepancy highlights the importance of developing eco-friendly methods for its degradation, such as using photocatalytic NPs [5]. Table 1 presents the photocatalytic activity of ZnO NPs, which generate reactive oxygen species (ROS) upon exposure to ultraviolet light [6], achieving degradation efficiencies for various dyes, ranging from 75% to over 98%, with treatment times between 50 and 180 min. Most dyes were effectively degraded within 60 to 130 min, with some achieving complete removal in just 60 min. The dyes assessed include methylene blue, methyl orange, Congo red, rhodamine B, eriochrome black-T, and Sunfix red, highlighting the potential of photocatalysis for treating various wastewater dyes.

In this study, we highlight for the first time the use of green synthesis of ZnO NPs utilizing the waste from endemic blueberries. Furthermore, by varying the NaOH concentration, we produce flower-like morphologies that can degrade MG dye using sunlight in less time than other studies. These new functionalities open up possibilities for implementing their photocatalytic properties. Additionally, these nanoparticles have potential applications in the health sector, electronics, and bioremediation, expanding the range of applications across different industries.

Table 1. Comparative photocatalytic activity of ZnO NPs synthesized from different green sources for hazard dyes degradation.

Extract Precursor	Dye	Light Source	Degradation Efficiency/Time	Reference
Cassia fistula	Methylene blue	UV light and sunlight	~98%/120 min	[7]
Canna indica		Sunlight	94.23%/150 min	[8]
Brassica oleracea var. botrytis			80%/75 min	[9]
Pyrus pyrifolia		UV light	88%/60 min	[10]

Table 1. *Cont.*

Extract Precursor	Dye	Light Source	Degradation Efficiency/Time	Reference
Carissa edulis	Congo red	UV light	97%/130 min	[11]
Artocarpus heterophyllus			90%/60 min	[12]
	Rose Bengal		>80%/60 min	[13]
Eucalyptus globulus	Methyl orange		96.6%/60 min	[14]
	Methylene blue		98.2%/50 min	
Thymus vulgaris		Sunlight	91%/60 min	[15]
	Sunfix red		75%/120 min	
	Real textile wastewater		79%/90 min	
Ziziphus jujube	Eriochrome black-T		86% 5 h	[16]
	Methylene blue		92%	
Abelmoschus esculentus			Completely removed/60 min	[17]
Aerva lanata	Rhodamine B	UV light	96% 180 min	[18]
Aerva javanica			85%	

2. Materials and Methods

2.1. Materials

Zinc chloride ($ZnCl_2$) was the precursor for synthesizing ZnO NPs, and sodium hydroxide (NaOH) controlled the pH and acted as a precipitating agent. MG dye served as a model of water pollutants. All reagents were chemical-grade and obtained from Sigma-Aldrich Chemicals (St. Louis, MO, USA).

2.2. Preparation of the Blueberry Extract

To prepare the green extract, 1 g of freeze-dried blueberry waste was boiled in 100 mL of water at 90 °C for 15 min with constant stirring. After cooling, the mixture was centrifuged at 10,000 rpm for 15 min. The supernatant was filtered and stored at 4 °C.

2.3. Green Synthesis of ZnO NPs

As depicted in Figure 1, the green synthesis of ZnO NPs was carried out using 90 mL of a 1×10^{-3} M $ZnCl_2$ solution stirred for 20 h at room temperature. After this time, the pH of the solution was adjusted to 8 using 1 M NaOH solution. Next, 10 mL of blueberry extract was added dropwise to the solution, and the mixture was heated to 60 °C for 1 h while stirring continuously. The resulting precipitate was collected by centrifugation at 10,000 rpm for 15 min. The product was then calcined at 600 °C. This synthesis protocol was also repeated with pH adjustments at 12 using 7 M and 14 M NaOH solutions.

2.4. Characterization of ZnO NPs

The surface characterization of the synthesized ZnO NPs was conducted using various solid-state techniques. The crystallinity of each calcined precursor was carried out through X-ray diffraction (XRD) with an Empyrean diffractometer (PANalytical, Westborough, MA, USA). The Fourier-transform infrared (FTIR) spectrum was recorded in the 4000–400 cm^{-1} range using a 400 FTIR spectrometer (Perkin Elmer, Waltham, MA, USA). The morphology of the calcined powders was analyzed by transmission electron microscopy (TEM) on a JEM 1010 microscope (Jeol, Tokyo, Japan) and by scanning electron microscopy

(SEM) carried out using a JEOL 7610F microscope (Jeol, Tokyo, Japan), with samples coated with a thin Au-Pd layer using a Denton Vacuum Desk V sputter coater under an argon atmosphere. Chemical species present in the ZnO NPs powders were analyzed by X-ray photoelectron spectroscopy (XPS) with a Sigma2 XPS and a Theta-probe (Thermo Fisher Scientific, East Grinstead, UK). Absorbance measurements were obtained using an Ultraviolet-visible spectrophotometer (Cary-300, Varian, Mulgrave Victoria, Australia). The Raman spectroscopy was performed using an alpha300 RA confocal microscope (WITec-Oxford Instrument, Ulm, Germany) with a 532 nm excitation laser at a 5–6 mW power and a 50× objective. UV also studied the optical properties of the ZnO NPs–Vis diffuse reflectance spectroscopy (Varian, Cary 300 model) in the wavelength range from 190 to 800 nm. The band gap energy (E_g) values were determined using the Kubelka–Munk function and the relation between the absorption edge and the absorbance, as described in Equation (1).

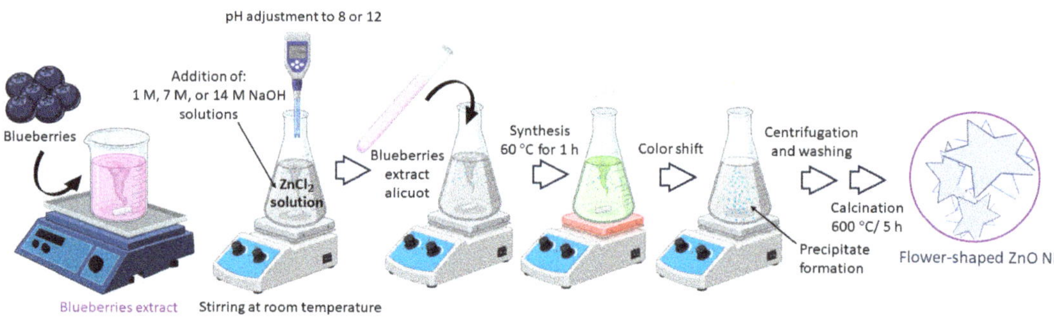

Figure 1. Schematic representation of the green synthesis protocol for ZnO NPs using blueberry (*Vaccinium corymbosum*) extract.

$$[F(R)h\nu]^2 = B(h\nu - E_g) \tag{1}$$

where F(R) is the Kubelka–Munk function, hν is the photon energy, E_g is the band gap energy, and B is a constant characteristic of the material.

2.5. Photocatalytic Dye Degradation

Photocatalytic experiments were carried out on sunny summer days between 11 a.m. and 2 p.m. on the rooftop of the Molecular Biology Research Laboratory at Universidad de Guadalajara, Mexico (20°39′25″ N, 103°19′31.2″ W). The optical irradiance of sunlight was 117.6 mW/cm^2, measured using a Thorlabs power meter at a wavelength of 365 nm. The photocatalytic activity of ZnO NPs was evaluated using 5 mg of each powder mixed with 20 mL of a 1×10^{-4} M MG dye solution. Subsequently, aliquots were extracted from each solution at different intervals. Thus, the absorbance spectra of each aliquot were obtained using a NanoDrop 2000 spectrophotometer (Thermo Scientific, Waltham, MA, USA). Based on the data recorded from each UV-Vis spectrum, the degradation percentage was determined using Equation (2).

$$\text{Degradation}(\%) = \left(\frac{A_0 - A}{A_0}\right) * 100 \tag{2}$$

where A_0 and A correspond to the absorbance values of the solution with the dye before and after exposure to sunlight, respectively.

3. Results and Discussion

3.1. XRD Analysis

Figure 2 shows the diffraction patterns of the precursor powders calcined at 600 °C using blueberry extract and various concentrations of NaOH. The XRD analysis was conducted in the 2θ range of 20° to 70°. All samples exhibit diffraction peaks at 2θ values around 31.7°, 34.4°, 36.2°, 47.5°, 56.6°, 62.8°, 66.3°, 67.9°, and 69.0° associated with (100), (002), (101), (102), (110), (103), (200), (112), and (201) planes of ZnO, respectively. According to the JCPDF #36-1451 file, all diffraction peaks were identified and linked to the hexagonal (wurtzite) phase of ZnO. It is important to note that no impurities or secondary phases were observed.

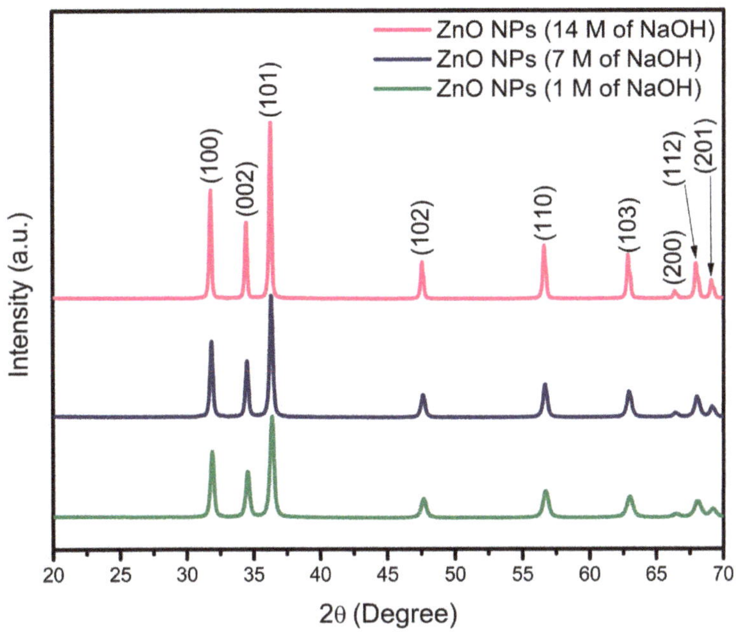

Figure 2. XRD patterns of the precursor powders calcined at 600 °C.

The data recorded in the diffractograms were used to assess the crystallinity and crystallite size of each ZnO NPs sample. The results are presented in Table 2. The crystallite size was determined using the Scherrer formula (Equation (3)) where d is the crystalline size, λ denotes the X-ray wavelength, β represents the full width at half maximum (FWHM) of the peak, and θ is the Bragg angle. The crystallinity was evaluated by Equation (4).

$$d = \frac{(0.95)(\lambda)}{\beta(\cos\theta)} \quad (3)$$

$$\text{Crystallinity} = \frac{\text{Area}_{\text{Crystalline peaks}}}{\text{Area}_{\text{All peaks}}} \quad (4)$$

As observed, crystallite size increases with higher concentrations of NaOH. These results demonstrate a relationship between sodium hydroxide concentration and crystal size, indicating that a higher NaOH concentration leads to particle coalescence, resulting in larger crystallites [19].

Table 2. Evaluation of the crystallinity and crystal size of the different ZnO NPs powders calcined at 600 °C.

ZnO NPs (Synthesis)	Crystallinity (%)	Crystallite Size (nm)
1 M NaOH	84.7	24.6
7 M NaOH	88.1	30.8
14 M NaOH	95.4	40.1

3.2. UV-Vis Spectroscopy

Figure 3 illustrates the UV-Vis spectra and Tauc plots of the ZnO NPs powders synthesized by 1 M NaOH and those produced using solutions of 7 M and 14 M NaOH. All ZnO NPs powders exhibit a broad absorption band in the 380–240 nm (Figure 3a). However, the ZnO NPs powders synthesized by 1 M NaOH display the highest reflectance, whereas those obtained with a concentration of 14 M NaOH show the lowest reflectance. These findings indicate a relationship between the concentration of NaOH used in each synthesis and the reflectance rate. Furthermore, according to Table 2, the powders obtained via the 1 M NaOH method also exhibit the smallest particle size (24.6 nm). In comparison, those synthesized with a concentration of 14 M NaOH demonstrate the largest particle size (40.1 nm). Based on this information, a potential explanation linking reflectance to particle size is grounded in the Kubelka–Munk theory, which suggests that penetration depth and absorption also increase as particle size increases, resulting in a decrease in reflectance. In summary, the reflectance of ZnO NPs increases as particle size decreases.

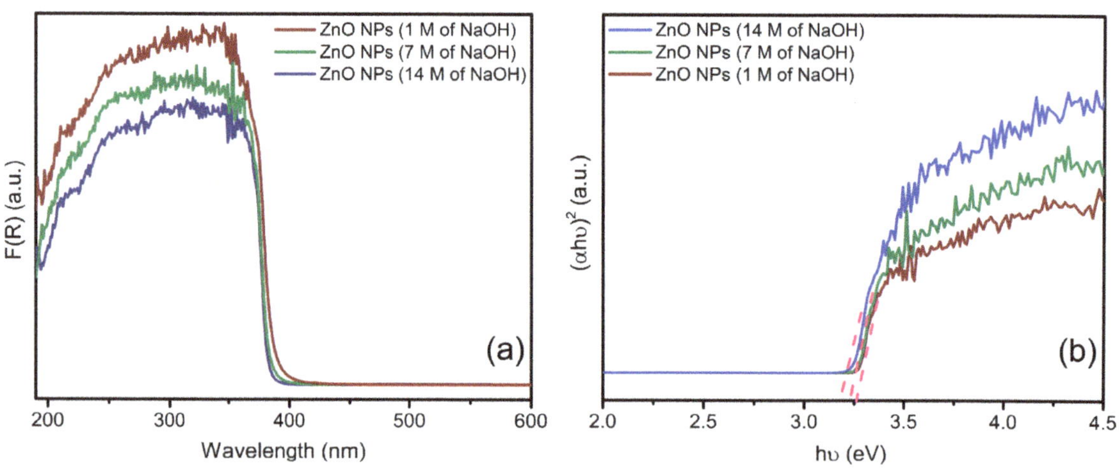

Figure 3. (**a**) UV-Vis absorbance spectra and (**b**) Tauc plots for ZnO NPs.

Conversely, the Tauc plots (Figure 3b) indicate a decrease in the E_g energy as the concentration of NaOH increases, measuring 3.27, 3.22, and 3.17 eV for the powders obtained with 1 M, 7 M, and 14 M of NaOH, respectively. A plausible explanation for the decrease in the E_g alongside the rising NaOH concentration could be as follows: the elevation in NaOH concentration also leads to an increase in defect formation, which results in the delocalization of molecular orbitals at the conduction band edge, ultimately causing a reduction in E_g [6,19].

3.3. FTIR Analysis

Figure 4 shows the FTIR spectra of the ZnO NPs synthesized by the three methods: 1 M, 7 M, and 14 M. In all samples, surface hydroxyl groups (OH^-) are indicated by a broad absorption band around 3400 cm^{-1}, reflecting O–H stretching vibrations. Furthermore, the ZnO NPs synthesized at 14 M exhibited the highest absorbance in this region, suggesting a greater O–H density than the ZnO NPs produced via 1 M NaOH and 7 M. The presence of O–H groups on the surface of ZnO NPs can be attributed to two main factors: first, the hygroscopic nature of this material, and second, the formation of defects, particularly the existence of oxygen vacancies [20].

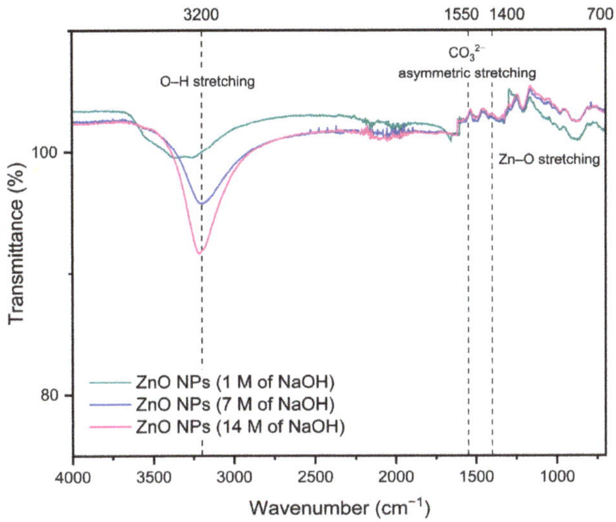

Figure 4. FTIR spectra of ZnO NPs.

Notably, the three synthesized powders showed no impurities related to organic matter. The peaks in the 1250–1550 cm^{-1} range may correlate with the adsorption of atmospheric CO_2 occurring during FTIR analysis. In contrast, all samples exhibit a consistent band at 700 cm^{-1}, indicating Zn–O stretching vibrations tied to the hexagonal structure of ZnO. However, variations in transmittance in other regions suggest differences in surface characteristics and particle shape.

3.4. XPS Analysis

The O 1s core-level spectra of each ZnO NPs were analyzed, as shown in Figure 5a–c, to propose a photodegradation mechanism. In these measurements, the C 1s, located at 284.5 eV was used as a reference. Additionally, each spectrum was deconvoluted, resulting in two Gaussians centered at 529.5 ± 0.2 eV (O_{I}) and 531.2 ± 0.1 eV (O_{II}). The Gaussians identified as O_{I} were associated with lattice oxygen (O–Zn), while the second (O_{II}) were attributed to the formation of oxygen vacancy (V_o) [21]. Although the contribution of V_o to photocatalytic activity in metallic oxides is still unclear, its presence promotes the formation of OH^- and hydroxyl radicals ($OH^\bullet$).

On the other hand, the percentage of the area associated with V_o was determined. The ZnO NPs synthesized by 1, 7, and 14 M of NaOH were 49.4%, 50.1%, and 51.9%, respectively. The results reveal an increase in the formation of V_o when the concentration of NaOH increases. Based on these results, it is probable that the ZnO NPs synthesized using a concentration of 14 M NaOH present the highest photocatalytic efficiency.

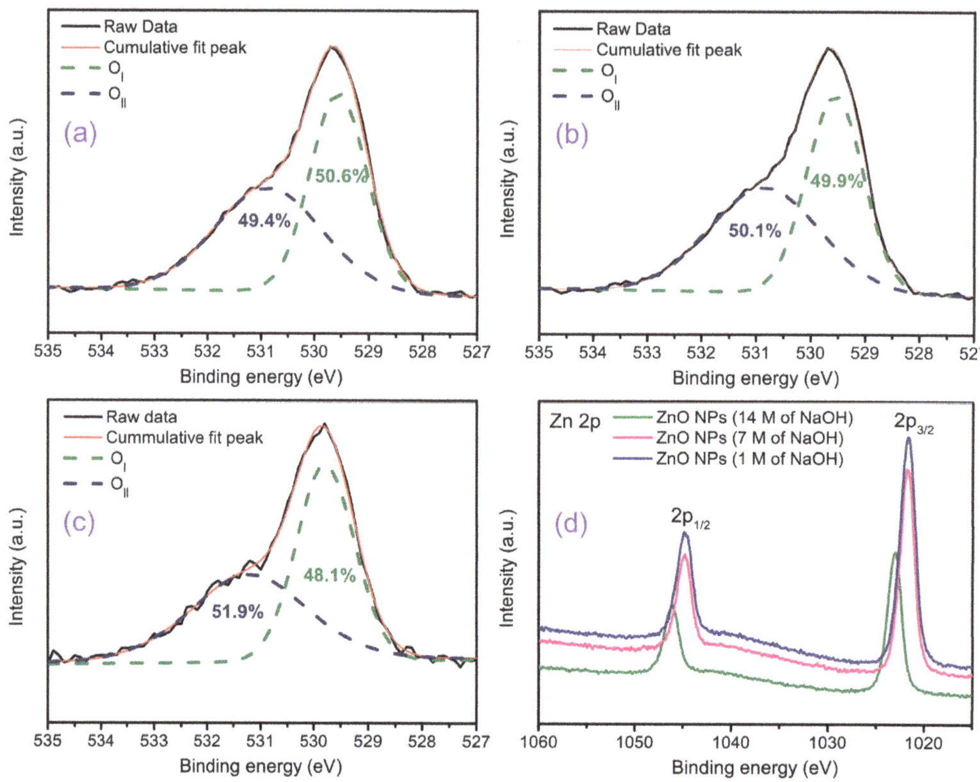

Figure 5. XPS analysis of O 1s in ZnO NPs synthesized using (**a**) 1 M, (**b**) 7 M, and (**c**) 14 M of NaOH, and (**d**) Zn 2p spectra.

The XPS spectra of each synthesized powder's Zn 2p core-level (Figure 5d) were also obtained. This core-level is characterized by two peaks, identified as $2p_{3/2}$ and $2p_{1/2}$. In the case of the powders synthesized by the 1 M and 7 M methods, the $2p_{3/2}$ peak presents a binding energy of approximately 1021.6 eV, whereas the powders synthesized with the solution at 14 M were 1022.5 eV. In the three cases, the binding energy range described above was associated with the Zn^{2+} of ZnO. However, the ZnO NPs powders synthesized using the 14 M NaOH solution show a shift of around 0.9 eV towards higher binding energy. This shift has already been observed in ZnO NPs and can be attributed to a hydroxylation process associated with oxygen vacancies [22].

3.5. Morphological Analysis of ZnO NPs Powders

Figure 6 presents the SEM micrographs of the ZnO NPs powders obtained through 1 M NaOH (Figure 6a), along with those produced using NaOH at concentrations of 7 M (Figure 6b) and 14 M (Figure 6c). The image of the powders synthesized by 1 M NaOH illustrates the formation of nanostructures with pseudo-spherical shapes, featuring an average diameter of 337 nm, with the smallest measuring 73 nm. In contrast, the micrographs of the powders synthesized with NaOH (7 and 14 M) display the development of flower-like nanostructures (Figure 6b,c), which are larger than those obtained through 1 M of NaOH synthesis.

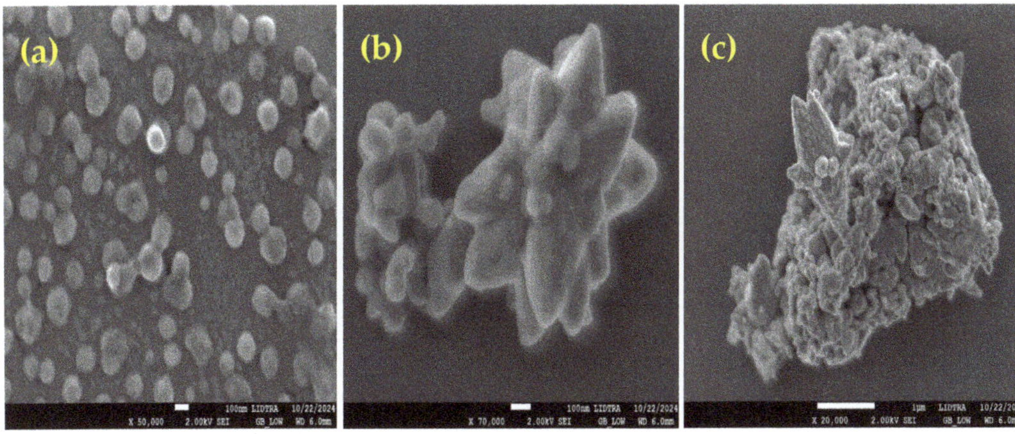

Figure 6. SEM micrographs of ZnO NPs powders synthesized by (**a**) 1 M of NaOH and using concentrations of (**b**) 7 M and (**c**) 14 M NaOH.

The synthesized ZnO powders were also analyzed using TEM (Figure 7a–c). The observed morphologies aligned with those previously described. Based on the TEM micrographs of the powders where NaOH was applied (Figure 7b,c), the average sizes of ZnO NPs synthesized with 7 M and 14 M NaOH were 1298 nm and 1455 nm in length, respectively. The width of the petals was also measured, with values of 365 nm and 650 nm for the powders synthesized with 7 M and 14 M NaOH, respectively.

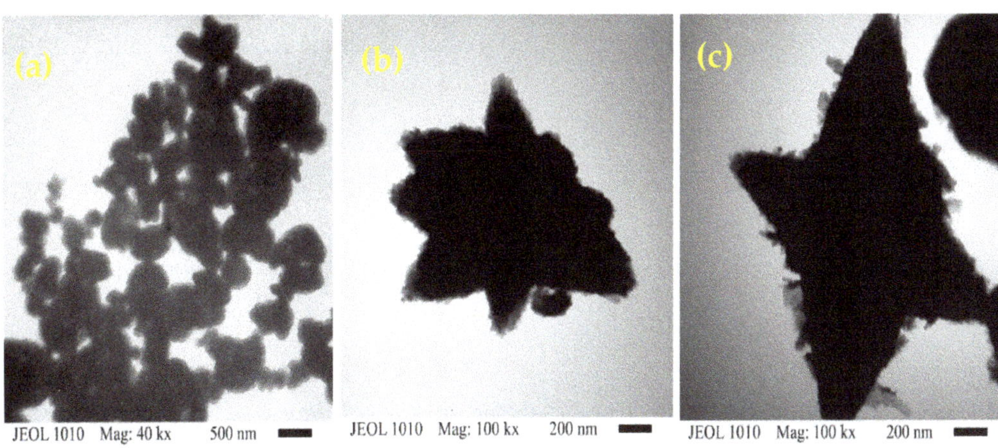

Figure 7. TEM micrographs of ZnO NPs powders synthesized by (**a**) 1 M and using concentrations of (**b**) 7 M and (**c**) 14 M NaOH.

According to the micrographs obtained through SEM and TEM, the presence of OH^- substantially modifies the morphology of the ZnO NP powders, influencing the formation of flower-like nanostructures. Furthermore, these ions promote the growth of nanocrystallites, causing the nanostructures to increase in size as the concentration of NaOH rises.

Regarding photocatalysis, this flower-like nanostructure may enhance the photocatalytic activity of ZnO NPs. Its open petal-like morphology potentially increases the surface area, making these ZnO NPs especially well-suited for photocatalytic applications where interaction with dye molecules is crucial, along with improved light refraction and dye molecule absorption.

3.6. Photocatalytic MG Dye Degradation

Figure 8 shows the UV-Vis absorbance spectra of the MG dye aliquots extracted at different intervals. The absorbance peak of the MG dye located at 617 nm was selected to evaluate the photocatalytic efficiency of the ZnO NPs.

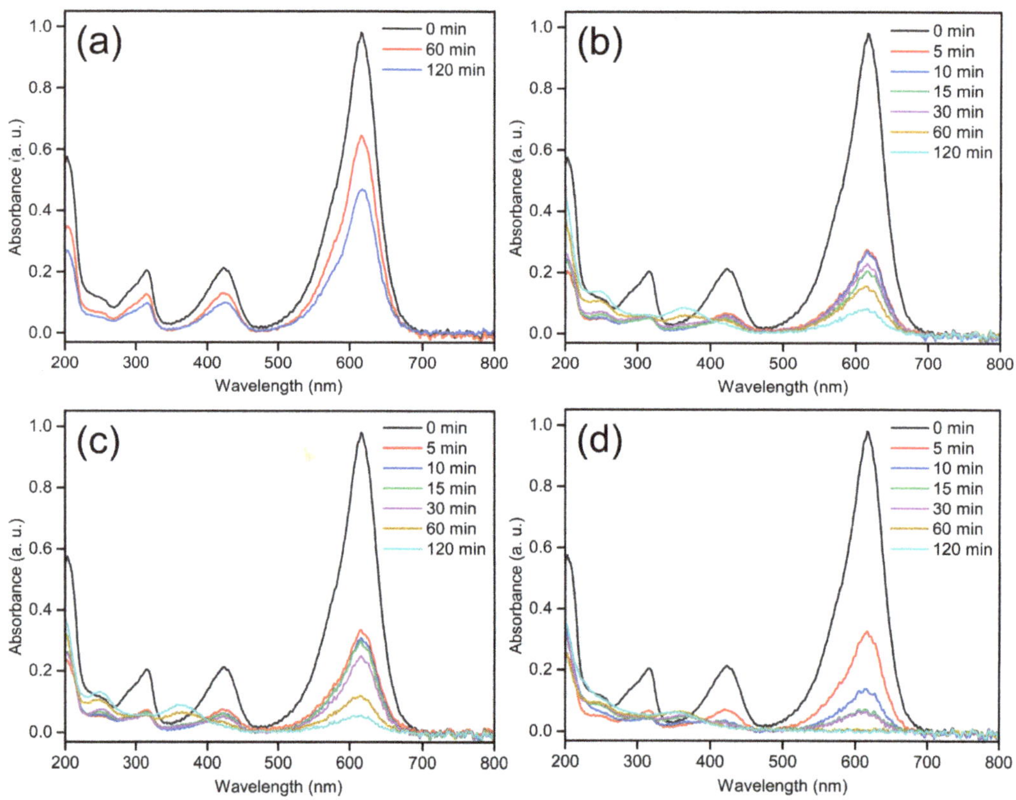

Figure 8. UV-Vis absorbance spectra of malachite green dye solutions being exposed to sunlight under the following conditions: (**a**) without ZnO NPs, (**b**) ZnO NPs 1 M, (**c**) ZnO NPs 7 M, and (**d**) ZnO NPs 14 M of NaOH.

At the start of the experiments, the role of sunlight in the degradation of the dye in the absence of a photocatalyst was assessed (Figure 8a). As noted, after 120 min of sunlight exposure, the dye degrades by 34%. Conversely, the photocatalytic activity of the ZnO NPs synthesized by each method was also evaluated. Accordingly, 5 mg of ZnO NPs was added to three solutions of MG dye. Figure 8b–d present the UV-Vis absorbance spectra of the dye solutions containing ZnO NPs exposed to sunlight for 120 min. In particular, after 5 min of sunlight exposure, the ZnO NPs synthesized using the 1 M, 7 M, and 14 M NaOH methods demonstrate photocatalytic efficiencies of 40%, 54%, and 60%, respectively (Figure 9). Furthermore, after 30 min, all ZnO NPs achieve a degradation greater than 80%.

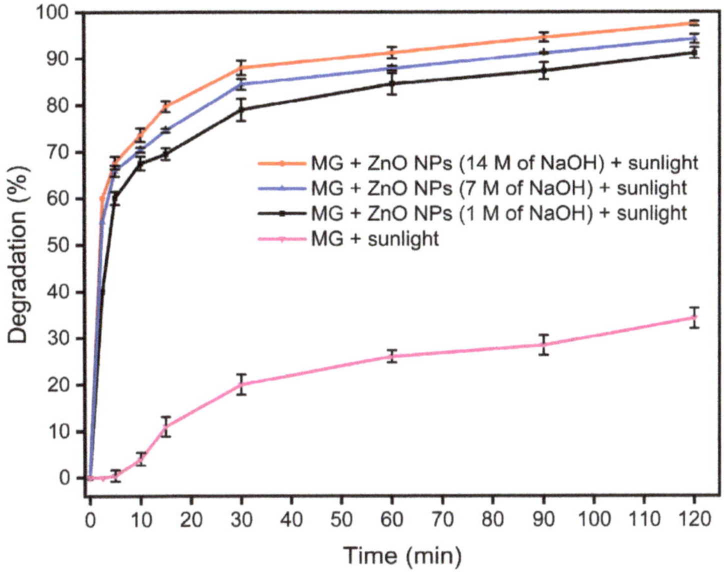

Figure 9. Percentage degradation of malachite green dye over time compared to sunlight alone.

As can be observed, in all the UV-Vis absorbance spectra obtained during the degradation of the MG dye, three typical peaks at 617 nm, 422 nm, and 316 nm appear. These peaks were identified and associated with C-N bonds of the chromophore conjugate structure of the dye [23]. However, after 5 min, a considerable decrease in the intensity of its main peak (617 nm) and a shift towards lower wavelength values occurs. This anomaly observed during the degradation process of the dye has already been recorded in other research, in which it was attributed to N-demethylation reactions [23,24]. Furthermore, new absorbance peaks at 250 and 360 nm appeared during the dye degradation. According to other research, the first peak (250 nm) was identified and attributed to the formation of 4-(N-ethylamino)benzoic acid [24], whereas the second peak (360 nm) was associated with the presence of 4-dimethylaminobenzophenone (DLBP) [25]. The formation of both adducts described above can be related to the cleavage of the central carbon in the dye molecule [26].

Based on these results, a possible mechanism describing the degradation process of the MG dye using ZnO NPs consists of the decomposition of the conjugated structure of the dye when ROS attacks its central carbon, followed by N-demethylation reactions.

These results reveal how the synthesis method affects the photocatalytic efficiency of ZnO NPs. Specifically, the ZnO NPs synthesized using NaOH demonstrate greater efficiency in degrading the MG dye than those produced by the 1 M NaOH method. This can be attributed to V_o concentration in the ZnO NPs. According to the previously described XPS results, the ZnO NPs synthesized with 7 M and 14 M NaOH contained the highest V_o concentrations. As is well known, these groups facilitate the formation of $OH^{\bullet}$, which allows for more efficient degradation of the MG dye.

Additionally, the photocatalytic efficiency achieved with the ZnO NPs synthesized in this study was compared to the results reported in other research on the degradation of the MG dye. Table 3 presents the findings.

Table 3. Photocatalytic degradation of green malachite using ZnO NPs by green synthesis.

Nanoparticle	Precursor	SHAPE/SIZE	Nanoparticle/ Dye Proportion	Conditions	Degradation Efficiency/Time	Reference
ZnO	Zinc nitrate hexahydrate	Flower-like/17.72 and 25.0 nm	60 mg ZnO NPs in 250 mL of 5.48×10^{-5} M	UV light irradiation	92.3%/60 min	[27]
		Sponge-like array/ ~52–76 nm	20 mg ZnO NPs 1.37×10^{-5} M		~95%/150 min	[28]
		Hexagonal/30 to 85 nm	200 mg ZnO NPs 2.74×10^{-5} M		89%/180 min	[29]
		Tetrameric structured/ ~50 nm	50 mg ZnO NPs in 50 mL of 5×10^{-5} M		38.1%/60 min	[30]
		Spindle and spherical/23.5 nm	12 mg ZnO NPs in 50 mL of 5.48×10^{-5} M and pH 9		90%/60 min	[31]
		Flower-like/ ~20.17 nm	25 mg ZnO NPs in 50 mL of 2.74×10^{-5} M		89%/180 min	[32]
	Zinc acetate dihydrate	Sphere-like/ 8–11 nm	75 mg ZnO NPs in 50 mL of 5.48×10^{-5} M and pH 10		96.5%/60 min	[33]
		Spherical/~65.9 nm	2 mg of ZnO NPs in 10 mL of 2.74×10^{-5} M		89.5%/120 min	[34]
		Spherical/~18.7 nm	5 mg of ZnO NPs in 40 mL of 6.86×10^{-2} M		23.31%/35 min	[35]
	Zinc nitrate hexahydrate	Foam-like bunch/ 6–11 nm	50 mg ZnO NPs in 100 mL of the 1.37×10^{-5} M	Sunlight/ UV light irradiation	~95%/150 min	[36]
		Spherical/21 nm	50 mg ZnO NPs in 50 mL of 2.74×10^{-5} M	Sunlight irradiation	~99%/180 min	[37]
	Zinc Sulfate	Rod shaped/ 40–50 nm	20 mg ZnO NPs in 100 mL of 2.74×10^{-5} M		92%/150 min	[6]
	Zinc chloride	Fower-like/~34 nm	8 mg ZnO NPs in 4.11×10^{-6} M		92%/60 min	[38]
		Flower-shaped	5 mg ZnO NPs in 20 mL of 1×10^{-4} M		97.5%/120 min	This study

In our work, we achieved a degradation of 97% in 120 min by adding 5 mg of the synthesized ZnO NPs to the MG dye solution. Comparing this degradation capability with those recorded in Table 3, we observe greater efficiency. Furthermore, it is noted that the photocatalytic efficiencies achieved in this work were under sunlight, whereas in most recorded cases, UV light was used. It is also worth mentioning that the concentration of the MG dye used in the photodegradation experiments is higher than that in the research proposed in Table 3.

3.7. Kinetics of the Photocatalytic Process

Additionally, a kinetic model that followed the photodegradation process of the MG dye using the ZnO NPs was selected. Four main kinetic models were evaluated: first-order (Equation (5)), second-order (Equation (6)), modified Freundlich (Equation (7)), and parabolic-diffusion (Equation (8)) models.

$$\ln\left(\frac{C}{C_0}\right) = -kt \quad (5)$$

$$\frac{1}{C} - \frac{1}{C_0} = kt \tag{6}$$

$$C = k_F t^n \tag{7}$$

$$\frac{1 - \frac{C}{C_0}}{t} = kt^{-0.5} \tag{8}$$

The absorbance values recorded in the UV-Vis absorbance spectra (Figure 10) were considered to evaluate each kinetic model. Subsequently, a linear fit was applied for each model, and the correlation coefficient (R^2) was determined. Table 4 shows the results obtained.

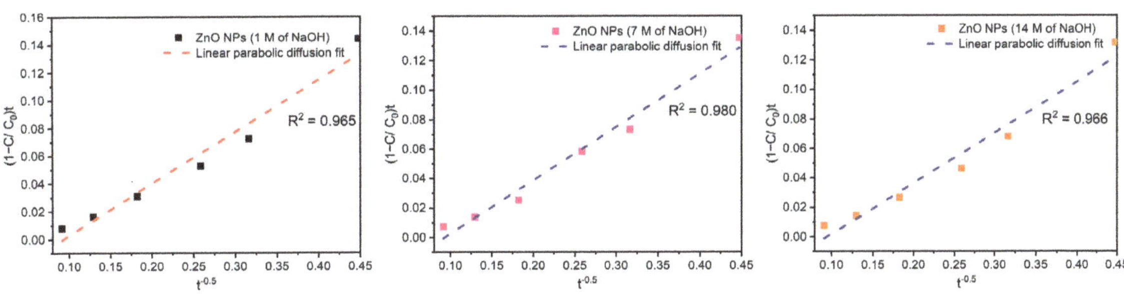

(a) (b) (c)

Figure 10. Graphic results of the linear fit applied to the parabolic diffusion kinetics model in the ZnO NPs synthesized by (**a**) 1 M, (**b**) 7 M, and (**c**) 14 M of NaOH.

Table 4. R^2 values for various kinetic models applied to the photocatalytic degradation of dye using ZnO NPs synthesized under different conditions.

ZnO NPs (Synthesis)	First-Order	Second-Order	Parabolic-Diffusion	Modified-Freundlich
	R^2			
1 M NaOH	0.8751	0.9635	0.9655	0.9052
7 M NaOH	0.5113	0.7091	0.9807	0.6489
14 M NaOH	0.8642	0.9625	0.9652	0.9674

From Table 4, the parabolic diffusion model best represents the degradation process of the MG dye using the ZnO NPs synthesized through the three different methods. This model yielded R^2 values of 0.965, 0.980, and 0.966 for the ZnO NPs synthesized by 1 M, 7 M, and 14 M NaOH, respectively. Figure 10 presents the graphical results of the linear fit applied to the parabolic diffusion kinetics model. This model is particularly noteworthy as it shows R^2 values closest to 1. In this model, the degradation process is characterized by the dye's adsorption onto the active sites of the photocatalyst [10].

The adsorption kinetics of dye onto ZnO NPs synthesized under different conditions reveal distinct trends in Figure 11. Within the first 20 min, a rapid decrease in C/C_2 indicates a fast adsorption. This suggests strong dye–ZnO NPs interactions, likely driven by electrostatic forces or hydrogen bonding [39]. As time progresses beyond 20 min, the adsorption rate slows, indicating a transition to a slower equilibrium phase influenced by diffusion limitations and saturation of available surface sites. Among the different ZnO NPs, those synthesized with 14 M NaOH exhibited the highest adsorption efficiency, followed by the 7 M NaOH sample. At the same time, the 1 M of NaOH sample reveals the lowest adsorption capability.

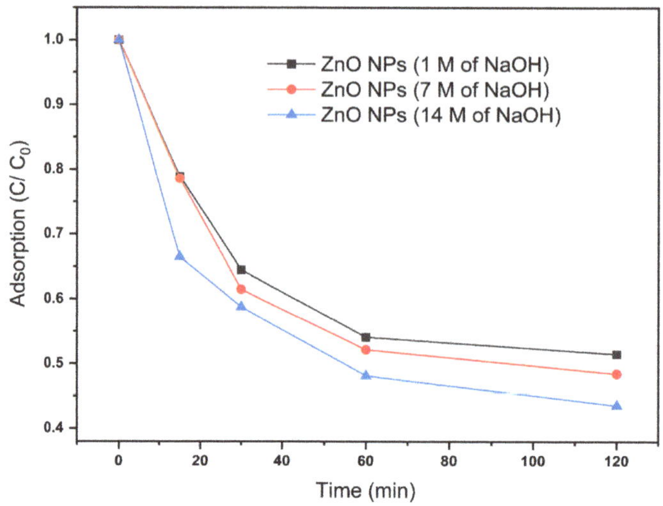

Figure 11. Adsorption kinetics of dye onto ZnO NPs synthesized under different conditions.

3.8. Reusability Test

The photocatalytic degradation of MG under sunlight was evaluated across five cycles using ZnO NPs synthesized with varying NaOH molarities (1 M, 7 M, and 14 M), and structural changes were assessed using Raman spectroscopy. To preserve the same amounts of ZnO NPs during the reusability tests, the recipients containing the ZnO NPs were heated to 120 °C. This causes the residual water to evaporate, preserving the same amount of ZnO NPs in subsequent cycles. Figure 12a, representing the ZnO NPs synthesized via 1 M NaOH, shows that the catalyst demonstrated high initial performance, achieving over 90% degradation within the first 120 min. However, by the fifth cycle, the efficiency decreased by less than 10%, likely due to surface fouling or ZnO NP agglomeration. In contrast, Figure 12b shows ZnO NPs synthesized with 7 M NaOH, where the first cycle also reached about 90% degradation. By the fifth cycle, however, performance dropped by approximately 10%, suggesting that the 7 M NaOH formulation resulted in a less stable catalyst, possibly due to the accumulation of byproducts or ineffective recovery. Figure 12c, representing ZnO NPs synthesized with 14 M of NaOH, displayed the most consistent performance, with the first cycle achieving over 95% degradation and only a slight decrease of less than 10% by the fifth cycle. This suggests that higher NaOH molarity results in a more stable photocatalyst capable of maintaining high degradation rates over multiple cycles [40–42]. These findings highlight the significant role of synthesis conditions in optimizing the stability and long-term effectiveness of ZnO-based photocatalysts.

The gradual decline in photocatalytic activity observed in the ZnO NPs over five cycles align with previous studies on reusing green-synthesized ZnO NPs. The ZnO NPs synthesized using *Passiflora foetida* demonstrated excellent reusability in degrading methylene blue and rhodamine B dyes, with efficiency decreasing from 93.25% to 86.63% for methylene blue and from 91.06% to 83.61% for rhodamine B over five cycles [36]. Similarly, ZnO NPs synthesized with *Aegle marmelos* showed a slight 3% reduction in degradation efficiency per cycle. Despite an overall 15% decline over five cycles, the photocatalytic performance remained relatively stable [43]. ZnO NPs synthesized using an aqueous *Punica granatum* extract also showed a gradual decrease in methylene blue photodegradation efficiency, from 92.3% in the first cycle to 88.2% after the fifth cycle [44]. In contrast, ZnO NPs fabricated with a leaf extract of *Rosmarinus officinalis* demonstrated a reduction in methylene blue and crystal violet degradation efficiency, from 99.9% and 99.6%

in the first cycle to 92.5% and 91.2% after three cycles, with a 7–8% efficiency reduction [45]. For ZnO NPs synthesized using *Nauclea latifolia* for methyl green dye removal, a similar decline in efficiency was observed. Residual dye molecules from previous cycles can occupy adsorption sites, reducing the number of active sites available. After the third cycle, efficiencies were recorded at approximately 60% [40]. Nonetheless, good efficiencies were still achieved after three cycles, highlighting the potential of NPs as efficient and cost-effective adsorbents for dye removal from aqueous solutions. This reusability is especially advantageous for lowering practical application costs.

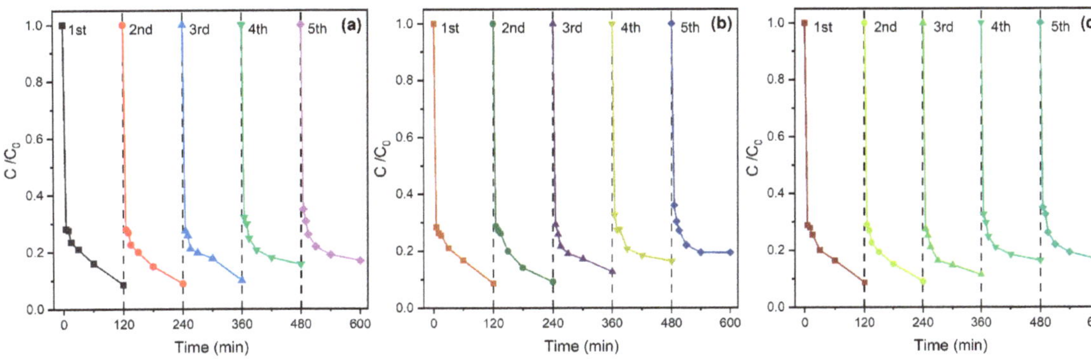

Figure 12. Cycling test of the photocatalytic degradation under sunlight irradiation of a malachite green aqueous solution in the presence of the ZnO NPs synthesized by (**a**) 1 M, (**b**) 7 M, and (**c**) 14 M of NaOH.

3.9. Raman Analysis

The ZnO NPs were also analyzed by Raman spectroscopy before and after being used as a photocatalyst, as shown in Figures 13a and 13b respectively. Both spectra reveal the typical ZnO wurtzite crystal structure in the P63mc space group. This is a hexagonal lattice characterized by two interpenetrating sub-lattices of Zn^{2+} and O^{2-} ions, such that if the crystal has no imperfections, each Zn^{2+} ion will be surrounded by a tetrahedron of oxygen ions and vice versa [46]. This leads to polar symmetry along the vertical axes of a hexagon (c-axis). This structure exhibits first and second-order peaks in the Raman spectrum, corresponding to E_2^{low}, $E_1^{high}-E_2^{low}$, A_1 transverse optical (TO), E_2^{high}, E_1 longitudinal optical (LO), E_1 (TO), $2A_1$ (LO), and $2E_2$ (low). The range of assignments can be observed in Table 5.

On the other hand, the intensity of the assigned signals largely depends on the synthesis method. In particular, E_2^{high}, $2A_1$ (LO), and E_1 (LO) are the most sensitive for determining defects or disorders in the ZnO lattice associated with changes in oxygen vacancies and interstitial Zn. The E_2 (LO) mode is associated solely with the vibrations of the Zn atoms, while the E_2^{high} mode is associated only with the vibration of the oxygen atoms in the Zn–O interaction [47]. In our particular case, the E_2^{high} mode was located at 437–439 cm^{-1}, while E_1 (LO) was located between 591–594 cm^{-1}. Otherwise, the spectra were normalized to the intensity corresponding to E_2^{low} for each sample to determine possible defects associated with vacancies in the ZnO lattice. Once the spectra were normalized and deconvoluted, the E_2^{high}/E_1 (LO) ratio was calculated (Table 6). These results reveal that the crystalline structure of ZnO is maintained before and after the photocatalytic reaction, and the sample that exhibited the highest concentration of defects was the one containing the most NaOH. At the same time, the one with the least was the 1 M NaOH. These results are consistent with those obtained from XPS and XRD.

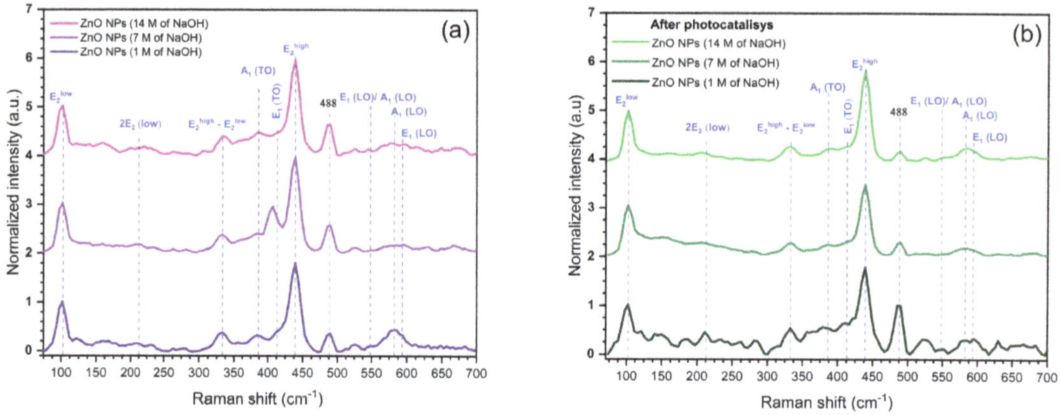

Figure 13. Raman spectrum for the ZnO NPs (**a**) before and (**b**) after being used as a photocatalyst.

Table 5. Assigned first and second-order Raman mode for ZnO.

Process	Raman Shift (cm^{-1})
A_1 (TO)	378–380
E_1 (TO)	409–413
A_1 (LO)	574–579
E_1 (LO)	583–594
E_2^{low}	99–102
E_2^{high}	437–439
$E_2^{high}-E_2^{low}$	330–333
$2A_1$ (LO)	1146–1158
$2E_2^{low}$	201–205

Table 6. Changes in the E_2^{high}/E_1 (LO) ratio in photocatalytic treatment.

ZnO NPs (Synthesis)	Before			After		
	E_2^{high}	E_1 (LO)	E_2^{high}/E_1 (LO)	E_2^{high}	E_1 (LO)	E_2^{high}/E_1 (LO)
1 M NaOH	1.81	0.34	5.3	1.77	0.31	5.6
7 M NaOH	1.93	0.22	8.7	1.82	0.22	8.2
14 M NaOH	1.94	0.15	12.5	1.36	0.11	12.3

3.10. Scavengers Test

To identify the contribution of the main reactive oxygen species (h^+, O_2^-, $OH^\bullet$) involved in the degradation process of MG dye using ZnO NPs, three different scavengers were used: disodium ethylenediamine tetraacetic acid (EDTA) for holes (h^+), p-benzoquinone for superoxide radicals (O_2^-), and isopropyl alcohol for $OH^\bullet$ [48]. In this setup, 5×10^{-3} M of each scavenger was added to three separate 20 mL of MG dye, each containing 5 mg of ZnO NPs as the photocatalyst. All other experimental conditions remained the same as those used in the previous photocatalytic experiments. Figure 14 shows the graphical results of this test. As observed, in all three cases, the main ROS involved in the photodegradation process were h^+ and $OH^\bullet$, while the least involved was O_2^-.

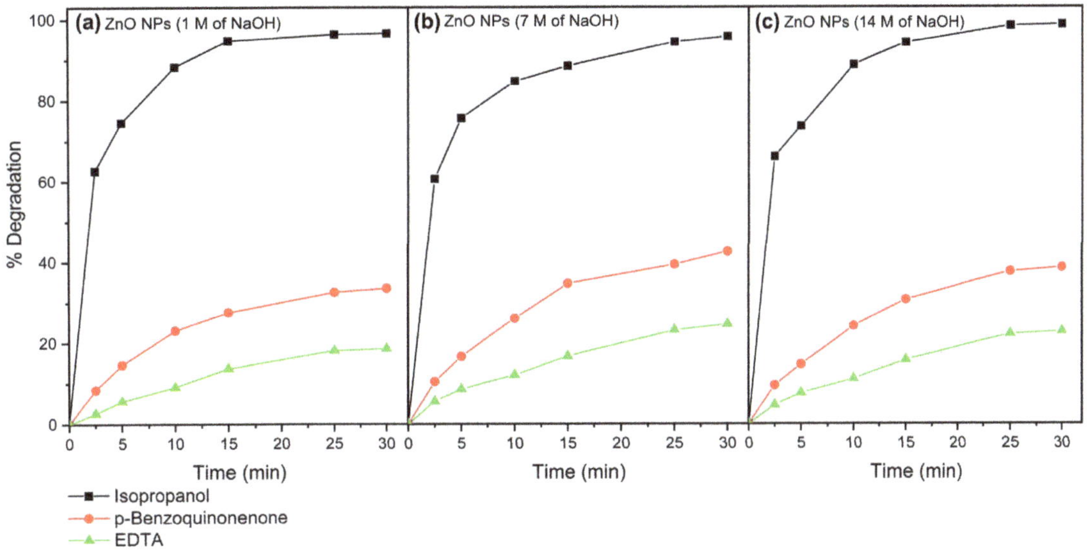

Figure 14. Degradation percentage using different scavengers under sunlight.

Scavengers test using isopropanol, EDTA, and p-benzoquinone revealed strong evidence for the dominant role of $OH^{\bullet}$ in the photocatalytic degradation of organic dyes using green-synthesized ZnO NPs. These NPs, derived from plant extracts including *Lupinus albus* [49], *Padina pavonica* [50], *P. granatum* [44], *P. foetida* [51], and *Hevea brasiliensis* [52], were tested against dyes such as MG, methylene blue, and rhodamine B. The findings consistently indicate that $OH^{\bullet}$ is the main ROS responsible for dye degradation, while O_2^{-} and h^{+} appear to play a secondary or indirect role.

3.11. Mechanism of Photodegradation

Based on the results obtained from the tests with scavengers, a mechanism for the photodegradation of the MG dye was proposed (Figure 15). This mechanism involves several chemical reactions and is based on generating an e^{-}/h^{+} pair. The proposed mechanism is discussed in detail below:

$$\text{ZnONPs}_{(\text{surface})} + h\upsilon_{(\text{photon})} \rightarrow e^{-}_{(\text{CB})} + h^{+}_{(\text{VB})} \quad (9)$$

$$O_2 + e^{-}_{(\text{CB})} \rightarrow O_2^{-} \quad (10)$$

$$H_2O + h^{+}_{(\text{VB})} \rightarrow OH^{\bullet} + H^{+} \quad (11)$$

$$O_2^{-} + H^{+} \rightarrow OOH^{\bullet} \quad (12)$$

$$2(OOH^{\bullet}) \rightarrow O_2 + H_2O_2 \quad (13)$$

$$H_2O_2 + O_2^{-} \rightarrow OH^{-} + OH^{\bullet} + O_2 \quad (14)$$

$$OH^{-} + h^{+}_{VB} \rightarrow OH^{\bullet} \quad (15)$$

$$OH^{\bullet} + MG_{\text{dye}} \rightarrow \text{Degradation products} \quad (16)$$

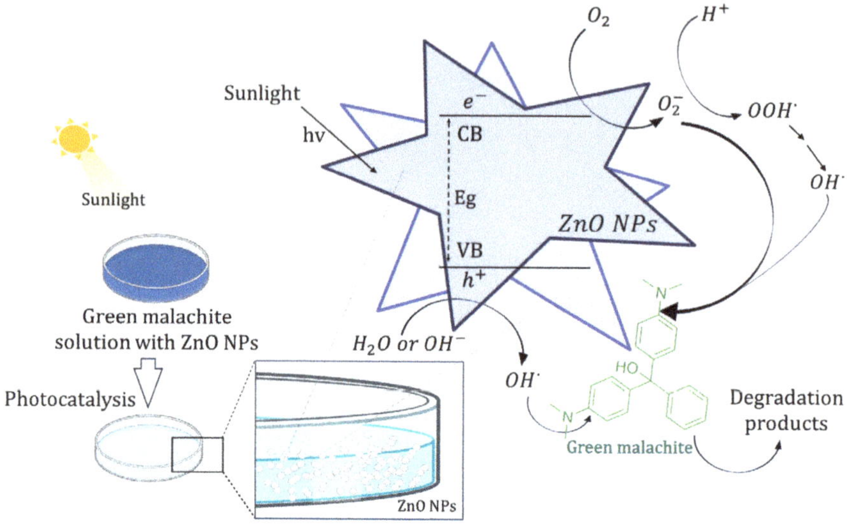

Figure 15. The photocatalytic degradation process of the malachite green dye using ZnO NPs as the photocatalyst exposed to sunlight.

When incident photons (hν) with energy greater than the E_g energy of ZnO are absorbed on its surface, electrons from the valence band (VB) are excited to the conduction band (CB), promoting the separation of charge carriers (e^- / h^+) (Equation (9)). The e^- in the CB can react with molecular oxygen to form $O_2{}^-$ (Equation (10)). In contrast, the h^+ in the VB can react with water and the OH^- to produce $OH^\bullet$ and protons (H^+) (Equations (11) and (15)). On the other hand, electrons in the CB also promote the formation of additional oxygen species, particularly hydroperoxyl ($OOH^\bullet$) and hydrogen peroxide (H_2O_2) (Equations (12)–(14)), which contribute to the degradation of the MG dye, albeit to a lesser extent. Finally, $OH^\bullet$, being the main ROS involved, contributes mainly to the degradation of this dye (Equation (15)).

4. Conclusions

Green-synthesized ZnO NPs were successfully synthesized using a low-cost and eco-friendly method involving aqueous *V. corymbosum* extract as efficient photocatalysts for environmental remediation. Controlling NaOH levels yielded ZnO NPs with a stable, flower-like morphology confirmed by XRD, TEM, and SEM. The results obtained from the photodegradation process of the MG dye using the ZnO NPs synthesized by the three different methods showed a photocatalytic efficiency greater than 80% in 30 min under sunlight. Additionally, a relationship was observed between the NaOH concentration used in each synthesis and the photocatalytic efficiency achieved. Specifically, 40%, 54%, and 60% of photocatalytic efficiencies were achieved in 5 min for the ZnO NPs synthesized by 1 M NaOH and using NaOH concentrations of 7 M and 14 M, respectively.

On the other hand, the concentration of NaOH also influenced the E_g of the synthesized materials. Specifically, the ZnO NPs demonstrated decreased E_g with an increased NaOH concentration. The ZnO NPs produced by infusion (where 1 M NaOH was used) had an E_g of 3.27 eV, while those prepared with a NaOH concentration of 14 M had an E_g of 3.17 eV. This decrease was attributed to the formation of defects, mainly due to oxygen vacancies. The morphology of each ZnO NP also underwent significant changes due to the concentration of NaOH. The ZnO NPs synthesized with a lower concentration of NaOH exhibited a pseudo-spherical shape with an average diameter of 180 nm. Conversely, a flower-like morphology was observed with increased NaOH usage. Additionally, the size of the flower-like nanostructures was noted to increase as the concentration of NaOH rose.

In summary, the ZnO NPs synthesized using the three different methods present a promising solution for wastewater treatment, demonstrating the potential for effectively and affordably reducing environmental pollutants.

Author Contributions: Conceptualization, M.Á.L.-Á., J.M.S.-J. and M.E.M.-R.; methodology, I.B.-L., O.F.-S., A.M.-G. and P.O.-G.; software, P.O.-G. and I.B.-L.; validation, R.I.L.-R. and P.O.-G.; formal analysis, I.B.-L., M.Á.L.-Á., M.E.M.-R. and J.M.S.-J.; investigation, I.B.-L., M.Á.L.-Á. and J.M.S.-J.; resources, M.Á.L.-Á., M.E.M.-R. and J.M.S.-J.; data curation, P.O.-G., O.F.-S., A.M.-G. and R.I.L.-R.; writing—original draft preparation, I.B.-L., M.Á.L.-Á. and J.M.S.-J.; writing—review and editing, J.M.S.-J. and M.E.M.-R.; visualization, M.Á.L.-Á. and J.M.S.-J.; supervision, J.M.S.-J. and M.E.M.-R. All authors have read and agreed to the published version of the manuscript.

Funding: This research received no external funding.

Institutional Review Board Statement: Not applicable.

Informed Consent Statement: Not applicable.

Data Availability Statement: The original contributions presented in this study are included in the article. Further inquiries can be directed to the corresponding authors.

Acknowledgments: This work was financially supported by Centro Universitario de Ciencias Exactas e Ingenierías, Universidad de Guadalajara, Mexico, and Consejo Nacional de Humanidades, Ciencias y Tecnologías (CONHACYT) for the first author's postdoctoral scholarship, CVU 588976. We thank Jose Antonio Rivera Mayorga (Universidad de Guadalajara) for his assistance with XPS measurements.

Conflicts of Interest: The authors declare no conflicts of interest.

References

1. Ong, C.B.; Ng, L.Y.; Mohammad, A.W. A Review of ZnO Nanoparticles as Solar Photocatalysts: Synthesis, Mechanisms and Applications. *Renew. Sustain. Energy Rev.* **2018**, *81*, 536–551. [CrossRef]
2. Agarwal, H.; Venkat Kumar, S.; Rajeshkumar, S. A Review on Green Synthesis of Zinc Oxide Nanoparticles—An Eco-Friendly Approach. *Resour. Effic. Technol.* **2017**, *3*, 406–413. [CrossRef]
3. Madkour, L.H. Ecofriendly Green Biosynthesized of Metallic Nanoparticles: Bio-Reduction Mechanism, Characterization and Pharmaceutical Applications in Biotechnology Industry. *Drugs Ther.* **2017**, *3*, 1–11. [CrossRef]
4. Varadavenkatesan, T.; Lyubchik, E.; Pai, S.; Pugazhendhi, A.; Vinayagam, R.; Selvaraj, R. Photocatalytic Degradation of Rhodamine B by Zinc Oxide Nanoparticles Synthesized Using the Leaf Extract of Cyanometra Ramiflora. *J. Photochem. Photobiol. B* **2019**, *199*, 111621. [CrossRef] [PubMed]
5. Tiwari, N.; Chakrabortty, S.; Samal, K.; Moulick, S.; Mohapatra, B.G.; Samanta, S.; Mohapatra, P.K.; Sanjay, K.; Nayak, J.; Banerjee, S.; et al. Photocatalytic Degradation of Malachite Green Using TiO_2 and ZnO Impregnated on Fecal Sludge Derived Biochar. *J. Taiwan. Inst. Chem. Eng.* **2023**, *145*, 104800. [CrossRef]
6. Vasantharaj, S.; Sathiyavimal, S.; Senthilkumar, P.; Kalpana, V.N.; Rajalakshmi, G.; Alsehli, M.; Elfasakhany, A.; Pugazhendhi, A. Enhanced Photocatalytic Degradation of Water Pollutants Using Bio-Green Synthesis of Zinc Oxide Nanoparticles (ZnO NPs). *J. Environ. Chem. Eng.* **2021**, *9*, 105772. [CrossRef]
7. Suresh, D.; Nethravathi, P.C.; Udayabhanu; Rajanaika, H.; Nagabhushana, H.; Sharma, S.C. Green Synthesis of Multifunctional Zinc Oxide (ZnO) Nanoparticles Using Cassia Fistula Plant Extract and Their Photodegradative, Antioxidant and Antibacterial Activities. *Mater. Sci. Semicond. Process* **2015**, *31*, 446–454. [CrossRef]
8. Nguyen, D.T.C.; Le, H.T.N.; Nguyen, T.T.; Nguyen, T.T.T.; Bach, L.G.; Nguyen, T.D.; Tran, T. Van Multifunctional ZnO Nanoparticles Bio-Fabricated from Canna Indica L. Flowers for Seed Germination, Adsorption, and Photocatalytic Degradation of Organic Dyes. *J. Hazard. Mater.* **2021**, *420*, 126586. [CrossRef] [PubMed]
9. Manojkumar, U.; Kaliannan, D.; Srinivasan, V.; Balasubramanian, B.; Kamyab, H.; Mussa, Z.H.; Palaniyappan, J.; Mesbah, M.; Chelliapan, S.; Palaninaicker, S. Green Synthesis of Zinc Oxide Nanoparticles Using *Brassica oleracea* Var. Botrytis Leaf Extract: Photocatalytic, Antimicrobial and Larvicidal Activity. *Chemosphere* **2023**, *323*, 138263. [CrossRef] [PubMed]

10. Ali, Z.A.; Shudirman, I.; Yahya, R.; Venkatraman, G.; Hirad, A.H.; Ansari, S.A. Green Synthesis of ZnO Nanostructures Using Pyrus Pyrifolia: Antimicrobial, Photocatalytic and Dielectric Properties. *Crystals* **2022**, *12*, 1808. [CrossRef]
11. Fowsiya, J.; Madhumitha, G.; Al-Dhabi, N.A.; Arasu, M.V. Photocatalytic Degradation of Congo Red Using *Carissa edulis* Extract Capped Zinc Oxide Nanoparticles. *J. Photochem. Photobiol. B* **2016**, *162*, 395–401. [CrossRef]
12. Vidya, C.; Manjunatha, C.; Chandraprabha, M.N.; Rajshekar, M.; Raj, M.A.L.A. Hazard Free Green Synthesis of ZnO Nano-Photo-Catalyst Using Artocarpus Heterophyllus Leaf Extract for the Degradation of Congo Red Dye in Water Treatment Applications. *J. Environ. Chem. Eng.* **2017**, *5*, 3172–3180. [CrossRef]
13. Vidya, C.; Prabha, M.N.C.; Raj, M.A.L.A. Green Mediated Synthesis of Zinc Oxide Nanoparticles for the Photocatalytic Degradation of Rose Bengal Dye. *Environ. Nanotechnol. Monit. Manag.* **2016**, *6*, 134–138. [CrossRef]
14. Siripireddy, B.; Mandal, B.K. Facile Green Synthesis of Zinc Oxide Nanoparticles by *Eucalyptus globulus* and Their Photocatalytic and Antioxidant Activity. *Adv. Powder Technol.* **2017**, *28*, 785–797. [CrossRef]
15. Haspulat Taymaz, B.; Demir, M.; Kamış, H.; Orhan, H.; Aydoğan, Z.; Akıllı, A. Facile and Green Synthesis of ZnO Nanoparticles for Effective Photocatalytic Degradation of Organic Dyes and Real Textile Wastewater. *Int. J. Phytoremediat.* **2023**, *25*, 1306–1317. [CrossRef] [PubMed]
16. Golmohammadi, M.; Honarmand, M.; Ghanbari, S. A Green Approach to Synthesis of ZnO Nanoparticles Using Jujube Fruit Extract and Their Application in Photocatalytic Degradation of Organic Dyes. *Spectrochim. Acta A Mol. Biomol. Spectrosc.* **2020**, *229*, 117961. [CrossRef] [PubMed]
17. Prasad, A.R.; Garvasis, J.; Oruvil, S.K.; Joseph, A. Bio-Inspired Green Synthesis of Zinc Oxide Nanoparticles Using *abelmoschus Esculentus* Mucilage and Selective Degradation of Cationic Dye Pollutants. *J. Phys. Chem. Solids* **2019**, *127*, 265–274. [CrossRef]
18. Duraimurugan, J.; Shanavas, S.; Ramesh, R.; Acevedo, R.; Anbarasan, P.M.; Maadeswaran, P. Hydrothermal Assisted Phytofabrication of Zinc Oxide Nanoparticles with Different Nanoscale Characteristics for the Photocatlytic Degradation of Rhodamine B. *Optik* **2020**, *202*, 163607. [CrossRef]
19. Katiyar, A.; Kumar, N.; Shukla, R.K.; Srivastava, A. Influence of Alkali Hydroxides on Synthesis, Physico-Chemical and Photoluminescence Properties of Zinc Oxide Nanoparticles. *Mater. Today Proc.* **2020**, *29*, 885–889. [CrossRef]
20. Xiong, G.; Pal, U.; Serrano, J.G.; Ucer, K.B.; Williams, R.T. Photoluminesence and FTIR Study of ZnO Nanoparticles: The Impurity and Defect Perspective. *Phys. Status Solidi C* **2006**, *3*, 3577–3581. [CrossRef]
21. Peng, G.; Chou, N.-N.; Lin, Y.-S.; Yang, C.-F.; Meen, T.-H. Comparison of the Degradation Effect of Methylene Blue for ZnO Nanorods Synthesized on Silicon and Indium Tin Oxide Substrates. *Materials* **2023**, *16*, 4275. [CrossRef] [PubMed]
22. Guye, M.E.; Egualle, F.K.; Appiah-Ntiamoah, R.; Kassahun, S.K.; Kim, H. Synergetic Effect of Optimized β-Zn(OH)2/ZnO Heterostructure towards Electrochemical Dopamine Detection. *J. Alloys Compd.* **2024**, *1002*, 175184. [CrossRef]
23. Ju, Y.; Yang, S.; Ding, Y.; Sun, C.; Zhang, A.; Wang, L. Microwave-Assisted Rapid Photocatalytic Degradation of Malachite Green in TiO2 Suspensions: Mechanism and Pathways. *J. Phys. Chem. A* **2008**, *112*, 11172–11177. [CrossRef]
24. Lu, C.-S.; Chen, C.-C.; Mai, F.-D.; Wu, Y.-C. Photocatalytic Degradation of Michler's Ethyl Ketone in Titanium Dioxide Dispersions under UV Irradiation. *J. Photochem. Photobiol. A Chem.* **2007**, *187*, 167–176. [CrossRef]
25. Ju, Y.; Yang, S.; Ding, Y.; Sun, C.; Gu, C.; He, Z.; Qin, C.; He, H.; Xu, B. Microwave-Enhanced H_2O_2-Based Process for Treating Aqueous Malachite Green Solutions: Intermediates and Degradation Mechanism. *J. Hazard. Mater.* **2009**, *171*, 123–132. [CrossRef] [PubMed]
26. Ju, Y.; Qiao, J.; Peng, X.; Xu, Z.; Fang, J.; Yang, S.; Sun, C. Photodegradation of Malachite Green Using UV–Vis Light from Two Microwave-Powered Electrodeless Discharge Lamps (MPEDL-2): Further Investigation on Products, Dominant Routes and Mechanism. *Chem. Eng. J.* **2013**, *221*, 353–362. [CrossRef]
27. Kumar, M.R.A.; Ravikumar, C.R.; Nagaswarupa, H.P.; Purshotam, B.; Gonfa, B.; Murthy, H.C.A.; Sabir, F.K.; Tadesse, S. Evaluation of Bi-Functional Applications of ZnO Nanoparticles Prepared by Green and Chemical Methods. *J. Environ. Chem. Eng.* **2019**, *7*, 103468. [CrossRef]
28. Pavan Kumar, M.A.; Suresh, D.; Nagabhushana, H.; Sharma, S.C. Beta Vulgaris Aided Green Synthesis of ZnO Nanoparticles and Their Luminescence, Photocatalytic and Antioxidant Properties. *Eur. Phys. J. Plus* **2015**, *130*, 109. [CrossRef]
29. Park, J.K.; Rupa, E.J.; Arif, M.H.; Li, J.F.; Anandapadmanaban, G.; Kang, J.P.; Kim, M.; Ahn, J.C.; Akter, R.; Yang, D.C.; et al. Synthesis of Zinc Oxide Nanoparticles from Gynostemma Pentaphyllum Extracts and Assessment of Photocatalytic Properties through Malachite Green Dye Decolorization under UV Illumination—A Green Approach. *Optik* **2021**, *239*, 166249. [CrossRef]
30. Liu, Y.C.; Li, J.; Ahn, J.; Pu, J.; Rupa, E.J.; Huo, Y.; Yang, D.C. Biosynthesis of Zinc Oxide Nanoparticles by One-Pot Green Synthesis Using Fruit Extract of *Amomum longiligulare* and Its Activity as a Photocatalyst. *Optik* **2020**, *218*, 165245. [CrossRef]
31. Devi, S.A.; Singh, K.J.; Devi, K.N. A Comparative Study on the Photocatalytic Activity of Eucalyptus Leaf Assisted Green Synthesized ZnO and Chemically Synthesized ZnO towards the Degradation of Malachite Green Dye. *Integr. Ferroelectr.* **2020**, *205*, 38–51. [CrossRef]

32. Rupa, E.J.; Anandapadmanaban, G.; Chokkalingam, M.; Li, J.F.; Markus, J.; Soshnikova, V.; Perez, Z.E.J.; Yang, D.-C. Cationic and Anionic Dye Degradation Activity of Zinc Oxide Nanoparticles from Hippophae Rhamnoides Leaves as Potential Water Treatment Resource. *Optik* **2019**, *181*, 1091–1098. [CrossRef]
33. Hasan, I.M.A.; Tawfik, A.R.; Assaf, F.H. GC/MS Screening of Buckthorn Phytochemicals and Their Use to Synthesize ZnO Nanoparticles for Photocatalytic Degradation of Malachite Green Dye in Water. *Water Sci. Technol.* **2021**, *85*, 664–684. [CrossRef]
34. Rajapriya, M.; Sharmili, S.A.; Baskar, R.; Balaji, R.; Alharbi, N.S.; Kadaikunnan, S.; Khaled, J.M.; Alanzi, K.F.; Vaseeharan, B. Synthesis and Characterization of Zinc Oxide Nanoparticles Using *Cynara scolymus* Leaves: Enhanced Hemolytic, Antimicrobial, Antiproliferative, and Photocatalytic Activity. *J. Clust. Sci.* **2020**, *31*, 791–801. [CrossRef]
35. Singh, A.K.; Nakate, U.T. Photocatalytic Properties of Microwave-Synthesized TiO_2 and ZnO Nanoparticles Using Malachite Green Dye. *J. Nanoparticles* **2013**, *2013*, 310809. [CrossRef]
36. Suresh, D.; Nethravathi, P.C.; Udayabhanu; Pavan Kumar, M.A.; Raja Naika, H.; Nagabhushana, H.; Sharma, S.C. Chironji Mediated Facile Green Synthesis of ZnO Nanoparticles and Their Photoluminescence, Photodegradative, Antimicrobial and Antioxidant Activities. *Mater. Sci. Semicond. Process* **2015**, *40*, 759–765. [CrossRef]
37. Aminuzzaman, M.; Ying, L.P.; Goh, W.-S.; Watanabe, A. Green Synthesis of Zinc Oxide Nanoparticles Using Aqueous Extract of Garcinia Mangostana Fruit Pericarp and Their Photocatalytic Activity. *Bull. Mater. Sci.* **2018**, *41*, 50. [CrossRef]
38. Ekennia, A.C.; Uduagwu, D.N.; Nwaji, N.N.; Oje, O.O.; Emma-Uba, C.O.; Mgbii, S.I.; Olowo, O.J.; Nwanji, O.L. Green Synthesis of Biogenic Zinc Oxide Nanoflower as Dual Agent for Photodegradation of an Organic Dye and Tyrosinase Inhibitor. *J. Inorg. Organomet. Polym. Mater.* **2021**, *31*, 886–897. [CrossRef]
39. Abegunde, S.M.; Adebayo, M.A.; Olasehinde, E.F. Green Synthesis of ZnO Nanoparticles and Its Application for Methyl Green Dye Adsorption. *Green Energy Resour.* **2024**, *2*, 100073. [CrossRef]
40. Elavarasan, N.; Kokila, K.; Inbasekar, G.; Sujatha, V. Evaluation of Photocatalytic Activity, Antibacterial and Cytotoxic Effects of Green Synthesized ZnO Nanoparticles by Sechium Edule Leaf Extract. *Res. Chem. Intermed.* **2017**, *43*, 3361–3376. [CrossRef]
41. Abomuti, M.A.; Danish, E.Y.; Firoz, A.; Hasan, N.; Malik, M.A. Green Synthesis of Zinc Oxide Nanoparticles Using Salvia Officinalis Leaf Extract and Their Photocatalytic and Antifungal Activities. *Biology* **2021**, *10*, 1075. [CrossRef]
42. Surendra, B.S.; Mallikarjunaswamy, C.; Pramila, S.; Rekha, N.D. Bio-Mediated Synthesis of ZnO Nanoparticles Using Lantana Camara Flower Extract: Its Characterizations, Photocatalytic, Electrochemical and Anti-Inflammatory Applications. *Environ. Nanotechnol. Monit. Manag.* **2021**, *15*, 100442. [CrossRef]
43. Mallikarjunaswamy, C.; Lakshmi Ranganatha, V.; Ramu, R.; Udayabhanu; Nagaraju, G. Facile Microwave-Assisted Green Synthesis of ZnO Nanoparticles: Application to Photodegradation, Antibacterial and Antioxidant. *J. Mater. Sci. Mater. Electron.* **2020**, *31*, 1004–1021. [CrossRef]
44. Fouda, A.; Saied, E.; Eid, A.M.; Kouadri, F.; Alemam, A.M.; Hamza, M.F.; Alharbi, M.; Elkelish, A.; Hassan, S.E.-D. Green Synthesis of Zinc Oxide Nanoparticles Using an Aqueous Extract of Punica Granatum for Antimicrobial and Catalytic Activity. *J. Funct. Biomater.* **2023**, *14*, 205. [CrossRef]
45. Saad Algarni, T.; Abduh, N.A.Y.; Al Kahtani, A.; Aouissi, A. Photocatalytic Degradation of Some Dyes under Solar Light Irradiation Using ZnO Nanoparticles Synthesized from Rosmarinus Officinalis Extract. *Green Chem. Lett. Rev.* **2022**, *15*, 460–473. [CrossRef]
46. Calizo, I.G.; Alim, K.A.; Fonoberov, V.A.; Krishnakumar, S.K.M.; Shamsa, M.; Balandin, A.A.; Kurtz, R.M. Micro-Raman Spectroscopic Characterization of ZnO Quantum Dots, Nanocrystals, and Nanowires. In Proceedings of the SPIE OPTO Integrated Optoelectronic Devices, San Jose, CA, USA, 20–25 January 2007.
47. Momot, A.; Amini, M.N.; Reekmans, G.; Lamoen, D.; Partoens, B.; Slocombe, D.R.; Elen, K.; Adriaensens, P.; Hardy, A.; Van Bael, M.K. A Novel Explanation for the Increased Conductivity in Annealed Al-Doped ZnO: An Insight into Migration of Aluminum and Displacement of Zinc. *Phys. Chem. Chem. Phys.* **2017**, *19*, 27866–27877. [CrossRef]
48. Selvaraj, S.; Vangari, G.A.; Mohan, M.K.; Ponnusamy, S.; Muthamizchelvan, C. Facile Synthesis of Sm Doped ZnO Nanoflowers by Co-Precipitation Method for Enhanced Photocatalytic Degradation of MB Dye under Sunlight Irradiation. *Ceram. Int.* **2022**, *48*, 29049–29058. [CrossRef]
49. Bizuayen Adamu, T.; Melese Mengesha, A.; Assefa Kebede, M.; Lake Bogale, B.; Walle Kassa, T. Facile Biosynthesis of Zinc Oxide Nanoparticles (ZnO NPs) Using *Lupinus albus* L. (Gibto) Seed Extract for Antibacterial and Photocatalytic Applications. *Results Chem.* **2024**, *10*, 101724. [CrossRef]
50. Alprol, A.E.; Eleryan, A.; Abouelwafa, A.; Gad, A.M.; Hamad, T.M. Green Synthesis of Zinc Oxide Nanoparticles Using Padina Pavonica Extract for Efficient Photocatalytic Removal of Methylene Blue. *Sci. Rep.* **2024**, *14*, 32160. [CrossRef] [PubMed]

51. Khan, M.; Ware, P.; Shimpi, N. Synthesis of ZnO Nanoparticles Using Peels of Passiflora Foetida and Study of Its Activity as an Efficient Catalyst for the Degradation of Hazardous Organic Dye. *SN Appl. Sci.* **2021**, *3*, 528. [CrossRef]
52. Maijan, P.; Waen-Ngoen, T.; Suwanboon, S.; Chantarak, S.; Voravuthikunchai, S.P. Sustainable Environmental-Based Synthesis of Zinc Oxide Nanoparticles Using Para Rubber Leaf Extract for Photocatalytic Degradation of Organic Pollutants and Microbial Control in Wastewater Treatment. *Inorg. Chem. Commun.* **2024**, *162*, 112086. [CrossRef]

ceramics

MDPI

Article

Tracking Secondary Raw Material Operational Framework—DataOps Case Study

Gabriel Pestana [1,2,*], Marisa Almeida [3] and Nelson Martins [4]

1 INOV—Instituto de Engenharia de Sistemas e Computadores Inovação, 1000-029 Lisbon, Portugal
2 School of Technology, Setubal Polytechnic University, 2910-761 Setubal, Portugal
3 CTCV—Centro Tecnológico da Cerâmica e do Vidro, 3025-307 Coimbra, Portugal; marisa@ctcv.pt
4 APICER—Associação Portuguesa da Indústria Cerâmica e da Cristalaria, 3025-307 Coimbra, Portugal; nmartins@apicer.pt
* Correspondence: gabriel.pestana@inov.pt

Abstract: The ceramic and glass industries, integral to the EU Emissions Trading System (EU ETS), face significant challenges in achieving decarbonization despite advancements in energy efficiency. The circular economy offers a promising pathway, emphasizing the reuse and recycling of waste materials into secondary raw materials (SRMs) to reduce resource consumption and emissions. This study investigates a standardized waste supply chain framework, developed collaboratively with stakeholders, tailored for the ceramic sector. The Waste Resource Platform (WRP) integrates Industry 4.0 paradigms, utilizing a modular, layered architecture and a process-centric design. The framework includes experimental tests and co-creation methodologies to refine a digital marketplace that connects stakeholders, facilitates SRM exchange, and fosters industrial symbiosis. The WRP demonstrates the potential for SRMs to replace virgin materials, reducing environmental impacts and production costs. It enhances supply chain transparency through digital traceability, promotes predictive material sourcing, and streamlines logistics via algorithmic optimization. Challenges such as regulatory gaps and quality standards are addressed through standardized processes, open data governance, and innovative algorithms. The WRP project advances circular economy goals in the ceramic sector, promoting waste reuse, industrial symbiosis, and supply chain resilience. Its standardized, open-access platform offers a scalable model for other industries, fostering sustainable practices and resource efficiency while addressing global climate targets.

Keywords: data governance; supply chain digital evidence items; situational awareness; system architecture; data-centric

1. Introduction

The ceramic and glass industries are highly energy-intensive and covered mainly by the EU Emissions Trading System (EU ETS), a cap-and-trade mechanism designed to curb greenhouse gas emissions [1]. The EU ETS allocates or sells emission allowances that companies can trade, incentivizing emission reduction and supporting the shift to a low-carbon economy. Although these industries have enhanced energy efficiency by adopting best practices, achieving full decarbonization remains a significant challenge. Existing studies suggest that incremental changes alone are insufficient; disruptive innovation and circular economy principles should be pursued as complementary strategies.

The circular economy model emphasizes reusing and recycling materials traditionally viewed as waste, transforming them into resources or secondary raw materials (SRMs).

According to EU and national policies [2], the circular economy aims to preserve product value, conserve resources, and shift away from the linear "take-make-dispose" model. The EU's Circular Economy Action Plan, part of the European Green Deal, promotes resource efficiency and waste reduction. Key policies focus on waste management, extended producer responsibility (EPR), and incentives for sustainable product lifecycles. This initiative targets sectors like electronics, textiles, and construction, encouraging design practices that support repairability and sustainability. Circular economy efforts are a path for reaching climate neutrality by 2050, as a large portion of greenhouse gas emissions results from resource extraction and processing [3].

To tackle logistical and market access challenges, particularly for small and medium-sized enterprises (SMEs) in remote areas, an omnichannel marketplace strategy can establish a digital value chain that supports circularity. The proposed platform offers SMEs a robust marketplace for trading SRM, granting them integrated benefits typically accessible only to large brands—at no extra cost. Acting as a global mediation gateway, it fosters cross-sector collaboration, advancing the circular economy by enabling SMEs to access secondary resources and forge sustainable partnerships. The final pilot seeks to engage consumers directly in shaping product offerings and customizing packages of eco-sustainable products they value.

With a focus on market needs, the Ceramic and Glassware Industry sector is examined using Portugal as a case study. Developed in close cooperation with stakeholders, such as suppliers, consumers, and regulatory bodies, these partners are crucial in refining the platform's features and participating in validation stages. The following two stakeholders are actively engaged in the co-creation cycle of the proposed solution:

- APICER: Associação Portuguesa das Indústrias de Cerâmica e de Cristalaria (www.apicer.pt) is a nationwide association for the ceramics and glassware sector. Its mission is to represent its members, promoting and defending their business interests to enhance the sector's competitiveness and growth. APICER is responsible for managing international promotion activities, analyzing and measuring the impact of industry developments, and running awareness campaigns.
- CTCV: Centro Tecnológico da Cerâmica e do Vidro (www.ctcv.pt) is a technological center focused on research, innovation, consulting services, and technology transfer in the ceramics and glass sectors. CTCV has experience in developing new materials and processes for these industries. Their work includes developing new combustion solutions, decarbonization initiatives, promoting raw material circularity, creating low-temperature sintering processes, and facilitating the digital transition.

The collaboration with APICER and CTCV has resulted in developing technical specifications for an open-access platform to improve communication flows between ceramic companies, especially SMEs, regarding information sharing and the use of SRMs, aiming to establish a new paradigm for managing circularity in the ceramic sector. This research cooperation is conducted as part of the Ecocerâmica e Cristalaria de Portugal (ECP), https://agendaecp.pt/en, accessed 16 December 2024, DecGreen Agenda for Business Innovation, financed by the RRP—Recovery and Resilience Plan, within the scope of the European Union's Next Generation EU | Project nº 76) project, focusing on potential standardization aspects to establish a marketplace for circularity within the ceramic sector.

PRR Plano de Recuperação e Resiliência | REPÚBLICA PORTUGUESA | Financiado pela União Europeia NextGenerationEU

The methodology included initial studies of potential technologies, experimental tests to validate solutions, co-creation with key stakeholders in the ceramics sector, and the implementation of the Waste Resource Platform (WRP) prototype for early-stage laboratory tests.

This will be followed by deployment in a relevant environment to assess the platform's impact and interoperability with existing systems. The study outlines processes to support system development and facilitate information exchange among acquirers, suppliers, and other stakeholders involved in the system's lifecycle. The goal is to develop an efficient process for collecting, tracking, and revalorizing ceramic waste, promoting circular economy actions within the ceramic sector by integrating information about the sector, waste typology, and use/valorization as secondary resources.

The WRP aims to contribute to eco-innovation and sustainability by promoting industrial symbioses for eco-efficiency and economic resilience. It seeks to create a standard data governance approach for SRMs, minimizing the consumption of virgin raw materials, increasing the economy's circularity, and developing innovative products from a life cycle perspective. The main objectives of the WRP are (1) to design and develop an interoperable platform that can act as a marketplace reference with services to identify potential raw materials from subproducts within the value chain, aggregating and sharing information about the availability and provenance of SRMs, and (2) to present a standardized open architectural approach for reusing and recycling waste from the ceramics sector and integrating waste from other sectors. This framework ensures comprehensive traceability of digital evidence throughout the process by documenting all actions within the digital value chain. This enhances transparency and accountability, fostering consistent practices among stakeholders across EU member states. Traceability is achieved through a non-repudiation event log supported by a matching algorithm that identifies the optimal alignment between SRM supply offers and reported consumption needs. The log employs a metadata structure to capture the status of the service agreements between the initiating entity and collaborating entities, with each collaboration mapped to a specific role in the value chain. While solutions like blockchain were excluded due to computational demands, the framework was designed with the flexibility to integrate similar secure transaction-linking mechanisms for SRM events where needed.

The log captures observable information stored on an immutable record accessible only to permissioned members, providing a shared, single source of truth for all agreed service details. Once recorded in the SRM log, information cannot be altered or tampered with by any participant. The log forms a secure data chain that tracks an asset (e.g., SRM) as it changes location or ownership, recording the precise time and sequence of actions to prevent data manipulation or unauthorized insertion.

The solution integrates an open-access platform for message streaming across the supply chain, connecting SRM suppliers and consumers while involving intermediaries as needed to facilitate service delivery. The platform fosters collaboration through high-value-added services and supports service-level agreements (SLAs) with selectable or confirmable service options. The informational workflow leverages operational artifacts that define the data governance framework, accommodating the dynamic needs of modern manufacturing and promoting cross-sector cooperation critical for advancing a digital circular economy.

The paper's structure is as follows: Section 2 reviews relevant work in this field. Section 3 presents the problem description and introduces and derives the proposed architectural approach. Section 4 showcases extensive experimental results and analysis. Conclusions drawn from the experiments are summarized in Section 5.

2. Related Work—The Path Towards Digital Transformation

The ceramic industry has been working and investing in innovations within the scope of the circular economy [4], and various ceramic sectors can innovate and incorporate waste/by-products from their production or generated by other industrial sectors, promoting circular economy strategies. In this domain, the digital transformation of the ceramic sector, framed by the Industry 4.0 paradigm, is one of the paths for enhancing production efficiency, fostering circular economy practices, and ensuring sustainable development [5]. The sector can overcome current challenges and build a resilient, future-ready industry by adopting advanced technologies and innovative design methodologies.

The digital transition within the manufacturing industry is revolutionizing production environments with more connected devices and automated production processes. By leveraging technologies such as the Internet of Things (IoT), Cyber-Physical Systems, and Big Data Analytics, the ceramic sector can harness the generated data to gain deeper insights into production processes, contributing to their optimization and sustainability. By adopting advanced digital technologies, the ceramic industry can reduce inefficiencies, lower costs, and increase productivity [6]. However, to obtain the maximum value from this myriad of novel technologies and the generated information, leveraging a real disruption in the manufacturing industry, it will be necessary to have a comprehensive knowledge of all processes within the value chain, encompassing information from different agents and integrating a wide diversity of data sources [7]. Only then will it be possible to deliver effective horizontal services, e.g., to achieve product traceability, from raw material to recycling, or to evaluate and minimize its ecological footprint, monitoring greenhouse gas emissions and fostering the circular economy through an industrial symbiosis between the various players in the value chain.

New methodologies in design projects, particularly design for circularity, are decisive for transitioning from a linear economy model to a circular economy framework. This shift emphasizes rethinking product lifecycles and incorporating sustainable practices from the initial design phase to end-of-life recycling. To ensure that ceramic production is sustainable, it must adhere to two fundamental rules: (i) Incorporate the concept of circular economy through the use of waste materials in product formulations, and (ii) reduce emissions by lowering firing temperatures and the total embodied energy. The ceramic industry can minimize environmental impacts and promote long-term sustainability by following these principles.

2.1. The Ceramic Industry in Numbers

To grasp the scale of the proposed challenge, Table 1 provides an overview of Portugal's Ceramic Industry. In 2023, Portuguese ceramic exports reached 162 international markets. France was the leading destination for Portuguese ceramics, followed by Spain, the United States, Germany, and the United Kingdom. Domestically, ceramic exports represented 1.20% of Portugal's total goods exports for the year. Among European countries, Portugal ranked as the 1st largest exporter within the EU on non-porcelain tableware, the 4th in sanitaryware, and the 5th of the EU on flooring and wall tiles. In 2023, the ceramics industry comprised 1214 companies. The turnover generated by the ceramics industry amounted to EUR 1492 million, with particular emphasis on flooring, wall tiles, and utilitarian and decorative ceramics.

Despite the year 2020, where the effects of the global pandemic significantly influenced the production and exports of the ceramic industry in Portugal, the GVA (Gross Value added) and productivity increased year after year until the year 2022. However, in 2023, there was a slight decrease in national production and exports, which can be explained by a contraction of the economy and international markets due to European and global inflation. At the same time, energy consumption is significantly decreasing, which is explained by the huge efforts and investments this industry is making in energy and process efficiency and implementing the circular economy model.

Table 1. Ceramic Sector main indicators in Portugal.

Ceramics	2019	2020	2021	2022	2023
Total Number of Companies (*)	1168	1092	1120	1162	1214
Companies with 10 or more employees (*)	179	175	175	180	184
Number of Employees (*)	18,458	18,196	18,591	19,505	18,996
Production (EUR) (*)	1,101,412,441	983,214,244	1,218,858,985	1,501,171,987	1,386,941,734
Turnover (EUR) (*)	1,187,822,504	1,101,714,459	1,322,381,279	1,564,680,787	1,491,648,223
GVA (Gross Value-added) (EUR) (*)	459,091,229	430,610,114	530,354,186	551,891,471	577,663,530
Apparent Labor Productivity (EUR)	24,872	23,665	28,527	28,295	30,410
Exports (EUR) (**)	707,743,607	661,424,935	813,181,081	964,123,218	879,506,794
Imports (EUR) (**)	202,262,289	196,149,511	226,946,672	290,537,330	293,167,901
Coverage Ratio of Imports by Exports	350%	337%	358%	332%	300%
Number of Export Markets (**)	163	159	158	170	162
Energy Consumption (toe—ton of oil equivalent) (***)	299,709	276,441	296,157	282,583	259,187

(*): Source: INE (National Statistical Institute), Integrated Business Accounts System (final data); (**): Source: INE (National Statistical Institute), International Trade in Goods Statistics (final data); (***): Source: Directorate-General for Energy and Geology | Energy Balances.

2.2. Waste Management and Circularity in the Ceramic Industry in Portugal

The ceramics industry generates various types of waste, the majority of which is inert or non-hazardous, with only a tiny portion classified as hazardous, typically related to maintenance or emission treatment processes. Decree-Law No. 102-D/2020, in line with EU Directives 2018/849, 2018/850, 2018/851, and 2018/852, regulates waste management and establishes criteria for specific substances or objects from production processes to be classified as by-products rather than waste. To be recognized as by-products, applications must be submitted to the National Waste Authority (ANR), overseen by the Portuguese Environment Agency [8]. By-products are substances resulting from production processes where their creation is not the primary aim. To qualify these materials as by-products, it was essential to demonstrate their inert nature. CTCV conducted leaching tests on various ceramic waste materials to verify this. Inert waste is defined as waste that does not undergo significant transformations and is neither soluble, flammable, nor reactive. It is not biodegradable and does not adversely impact other substances, preventing environmental pollution or health hazards. Its leachability, pollutant levels, and ecotoxicity are minimal, posing no risk to surface or groundwater quality.

Additionally, potential reuse methods needed to be proven to facilitate future utilization of these by-products. Ceramic waste offers multiple valorization opportunities, with several industries willing to incorporate it into their production processes. For example, ceramic waste can replace sand in ceramic mixtures or serve as ground-inert material for paving extraction sites, roadways, construction fillings, or tennis courts. It may also be used as a by-product in the cement industry, showcasing its versatility in recycling applications.

Considering the general criteria defined for classification as a by-product [9], classifying specific types of ceramic manufacturing waste as by-products under Article 44-A of the General Waste Management Regime (RGR) started to be regulated. This classification applies to waste from stages such as ceramic mixture preparation (e.g., before the thermal process), dust from dedusting systems, and non-conforming ceramic pieces post-thermal process. Following this declaration, APICER collected data from member companies benefiting from the reclassification, real data from other ceramic industries, data from APA, and calculations by CTCV based on real data. As shown in Table 2, the amount of waste reused as by-products is significant when compared with the dispatched as waste, highlighting the ceramics industry's potential for waste reintegration and promoting circular economy practices.

Table 2. Ceramic waste routed as by-products (tons)—APICER members (benefiting from the by-product reclassification decision) and other ceramic industries.

YEAR	2019		2020		2021		2022		2023	
Nº of Companies	19 ǀ > 150		20 ǀ > 150		23 ǀ > 150		24 ǀ > 150		24 ǀ > 150	
	Quantity Routed as By-Product	Quantity Routed as Waste	Quantity Routed as By-Product	Quantity Routed as Waste	Quantity Routed as By-Product	Quantity Routed as Waste	Quantity Routed as By-Product	Quantity Routed as Waste	Quantity Routed as By-Product	Quantity Routed as Waste
LER 10 12 08	30,576	65,000	18,117	58,000	35,968	68,263	11,343	63,181	14,945	55,758
LER 10 12 03	3859	10,500	2502	5200	6057	11,320	2465	9784	660	11,414
LER 10 12 01	7466	15,250	9472	2754	9387	29,336	8643	23,003	4044	19,889
LER 10 12 13		76,350		35,500		73,835		65,004		53,081
Total (ton)	41,901	167,100	30,091	101,454	51,411	182,754	22,451	160,972	19,648	140,142

LER 101208—Waste from the manufacture of ceramic pieces, bricks, tiles, and construction products (after the thermal process)—fired Waste; LER 101203—Waste from the manufacture of ceramic pieces, bricks, tiles, and construction products—Particles and Dust; LER 101201—Waste from the preparation of the mixture (before the thermal process)—Greenware; LER 101213—Sludge from on-site effluent treatment.

Following the change in the mode of communication of the waste produced, as of 2022, companies must directly communicate the annual waste production report to the Portuguese Environment Agency (APA). The data obtained and collected from that year may not be reliable. The Waste Resource Platform (WRP) that we are developing in this work can also serve to monitor the waste data companies communicate more trustfully and reliably. This leads to improved waste management and efficient support for the sustainability indicators for implementing the circular economy in the ceramics industry.

The ceramic and glass sectors have a distinct advantage in their ability to reuse by-products, waste, and residues from other industries, supporting industrial symbiosis. This often involves pre-processing steps like grinding, heat treatment, and developing tailored product formulations to preserve ceramic properties. Boosting productive efficiency through digital transition contributes to both internal processes within companies and the logistics across the ceramic sector's entire supply chain [10]. Studies show that digital transformation is essential for the ceramic and glassware industries to modernize and remain competitive, overcoming low levels of automation and limited cross-sector communication in these industries [11,12]. Integrating digital solutions can enhance resource management, elevate product quality, and drive innovation. Figure 1 depicts the concepts of endogenous and exogenous circularity within the ceramic industry. Endogenous circularity focuses on the internal reuse of materials within the ceramic production cycle. Examples include recycling broken ceramic tiles as raw materials for new tile or brick production or reintegrating ceramic dust from manufacturing lines into the clay mixture. Conversely, exogenous circularity emphasizes incorporating waste materials from other industries into ceramic production. This approach includes utilizing fly ash from power plants as a substitute for traditional clay or sand or incorporating glass cullet and waste aggregates from construction demolition into ceramic glazes or mixtures.

Endogenous circularity refers to a system's internal, self-sustaining processes that promote resource reuse, recycling, and efficient management. It emphasizes how organizations, industries, or even ecosystems can develop closed-loop systems that minimize waste and maximize resource efficiency from within. This concept is particularly relevant in sectors where waste materials or by-products can be reintegrated into production, reducing the need for external inputs and fostering sustainability.

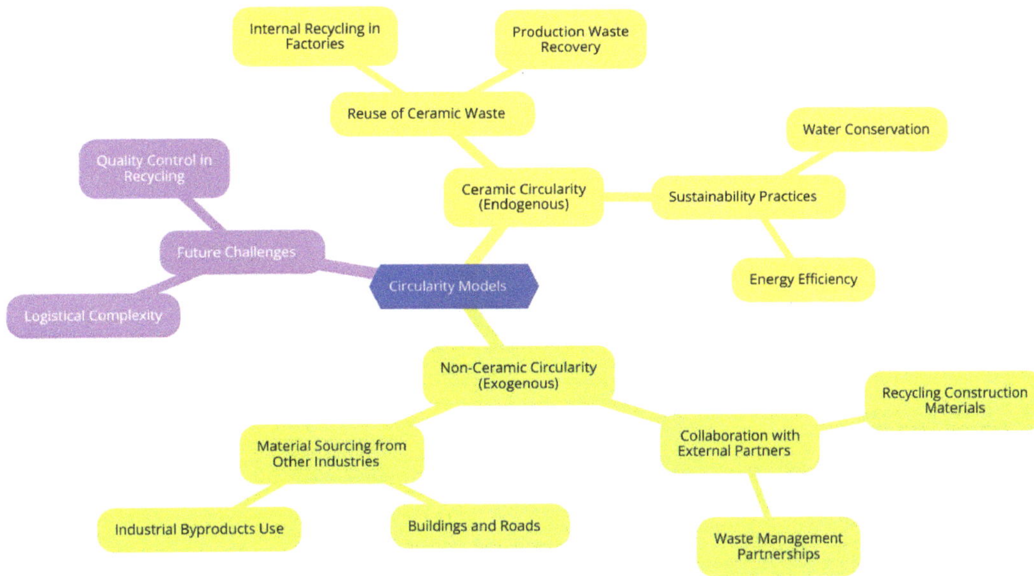

Figure 1. A mindmap illustrating ceramic vs. non-ceramic circularity.

Exogenous circularity involves incorporating external materials, processes, or systems into a circular economy framework. Unlike endogenous circularity, which focuses on internal resource reuse, exogenous circularity leverages resources from outside an organization's operations. This may include sourcing recycled materials from other industries, collaborating with external partners for waste management, or engaging in shared resource networks, including energy savings. For instance, a glass manufacturer could use recycled glass sourced from construction waste, while the ceramic industry might incorporate sand waste from other sectors, like foundries. Exogenous circularity highlights cross-sector collaboration, where different industries benefit from each other's waste streams, creating synergies and reducing environmental impact. This approach extends circularity beyond individual companies, building a broader sustainability network. By utilizing external resources, businesses gain access to new materials, reduce production costs, have possible energy savings (and CO_2 emissions), and contribute to a larger circular economy, where one industry's waste serves as another's raw material.

However, the challenges to implementing circular economy strategies and industrial symbioses in the creation of Eco-Innovative solutions (raw materials, ceramic pastes, and glasses) at an industry level are huge and require incremental actions to acquire the knowledge to cover all the needed aspects (holistic approach) to design proper solutions.

3. Introduction to the Data Governance Framework

This section presents research findings on implementing a data governance infrastructure to keep stakeholders and decision-makers informed on current and future material needs. This approach balances production demands with reusable and recycled materials, ensuring alignment with consumer expectations. The proposed solution adopts a process-centered approach, utilizing global tracking metrics (e.g., business key indicators) to ensure interoperability and promote industrial symbiosis. It employs a modular architecture with an isolation layer orchestrating communication between the software modules in the presentation layer and the application layer subsystems. The solution incorporates an open-access platform for message streaming across the supply chain, connecting suppliers and consumers of SRM and involving intermediaries to facilitate service delivery. It aims to enhance collaboration with high-value-added services, supporting service-level agreements (SLAs) through selectable or confirmable service options. The following sections delve into the operational artifacts that shape the data governance framework, supporting the dynamic nature of modern manufacturing and cross-sector cooperation necessary for a digital transition to a circular economy.

The architectural approach aligns with ISO42010 [13], emphasizing that stakeholder concerns should drive architectural descriptions. Although the current standard version uses the term "entity-of-interest" rather than "system-of-interest" (SoI), this paper retains SoI for simplicity, avoiding deeper technicalities beyond the paper's scope. Here, the SoI represents the SRM supply chain workflow, addressing stakeholder interests and concerns through various viewpoints. Each stakeholder focuses their interaction with the system based on their unique perspective, interests, and expertise.

3.1. System of Interest—Context Diagram

A context diagram in software engineering projects provides an overview of interactions between external entities and the system of interest (SoI). Also called a level 0 data flow diagram, this diagram presents a broad, easy-to-understand view of the system's boundaries and its relationships with external entities. It highlights key components of the system, external actors (like users or other systems), and the data flow between them, clearly delineating the boundaries and relationships between the system and its environment.

The context diagram identifies the actors, system scope, and boundaries, serving as a communication tool to help stakeholders understand and define how the system integrates within an organization's broader processes and infrastructure. By mapping these interactions, stakeholders—such as developers, project managers, and clients—achieve a shared understanding of the system's operational context. This common perspective helps ensure that requirements and dependencies are identified and managed throughout the software development lifecycle. Additionally, context diagrams document interactions and interfaces with different data flow or distinct sources. Figure 2 illustrates the information workflow for operationalizing the SRM supply chain, highlighting interactions among various stakeholders to encourage circular economy practices through digitalized supply chain logistics. It aims to establish traceability to track each order from placement to delivery—a critical requirement across companies in this sector.

Figure 2. Context diagram of the supply chain for secondary raw materials.

Some stakeholders are directly interested in the SoI, and others are impacted by the opportunities the system can provide. When stakeholders are included in the concept of a SoI, the following concepts emerge: (1) their skills, knowledge, and experience related to the SoI, (2) their perceptions of the current and future state of the system and any systemic problems, (3) their values and beliefs about the system that may shape the purpose and outcomes, (4) their understanding of the system based upon their role when interacting with the SoI. Table 3 briefly introduces the entities (i.e., stakeholders) participating in the digital marketplace deployed by the WRP system. The table also outlines the stakeholders' interests, expressed as concerns about the SoI or the architecture they know.

Table 3. List of stakeholders interacting with the SoI.

Stakeholder	Entity Role and Type of Interest in the System
Industry	Role: When reporting an SRM service, an entity can either report a specific need (C—consumption) or provide an availability (S—supply). It can be involved in establishing an SLA. Type of interest: Direct interest and possible promoter of actions to establish a commercial partnership.
Carrier	Role: An entity that operates at the level of SRM transport and may simultaneously be an industrial partner. It can be involved in establishing an SLA. Type of interest: Indirect interest
Processor	Role: An entity that processes SRM materials by grinding them to the necessary/specified granulometry requested by the consumer entity. It can be involved in establishing an SLA. Type of interest: Indirect interest
Intermediate Depositor	Role: An entity that operates solely as a temporary storage for SRM (without processing capacity). It can be involved in establishing an SLA. In some cases (e.g., when the SRM is classified as waste requiring a specific treatment), the entity might need to be a specialized waste treatment eco-recycling center. Type of interest: Indirect interest
Quality Analysis Laboratory	Role: An entity that performs analyses on the composition of SRM, identifying potential risk substances that may compromise the quality of the by-product or its intended use. Typically, this is a service contracted by an industry entity. Type of interest: Indirect interest
Owner	Role: An entity that manages the WRP platform, ensuring the integrity of membership applications. Responsible for providing community members to use the platform as intended, avoiding situations that could compromise transparency or business requirements relevant to the sector. Has the authority to change the registration status of an entity if non-compliance is detected. Type of interest: Indirect interest, but regulator of ethical behavior compliance within the community

Concerns arise from stakeholders' perspectives shaped by domain knowledge, experience, training, roles, and authority. They reflect matters of interest or significance for one or more stakeholders, with some concerns shared across multiple parties, while a single stakeholder may have various concerns. These concerns span stakeholders' needs, architecture goals, expectations, responsibilities, requirements, design constraints, and assumptions and may also involve dependencies, quality attributes, architectural decisions, risks, or other issues. In designing the SoI architecture, adversarial concerns, such as political objections, can be addressed through negotiated solutions, while threats may require preventive measures. Concerns may surface at any stage of the SoI lifecycle—from conceptualization, design decisions, and construction to implementation, deployment, operation, and ownership transfer.

Stakeholders often develop distinct perspectives shaped by their roles, experiences, beliefs, and other characteristics. These perspectives may stem from domain expertise, professional experience, training, or involvement in various SoI lifecycle stages, such as design, development, manufacturing, supply, and operation. Influencing factors may also include personality traits, cultural background, peer influence, or affiliations. Perspectives represent how stakeholders view the SoI in a particular context, especially concerning their concerns, and are typically subjective. Each perspective may generate one or more concerns. Since concerns stem from these perspectives, architectural viewpoints are often organized according to stakeholder perspectives. Four perspectives (i.e., user profiles) were created within the project scope, each with a specific role in managing the information.

- Owner user (O-User): responsible for the governance of the data reported on the WRP platform, including auditing entity-reported data and validating/approving membership applications. O-Users have read-only (R) access to the platform data but can change the entity's registration status. An O-User can authorize the registration of new O-Users in the WRP.
- Administrator user (A-User): the individual who creates the entity's record is automatically assigned the administrator profile for that entity. Upon O-User approval of the membership application, the A-User has full permissions (CRUD) for entity data management services, including adding new users as A-User, M-User, and S-User. These users are automatically associated with the record of the entity created by the A-User.
- Master user (M-User): similar to the A-User profile but cannot change data of A-User profiles or access the Settings option. M-Users can add new users as an M-User or S-User.
- Standard user (S-User): Has read-only (R) access to the entity's data but has full permissions (CRUD) to manage data relating to the SRM record of the entity to which they are associated. S-Users cannot add new users but can edit some attributes of their own records.

The legitimacy and importance of a concern held by a stakeholder can be a consequence of the stakeholder's role (e.g., platform owner, SRM supplier or consumer, SRM carrier or processor), the importance of including these stakeholders in the understanding of the SoI will be shown when gaining a shared understanding of the WRP system.

3.2. Architecture Framework Description

This section presents a framework developed to formally represent the system architecture and establish a structured development process. This framework defines an open-access platform designed to enhance communication and foster partnerships among ceramic companies through information sharing. The architecture concept is a work product that embodies conventions, principles, and best practices relevant to architecture-related activities within a specific application domain or stakeholder community [13]. The architectural model captures the SoI's key concepts or properties. It maps the foundational elements of a system in its environment and sets governing principles for the system's implementation, evolution, and lifecycle processes [14]. The framework aims to produce a product that meets user and stakeholder requirements, supporting effective communication and collaboration among parties involved in creating, using, and managing modern systems, ensuring they function in an integrated, cohesive manner.

Figure 3 illustrates a block diagram of the modular architecture designed to establish a B2B infrastructure accessible to all ceramic sector companies. This infrastructure enhances communication, management, stock control, order tracking, and raw material quality assessment. The aim is to create a marketplace that supports data processing and acquisition throughout the supply chain, automating order fulfillment and delivery processes. An integrated system for identifying and tracking ceramic production needs within a digital circularity chain offers companies insights into production waste valorization, reducing costs through industrial symbiosis, where one company's waste becomes another's resource. As depicted in Figure 3, the WRP platform architecture follows a modular and layered approach, with human interaction limited to the presentation layer, while core business logic and services operate within the application layer.

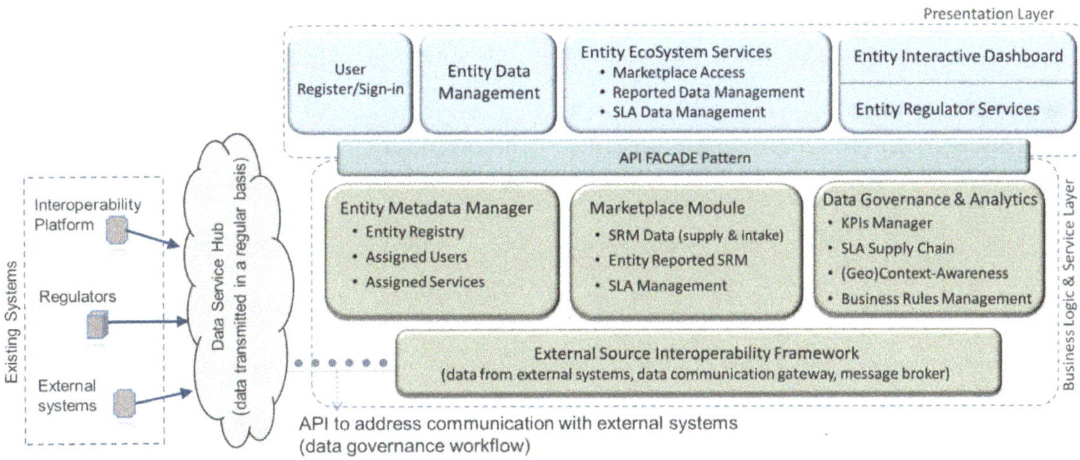

Figure 3. Block diagram of the architecture approach.

The software modules at the presentation layer provide a responsive web interface that adjusts dynamically based on the user's access profile and the characteristics of the device screen (e.g., 15″ screen or 8″ screen). This interconnected organization of software modules facilitates the system's understanding, maintenance, and evolution. Each module is represented as an independent functional unit, which reduces components' maintenance, scalability, and reuse, especially for analogous use cases, allowing for specific updates without globally affecting the system's behavior. Below, we briefly describe the software modules that make up the WRP platform block diagram.

- User Register/Sign-in: this module collects user information (e.g., access credentials or new entity registration). The main scenario addresses validating access to the WRP platform using user credentials. For usability and security reasons, i.e., compliance with ISO 25010 [15], users can recover their password (use case: Forgot Password) or request registration for a new entity (use case: Register New User) by filling out a form indicating whether the registration is for an individual or a collective entity. The system displays error messages for non-compliance situations, logging these events to ensure conformity with the quality and security requirements specified in ISO 25010. The Entity Metadata Manager module verifies and validates metadata associated with an entity's registration in the backend. For a "Collective Entity", it manages metadata characterizing users registered as entity members. It allows the registration status (e.g., Individual User Entity or Entity member) to be checked. An entity member is a user whose registration on the WRP platform is linked to that entity. This addresses functional integrity requirements, ensuring the system's behavior and compliance with security requirements concerning the AAA communication protocol (authentication, authorization, and auditing), mitigating risks such as Non-repudiation, Authenticity, and Accountability [16]. It also logs state change events (e.g., messages reported by O-User, A-User, M-User, and S-User) to enable the auditability of historical events.
- Entity EcoSystem Services: the user profile determines the level of service provided, with the most restrictive applied to an O-User and the most complete to an A-User. The base configuration allows access to data of the entity to which the user is associated (e.g., an entity registered as a collective person). It interacts with the Marketplace module at the backend to compute information about the availability and consumption of SRM Data (supply and intake). When an authenticated user reports an SRM, the registration can be classified as Supply (S—Supply) or Consumption/Intake

(C—Consumption). For supply (SRM Data), it presents a marketplace for viewing availabilities and consumptions. Authorized users can manually navigate the reported data to select records to establish an SLA. Registry management (Reported SRM) enables CRUD operations (Create, Update, Read, and Delete) on Marketplace-reported data. The SLA Management module provides services for managing established or prospective contracts, especially concerning collaboration with entities like processors, carriers, laboratories, and intermediate repositories.

- Interactive Dashboard: this interface displays information on SRM management indicators per entity and the information workflow for establishing an SLA. It features a Dashboard with indicators structured by scorecards and a metadata structure for indicator characterization and configuration. It also includes a Dendrogram with information on the custody chain of services contracted via SLA with entities registered on the WRP platform. The Dendrogram is a graph that can be managed using graphical elements. The focus is on Usability (ISO 25010), particularly Operability (ease of operation and control) and Interface Aesthetics (interface appeal).
- Entity Regulator Services: this web-based interface allows the O-User to analyze compliance with the requests for joining the WRP platform. During registration on the platform, the responsible user must indicate the type of entity, specifying whether the request is for an individual entity (i.e., an individual person) or a business entity (i.e., a registration concerning a collective). Once the membership request is approved, the A-User can register affiliated entities (e.g., a group of companies) and manage users reported as members of a specific company or group of companies.
- Data Governance and Analytics: this module processes information flows related to the georeferencing of SRM locations. It triggers contextual notifications based on the event's scope and location, encapsulated in the (geo)Context-Awareness submodule. It implements business rules for establishing SLAs, defines SRM registration criteria, and configures criteria for activating state machines that execute actions on the registration state of entities, entity members, and SRM records—encapsulated in the Business Rules Management submodule. The KPI Manager module orchestrates the governance of indicators related to the SLA process with different entities. The SLA Supply Chain submodule coordinates with the execution of the Matching algorithm based on user-defined selection criteria for establishing an SLA. It interacts with the Dashboard at the presentation layer to present a dendrogram-style graphical interface for visualizing involved entities and the SLA status with each.
- External Source Interoperability Framework: this module encapsulates system behavior regarding interfaces with external systems. It aggregates interaction requirements at the API level, such as interaction with Google Maps and the Interoperability Platform. The focus is on maintenance, which includes the system's ability to be updated and adapted to changing requirements and environments. It aims to enhance test effectiveness (Testability) and system operability in response to changes or adjustments during the development cycle (Modifiability).

The solution employs a layered and modular architecture with an isolation layer between the presentation and application subsystems and integrates software components from other consortium projects. The Facade design pattern [17] is adopted to offer a structural interface that simplifies interactions with complex subsystems, allowing clients to use a unified interface without needing to understand intricate underlying details. The Facade pattern is especially beneficial in systems with multiple interdependent classes, as it decouples clients from subsystems, minimizing dependencies and boosting modularity and maintainability.

The Facade pattern is equally effective in a microservices architecture where independent services handle distinct functions. Although microservices operate independently, they often need to communicate to preserve the system's modularity and support updates or replacements. In this case, the Facade can act as a gateway for inter-microservice communication by providing a unified API, managing calls across services, and delivering consolidated responses to clients [18]. This streamlines client-side code and enhances scalability and maintainability by isolating clients from communication complexities among services. Additionally, the Facade can manage cross-cutting concerns such as logging, authentication, and rate limiting, reducing redundancy and increasing system robustness. However, facades should be used cautiously, as excessive use can add complexity. They are most effective for simplifying complex systems or offering specific, streamlined views of subsystems.

3.3. Technical Overview of the Domain Model

A Domain Model is a conceptual framework representing the data structures and relationships within the SoI, abstracted from specific database storage or management details [19]. Unlike the data storage layer, the Domain Model focuses on the relationships and connections among data elements within the business context. This separation of concerns supports system flexibility and adaptability, as changes in data storage do not impact the domain representation, thereby enhancing system scalability and maintainability.

In a typical Domain Model, entities are depicted as objects or classes, with attributes that define their properties and associations representing relationships between them. This structured approach aids in system analysis and design, facilitating the identification of redundancies, inconsistencies, and opportunities for optimization [20]. The Domain Model details informational entities like SRM reports or established SLAs and their structured interconnections, often called metadata. These data elements serve as inputs for defining the system's behavior, or "function", framing the system's environment and capturing factors that affect system properties and performance while remaining outside its direct control. For instance, consumers and suppliers form part of a company's transactional environment, which influences system behavior as a series of events over time, highlighting the system's function as a core determinant of its behavior.

As illustrated in Figure 4, the Domain Model acts as a blueprint for the system's data architecture, offering a high-level view that enhances communication among stakeholders, such as business analysts, developers, and database administrators [21]. By centering on the domain's fundamental aspects, the Domain Model ensures all team members have a shared understanding of core business concepts and their interactions, which is essential for designing and implementing coherent, efficient systems.

Stakeholders with profiles like A-User, O-User, M-User, and S-User have distinct perspectives and data access restrictions. This structured information workflow, governed by architecture viewpoints, separates concerns among stakeholders while offering an integrated view of the whole entity, fundamental to the system architecture. Each stakeholder perspective represents an aggregation of concerns, and the proposed solution associates each perspective with the corresponding stakeholders. The architecting effort should identify current or future stakeholder perspectives relevant to the SoI.

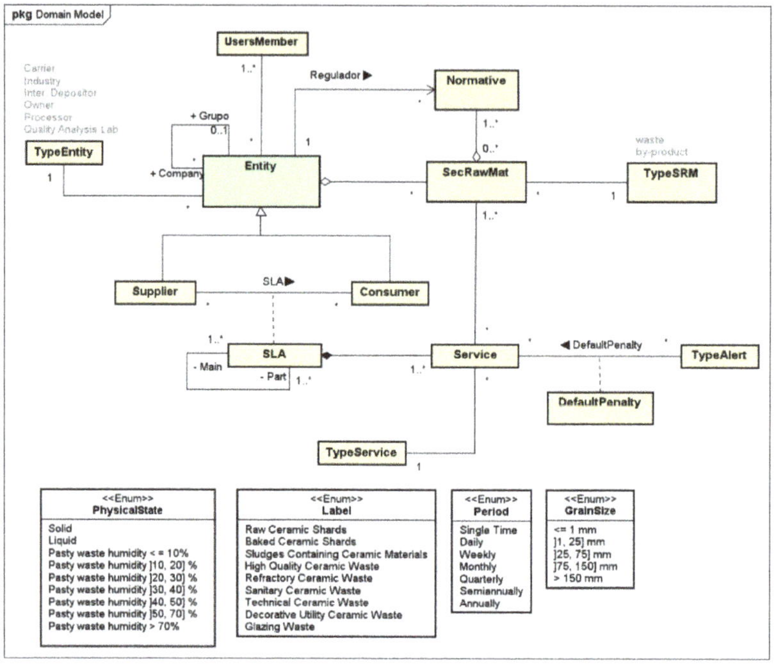

Figure 4. The domain model of the WRP Platform—collapsed view.

In this domain, DataOps and Data Governance Frameworks closely align with the principles of the Domain Model. DataOps emphasizes the continuous delivery and automation of data workflows [22]. It benefits from a well-defined Domain Model by ensuring that data pipelines and processes are built on a clear understanding of business entities and their relationships. A robust Domain Model provides a stable foundation for DataOps practices [23], facilitating efficient data integration, quality assurance, and deployment. It allows data engineers and operations teams to automate and manage data flow more effectively, ensuring that the correct data are delivered to the right place at the right time, adhering to predefined business rules and relationships.

A Domain Model is also integral to a comprehensive Data Governance Framework. Data Governance involves managing data availability, usability, integrity, and security across an organization [24]. A clear and well-structured Domain Model aids in defining data ownership, establishing data quality standards, and enforcing data access and usage policies. By mapping out essential domain concepts and their interactions, the Domain Model provides a reference for governance policies, ensuring that data management practices align with business objectives and regulatory requirements. This alignment is crucial for maintaining data integrity, reducing risks, and ensuring compliance, ultimately enhancing the reliability and trustworthiness of organizational data.

4. Discussion

Applying DataOps to develop a data-centric prototype has profound implications, primarily by promoting a culture of continuous delivery and automation of data workflow, secondly because the prototype can achieve a higher level of data quality and consistency, and thirdly because it facilitates the integration and optimization of data pipelines, ensuring that data are collected, processed, and analyzed efficiently and accurately, contributing to improving an informed decision-making process. DataOps' focus on collaboration and communication among data professionals ensures that the system can adapt to changing requirements and continuously improve its performance. These implications are essential for stimulating circular economy practices, enabling real-time monitoring and management of resources, optimizing their use, and minimizing waste.

DataOps also helps define proficient use cases by enabling rapid iteration and refinement of data-driven applications. Use cases can be developed to address specific needs within the supply chain, such as tracking the availability and usage of SRM. By leveraging automated data processing and real-time analytics or by triggering situational awareness whenever data inconsistencies are detected, these use cases can provide actionable insights, supporting faster decision-making. For instance, a use case might involve predicting demand for recycled materials, allowing companies to adjust their production schedules and inventory levels accordingly. This agility was explored within the case study for maintaining a circular economy, where the goal is to minimize waste and maximize resource efficiency through continual reuse and recycling.

Establishing laboratory data analysis routines and technical protocols is another area where the case study benefits from the DataOps principles. By standardizing the procedures for evaluating and characterizing ceramic by-products and wastes, DataOps ensures that the data collected is reliable and comparable across different batches and sources. This standardization demonstrates the technical feasibility of using the reported information about SRM as open data, streamlining the data sharing between existing systems. Open data within the scope of the study (i.e., ceramic sector) refers to data available to be accessed, used, and shared without restrictions within the WRP community. In the context of the ceramic industry, implementing open data principles for SRM involves creating a core metadata structure that standardizes how data are described, cataloged, and exchanged.

4.1. Marketplace Business Scenario

Figure 5 showcases a screenshot of the Marketplace layout developed as part of the ECP project. Due to data privacy concerns, the information presented in this study is based on dummy data. The emphasis is on the implemented functionalities and, more importantly, the metadata structure rather than the actual data content. The list of metadata was co-created with stakeholders, including the specification of a standardized procedure to map the digital logistics chain information workflow for SRM reuse in the ceramic sector. Each reported record has a unique identifier (i.e., attribute "Ref. ID" in the form), and a set of icons indicates its status (see Table 4 for more information). Users can expand a specific item to obtain an overview of the reported data. Interested users can access a brief overview of a particular SRM record by expanding it. To access detailed information, a double-click opens a full-screen form containing technical data for proper assessment. This includes details about the reporting entity, its classification, and comments from other users who have interacted with the entity in various roles within an SLA business interaction.

To streamline user interactions, the Marketplace tab offers standard features such as reordering the list by multiple criteria, filtering by specific keywords, selecting a particular raw material, and switching to a map view for geo-visualization and map-based analysis. The status of the raw data is also displayed, providing visual cues about the availability of the reported information. Table 4 explains the meaning of each status icon. Users can activate the checkbox next to each raw material to mark it. Upon completing the search, clicking the Submit button sends an invitation to the corresponding selected entities, indicating interest in establishing an SLA.

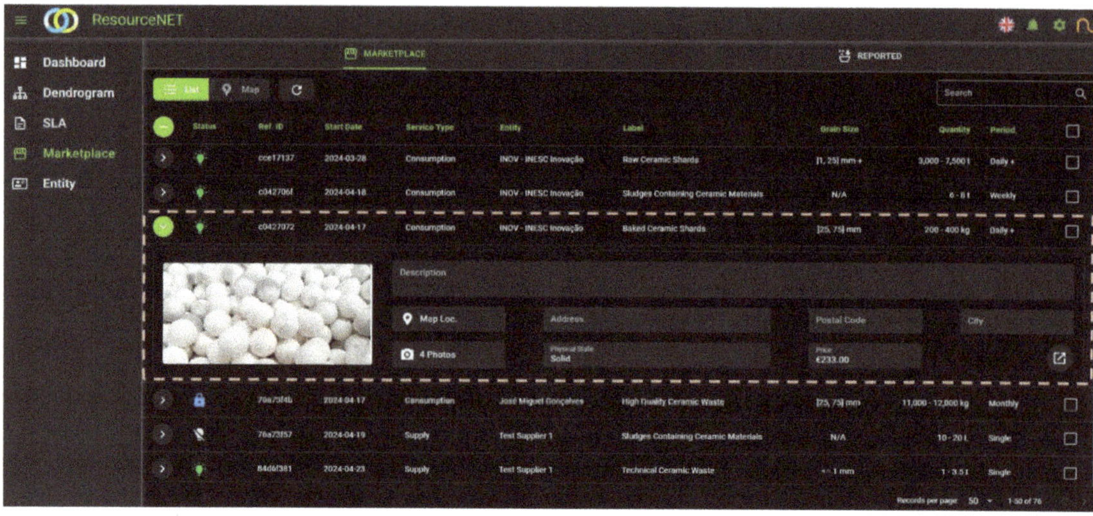

Figure 5. Marketplace tab with a specific SRM record overview.

Table 4. List of icons characterizing the status of an SRM record.

Icon	Description of the Record Status
	This status indicates that the reported SRM record is available for analysis and consideration by interested market players (e.g., consumers or suppliers) for establishing an SLA. A—Active (icon in green)
	This status indicates that the reported information is not valid. However, interested users can still analyze the data and contact the corresponding entity for a potential B2B operation. This icon maps the following possibilities (icon in grey): C—Canceled, the entity that reported the SRM record has canceled the reported data. E—Expired, the due date was exceeded; therefore, the SRM record validation has expired. U—Unavailable, the SRM is temporarily unavailable.
	This status indicates that the O-User has classified the SRM record as inconsistent, making it unsuitable for B2B operations. If the inconsistency persists, there is a high likelihood that the entity will be suspended. S—Suspended (icon in red)
	This status indicates that the reported SRM is no longer available, as it has already been assigned to an SLA. Due to commercial confidentiality concerns, the entity that reported the SRM record can choose to either hide the record from the marketplace or keep it visible for contact by interested market players. L—Locked (icon in blue)

Although only raw materials with an Active status are available from a B2B perspective, displaying information in the marketplace can foster interactions between market players interested in specific SRMs. Due to business data constraints, information on locked status can be omitted. In this case, the record owner must deselect the visibility option. The open nature of the marketplace also ensures transparency regarding entities' compliance with policies or market rules for B2B operations involving SRMs (e.g., compliance with the legislation on Waste Declassification or legal criteria for End-of-Waste classification). Therefore, transparent information is provided whenever the O-User suspends entity-reported information. Additionally, it is possible to check comments from market players who have interacted with the entity commercially.

The marketplace enables interested entities to request analyses from laboratories to ensure SRMs are free of contaminants or any other agents that could compromise quality or intended use. Through continuous monitoring and automated reporting, the DataOps approach helps laboratories maintain high-quality data and quickly report deviations, enhancing WRP's situational awareness and supporting the adoption of recycled materials in industrial applications.

DataOps also improves the identification and traceability of recycled materials and related products. By integrating with existing systems like quality control systems, DataOps enables comprehensive tracking throughout the supply chain, ensuring quality metrics are available at every stage, from initial material acquisition to final product delivery. Customers and stakeholders can verify the order status and raw material quality in real-time, fostering transparency and trust. Moreover, by maintaining detailed records of material origins and processing histories, DataOps supports regulatory compliance and helps companies demonstrate their commitment to sustainable practices. This holistic approach improves operational efficiency and strengthens supply chain integrity, promoting the principles of a circular economy.

4.2. Marketplace—Reported Data

Each entity member can access their data by clicking the Reported tab, which filters the information to show only records related to their entity (see Figure 6). The WRP system is designed to comply with the ISO 25010 standard for user interaction, ensuring that the solution is user-friendly and efficient for specified users to exchange information and complete tasks within a coherent context. This includes attributes such as appropriateness, recognizability, and learnability, which make the system easy to operate and control, and a standardized layout with graphical elements that help prevent user errors. These features encourage continued interaction and user engagement.

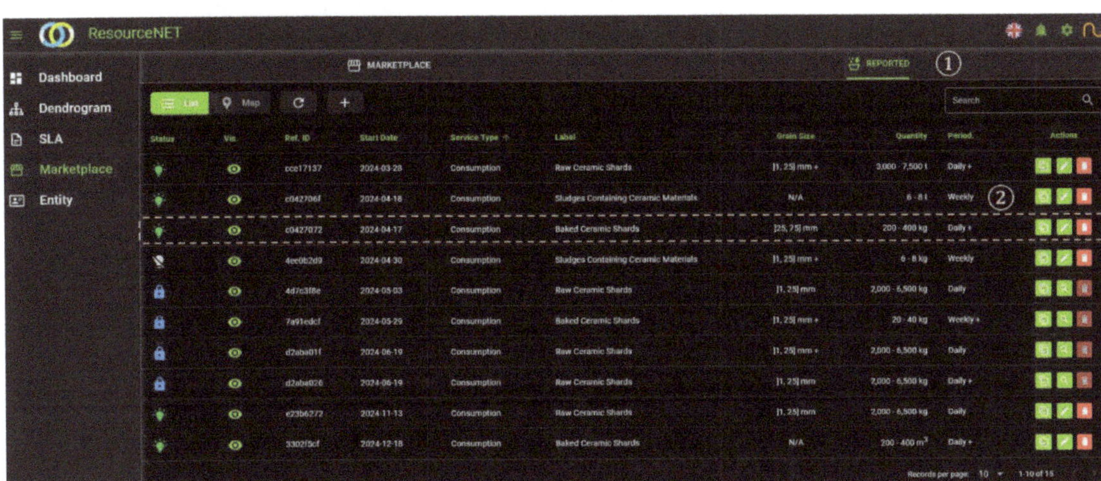

Figure 6. The list of SRM data reported by an entity. (1) the Reported tab filters the information to list only the records related to the entity to which the logged-in user belongs; (2) the available actions allow the user to delete, edit, or clone a record.

Self-descriptiveness is another relevant aspect, providing users with the necessary information to make the system's capabilities and use immediately apparent without excessive interactions or the need for extensive training. The interface is designed to cater to a wide range of stakeholders, including developers, operational users, and those influenced by regulations and legal procedures. The standardized layout promotes user assistance, making the system accessible to a broad range of users to achieve specified goals. The system's availability and intuitiveness ensure it is operational and accessible when required. The security dimension is of significant concern to ensure that data are accessible only to authorized users, that actions are verifiable, and that entities can be uniquely traced. This ensures that actions and events are non-repudiable and that the identity of entities can be accurately confirmed. Finally, the system is designed for interoperability, allowing it to exchange information with other products and use the exchanged information effectively.

Combining DataOps with service design thinking to create a standardized interface yields numerous benefits. This approach ensures that data workflows are efficient and automated while fostering an intuitive and user-centric design. By aggregating and segmenting multiple information into specific sections, the interface promotes self-descriptiveness, making it immediately clear to users how to interact with the system. The design thinking process emphasizes user engagement, ensuring that the layout is intuitive and that users can quickly achieve their specified goals within the context of use. Figure 7 illustrates a holistic approach that improves operational efficiency, data quality, and user engagement, fostering continuous interaction with the platform. The SRM section aggregates metadata to characterize the reported SRM, organizing technical attributes to guide the user effectively, particularly in high-pressure scenarios. To engage industrial stakeholders like suppliers and consumers, the platform offers dynamic search functionalities that align marketplace information with user interests. A matching algorithm executes data analytic models based on user-defined criteria to identify and cluster relevant records. For instance, to reduce transportation costs, consumers can specify a supplier search radius around the delivery location, with suggestions ranked by economic advantage. Distance can be calculated as a straight radius or driving distance in kilometers.

In the scenario depicted in Figure 7, a consumer seeks a specific SRM quantity with a particular grain size but is open to two size options. Users can indicate if the processing is required to reduce grain size from [25, 75] mm to [1, 25] mm and request carrier services to transport the SRM to an intermediate processor or destination. Additionally, laboratory services can be requested to collect and analyze an SRM sample. By clicking "Search", the matching algorithm processes the specified criteria, presenting results in the "Matching Results" section. These results are displayed as grouped combinations of entities, facilitating informed decision-making and promoting efficient collaboration.

This core metadata structure ensures consistency and interoperability across different systems and stakeholders within the industry. By adopting a standard metadata framework, companies can more easily share information about SRM sources, properties, and availability, facilitating collaboration and innovation. Such standardization enhances transparency and efficiency in the supply chain, supports regulatory compliance, and promotes the broader adoption of circular economy practices by making valuable data widely accessible and usable.

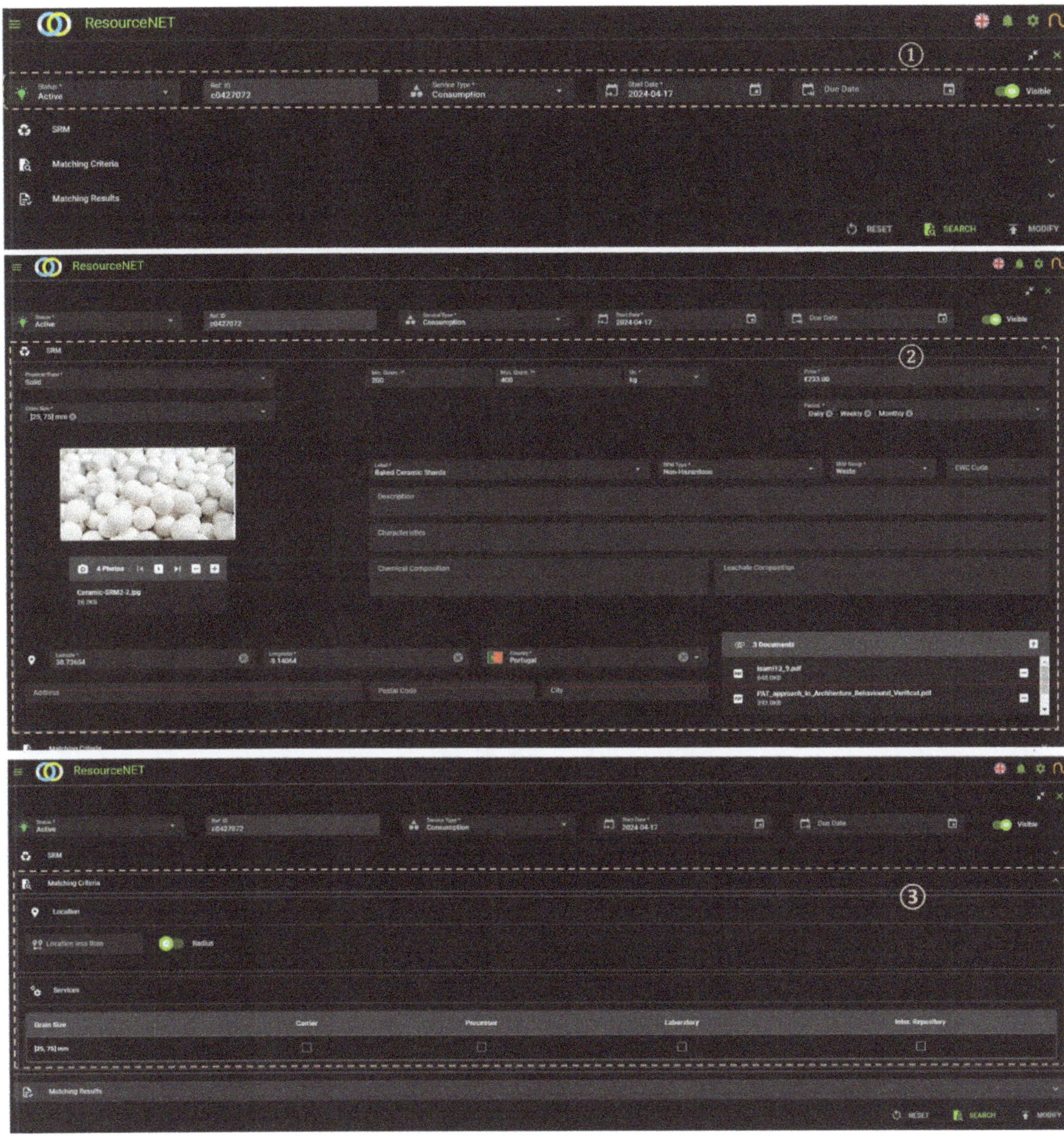

Figure 7. The layout for adding a new record to the marketplace. (1) This section presents the collapsed view of the SRM form, displaying attributes such as status, identifier, type of service, start/due date, and SRM record visibility; (2) this section provides a detailed view of the metadata used to characterize an SRM record; (3) this section outlines the elements that users can configure to set preferences for the matching algorithm to use as filtering criteria.

4.3. The Entity Status Informational Workflow

The WRP platform adheres to a standardized open architectural approach designed to reuse and recycle waste within the ceramics sector while integrating waste from other industries. This framework allows data to be accessed and shared freely within the WRP community, though under a supervised context. This supervision ensures that a qualified entity reviews new member applications and monitors community behavior to prevent deviations from established procedures. Additionally, it ensures that all published information complies with legal requirements and best practices for SRM services reported by WRP members.

The workflow in Figure 8 outlines the process triggered when a new application is submitted. The application is marked as Pending (P) while awaiting O-User review, which may request additional documentation, mainly if the applicant is unknown or not from the Ceramic sector. For instance, when the entity Portuguese Classification of Economic Activities is not from the Ceramic sector, the O-User might request additional documentation/certifications before being able to analyze the request.

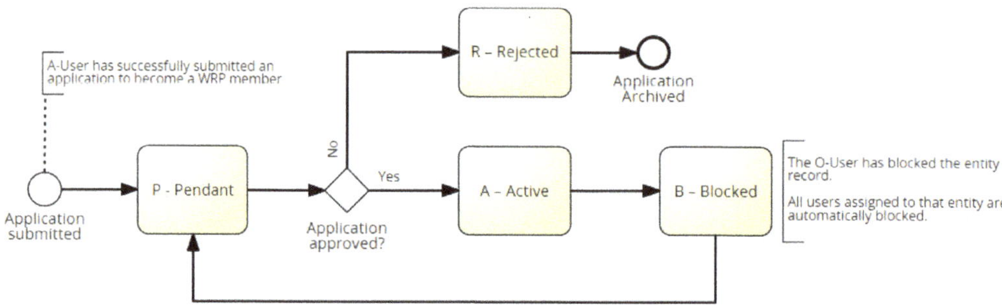

Figure 8. Diagram showing the sequence of states that an entity's record can assume.

If the application is rejected, the entity is notified, and its data are archived. If approved, the entity's status is updated to Active (A), allowing the A-User to add new users. For security, the platform sends an email requesting users to acknowledge their invitations and update their credentials. Once access is confirmed, their status changes to Active, granting full SRM record management capabilities. All Create, Update, and Delete (CUD) operations are logged for historical analysis, ensuring accountability.

A-Users and M-Users can block users within their entity, while O-Users can block any user across the platform. If an entity is blocked, all associated users are also blocked. Justification for blocking must be provided so affected users can address the issue. This procedure allows the supervisor to monitor user behavior and take action if misconduct is detected. The transition from a Blocked to Pending state (B-to-P) may be necessary if the user or entity needs to perform specific actions, such as uploading required certifications, which are then verified by credentialed users like O-Users, A-Users, or M-Users.

5. Conclusions

The case study outlines an architectural framework for a standardized waste supply chain solution featuring online services for exchanging and managing secondary resources and waste. This approach facilitates faster decision-making and promotes a circular economy. The study employs an innovative, process-centric design methodology that supports the planning, managing, and enhancing lifecycle activities within organizations or projects. It utilizes a context diagram to map out the system's key components, external actors (such as users or other systems), and the data flow between them, clearly defining how the system interacts with its environment. The study also details processes that support system development and enable information exchange among stakeholders, including acquirers, suppliers, and other parties involved in the system's lifecycle. The objective is to create an efficient process for collecting, tracking, and revalorizing ceramic waste, thereby advancing circular economy practices within the ceramic industry. This is achieved by integrating sector-specific information, waste typology, and secondary resource utilization.

SRMs, such as waste or recycled materials, can replace natural raw materials in manufacturing, offering benefits such as reduced resource use, lower environmental impact, and cost savings. However, challenges persist, including the lack of EU-wide quality standards for certain materials, regulatory issues, market limitations, and the risk of contaminants in recycled inputs. The proposed approach incorporates the Industry 4.0 paradigm to introduce innovations in data governance and algorithmic models, optimizing supply and demand matching within the logistics model. By implementing disintermediation and process automation strategies, the modular architecture enhances the procurement of recycled materials and related products. As system interaction evolves, the understanding of production patterns improves, enabling the use of predictive methods to streamline material sourcing and ordering. This integration simplifies logistics and improves the overall efficiency and sustainability of the ceramic sector supply chain.

The WRP project has made significant progress in advancing the circular economy, particularly within the ceramics sector, by improving the exchange and management of SRM. It addresses key challenges in waste reuse and recycling standardization, aiming to build a robust SRM supply chain logistics framework. This initiative fosters industrial symbiosis—turning one company's waste into another's resource—and opens new business opportunities. A significant achievement of the project is the development of a standardized, open architectural approach that facilitates the integration of waste from various industries into ceramics production. By aggregating and sharing data on SRM availability and origin, the project strengthens supply chain sustainability and resilience while promoting more efficient resource use, social responsibility, and redundancy in times of crisis.

Central to the WRP project is the creation of an open-access platform with a modular, layered architecture that simplifies complex interactions. Using design patterns like the Facade, the platform enhances system usability, adaptability, and scalability, particularly within a microservices environment. The project's collaborative approach, involving stakeholders such as suppliers, consumers, and regulatory bodies, ensures that the platform aligns with user needs and expectations. Ultimately, the WRP project demonstrates the potential of standardized waste reuse strategies to support a circular economy within ceramics and across various industries. Focusing on stakeholder collaboration, open data, and sustainable practices serves as a model for industries seeking to adopt circular economy principles.

Author Contributions: Conceptualization, G.P.; methodology, G.P.; software, G.P.; validation and formal analysis, G.P., M.A. and N.M.; investigation, resources, and data curation, G.P., M.A. and N.M.; writing—original draft preparation, G.P.; writing—review and editing, M.A. and N.M. All authors have read and agreed to the published version of the manuscript.

Funding: Green Agenda for Business Innovation, financed by the RRP—Recovery and Resilience Plan, within the scope of the European Union's Next Generation EU | Project nº 76.

Institutional Review Board Statement: Not applicable.

Informed Consent Statement: Not applicable.

Data Availability Statement: The data presented in this study are available in this article.

Conflicts of Interest: The authors declare no conflict of interest.

References

1. Official Journal of the European Union. Directive-2023/959-EN-EUR-Lex. 2023. Available online: http://data.europa.eu/eli/dir/2023/959/oj (accessed on 17 November 2024).
2. European Commission. A New Circular Economy Action Plan, EUR-Lex-52020DC0098, 2020. Available online: https://eur-lex.europa.eu/legal-content/EN/TXT/?uri=CELEX:52020DC0098 (accessed on 17 November 2024).

3. European Comission. Circular Economy Action Plan. Energy, Climate Change, Environment, 2023. Available online: https://environment.ec.europa.eu/strategy/circular-economy-action-plan_en (accessed on 17 November 2024).
4. Vasić, M.V.; Velasco, P.M.; Bueno-Rodríguez, S.; Grubeša, I.N.; Dondi, M.; Villarejo, L.P.; Eliche-Quesada, D.; Zanelli, C. State and perspectives of sustainable production of traditional silicate ceramics. In *Open Ceramics*; Elsevier: Amsterdam, The Netherlands, 2024. [CrossRef]
5. Vacchi, M.; Siligardi, C.; Cedillo-González, E.I.; Ferrari, A.M.; Settembre-Blundo, D. Industry 4.0 and Smart Data as Enablers of the Circular Economy in Manufacturing: Product Re-Engineering with Circular Eco-Design. *Sustainability* **2021**, *13*, 10366. [CrossRef]
6. Bueno, R.E.; Pohlmann, M.N.; dos Santos, H.A.; Gonçalves, R.F. The Procurement 4.0 Contributions to Circular Economy. *Sustainability* **2024**, *16*, 5838. [CrossRef]
7. Pestana, G.; Antunes, W.; Carvalho, J. Digital Chain of Custody Operational Framework. In Proceedings of the 2023 IEEE International Workshop on Technologies for Defense and Security (TechDefense), Rome, Italy, 20–22 November 2023. [CrossRef]
8. Portuguese Environment Agency. *Portuguese National Inventory Report on Greenhouse Gases, 1990–2021*; Portuguese Environment Agency: Amadora, Portugal, 2023.
9. Official Journal of the European Union. Document 32014D0955: Amending Decision 2000/532/EC on the list of waste pursuant to Directive 2008/98/EC. *Off. J. Eur. Union* **2014**.
10. Silva, J.d.S.; Leite, M.S.A. Analysis of a supply chain in the ceramic sector: A look at business processes. *Gest. Prod.* **2024**, *31*, e12021. [CrossRef]
11. Hsu, C.-H.; Li, Z.-H.; Zhuo, H.-J.; Zhang, T.-Y. Enabling Industry 5.0-Driven Circular Economy Transformation: A Strategic Roadmap. *Sustainability* **2024**, *16*, 9954. [CrossRef]
12. Oyungerel, T.; Bayarsaikhan, B.; Sainbileg, G. Study of the gap in the circular economy practices. *Mong. J. Econ. Rev.* **2024**, *28*, 64–77. [CrossRef]
13. *ISO/IEC/IEEE 42010*; 2022 Software, Systems and Enterprise—Architecture Description. ISO/IEC/IEEE: Geneva, Switzerland, 2022.
14. Martin, J.N. Overview of the Revised Standard on Architecture Description–ISO/IEC 42010. *INCOSE Int. Symp.* **2021**, *31*, 1363–1376. [CrossRef]
15. Islam, M.; Imran, R.; Hosain, S. The Evaluation of Enterprise Resource Planning using ISO 25010 Based Quality Model. In Proceedings of the 2021 2nd International Informatics and Software Engineering Conference (IISEC), Ankara, Turkey, 16–17 December 2021. [CrossRef]
16. Pestana, G.; Sofou, S. Data Governance to Counter Hybrid Threats against Critical Infrastructures. *Smart Cities* **2024**, *7*, 1857–1877. [CrossRef]
17. Kulkarni, N.D.; Bansal, S. Utilizing the Facade Design Pattern in Practical Phone Application Scenario. *J. Artif. Intell. Cloud Comput.* **2022**. [CrossRef] [PubMed]
18. Lercher, A.; Glock, J.; Macho, C.; Pinzger, M. Microservice API Evolution in Practice: A Study on Strategies and Challenges. *J. Syst. Softw.* **2024**, *215*, 112110. [CrossRef]
19. Shvets, O.; Murtazin, K.; Meeter, M.; Piho, G. Towards a Domain Model for Learning and Teaching. In Proceedings of the 12th International Conference on Model-Based Software and Systems Engineering, Rome, Italy, 21–23 February 2024. [CrossRef]
20. Deckers, R.; Lago, P. Systematic literature review of domain-oriented specification techniques. *J. Syst. Softw.* **2022**, *192*, 111415. [CrossRef]
21. Zafar, A.; Azam, F.; Latif, A.; Anwar, M.W.; Safdar, A. Exploring the Effectiveness and Trends of Domain-Specific Model Driven Engineering: A Systematic Literature Review (SLR). *IEEE Access* **2024**. [CrossRef]
22. Nadal, S.; Jovanovic, P.; Bilalli, B.; Romero, O. Operationalizing and automating Data Governance. *J. Big Data* **2022**, *9*, 117. [CrossRef]
23. Rodriguez, M.; de Araújo, L.J.P.; Mazzara, M. Good practices for the adoption of DataOps in the software industry. *J. Phys. Conf. Ser.* **2020**, *1694*, 012032. [CrossRef]
24. Torre-Bastida, A.I.; Gil, G.; Miñón, R.; Díaz-De-Arcaya, J. Technological Perspective of Data Governance in Data Space Ecosystems. In *Data Spaces*; Springer International Publishing: Cham, Switzerland, 2022. [CrossRef]

ceramics

MDPI

Article

Improving the Mechanical, Corrosion Resistance, Microstructural and Environmental Performance of Recycled Aggregate Concrete Using Ceramic Waste Powder as an Alternative to Cement

Ansam Ali Hashim [1,2], Rana Anaee [2] and Mohammed Salah Nasr [3,*]

1 Technical Institute of Babylon, Al-Furat Al-Awsat Technical University, Najaf 54001, Iraq; ansamly2@atu.edu.iq
2 Department of Materials Engineering, University of Technology, Baghdad 10066, Iraq; 130033@uotechnology.edu.iq
3 College of Engineering, University of Babylon, Babylon 51002, Iraq
* Correspondence: eng511.mohammed.nasr@uobabylon.edu.iq

Abstract: This study investigates the effectiveness of replacing the cement with 0, 5, 10, 15, and 20 wt.% of ceramic waste powder (HCCP) to improve the performance of recycled aggregate concrete (RCA) prepared using 25 wt.% wall tile ceramic coarse aggregates. The slump, initial and final setting time, compressive strength, splitting tensile strength, flexural strength, electrical resistivity, bulk density, porosity, total and surface water absorption, pH level, ultrasonic pulse velocity, dynamic elastic modulus, chloride ion diffusion coefficient, chloride penetration depth, microstructure analysis, and environmental assessment properties were investigated. The results showed that replacing cement with HCCP by 5 to 20 wt.% prolonged the setting time and improved all hardened properties. The highest improvements in mechanical properties were observed at 5 wt.% HCCP, with increasing rates of 26.5%, 22%, and 22.4% at 90 days for compressive strength, tensile strength, and flexural strength, respectively. On the other hand, the optimum enhancement for the durability, microstructural, and environmental efficiency properties was recorded at a 20 wt.% HCCP replacement rate. However, the strength at this ratio tended to decrease but remained higher than that of the control RAC. For instance, the total water absorption, surface water absorption, void ratio, chloride penetration depth, and migration coefficient were reduced by 47%, 45%, 38%, 62.3%, and 55.52%, respectively, compared to the reference sample.

Keywords: high calcium ceramic waste powder; recycled ceramic coarse aggregates; mechanical and durability properties; microstructure analysis; CO_2 emission

1. Introduction

Concrete is a highly prevalent building material globally because of its reliability, strength, durability, and convenient construction [1,2]. With its outstanding qualities and cost-effectiveness, the employment of concrete is continually increasing [3]. An estimation was made that the concrete construction sector requires over 1.5 billion tons of cement, 10–20 billion tons of aggregates, and around 1 billion tons of water per year for the manufacturing of concrete [4]. Furthermore, it has been noted that the yearly utilization of concrete exceeds 10 billion tons in contemporary industrialized societies [5,6]. With the growth of the global population, it is projected that the annual consumption of concrete will reach around 18 billion tons [5]. Concrete production has been responsible for a range of severe environmental issues, including the substantial release of carbon dioxide (CO_2) in the manufacturing of cement and the extensive use of naturally occurring aggregates (both coarse and fine) obtained through earth extraction. According to Chen et al. [7], cement manufacturing accounts for around 8% of worldwide carbon dioxide (CO_2) emissions. Additionally, the energy consumption regarding the manufacture of cement is measured at 118 kWh per ton [8,9]. It is possible to decrease the ecological impact of the concrete industry, through reduced raw material consumption and CO_2 emissions [10]. Hence, utilizing alternative materials like aggregates derived from byproducts of other sources may significantly contribute to the efficient manufacture of concrete [11,12]. Green concrete construction, which focuses on efficiently utilizing various agricultural or industrial waste materials in the manufacturing of concrete, has garnered significant research interest. Several studies have utilized recycled materials to substitute aggregates or cement in the manufacturing of environmentally friendly and long-lasting concrete [13–21]. Ceramic waste is a type of industrial waste that might potentially be utilized as a replacement material for concrete. Ceramic tiles are manufactured by submitting naturally occurring geological materials, such as clay, feldspar, and quartz, to elevated heat. Factories typically create two categories of tiles: floor tiles and wall tiles. An ordinary floor tile composition comprises around 50 wt.% clay, 40 wt.% feldspar, and 10 wt.% quartz, and requires a firing temperature of 1180 °C. Nevertheless, the recipe for wall tiles exhibits certain distinctions in comparison to those of floor tiles. Feldspar is substituted for calcite in the wall tile formula, and it requires a firing temperature of 1080 °C [22]. Recycling ceramic waste is challenging, resulting in a significant portion being disposed of in landfills. This practice greatly strains land resources, environment, and groundwater reserves [23]. While using ceramics in concrete offers an eco-friendly recycling technique, certain types of ceramics also have components that can enhance concrete qualities such as mechanical strength, as well as durability properties [4]. Moreover, ceramics can augment concrete's durability by increasing its resistance to chemical agents present in the surrounding environment. The process of incorporating recyclable ceramic material into concrete could have been mostly accomplished by substituting either the coarse aggregate or fine aggregate, or as a mineral addition [24]. Waste ceramic tile aggregates are durable and possess a high specific gravity. They also have increased voids, stronger crushing value, impact value, and abrasion value when in comparison with natural crushed stone [23]. Consequently, academics have lately shown significant interest in using recycled ceramic aggregates (RCA) in concrete. Three distinct types of ceramic tile materials—unused ceramic floor tiles, unused ceramic wall tiles, and waste ceramic wall tiles obtained from a demolition site —were utilized to replace a portion of the larger-sized aggregates in concrete. The concrete-mechanical characteristics obtained were found to be comparable to those of conventional concrete, as per the study conducted by Anderson et al. [24]. Medina et al. [25] studied the effect of replacement natural coarse aggregate with 20 and 25 wt.% of sanitary ceramic wastes on the electrical conductivity and durability properties of the resultant concrete. The findings demonstrated a notable improvement in the electrical conductivity and durability properties of concrete, including sanitary ceramic wastes. Ceramic waste was used by Khalid et al. [26] as gravel in producing concrete. The researchers discovered that the concrete mixture containing 35 wt.% ceramic waste exhibited superior characteristics when compared to standard concrete. Gharibi et al. [27] substituted natural sand and gravel (separately) with crushed ceramic electrical insulator waste at rates of 25%, 50%, 75%, and 100%. They also examined the simultaneous substitution of coarse and fine insulator waste categories at 50% and 100%. The outcomes indicate that substituting natural aggregate with ceramic insulator wastes at all levels led to an improvement in compressive strength and a decrease in the absorption of water, in comparison to the specimens in the untreated group. Klimek et al. [28] found that substituting natural fine aggregate in a mortar with aggregate from sanitary ceramic waste resulted in a positive impact on the strength characteristics of the mortars, while also reducing absorbability and the action of the capillary. However, Giridhar et al. [29] found a decrease in strength as the amount (wt.%) of recycled ceramic used increased. The study revealed that the compressive and split tensile strength of concrete decreased as the proportion of ceramic tile aggregate in the mixture increased. Other researchers have reached similar results as well [30–33]. The significant variation in strength outcomes primarily stems from the inconsistency in the ceramic material used as the source, which encompasses sanitary ware, kitchenware, electric insulators, bricks, and tiles, commonly employed in concrete [34]. The building industry is limited in its usage of RCA due to its significant absorption capacity. The difference in water absorption between ceramic and natural aggregates poses a significant challenge when utilizing ceramic aggregates in concrete manufacturing. Consequently, this leads to reductions in strength, workability, and durability [35,36]. To overcome the limitations of RCA, various pre-treatment procedures have been used to improve the strength and durability of RCA, such as nanomaterial modification [37], accelerated carbonation method [38], and microbial carbonate precipitation [39]. Nevertheless, the complex treatment procedure and additional expenses make the pre-treatment approaches unsuitable for civil engineering applications. Incorporating supplementary cementitious materials (SCMs) directly into recycled aggregate concrete (RAC) is the most favorable choice to avoid the complex alteration of the raw material, such as adding silica fume (SF), ground granulated blast furnace slag, metakaolin, fly ash, etc. This approach is considered effective in reducing additional expenses and improving the strength and longevity of RAC for engineering purposes [40–45]. Also, fired clay [46], bricks [47], tiles [48], and sanitary ware wastes [49] have been proposed as potential pozzolanic additives to cement. The pozzolanic reactivity of the material is derived from the presence of reactive SiO_2 and Al_2O_3. The reactive silica content in certain ceramic powders exceeds 25 wt.% [50]. The ceramic powder does not exhibit any discernible beneficial effect regarding the initial strength progress of concrete, primarily serving as a micro-filler [51]. The strength of concrete that contains ceramic powder is improved in the later ages of its development because of the pozzolanic reaction of the ceramic powder [52,53]. Moreover, the inclusion of ceramic powder can significantly enhance the process of refining the pore structure in concrete [54].

Based on the existing literature, few studies have focused on replacing natural coarse aggregate with wall ceramic tile waste in RAC. Additionally, most published papers have used low-calcium ceramic waste powder (CaO less than 10 wt.%) as a substitute for cement. To clarify, very few studies have utilized high-calcium ceramic waste powder (HCCP) with a CaO content of more than 30 wt.% as a replacement for cement. Furthermore, there have been very few, if any, studies that have enhanced the performance of wall ceramic waste-based RAC with HCCP. Thus, the research's main objective is to utilize HCCP as a substitute material for cement in RCA concrete. This study utilized crushed wall ceramic tile waste as a substitute for 25 wt.%. of the coarse aggregate, and ground wall ceramic tiles (HCCPs) as a substitution for 5, 10, 15, and 20 (wt.%). of cement in the creation of sustainable RCA. An investigation was conducted to examine the properties of the resulting RCA–HCCP concrete, including its fresh, physical characteristics, mechanical strength, durability, microstructure, and environmental impact. Using ceramic waste in RCA manufacturing reduces the amount of construction materials required, reduces the need for landfill space, and helps preserve the environment by mitigating potential pollution effects.

2. Materials and Methods

2.1. Materials Used

2.1.1. Cement

The Ordinary Portland Cement (OPC) (CEMI- 42.5R, Mass cement factory, Bazyan, Al Sulaymaniah, Iraq) under the trade name "Al MASS" was used in this study and classified as "Type I" by ASTM C 150 [55]. The specific gravity and fineness of the cement used were measured, and their values were 3.15 and 358 m^2/kg. The chemical composition of the cement is illustrated in Table 1, which was obtained through the chemical analysis conducted in the Environmental Laboratory of the University of Babylon (according to BE EN 196-2 [56]). The properties of cement meet the requirements of the Iraqi Standard Specification (IQS No. 5/2019) [57]. The distribution of cement particle sizes shown in Figure 1 was obtained through a laser particle analyzer (Malvern MS3000, Malvern Panalytical, Worcestershire, United Kingdom) device.

Table 1. The chemical composition of OPC and HCCP.

Oxide	HCCP (wt.%)	OPC (wt.%)
CaO	33.72	61.41
Al_2O_3	5.45	4.72
SiO_2	33.27	21.84
Fe_2O_3	18.67	3.60
MgO	0.79	2.97
K_2O	3.09	0.65
Na_2O	1.40	0.29
SO_3	0.18	2.26
TiO_2	3.12	1.69

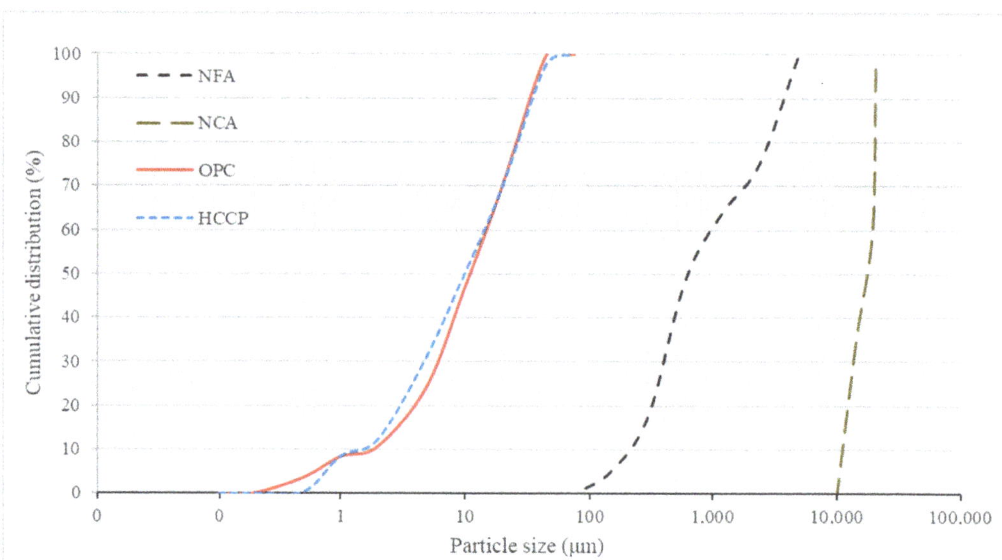

Figure 1. Particle size pattern of HCCP, OPC, fine aggregate (NFA), and coarse aggregates (NCA).

2.1.2. HCCP and RCA

Broken pieces of Chinese wall ceramic tiles provided by ceramic supply companies (Al-Mustafa Company, Babil Governorate, Iraq) were used to obtain RCA and HCCP. The gathered tiles were physically fractured (with a hammer) into typical aggregate dimensions (<19 mm). Figure 2 illustrates the laboratory procedure for acquiring HCCP. Due to the crushing action and the fragile characteristics of the ceramic wall tile, they had an angular form and a rougher surface compared to natural coarse aggregate (NCA). Table 2 depicts the physical characteristics of RCA and natural aggregates.

Figure 2. Schematic diagram of production of RCA and HCCP from discarded wall ceramics.

Table 2. The physical properties of natural and recycled aggregates.

Property	Fine Aggregate (Sand)	Natural Coarse Aggregate (Gravel)	Recycled Ceramic Coarse Aggregate
Fineness Modulus	2.91	6.2	6.9
Maximum Aggregate Size (mm)	4.75	19	19
Unit Weight, kg/m^3 (as per ASTM C29 [58])	1660	1640	1038
Specific Gravity (as per ASTM C127 [59])	2.65	2.58	2.46
Water Absorption, wt.% (as per ASTM C127 [59])	0.93	0.47	16

The ceramic powder was obtained from ceramic wall tiles that were deliberately fractured with a hammer, and then, the fragmented tiles were put into a Los Angeles Grinder (OSAKA Company, Japan, Tokyo) for 15 min to reduce the size to 300 µm. Then, the grinding process was completed using an electric steel grinder for a grinding period of 10 min. Figure 2 illustrates the procedure employed to obtain HCCP. The chemical components of HCCP, as determined using X-ray fluorescence (XRF), are shown in Table 1. The primary oxides present in the HCCP are CaO, SiO_2, Fe_2O_3, and Al_2O_3. The CaO percentage is 33.72 wt.%. Figure 1 displays the particle size pattern of OPC and HCCP as measured by a laser particle analyzer (Malvern MS3000, Malvern Panalytical, Worcestershire, United Kingdom). The average particle size of HCCP and OPC was 9.941 μm and 10.96 μm, respectively. A fineness test was also carried out using a Blaine device (Safety Energy Test Services Ltd., Nanjing, China), and the specific density of the resulting powder (HCCP) was measured by Chinese made Le Chatelier Flask, and its values were 367 kg/m^2 and 2.55, respectively.

2.1.3. Normal Aggregate

Crushed gravel with 2.58 specific gravity has been used as a coarse aggregate in all mixes. The size was limited to a maximum of 20 mm and was acquired from quarries in Al-Nibaai, Iraq. The sieving method determined the particle size distribution (PSD) of aggregates, which is also included in Figure 1. The physical characteristics of the crushed gravel were following the Iraqi standard (IQS No. 45) [60] and are shown in Table 2.

The fine aggregates (natural sand) were imported from a neighborhood in the Karbala region (middle of Iraq) known as AL-Akhaidhir, possessing a maximum standard size of 4.75 mm and a saturated surface dry (SSD) density of 1660 kg/m^3. The findings of the sand's gradation are depicted in Figure 1. This sand was graded according to Iraqi standards No. 45, specifically for the second zone [60]. The physical characteristics of natural fine aggregates are shown in Table 2.

2.1.4. Superplasticizer (SP)

In all concrete mixes, hyperplastic PC200, a high-range water reducer (superplasticizer), was utilized by combining with water to enhance the workability of fresh RAC mixtures. It was employed at the suggested doses of 0.2–0.4% by using the weight of the binder to produce acceptable workability. The manufacturer recommended using 0.75–2 liters per 100 kg of cementitious material. This superplasticizer satisfies the specifications of ASTM C494 Type F [61].

2.1.5. Water

For the preparation of all RAC mixes, tap water with a pH level of 7.5 was utilized.

2.2. *Mix Proportion*

The primary aim of the research is to investigate the impact of HCCP derived from wall ceramic waste on the characteristics and durability of RAC; several HCCP replacement percentages by mass of cement were utilized. The five concrete mixes, namely RCA–HCCP5, RCA–HCCP10, RCA–HCCP15, and RCA–HCCP20, were created by substituting cement with 5%, 10%, 15%, and 20% of HCCP, respectively. All of these mixes included 25% of RCA as partial substitution (by weight) for the natural coarse aggregate (NCA). All mixes, comprising the reference mix, were made using a constant water-to-cement ratio of 0.5. In all mixes, a superplasticizer admixture of 1.6% of the cementitious materials was utilized to achieve concrete with satisfactory workability and a target mean strength of 30 MPa. Table 3 provides a concise summary of the mixing proportions for concrete.

Table 3. Mix proportions (kg/m^3) clarity.

Mix Name	Cement	HCCP	Coarse Aggregate		NFA	Water	SP
			NCA	RCA			
RCA–HCCP0	375	0	750	250	750	0.5	1.6
RCA–HCCP5	356.25	18.75	750	250	750	0.5	1.6
RCA–HCCP10	337.5	37.5	750	250	750	0.5	1.6
RCA–HCCP15	318.75	56.25	750	250	750	0.5	1.6
RCA–HCCP20	300	75	750	250	750	0.5	1.6

RCA, recycled ceramic coarse aggregate concrete; HCCP, high calcium ceramic waste powder; NCA, natural coarse aggregate; NFA, natural fine aggregate; and SP, superplasticizer.

2.3. Mixing, Curing, and Preparation of Specimens

A laboratory mixing machine with a pan-type design and a capacity of 250–300 kg, was employed to blend concrete components. NCA, RCA, and fine aggregates were mixed in the mixer for 1 min. This step ensures a uniform distribution of the aggregates before adding other components. After that, 50% of the required mixing water was combined with the aggregates and mixed for another minute. The purpose of this step is to moisten the aggregates and achieve initial hydration. Once the water and aggregates were mixed, cement and HCCP were added to the mixer. The mixer ran for 1 min to thoroughly mix the cementitious materials with the moistened aggregates. This step ensures the uniform distribution of cement and HCCP throughout the mixture. Next, the superplasticizer (a chemical admixture used to improve workability) and the residual water were incorporated into the mixer, and the mixer operated for 1 min. The superplasticizer helps in achieving the desired fluidity or workability of the concrete. The mixer operated for 1 min.

Standard molds are used to shape the concrete samples according to BS EN 12390-1 [62]. To prevent the concrete from sticking to the molds after it has solidified, the inner surfaces of the molds are lubricated with oil. The concrete mix is poured into the molds, and then the samples are compacted using an electric vibrating device to remove any gaps or voids in the mixture. This ensures proper compaction and eliminates air pockets. In this experiment, different types of samples are cast using specific molds, cubes, cylinders, and prisms. After 24 h, the specimens were extracted. Afterward, each of the RCA–HCCP concrete samples was set up in reservoirs of water and exposed to ambient temperature for 28 and 90 days.

2.4. Program of Testing

Multiple experiments were conducted in this investigation, as outlined in the subsequent sections:

2.4.1. Fresh Tests (Workability and Setting Times)

The slump test is performed on fresh concrete for all mixtures. The test is conducted on all RCA–HCCP concrete mixes under the ASTM-C 143 standard [63], while setting time tests are conducted with the Vicat apparatus by ASTM C191 [64].

2.4.2. Hardened Tests

Compressive Strength

The test is conducted following the specifications given in BS 1881-Part 116 [65]. A series of thirty concrete cubes, each measuring 100 mm in size, was created to study the progression of compressive strength in different concrete mixes over time. The ages under investigation are 28 and 90 days. Three cubes are produced for each mixture at the designated age.

Tensile Strength

The ASTM C496 standard is used to conduct the tensile strength test on concrete cylinders [66]. The molds utilized are cylindrical, measuring 20 × 10 cm^2. Three specimens are tested for all mixes at each age (28 and 90 days of curing).

Flexural Strength

The flexural strength tests, by ASTM C293 [67], were conducted on 100 × 100 × 400 mm^3 prism specimens. Three specimens are tested for each mix at the ages of 28 and 90 days.

Bulk Density

The test was conducted on the same cubes that were utilized for compression testing, during which the cubes were weighed, and their precise measurements were quantified. The bulk density was determined by dividing the weight of each cube by its volume [68]. Each result was measured three times, and the average of three measurements was utilized to symbolize the outcome.

Ultrasonic Pulse Velocity (UPV) and Dynamic Elastic Modulus (Ed)

The PUNDIT device (MATEST S.p.A., Treviolo, Italy) was used to measure the UPV and Ed by ASTM C597 [69]. Their measurements were taken using a 100 mm^3 cube when they were 28 days of age. Each result was obtained by taking an average of three readings. The Ed was determined using Equation (1) [69]:

$$v = \sqrt{\frac{Ed(1-\mu)}{\rho(1+\mu)(1-2\mu)}} \tag{1}$$

The variables in question are ρ, denoting density, and μ, denoting dynamic Poisson's ratio.

Equation (2) can be reformulated by substituting the value of Poisson's ratio (0.2) as follows:

$$Ed = 0.9\ \rho\ v2 \tag{2}$$

Permeable Voids, Total Water Absorption, and Surface Water Absorption

The samples used for these tests are the portions resulting from the flexural strength test specimen (prisms 100 × 100 × 400 mm^3) after they have been broken in half in the testing apparatus. First, the specimens underwent a drying procedure in a furnace at a temperature of 110 °C for a minimum period of 24 h until the weight was constant. Once the specimen is taken out of the oven, allow it to cool in dry air until it reaches a temperature of 20–25 °C, and the dry mass (Md) is obtained. The rate of surface absorption of RCA–HCCP concrete was measured for the hardened state as presented in ASTM C1585-2013 [70]. The dried specimen was exposed to water at 5 mm depth for 24 h on one of their faces, and in this case, it was the undamaged face resulting from the flexural test, it was weighed (Mc), and the surface absorption (%) was determined using the following equation:

$$Surface\ absorption(S\%) = \frac{Mc - Md}{A} \times 100 \tag{3}$$

The percentage of permeable pores and total water absorption tests were executed using the methodology outlined in ASTM C642 [71]. The samples after drying are fully flooded in tap water and left not less than 48 h (until the constant weight). Next, the specimen is submerged in tap water, and subjected to boiling for 5 h, leaving to cool gradually for a minimum of 14 h until it reaches a final temperature between 20 and 25 °C. A cloth was used to eliminate any moisture from the surface, and the mass of the specimen (Ms) was measured. Subsequently, the weight of the specimen was measured while it was fully submerged in water (Mb). The total water absorption and percentage of permeable pores calculation are determined using the following equation:

$$\textit{Total water absorption } (TA\%) = \frac{\mathrm{Ms} - \mathrm{Md}}{\mathrm{Md}} \times 100 \quad (4)$$

$$\text{Permeable voids}(\%) = \frac{\mathrm{Ms} - \mathrm{Md}}{\mathrm{Ms} - \mathrm{Mb}} \times 100 \quad (5)$$

Electrical Resistivity (ER)

A LCR meter (DER EE Electrical Instrument Co., Ltd., New Taipei City, Taiwan), was utilized to measure the impedance at a frequency of 1000 Hz. Saturated 100 mm^3 cube samples were examined utilizing the two-metal plate technique [72]. To conduct this experiment, a 100 mm^3 cube was placed between two plates and a moist sponge was included to ensure electrical conductivity. The electrical resistivity of the concrete was subsequently calculated using Equation (6). The equation necessitates the measurement of the electrical impedance value of the sample utilizing the designated instrument. The result is multiplied by the area of the sample and then divided by its height. Each result was documented by taking an average of three measurements.

$$\mathrm{ER} = \frac{\mathrm{A}}{\mathrm{L}} \times \mathrm{R} \quad (6)$$

The equation utilizes the following variables: R denotes the impedance, L denotes the height of the specimen in centimeters, ER denotes the electrical resistivity in kilo-ohms per centimeter, and A denotes the cross-sectional area of the specimen in square centimeters.

Test for Chloride Migration

In this test, the chloride ions penetration in concrete is accelerated, and the test is conducted according to the Nordtest method [73]. For each mix, three concrete specimens (Ø100 mm × 50 mm height) were cast and moist-cured for 28 days. After that, the specimens were positioned within a vacuum chamber. Both end surfaces of the sample were exposed, and absolute pressures in the vacuum vessel were increased to a pressure range of 1–5 kPa within several minutes, and the vacuum specimens remained for 3 h. The pump vacuum is then kept running to fill the vessel with a saturated solution of $Ca(OH)_2$ by different pressure between the vacuum container and the other container containing $Ca(OH)_2$ to immerse all the specimens and to avoid the leaching of $Ca(OH)_2$. The vacuum was maintained for an hour before releasing the air back into the container. After soaking the specimens in lime solution for 18 ± 2 h, the migration of chloride penetration was accelerated by using an electrical circle on solutions of NaCl and NaOH, following the Nordtest method [73]. Each specimen was drilled into a cylinder rubber tube and fitted with two clamps. The rubber tube was filled with 300 mg solution of sodium hydroxide (0.3 N NaOH) and fixed in a tank containing sodium chloride solution (10% NaCl by mass), where the electrode in NaOH solution becomes an anode, and the electrode in NaCl solution acts as a cathode. The chloride ion depth, xd, was then measured by spraying a silver nitrate ($AgNO_3$) solution with a certain concentration of 0.1 N onto the split cylinder sample and measuring the chloride penetration (xd) caused by the reaction between Cl and Ag^+ to find the chloride penetration depth of the concrete prism. Then, the chloride penetration depth in the short cylinder was employed to calculate the diffusion coefficient for non-steady state migration (Dnssm), as shown in Equation (7).

$$\text{Dnssm} = \frac{0.0239(273 + \text{T})\text{l}}{(\text{Uv} - 2)\text{t}}(\text{xd} - 0.0238(\sqrt{\frac{(273 + \text{T})\text{l xd}}{\text{Uv} - 2}})) \quad (7)$$

The symbol Dnssm denotes the product of the non-steady state migration coefficient $\times 10^{-12}$ square meters per second. U denotes the magnitude of the applied potential, measured in volts. T reflects the mean value of the beginning and final temperatures, measured in degrees Celsius, of the anolyte solution. L denotes the magnitude of the specimen's thickness, measured in millimeters. The variable "xd" denotes the mean measurement of the depth of chloride penetration, expressed in millimeters. Finally, t denotes the length of the test in hours.

pH Test

Small portions were obtained from broken concrete cubes under the compressive strength test and used for measuring the alkalinity (pH) of concrete mixes [74]. These parts are crushed and then ground to a powder for ten minutes using electric steel grinding (the same machine that was used to grind the ceramic powder) and then sieved using a 300-micrometer sieve. The total weight of the prepared sample was about 75 g. As a reference, the powder suspension method depends on this test [75]. A Chinese-made TRANS Meter is used in the current test. Prior to the test, the pH meter underwent calibration with a pH value of 10. A total of 50 ± 0.1 g of the specimens were combined with 50 ± 0.1 g of fresh distilled water in a 100 mL self-standing centrifuge tube. A magnetic stir bar was then added, and the screw cap on the tube was tightened. The tube was positioned on the magnetic stirrer to agitate without applying heat for 5 min during leaching. The suspension was passed through a Grade 40 quantitative filter paper using a Buchner funnel filter to collect a minimum of 10 mL of the solution. Subsequently, pH measurements were recorded. The projected timeline for the experiment, after dividing the specimens into their respective sizes, was as follows: the production of the suspension was expected to take 5 min, stirring and leaching would take another 5 min, filtering would require 10 min, and measuring the pH would take 10 min. The pH of the ceramic powder was tested using the same procedure described earlier.

Microstructural Analysis

The microstructure examination was conducted using a scanning electron microscope (SEM, TESCAN, Brno, Czech Republic). For this examination, a tiny concrete sample of approximately 5×5 mm^2 was removed from each of the specimens. The sample was dehydrated in an oven at a temperature of 105 °C for 24 h. Subsequently, its surface was polished by applying diamond paste. The specimen was light-covered with copper to impart electrical charge, and subsequently positioned on an SEM stem.

Environmental Assessment

Utilizing solid waste as a supplementary cementitious material in cement or concrete is a highly efficient method for decreasing the expenses associated with cement or concrete. The concrete industry has emerged as a highly efficient solution for managing diverse types of solid waste. Notably, China has made significant advancements in solid waste utilization technology, largely due to market competition [76]. Unfortunately, the presence of a large amount of solid waste in cement or concrete poses a risk to the integrity of the cement or concrete. While the carbon emissions from the cement industry may decrease, the carbon emissions from concrete may not necessarily decrease. Therefore, a comprehensive evaluation and solution are required to effectively utilize solid waste in cement or concrete. The scientific method used to assess the ecological consequences of a product takes into account not only its quantity but also its quality or performance. Scientists have suggested using an index, like the one created by Filho et al. [54], to measure the level of Portland cement (PC intensity) and carbon dioxide emissions (CO_2 intensity) to assess the environmental efficiency of cement usage. The OPC intensity index, as defined in Equation (8), quantifies the quantity of Portland cement required per unit volume of concrete to achieve a mechanical strength of 1.0 MPa at a specific age 'j' day (28 and 90 days).

$$\text{OPC intensity} = \frac{\text{portland cement consumption}}{\text{compessive strength at (J days)}} \tag{8}$$

$$CO_2 \text{ intensity} = \frac{\text{carbon dioxide immesion}}{\text{compessive strength at (J days)}} \tag{9}$$

The CO_2 intensity index ($kg \cdot m^3 \cdot MPa^{-1}$) quantifies the global warming impact of a cementitious formulation by measuring the amount of carbon dioxide released by the binder (cement and HCCP) per unit volume of the mixture (in kg/m^3) when achieving a mechanical strength of 1.0 MPa at a specific age (28 and 90 days).

CO_2 emission is the total greenhouse gas emission (kg CO_2eq) per kg of material produced. The carbon emissions from the materials used in the formation of the RCA–HCCP mixes formulated in this research were obtained by surveying the previous literature and adopting the values mentioned therein, as shown in Table 4.

Table 4. Environmental impact data for the production of RAC ingredients.

Material	kg CO_2eq	Author
OPC (1 kg)	9.27×10^{-1}	Kurda et al. [77]
NFA (1 kg)	9.87×10^{-3}	Kurda et al. [77]
NCA (1 kg)	2.44×10^{-2}	Majhi and Nayak [78]
RCA (1 kg)	2.94×10^{-2}	Park et al. [79]
HCCP (1 kg)	4.5×10^{-2}	Hussein et al. [80]
Water (1 kg)	1.33×10^{-4}	Kurda et al. [77]

3. Results and Discussion

3.1. Initial and Final Setting Time

Figure 3 illustrates the outcomes of the initial and final setting times of RCA–HCCP, which was made using different amounts of HCCP as a substitute for OPC as a binder. The data indicated that the initial and final setting times varied between 120 and 132 min and 170 and 192 min, respectively. The control sample, which consisted of 100% cement, had the lowest values for both initial and final setting times. The substitution of OPC with HCCP resulted in a slightly prolonged period at which the pastes began to lose their plasticity. The greatest difference was observed in the pastes containing 20 wt.% HCCP, with a 12 min later than the reference paste (RCA–HCCP0). The slightly extended initial setting time (IST) enhances the efficiency of transporting, laying, and compacting the concrete when using this particular type of blended cement. Nevertheless, the rise in HCCP concentrations resulted in a modest increase in both the initial and final setting times of all pastes. The HCCP is impacting the process of hydration and solidification. The disparity in variability between initial and final setting time is small. Nevertheless, the disparity in the setting durations became more pronounced as the HCCP contents in RCA–HCCP samples increased. As the HCCP concentration reached 20 wt.%, the SiO_2 content also increased, causing a slower chemical reaction rate and longer setting time [81]. The outcomes of the RCA–HCCP blended systems align with the results previously documented by Pitarch et al. [52], that discovered that substituting a portion of OPC with ceramic powder (15 wt.% to 50 wt.%) altered the flow characteristics of the pastes and slowed down their hydration process. These results confirmed that the inclusion of HCCP in the concrete matrix effectively prolonged the setting time of concretes under normal environmental conditions.

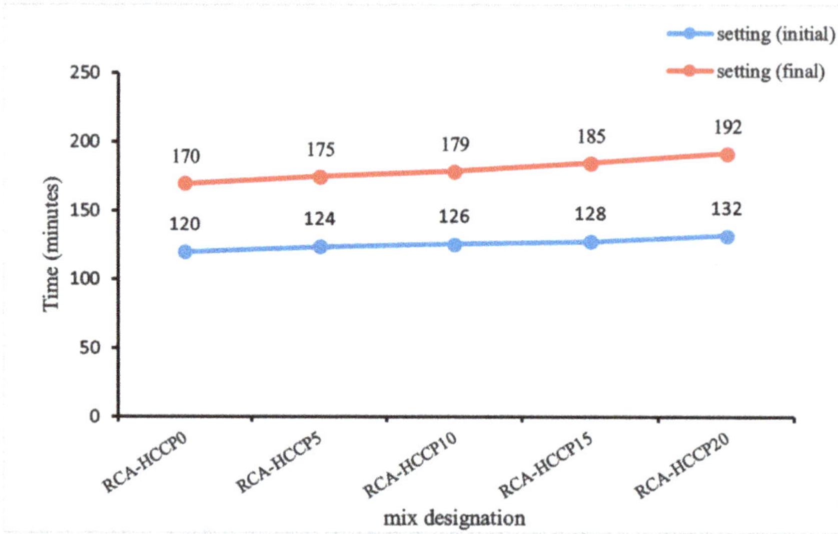

Figure 3. The initial and final setting time for RCA mixtures with and without HCCP.

3.2. Slump

The slump of both the control concrete mix and the mixes with partial replacement of HCCP were tested, as indicated in Figure 4. Immediately after, the water was added and well-blended. The slump measurement of RCA–HCCP0 concrete was determined to be 100 mm. Subsequently, upon beginning the replacement process, there was a significant and rapid increase in the slump value, suggesting that HCCP is functioning as a plasticizing agent consistent with the findings reported by Nalli & Vysyaraju [82]. It is observed that the RCA–HCCP20 mixture achieved the maximum slump value of 120 mm. On the other hand, the RCA–HCCP0 mixture demonstrated the smallest slump measurement, measuring 100 mm. Hence, it is feasible to reduce the water-to-cement ratio (w/c) by employing HCCP to achieve equivalent workability to that of the control mix. The improvement in workability of RAC with the presence of ceramic powder is due to several reasons, including the smoothness of the surface of the HCCP particles and the presence of a wide range of particle sizes, which reduces friction and facilitates particle movement [83]. In addition, the slow hydraulic reaction of HCCP results in more water being available than pure cement at the time of pouring, which increases the plasticity of the concrete mixture [83,84].

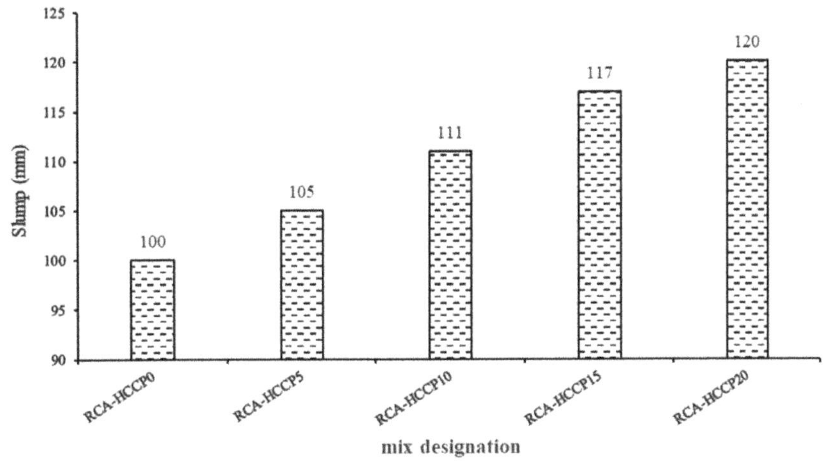

Figure 4. Slump value from RCA mixtures with and without HCCP.

3.3. Compressive Strength

Figure 5 illustrates the outcomes of the compressive strength test of cubic specimens ($10 \times 10 \times 10$ cm^3) at the 28- and 90-day ages which were subjected to testing at a loading rate of 0.35 MPa/s, each result shown was calculated from the average of three readings. The findings revealed that at 28 days of age, compression test values for RCA–HCCP5, RCA–HCCP10, and RCA–HCCP15 recorded compressive strength higher than of the RCA–HCCP0 mix by 11.76%, 3.90% and 2.00%, respectively. This was attributed to the effect of the high calcium ceramic particles filling the gaps, as they had a smaller particle size compared to the OPC grains (9.941 µm vs. 10.96 µm), based on the obtained particle size results. The HCCP grains can occupy the empty spaces between the aggregate and the cement paste, resulting in a denser microstructure and ultimately leading to an increase in compressive strength [85]. While incorporating 20 wt.% HCCP reduced the compressive strength by 9.70%. This might be due to the dilution effect that slows down the strength development [86]. The micro-filling impact of HCCP was insufficient to compensate for the decrease in cement content. Nevertheless, when the RCA–HCCP mixes reached 90 days of age, all the RCA–HCCP mixes exhibited sufficient levels of compressive strength, exceeding that of the reference mixture. Regarding the recorded results, extending the time of curing improves the compressive strength values of the RCA–HCCP mixes in comparison to the strength levels at 28 days of age. For example, at 90 days, the RAC with 5, 10, 15, and 20 wt.% HCCP achieved compressive strengths of 50.17, 46.67, 44.50, and 41.83 MPa, respectively. These values represent increases of approximately 26.5%, 16.9%, 11.4%, and 4.8% compared to the 39.63 MPa recorded for the RCA–HCCP0 mix at the same age. The effect of delayed setting caused by the ceramic powder becomes negligible at 90 days [52]. The reason for this is the continual process of cement hydration and the subsequent production of more calcium hydroxide, which undergoes a chemical reaction with the ceramic powder, resulting in the formation of more calcium silicate hydrate (CSH) gel. As a result, the compressive strength was enhanced [87]. Ebrahimi et al. [88], Mohammadhosseini et al. [89] and Rashid et al. [90] achieved similar results.

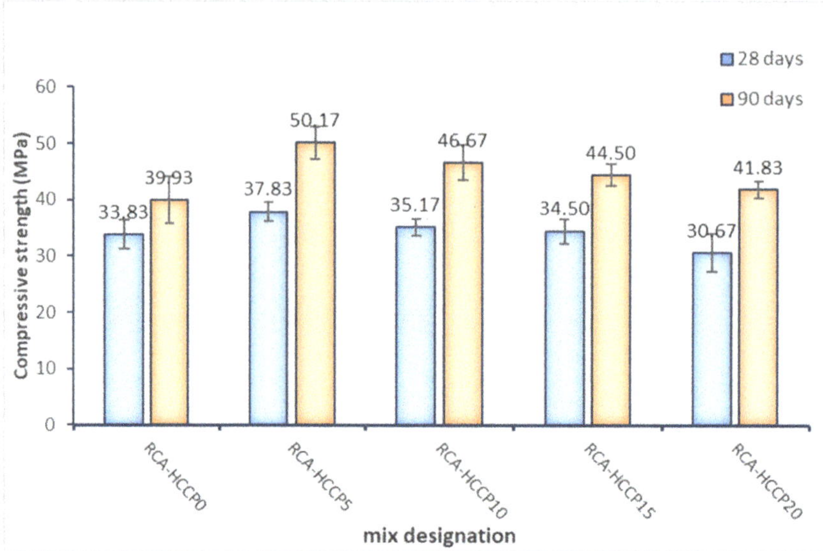

Figure 5. Compressive strength from RCA concrete at 28 and 90 days with and without HCCP.

3.4. Splitting Tensile Strength

Figure 6 illustrates the outcomes of the tensile strength test of cylindrical specimens (10×20 cm^2) at the 28- and 90-day ages, which were subjected to testing at a loading rate of 0.5 MPa/s; the tensile strength outcomes were determined by using the average value of three samples. An analogous pattern to that seen in the compressive strength was identified in the tensile strength of RCA–HCCP mixes employing HCCP. The splitting tensile strength of the RCA–HCCP concrete mixes was seen to rise as the curing time increased. The observed rise in strength values can be attributed to the pozzolanic effect of the HCCP and the micro-filling effect. The HCCP particles functioned as sites of nucleation for cement granules and the products of hydration, resulting in a more compact microstructure [91]. At the longer curing times (90 days), the splitting tensile strength of all mixes, including HCCP, exhibited an increase in comparison to the RCA–HCCP0 mixture. At 90 days of age, the tensile strength of RCA–HCCP5, RCA–HCCP10, RCA–HCCP15, and RCA–HCCP20 was found as 4.85, 4.32, 4.05, and 4.32 MPa, which is about 23.5%, 9.9%, 3.1% and 9.9% greater than the value of RCA–HCCP0 mix. The results obtained in this research were consistent with the outcomes of Ebrahimi et al. [88], who examined the impact of ceramic powder on the tensile strength of concrete. Their findings demonstrated that replacing OPC with ceramic powder revealed good tensile strength development after 90 days.

Figure 7 illustrates the correlation between the compressive strength and tensile strength of RCA–HCCP samples. Specimens were produced using varying amounts of HCCP instead of OPC and were tested at 28 and 90 days. A strong exponential correlation was found between compressive strength and tensile strength, with R^2 values of 0.9233 and 0.9425 at 28 and 90 days. This indicates a high level of confidence in the correlations.

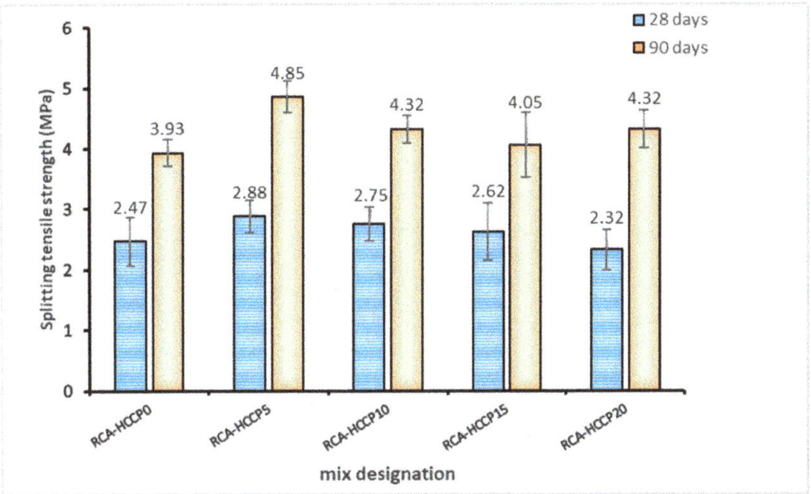

Figure 6. The tensile strength of RCA concrete at 28 and 90 days with and without HCCP.

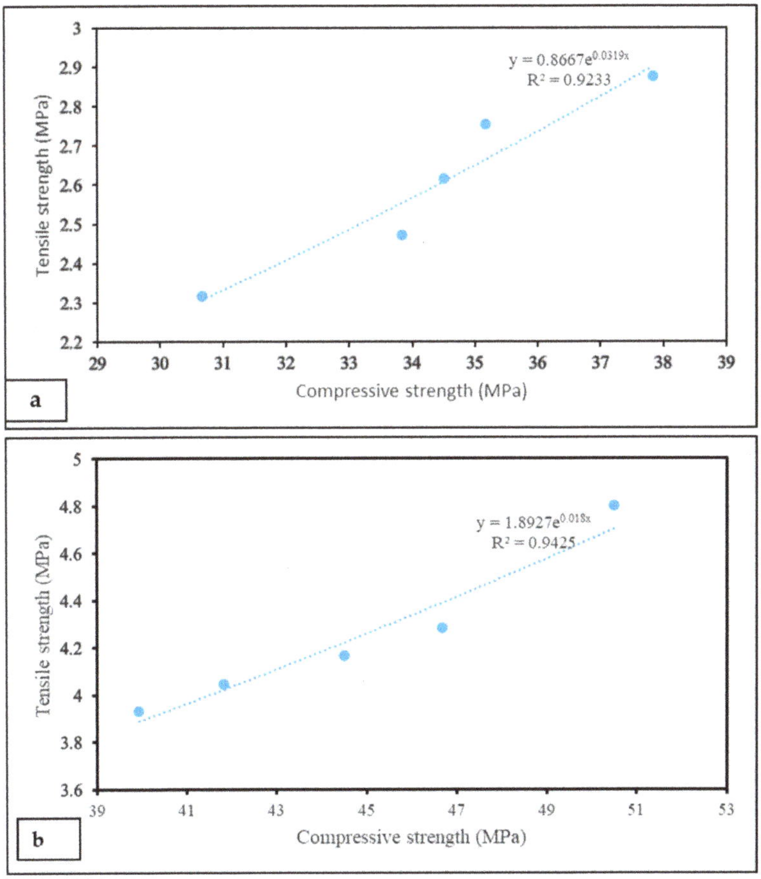

Figure 7. The correlation between compressive strength and tensile strength (**a**) at 28 days and (**b**) at 90 days of age. The dotted line represents the trendline used to find a mathematical relationship between the datapoints.

3.5. Flexural Strength

The outcomes of the flexural strength obtained at 28 days of curing for prism specimens ($10 \times 10 \times 40$ cm^3) of concrete were subjected to a loading rate of 1 MPa/min during testing, which is presented in Figure 8. The flexural strength of RCA–HCCP mixes displayed a comparable pattern to the compressive and tensile strength of those same RCA–HCCP mixes. Based on the outcomes shown in Figure 8, the RCA–HCCP5, RCA–HCCP10, RCA–HCCP15, and RCA–HCCP20 mixtures recorded flexural strength values of (4.7 MPa, 4.27 MPa, 4.19 MPa, and 3.8 MPa) at 28 days, respectively. Moreover, the flexural strength value of 4.10 MPa was recorded for RCA–HCCP0mix. Concerning the results recorded, the flexural strength of all the RCA–HCCP mixtures incorporating HCCP, increased with increasing time of hydration. This points to a higher impact when curing times with greater amounts of HCCP. At the age of 90 days, the RCA with varying percentages of HCCP (5, 10, 15, and 20 wt.%) achieved flexural strengths of 5.83 MPa, 5.57 MPa, 5.27 MPa, and 5.03 MPa, respectively. These values are 22.4%, 16.8%, 10.5%, and 5.6% greater than the control mixture's stated strength of 4.77 MPa. The RCA–HCCP5 concrete mix demonstrates superior flexural strength with a value of 5.83 MPa, surpassing that of the RCA–HCCP0 (control mix). These data suggest that the ideal replacement amount for producing RCA concrete is around 5%. In other studies, Lasseuguette et al. [53] found that approximately 15 wt.% of ceramic powder exhibited the highest mechanical strength, while the 20 wt.% content was necessary to attain mechanical qualities similar to those of OPC. Mohit and Sharif [92] found that the addition of 10 wt.% ceramic powder gave the best flexural strength. This improvement in the strength of the RCA incorporating HCCP is related to the physical–chemical characteristics of HCCP and its interactions with hydrated cement. The additional surface area of ceramic powder, similar to silica fume [93], provides more locations for the generation of hydration products. The even distribution of HCCP granules also decreases the path that ions need to travel from the active minerals to the sites of nucleation. These effects collectively promote higher nucleation rates and accelerate cement hydration processes. Physically, HCCP serves as a filler that enhances the density of the cement matrix by occupying the empty spaces between cement particles [94]. Chemically, the pozzolanic reactions of ceramic powder result in the formation of additional bonding phases, which strengthen the contacts between the ceramic powder particles and the dehydrated cement [95]. Together, the chemical and physical influences have a combined effect on the mechanical behavior of the RCA–HCCP concrete mix [95], consequently resulting in higher strength with longer curing periods [96]. Figure 9 illustrates the correlation between the compressive strength and flexural strength of the RCA–HCCP sample which reveals a strong exponential correlation between compressive strength and flexural strength, with an R^2 value of 0.9849, and 0.9783 at 28, and 90 days. This demonstrates a high level of confidence in the relations.

3.6. Bulk Density Results

The outcomes for bulk density are displayed in Figure 10. The bulk density of hardened mortar or concrete is determined by many factors, including the presence of trapped or incorporated air spaces, the quantity and density of the aggregate, and the cement and water percentage [97]. The outcomes showed two trends. First, an increase in density as a function of age. This indicates a gradual increase in the pozzolanic effect as time progresses [79]. Second, a slight reduction in the density when HCCP is used as a substitute comes from reducing the cement content, which has a higher specific gravity (the specific gravity for cement is 3.15 compared to 2.55 for HCCP). After the 28-day curing period, the density values for the RCA samples containing 5, 10, 15, and 20 wt.% HCCP were 2327, 2317, 2290, and 2277 Kg/m^3 and increased to 2330, 2322, 2297, and 2284 Kg/m^3 at

90 days of age. The change in density values did not surpass 2.6% and 2.5%, as observed in the RCA–HCCP20 mixture at 28 and 90 days of age. Thus, replacing cement with HCCP, ranging from 5 wt.% to 20 wt.%, had minimal impact on the density of the hardened concrete despite HCCP having a lower specific gravity than cement, as has also been reported in past studies [88,98]. These findings comply with the work by Nasr et al. [87], which pointed out that the bulk density decreases slightly as the cement replacement with ceramic powder increases. This phenomenon might be attributed to the tiny particles of HCCP, which improved the compatibility of the RCA by filling in the gaps. As a result, the microstructure of the concrete became denser [99]. The chemical composition of the HCCP contains silica, lime, and alumina, which create a pozzolanic microenvironment inside the concrete matrix. The calcium silicates and aluminates ensuing from the pozzolanic reactions of HCCP have enhanced the integrity of the concrete. Additionally, the water that ceramic particles release has an internal curing effect that aids in forming a denser CSH (calcium silicate hydrate) gel. The appropriate dispersion of CSH gel and the generation of more CSH gel is due to the better performance of the RCA–HCCP concrete mix. This effect countered the decrease in density caused by the inclusion of HCCP [100].

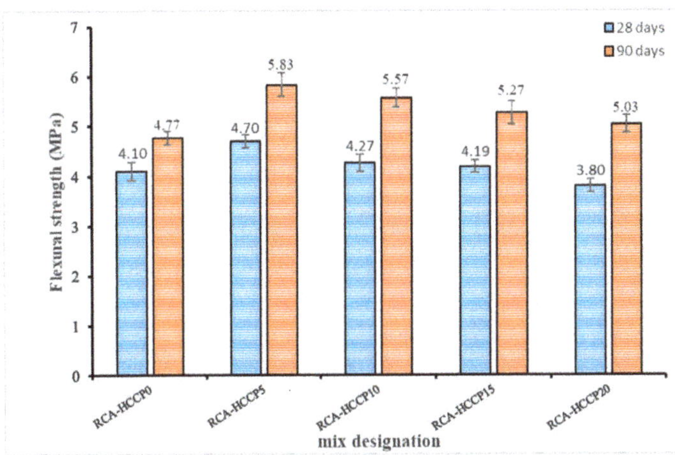

Figure 8. The flexural strength of RCA concrete at 28 and 90 days with and without HCCP.

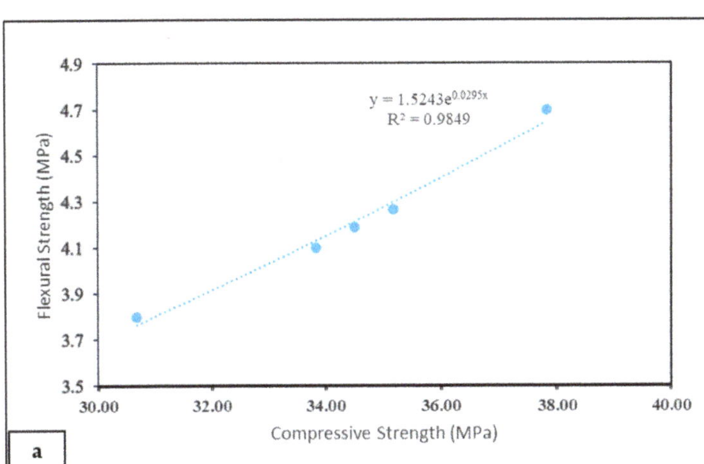

Figure 9. *Cont.*

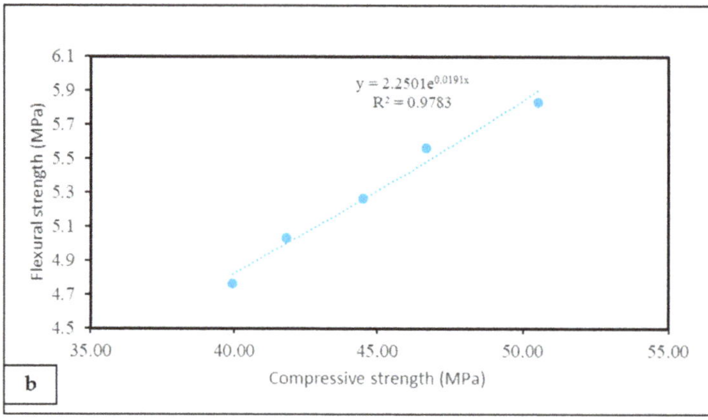

Figure 9. The correlation between compressive strength and flexural strength, (**a**) at 28, (**b**) at 90 days of age.

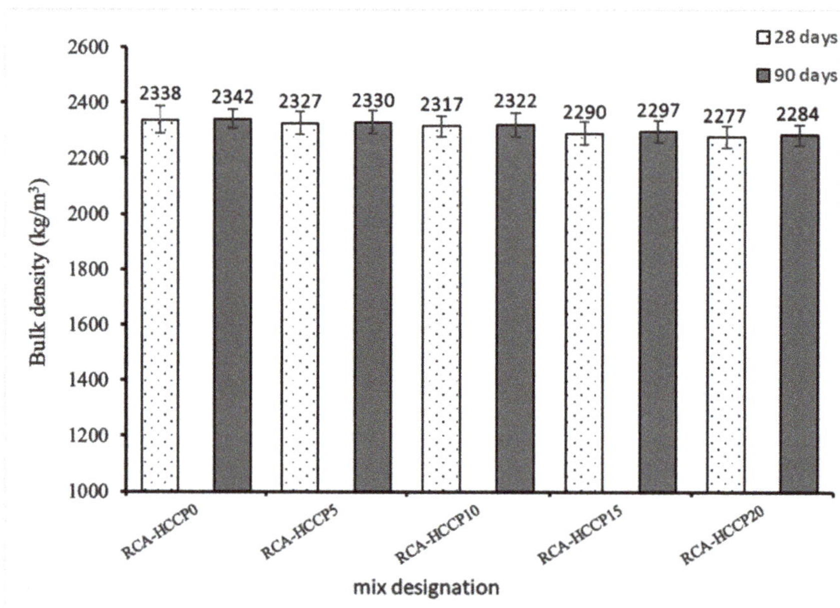

Figure 10. Bulk density results from RCA with and without HCCP.

3.7. Ultrasonic Pulse Velocity (UPV)

The UPV results are depicted in Figure 11. The UPV approach is very efficient, potent, and adaptable, enabling thorough examination of material uniformity. UPV may be utilized to ascertain the consistency of concrete, regulate its quality, monitor its degradation, and detect internal defects and empty spaces [101,102]. The UPV values for RAC mixes that included HCCP as a cement substitute ranged from 4397 to 4373 m/s. The highest reduction in the velocity (1.14%) was found when the cement was replaced with 20 wt.% HCCP. There is no significant change in UPV results after replacing cement with HCCP. Since UPV depends on concrete density [103], this argument is substantiated by the density findings of this research. However, UPV m/s in the range of less than 2000 m/s, 2000–3000, 3000–3500, 3500–4500 m/s, to more than 4500 m/s are classified into very poor, poor, medium, good, and excellent in quality of concrete, respectively [104]. The outcomes indicate that all RCA–HCCP concrete mixes exhibit satisfactory quality when the UPV is measured. The coefficient variance of the test results of UPV is less than 4%.

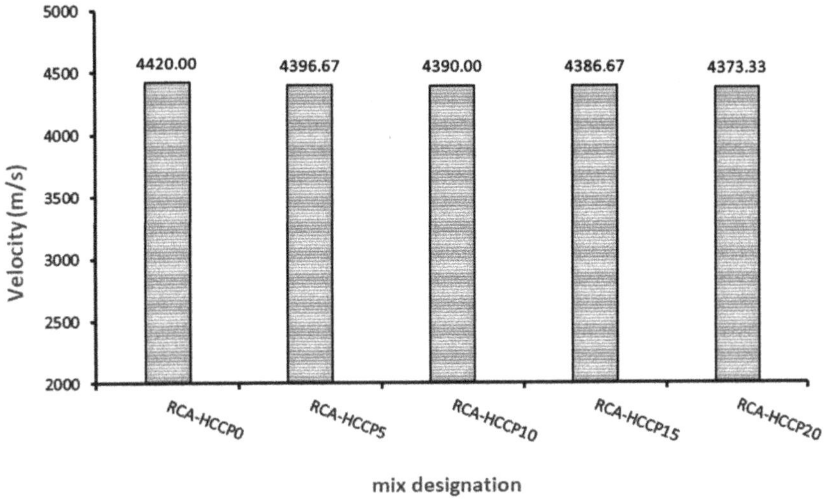

Figure 11. UPV results from RCA with and without HCCP.

3.8. Dynamic Modulus of Elasticity

The modulus of elasticity is a substantial mechanical property of concrete that significantly influences the construction of concrete structures. This property, along with compressive strength, directly controls the proper timing for formwork removal and pre-stress application on concrete members. If concrete is mishandled before it develops the desired modulus of elasticity, it can lead to large deformations, crushing, or even catastrophic failures. Therefore, understanding the early-age elastic modulus of concrete is crucial to ascertaining the sustained effectiveness and safety of the construction [105]. Figure 12 displays the outcomes of the dynamic elastic modulus investigation. It was found that the modulus of elasticity values of the RCA–HCCP0 mix and the mixture containing HCCP as an alternative to cement were either similar to or slightly less than the RCA–HCCP0 mix. A similar result was obtained by Mohammadhosseini et al. [89], who stated that the incorporation of 20 wt.% ceramic powder as a cement substitute led to a minor, approximately within a 6% range decrease in the modulus of elasticity of the concrete. The elastic modulus measurements for all RCA–HCCP mixes were between 41.17 and 39.22 GPa. The RCA–HCCP20 mixture had the lowest modulus of elasticity, 4.7% below the RCA–HCCP0 mix. The elastic modulus measurements of all RCA–HCCP mixes decreased as the concentration of HCCP increased, with the drop falling within the range of 1.7% to 4.7%. Although there was a 20 wt.% drop in cement, the modulus of elasticity did not see a substantial fall, despite the dilution impact. This may be due to the reaction of Al_2O_3 and SiO_2 in HCCP with cement hydration products, forming a more compact calcium aluminate and hydrated C-S-H gel. These cementitious materials are better filled with reactivated coagulation so that the adhesion between the aggregate and the cement paste is improved [106].

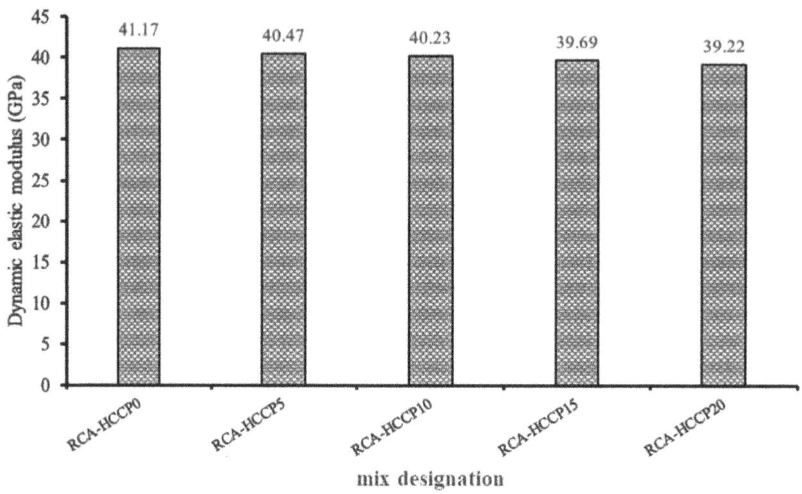

Figure 12. The dynamic of elasticity results from RCA concrete with and without HCCP.

3.9. Percentage of Permeable Voids

The outcomes of the permeable pores at 28 and 90 days of age are depicted in Figure 13. The results showed that the percentage of permeable voids decreased at 28 days as the substitute level of HCCP rose to 10 wt.%. Including HCCP as a substitute for a portion of the cement (5 wt.% and 10 wt.%) significantly reduced the total porosity by 24.2% and 9.2% compared to RCA–HCCP0. The reduction in the porosity can be explained according to El-Dieb and Kanaan [86], who reported that the decrease in the porosity could be attributed to the physical micro filling effect of ceramic waste powder (i.e., fine particles with high S.S.A. which is about 367 m^2/kg), which changed the arrangement of the pores and decreased the size of open pores by enhancing the arrangement of the particles, particularly at the interface between aggregates and paste (ITZ). The porosity value observed in the RCA–HCCP15 mix was like the one in the RCA–HCCP0 mix, and the value of RCA–HCCP20 was higher than that in the RCA–HCCP0 mix, as its porosity value was 9.6%. As the duration of the curing process extends from 28 days to 90 days, the percentage of permeable voids experiences a significant decrease. The percentage of permeable voids in RCA–HCCP5, RCA–HCCP10, RCA–HCCP15, and RCA–HCCP20 mixes decreased by 17%, 21%, 31%, and 38% compared to what was in the reference mix. Prolonging the curing time in the existence of HCCP enhances the pore structure of concrete and reduces the percentage of permeable pores through the pozzolanic reaction. In addition to their ability to act as fillers within capillary pore structures, HCCP particles also can act as such. The pozzolanic reaction involves the dissolution of reactive alumina and silica in the ceramic granules into the pore fluid. As a result, the ceramic granules move outward and react with active ions (such as Ca^{2+} and OH^-) to produce newly generated calcium silicate hydrates (C-S-A-H) [94]. Jerônimo et al. [107] found similar outcomes when investigating the impact of substituting cement with ground clay brick waste (GCBW) on the characteristics of paste and concrete.

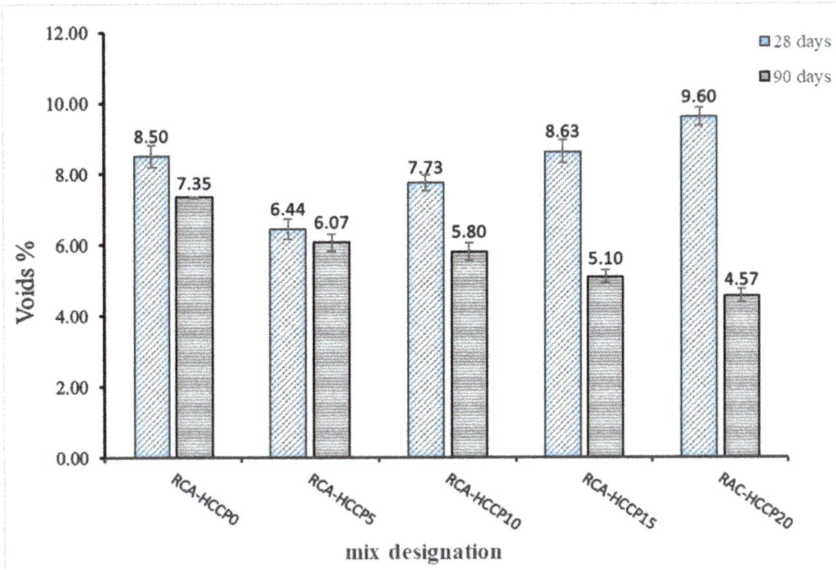

Figure 13. Percentage of permeable void results from RCA with and without HCCP.

3.10. Total and Surface Water Absorption

Figures 14 and 15 display the total and surface water absorption (TW and SW) results for varying levels of HCCP-containing RCA at 28 and 90 days. The water absorption test yields data on the capillary suction. Furthermore, the durability of concrete is significantly impacted by the flow of water within it [108]. As noted, water absorption is directly affected by HCCP content and curing time. The results show that at 28 days, the use of HCCP decreases the porosity, and the water absorption of RCA concretes up to a certain point. A similar pattern to that seen in the permeable void in TW and SW was identified for RCA–HCCP mixtures using HCCP (see Figure 13). The micro-filler effect seen in this case is a result of the smaller particle size of ceramic powder in comparison to cement. This phenomenon leads to sealing the porous openings inside the concrete, enhancing its ability to resist water absorption [109]. Similar results were observed by Mohit and Sharif [92] when they investigated the impact of substituting cement with ceramic waste powder on mortar properties. Replacing cement with 15 wt.% and 20 wt.% ceramic powder achieved total and surface water absorption of 3.47 and 4.63 g/cm^2 and, 0.40 and 0.45 g/cm^2 respectively, which is higher than that for the RCA–HCCP0 mixture by 1.3% and 35.3%, and 4.4% and 19.3%, respectively. The findings were ascribed to the increase in HCCP content at 28 days, which affects the partly and non-reacted particles. This leads to the formation of a more porous structure and increases water absorption in RCA–HCCP specimens [110]. The water absorption ratios of all mixes after 90 days are lower than those after 28 days. At 90 days of age, all RAC mixes containing HCCP showed lower water absorption than the reference mixtures. The surface and total water absorption values of the RCA mixture containing 20 wt.% HCCP were recorded to be much lower than those at 28 days, as the surface and total absorption values were lower than the RCA–HCCP0 mixture by 45% and 47%, respectively. Reduced water absorptions indicated few porous structures in the mixtures [92]. This decrease in TW and SW is attributable to the pozzolanic activity of HCCP, which led to an increased quantity of C-S-H that transformed linked holes into disconnected pores [111]. Moreover, the presence of Fe_2O_3 at 18.67 wt.% in HCCP can enhance the compaction of cement and increase the density of the concrete microstructure [112]. Furthermore, as stated in the literature [109,112], mortar or concrete is deemed durable if the absorption values are below 10%. This study's TW and SW values were less than 10%. Therefore, RCA–HCCP containing all the mentioned ratios is considered durable.

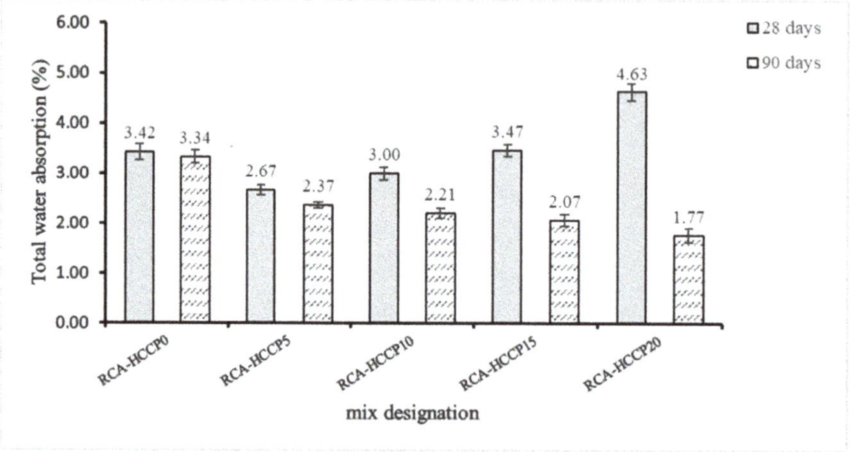

Figure 14. Total water absorption results from RCA with and without HCCP.

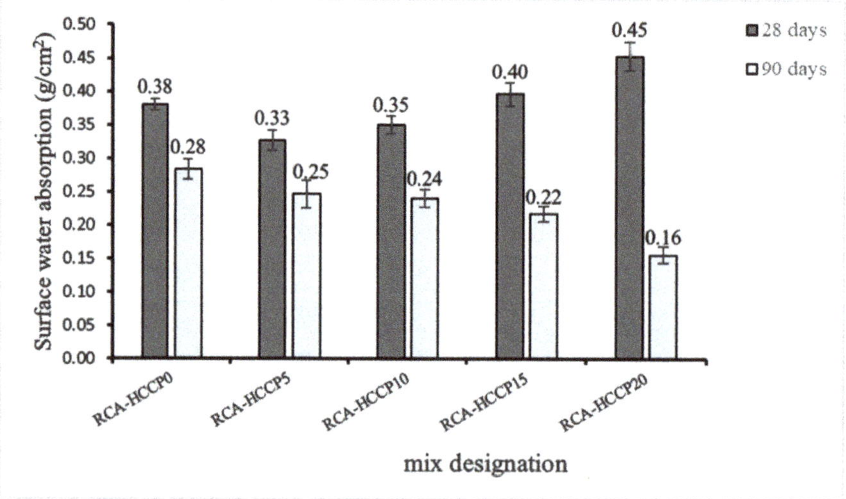

Figure 15. Surface water absorption results from RCA with and without HCCP.

3.11. Electrical Resistivity

Electrical resistivity evaluates the capacity of concrete to prevent the corrosion of reinforcing steel as per the ACI 222R-01 [113] committee. The resistivity measurements for the 28- and 90-day ages are presented in Figure 16. At 28 days, the resistivity results for the RCA–HCCP5, RCA–HCCP10, and RCA–HCCP15 mixes showed values greater than the control mixture (11.9 KΩ·cm). On the other hand, the RCA–HCCP20 mixture had an electrical resistivity of 11.53 KΩ·cm less than the control mixture by no more than 3.1%. These values are categorized by ACI 222R-01 [113] as having corrosion protection levels that range from "high" to "very high". It was noted that the rise in resistivity was directly related to the degree of HCCP replacement. At 90 days, utilizing HCCP at levels of 5, 10,

15, and 20 wt.%, there was a notable rise in the electrical resistivity values compared to the RCA–HCCP0 mix, the increases were 5.3%, 8.8%, 10.5%, and 14.7%, respectively. These outcomes demonstrated that the electrical resistivity was not adversely affected when replacing cement with HCCP, despite the reduction in cement content. Including HCCP resulted in a decrease in the linked pore network, leading to a drop in the conductivity of the concrete. Another thing is that HCCP pozzolanic activity helps to smooth out the pores and microstructure of concrete, making it even harder for ions to move, lowering the concrete's conductivity [114]. Moreover, the degree of corrosion protection is enhanced with an increase in resistivity value. The electrical resistivity results at 90 days for all the RAC mixtures containing HCCP are higher than 13.6 kΩ·cm. ACI 222R-01 [113] classified this as "very high" corrosion protection levels. Similar results were obtained by Kannan et al. [114], when they assessed the inclusion of ceramic powder, ranging from 10 wt.% to 40 wt.%, as a substitute for OPC.

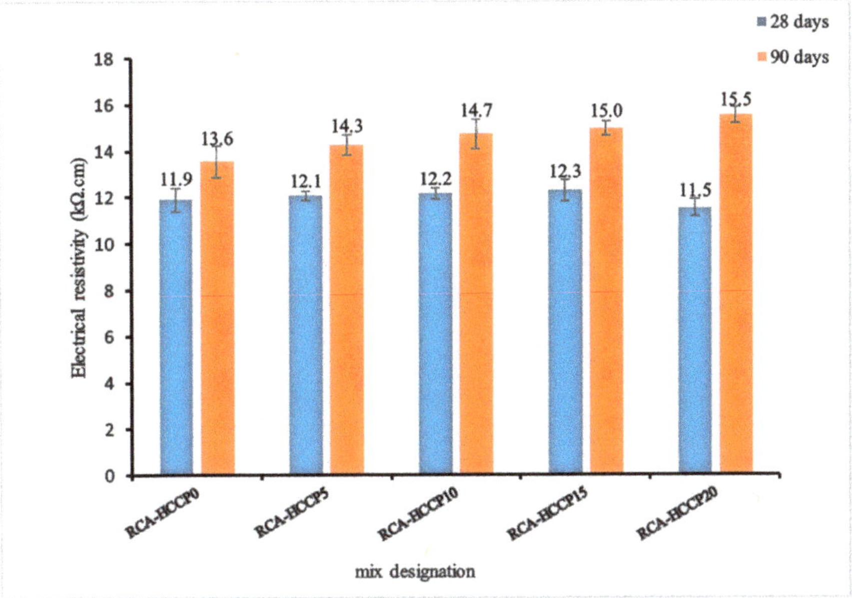

Figure 16. Electrical resistivity results from concrete with and without HCCP.

3.12. *Depth of Chloride Ion Penetration*

A migration test was conducted at 28 and 90 days of age to determine the depth of chloride ion penetration. This was conducted by spraying the new fracture surfaces of the specimen with a 0.1 N solution of silver nitrate. Figure 17 displays the cross-section of the concrete samples that were treated with the $AgNO_3$ solution to measure the depth of chloride ion penetration. The concrete specimens had a lighter tint on their surface due to the presence of silver chloride, which is a white deposit indicating the depth of chloride penetration. In contrast, the inner parts of the specimens exhibited a darker color because of the formation of silver hydroxide. Figure 18 depicts the depth of chloride ion penetration into RCA–HCCP concrete. At 28 days, all mixtures containing HCCP exhibited a shallower depth of chloride penetration than that of RCA–HCCP0, and at 90 days of age, all the concrete mixes containing ceramic powder exhibited a significantly lower depth of chloride ion penetration compared to the reference mix. Based on the outcomes of chloride ion penetration into RCA, as depicted in Figure 19, the depths of chloride ion penetration for the RCA–HCCP5, RCA–HCCP10, RCA–HCCP15, and RCA–HCCP20 concretes (HCCP replacement levels of 5, 10, 15, and 20, respectively) were 23.72, 23.46, 20.57, and 19.05 mm at the 28 days, which was around 14.7, 19.5, 28.6, and 35.6% lower than that of the RCA–HCCP0 (control mixture), respectively, and decreased to 13.76, 11.82,10.43, and 7 mm at the 90 days, which was around 10.9, 32.6, 38.4, and 62.3% less than that of the RCA–HCCP0 mix, respectively. The increase in HCCP substitute levels led to an elevation in the resistance of RCA against chloride ion penetration. The substitution of OPC with HCCP as a binding material had a distinct and beneficial impact on the durable nature of RCA, specifically, its ability to withstand the penetration of chloride ions. This outcome yielded identical results to concretes that contained ground bagasse ash [115]. Moreover, when cement was replaced by 15 and 20 wt.% HCCP, the penetration extent of chloride ions into RCA–HCCP15 and RCA–HCCP20 concrete was reduced than that into RCA–HCCP5 and RCA–HCCP10 concrete, although the compressive strengths of RCA–HCCP15 and RCA–HCCP20 were lower than those of RCA–HCCP5 and RCA–HCCP10 concrete at the same age. This study confirmed that substituting OPC with HCCP is more beneficial than increasing compressive strength in enhancing the durability of RAC in terms of chloride resistance. The outcomes align with the results made by Mohammadhosseini et al. [89] when they used ceramic powder as a substitute for cement to enhance the durability characteristics of recycled ceramic mortar. The reason for this phenomenon is the filler effect, where tiny ceramic powder particles fill the gaps between the cement particles, leading to a significantly more compact binder. Furthermore, the ceramic's pozzolanic products (CSH) and calcium aluminate hydrate (CAH) gel played a role in physically binding, allowing the adsorption of chloride ions onto the surfaces of these products [116,117].

Figure 17. Depth of chloride penetration into RCA with and without HCCP: (**A**) 28 days, (**B**): 90 days.

3.13. Non-Steady-State Diffusion Coefficients of Chloride Ions (Dnssm)

Chloride ions are the leading cause of steel bar corrosion, and they diminish the longevity of concrete buildings. The chloride ion permeability is a crucial factor in assessing the concrete's resistance to chloride corrosion. The corrosion of steel bars is directly linked to concrete cracking and spalling, particularly in buildings exposed to maritime environments and roads [118]. The rapid chloride migration coefficient test assessed the resistance to chloride ion penetration of RCA–HCCP. Therefore, the chloride migration coefficient of RCA–HCCP at the 28 and 90 days could be determined by the corresponding chloride migration coefficient according to Equation (7). Figure 19 displays the test outcomes. From the figure, the chloride migration coefficients at 28 days for the control mixture and RCA concrete mixture containing HCCP were 15.52×10^{-12} m^2/s, 13.23×10^{-12} m^2/s, 12.49×10^{-12} m^2/s, 11.07×10^{-12} m^2/s, and 9.94×10^{-12} m^2/s for RCA–HCCP5, RCA–HCCP10, RCA–HCCP15, and RCA–HCCP20, respectively. At 90 days, the RCA–HCCP0 mix and the RCA concrete mixture containing HCCP at 5, 10, 15, and 20 wt.%(had chloride migration coefficients of 8.56×10^{-12} m^2/s, 7.63×10^{-12} m^2/s, 5.77×10^{-12} m^2/s, 5.27×10^{-12} m^2/s, and 3.23×10^{-12} m^2/s, respectively. As can be noticed, adding the ceramic powder to RCA led to a reduction in the Dnssm compared to what it is in the RCA–HCCP0 mix. The result shows that using the HCCP as a cement replacement in RCA concrete looks promising. Moreover, the presence of Si^{4+} ions derived from SiO_2, the main chemical component of HCCP, which is 33.27 wt.% of HCCP, can effectively hinder the migration of chloride ions, overcoming the limitation observed in RCA where HCCP does not replace OPC. In addition, the creation of calcium silicate hydrate (CSH) can absorb and obstruct the pathway through which chloride ions enter the RCA [119,120]. The findings were consistent with those of Rattanachu et al. [115]. The researcher determined that increasing the quantity of ground bagasse ash up to 50 wt.% in high-strength RCA resulted in higher porosity, but also improved durability in terms of the capacity to withstand the penetration of chloride ions. Furthermore, Wongkeo et al. [121] showed that the inclusion of fly ash in concrete enhanced its resistance to chloride, and this effect was observed to be more pronounced with higher fly ash content. However, it should be noted that the porosity of the concrete was also increased. Nevertheless, the levels at which the HCCP was replaced in RCA significantly affected the resistance of chloride ions, when compared to compressive strength. The compressive strength of the RCA–HCCP5 mix at 90 days was 50.5 MPa, and its chloride migration coefficient was 8.6×10^{-12} m^2/s. In comparison, the compressive strengths of the RCA–HCCP15 and RCA–HCCP20 mix were 44.50 and 41.83 MPa, respectively. The chloride migration coefficients for this concrete were 5.3×10^{-12} and 3.2×10^{-12} m^2/s, respectively.

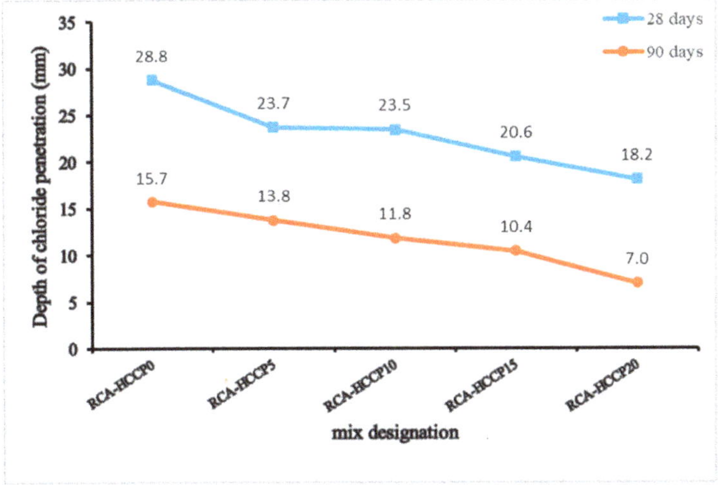

Figure 18. Relationship between the chloride penetration depth and the HCCP replacement level for RCAs.

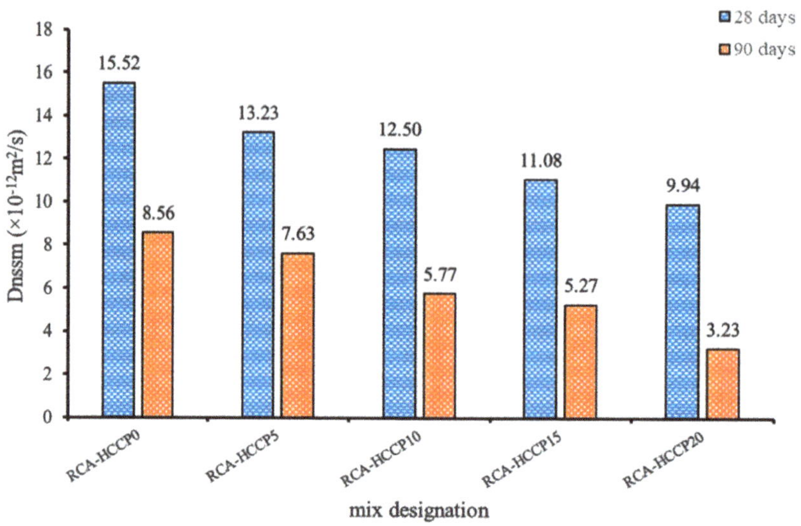

Figure 19. Chloride migration coefficient from RCA with and without HCCP.

3.14. pH Level

Figure 20 displays the pH test outcomes at 28 days. The pH values for all RCA–HCCP mixtures were between 13 and 12.6. It is evident that, overall, increasing HCCP results in a reduction in the pH of all the RCA–HCCP mixes. The RCA–HCCP20 mix had the lowest pH value, measuring 12.6. This is mainly attributed to an increase in the intake of calcium hydroxide, which arises from the process of creating a greater amount of C-S-H gel during the pozzolanic processes [88]. Moreover, the pH value of HCCP (12.4) is slightly lower than that of the RCA–HCCP0 blend (13), which may cause a decrease in the pH value of RCA–HCCP blends at a high substitution ratio with HCCP. However, the pH for all RCA–HCCP mixes incorporating HCCP is higher than (12.5) indicating that the pH value of RCA–HCCP was not affected by the substitution. Consequently, steel remains protected from corrosion when it is contained within RCA–HCCP concrete [122]. Vejmelková et al. [123] achieved a comparable outcome, where they found that the pH values slightly decreased with a rising proportion of finely milled ceramics in the concrete mixes.

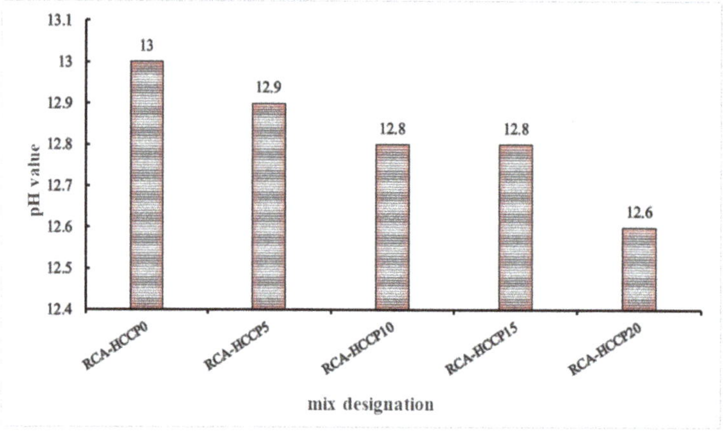

Figure 20. pH values from RCA specimens with and without HCCP at 28 days of age.

3.15. Environmental Assessment

Figure 21 displays the OPC intensity of RCA–HCCP mixtures. The OPC intensity of RCA–HCCP0 at 28 days is greater than that of the RCA mixture produced with HCCP. The OPC intensity of RCA–HCCP5, RCA–HCCP10, RCA–HCCP15, and RCA–HCCP20 at 28 days were 9.42, 9.60, 9.24, and 9.78 kg·m^{-3}·MPa^{-1}, respectively, which was approximately 14.98, 13.35, 16.60, and 11.73 % lower than that (11.083 kg·m^{-3}·MPa^{-1}) of the RCA– HCCP0 (control mixture). Moreover, at 90 days, they decreased to 7.05, 7.23, 7.16, and 7.17 kg·m^{-3}·MPa^{-1}, which was around 24.92, 23, 23.24, and 23.64% lower than that (9.39 kg·m^{-3}·MPa^{-1}) of the RCA–HCCP0 mixture, respectively. Utilizing HCCP resulted in a reduction in the amount of Portland cement required. At 28 days, HCCP inclusion leads to increases in the compressive strength of RCA up to 15 wt.% replacement and increases the compressive strength of all RCA–HCCP mixtures containing HCCP at 90 days of age. The compressive strength of the RCA–HCCP20 mix can be re-established at 90 days of age due to the HCCP's pozzolanic action; this leads to a favorable effect in terms of the intensity of OPC. Thus, the utilization of HCCP leads to a decrease in the quantity of OPC required to obtain a specific compressive strength. This trend improves the environmental efficiency of RCA manufactured using HCCP.

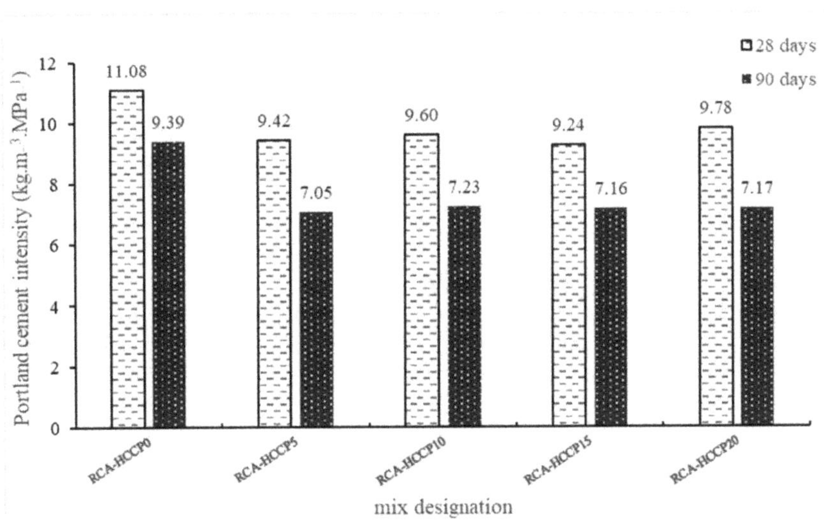

Figure 21. Portland cement intensity from RCA with and without HCCP at 28 and 90 days of age.

The carbon dioxide intensity of RCA–HCCP mixtures is shown in Figure 22. The substitution of cement for HCCP led to a reduction in the amount of carbon dioxide released per unit of compressive strength in all the mixes containing HCCP, in comparison to the RCA–HCCP0 mix, and the largest decline in the CO_2 intensity was seen at the age of 90 days in the RCA mixture containing 5 wt.% of the HCCP, which was 25% less than the RCA–HCCP0. Therefore, the use of HCCP in RCA decreased the environmental consequences arising from inappropriate disposal, as well as alleviated greenhouse gas emissions from the building sector. The outcomes depicted in Figures 21 and 22 suggest that the altered binder, with a 5 wt.% substitution of Portland cement by HCCP, exhibits superior environmental efficiency in terms of cement usage and carbon dioxide emissions.

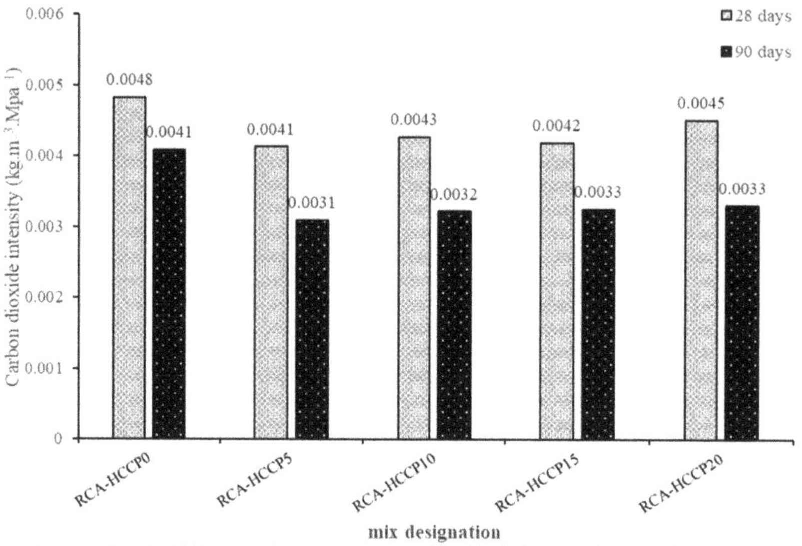

Figure 22. Carbon dioxide intensity for RCA with and without HCCP at 28 and 90 days of age.

3.16. Microstructural Analysis

To fully understand the origin, speed, and process of deterioration, as well as how to enhance certain characteristics of concrete, a comprehension of the microstructure of hardened concrete is important. Moreover, the microstructure of concrete frequently influences its mechanical properties. This research identifies two distinct phases in concrete: the cement paste phase and the aggregate phase. Another phase that is seen during microstructure investigation is known as the Inter-facial Transition Zone (ITZ) [124]. The ITZ phase has lower strength compared to the other two phases that have been seen. The assessment was conducted on the specimens exhibiting the highest mechanical strength after 28 days of curing (RCA–HCCP5) as well as control specimens (RCA–HCCP0). Figure 23 (A and B) displays the SEM images at 1000 magnification power of RCA–HCCP0 and RCA–HCCP5 mixes, respectively. The SEM pictures of RCA–HCCP0 exhibit a deficient morphology characterized by the presence of expanded voids, porosity, and cracking in the ITZ between the aggregate and the cement paste. This is due to the lower hydration of the RCA–HCCP0 mix, which is not enough to fill the pores of the specimen. The RCA–HCCP5 mix shows a dense microstructure of ITZ and a restricted amount of observable empty spaces as a result of the existence of ceramic powder particles. The particles are completely enclosed and evenly distributed within the cement, allowing for a strong connection with the aggregate in concrete due to their small dimensions [112]. Furthermore, the ceramic powder undergoes a pozzolanic reaction that consumes $Ca(OH)_2$ and generates further hydration products. As a result, the RCA–HCCP5 specimen exhibits a more compact microstructure with reduced pore content. This discovery is consistent with the results of the mechanical testing, density, UPV, and Dynamic modulus of elasticity.

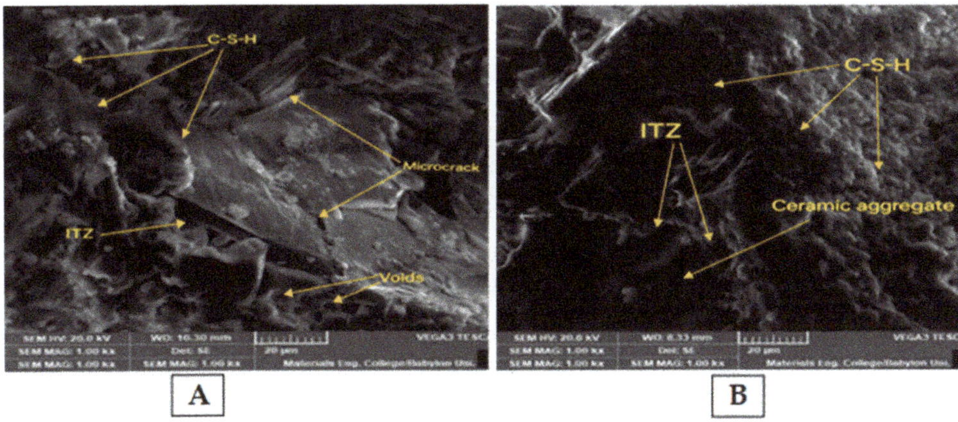

Figure 23. SEM micrographs from the (**A**) control mixture RCA–HCCP0 at 1000 magnification power and (**B**) the RCA mixture containing 5 wt.% HCCP at 1000 magnification power.

4. Conclusions

The present study examines the strength, durability, and microstructural properties of concrete that incorporates ceramic wall waste as a partial replacement for cement and coarse aggregate. Additionally, the environmental impact of the RCA produced was investigated. In this study, ceramic waste powder was used to replace cement in proportions of 5%, 10%, 15%, and 20% by weight. Furthermore, 25 wt.% of the natural coarse aggregate was replaced with aggregate produced from crushed ceramic wall waste. The following conclusions were determined:

1. Raising the substitution rate of OPC with HCCP increases the slump value and enhances the workability. The greatest enhancement in slump value was observed at 20 wt.% HCCP content. Moreover, both initial and final setting times of concrete are prolonged as the proportion of OPC substitution with HCCP increases.
2. Substituting cement with 5 to 20 wt.% HCCP improved the mechanical properties of RCA in the same pattern. The most significant improvement was recorded at 5 wt.% HCCP content. The compressive strength, tensile strength, and flexural strength increased by 11.8%, 16.3%, and 14.6% at 28 days and by 26.5%, 22%, and 22.4% at 90 days over the reference mixture. Furthermore, up to 20 wt.% replacement level, the mechanical properties remained higher than that of the RCA–HCCP0 (control) mix.
3. The chloride migration coefficient of RCA was significantly reduced after cement was replaced with HCCP. In addition, incorporating HCCP in RCA concrete greatly enhanced its resistance to chloride ion penetration. These parameters improved proportionally with the increasing substituting rate. At 28 days, using a 20 wt.% replacement ratio, the obtained values of chloride penetration depth and migration coefficient were 37.94% and 36.88% less than that for the reference RCA. After 90 days, the reduction rates were quite substantial, with 62.3% and 55.52% lower values than that of the reference concrete.
4. Substituting cement with HCCP resulted in significantly better environmental performance in terms of cement usage and CO_2 emissions. After 90 days, the Portland cement intensities of RCA–HCCP5, RCA–HCCP10, RCA–HCCP15, and RCA–HCCP20 decreased by 24.92%, 23%, 23.24%, and 23.64%, respectively, compared to the RCA–HCCP0 mixture.

5. The examination of the microstructure using SEM showed that substituting cement with HCCP enhanced the microstructure of RCA and raised the density of the interfacial transition zone (ITZ).
6. HCCP, as an alternative cementitious material, has enhanced the properties of cement binders and, in turn, improved the resulting concrete microstructure. Additionally, it has positively affected the characteristics of concrete made with porous recycled aggregates due to its ability to retain water, which supports continuous hydration over time. Based on the results of all the tests conducted, it has been proven feasible to manufacture eco-friendly RCA with 25 wt.% recycled coarse ceramic aggregate and 20 wt.% HCCP. This composition shows significant improvements in mechanical properties, and durability characteristics as well as a notable reduction in carbon dioxide emissions.

Author Contributions: A.A.H.: Data curation, conceptualization, methodology, investigation, writing—original draft. R.A.: Writing—review and editing, writing—original draft, supervision. M.S.N.: Writing—review and editing, writing—original draft, supervision. All authors have read and agreed to the published version of the manuscript.

Funding: This research received no external funding.

Institutional Review Board Statement: Not applicable.

Informed Consent Statement: Not applicable.

Data Availability Statement: The original contributions presented in this study are included in the article.

Conflicts of Interest: The authors declare no conflicts of interest.

References

1. Gautam, L.; Jain, J.K.; Kalla, P.; Danish, M. Sustainable utilization of granite waste in the production of green construction products: A review. *Mater. Today Proc.* **2021**, *44*, 4196–4203. [CrossRef]
2. Azeez, A.T.; Hassan, M.S.; Attiyah, A.A.H. Modelling the complementary role of inert slag aggregate and waste glass on concrete mechanical strengths using the response surface and design-of-experiment approach. *Constr. Build. Mater.* **2023**, *409*, 134199. [CrossRef]
3. Kalakada, Z.; Doh, J.-H.; Chowdhury, S. Glass powder as replacement of cement for concrete–an investigative study. *Eur. J. Environ. Civ. Eng.* **2022**, *26*, 1046–1063. [CrossRef]
4. Aslam, M.; Shafigh, P.; Jumaat, M.Z. Oil-palm by-products as lightweight aggregate in concrete mixture: A review. *J. Clean. Prod.* **2016**, *126*, 56–73. [CrossRef]
5. Aprianti, E.; Shafigh, P.; Bahri, S.; Farahani, J.N. Supplementary cementitious materials origin from agricultural wastes–A review. *Constr. Build. Mater.* **2015**, *74*, 176–187. [CrossRef]
6. Bhargav, M.; Kansal, R. Experimental investigation to substitute of cement with ceramic tiles powder in concrete. *Int. J. Res. Appl. Sci. Eng. Technol.* **2020**, *8*, 302–307. [CrossRef]
7. Chen, X.; Zhang, D.; Cheng, S.; Xu, X.; Zhao, C.; Wang, X.; Wu, Q.; Bai, X. Sustainable reuse of ceramic waste powder as a supplementary cementitious material in recycled aggregate concrete: Mechanical properties, durability and microstructure assessment. *J. Build. Eng.* **2022**, *52*, 104418. [CrossRef]
8. Azeeza, A.T.; Hassanb, M.S.; Atiyaha, A.A. Evaluation of the performance of steel slag and waste glass as a cement replacement. *Eng. Technol. J.* **2023**, *41*, 1567–1577. [CrossRef]
9. Tobeia, S.B.; Assi, N.S.; Abbas, N.S. Mechanical Properties Prediction of Normal and High Strength Geopolymer Concrete. *Eng. Technol. J.* **2021**, *39*, 1781–1788. [CrossRef]
10. Akhtar, A.; Sarmah, A.K. Construction and demolition waste generation and properties of recycled aggregate concrete: A global perspective. *J. Clean. Prod.* **2018**, *186*, 262–281. [CrossRef]
11. Nie, Q.; Hu, W.; Huang, B.; Shu, X.; He, Q. Synergistic utilization of red mud for flue-gas desulfurization and fly ash-based geopolymer preparation. *J. Hazard. Mater.* **2019**, *369*, 503–511. [CrossRef]
12. Hasan, Z.A.; Abed, M.K.; Nasr, M.S. Studying the Mechanical Properties of Mortar Containing Different Waste Materials as a Partial Replacement for Aggregate. *Int. Rev. Civ. Eng.* **2019**, *10*, 155. [CrossRef]

13. Heniegal, A.M.; Maaty, A.A.E.S.; Agwa, I.S. Simulation of the behavior of pressurized underwater concrete. *Alex. Eng. J.* **2015**, *54*, 183–195. [CrossRef]
14. Heniegal, A.M.; Omar, O.M.; Agwa, I.S.; Youssef, K.G. Effect of Forced casting joints on the mechanical properties of self-compacting concrete incorporating cement waste. *IJEIT* **2017**, *7*, 1–6.
15. Agwa, I.S.; Ibrahim, O.M.O. Fresh and hardened properties of self-compacting concrete containing cement kiln dust. *Chall. J. Concr. Res. Lett.* **2019**, *5*, 13–19. [CrossRef]
16. Zeyad, A.M.; Johari, M.A.M.; Tayeh, B.A.; Yusuf, M.O. Pozzolanic reactivity of ultrafine palm oil fuel ash waste on strength and durability performances of high strength concrete. *J. Clean. Prod.* **2017**, *144*, 511–522. [CrossRef]
17. Xiao, R.; Ma, Y.; Jiang, X.; Zhang, M.; Zhang, Y.; Wang, Y.; Huang, B.; He, Q. Strength, microstructure, efflorescence behavior and environmental impacts of waste glass geopolymers cured at ambient temperature. *J. Clean. Prod.* **2020**, *252*, 119610. [CrossRef]
18. Fahmy, W.S.; Heneidy, E.A.-L.; Ali, E.E.; Agwa, I.S. Performance of concrete containing crushed waste concrete exposed to different fire temperatures. *Int. J. Civ. Struct. Eng.* **2012**, *3*, 9–22.
19. Nuaklong, P.; Jongvivatsakul, P.; Pothisiri, T.; Sata, V.; Chindaprasirt, P. Influence of rice husk ash on mechanical properties and fire resistance of recycled aggregate high-calcium fly ash geopolymer concrete. *J. Clean. Prod.* **2020**, *252*, 119797. [CrossRef]
20. Turkia, S.K.; Ibrahima, S.I.; Al Maamorib, M.H.D. Comparison of locally manufactured composite tile adhesive using waste material with imported tile adhesive. *Eng. Technol. J.* **2024**, *42*, 569–581. [CrossRef]
21. Hashim, A.A.; Rodhan, Z.K.; Abbas, S.J. Fresh and hardened properties of self-compacting high performance concrete containing nano-metakaolin as a partial replacement. In Proceedings of the IOP Conference Series: Materials Science and Engineering, Thi Qar, Iraq, 15–16 July 2020; IOP Publishing: Bristol, UK; Volume 928, p. 22036.
22. Elçi, H. Utilisation of crushed floor and wall tile wastes as aggregate in concrete production. *J. Clean. Prod.* **2016**, *112*, 742–752. [CrossRef]
23. Ray, S.; Haque, M.; Sakib, M.N.; Mita, A.F.; Rahman, M.D.M.; Tanmoy, B.B. Use of ceramic wastes as aggregates in concrete production: A review. *J. Build. Eng.* **2021**, *43*, 102567. [CrossRef]
24. Anderson, D.J.; Smith, S.T.; Au, F.T.K. Mechanical properties of concrete utilising waste ceramic as coarse aggregate. *Constr. Build. Mater.* **2016**, *117*, 20–28. [CrossRef]
25. Medina, C.; De Rojas, M.I.S.; Thomas, C.; Polanco, J.A.; Frías, M. Durability of recycled concrete made with recycled ceramic sanitary ware aggregate. Inter-indicator relationships. *Constr. Build. Mater.* **2016**, *105*, 480–486. [CrossRef]
26. Khalid, F.S.; Azmi, N.B.; Sumandi, K.A.S.M.; Mazenan, P.N. Mechanical properties of concrete containing recycled concrete aggregate (RCA) and ceramic waste as coarse aggregate replacement. In Proceedings of the AIP Conference Proceedings, Varna, Bulgaria, 16 January 2025; AIP Publishing: Melville, NY, USA; Volume 1891.
27. Gharibi, H.; Mostofinejad, D.; Bahmani, H.; Hadadzadeh, H. Improving thermal and mechanical properties of concrete by using ceramic electrical insulator waste as aggregates. *Constr. Build. Mater.* **2022**, *338*, 127647. [CrossRef]
28. Klimek, B.; Szulej, J.; Ogrodnik, P. The effect of replacing sand with aggregate from sanitary ceramic waste on the durability of stucco mortars. *Clean Technol. Environ. Policy* **2020**, *22*, 1929–1941. [CrossRef]
29. Giridhar, V.; Rao, H.S.; Kumar, P.S.P. Influence of ceramic waste aggregate properties on strength of ceramic waste aggregate concrete. *IJRET* **2015**, *4*, 15–24.
30. Ch, H.K.; Ramakrishna, K.A.; Babu, K.S.; Guravaiah, T.; Naveen, N.; Sk, J. Effect of waste ceramic tiles in partial replacement of coarse and fine aggregate of concrete. *Int. Adv. Res. J. Sci. Eng. Technol.* **2015**, *2*, 13–16.
31. Javed, A.; Siddique, S.; Prasad, V.R. Investigation on ceramic waste and stone dust as aggregate replacement in concrete. *Int. J. Eng. Technol. Manag. Appl. Sci.* **2015**, *3*, 127–130.
32. Umapathy, U.; Mala, C.; Siva, K. Assessment of concrete strength using partial replacement of coarse aggregate for waste tiles and cement for rice husk ash in concrete. *Int. J. Eng. Res. Appl.* **2014**, *4*, 72–76.
33. Nepomuceno, M.C.S.; Isidoro, R.A.S.; Catarino, J.P.G. Mechanical performance evaluation of concrete made with recycled ceramic coarse aggregates from industrial brick waste. *Constr. Build. Mater.* **2018**, *165*, 284–294. [CrossRef]
34. Keshavarz, Z.; Mostofinejad, D. Porcelain and red ceramic wastes used as replacements for coarse aggregate in concrete. *Constr. Build. Mater.* **2019**, *195*, 218–230. [CrossRef]
35. Etxeberria, M.; Gonzalez-Corominas, A. The assessment of ceramic and mixed recycled aggregates for high strength and low shrinkage concretes. *Mater. Struct.* **2018**, *51*, 129. [CrossRef]
36. Silva, R.V.; De Brito, J.; Dhir, R.K. Properties and composition of recycled aggregates from construction and demolition waste suitable for concrete production. *Constr. Build. Mater.* **2014**, *65*, 201–217. [CrossRef]
37. Li, L.; Xuan, D.; Sojobi, A.O.; Liu, S.; Chu, S.H.; Poon, C.S. Development of nano-silica treatment methods to enhance recycled aggregate concrete. *Cem. Concr. Compos.* **2021**, *118*, 103963. [CrossRef]
38. Zhan, B.J.; Xuan, D.X.; Zeng, W.; Poon, C.S. Carbonation treatment of recycled concrete aggregate: Effect on transport properties and steel corrosion of recycled aggregate concrete. *Cem. Concr. Compos.* **2019**, *104*, 103360. [CrossRef]

39. Wu, C.-R.; Zhu, Y.-G.; Zhang, X.-T.; Kou, S.-C. Improving the properties of recycled concrete aggregate with bio-deposition approach. *Cem. Concr. Compos.* **2018**, *94*, 248–254. [CrossRef]
40. Muduli, R.; Mukharjee, B.B. Effect of incorporation of metakaolin and recycled coarse aggregate on properties of concrete. *J. Clean. Prod.* **2019**, *209*, 398–414. [CrossRef]
41. Bui, N.K.; Satomi, T.; Takahashi, H. Effect of mineral admixtures on properties of recycled aggregate concrete at high temperature. *Constr. Build. Mater.* **2018**, *184*, 361–373. [CrossRef]
42. Tran, D.L.; Mouret, M.; Cassagnabère, F.; Phung, Q.T. Effects of intrinsic granular porosity and mineral admixtures on durability and transport properties of recycled aggregate concretes. *Mater. Today Commun.* **2022**, *33*, 104709. [CrossRef]
43. Fernando, A.; Selvaranjan, K.; Srikanth, G.; Gamage, J. Development of high strength recycled aggregate concrete-composite effects of fly ash, silica fume and rice husk ash as pozzolans. *Mater. Struct.* **2022**, *55*, 185. [CrossRef]
44. Saleem, M.H.; Abdul-Hamead, A.A.; Othman, F.M. Performance of waste materials and carbon nanotube in concrete incorporated with carbon fiber. In Proceedings of the AIP Conference Proceedings, Athens, Greece, 17 August 2022; AIP Publishing: Melville, NY, USA; Volume 2437.
45. Trník, A.; Scheinherrová, L.; Kulovaná, T.; Reiterman, P.; Vejmelková, E.; Černý, R. Thermal analysis of high-performance mortar containing burnt clay shale as a partial Portland cement replacement in the temperature range up to 1000 C. *Fire Mater.* **2017**, *41*, 54–64. [CrossRef]
46. de Rojas, M.I.S.; Marín, F.P.; Frías, M.; Rivera, J. Properties and performances of concrete tiles containing waste fired clay materials. *J. Am. Ceram. Soc.* **2007**, *90*, 3559–3565. [CrossRef]
47. Arif, R.; Khitab, A.; Kırgız, M.S.; Khan, R.B.N.; Tayyab, S.; Khan, R.A.; Anwar, W.; Arshad, M.T. Experimental analysis on partial replacement of cement with brick powder in concrete. *Case Stud. Constr. Mater.* **2021**, *15*, e00749. [CrossRef]
48. Taher, M.J.; Abed, E.H.; Hashim, M.S. Using ceramic waste tile powder as a sustainable and eco-friendly partial cement replacement in concrete production. *Mater. Today Proc.* 2023, *in press*. [CrossRef]
49. Ahmad, S.; Khan, R.A.; Shamim, S.; Chandra, U. Effect of waste ceramic sanitary ware as partial replacement of aggregates and cement in concrete. *Innov. Infrastruct. Solut.* **2023**, *8*, 205. [CrossRef]
50. de Rojas, M.I.S.; Frías, M.; Sabador, E.; Asensio, E.; Rivera, J.; Medina, C. Use of ceramic industry milling and glazing waste as an active addition in cement. *J. Am. Ceram. Soc.* **2018**, *101*, 2028–2037. [CrossRef]
51. Zhang, P.; Zhang, P.; Wu, J.; Guo, Z.; Zhang, Y.; Zheng, Y. Mechanical Properties and Durability of Sustainable Concrete Manufactured Using Ceramic Waste: A Review. *J. Renew. Mater.* **2023**, *11*, 937–974. [CrossRef]
52. Pitarch, A.M.; Reig, L.; Tomás, A.E.; Forcada, G.; Soriano, L.; Borrachero, M.V.; Paya, J.; Monzó, J.M. Pozzolanic activity of tiles, bricks and ceramic sanitary-ware in eco-friendly Portland blended cements. *J. Clean. Prod.* **2021**, *279*, 123713. [CrossRef]
53. Lasseuguette, E.; Burns, S.; Simmons, D.; Francis, E.; Chai, H.K.; Koutsos, V.; Huang, Y. Chemical, microstructural and mechanical properties of ceramic waste blended cementitious systems. *J. Clean. Prod.* **2019**, *211*, 1228–1238. [CrossRef]
54. Hoppe Filho, J.; Pires, C.A.; Leite, O.D.; Garcez, M.R.; de Medeiros, M.H.F. Red ceramic waste as supplementary cementitious material: Microstructure and mechanical properties. *Constr. Build. Mater.* **2021**, *296*, 123653. [CrossRef]
55. *ASTM C150/C150M*; Standard Specification for Portland Cement. ASTM International: West Conshohocken, PA, USA, 2015.
56. *BS EN 196-2*; Method of Testing Cement Part 2: Chemical Analysis of Cement. British Standards Institution: London, UK, 2013.
57. Al-Jiboory, Y.M.; Al-Hazaa, S.H. Assessment of Altawseea Ordinary Portland Cement Northern Iraq: Mineralogy, Microstructure, and Hydration. *Iraqi Geol. J.* **2022**, *55*, 198–208. [CrossRef]
58. *ASTM C29*; Standard Test Method for Bulk Density ("Unit Weight") and Voids in Aggregate. ASTM International: West Conshohocken, PA, USA, 2009.
59. *ASTM C127*; Standard Test Method for Relative Density (Specific Gravity) and Absorption of Coarse Aggregate. ASTM International: West Conshohocken, PA, USA, 2015.
60. Iraqi Specifications No. (45). In *Aggregate from Natural Sources for Concrete and Construction*; Central Organization for Standardization & Quality Control (COSQC): Baghdad, Iraq, 1984; p. 11.
61. *ASTM C 494/C 494M-08*; Standard Specification for Chemical Admixtures for Concrete. ASTM: West Conshohocken, PA, USA, 2008.
62. *BS EN 12390-1*; Testing Hardened Concrete—Part 1: Shape, Dimensions and Other Requirements for Specimens and Moulds. British Standard Institution (BSI): London, UK, 2021.
63. *ASTM C 143/C 143M-03*; Standard Test Method for Slump of Hydraulic-Cement Concrete. ASTM: West Conshohocken, PA, USA, 2003; pp. 1–4.
64. *ASTMC191-08*; Standard Test Methods for Time of Setting of Hydraulic Cement by Vicat Needle. ASTM International: West Conshohocken, PA, USA, 2009; Volume 4, pp. 1–8.
65. *BS 1881-116*; Testing Concrete—Part 116: Method for Determination of Compressive Strength of Concrete Cubes. BSI: London, UK, 2003.

66. *ASTM C496*; Standard Test Method for Splitting Tensile Strength of Cylindrical Concrete Specimens. ASTM International: West Conshohocken, PA, USA, 2011.
67. *ASTM C293/C293M*; Standard Test Method for Flexual Strength of Concrete (Using Simple Beam with Center-Point Loading) 1. ASTM International: West Conshohocken, PA, USA, 2016.
68. Abdulridha, S.Q.; Nasr, M.S.; Al-Abbas, B.H.; Hasan, Z.A. Mechanical and structural properties of waste rope fibers-based concrete: An experimental study. *Case Stud. Constr. Mater.* **2022**, *16*, e00964. [CrossRef]
69. *ASTM C597*; Standard Test Method for Pulse Velocity Through Concrete. ASTM International: West Conshohocken, PA, USA, 2009.
70. *ASTM C1585*; Standard Test Method for Measurement of Rate of Absorption of Water by Hydraulic—Cement Concretes. ASTM: West Conshohocken, PA, USA, 2013.
71. *ASTM C642*; Standard Test Method for Density, Absorption, and Voids in Hardened Concrete. ASTM International: West Conshohocken, PA, USA, 2013.
72. Gopalakrishnan, R.; Nithiyanantham, S. Microstructural, mechanical, and electrical properties of copper slag admixtured cement mortar. *J. Build. Eng.* **2020**, *31*, 101375. [CrossRef]
73. *N.B. 492*; Concrete, Mortar and Cement-Based Repair Materials: Chloride Migration Coefficient from Non-Steady-State Migration Experiments. Nordtest: Taastrupl, Denmark, 1999.
74. Nasr, M.S.; Salih, S.A.; Hassan, M.S. Some Durability Characteristics of Micro Silica and Nano Silica Contained Concrete. *J. Univ. Babylon* **2016**, *24*, 980–990.
75. Wang, W.-C.; Huang, W.-H.; Lee, M.-Y.; Duong, H.T.H.; Chang, Y.-H. Standardized procedure of measuring the pH value of cement matrix material by ex-situ leaching method (ESL). *Crystals* **2021**, *11*, 436. [CrossRef]
76. Shen, W.; Liu, Y.; Yan, B.; Wang, J.; He, P.; Zhou, C.; Huo, X.; Zhang, W.; Xu, G.; Ding, Q. Cement industry of China: Driving force, environment impact and sustainable development. *Renew. Sustain. Energy Rev.* **2017**, *75*, 618–628. [CrossRef]
77. Kurda, R.; Silvestre, J.D.; de Brito, J. Toxicity and environmental and economic performance of fly ash and recycled concrete aggregates use in concrete: A review. *Heliyon* **2018**, *4*, e00611. [CrossRef] [PubMed]
78. Majhi, R.K.; Nayak, A.N. Production of sustainable concrete utilising high-volume blast furnace slag and recycled aggregate with lime activator. *J. Clean. Prod.* **2020**, *255*, 120188. [CrossRef]
79. Park, W.-J.; Kim, T.; Roh, S.; Kim, R. Analysis of life cycle environmental impact of recycled aggregate. *Appl. Sci.* **2019**, *9*, 1021. [CrossRef]
80. Huseien, G.F.; Sam, A.R.M.; Shah, K.W.; Mirza, J.; Tahir, M.M. Evaluation of alkali-activated mortars containing high volume waste ceramic powder and fly ash replacing GBFS. *Constr. Build. Mater.* **2019**, *210*, 78–92. [CrossRef]
81. Nath, P.; Sarker, P.K. Effect of GGBFS on setting, workability and early strength properties of fly ash geopolymer concrete cured in ambient condition. *Constr. Build. Mater.* **2014**, *66*, 163–171. [CrossRef]
82. Nalli, B.R.; Vysyaraju, P. Utilization of ceramic waste powder and rice husk ash as a partial replacement of cement in concrete. In Proceedings of the IOP Conference Series: Earth and Environmental Science, Bapatla, India, 24–25 September 2022; IOP Publishing: Bristo, UK; Volume 982, p. 12003.
83. Basak, S.; Haque, M.R.; Azad, A.K.; Rahman, M.M. Performance of Ceramic Tiles Waste as a Partial Replacement of Brick Aggregate on Mechanical and Durability Properties of Concrete. *J. Eng. Adv.* **2024**, *5*, 9–13. [CrossRef]
84. Abubakr, M.; Khitab, A.; Sadiq, S.; Anwar, W.; Tayyeb, S. Evaluation of ordinary concrete having ceramic waste powder as partial replacement of cement. In Proceedings of the International Conference on Sustainable Development, Sindh, Pakistan, 5–7 December 2019; Mehran University of Engineering & Technology: Jamshoro, Pakistan; pp. 1–4.
85. Mohit, M.; Haftbaradaran, H.; Riahi, H.T. Investigating the ternary cement containing Portland cement, ceramic waste powder, and limestone. *Constr. Build. Mater.* **2023**, *369*, 130596. [CrossRef]
86. El-Dieb, A.S.; Kanaan, D.M. Ceramic waste powder an alternative cement replacement–characterization and evaluation. *Sustain. Mater. Technol.* **2018**, *17*, e00063. [CrossRef]
87. Nasr, M.S.; Salih, M.A.; Shubbar, A.; Falah, M.W.; Abadel, A.A. Influence of mechanical activation on the behavior of green high-strength mortar including ceramic waste. *Mater. Sci.* **2023**, *41*, 41–56. [CrossRef]
88. Ebrahimi, M.; Eslami, A.; Hajirasouliha, I.; Ramezanpour, M.; Pilakoutas, K. Effect of ceramic waste powder as a binder replacement on the properties of cement- and lime-based mortars. *Constr. Build. Mater.* **2023**, *379*, 131146. [CrossRef]
89. Mohammadhosseini, H.; Lim, N.H.A.S.; Tahir, M.M.; Alyousef, R.; Samadi, M.; Alabduljabbar, H.; Mohamed, A.M. Effects of waste ceramic as cement and fine aggregate on durability performance of sustainable mortar. *Arab. J. Sci. Eng.* **2020**, *45*, 3623–3634. [CrossRef]
90. Rashid, K.; Razzaq, A.; Ahmad, M.; Rashid, T.; Tariq, S. Experimental and analytical selection of sustainable recycled concrete with ceramic waste aggregate. *Constr. Build. Mater.* **2017**, *154*, 829–840. [CrossRef]
91. Attaelmanan, M.; Kambal, M.E.M.; Mansour, M.I. Study the effect of using ceramic waste powder as partial replacement for cement on concrete properties. *J. Karary Univ. Eng. Sci.* 2021. [CrossRef]

92. Mohit, M.; Sharifi, Y. Ceramic waste powder as alternative mortar-based cementitious material. *ACI Mater. J.* **2019**, *116*, 107–116. [CrossRef]
93. Lü, Q.; Qiu, Q.; Zheng, J.; Wang, J.; Zeng, Q. Fractal dimension of concrete incorporating silica fume and its correlations to pore structure, strength and permeability. *Constr. Build. Mater.* **2019**, *228*, 116986. [CrossRef]
94. Nasr, M.S.; Hussain, T.H.; Najim, W.N. Properties of Cement Mortar Containing Biomass Bottom Ash and Sanitary Ceramic Wastes as a Partial Replacement of Cement. *Int. J. Civ. Eng. Technol.* **2018**, *9*, 153–165.
95. Li, L.; Liu, W.; You, Q.; Chen, M.; Zeng, Q. Waste ceramic powder as a pozzolanic supplementary filler of cement for developing sustainable building materials. *J. Clean. Prod.* **2020**, *259*, 120853. [CrossRef]
96. Siddique, S.; Shrivastava, S.; Chaudhary, S. Lateral force microscopic examination of interfacial transition zone in ceramic concrete. *Constr. Build. Mater.* **2017**, *155*, 688–725. [CrossRef]
97. Nasr, M.S.; Ali, I.M.; Hussein, A.M.; Shubbar, A.A.; Kareem, Q.T.; AbdulAmeer, A.T. Utilization of locally produced waste in the production of sustainable mortar. *Case Stud. Constr. Mater.* **2020**, *13*, e00464. [CrossRef]
98. Mohit, M.; Sharifi, Y. Thermal and microstructure properties of cement mortar containing ceramic waste powder as alternative cementitious materials. *Constr. Build. Mater.* **2019**, *223*, 643–656. [CrossRef]
99. Rashad, A.M.; Essa, G.M.F. Effect of ceramic waste powder on alkali-activated slag pastes cured in hot weather after exposure to elevated temperature. *Cem. Concr. Compos.* **2020**, *111*, 103617. [CrossRef]
100. Siddique, S.; Shrivastava, S.; Chaudhary, S.; Gupta, T. Strength and impact resistance properties of concrete containing fine bone china ceramic aggregate. *Constr. Build. Mater.* **2018**, *169*, 289–298. [CrossRef]
101. Ghosh, R.; Sagar, S.P.; Kumar, A.; Gupta, S.K.; Kumar, S. Estimation of geopolymer concrete strength from ultrasonic pulse velocity (UPV) using high power pulser. *J. Build. Eng.* **2018**, *16*, 39–44. [CrossRef]
102. Rao, A.U.; Shetty, P.P.; Bhandary, P.R.; Tantri, A.; Yaragal, S.C. Assessment of fly ash and ceramic powder incorporated concrete with steam-treated recycled concrete aggregates prioritising nano-silica. *Emergent Mater.* **2024**, *7*, 443–472. [CrossRef]
103. Sasanipour, H.; Aslani, F. Effect of specimen shape, silica fume, and curing age on durability properties of self-compacting concrete incorporating coarse recycled concrete aggregates. *Constr. Build. Mater.* **2019**, *228*, 117054. [CrossRef]
104. Mendes, S.E.S.; Oliveira, R.L.N.; Cremonez, C.; Pereira, E.; Pereira, E.; Medeiros-Junior, R.A. Mixture design of concrete using ultrasonic pulse velocity. *Int. J. Civ. Eng.* **2020**, *18*, 113–122. [CrossRef]
105. Zhou, Y.; Gao, J.; Sun, Z.; Qu, W. A fundamental study on compressive strength, static and dynamic elastic moduli of young concrete. *Constr. Build. Mater.* **2015**, *98*, 137–145. [CrossRef]
106. Kuan, P.; Hongxia, Q.; Kefan, C. Reliability analysis of freeze–thaw damage of recycled ceramic powder concrete. *J. Mater. Civ. Eng.* **2020**, *32*, 5020008. [CrossRef]
107. Jerônimo, V.L.; Meira, G.R.; Filho, L.C.P.d.S. Performance of self-compacting concretes with wastes from heavy ceramic industry against corrosion by chlorides. *Constr. Build. Mater.* **2018**, *169*, 900–910. [CrossRef]
108. Bameri, M.; Rashidi, S.; Mohammadhasani, M.; Maghsoudi, M.; Madani, H.; Rahmani, F. Evaluation of Mechanical and Durability Properties of Eco-Friendly Concrete Containing Silica Fume, Waste Glass Powder, and Ground Granulated Blast Furnace Slag. *Adv. Mater. Sci. Eng.* **2022**, *2022*, 2730391. [CrossRef]
109. Abadel, A.A.; Nasr, M.S.; Shubbar, A.; Hashim, T.M.; Tuladhar, R. Potential Use of Rendering Mortar Waste Powder as a Cement Replacement Material: Fresh, Mechanical, Durability and Microstructural Properties. *Sustainability* **2023**, *15*, 11659. [CrossRef]
110. Huseien, G.; Sam, A.R.M.; Shah, K.W.; Mirza, J. Effects of ceramic tile powder waste on properties of self-compacted alkali-activated concrete. *Constr. Build. Mater.* **2020**, *236*, 117574. [CrossRef]
111. Yusuf, M.O.; Al-Sodani, K.A.A.; AlAteah, A.H.; Al-Tholaia, M.M.H.; Adewumi, A.A.; Bakare, A.O.; Usman, A.K.; Momohjimoh, I. Performances of the synergy of silica fume and waste glass powder in ternary blended concrete. *Appl. Sci.* **2022**, *12*, 6637. [CrossRef]
112. Li, L.; Joseph, P.; Zhang, X.; Zhang, L. A study of some relevant properties of concrete incorporating waste ceramic powder as a cement replacement agent. *J. Build. Eng.* **2024**, *87*, 109106. [CrossRef]
113. *ACI 222R-01*; Protection of Metals in Concrete Against Corrosion. American Concrete Institute Farmington Hills: Farmington Hills, MI, USA, 2009.
114. Kannan, D.M.; Aboubakr, S.H.; El-Dieb, A.S.; Taha, M.M.R. High performance concrete incorporating ceramic waste powder as large partial replacement of Portland cement. *Constr. Build. Mater.* **2017**, *144*, 35–41. [CrossRef]
115. Rattanachu, P.; Karntong, I.; Tangchirapat, W.; Jaturapitakkul, C.; Chindaprasirt, P. Influence of bagasse ash and recycled concrete aggregate on hardened properties of high-strength concrete. *Mater. Construcción* **2018**, *68*, e158. [CrossRef]
116. Huseien, G.F.; Sam, A.R.M.; Shah, K.W.; Budiea, A.M.A.; Mirza, J. Utilizing spend garnets as sand replacement in alkali-activated mortars containing fly ash and GBFS. *Constr. Build. Mater.* **2019**, *225*, 132–145. [CrossRef]
117. Rattanachu, P.; Toolkasikorn, P.; Tangchirapat, W.; Chindaprasirt, P.; Jaturapitakkul, C. Performance of recycled aggregate concrete with rice husk ash as cement binder. *Cem. Concr. Compos.* **2020**, *108*, 103533. [CrossRef]

118. Wang, J.; Liu, E. The relationship between steady-state chloride diffusion and migration coefficients in cementitious materials. *Mag. Concr. Res.* **2020**, *72*, 1016–1026. [CrossRef]
119. Chopra, D.; Siddique, R. Strength, permeability and microstructure of self-compacting concrete containing rice husk ash. *Biosyst. Eng.* **2015**, *130*, 72–80. [CrossRef]
120. Pradhan, B. Corrosion behavior of steel reinforcement in concrete exposed to composite chloride–sulfate environment. *Constr. Build. Mater.* **2014**, *72*, 398–410. [CrossRef]
121. Wongkeo, W.; Thongsanitgarn, P.; Ngamjarurojana, A.; Chaipanich, A. Compressive strength and chloride resistance of self-compacting concrete containing high level fly ash and silica fume. *Mater. Des.* **2014**, *64*, 261–269. [CrossRef]
122. Wan, X.; Wittmann, F.H.; Zhao, T.; Fan, H. Chloride content and pH value in the pore solution of concrete under carbonation. *J. Zhejiang Univ. Sci. A* **2013**, *14*, 71–78. [CrossRef]
123. Vejmelková, E.; Keppert, M.; Rovnaníková, P.; Ondráček, M.; Keršner, Z.; Černý, R. Properties of high performance concrete containing fine-ground ceramics as supplementary cementitious material. *Cem. Concr. Compos.* **2012**, *34*, 55–61. [CrossRef]
124. Jain, K.L.; Sancheti, G.; Gupta, L.K. Durability performance of waste granite and glass powder added concrete. *Constr. Build. Mater.* **2020**, *252*, 119075. [CrossRef]

ceramics

MDPI

Article

Effect of Alkaline Activated Cashew Nut Shell Ash in the Stabilization of Weak Clayey Soil—An Experimental Study

Parthiban Devarajan [1], Dhanasingh Sivalinga Vijayan [1,*], Shanmuga Sundar Saravanabhavan [2], Arvindan Sivasuriyan [3], Naveen Sankaran [4], Theodoros Chrysanidis [5] and Zineb Moujoud [6]

1 Department of Civil Engineering, Aarupadai Veedu Institute of Technology, Vinayaka Missions Research Foundation, Chennai Campus, Tamil Nadu 603104, India; parthi92bhde@gmail.com
2 Department of Biotechnology, Aarupadai Veedu Institute of Technology, Vinayaka Missions Research Foundation, Chennai Campus, Tamil Nadu 603104, India; sundarannauniv85@gmail.com
3 Institute of Civil Engineering, Warsaw University of Life Sciences, 02-787 Warsaw, Poland; arvindan_sivasuriyan@sggw.edu.pl
4 DACK Consulting Solutions, Inc., White Plains, NY 10601, USA; naveensankaran24@gmail.com
5 Department of Civil Engineering, Aristotle University of Thessaloniki, 541 24 Thessaloniki, Greece; thchrysa@arch.auth.gr
6 Faculty of Sciences Ben M'Sick, Hassan II University of Casablanca, Casablanca 9167, Morocco; zineb.moujoud-etu@etu.univh2c.ma
* Correspondence: vijayan@avit.ac.in

Abstract: Weak clayey soils in construction are considered problematic due to their high compressibility and low bearing capacity. This study proposes an environmentally friendly replacement for conventional soil stabilizers through the use of geopolymer (GP) containing Cashew Nut Shell Ash (CNSA) to improve soil characteristics. In this study, the CNSAGP was compared with lime-stabilized soil for unconfined compressive strength (UCS), durability, and improved microstructure. The experimental outcomes showed that 9 M + CNSAGP with 4% CNSA provided a UCS of 1900 kPa, which was higher than the lime-stabilized soil (6% lime with 4% CNSA) at 1400 kPa. Durability test results revealed that the CNSAGP-treated sample had better protection against water damage with a strength loss of about 18%, while the lime-treated sample had a strength loss of about 25%. Thermal stability analysis showed that CNSAGP had lower LOI values compared to lime-stabilized samples (0.17% at 900 °C), which indicates CNSAGP's heat resistance. Microstructure analysis revealed that CNSAGP-stabilized soil was less porous, the microstructure being denser because of reactions of aluminosilicate and pozzolanic activity. Moreover, it affected the soil's alkalinity, making it better, and improved Atterberg limits, which affected the plasticity and workability. These findings show that CNSAGP is a long-lasting and eco-friendly means of soil stabilization with higher strength, thermal stability, and durability than traditional methods and can be used in engineering.

Keywords: soil stabilization; cashew nut shell ash; geopolymer; molarity; Atterberg limit; microstructure

1. Introduction

Clay soils are generally considered problematic in construction activities, potentially causing significant damage and uneven settlement of structures such as pavements and runways owing to their high plasticity and sudden swelling and shrinkage behavior in both wet and dry conditions [1]. Highly plastic clayey soils naturally exhibit poor bearing strength, rendering them unsuitable for constructing high-rise and heavily loaded structures without treatment [2]. Clay soils are characterized by low permeability, resulting in prolonged water retention and an increased risk of waterlogging and structural-related issues during construction [3]. During the construction of underground structures, such as retaining walls and basements, these soils pose potential damage owing to the pressure of the retained water [4]. Clay soils typically contain a high proportion of fine particles that are susceptible to erosion in running water, potentially leading to foundation failure and unstable slopes, thereby increasing the risk of landslides.

Prior to construction, the soil should be stabilized to enhance the performance of clayey soil, particularly its mechanical and physical characteristics, making it suitable for building purposes [5]. During the mechanical stabilization process, the strength of the clayey soil improves considerably, enabling it to support heavy structural and vehicular loads. This technique addresses issues related to changes in soil volume by regulating soil plasticity [6]. Stabilized clay soil demonstrates increased resistance to erosion over time and simplifies the construction process because it directly extends the lifespan of the structure. Numerous studies have demonstrated the importance of soil stabilization in improving soil properties [7]. This method involves combining soil with conventional stabilizing agents such as cement, lime, or fly ash to enhance its strength, durability, and load-bearing capacity [8]. In conventional cement stabilization, the Portland cement was mixed with soil for stabilization, creating a compact soil-cement matrix that decreases plasticity and provides moisture resistance. In comparison, lime-stabilized soil has been shown to improve stability and reduce plasticity [9]. The addition of lime to soil alters its physical behavior and diminishes its swelling potential, resulting in an increased load-bearing capacity.

Additionally, sulfate reactions in soils stabilized with cement and lime can result in swelling and road surface failures, complicating the selection of appropriate soil treatment methods and stabilizing agents. The use of fly ash, either alone or in combination with lime, has demonstrated advantages comparable to those of conventional techniques while offering environmental benefits [10]. Nevertheless, recent research indicates that despite the impressive results of using cement and lime for soil stabilization, these methods have considerable environmental drawbacks, primarily owing to their carbon dioxide emissions during manufacturing [11]. The ongoing extraction of raw materials for lime production has led to resource depletion and habitat disruption. Furthermore, the production and application of these traditional stabilizers can generate dust and aerosols, potentially causing respiratory issues and skin irritation [12]. The expense and transportation of raw materials also increase the overall project costs, as most of these materials are not locally available. Moreover, conventional stabilizers are not suitable for soils with high levels of organic matter and sulfates.

Finding alternative stabilization materials is crucial for mitigating environmental impacts, addressing the negative effects of traditional stabilizers, and achieving sustainability, safety, cost efficiency, and improved performance. Environmental scientist Raymond Pierrehumbert emphasizes that the fundamental goal consists of the complete decarbonization of human activities that produce CO_2, such as the burning of fossil fuels, the creation of cement, and deforestation. What he states is that the time needed to get to zero can be measured by the degree of global warming. Pierrehumbert disappointedly describes the slow pace of decarbonization, and as a result, it has opened up enthusiasm for technological solutions, including geoengineering. But he discouraged any country from relying on such weak narratives and urged everyone to take definite and ongoing actions on emission reductions via the shift to renewable energy and the enforcement of policies [13]. This can be accomplished by recycling and utilizing eco-friendly waste materials as additives or by implementing novel methods such as geopolymers with or without industrial-based admixtures [14].

Geopolymers are synthetic materials created through the reaction of alkaline-activated silicates or hydroxides with binding agents [15]. These materials have found widespread industrial use because of their interconnected networks and chains of mineral compounds, which are held together by covalent bonds formed during alkali activation reactions. These reactions involve the exchange of oxygen and negatively charged particles between Al^{3+} ions [16]. The main components of geopolymers are aluminum silicate compounds along with calcium-rich substances that aid the formation of tobermorite, a specific compound [17]. The strength of geopolymer materials is determined by their polymerization levels [18]. Geopolymerization typically involves the use of alkali-silica cations combined with alkaline activators and is generally classified into four categories [19]. Many researchers are investigating geopolymer materials because of their environmentally friendly raw materials and alkali-activation reactions, which help reduce their environmental impact [20].

In this study, a geopolymer matrix was developed using CNSA, which was introduced between soil particles. An alkaline activator solution was created using sodium hydroxide (NaOH) and sodium silicate (Na_2SiO_3) to activate the CNSA for stabilization. This research aims to enhance the strength and durability of structures by improving the mechanical properties, durability, and microstructure of clayey soil treated with cashew shell ash-based geopolymer (CNSAGP). The goal is to reduce or control soil shrinkage and swelling behavior and increase the soil's resistance to environmental factors during wet-dry cycles. This study evaluated the performance of CNSA-based geopolymer-stabilized clayey soil in comparison with lime-treated soil. Initially, the research identified problematic soil with high plasticity and examined its parameters to determine the composition of the CNSA geopolymer matrix and the optimal percentage of lime required for effective stabilization. After stabilization, the stabilized soil samples' characteristics were evaluated under different CNSA ratios, curing temperatures, and conditions, and they were compared with soils that had been treated with lime. The engineering behavior of the soil was evaluated using strength and durability parameters, whereas the microstructure was examined using scanning electron microscopy (SEM) analysis. Finally, the results of lime- and CNSAGP-stabilized specimens were compared to address the identified research gap, evaluate the performance of the novel CNAGP stabilizers, and offer a sustainable and innovative solution for the construction industry.

2. Materials and Methods

2.1. Clayey Soil

This study conducted an investigation by extracting clayey soil from an abandoned dry lake near Paiynoor Village, Chennai, India. The soil was obtained at a depth of one meter beneath the surface and exhibited dark brown coloration. Sufficient samples were collected for laboratory examination and carefully stored in sealed plastic containers to preserve moisture content. The collected materials, including weeds, were extracted using a trowel. A notable characteristic observed at the site was the soil's porous nature and its susceptibility to compression, setting it apart from inorganic soils, such as clay and sand, which are primarily composed of solid silicate particles. The soil underwent a drying process, first in sunlight and then in an oven at a controlled temperature of 110 °C for 24 h. Subsequently, the physical properties of the pre-stabilized soil were analyzed, and the results are presented in Table 1.

Table 1. Pre-stabilization property of soil.

Soil Behaviour	Values
Liquid Limit (LL)	59.25%
Plastic Limit (PL)	24.25%
Plasticity Index (PI)	40%
Specific Gravity @ 27 °C (G)	2.48
pH value of Soil	6.3
Maximum Dry Density (MDD)	1570 kg/m^3
Optimum Moisture content (OMC)	25.40%
UCS Strength (kPa) (28 days curing)	256.19

The graphs below illustrate the particle size distribution composition of the soil material using sieve and hydrometer analyses. Figure 1a shows the cumulative percentage of finer particles against particle size and reveals that most particles are smaller than 2 mm, with 90.9% of the sample passing through this size and nearly all particles passing through a 4.75 mm gradation. This shows a gradual buildup of smaller particles. Figure 1b, showing particle size distribution, indicates that the material is dominated by silt and clay at 88.38%, and fine sand was 11.26% with small portions of medium sand, coarse sand, and gravel. In this analysis, the high level of the material's fineness was emphasized, the presence of which is observed from the dominance of silt and clay fractions.

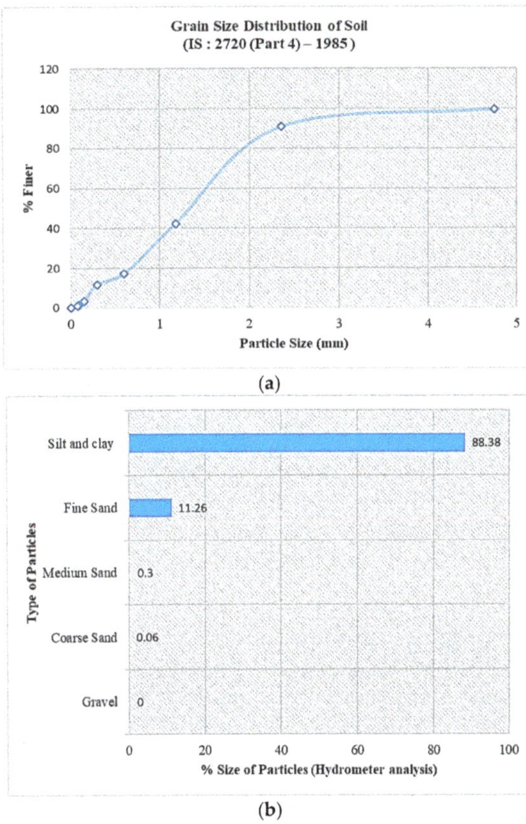

Figure 1. (**a**). Particle size distribution of Collected Soil. (**b**). % Particle size of Collected Soil.

2.2. Cashew Nut Shell Ash

Cashew nut shell ash (CNSA) was utilized in this study to synthesize the industrial-based geopolymer collected from the cashew refinery and oil extract industry, Panruti, Tamil Nadu, India. CNSA was produced by incinerating the shells at 750 °C, and the burnt ash was utilized as a supplementary cementitious material in combination with a geopolymer in this study, as shown in Figure 2.

Literature study reveals that CNSA contains high concentrations of silica and minerals, which contribute to its enhanced compressive strength and longevity. As usual, the chemically processed cashew nut shells exhibit high adsorption behavior toward harmful substances and heavy metals, thereby offering an economical and eco-friendly solution [21]. Shells treated with H_2SO_4 showed the highest adsorption capacity through monolayer physisorption, as evidenced by the pseudo-first-order and Langmuir model results [22]. For an extended period, cashew nut shell ash has been employed as a sustainable cementitious material owing to its potential pozzolanic properties [23].

Cashew nut shell ash (CNSA) is an industrial by-product created from the remnants of cashew nut processing. The process involves separating the nut from its shell, extracting oil, and burning the leftover material. CNSA serves as an alternative to traditional cement stabilizers [24]. The chemical composition of CNSA, particularly its CaO and SiO_2 contents, plays a crucial role in determining its rheological and mechanical properties, which in turn affect its compatibility and flow characteristics [25]. Research has shown that CNSA can enhance the early strength of stabilized soil composites within a week of curing, which is comparable to that of other agricultural waste-derived additives. Furthermore, when used in combination with lime, CNSA significantly increased the unconfined compressive strength of the soil. In concrete applications, the incorporation of cashew nut shell ash (CNSA) improves thermal insulation, enhances overall performance, and reduces environmental impact [26]. Additionally, it increases the compressive strength in a sustainable way, with precise regression models capable of predicting the thermophysical properties of the resulting concrete.

Figure 2. Cashew Nut Shell ash.

Researchers have investigated the use of cashew nut shell ash (CNSA) as an eco-friendly and economical substitute for conventional cement in the cement industry [18]. Various literature studies state that incorporating cashew nut shell ash (CNSA) in soil stabilization results in beneficial pozzolanic and chemical compositions within the soil matrix, rendering it an excellent material for stabilizing clay-rich soils. In addition to their economic and environmental benefits, CNSA materials have emerged as promising substitutes for traditional stabilizers. It is recommended that ongoing research be conducted to further investigate CNSA's potential, and more extensive applications in the construction sector are necessary. Figure 3 illustrates the energy-dispersive X-ray spectroscopy (EDS) analysis used to identify the percentage of individual elements present in CNSA. The analysis revealed that the presence of the following elements in the cashew nut shell ash, as shown in Table 2, was used in this study.

Table 2. CNSA Compounds (EDS) and physical parameters.

Properties/Compounds	Proportion
N (Nitrogen)	0.01%
O (Oxygen)	20.88%
Na (Sodium)	0.09%
Mg (Magnesium)	0.17%
Al (Aluminum)	0.57%
Si (Silicon)	0.19%
Cl (Chlorine)	0.12%
K (Potassium)	1.97%
Ca (Calcium)	0.56%
Ti (Titanium)	0.10%
Fe (Iron)	0.47%
Cu (Copper)	0.26%
C (Carbon)	78%

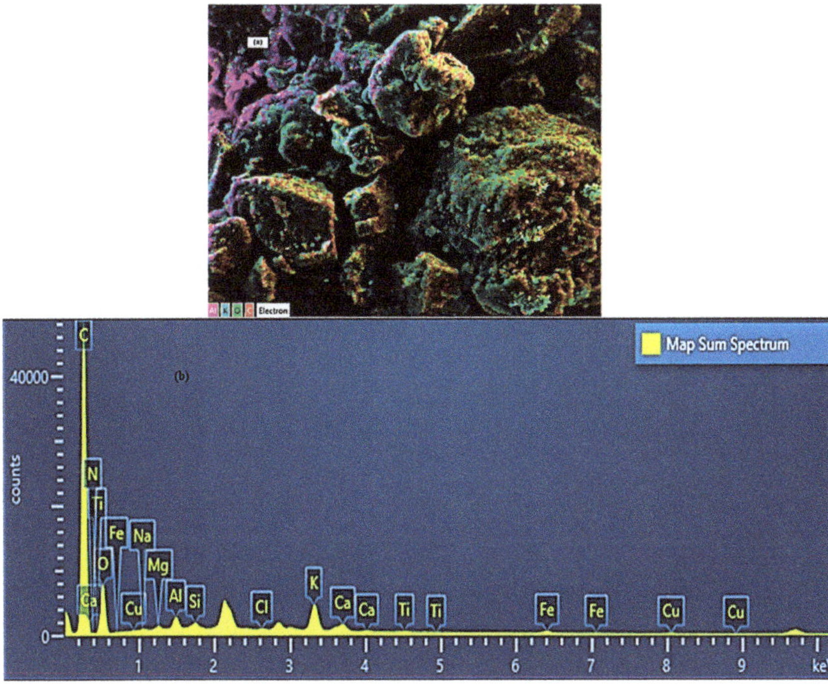

Figure 3. EDS of Cashew Nut Shell ash.

2.3. Geopolymer Syntheziation

Alkali activators such as highly concentrated sodium hydroxide reacted with solid aluminosilicate to form a "geopolymer", which is an inorganic binding material [27]. The geopolymer was considered beneficial to the environment because of its manufacturing process, which involves the use of byproducts of industries including fly ash, ground granulated blast furnace slag (GGBS), metakaolin, and metal slag [28]. Geopolymers have been discussed to present several advantages over traditional construction binding materials in the construction industry; primarily, they display great durability and strength, less shrinkage, and are resistant to fire and corrosion [29]. The synthesis of a geopolymer entails a chemical reaction between the alkaline solution and aluminosilicate called "alkaline activation", which forms a three-dimensional polymer chain of geopolymer similar to that of natural zeolite [30]. Usually, alumino silicate materials contain high content of alumina (Al_2O_3) and silica (SiO_2), which are obtained from fly ash, GGBS, and metakaolin [31]. An alkaline activator can be made up of potassium (KOH), sodium hydroxide (NaOH), potassium (K_2SiO_3), and sodium silicate (Na_2SiO_3) [32]. The preparation of geopolymers for different uses commonly involves the dissolution, condensation, gelation, and polycondensation of materials [33]. To assess the strength, UCS specimens were prepared by combining a calculated amount of sodium-based hydroxide solid for a 1 L solution, with the quantity determined based on the concentration in terms of molarity. Sodium-based silicates, also known as water glass, were obtained in powder form from a local chemist in Chennai, India. It had a SiO_2/Na_2O molar ratio of 3 and contained 11% Na_2O and 31% SiO_2 by weight.

As illustrated in Figure 4, sodium hydroxide solutions with molarities of 3 M, 6 M, and 9 M were prepared and mixed with a sodium silicate under a solution binder ratio of 1:2 (V:V) with distilled water. The combination of sodium hydroxide and sodium silicate produces an exothermic reaction, necessitating the preparation of an alkaline activator solution one day prior to use.

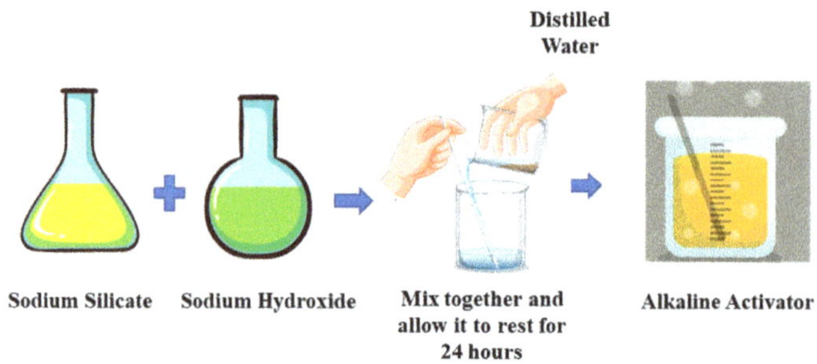

Figure 4. Preparation of Alkaline activator.

In the geopolymerization process, the aluminosilicate precursors like fly ash/clay transform into a long-lived 3D geopolymer gel through the medium of an alkaline activator [20]. This sequence starts by dissolving silica (Si) and alumina (Al) from precursors in highly concentrated sodium hydroxide (NaOH) and sodium silicate (Na_2SiO_3) [34]. Based on the alkaline activator, this dissolution leads to the release of $(SiO_4)^{4-}$ and $(AlO_4)^{5-}$ tetrahedra, which undergo condensation reactions to form oligomers as shown in the below equation.

$$Si(OH)_4 + Al(OH)_4^- \rightarrow (Si - O - Al) + 4H_2O$$

These oligomers polymerize into a three-dimensional gel network containing Si–O–Si and Si–O–Al bonds with the support of alkali ions such as Na^+ or K^+ ions. Water also has an important function of improving ion conductivity and increasing the density of the structure during the curing process [35]. Curing at temperatures varying between 40 and 100 °C is preferred, as it increases the rate of the reaction and the strength of the cement [36]. The synthesized geopolymer has features of high compressive strength, enhanced durability, and resistance to chemicals and heat and can be utilized for construction, environmental remediation, and fire resistance [37]. This eco-friendly process also lowers CO_2 emissions by using industrial waste and serves as a better option than regular cement.

2.4. Preparation of Soil Specimen

In this study, the specimens were prepared by using manual hand mixing to maintain homogeneity. Also, a pre-process was carried out before the introduction of the alkaline solution into the reaction mixture. Before the hydroponic experiment, the soil was first sieved in order to remove debris and to have a uniform grain size of 425 microns. The sieved soil was then dried in an oven at 110 °C to remove any moisture left on the soil because moisture could alter the stability of the mixture and also interfere with the chemical reactions that take place during stabilization. These pretreatment procedures were important to enhance the reactivity of the soil with the alkaline solution and to obtain results with a high degree of repeatability. Unconfined compression (UCS) cylindrical specimens were cast using a prepared clayey soil with traditional stabilizer laboratory-grade lime and novel geopolymer as a stabilizer with or without CNSA. To maintain sample integrity, the soil was examined for contaminants before specimen formation. Cylindrical samples 38 mm in diameter and 76 mm in length were fabricated using various ratios of lime and CNSA-based geopolymers, as outlined in Table 3. The water content was maintained constant at the optimum moisture content (OMC) value of the soil throughout the sample preparation process.

Table 3. Mix proportion of Materials.

Mix ID	Weight of Soil (g)	Weight of Lime (g)	Weight of CNSA (g)	Geopolymer (GP) Molarity	Volume of Water (mL)	Volume of GP Solution (mL)
PS	153	-	-	-	24	-
PS + 2L	149.94	3.06	-	-	24	-
PS + 4L	146.88	6.12	-	-	24	-
PS + 6L	143.82	9.18	-	-	24	-
PS + 2 L + 2 CNS	146.88	3.06	3.06	-	24	-
PS + 2 L + 4 CNS	143.82	3.06	6.12	-	24	-
PS + 2 L + 6 CNS	140.76	3.06	9.18	-	24	-
PS + 4 L + 2 CNS	143.82	6.12	3.06	-	24	-
PS + 4 L + 4 CNS	140.76	6.12	6.12	-	24	-
PS + 4 L + 6 CNS	137.7	6.12	9.18	-	24	-
PS + 6 L + 2 CNS	140.76	9.18	3.06	-	24	-
PS + 6 L + 4 CNS	137.7	9.18	6.12	-	24	-
PS + 6 L + 6 CNS	134.64	9.18	9.18	-	24	-
PS + 2 CNS +2 M GP	149.94	-	3.06	2 M	-	24
PS + 4 CNS + 2 M GP	146.88	-	6.12	2 M	-	24
PS + 6 CNS + 2 M GP	143.82	-	9.18	2 M	-	24
PS + 2 CNS + 4 M GP	149.94	-	3.06	4 M	-	24
PS + 4 CNS + 4 M GP	146.88	-	6.12	4 M	-	24
PS + 6 CNS + 4 M GP	143.82	-	9.18	4 M	-	24
PS + 2 CNS + 6 M GP	149.94	-	3.06	6 M	-	24
PS + 4 CNS + 6 M GP	146.88	-	6.12	6 M	-	24
PS + 6 CNS + 6 M GP	143.82	-	9.18	6 M	-	24

Soil samples, sieved through a 4.75 mm sieve, were cast and molded the unconfined compression (UCC) using a 30 kN hydraulic jack compressor, as shown in Figure 5. Cast samples were stored in sealed Ziplock bags for 7, 14, and 28 days. For traditional stabilization, soil samples were mixed with lime at 2%, 4%, and 6% of the total soil weight and cured at room temperature for 28 days. Conversely, CNSA geopolymer specimens used an alkaline activator with NaOH molarity levels of 2 M, 4 M, and 6 M, maintaining an alkaline binder-to-CNSA ratio of 1:2, with CNSA at 2%, 4%, and 6% of the total soil weight. These samples were cured under different conditions, including ambient room temperature and in an oven at 50 °C, 75 °C, 100 °C, and 125 °C for a period of 48 h.

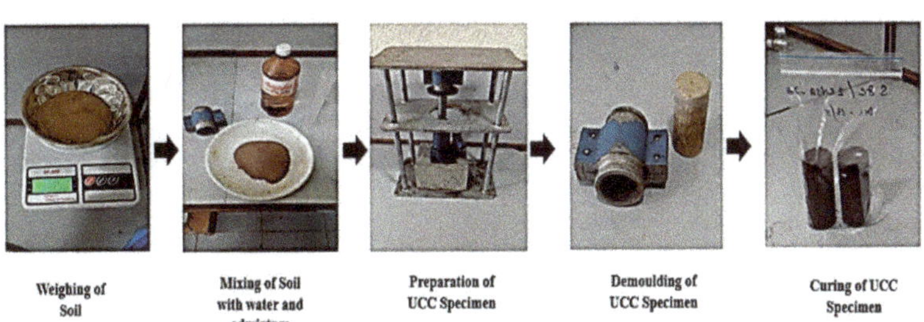

Figure 5. Preparation of Soil Specimen.

3. Results and Discussion

3.1. Engineering Behavior of Stabilized Soil

This study investigated the engineering properties of stabilized soils by evaluating their strength characteristics through unconfined compression strength (UCS) testing and durability via capillary soaking. Strength tests were performed using the UCS apparatus after 7, 14, and 28 days of curing. Simultaneously, soil specimen durability was evaluated by measuring the percentage decrease in the strength of the UCS. This was achieved by immersing a 28-day cured stabilized UCS specimen in a basin containing distilled water at a standard depth of 1.5 cm for 24 h. The soaked specimens were then tested using a UCS apparatus, and the results were compared to those of the unsoaked UCS specimens. This section analyzes the strength and durability characteristics of soil stabilized with lime and CNSA geopolymer.

3.1.1. Strength Assessment of Stabilized Soil Using Lime with CNSA

Figure 6 below shows the changes in the 28-day unconfined compressive strength (UCS) of stabilized soil with different percentages of lime alone and lime mixed with CNSA, as shown in Figure 6. This is presented using a two-tailed, unpaired Student's *t*-test from the experimental replicates.

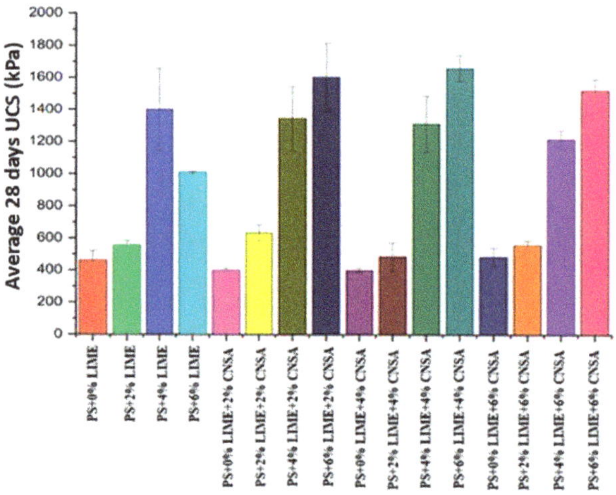

Figure 6. Lime composition vs. 28 days average UCC Strength (kPa) of Stabilized Soil with and without CNSA.

It was also observed that the unconfined compression (UCC strength gradually increased from 400 kPa at 2% lime to 1600 kPa at 6% lime, which revealed that lime improves the strength of the soil through pozzolanic reactions and soil aggregation. When CNSA is added, the value of UCS further increases, whereby the combination of 6% lime and 4% CNSA gave the highest UCS value of nearly 1750 kPa. This signifies a synergistic relationship whereby CNSA contributes extra silica and alumina, which is used to react with lime to produce more durable cementitious materials. However, when only CNSA is used, it can enhance the UCC strength, but to a lesser extent, with a maximum of around 1200 kPa at 6% CNSA, relative to the outcome attained with lime. Furthermore, these outcomes corroborate that incorporating CNSA with lime enhances the dependency and adjustment of reactive compounds within the soil, hence adding strength as per the study. This means that the observed strength variation is a result of increased binding due to pozzolanic reactions and the right proportions of lime/CNSA.

3.1.2. Unconfined Compression Strength of GP Treated Soil Under Varied Molarity with and Without CNSA

Figure 7 shows the fluctuation in unconfined compressive strength (UCS) of CNSAGP with different molarities of solution 3M, 6M, and 9M and curing conditions and without Cashew Nut Shell Ash (CNSA and without CNSA) presented using a two-tailed, unpaired Student's *t*-test from the experimental replicates. It illustrates the impact of curing temperature (ambient, 50 °C, 75 °C, 100 °C, and 125 °C) on the UCC strength. For soil stabilized at 3 M molarity, the UCC strength improves from about 400 kPa under normal curing conditions to possess a value of 800 kPa at 125 °C, although 4% CNSA develops a higher strength than 2% CNSA. UCC strength increases significantly at 6 M molarity, reaching up to 1400 kPa with 4% CNSA at 125 °C, suggesting that further improvements to molarity and heat curing could lead to increased stabilization. The 9 M molarity treatments demonstrate the highest strength improvements of up to 1800 kPa at 125 °C with 4% CNSA. It could be deduced that the incorporation of higher molarity and higher curing temperatures promotes the formation of the pozzolanic reactions and geopolymerization, leading to better binding. Curing at ambient temperature always results in the lowest values of compressive strength regardless of the curing condition, while curing at 100 °C and 125 °C have higher values of compressive strength. This synergistic effect between CNSA, higher molarity, and elevated curing temperatures enhances soil stabilization as a result of better geopolymer binder formation.

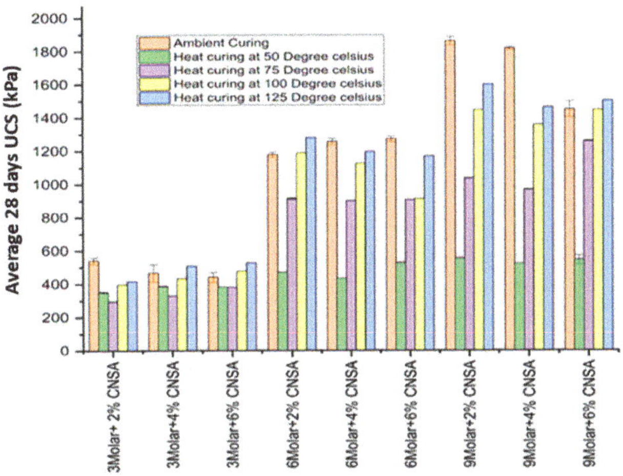

Figure 7. UCS of GP Stabilized soil under different Molarity with and without CNSA.

3.1.3. UCS Behavior of Lime vs. CNSAGP Stabilized Soil

Figure 8 shows the average 28-day UCS of soil stabilized with different proportions of lime and CNSAGP, with and without CNSA. As the molarity and CNSA content rise, UCC strength significantly rises for CNSAGP-stabilized soil at varying molarities. The maximum strength is observed with 9 M + 4% CNSA, which gives about 1900 kPa, showing that the optimum geopolymer reactions are highly favored where the molarity is high and the CNSA content is also high. At lower molarities, for example, at 3 M, the UCC strength is considerably lower and ranges from about 700 to 800 kPa at 4% CNSA, which reflects the poor stabilization attributable to weak geopolymerization reactions.

On the other hand, for the lime-stabilized soil (PS without CNSAGP), the UCC strength rises moderately with lime content. For example, 6% lime at a PS+ level provides a maximum of 800 kPa, which is significantly lower than that of stabilized soils with CNSAGP. Lime and CNSA show the aspect of synergy whereby their combined influence on the strength of the composite surpasses the sum of the individual effects, with the PS+6% lime + 4% CNSA producing strength of about 1400 kPa. These changes can be caused by the increase in pozzolanic and geopolymer reactions in CNSAGP-stabilized soils, as well as the effect of molarity and CNSA content on cementitious bonding strength compared to lime-stabilized soils. The findings suggest that CNSAGP, when combined with CNSA at their optimal blends, provides better stabilization of soil than lime alone.

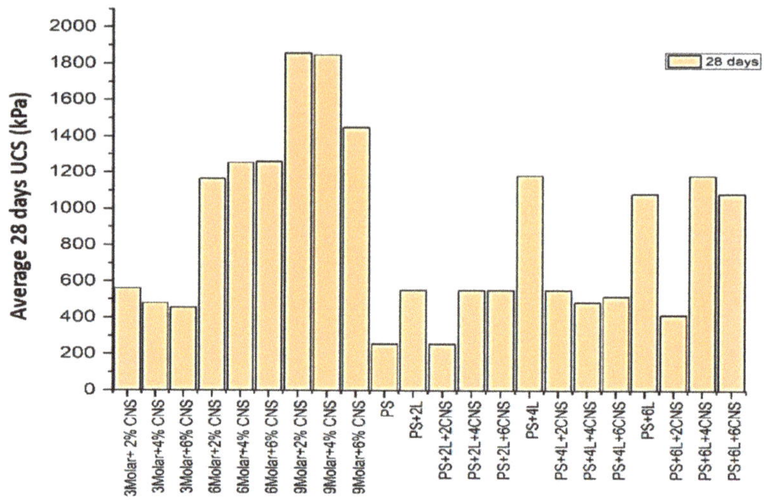

Figure 8. Strength of Lime vs. CNSAGP stabilized soil.

3.1.4. Average Stress–Strain Behavior of Lime and CNSAGP Stabilized Soil

Figure 9 soil stress–strain graph indicates the performance of soil treated with lime and geopolymer with CNAs and without CNAs. The detailed investigation suggests that the samples that incorporated lime and CNSA have higher peak stress than the other samples containing lime only, this leads to the improvement of mechanical characteristics through improved pozzolanic reactivity of CNSA and lime.

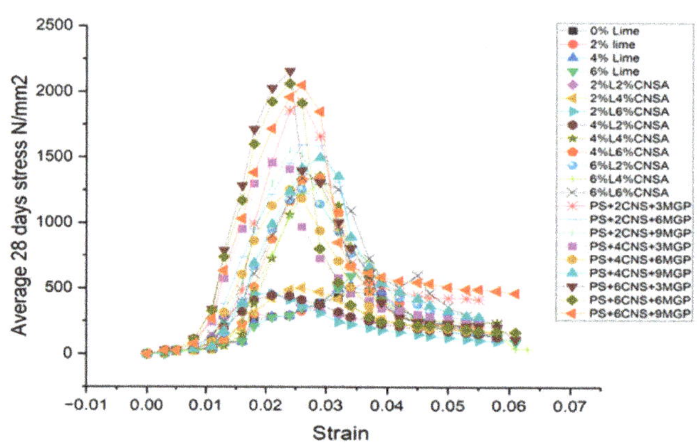

Figure 9. Stress–Strain Behavior of Stabilized soil specimen.

Further performance enhancement is achieved by the integration of geopolymer, where the geopolymer-stabilized samples possess a steeper stress–strain curve and enhanced peak stress that leads to an enhanced elastic modulus and, therefore, higher load-bearing capability. A steep slope, which is actually the stress–strain ratio, is shown by curve fits of the elastic regions using graphs. This suggests that the geopolymer-CNSA combinations exhibit an enhanced modulus of elasticity. The pozzolanic reactions that occur between lime, CNSA, and geopolymer may be the reason for this change. These reactions lead to the formation of a harder matrix, reduced porosity, and strengthened bonds. These properties are two-fold, revealed in the enhanced elastic and plasticity performance, while CNSA serves as the best admixture to enhance stabilization.

3.1.5. Durability Assessment

Figure 10 shows the percentage loss in the UCS of the soil stabilized with lime and geopolymer and with the addition of CNSA and after capillary soaking for a period of 48 h over the saturated sand layer. The results reveal that untreated soils and those treated with CNSA have a relatively higher UCS, and the lime-treated soils without CNSA have lower UCS reductions of more than 25% for 0% and 2% lime treatments. This suggests that the proposed technique offers little resistance to capillary soaking, as lime alone cannot effectively address moisture penetration. On incorporation of CNSA, the percentage reduction in UCS is lesser and ranges between 20 and 23% due to the durability aspect where CNSA reacts with lime to form a denser pozzolanic gel. CNSA has been proven to enhance the geopolymer-stabilized soils, and this is evidenced from the development shown above. All of these combinations result in lower UCS values below 20%, with the PS + 6% CNS + 6 M GP combination having the lowest at approximately 18%. This is compared to the improved water resistance that geopolymer stabilization offers to form a tighter matrix that is less permeable. The variation can be explained on the basis of the higher bonding material and water resistance of the geopolymer as well as the presence of CNSA, which also acts as a pozzolanic material. It eliminates the negative impacts of capillary soaking and therefore enhances the performance of the stabilized soil in the long run.

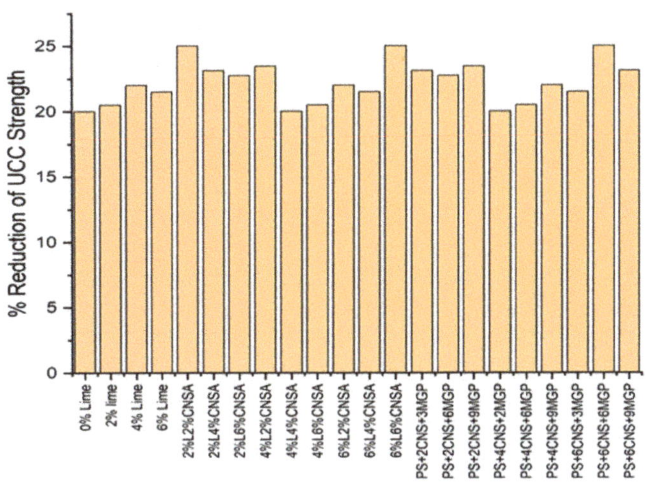

Figure 10. Reduction strength due to capillary soaking.

3.2. Physical Characteristics of Stabilized Soil

3.2.1. pH Value of Stabilized Soil

Figure 11 shows the pH ranges of the stabilized soils based on the additions of lime and geopolymer with/without CNSA, presented with a 5% error bar from the experimental replicates. It is also observed that there is a general trend of increase in the pH values with the increase in stabilizer content and geopolymer molarity. Lime-treated soils demonstrate a slow but steady process of pH increase; the 2% lime-treating program yields soils with a pH of 6.5, while the 6% lime-treating program results in 7.0. Lime-SAA has a pH of approximately 7.2–7.4 and thus implies an increase in alkalinity due to the products of the reaction that include C-S-H gel and other strongly alkaline products.

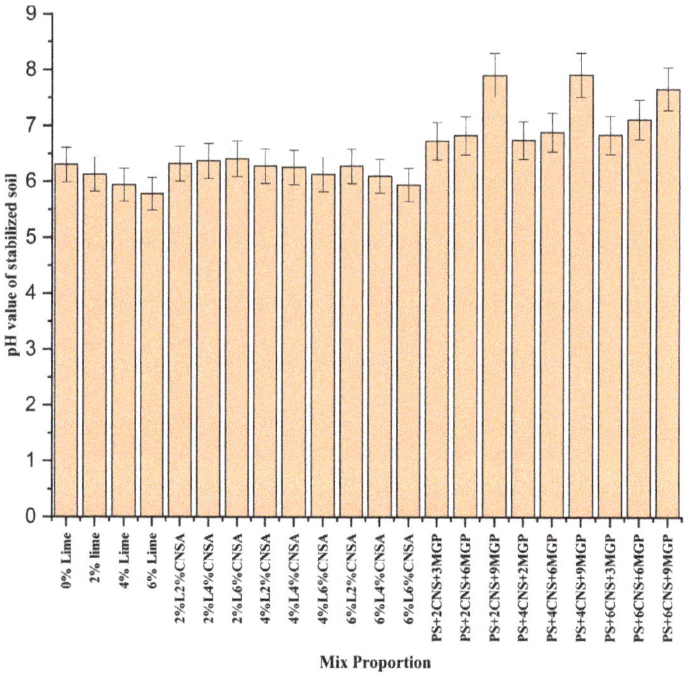

Figure 11. pH value of stabilized soil.

Similar to the case of untreated soils, the pH values of geopolymer-treated soils are higher, most notably when CNSA is included. For instance, geopolymer with CNSA and higher molarities, such as the combination of PS + 6% CNSA + 9 M GP, yielded the highest pH of approximately 8.5–8.8 and supported the strongly alkaline environment of the geopolymer matrix. This is caused by the enhancement of aluminosilicate phases in CNSA through the highly alkaline geopolymer solution to form a strong and interconnected matrix. The swing in the pH levels demonstrates the efficiency of CNSA through chemically stabilizing the soils that have been worked on. The higher pH in the geopolymer-stabilized soils with CNSA offers better setting and proper cementation/hardening, in which the UCS is beneficial. The outcome of the research proves that the integration of geopolymer and CNSA not only improves the mechanical properties of the soil but also alters the chemical properties of the soil in a way that could be more suitable for long-term stability and durability.

3.2.2. Atterberg Limit

Figure 12 shows the change in the Atterberg limits, load limit (LL), plastic limit (PL), and plasticity index (PI) depending on the stabilized type and the use of the Cashew nut Shell Ash (CNSA) additives, lime stabilization, and geopolymer stabilization, which were tested after 28 days of curing, with a 5% error bar from the experimental replicates. It was found that the use of lime, CNSA, and geopolymer has been associated with a decrease in the liquid limit and the plasticity index while the value of the plastic limit increases, thus increasing the workability and stability of the soil. As for its liquidity limit, it decreases gradually with an increase in lime content and reaches the minimum at 6% of lime content equal from 59% of untreated soil to about 56%. There is a slight rise in the plastic limit from about 24% to around 27% and a decrease in the plasticity index from around 35% to roughly 29%, suggesting the change to less swelling-shrinking or better workable soil.

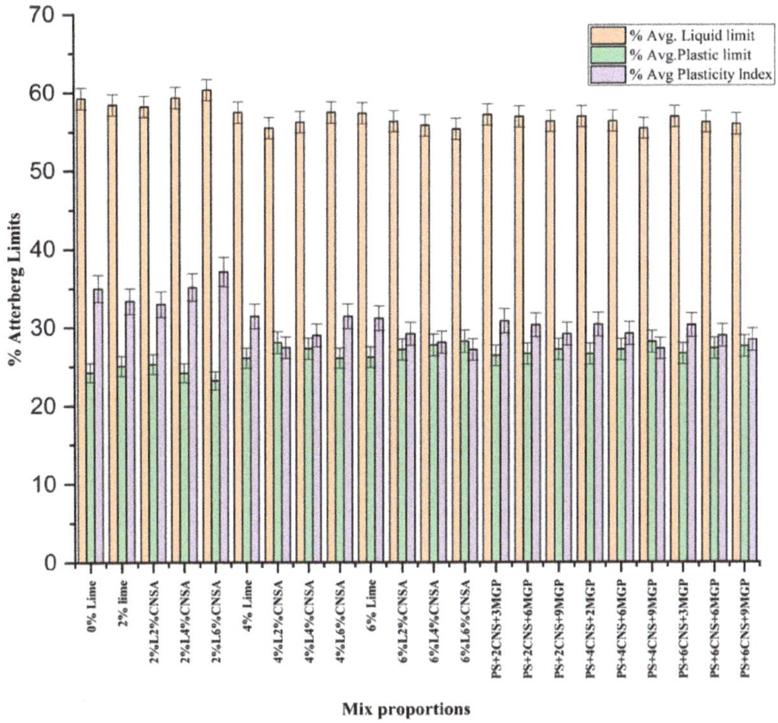

Figure 12. % Atterberg Limits of Stabilized Soil.

Incorporation of CNSA doubles these effects. For example, the soils containing 6% lime and 4% CNSA have a liquid limit of 55%, a plastic limit of 28%, and a plasticity index of 27%. This is attributed to the pozzolanic reactions between the lime and CNSA, which cause the enhancement of the soil particles binding and reduction of water trapped. The pertaining analyses also indicate that the combined use of geopolymer with CNSA applied on the soils shows the maximum enhancement. For instance, the blend produced using 9 M geopolymer and 6% CNSA has a liquid limit of around 52%, a plastic limit of around 30%, and a plasticity index of around 22%. This goes a long way to explaining how geopolymerization is able to create a denser soil structure with less plasticity and higher resistance to deformation. These variations can be explained on the basis of the chemical and physical activity of lime, CNSA, and geopolymer with soil particles to affect the availability of free water for swelling and to enhance the mechanical strength and structural stability of the soil. The outcomes support that geopolymer together with CNSA is the best stabilizer for improving the characteristics of the soil.

3.2.3. Loss of Ignition

The percentage of weight reduction in the soil at elevated temperatures directly correlates with the strength and durability of the stabilized soil. The loss of Ignition (LOI) test is commonly used to measure curing, which involves the emission of volatile compounds and moisture, burning of organic matter, and breakdown of carbonate substances in the stabilized soil. The key factors influencing the percentage weight reduction during high-temperature curing include the amount of binding agent added, initial organic content, and curing duration. In this study, we subjected the stabilized samples to high temperatures between 600 °C, 750 °C, and 900 °C for a period of 2 h in a muffle furnace with an initial LOI value of 0.12. Loss of ignition (LOI) was determined by comparing the initial and post-burn weights of the soil samples.

Figure 13 illustrates the assessment of the percentage loss of organic matter and shows the % LOI of lime-stabilized and geopolymer-stabilized soil with and without CNSA at 600 °C, 750 °C, and 900 °C. The LOI values in general decrease with increasing temperature of the treatment, suggesting enhanced thermal stability. The percentage LOI of the lime-treated soils without CNSA is approximately 0.25 at 600 °C, indicating lower thermal stability owing to the decomposition of lime and the absence of any considerable pozzolanic reaction. The involvement of CNSA brings the LOI percentage down to around 0.22–0.23% due to pozzolanic chemistry that infused a more compact and sound matrix. However, the percentage LOI of the geopolymer-stabilized samples is lower and more stable, ranging from around 0.21% at 600 °C to 0.17% at 900 °C in PS + CNS + 6 M GP. This indicates that the inorganic binder system of the geopolymer provides enhanced thermal stability. The lower percentage LOI in geopolymer-treated soils is due to the aluminosilicate framework formed during geopolymerization that is thermally stable, complemented by the CNSA that provides reactive silica and alumina. Although there is some enhancement in the CNSA and lime-stabilized samples, the thermal profile is poorer as compared to geopolymer-treated samples. These results explain the effectiveness of the integration of geopolymer and CNSA in applications where higher thermal stability is required. Furthermore, the decrease in LOI value in stabilized soil indicates the removal of the extent of organic content and volatile elements that evaporated during heat treatment. This reduction does not affect the strength of the soil in any way as a result of pozzolanic reactions that occur and the creation of a dense geopolymer matrix that offers enhanced strength in place of the lost weight. The stabilization process also improves the other properties of the composite, such as durability and the mechanical properties that provide stability.

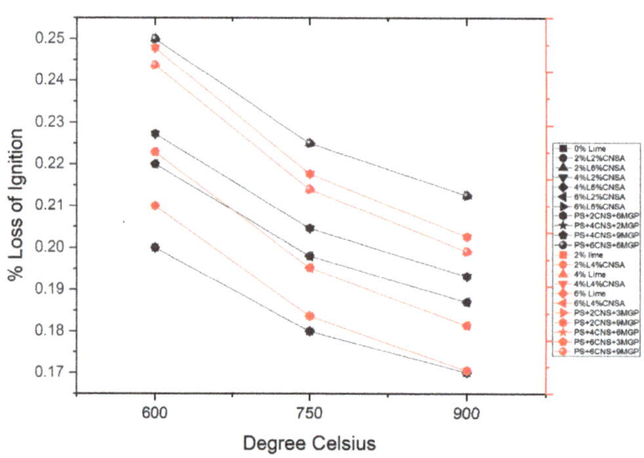

Figure 13. Loss of Ignition of Stabilized Soil.

3.3. *Microstructural Behavior of Stabilized Soil*

Figure 14 shows the FESEM images of the unconfined compressive strength (UCS) at 340× magnification to help explain the changes in the microstructure of lime-stabilized and geopolymer-stabilized soils with and without CNSA. Figure 14a, lime-treated clay soil without CNSA shows a good porous structure with numerous gaps and low inter-particle forces. This corresponds with the poor UCS. This structure does not contain compact reaction products, which makes it less capable of withstanding external forces. In the CNSA case, Figure 14b shows that lime-treated soil has low porosity and high connectivity of particles as a result of pozzolanic reactions that lead to the conversion of Lime–CNSA to C-S-H and C-A-H. This holds the key to explaining the observed moderate UCS increases due to the densification of the matrix and enhancing load transfer between particles. Figure 14c consists of soil that has been treated with geopolymer but does not contain CNSA and shows a more compact and stronger structure of the chemical bonds with the network of the aluminosilicate gel being formed. The microstructure also substantially decreases porosity and increases the UCS because the particle linkages through geopolymerization are strong and restrict microcracking. Lastly, Figure 14d presenting the geopolymer-treated soil with CNSA, has the least number of pores and evenly disseminated reaction products due to favorable compatibility between geopolymer and CNSA Si/Al phases. This structure provides the highest UCS and geometries that suggest enhanced mechanical performance and thermal insulation properties. The microstructural differences have been further amplified; a clear relation is evident between these changes and the mechanical enhancements, thus confirming the effectiveness of CNSA and geopolymer for the stabilization of soils.

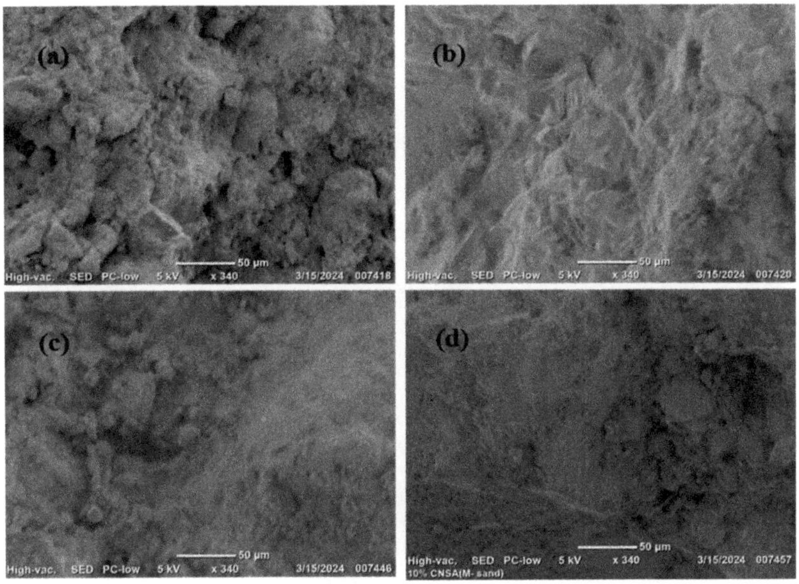

Figure 14. FESEM Analysis of Stabilized Soil (**a**) Plain Soil (**b**) Plain soil with 4% Lime (**c**) Plain soil with 6% Lime + 4% CNSA (**d**) Plain soil with 9 M GP + 2% CNSA (at 340×).

4. Conclusions

Based on the comprehensive experimental study, the following conclusions can be drawn:

1. For effective stabilization with CNSA, geopolymer-stabilized soils gained a significantly higher Unconfined Compressive Strength (UCS) than lime-stabilized soils. Consequently, soil treated with geopolymer with 9 M NaOH and 4% CNSA provided the highest UCS value, which is about 1900 kPa, compared to that of lime-stabilized soil, which was only 1400 kPa with 6% lime and 4% CNSA. This shows the enhanced chemistry between the geopolymer and CNSA system such as bonding and formation of new compounds.
2. Generally, soil samples that received geopolymer treatment have better proven resistance to degradation from water immersion, as evident from the approximately 18% reduction in strength for geopolymer with CNSA after the capillary soaking test as compared to an approximately 25% reduction in strength for lime-only-treated soil. Hence, there is an increase in the density of the created geopolymer matrix, and the permeability of the material is reduced due to CNSA pozzolanic reactions.

3. The thermal stability of the geopolymer-stabilized soils is further evident from the Loss of Ignition (LOI) results wherein the LOI values have decreased with temperature 0.17% at 900 °C for geopolymer + CNSA for geopolymer-stabilized soils, while the lime-treated CNSA-stabilized soil samples have much higher LOI values of 0.25% at 600 °C. The higher thermal resistances of the geopolymer-CNSA-treated soils can be explained by the progressive consolidation of the aluminosilicate framework of the geopolymer and the increased reactivity of CNSA.
4. FESEM analysis established that the soils treated with geopolymer-CNSA had a higher density and uniformity, free from pores, and a uniform distribution of reaction products. This was different from the lime-stabilized soils, which have shown porous and less compact structures. In addition, the improvement of the microstructure of geopolymer-treated soils explains the increased mechanical and thermal performance.
5. The incorporation of CNSA and geopolymer has led to a decrease in the value of the plasticity index and liquid limit and an increase in that of the plastic limit, which makes the soil more suitable for construction as its volume does not expand or shrink severely when it is being compacted. This characteristic is useful for engineering applications where the dimensional stability of the material is desired.

Therefore, CNSA, when used as a pozzolanic material alongside a geopolymer shows a more environmentally friendly and sustainable method of operation than the regular lime and cement stabilizers. This also reduces carbon emissions while putting industrial by-products into proper use. Therefore, the integration of a geopolymer and CNSA in the stabilization of soils is an effective and sustainable solution that can provide additional strength, durability, and thermal stability and is eco-friendly when compared with lime treatment. This study reveals the possibility of employing CNSA-based geopolymers in the civil engineering and construction industries.

Author Contributions: Conceptualization, P.D. and D.S.V.; methodology, P.D. and D.S.V.; validation, S.S.S., A.S. and T.C.; formal analysis, P.D. and S.S.S.; investigation, P.D.; resources, N.S.; data curation, P.D., A.S. and Z.M.; writing—original draft preparation, P.D.; writing—review and editing, P.D., S.S.S., N.S. and Z.M.; visualization, P.D. and Z.M.; supervision, D.S.V., S.S.S. and T.C.; project administration, D.S.V. All authors have read and agreed to the published version of the manuscript.

Funding: This research received no external funding.

Institutional Review Board Statement: Not applicable.

Informed Consent Statement: Not applicable.

Data Availability Statement: Data are contained within the article.

Conflicts of Interest: Naveen Sankaran is employed by the company DACK Consulting Solutions Inc. The remaining authors declare that the research was conducted in the absence of any commercial or financial relationships that could be construed as a potential conflict of interest.

References

1. Safi, W.; Singh, S. Efficient & effective improvement and stabilization of clay soil with waste materials. *Mater. Today Proc.* **2022**, *51*, 947–955. [CrossRef]
2. Ta'negonbadi, B.; Noorzad, R. Stabilization of clayey soil using lignosulfonate. *Transp. Geotech.* **2017**, *12*, 45–55. [CrossRef]
3. Onyelowe, K.C.; Onyia, M.E.; Bui Van, D.; Baykara, H.; Ugwu, H.U. Pozzolanic Reaction in Clayey Soils for Stabilization Purposes: A Classical Overview of Sustainable Transport Geotechnics. *Adv. Mater. Sci. Eng.* **2021**, *2021*, 6632171. [CrossRef]
4. Hung, C. Reutilization of solid wastes to improve the hydromechanical and mechanical behaviors of soils—A state-of-the-art review. *Sustain. Environ. Res.* **2023**, *33*, 17. [CrossRef]
5. Brand, A.S.; Singhvi, P.; Fanijo, E.O.; Tutumluer, E. Stabilization of a Clayey Soil with Ladle Metallurgy Furnace Slag Fines. *Materials* **2020**, *13*, 4251. [CrossRef] [PubMed]
6. Debnath, A.; Saha, S.; Chattaraj, R. Stabilization of Clayey Soil with Marble Dust Recent Developments in Sustainable Infrastructure. *Lect. Notes Civ. Eng.* **2020**, *75*, 175–183.
7. Subbarao, R. Industrial Wastes in Soil Improvement. *ISRN Civil Eng.* **2011**, *2011*, 138149. [CrossRef]

8. Bandara, N.; Hettiarachchi, H.; Jensen, E.; Binoy, T.H. Upcycling Potential of Industrial Waste in Soil Stabilization: Use of Kiln Dust and Fly Ash to Improve Weak Pavement Subgrades Encountered in Michigan, USA. *Sustainability* **2020**, *12*, 7226. [CrossRef]
9. Osinubi, K.; Nwaiwu, C. Compaction Delay Effects on Properties of Lime-Treated Soil. *J. Mater. Civ. Eng.* **2006**, *18*, 250–258. [CrossRef]
10. Anupam, A.K.; Kumar, P.; Ransinchung, G.D. Use of Various Agricultural and In-dustrial Waste Materials in Road Construction. *Procedia—Soc. Social. Behav. Sci.* **2013**, *104*, 264–273. [CrossRef]
11. Waciński, W.; Kuligowski, K.; Olejarczyk, M.; Zając, M.; Urbaniak, W.; Cyske, W.; Kazimierski, P.; Tylingo, R.; Mania, S.; Cenian, A. Recycling of Industrial Waste as Soil Binding Additives—Effects on Soil Mechanical and Hydraulic Properties during Its Stabilisation before Road Construction. *Materials* **2024**, *17*, 2000. [CrossRef] [PubMed]
12. Kharade, A.S.; Suryavanshi, V.V.; Gujar, B.S.; Deshmukh, R.R. Waste product "bagasse ash" from sugar industry can be used as stabilizing material for expansive soils. *Int. J. Res. Eng. Technol.* **2014**, *3*, 506–512.
13. Pierrehumbert, R. There Is No Plan B for Dealing with the Climate Crisis. *Bull. At. Sci.* **2019**, *75*, 215–221. [CrossRef]
14. Pourakbar, S.; Asadi, A.; Huat, B.B.; Fasihnikoutalab, M.H. Stabilization of clayey soil using ultrafine palm oil fuel ash (POFA) and cement. *Transp. Geotech.* **2015**, *3*, 24–35. [CrossRef]
15. Vizureanu, P.; Burduhos-Nergis, D.D. *Geopolymers: Properties and Application*, 1st ed.; CRC Press: Boca Raton, FL, USA, 2021. [CrossRef]
16. Abdullah, H.H.; Shahin, M.A. Geomechanical Properties of Clay Stabilised with Fly Ash-Based Geopolymer Subjected to Long-Term Sulfate Attack. *Int. J. Geosynth. Ground Eng.* **2023**, *9*, 75. [CrossRef]
17. Hasan, M.; Zaini, M.S.I.; Yie, L.S.; Masri, K.A.; Jaya, R.P.; Hyodo, M.; Winter, M.J. Effect of optimum utilization of silica fume and eggshell ash to the engineering properties of expansive soil. *J. Mater. Res. Technol.* **2021**, *14*, 1401–1418. [CrossRef]
18. Parthiban, D.; Vijayan, D.S.; Koda, E.; Vaverkova, M.D.; Piechowicz, K.; Osinski, P.; Van Duc, B. Role of industrial based precursors in the stabilization of weak soils with geopolymer—A review. *Case Stud. Constr. Mater.* **2022**, *16*, e00886. [CrossRef]
19. Geng, K.; Chai, J.; Qin, Y.; Xu, Z.; Cao, J.; Zhou, H.; Zhang, X. Damage evolution, brittleness and solidification mechanism of cement soil and alkali-activated slag soil. *J. Mater. Res. Technol.* **2023**, *25*, 6039–6060. [CrossRef]
20. Khadka, S.D.; Jayawickrama, P.W.; Senadheera, S. Strength and Shrink/Swell Behavior of Highly Plastic Clay Treated with Geopolymer. *Transp. Res. Rec.* **2018**, *2672*, 174–184. [CrossRef]
21. Nyirenda, J.; Zombe, K.; Kalaba, G.; Siabbamba, C.; Mukela, I. Exhaustive valorization of cashew nut shell waste as a potential bioresource material. *Sci. Rep.* **2021**, *11*, 11986. [CrossRef]
22. Nuithitikul, K.; Phromrak, R.; Saengngoen, W. Utilization of chemically treated cashew-nut shell as potential adsorbent for removal of Pb(II) ions from aqueous solution. *Sci. Rep.* **2020**, *10*, 3343. [CrossRef]
23. Oyebisi, S.; Igba, T.; Oniyide, D. Performance evaluation of cashew nutshell ash as a binder in concrete production. *Case Stud. Constr. Mater.* **2019**, *11*, e00293. [CrossRef]
24. Tantri, A.; Nayak, G.; Kamath, M.; Shenoy, A.; Shetty, K.K. Utilization of cashew nut-shell ash as a cementitious material for the development of reclaimed asphalt pavement incorporated self compacting concrete. *Constr. Build. Mater.* **2021**, *301*, 124197. [CrossRef]
25. James, J.; Roshna, R.; Santhiya, S. Cashew nut shell ash as a supplementary additive in lime stabilized expansive soil composites. *Mater. Today Proc.* **2022**, *62*, 644–649. [CrossRef]
26. Little, D. *Evaluation of Structural Properties of Lime Stabilized Soils and Aggregates: Mixture Design and Testing Protocol for Lime Stabilized Soils*; National Lime Association: Arlington, VA, USA, 2000.
27. Sargent, P. The Development of Alkali-Activated Mixtures for Soil Stabilisation. In *Handbook of Alkali-Activated Cements, Mortars and Concretes*; Elsevier: Amsterdam, The Netherlands, 2015; pp. 555–604. [CrossRef]
28. Deventer. Introduction to Geopolymers. In *Geopolymers*; Elsevier: Amsterdam, The Netherlands; pp. 1–11. [CrossRef]
29. Nagaraju, T.V. Geopolymer-stabilized soils: Influencing factors, strength development mechanism and sustainability. *J. Taibah Univ. Sci.* **2023**, *17*, 2248651. [CrossRef]
30. Abdila, S.R.; Abdullah, M.M.A.B.; Ahmad, R.; Burduhos Nergis, D.D.; Rahim, S.Z.A.; Omar, M.F.; Sandu, A.V.; Vizureanu, P.; Syafwandi. Potential of Soil Stabilization Using Ground Granulated Blast Furnace Slag (GGBFS) and Fly Ash via Geopolymerization Method: A Review. *Materials* **2022**, *15*, 375. [CrossRef]
31. Yao, J.; Qiu, H.; He, H.; Chen, X.; Hao, G. Application of a Soft Soil Stabilized by Composite Geopolymer. *J. Perform. Constr. Facil.* **2021**, *35*, 04021018. [CrossRef]
32. Wu, D.; Zhang, Z.; Chen, K.; Xia, L. Experimental Investigation and Mechanism of Fly Ash/Slag-Based Geopolymer-Stabilized Soft Soil. *Appl. Sci.* **2022**, *12*, 7438. [CrossRef]
33. Ramezani, S.J.; Toufigh, M.M.; Toufigh, V. Utilization of Glass Powder and Silica Fume in Sugarcane Bagasse Ash-Based Geopolymer for Soil Stabilization. *J. Mater. Civ. Eng.* **2023**, *35*, 04023042. [CrossRef]
34. Wassie, A. Influence of Curing Time and Initial Moisture Content on Metakaolin-Based Geopolymer-Stabilized Soft Soil. *Adv. Civ. Civil. Eng.* **2023**, *2023*, 6673716. [CrossRef]
35. Okagbue, C.O. Stabilization of Clay Using Woodash. *J. Mater. Civ. Eng.* **2007**, *19*, 14–18. [CrossRef]

36. Aydin, K.; Sivrikaya, O.; Uysal, F. Effects of curing time and freeze–thaw cycle on strength of soils with high plasticity stabilized by waste marble powder. *J. Mater. Cycles Waste Manag.* **2020**, *22*, 1459–1474. [CrossRef]
37. Al-Rkaby, A.H.; Odeh, N.A.; Sabih, A.; Odah, H. Geotechnical characterization of sus-tainable geopolymer improved soil. *J. Mech. Behav. Mater.* **2022**, *31*, 484–491. [CrossRef]

ceramics

MDPI

Article

Evolution of CO_2 Uptake Degree of Ordinary Portland Cement During Accelerated Aqueous Mineralisation

Giuseppe Ferrara [1,*], Pedro Humbert [2], Davide Garufi [2] and Paola Palmero [1]

[1] INSTM R.U. PoliTO-LINCE, Department of Applied Science and Technology, Politecnico di Torino, Corso Duca Degli Abruzzi 24, 10129 Torino, Italy; paola.palmero@polito.it

[2] Innovation Centre for Sustainable Construction, CRH, De Klencke 10, 1083 HL Amsterdam, The Netherlands; phumbert@crh.com (P.H.); dgarufi@crh.com (D.G.)

* Correspondence: giuseppe.ferrara@polito.it

Abstract: The utilisation of carbonation treatments to produce building materials is emerging as a valuable strategy to reduce CO_2 emissions in the construction sector. It is of great importance to regulate the degree of carbonation when the mineralisation process is combined with hydration, as a high CO_2 uptake may impede the development of adequate strength. A significant number of studies focus on attaining the maximum carbonation degree, with minimal attention paid to the examination of the evolution of CO_2 uptake over the initial stages of the process. In this context, the present study aims to investigate the evolution of CO_2 uptake over time during carbonation. Ordinary Portland Cement (OPC) is employed as material, with aqueous carbonation selected as the mineralisation process. This investigation encompasses a range of carbonation durations, spanning from 5 to 40 min. The analysis of the evolution of the mineral composition with time demonstrated that the rate of the carbonation reaction accelerates in the initial minutes, resulting in the conversion of all the portlandite produced during the hydration process in the initial 10 min. Quantitative analysis of the carbonation degree indicated that the CO_2 uptake at 40 min is equal to 19.1%, which is estimated to be approximately 70% of the maximum achievable value. By contributing to the understanding of the early carbonation mechanisms in aqueous conditions of OPC, this study provides valuable support for further investigation focused on the use of cement mineralisation processes to produce building materials.

Keywords: cement; carbon capture and storage; accelerated carbonation; calcium carbonate; CO_2 uptake

1. Introduction

It is estimated that approximately 40% of the planet's energy resources are attributed to the construction sector [1]. Cement is responsible for approximately 7% of total greenhouse gas emissions due to its embodied energy and its extensive use, which is driven by the ever increasing global demand for concrete production [2]. In order to comply with the European Union's climate policy, it is imperative that urgent action be taken to achieve net zero emissions in the cement industry by 2050 [3].

The utilisation of natural raw products [4–7] or recycled materials [8] as a potential solution to reduce cement use is an emerging trend. Nevertheless, the complete replacement of cement remains impractical due to its mechanical properties and extensive availability. Recently, the utilisation of Carbon Capture and Utilisation (CCU) technologies has emerged as a potential strategy for significantly reducing CO_2 emissions in the building materials sector [2]. The reduction of CO_2 is achieved by inducing accelerated mineralisation of alkaline products, which results in permanent storage of CO_2 through the formation of stable calcium (or magnesium) carbonates. In addition to cement, a range of alkaline wastes and industrial byproducts can be employed in the carbonation processes, including steel slag, cement bypass dust and construction and demolition waste [9]. The mineralisation process can be induced to produce either precast building blocks [10,11] or carbonated powder for use in a cement blend [12,13]. The aqueous carbonation method is a technique that is widely employed for the production of carbonated powder. The process entails the injection of CO_2 gas into an aqueous suspension comprising alkaline powder, with a liquid-to-solid ratio between 3 and 10. The process occurs by stirring the suspension until the carbonation is completed [14]. Subsequently, the slurry is subjected to centrifugation and drying, thereby obtaining the carbonated powder. Despite the promising results in terms of the carbonation degree, concerns have been raised regarding the mechanical performance of these powders when used as supplementary cementitious material [15,16]. An alternative approach involves carbonating fresh mixtures immediately prior to casting [17,18]. Cement powder, which is highly alkaline and rich in portlandite when hydrated, is an appropriate feedstock for aqueous carbonation. Monkman et al. (2017) proposed a methodology wherein CO_2 is injected during the mixing of fresh concrete, thereby facilitating the formation of nucleation sites for the development of hydrated calcium silicates [18]. This process results in enhanced mechanical strength of the concrete. Nevertheless, while elevated levels of CO_2 injection result in enhanced carbonation, exceeding a specific threshold can reverse this beneficial effect. An excess of calcium carbonate impedes the correct hydration process, which ultimately results in a reduction in the mechanical strength of the material [17]. These findings emphasise the importance of achieving a specific carbonation degree, which emerges as a crucial factor in defining the properties of the end product. This aspect is also emphasised by Matschei et al. (2007), who investigated the role of calcium carbonate in cement hydration. Their findings indicate that calcite in cement paste added as limestone acts both as an inert filler and an active participant in the hydration process [19]. In particular, it is assumed that all the calcite is reactive with cement up to a 5% content.

In case calcium carbonate is introduced in the form of limestone or pure $CaCO_3$, it is relatively straightforward to regulate its quantity within a given mixture. Conversely, when the presence of carbonates is derived from an accelerated carbonation process, it is challenging to obtain the desired amount, as it depends on the parameters of the mineralisation process. Specifically, once the water content of the mixture and the CO_2 flow have been fixed, it is of the utmost importance to determine the requisite time of injection in order to achieve the desired carbonation degree. Fu et al. (2024) proposed an innovative approach in which aqueous carbonated cement slurry is mixed with dry cement and aggregates to create concrete. They demonstrated that the duration of the mineralisation process affects the carbonation degree and, consequently, the mechanical properties of the end product, confirming the importance of controlling the carbonation degree [20].

The mechanisms of OPC-enforced carbonation have been extensively studied in the literature with respect to the kinetics of the clinker reaction and phase analysis of carbonation products [21]. Nevertheless, the investigation of the evolution of the CO_2 uptake evolution over time during OPC carbonation, evaluated through established testing procedures documented in the literature [22–24] and currently under investigation by an international scientific committee [25], remains a gap in the field. Indeed, the majority of studies examining the carbon capture of cement-based materials in aqueous conditions utilise pulverised hydrated cement paste, either created in a laboratory setting or derived from concrete demolition waste [26–30]. The use of raw cement powder is less prevalent.

In order to address this knowledge gap, the present study aims to investigate the evolution of the carbonation degree during the aqueous mineralisation of OPC powder, adopting a similar approach to that used for other cement-based materials. This implies the rigorous quantification of the stored CO_2, which is a pivotal element in the formulation of a comprehensive cost-benefit analysis with a view to achieving zero net CO_2 emissions in the cement value chain [31,32]. To this purpose, a number of different timescales for the carbonation process were considered. The objective of the investigation of mineral composition is twofold: firstly, to identify the compounds involved in the process and secondly, to qualitatively observe the development of calcium carbonate with mineralisation time. Furthermore, quantitative analysis, such as thermogravimetric and thermal decomposition in a muffle furnace, was employed to provide quantitative information concerning carbonate content. The CO_2 uptake and carbonation efficiency were evaluated in accordance with established equations from the literature [22–25]. Ultimately, to provide a comprehensive analysis of the efficiency of cement powder aqueous carbonation, the results are compared with those of analogous studies that have employed cement-based materials in aqueous carbonation processes.

2. Materials and Methods

2.1. Cement Characterisation

The raw material selected for use was Ordinary Portland Cement (CEM I 52.5 N). The as-received powder was subjected to physical and chemical characterisation. The particle size distribution was assessed by means of laser granulometry (Malvern Mastersizer 3000E laser diffraction particle size analyser, Worcestershire, UK), and the results are presented in Figure 1, which shows the frequency and cumulative volume distribution. The percentile values, d90, d50, and d10, assume values of 37.00 μm, 11.70 μm and 1.32 μm, respectively. An X-ray fluorescence analysis (XRF, Rigaku Supermini 2000, Tokyo, Japan) was conducted to evaluate the elemental composition in terms of oxides (Table 1).

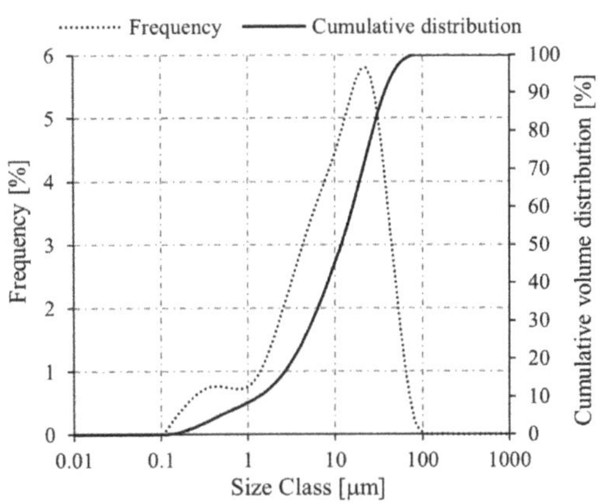

Figure 1. Particle size distribution of the as-received cement powder.

Table 1. XRF chemical composition of cement raw powder (wt.%).

CaO	SiO_2	SO_3	Al_2O_3	Fe_2O_3	MgO	K_2O	Na_2O	TiO_2	P_2O_5	SrO	MnO	LOI
64.5	17.9	4.3	4.2	3.5	1.3	0.9	0.4	0.4	0.1	0.1	0.1	2.1

The theoretical CO_2 uptake potential was evaluated based on the XRF chemical composition through the application of the modified Steinour equation [22], with an assumed value of 48.5%.

The chemical composition of the material was investigated by means of thermogravimetric differential thermal analysis (TG-DTA, LabSys evo machine, Setaram, Caluire-et-Cuire, France), carried out up to 1050 °C with a heating rate of 10 °C/min, with N_2 as carrier gas and on samples with a mass of 50 mg. X-ray diffraction analysis (XRD, Malvern Panalytical Empyrean, Worcestershire, United Kingdom) was carried out in a Bragg–Brentano configuration. The Cu-Kα wavelength radiation (0.15406 nm) was operated at 40 kV and 40 mA, with the 2θ spanning from 5° to 70° and an angular step of 0.006° held for 23 s. The diffractometer was equipped with Soller slits (0.04 rad), an anti-scatter slit (P7.5), a beam mask (10 mm), a divergence slit (¼°) and an axial slit (½°). XRD analysis was conducted to identify the crystalline phases present. Field emission

Scanning Electron Microscopy (FE-SEM, Supra 40 Zeiss Merlin, Oberkochen, Germany) was employed to analyse the microstructure of the materials after the carbonation.

2.2. Aqueous CO_2 Mineralisation

The aqueous carbonation was conducted by stirring a suspension comprising cement powder while injecting CO_2 as a gas (Figure 2). The suspension, with a liquid-to-solid ratio of 3, comprised 240 g of demineralised water and 80 g of cement. The carbonation process was conducted under room pressure and temperature conditions, with a stirring rate of 1500 rpm. The CO_2 gas was provided by a cylinder with a flow of 200 L/h, with a purity of 99.9%. The high flow rate is essential to ensure the adequate dissolution of CO_2 in the suspension under open laboratory conditions, which allows for the significant release of the injected gas into the atmosphere. Potential industrial applications of the mineralisation process may include the recirculation of gas to prevent dispersion in the atmosphere and facilitate the utilisation of reduced CO_2 concentrations and/or flow rates.

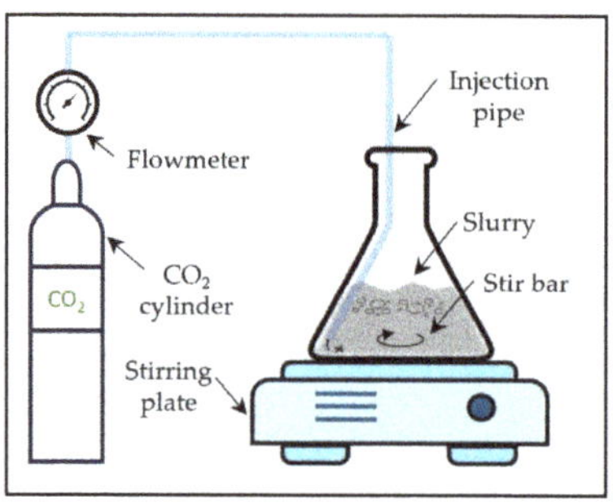

Figure 2. Setup of aqueous mineral carbonation process.

The duration of the carbonation process was varied, with experiments conducted for 5 min, 10 min, 20 min, 30 min and 40 min. Once the mineralisation process was complete, the carbonated cement powder was separated from the supernatant by means of centrifugation, followed by drying at 60 °C for 24 h. The efficiency of the mineralisation process was then assessed by quantifying the $CO_{2,uptake}$ of the processed powder. Two methods were employed to quantify the CO_2 content within the material: thermogravimetric analysis and thermal decomposition in a muffle furnace [23]. In accordance with the findings of several studies in this field [33–35], the release of CO_2 due to the decomposition of calcium carbonate decomposition is assumed to occur between 550 °C and 850 °C. The percentage of CO_2 content is calculated as the mass loss related to this temperature range, expressed as a percentage of the mass at 25 °C.

In the case of the thermogravimetric method, the mass values are obtained from the curves of the TG-DTA analysis. One sample, with an initial mass of 50 mg, was analysed for each carbonation test. The thermal decomposition in the muffle furnace consisted of a two-stage process: first, the sample was heated to 550 °C, and the weight was recorded; then, it was heated to 850 °C, and the weight was recorded. The percentage of CO_2 content was calculated as the difference between the two recorded weights, expressed as a percentage of the initial weight at 25 °C.

The initial mass of the samples subjected to thermal decomposition in a muffle furnace was approximately 2.5 g. Three replicates were conducted for each carbonation test, and the percentage of CO_2 content was assumed to be the average of the three values. The percentage of $CO_{2,uptake}$ was assessed using Equation (1):

$$\%CO_{2,uptake} = \frac{\%CO_{2,content} - \%CO_{2,initial}}{1 - \%CO_{2,content}} \quad (1)$$

XRD analysis was conducted on samples of the processed powders to examine the alteration in the mineral composition of the cement under investigation over the course of the mineralisation process and to identify the compounds involved in the carbonation reactions.

3. Results and Discussion

3.1. Mineral Composition

Figure 3 presents the XRD patterns of the powders subjected to varying carbonation times. For comparison, the XRD pattern of the raw cement powder (Cem as-received) is also included as a reference.

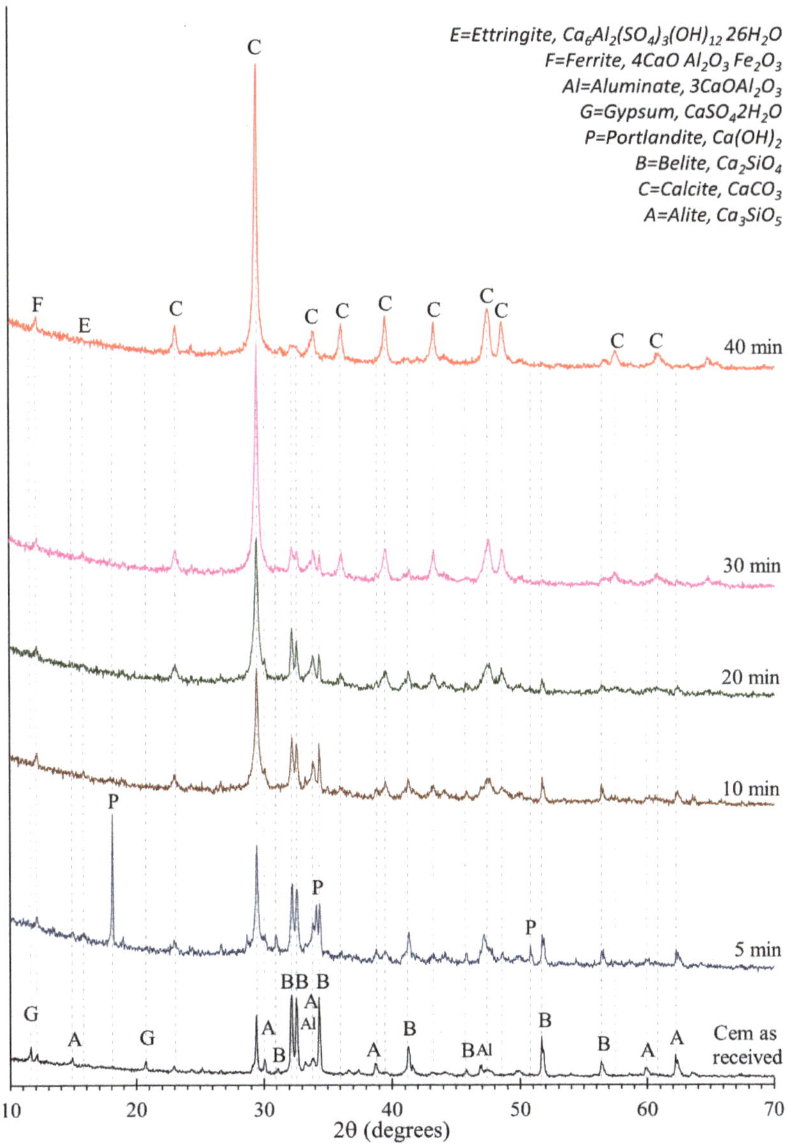

Figure 3. XRD pattern of as-received cement powder and powders at different times of carbonation.

The principal compounds identified in the raw material were calcium silicates, namely belite (C_2S) and alite (C_3S), calcium aluminates (C_3A), calcium ferroaluminates (C_4AF), and calcium sulphate (gypsum). Additionally, the presence of calcium carbonate was indicated by the identification of calcite peaks. At 5 min, the presence of portlandite is clearly identified, as evidenced by peaks at 2θ equal to 18.1° and 63.4°. The presence of $Ca(OH)_2$ is attributable to the commencement of the hydration process of calcium silicates, which results in the formation of calcium hydroxide.

Furthermore, the principal calcium carbonate peak, in the crystalline form of calcite, at 2θ equal to 29.4° is also markedly enhanced in comparison to the raw cement. This indicates that the carbonation process commences prior to the five minute mark. This finding is consistent with the literature, which indicates that the carbonation reaction of clinker in aqueous conditions occurs rapidly, with a reaction degree of alite phases equal to 15% after 0.5 min of carbonation [21]. However, the gypsum phase was not identified, while a slight signal attributed to the ettringite phase was observed. This is in line with expectations based on the reaction between calcium aluminates and calcium sulphate, which leads to ettringite formation.

The presence of other compounds already observed in the raw materials, alite and belite, can be still observed, indicating that the hydration process is not yet complete. Following a 10 min carbonation period, the portlandite peaks are no longer discernible, while the calcite peaks exhibit a notable intensification. It is also noteworthy that the alite signals exhibit a notable decline in intensity from the 10 min mark onward. Additionally, a reduction in the intensity of the belite phase can be observed as well, with a more pronounced decrease in the patterns starting from 30 min. These observations are consistent with the existing literature on the carbonation of cement, which associates this process with the formation of portlandite derived from the hydration of calcium silicates [36]. C_3S exhibits a rapid reaction, with a degree of reaction of approximately 80% after 10 min. C_2S, which commences the reaction process after 5 min, reaches a reaction degree of approximately 70% at 20 min [21].

Indeed, during the initial stages of the process, the dissolution of alite and precipitation of calcium carbonate result in a considerable consumption of CO_2, which is rapidly consumed at a rate that exceeds its solubility in the gas phase. Consequently, during this phase, the dissolution of CO_2 represents the limiting factor influencing the precipitation of calcium carbonate [21]. This explains the persistence of portlandite up to 5 min. Conversely, at longer carbonation times, the overall carbonation process is constrained by the dissolution of calcium silicates, as evidenced by the reduction of the intensity of their signal in the XRD spectrum and by the disappearance of portlandite.

Figure 4 illustrates the TG and DTA curves of the raw and mineralised cement at each specified carbonation time point. Regarding the raw powder, three main principal peaks in the DTA curve, corresponding to mass losses in the TG curve, were identified.

Peak I, occurring at approximately 140 °C, is associated with the dehydration of gypsum. Peak II, occurring at approximately 450 °C, is related to the dehydration of portlandite. Peak III, occurring at approximately 750 °C, is related to the decomposition of calcium carbonate. The mass loss associated with peak II is not substantial, indicating a relatively minor presence of portlandite. This conclusion is supported by the absence of a clear signal in the XRD pattern of the raw cement. The mass loss related to peak III, which is equal to 1.6%, represents the initial percentage of CO_2 content present in the raw material, $\%CO_{2,initial}$.

In accordance with the findings of the XRD analysis, the presence of portlandite (peak II) was identified at the 5 min carbonation stage, with a mass loss of 0.9% assessed via the tangent method [37]. Peak II is still discernible in the DTA curve at 10 min; however, no corresponding mass loss is evident in the TG curve, indicating a negligible amount of portlandite. No signals indicative of the presence of portlandite could be discerned at longer periods of carbonation.

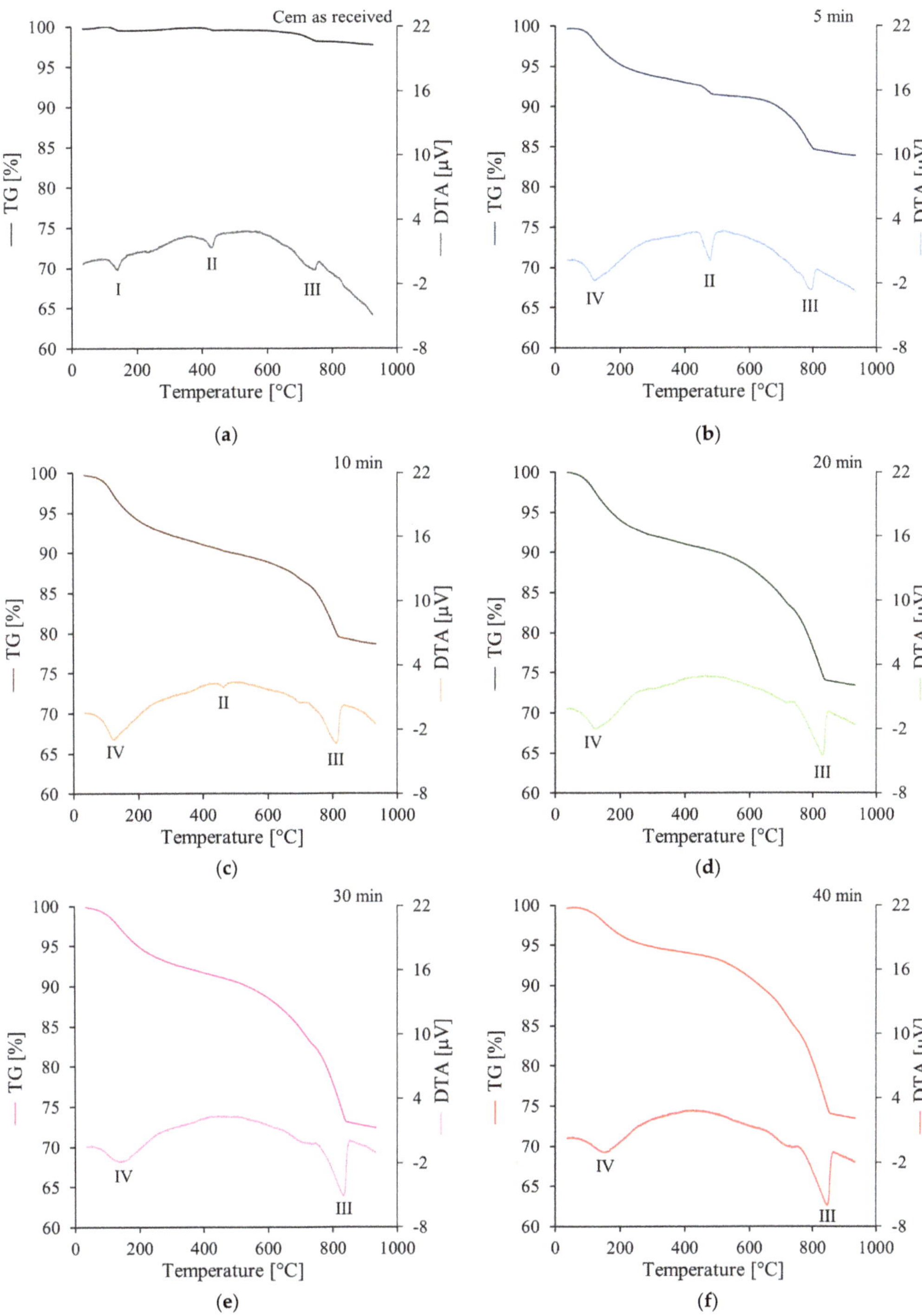

Figure 4. TG and DTA curves of (**a**) as-received cement powder and powders at times of carbonation of (**b**) 5 min; (**c**) 10 min; (**d**) 20 min; (**e**) 30 min; (**f**) 40 min.

The DTA curves of the carbonated samples display a peak between approximately 100 °C and 200 °C, which is indicative of the release of physically bonded water from the C-S-H gel. This confirms the occurrence of the silicate hydration process. The mass loss associated with peak III, resulting from the decomposition of calcium carbonate, is observed to increase with the process duration, indicating a higher degree of carbonation.

The analysis of mineral composition emphasises that in aqueous conditions, the process of hydration and carbonation occur concurrently, with carbonates formed through the conversion of portlandite derived from hydration. The presence of portlandite at a carbonation time of 5 min indicates that the dissolution of calcium silicate and the precipitation of calcite processes commence with a kinetic rate exceeding that of CO_2 dissolution, which does not permit the complete conversion of portlandite. At the 10 min mark, the absence of portlandite and the near total depletion of alite (as shown by a reduction in peak intensity), which reacts at a considerably faster rate than belite, may indicate that the limiting factor is the dissolution of silicates. Figure 5 depicts FE-SEM images of samples that have been carbonated for either 5 min or 40 min.

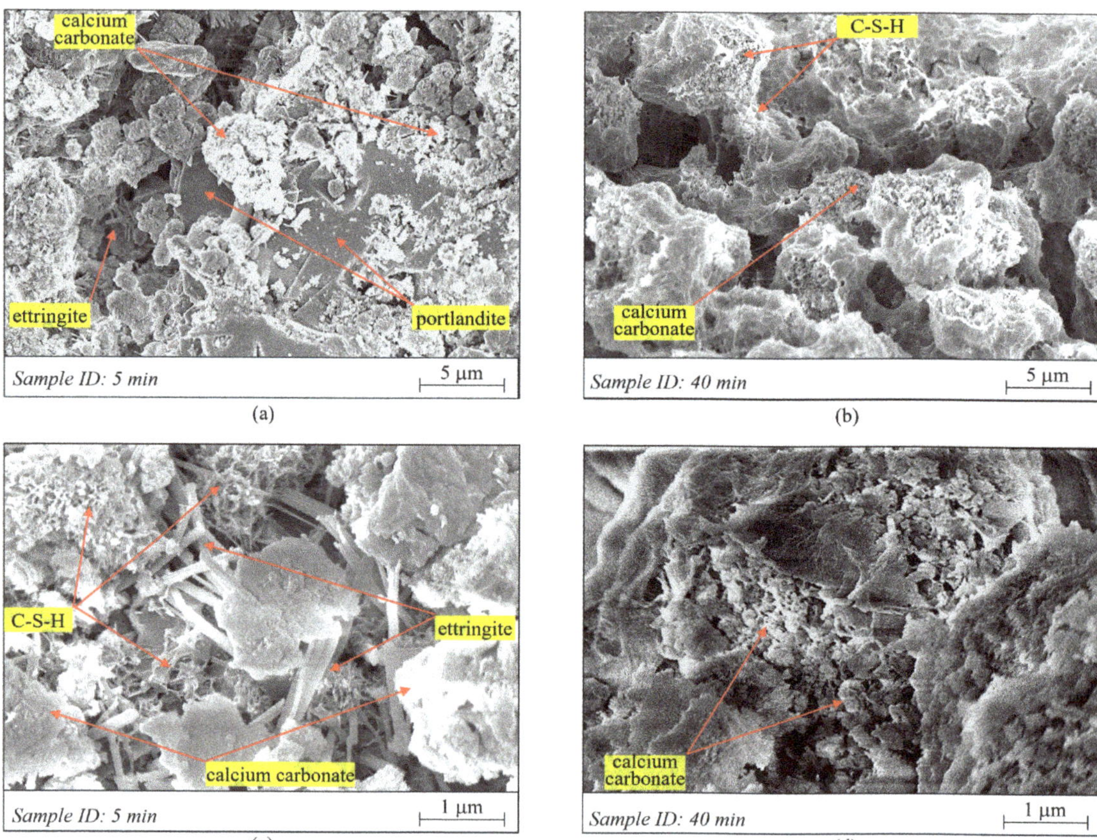

Figure 5. FE-SEM images of carbonated samples at (**a**) 5 min (Mag. 10,000×); (**b**) 40 min (Mag. 10,000×); (**c**) 5 min (Mag. 50,000×); (**d**) 40 min (Mag. 50,000×).

At 10,000× magnification, the sample carbonated for 5 min displays plate-like structures with hexagonal geometry, indicative of portlandite (Figure 5a). The elongated crystals, which are also discernible at 50,000× magnification (Figure 5c), indicate the presence of ettringite. Additionally, denser structures comprising grains that cover the portlandite crystals and needle-like structures can be observed. While the images do not permit precise identification of these formations, their structures may be associated with the presence of calcium carbonate (presumably calcite) and C-S-H gel [38]. This finding is corroborated by the results of the XRD and TG-DTA analyses. Furthermore, the presence of a needle-like structure indicative of hydrated calcium silicates at an early age is consistent with the characteristics of C-S-H gel.

Similarly, images of samples subjected to 40 min treatment demonstrate the presence of needle-like formations, which can be attributed to the presence of C-S-H gel (Figure 5b). Moreover, the presence of regular grains forming compact structures that extensively cover the surface of the sample and that may be associated with the presence of calcium carbonate (Figure 5d). The presence of portlandite is detected in the 40 min sample. The compounds identified via FE-SEM analysis align with the results from XRD and TG-DTA, indicating the presence of calcium hydroxide that has not yet undergone carbonation at the 5 min mark and a higher presence of calcium carbonate at the 40 min mark, where any portlandite is observed.

3.2. CO_2 Uptake

Three samples were subjected to thermal decomposition in a muffle furnace for each carbonation time. The CO_2 content, assessed by those decompositions for each test, along with the standard deviation, is presented in Table 2. The mean values derived from thermal decomposition in the muffle furnace, together with the values obtained from TG-DTA, are illustrated in Figure 6.

Table 2. $CO_{2,content}$ percentage (%) from Thermal Decomposition in a muffle furnace of as-received cement and carbonated samples; three replicates (I, II and III) and Standard Deviation (St. Dev.) for each carbonation test.

	$CO_{2,content}$ [%]			St. Dev. [%]
	I	II	III	
Cem as-received	1.6	1.6	1.6	0.01
5 min	7.8	7.9	7.9	0.04
10 min	10.2	10.1	10.1	0.04
20 min	13.5	13.6	13.6	0.08
30 min	16.0	16.1	16.2	0.08
40 min	16.8	16.6	16.6	0.14

Table 2 demonstrates that comparable outcomes are achieved with the two distinct methods. In greater detail, the measurements obtained via TG-DTA were found to be slightly higher than those obtained via thermal decomposition for times of carbonation exceeding 10 min. One possible explanation for this discrepancy is the difference in sample size between the two methods. The TG-DTA method was employed on a sample of approximately 50 mg, whereas the thermal decomposition method utilised samples of greater size, including three replicates of approximately 2.5 g each, which is more representative of the material. Alternatively, the lower value may be attributed to inadequate temperature control of the muffle furnace, with an actual temperature exceeding the set value of 550 °C, leading to an underestimation of CO_2 content.

Conversely, at a carbonation time of 5 min, the CO_2 content assessed via thermal decomposition in a muffle furnace was found to be higher than that estimated by TG-DTA. This discrepancy may be attributed to the operational parameters of the thermal decomposition method and the potential presence of portlandite. Following the initial heating phase at 550 °C, the sample is cooled, and the weight is recorded. It is then heated again up to 850 °C. During the interval between the two steps, portlandite may undergo partial carbonation due to the presence of CO_2 in the atmosphere, resulting in an overestimation.

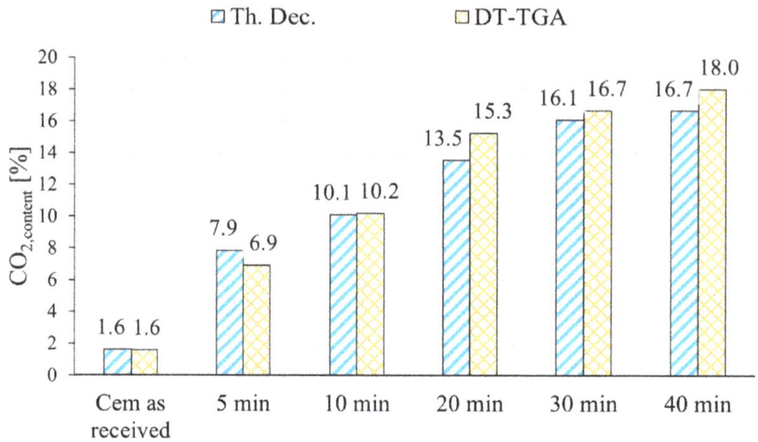

Figure 6. $CO_{2\,content}$ of as-received cement and carbonated samples; average values from Thermal Decomposition in a muffle furnace (Th. Dec.) and estimates from the Thermogravimetric analysis (DT-TGA).

Table 3 presents the percentage of $CO_{2,uptake}$ values assessed by means of Equation (1), considering the average values derived from thermal decomposition measurements and the values from TG-DTA.

Table 3. $CO_{2,uptake}$ percentage (%) from thermal decomposition in a muffle furnace (average of the three replicates) and from TG-DTA of carbonated samples.

	Th. Dec.	**TG-DTA**
	$CO_{2,uptake}$ [%]	**$CO_{2,uptake}$** [%]
5 min	6.8	5.7
10 min	9.5	9.6
20 min	13.8	16.1
30 min	17.3	18.1
40 min	18.1	20.1

The results presented in Table 3 are illustrated in Figure 7, which depicts the equation describing the CO_2 uptake as a function of time during carbonation. The chart displays the curve that approximates the experimental data with a standard error of the estimate, S, equal to 0.37 %. Additionally, the equation of the trend line is provided for reference.

As anticipated, the CO_2 uptake increases with the duration of carbonation. In greater detail, the slope of the curve decreases with time and tends toward an asymptotic behaviour, indicating that the kinetics of the carbonation process slow down over time. Indeed, during the initial minutes of the process, the rate is limited by the dissolution of the CO_2 in water. However, after approximately 10 min, the dissolution of calcium silicates becomes the limiting factor. The progressive conversion of the compounds to be carbonated results in a reduction in their availability within the suspension, which in turn gives rise to the observed asymptotic trend.

In accordance with the projected trajectory, the CO_2 uptake observed at the 40 min mark is estimated to be approximately 70 % of the maximum attainable value (asymptote). The high degree of carbonation observed after 40 min is in accordance with the findings of previous studies investigating the evolution of pH during carbonation. Initially, a notable increase in pH is caused by the dissolution of calcium silicates. However, after approximately ten min, there is a significant decline in pH as the degree of carbonation increases, reaching a nearly constant value after about 40 min [21]. Nevertheless, additional experimental tests at extended durations are necessary to substantiate the hypothesis, as other variables may influence the carbonation process.

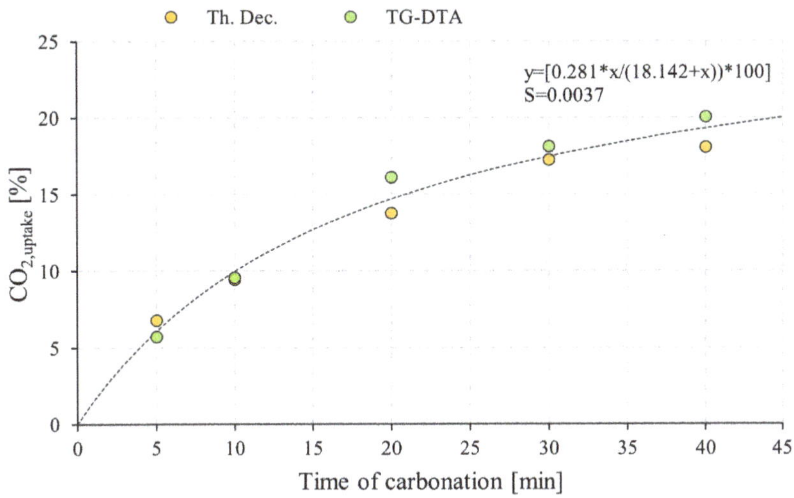

Figure 7. $CO_{2\ uptake}$ versus time of carbonation; values from Th. Dec. and TG-DTA; trendline with equation and standard error of the estimate (S).

According to the theoretical CO_2 uptake evaluated via the Steinour equation, the carbonation efficiency (ratio between experimental and theoretical values) reaches a value of 39.3% at 40 min.

4. Comparison with Other Cement-Based Materials Carbonated in Aqueous Conditions

The present study compared the carbonation efficiency of cement powder subjected to aqueous mineralisation with the findings of previous research investigating the accelerated carbonation of cementitious powders via the same process. The objective of the comparison is twofold: firstly, to verify that the results align with those of similar studies, and secondly, to identify the potential strategies for improving the efficiency of the process.

The comparison studies either involve cement powder as the raw material [20] or pulverised hydrated cementitious paste created in the laboratory [26,29,30] or from construction and demolition waste [27,28].

Table 4 presents a summary of the material types, operational parameters (liquid-to-solid ratio, CO_2 concentration and flow, pressure, temperature and time of carbonation), and the CO_2 uptake for all the considered studies. It should be noted that studies reporting pressure values higher than those in ambient conditions adopted closed systems in which the carbonation process occurred in an autoclave apparatus.

In a series of experiments conducted by Fu et al. (2024), cement powder was carbonated under similar operational conditions to those employed in the present study, including liquid-to-solid ratio, pressure and temperature. Additionally, the influence of carbonation time was investigated, extending up to 40 min [20]. Similarly, as in the present study, an increase in the degree of carbonation with time was observed. Nevertheless, the attained CO_2 uptake values were markedly lower. This may be attributed to the low CO_2 flow rate employed, which was one order of magnitude lower than that used in this study and represented a potential limiting factor of the process. A comparison of the rate at which the carbonation degree increases over time reveals that doubling the process duration from 10 to 20 min in both studies results in a CO_2 uptake approximately 1.5 times higher while increasing the time from 10 to 40 min in both studies leads to a CO_2 uptake that is approximately twice as high. This demonstrates that in both systems, the carbonation degree increased with time at comparable rates, irrespective of the disparate CO_2 flow conditions.

Teune et al. (2023) employed ambient operational conditions in a manner consistent with the present study, yet they permitted significantly longer periods of carbonation [26]. This may be attributed to the lower CO_2 flow and concentration employed. It is noteworthy that the two CO_2 uptake values, 20% and 32%, are associated with pastes produced, respectively, with cement of type CEM-III 42.5 N (70% blast furnace slag) and CEM-I 52.5 R (OPC). This result serves to highlight the enhanced reactivity of OPC in comparison to alternative types. The CO_2 uptake value attained by the CEM-I-related powder is comparable to the maximum value predicted by the trend line in Figure 7 (~28%), thereby confirming that the complete carbonation may occur at a high process duration.

Table 4. Comparison with other cement-based materials subjected to carbonation processes.

	Study	Material	Liquid-to-Solid Ratio	CO_2 Conc.	CO_2 Flow	Pressure	Temperature	Time	CO_2 Uptake
	This study	cement powder	3	100%	200 L/h	ambient	ambient	5 m 10 m 20 m 30 m 40 m	6.3% 9.5% 15.0% 17.7% 19.1%
[20]	Fu et al., 2024	cement powder	2	100%	12 L/h	ambient	ambient	10 m 20 m 40 m	3.6% 5.3% 8.2%
[26]	Teune et al., 2023	hydrated cement paste	10	15%	10 L/h	ambient	ambient	5 h ÷ 6 h	20% ÷ 32%
[27]	Ho et al., 2021	cement concrete waste	100	14%	240 L/h	-	ambient	1.5 h	4% ÷ 13%
[28]	Pasquier et al., 2018	cement concrete waste	10	18%	-	10.2 bar	ambient	15 min	7% ÷ 11%
[29]	Kong et al., 2022	hydrated cement paste *	2.5	100%	-	10 bar 20 bar 25 bar	60 °C	24 h	20.7% 20.8% 24.3%
[30]	** Hernandez-Rodriguez et al., 2021	hydrated cement paste	15	100%	-	11 bar	ambient	1 h 3 h 6 h 21 h 67 h 97 h 120 h	14% 22% 24% 23% 29% 21% 30%
		hydrated cement paste	15	100%	-	1 bar 11 bar 31 bar 51 bar	ambient	6 h	29% 28% 29% 28%

* hydrated and aged cement paste simulating cement paste from demolished concrete. ** CO_2 uptake was calculated by means of Equation (1) based on the CO_2 content data available in the article.

Ho et al. (2021) [27], who employed operational parameters analogous to those of the present study and a carbonation time with the same order of magnitude, demonstrated lower CO_2 uptake values. This can be attributed to the reduced reactivity of the waste concrete in comparison to the raw cement, as well as the higher initial CO_2 content present in the material. The results presented by Pasquier et al. (2018) [28] demonstrate that a comparable degree of carbonation can be achieved in a considerably shorter time frame and with significantly higher pressure. Significantly higher CO_2 uptake values were observed by Kong et al. (2022) [29], in which the carbonation was conducted at high pressure and temperature conditions, with a particular process including pretreatment of the powder via calcination.

Hernandez-Rodriguez et al. (2021) conducted an analysis to determine the influence of both pressure and time of carbonation [30]. Their study demonstrates that at the shorter time interval of 1 h, a CO_2 uptake value of approximately 14% is observed. In contrast, for all tests conducted over longer periods of time, the CO_2 uptake exhibits a notable increase, ranging from 21% to 30%. Additionally, it is noteworthy that when a prolonged carbonation time of 6 h is employed, the carbonation degree remains unaltered, irrespective of the high-pressure conditions.

The comparison with the literature shows that numerous studies utilising hydrated pastes have employed significantly longer periods of carbonation. Indeed, the carbonation of raw OPC powder (i.e., not pre-hydrated) is characterised by a higher rate and attains a similar carbonation degree in shorter times. This is a significant factor to be taken into account in the case of the utilisation of waste cement paste that has already undergone a prior hydration process. The data also demonstrate that high carbonation degrees are achieved when high temperature and pressure values are employed [29,30], thereby confirming the significance of these operational parameters in enhancing the efficiency of the aqueous carbonation process. In light of these findings, further investigation is required to quantify the beneficial effects associated with the increase in temperature and pressure while also considering the carbon footprint associated with the use of more severe operational parameters.

In conclusion, the comparison with the literature demonstrates that carbonation in aqueous conditions and the use of OPC powder results in CO_2 uptake values comparable to those of other studies despite the use of ambient temperature and pressure conditions. This is a pivotal consideration in the context of a cost-benefit analysis of potential applications where these processes could be deployed. The aqueous carbonation of cement powder can be employed to facilitate the formation of carbonates in fresh mixtures. In this regard, the utilisation of mild operational conditions (ambient temperature and pressure and relatively brief periods of carbonation) represents a favourable aspect that may facilitate the potential applications. Furthermore, the comparison demonstrates that the utilisation of reduced CO_2 flows, despite the retardation of carbonation reactions, still yields an effective process. This aspect should be considered in view of potential applications in which industrial flue gas may be used as a CO_2 supplier. In this regard, further investigation is required to explore the feasibility of integrating the mineralisation process with the management of CO_2-rich gases emitted by industrial plants.

5. Conclusions

This study investigates the evolution of the carbonation degree over time during the aqueous mineralisation process utilising cement powder. We present an analysis of both the carbonation degree and the mineral composition. Our principal findings are presented as follows:

- The XRD pattern analysis indicates that the intensity of peaks associated with calcium silicates diminishes as a consequence of the hydration and carbonation processes. Conversely, a significant increase in the intensity of calcium carbonate peaks with the carbonation time is observed.
- From both the TG-DTA and XRD analysis, the presence of portlandite at a carbonation time of 5 min is observed, while it is not shown from 10 min upward. This may indicate that the rate of CO_2 dissolution from the gas is the limiting factor of the carbonation process during the initial minutes. Conversely, at longer durations of the process, with C_3S (alite) almost fully reacted, and with C_2S (belite) characterised by a slower reaction rate, silicate dissolution represents the limiting factor of the process.
- The thermal decomposition in a muffle furnace and TG-DTA emerge as valuable methods for the assessment of CO_2 content in carbonated materials, with comparable results. The TG-DTA results to be more appropriate when portlandite is present in the material.
- The CO_2 uptake increases with the time of carbonation, following a trend of asymptotic behaviour. The trendline prediction curve indicates that CO_2 uptake, with a value of 19.1%, reaches approximately 70% of the maximum achievable value at 40 min.
- The comparison with the literature emphasises the attainment of significant CO_2 uptake values despite the use of mild and less energy-demanding operational parameters. Furthermore, it highlights the enhanced carbonation capacity of raw OPC in

comparison to hydrated cement pastes, including those derived from construction and demolition waste materials.

This article offers a valuable contribution to the field by defining the carbonation degree evolution of cement powder subjected to aqueous mineralisation. It provides a crucial reference point for the definition of the operational parameters of treatments in which the calcium carbonate content must be controlled. In this regard, a suitable application is represented by processes combining carbonation with the mixing process to produce fresh concrete. It is noteworthy that this study was conducted using a suspension with a liquid-to-solid ratio equal to 3. Future research should consider varying ratios in line with specific applications under investigation.

The equation predicting the evolution of the CO_2 uptake over the time of carbonation can be employed in studies aimed at optimising the mineralisation technique by balancing the emissions associated with the length of the process with the stored CO_2 to achieve an optimal balance between the two.

Finally, our study paves the way for further analysis aimed at defining the applicability of the mineralisation process to the production of building materials. In particular, the impact of carbonation degrees on the strength development should be investigated, with a specific focus on the hydration and/or pozzolanic reactions, as well as the potential nucleation effects that may occur when the processed powders are used in conjunction with traditional binders.

Author Contributions: Conceptualisation, G.F., D.G. and P.P.; Methodology, G.F. and P.H.; Validation, P.H. and D.G.; Investigation, G.F.; Resources, P.H.; Data curation, G.F.; Writing—original draft, G.F.; Writing—reviewing and editing, D.G. and P.P.; Supervision, D.G. and P.P.; Project administration, P.H. and D.G. All authors have read and agreed to the published version of the manuscript.

Funding: This research received no external funding.

Institutional Review Board Statement: Not applicable.

Informed Consent Statement: Not applicable.

Data Availability Statement: The original contributions presented in this study are included in the article; further inquiries can be directed to the corresponding author.

Acknowledgments: The authors gratefully acknowledge the Safety of Infrastructures and Constructions (SISCON) laboratory for providing the instrumentation for the thermal analysis.

Conflicts of Interest: Authors Pedro Humbert and Davide Garufi were employed by the company CRH. The remaining authors declare that the research was conducted in the absence of any commercial or financial relationships that could be construed as potential conflicts of interest.

References

1. Coppola, L.; Bellezze, T.; Belli, A.; Bianco, A.; Blasi, E.; Cappello, M.; Caputo, D.; Chougan, M.; Coffetti, D.; Coppola, B.; et al. New materials and technologies for durability and conservation of building heritage. *Materials* **2023**, *16*, 1190. [CrossRef]
2. Busch, P.; Kendall, A.; Murphy, C.W.; Miller, S.A. Literature review on policies to mitigate GHG emissions for cement and concrete. *Resour. Conserv. Recycl.* **2022**, *182*, 106278. [CrossRef]
3. CEMBUREAU. The European Cement Association 2021. In *Cementing the European Green Deal: Reaching Climate Neutrality Along the Cement and Concrete Value Chain by 2050*; CEMBUREAU: Brussels, Belgium, 2021.
4. Ferrara, G.; Martinelli, E. Tensile behaviour of Textile Reinforced Mortar composite systems with flax fibres. In Proceedings of the 12th Fib International PhD Symposium in Civil Engineering 2018, Prague, Czech Republic, 29–31 August 2018; pp. 863–869.
5. Pepe, M.; Lombardi, R.; Ferrara, G.; Agnetti, S.; Martinelli, E. Experimental Characterisation of Lime-Based Textile-Reinforced Mortar Systems Made of Either Jute or Flax Fabrics. *Materials* **2023**, *16*, 709. [CrossRef]
6. Reboul, N.; Saidi, M.; Mualla, S.; Homoro, O.; Amziane, S. Bond behaviour of Fibre Reinforced Polymers applied on masonry substrate: Analysis based on acoustic emission, digital image correlation and analytical modelling. *Constr. Build. Mater.* **2023**, *403*, 132921. [CrossRef]
7. Ferrara, G.; Michel, L.; Ferrier, E. Flexural behaviour of timber-concrete composite floor systems linearly supported at two edges. *Eng. Struct.* **2023**, *281*, 115782. [CrossRef]
8. Bolden, J.; Abu-Lebdeh, T.; Fini, E. Utilization of recycled and waste materials in various construction applications. *Am. J. Env. Sci.* **2013**, *9*, 14–24. [CrossRef]

9. Pan, S.Y.; Chang, E.E.; Chiang, P.C. CO_2 capture by accelerated carbonation of alkaline wastes: A review on its principles and applications. *AAQR* **2012**, *12*, 770–791. [CrossRef]
10. Ferrara, G.; Belli, A.; Keulen, A.; Razo, D.A.S.; Tulliani, J.; Palmero, P. Carbon sequestration in Steel Slag-based mortar: Investigation on chemical composition and mechanical properties. In Proceedings of the 2nd Fib Italy YMG Symposium on Concrete and Concrete Structures 2021, Rome, Italy, 18–19 November 2021; pp. 220–227.
11. Mahoutian, M.; Shao, Y. Production of cement-free construction blocks from industry wastes. *J. Clean. Prod.* **2016**, *137*, 1339–1346. [CrossRef]
12. Bonfante, F.; Ferrara, G.; Humbert, P.; Garufi, D.; Tulliani, J.M.C.; Palmero, P. CO_2 Sequestration Through Aqueous Carbonation of Electric Arc Furnace Slag. *JACT* **2024**, *22*, 207–218. [CrossRef]
13. Polettini, A.; Pomi, R.; Stramazzo, A. CO_2 sequestration through aqueous accelerated carbonation of BOF slag: A factorial study of parameters effects. *J. Environ. Manag.* **2016**, *167*, 185–195. [CrossRef]
14. Bonfante, F.; Humbert, P.; Tulliani, J.M.; Palmero, P.; Ferrara, G. CO_2 uptake of cement by-pass dust via direct aqueous carbonation: An experimental design for time and temperature optimisation. *Mater. Struct.* **2024**, *57*, 181. [CrossRef]
15. Ma, M.; Mehdizadeh, H.; Guo, M.Z.; Ling, T.C. Effect of direct carbonation routes of basic oxygen furnace slag (BOFS) on strength and hydration of blended cement paste. *Constr. Build. Mater.* **2021**, *304*, 124628. [CrossRef]
16. Fang, Y.; Su, W.; Zhang, Y.; Zhang, M.; Ding, X.; Wang, Q. Effect of accelerated precarbonation on hydration activity and volume stability of steel slag as a supplementary cementitious material. *J. Therm. Anal. Calorim.* **2022**, *147*, 6181–6191. [CrossRef]
17. Lippiatt, N.; Ling, T.C.; Eggermont, S. Combining hydration and carbonation of cement using super-saturated aqueous CO_2 solution. *Constr. Build. Mater.* **2019**, *229*, 116825. [CrossRef]
18. Monkman, S.; MacDonald, M.; Hooton, R.D.; Sandberg, P. Properties and durability of concrete produced using CO_2 as an accelerating admixture. *Cem. Concr. Compos.* **2016**, *74*, 218–224. [CrossRef]
19. Matschei, T.; Lothenbach, B.; Glasser, F.P. The role of calcium carbonate in cement hydration. *Cem. Concr. Res.* **2007**, *37*, 551–558. [CrossRef]
20. Fu, X.; Guerini, A.; Zampini, D.; Rotta Loria, A.F. Storing CO_2 while strengthening concrete by carbonating its cement in suspension. *Commun. Mater.* **2024**, *5*, 109. [CrossRef]
21. Zajac, M.; Lechevallier, A.; Durdzinski, P.; Bullerjahn, F.; Skibsted, J.; Haha, M.B. CO_2 mineralisation of Portland cement: Towards understanding the mechanisms of enforced carbonation. *J. CO_2 Util.* **2020**, *38*, 398–415. [CrossRef]
22. Huntzinger, D.N.; Gierke, J.S.; Kawatra, S.K.; Eisele, T.C.; Sutter, L.L. Carbon dioxide sequestration in cement kiln dust through mineral carbonation. *Environ. Sci. Technol.* **2009**, *43*, 1986–1992. [CrossRef]
23. Ferrara, G.; Belli, A.; Keulen, A.; Tulliani, J.M.; Palmero, P. Testing procedures for CO_2 uptake assessment of accelerated carbonation products: Experimental application on basic oxygen furnace steel slag samples. *Constr. Build. Mater.* **2023**, *406*, 133384. [CrossRef]
24. Librandi, P.; Nielsen, P.; Costa, G.; Snellings, R.; Quaghebeur, M.; Baciocchi, R. Mechanical and environmental properties of carbonated steel slag compacts as a function of mineralogy and CO_2 uptake. *J. CO_2 Util.* **2019**, *33*, 201–214. [CrossRef]
25. RILEM Technical Committee 309-MCP: Accelerated Mineral Carbonation for the Production of Construction Materials 2022. Chair: Snellings, R. Available online: https://www.rilem.net/groupe/309-mcp-accelerated-mineral-carbonation-for-the-production-of-construction-materials-442 (accessed on 27 October 2024).
26. Teune, I.E.; Schollbach, K.; Florea, M.V.A.; Brouwers, H.J.H. Carbonation of hydrated cement: The impact of carbonation conditions on CO_2 sequestration, phase formation, and reactivity. *J. Build. Eng.* **2023**, *79*, 107785. [CrossRef]
27. Ho, H.J.; Iizuka, A.; Shibata, E.; Tomita, H.; Takano, K.; Endo, T. Utilization of CO_2 in direct aqueous carbonation of concrete fines generated from aggregate recycling: Influences of the solid-liquid ratio and CO_2 concentration. *J. Clean. Prod.* **2021**, *312*, 127832. [CrossRef]
28. Pasquier, L.C.; Kemache, N.; Mocellin, J.; Blais, J.F.; Mercier, G. Waste concrete valorization; aggregates and mineral carbonation feedstock production. *Geosciences* **2018**, *8*, 342. [CrossRef]
29. Kong, Y.; Song, Y.; Weng, Y.; Kurumisawa, K.; Yan, D.; Zhou, X.; Wang, S.; Ruan, S. Influences of CO_2-cured cement powders on hydration of cement paste. *Greenh. Gases* **2022**, *12*, 249–262. [CrossRef]
30. Hernandez-Rodriguez, A.; Orlando, A.; Montegrossi, G.; Huet, B.; Virgili, G.; Vaselli, O. Experimental analysis on the carbonation rate of Portland cement at room temperature and CO_2 partial pressure from 1 to 51 bar. *Cem. Concr. Compos.* **2021**, *124*, 104271. [CrossRef]
31. Xi, F.; Davis, S.J.; Ciais, P.; Crawford-Brown, D.; Guan, D.; Pade, C.; Shi, T.; Syddall, M.; Lv, J.; Ji, L. Substantial global carbon uptake by cement carbonation. *Nat. Geosci.* **2016**, *9*, 880–883. [CrossRef]
32. Ashraf, W. Carbonation of cement-based materials: Challenges and opportunities. *Constr. Build. Mater.* **2016**, *120*, 558–570. [CrossRef]
33. Mo, L.; Zhang, F.; Deng, M. Mechanical performance and microstructure of the calcium carbonate binders produced by carbonating steel slag paste under CO_2 curing. *Cem. Concr. Res.* **2016**, *88*, 217–226. [CrossRef]
34. Moon, E.J.; Choi, Y.C. Development of carbon-capture binder using stainless steel argon oxygen decarburization slag activated by carbonation. *J. Clean. Prod.* **2018**, *180*, 642–654. [CrossRef]
35. Liu, G.; Schollbach, K.; van der Laan, S.; Tang, P.; Florea, M.V.; Brouwers, H.J.H. Recycling and utilization of high volume converter steel slag into CO_2 activated mortars–The role of slag particle size. *Resour. Conserv. Recycl.* **2020**, *160*, 104883. [CrossRef]

36. de Sena Costa, B.L.; de Oliveira Freitas, J.C.; Santos, P.H.S.; de Araújo Melo, D.M.; de Oliveira, Y.H. Effects of carbon dioxide in Portland cement: A relation between static sedimentation and carbonation. *Constr. Build. Mater.* **2017**, *150*, 450–458. [CrossRef]
37. Lothenbach, B.; Durdzinski, P.; De Weerdt, K. *Thermogravimetric Analysis. A Practical Guide to Microstructural Analysis of Cementitious Materials*; Crc Press: Boca Raton, FL, USA, 2016; Volume 1, pp. 177–211.
38. Zhang, D.; Wang, Y.; Ma, M.; Guo, X.; Zhao, S.; Zhang, S.; Yang, Q. Effect of equal volume replacement of fine aggregate with fly ash on carbonation resistance of concrete. *Materials* **2022**, *15*, 1550. [CrossRef]

ceramics

MDPI

Article

The Effect of Adding Banana Fibers on the Physical and Mechanical Properties of Mortar for Paving Block Applications

Ginan Al-Massri [1], Hassan Ghanem [1,*], Jamal Khatib [1,2,*], Samer El-Zahab [3] and Adel Elkordi [1,4]

1 Faculty of Engineering, Beirut Arab University, Beirut 12-5020, Lebanon; g.massri@bau.edu.lb (G.A.-M.); a.elkordi@bau.edu.lb (A.E.)
2 Faculty of Engineering, University of Wolverhampton, Wolverhampton WV1 1LY, UK
3 College of Engineering and Technology, American University of the Middle East, Egaila 54200, Kuwait; samer.el-zahab@aum.edu.kw
4 Faculty of Engineering, Alexandria University, Alexandria 5423021, Egypt
* Correspondence: h.ghanem@bau.edu.lb (H.G.); j.khatib@bau.edu.lb (J.K.)

Abstract: Paving blocks might encounter diverse environmental conditions during their lifespan. The durability of paving blocks is determined by their capacity to endure various exposure conditions. Synthetic fibers have been used in mortar and concrete to improve their properties. This research investigates the influence of including banana fiber (BF) on the physical and mechanical characteristics of mortar. Five different mortar mixes were developed, with varying amounts of BF ranging from 0 to 2% by volume. Testing included ultrasonic pulse velocity, compressive strength, flexural strength, total water absorption, and sorptivity. Specimens were cured for up to 90 days. The results indicate that using 0.5% BF resulted in an improvement in compressive and flexural strength compared to the control mix. There was an increase in total water absorption and the water absorption coefficient in the presence of fibers. There appeared to be good correlations between the compressive strength and the other properties examined.

Keywords: natural fibers; bio-fibers; mortar; strength; total water absorption; capillary water absorption

1. Introduction

The world's agricultural sector generates almost one billion tons of waste per year. Plant byproducts account for 80% of these pollutants [1]. Ecologists are concerned about these pollutants because of their negative impact on underground water sources and the ozone layer, two factors that strongly affect global warming and economic conditions [2]. Thus, part of these wastes was incorporated into cementitious-based materials used in building and paving construction. Natural fibers derived from plants are recognized as eco-friendly alternatives to synthetic fibers because of their abundance, affordability, longevity, and minimal energy consumption [3,4].

The mechanical properties of natural fibers are influenced by their chemical structure. As a result, the engineering features of cement matrices reinforced with natural fibers are improved to a level that is comparable to those of cement matrices reinforced with synthetic fibers [5]. The mechanical properties of concrete and mortar were investigated in previous studies addressing the effect of adding natural fibers to the mixture in terms of the total volume or total mass. For instance, Shah et al. [6] investigated the effect of varying the mass fractions of coconut and sisal fibers in high-strength concrete (HSC) from 0.5 to 1.5% of the cement content on the material's characteristics. At a fiber concentration of 1%, HSC's compressive strength was increased by 33% with coconut fibers and 24% with sisal fibers. According to Khan et al. [7], jute fiber boosts the mechanical properties of concrete, including its compressive strength, split tensile strength, and flexural strength. It was found that a ratio of 0.10% jute fiber is optimum for enhancing these features. The addition of coir fiber to concrete up to 0.5% by mass of mixture increased the material's strength, as reported by Prafulla et al. [8].

Regarding permeability, the addition of treated natural fibers to cement-based matrices can significantly reduce their total water absorption, contributing to the sustainability and durability of the material. As an illustration, Machaka et al. [9] reported that adding up to 1.5% of Phragmites Australis (PA) natural fibers to concrete can decrease water absorption by total immersion and capillary action by up to 45%, with a negligible decrease in concrete density and strength. This suggests that natural fibers can enhance the durability of concrete by reducing its susceptibility to water ingress [9]. Zouaoui et al. [10] showed that natural fibers such as palm stem and hemp fibers, when treated to reduce their hydrophilic nature, can be integrated into mortar to produce materials with low thermal conductivity and an excellent moisture-buffering capacity. Contradictorily, untreated natural fibers may absorb water due to their hydrophilic nature, which can be a challenge for their use in concrete. Li et al. [11] discussed the hydrophilic behavior of natural fibers, which can lead to reduced bonding between the concrete matrix and fibers, potentially affecting the overall performance of the concrete material. However, the chemical alteration of natural fibers demonstrates significant potential for enhancing their mechanical properties, boosting dimensional stability against moisture and biodeterioration, hence prolonging service life and expanding applicability. A multitude of reviews exists regarding the chemical modification of fibers using various substances. There are numerous approaches aimed at efficiently modifying the fiber's surface to render it hydrophobic. Alkaline treatment, acylation, and silylation represent the predominant chemical treatments [12,13]. Therefore, the use of treated natural fibers in concrete is a promising strategy for producing eco-friendly building materials with adequate mechanical and durability performance [9].

Worldwide, banana harvesting produces thousands of tons of agricultural waste [14]. Previous research showed that, up to a certain point, cementitious materials' mechanical characteristics can be improved by adding banana fiber (BF). As an example, Babar Ali et al. [15] investigated the impact of varying amounts of banana stem fiber BSF (0.25, 0.5, and 1% of the total mix volume) on the strength characteristics of concrete. The results showed that the compressive strength, splitting tensile strength, and flexural strength of the concrete were enhanced by 6, 40, and 10%, respectively, when incorporating BSF at 0.5% content. Logeshwar et al. [16] investigated the performance of composites made from concrete mixed with agricultural waste (bagasse ash (BA) and rice husk ash (RHA)) at different percentages of BF (0.2, 0.4, and 0.6%). Incorporating BF into concrete mixes with BA and RHA significantly enhanced their strength. When comparing several BF combinations, the one with 0.2% BF yielded the best results. Pathan et al. [17] examined the compressive and flexural strengths of high-strength concrete by varying the amount of BF from 0 to 2% of the weight of cement and observed that the mix of 1.5% BF generated the strongest results. The impact of different concentrations of BF of a 10 mm length on the mechanical properties of mortar was investigated by Lamichhane et al. [18]. Adding 0.25% BF to base mortar increased compressive strength by 18.7%, flexural strength by 29.9%, and splitting tensile strength by 41.1% after 28 days.

In terms of physical properties, the available sources do not explicitly quantify the specific effect of BF on the ultrasonic pulse velocity and total and capillary water absorption of mortar. To obtain a final verdict on this matter, additional research and experimental evidence are necessary. Considering the paucity of literature on the subject, this research sets out to fill that gap by investigating the physical and mechanical characteristics of mortar containing BF. In light of this, a sequence of experiments was carried out on mortar samples with different BF additions by total volume (0, 0.5, 1, 1.5, and 2%). Tests covered ultrasonic pulse velocity (UPV), compressive strength, flexural strength, total water absorption (TWA), and capillary water absorption (CWA). The experimental data in this study were thoroughly validated using a hyperbolic model to predict compressive strength and a capillary-diffusive model to estimate diffusion and sorptivity coefficients. These models were selected for their capacity to provide an extensive and comprehensive depiction of the underlying physical properties of the material under varying environmental conditions. This approach's innovation resides in the integration of these two models, which not only provides a more accurate prediction of mechanical properties and moisture-transport characteristics but also addresses traditional empirical approaches with advanced, physics-based simulations. This novel integration promotes a more profound comprehension of the relationship between material strength and moisture dynamics, providing insights that can substantially improve the design and efficacy of materials in practical applications. Furthermore, the correlations among several characteristics are examined in this study.

2. Materials and Methods

2.1. Materials

BF was harvested from banana pseudo-stems by means of a decortication machine in Cairo, Egypt. After slicing the BF into 1 cm pieces, it was immersed in a 6% NaOH solution for 24 h [19]. Next, BF was rinsed with clean water to neutralize the alkali, and lastly, it was sun-dried. Having been cleaned, dried, and then packed in plastic bags, the fibers were ready for use. Figure 1 represents the raw and cut-treated BF. The process of obtaining BF is shown in Figure 2. The bulk density of BF is 1350 kg/m^3. The stress–strain curve for BF is displayed in Figure 3. The curve shows the relationship between the applied stress and the resulting strain, revealing that the fibers can withstand increasing loads without breaking. The tensile strength is 53 MPa, and the elastic modulus is 3.5 GPa.

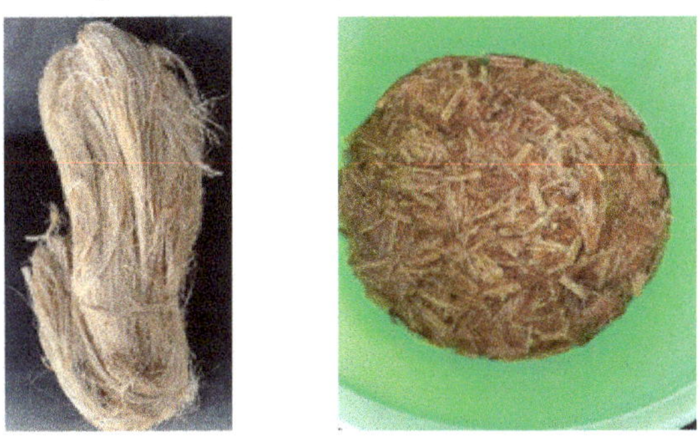

(a) (b)

Figure 1. (**a**) Raw bundled BF, (**b**) Cut-treated BF.

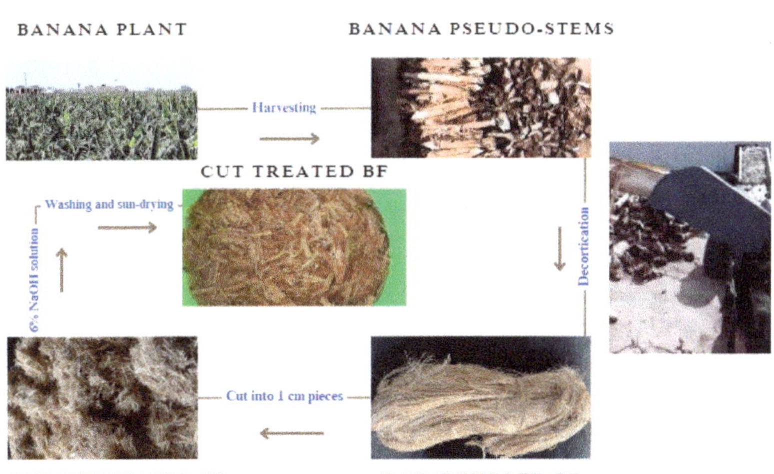

Figure 2. Preparation process of banana fiber.

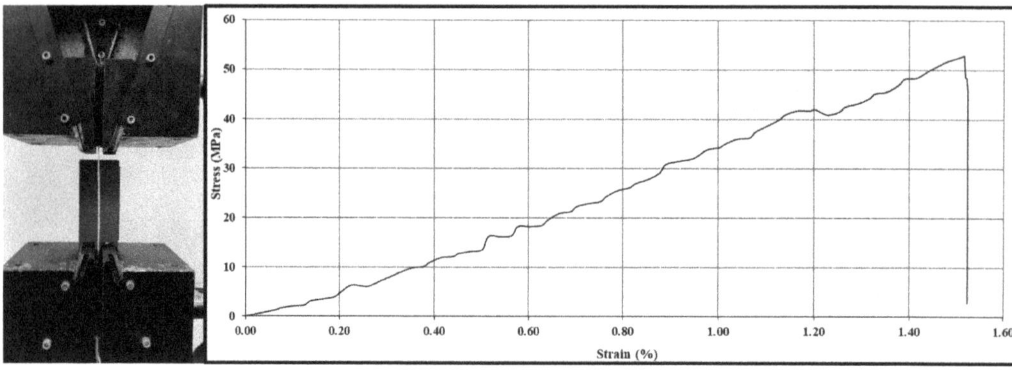

(a) (b)

Figure 3. (**a**) BF tensile strength test; (**b**) Stress–strain curve.

Type I Ordinary Portland Cement with 3.15 specific gravity and 1440 kg/m^3 density was employed. Siliceous sand with 2.8 fineness modulus, 3.5% water absorption, and specific gravity of 2.65 was used.

2.2. Mix Proportions

Five distinct percentages (0, 0.5, 1, 1.5, and 2% by volume) of BF were incorporated into the mortar mixes. For all mixes, the cement-to-water -to-sand (c:w:s) ratio remained constant at 1:0.45:2. Mix proportions are displayed in Table 1.

Table 1. Mortar mix proportions.

	Quantity (kg/m^3)						
Mortar Code	**Cement**	**Water**	**Sand**	**BF**	**BF (% by vol)**	**w/c**	**c/s**
M-0% BF	657	295.6	1314	0	0	0.45	0.5
M-0.5% BF	657	295.6	1314	6.75	0.5	0.45	0.5
M-1% BF	657	295.6	1314	13.5	1	0.45	0.5
M-1.5% BF	657	295.6	1314	20.25	1.5	0.45	0.5
M-2% BF	657	295.6	1314	27	2	0.45	0.5

2.3. Experimental Methodology

2.3.1. Mortar Sample Preparation

Initially, a dry mixture of cement, sand, and BF was prepared. Subsequently, water was introduced and agitated for a duration of two minutes. Following the mixing process, the mortar's consistency was assessed by conducting the flow test according to ASTM C 1437 [20]. The flows corresponding to M-0% BF, M-0.5% BF, M-1% BF, M-1.5% BF, and M-2% BF were 103, 96, 89, 81, and 77%, respectively. Flow table values exhibited a negative correlation with increasing BF%, as observed from the flow findings. Ultrasonic-pulse velocity and compressive strength tests were conducted on cubes with dimensions of 50 × 50 × 50 mm. Beams measuring 40 × 40 × 160 mm were used to determine flexural strength. For total water absorption and capillary water absorption, cubes measuring 100 × 100 × 100 mm were utilized.

2.3.2. Ultrasonic-Pulse Velocity (UPV)

One nondestructive method for evaluating concrete quality is UPV test. This method involves tracking the speed of an ultrasonic pulse wave as it travels through the concrete sample. The method is standardized by ASTM C597 [21]. UPV test was conducted at

1, 7, 28, and 90 days (Figure 4). Concrete quality is classified according to pulse velocity values, as presented in Table 2.

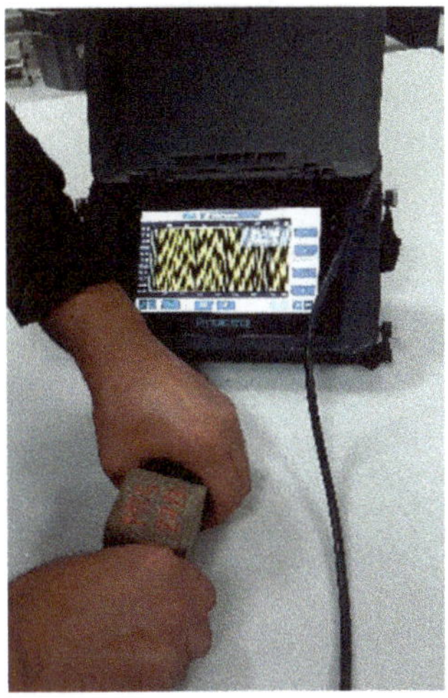

Figure 4. UPV test.

Table 2. Concrete quality classification based on UPV. Reproduced with permission from [22], Digital Commons @ BAU, 2024.

Quality of Concrete	Excellent	Good	Doubtful	Poor	Very poor
UPV (km/s)	>4.5	3.5-4.5	3.0-3.5	2.0-3.0	<2.0

2.3.3. Compressive Strength

The compressive strength test, considered as an indicator of the quality of the mortar, is affected by various parameters, including mix quantities, water-to-cement ratio, and curing methods. The ASTM C109 [23] was followed to measure the compressive strength at 1, 7, 28, and 90 days (Figure 5). The compressive strength value was taken as the average of three readings.

Figure 5. Compressive strength test.

2.3.4. Compressive Strength Modeling

Two variables, the initial rate of compressive strength (IRC) and the ultimate compressive strength (UC), are used to calculate compressive strength characteristics. As a measure of the material's early strength gain from moisture, the IRC indicates the rate of compressive strength during the initial phases of curing. On the other hand, the ultimate compressive strength (UC) defines the highest level of compressive strength, describing the long-term behavior of the material after reaching a constant moisture content. In order to examine these key variables, a hyperbolic model [24] (Figure 6) was generated as follows:

$$C = \frac{x}{\frac{1}{y} + \frac{x}{z}} \quad (1)$$

where the following apply:
C = modeled compressive strength;
x = curing period (days);
y = initial rate of compressive strength (IRC);
z = ultimate compressive strength (UC).

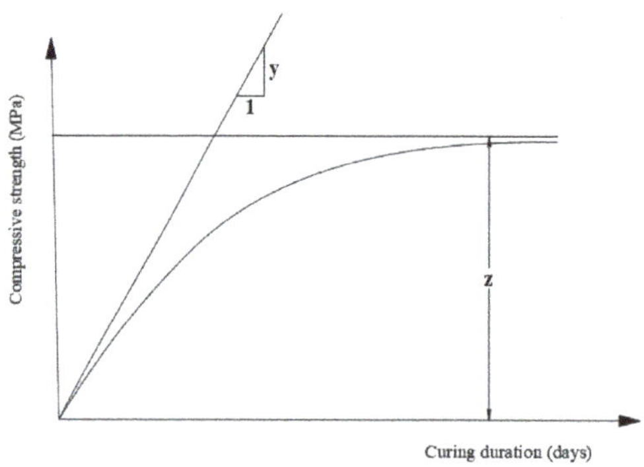

Figure 6. Compressive strength variables.

The aforementioned two variables were determined using the solver built in Microsoft Excel software version 2021.

2.3.5. Flexural Strength

The flexural strength test is an indirect test employed to quantify the tensile strength of mortar samples at days 1, 7, 28, and 90 following ASTM C348 [25] (Figure 7). The flexural strength value was the average of three samples.

Figure 7. Flexural strength test.

2.3.6. Total Water Absorption (TWA)

Mortar samples were subjected to TWA test measuring the amount of water absorbed after a particular duration according to ASTM C1585 [26] (Figure 8). After 1, 7, 28, and 90 days of curing, the mortar specimens were dried at 80 °C for 48 h until they reached a constant dry weight (A). The cubes were then submerged in water for 5 min, taken out, dried using a towel to remove surface water, and weighed again; at this point, the weight was recorded as the initial weight (B). At 10, 20, 60, 120, and 450 min intervals, this process was repeated. The percentage of TWA was calculated as follows:

$$TWA = \frac{B - A}{A} \times 100 \quad (2)$$

Figure 8. Total water absorption test.

2.3.7. Capillary Water Absorption (CWA)

After being dried at 80 °C until the sample reached a constant mass, mortar specimens were subjected to CWA test following ASTMC1585 [26]. To prevent the evaporation of water and to keep the uniaxial water flow constant during the test, a non-absorbent coating was applied to all four sides of the cubes. Both the top and bottom sides remained unsealed, and the weight of the sample was noted, W_1. As illustrated in Figure 9, the specimen was subsequently placed in a container and filled with water until it reached a depth of 5 mm, as measured from the bottom of the mortar cubes. Measurements of water absorption were taken at 3, 5, 10, 20, 30, 60, 120, and 240 min and 1, 2, and 3 days. The CWA was calculated as follows:

$$CWA = \frac{w_2 - w_1}{A \times d} \quad (3)$$

where the following apply:
$w_2 - w_1$ = cumulative water absorption (g);
A = surface area of the cube through which water penetrates (mm^2);
d = density of water (g/mm^3).

(a)

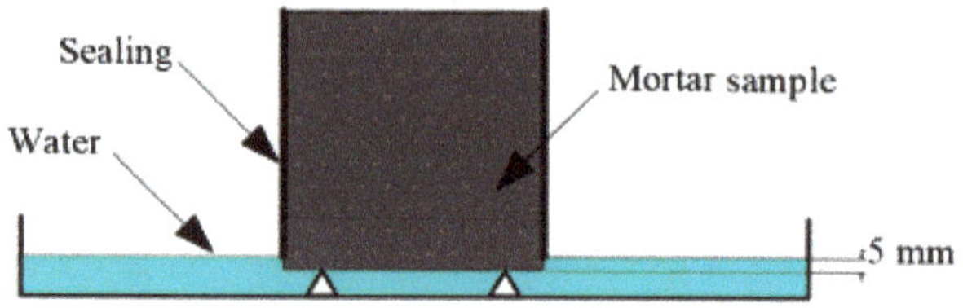

(b)

Figure 9. (**a**) CWA samples; (**b**) Experiment set-up.

2.3.8. Capillary-Diffusive Model

Water transport in mortar and concrete has been the subject of extensive research. During the first absorption period, the cumulative water absorption M/A is directly proportional to the square root of the elapsed wetting time t, when mortar or concrete surface is exposed to wetting by water. Once water has passed through all of the larger capillary pores, the initial rate of ingress shown reduces after a period of sorption. As time progresses, the significance of small pores grows, suggesting that sorptivity is now happening through the finer pores. As a result of gravitational forces, the rate of sorption is expected to decrease exponentially [27,28]. The transition from rapid to slow sorption rates is modeled using the following capillary-diffusive model [28]:

$$\frac{M}{A} = C\rho\left[1 - \exp\left[\frac{-s\sqrt{t}}{C\rho}\right]\right] + C_O L\left[1 - \frac{8}{\pi^2}\sum_{n=0}^{\infty}\frac{1}{(2n+1)^2}\exp\left[-\frac{(2n+1)^2\pi^2}{4L^2}Dt\right]\right] \quad (4)$$

where the following apply:
M/A = weight of water absorbed per unit of area (kg/m^2);
C = constant associated with the distance from the concrete surface where capillary pores regulate the initial sorption;
ρ = density of the water (kg/m^3);
S = sorptivity coefficient (kg/m^2.s$^{1/2}$);
t = time (s);
C_O = invariance of water concentration (kg/m^2);
L = depth of the sample = 0.1 m;
D = diffusion coefficient (m^2/s).

The aforementioned variables were calculated using the solver built in Microsoft Excel software.

3. Results and Discussion

3.1. Ultrasonic-Pulse Velocity (UPV)

The UPV results for the mortar mixes are shown in Figure 10. A minor reduction in the UPV was observed at all curing ages when 0.5, 1, 1.5, and 2% BF were integrated. As an illustration, the UPV at 90 days for the control mix was 3.88 Km/s. With the addition of 0.5, 1, 1.5, and 2% BF, the associated UPV values were 3.83, 3.71, 3.65, and 3.63 Km/s. The corresponding decreases in the UPV were 1.3, 4.38, 6, and 6.44%. This slight decrease may be due to the fact that fibers have the potential to generate heterogeneity and voids in the mortar's matrix. BF, like other organic fibers, alters matrix homogeneity, which may reduce internal structure continuity. This disturbance may impede ultrasonic pulse transmission, reducing the UPV. Fibers can cause local microstructure abnormalities such fiber-matrix bonding or fiber dispersion, reducing composite material density and homogeneity. These factors can slow ultrasonic waves, which are sensitive to material structural changes. The effects of increasing fiber volume on concrete homogenization and UPV have been demonstrated in earlier research using different organic fibers, such as coir fiber [29], cotton fiber, and wool fiber [30]. For example, using banana fibers in concrete reduced the UPV by 7% at a 1% fiber volume [15].

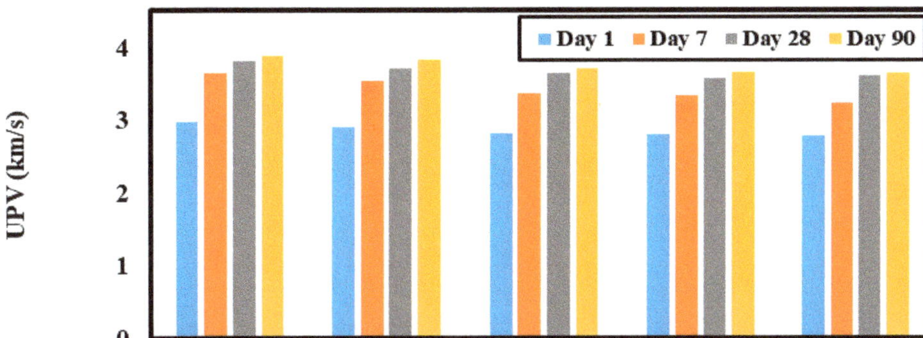

%BF	M-0% BF	M-0.5% BF	M-1% BF	M-1.5% BF	M-2% BF
Day 1	2.97	2.9	2.82	2.8	2.78
Day 7	3.64	3.54	3.36	3.32	3.23
Day 28	3.8	3.7	3.64	3.57	3.6
Day 90	3.88	3.83	3.71	3.65	3.63

Figure 10. UPV of mortar with different BF additions.

3.2. Compressive Strength

The compressive strength of several mortar mixes containing BF is presented in Figure 11. At 90 days, the results demonstrate that adding 0.5% BF resulted in a 7% improvement in compressive strength when compared to the control mix. A potential explanation is that the 1 cm fibers, when introduced to the matrix at a dosage of 0.5%, are distributed uniformly. The increased compressive strength as compared to plain mortar may be due to two factors: the alveolar structures on the surface of the fibers, which form the bond between the mortar and the fibers, and the tensile strength of the BF [31–33]. When compared to the control mix, the compressive strength is 20% lower for M-1% BF, 25% lower for M-1.5% BF, and 47% lower for M-2% BF. The primary chemical ingredient of BF, cellulose, may be responsible for this decrease. Cellulose, owing to its hydrophilic properties, can disrupt the cement-hydration process by competing for water, which is required for the hydration processes necessary for cement setting and strength development. The competition for water can diminish the quantity available for calcium silicate hydrate (C-S-H) gel formation, obstructing the hydration process and potentially compromising the mechanical qualities of the cement. Moreover, the contact between cellulose fibers and the cement matrix may result in the breaking of hydrogen bonds at the fiber–matrix interface, especially as water is absorbed during the curing process. This rupture weakens the bond between the fibers and the matrix, resulting in a decrease in the material's overall strength [34–37]. In line with this pattern is the finding of Lakawa [38], who demonstrated a 7.3% increase in compressive strength when cementitious materials were supplemented with palm fibers at dosages up to 0.5%.

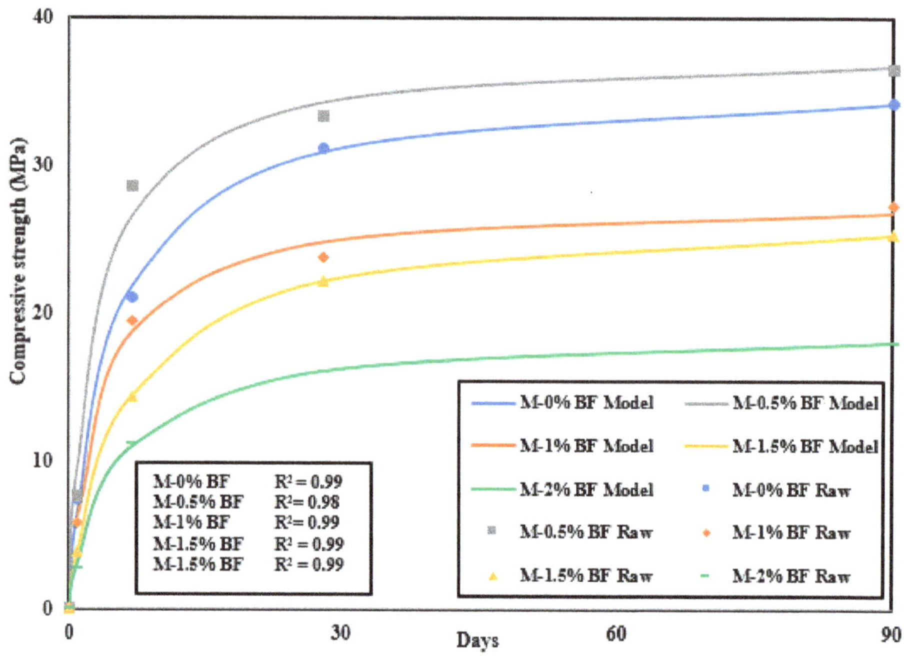

Figure 11. Modeling of the compressive strength of mortar with different BF additions.

A hyperbolic equation that attempted to simulate the mortars' compressive strength with time revealed promising outcomes, showing a significant agreement with the experimental results. R^2 values of 0.99, 0.98, 0.99, 0.99, and 0.99 for 0%, 0.5%, 1%, and 2% BF additions, respectively, indicated a high level of determination. The IRC and UC values for all mortar mixes are displayed in Figure 12. It is evident that these two variables followed an identical pattern. The mortar with 0.5% BF showed an increase in IRC and UC compared to the control mix by 53 and 6%, respectively. The increased values of IRC and UC observed for M-0.5% BF can be attributed to the presence of 0.5% BF, which promotes a uniform dispersion of fibers throughout the mortar matrix, resulting in a strong bonding structure. The regular distribution of fibers throughout the matrix enhances the efficiency of the hydration process during the initial curing stages, generating more consistent hydration rates. This consistency enhances microstructural development, leading to improved ultimate compressive strength of the composite material once a steady moisture content is attained. IRC experienced a gradual decrease of 4, 46, and 55% with 1, 1.5, and 2% BF, respectively, compared to the control mix. The decrease in the initial rate of strength can be related to the chemical consistency of cellulose, which hinders the hydration process of cement [31,34]. Similarly, UC values dropped progressively by 23, 25, and 47% in comparison to the control mix with the addition of 1, 1.5, and 2% BF. The drop in the ultimate strength may be due to the disruption of hydrogen bonds between the fibers and the matrix when the cement absorbs water during wet curing [31–34].

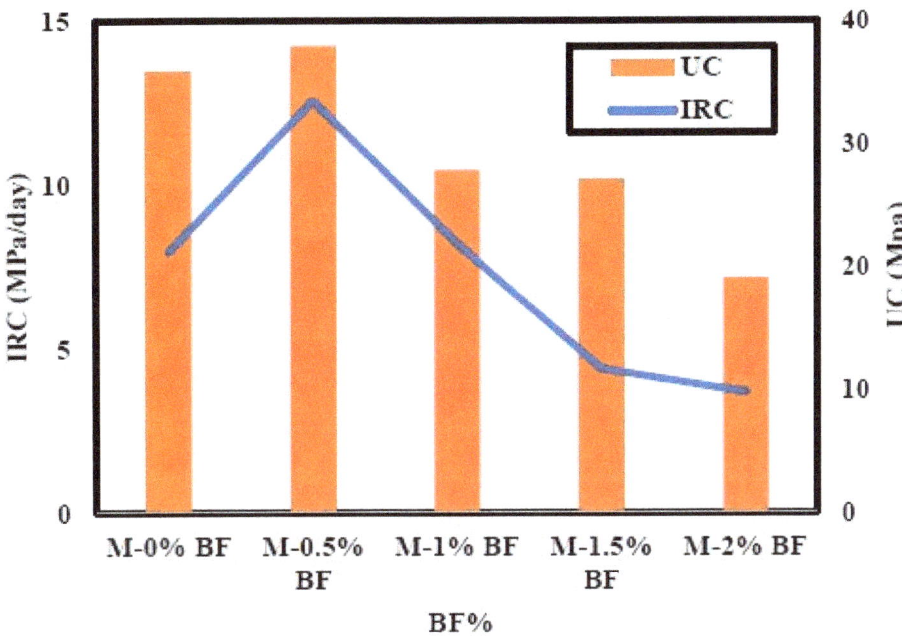

Figure 12. Compressive strength characteristics of mortar with different BF additions.

3.3. Flexural Strength

The flexural strength of different mortar mixes containing BF is shown in Figure 13. At 90 days, the results show that flexural strength was 5.1% higher with the addition of 0.5% BF compared to the control mix. Since BF is naturally resistant to cracking, it delays the beams' failure and improves their capacity to endure stress after peak loading, leading to an increase in the sample's flexural strength [39–42]. Increased fiber content may result in an unequal distribution of fibers inside the matrix, causing clumping. The non-uniform distribution of fibers into the matrix may lead to an ineffective interaction between fibers and mortar at higher fiber contents, leading to bonding issues. This may generate vulnerabilities that undermine the overall integrity of the mortar. Furthermore, the bridging effect produced by BF restricting fracture propagation [40] is responsible for this improvement. Above 0.5% BF addition, in comparison to fiber-free mortar, the flexural strength exhibited a 6.1, 14, and 18.3% drop for M-1% BF, M-1.5% BF, and M-2% BF, respectively, after 90 days. Mixing at 1, 1.5, and 2% BF causes the fibers to wrap and twist, which reduces their effectiveness in stress transfer throughout the total fiber length and, in turn, reduces flexural strength [43,44]. A further potential cause is that the porosity of the mixture is affected by an accumulation of fiber filaments caused by the balling effect [15]. In line with this finding, Lamichhane et al. [18] demonstrated that increasing the BF content increased its flexural strength up to a volumetric fiber content of 0.25%, after which strength began to decrease for higher fiber concentrations. According to another study investigating the fiber-reinforced recycled cement foundation material, the flexural strength of fiber-reinforced recycled-brick-powder cementitious composites increased initially and then declined with the concentration of polyvinyl alcohol (PVA) fiber. Fiber pull-out, fracture, and plastic degradation may be associated with the mechanism of strength drop at high fiber doses [45].

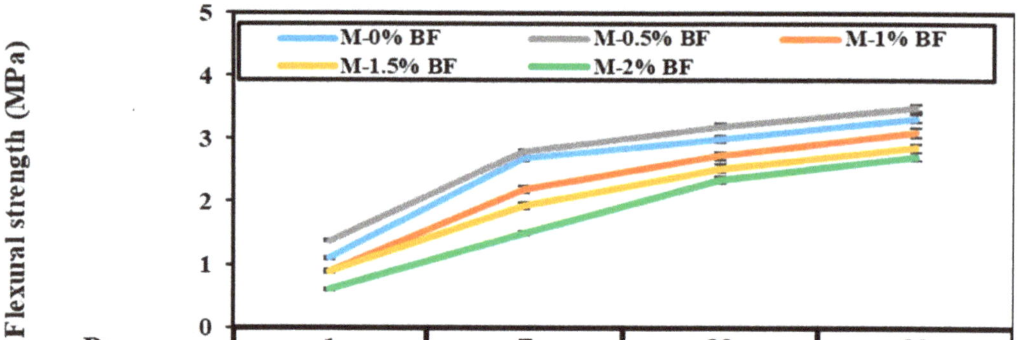

Days	1	7	28	90
M-0% BF	1.10	2.70	3.00	3.33
M-0.5% BF	1.38	2.80	3.20	3.50
M-1% BF	0.90	2.20	2.73	3.10
M-1.5% BF	0.89	1.93	2.53	2.86
M-2% BF	0.60	1.50	2.35	2.72

Figure 13. Flexural strength of mortar with different BF additions.

3.4. Total Water Absorption (TWA)

The TWA results of several mortar mixes containing BF at 1 and 28 days are shown in Figures 14a and 14b, respectively. As shown, there was an increase in TWA from 11.8% for M-0% BF to 15.6% for M-2% BF on day 1 and from 5.1% for M-0% BF to 7% for M-2% BF on day 28. The impact of curing time on the TWA for various mortar mixes is shown in Figure 15a. The % increase in TWA for different mortar mixes compared to the control mix at 1 day and 28 days is displayed in Figure 15b. The results show that as the curing duration increases, TWA drops significantly. For example, the M-1.5% BF TWA decreased from 15.5% at day 1 to 6.2% at day 28. Longer curing durations cause concrete's capillary pores to fill with water, which reduces water absorption over time [46]. Furthermore, it was noted that the TWA is unaffected by the addition of 0.5% BF. Beyond 0.5% BF, the TWA increased gradually with the addition of BF. For example, at 28 days, the TWA percentages for mixes containing 1, 1.5, and 2% BF are 5.7, 6.2 and 7%, respectively. When compared to the control mix, the equivalent increases are 11, 21.8, and 37.1%. This increase may be due to the formation of extra voids within the mortar matrix when the BF concentration surpasses 0.5%. The augmentation of fiber volume causes higher disruption of the matrix's structural integrity, leading to elevated porosity. Thus, the augmented porosity enables faster water penetration, resulting in increased overall water absorption [33,47].

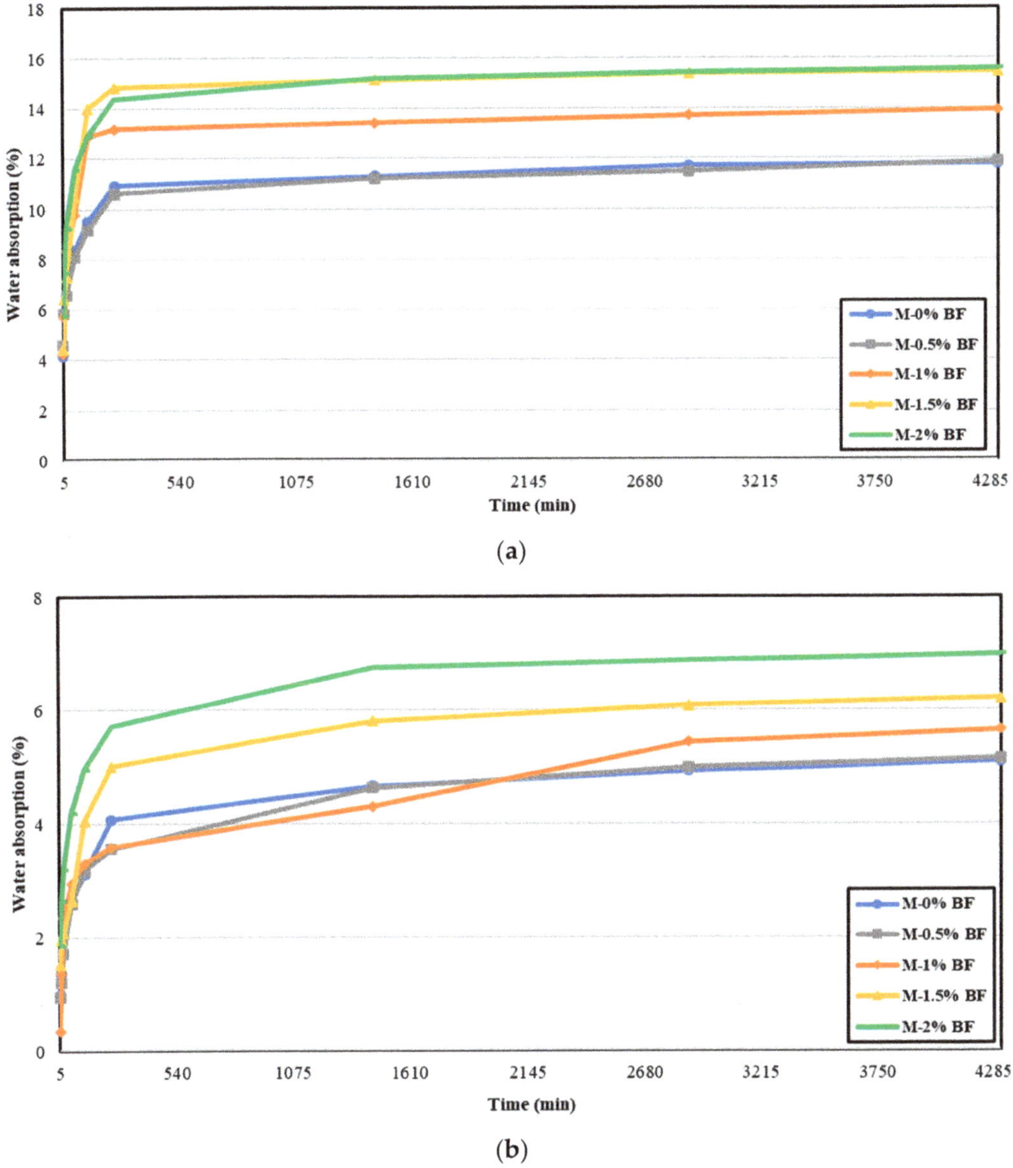

Figure 14. TWA of mortar with different BF additions: (**a**) 1 day; (**b**) 28 days.

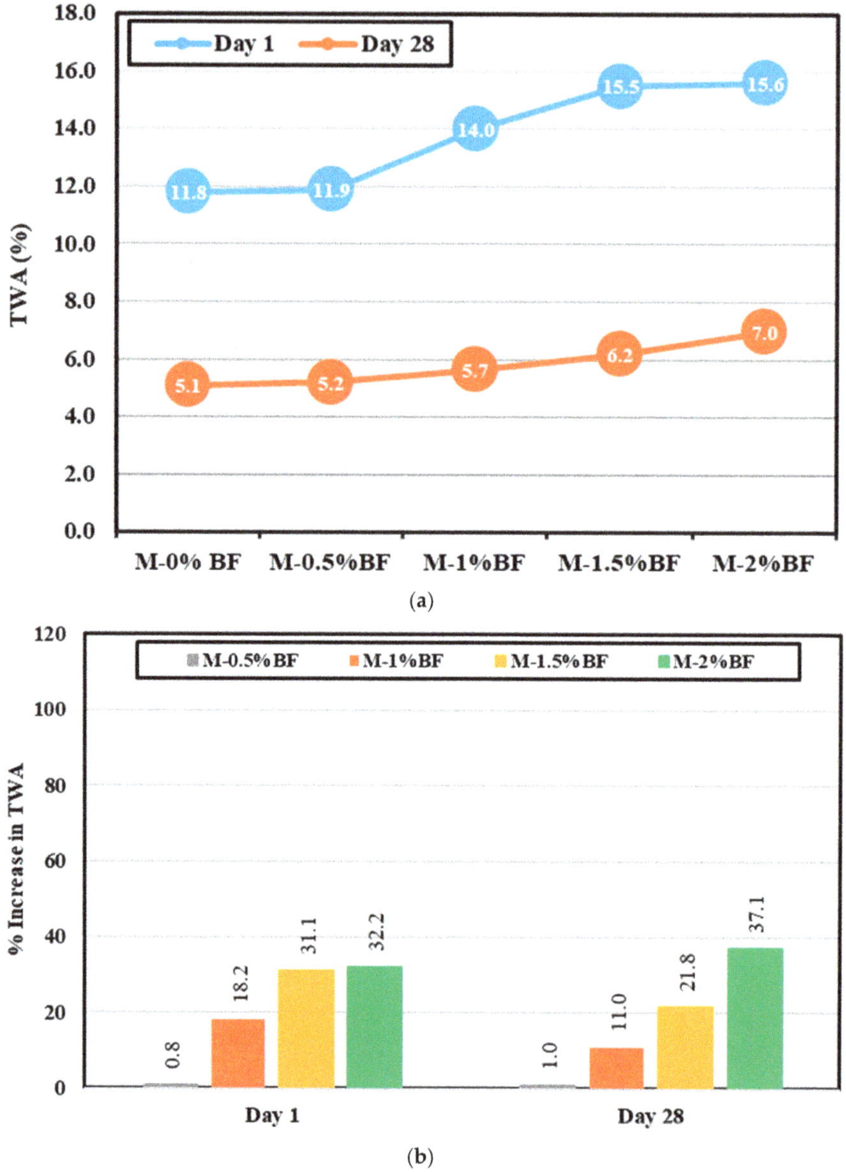

Figure 15. (**a**) Effect of curing duration on TWA coefficients; (**b**) % decrease in TWA at 1 day and 28 days.

3.5. Capillary-Diffusive Process

Figure 16a,b display the weight of absorbed water per area (M/A) results of various mortar mixes incorporating BF at days 1 and 28, respectively. Based on the data, M/A increased from 10 kg/m^2 for M-0% BF to 24.7 kg/m^2 for M-2% BF at day 1, and from 9 kg/m^2 for M-0% BF to 22.65 kg/m^2 for M-2% BF at day 28.

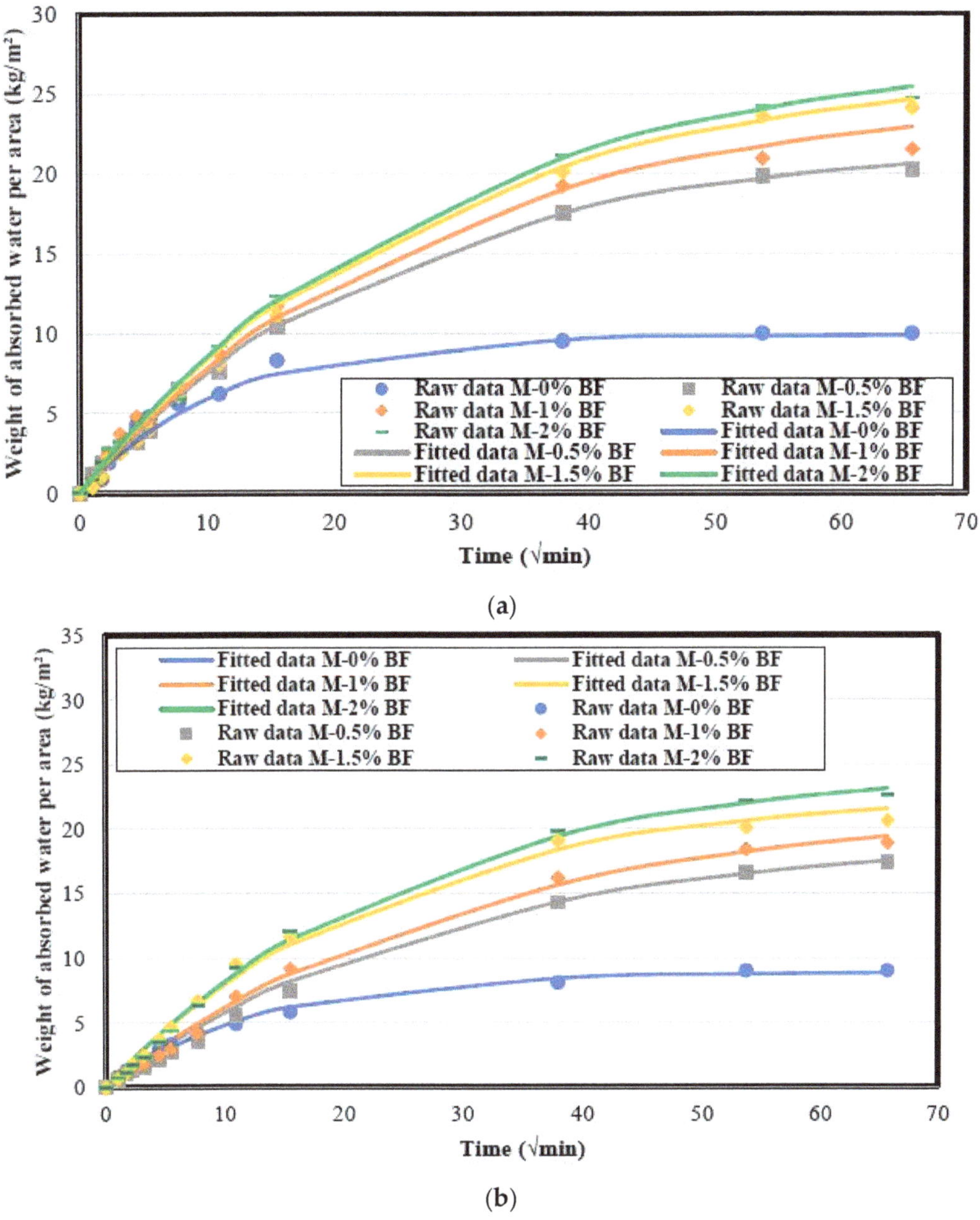

Figure 16. Modeling of capillary-diffusive process of mortar with different BF additions: (**a**) 1 day; (**b**) 28 days.

A capillary-diffusive model was used to estimate the mortars' sorptivity and diffusion coefficients, and the results demonstrated a strong fitting with the experimental data ($R^2 > 0.99$). The values of S, D, and R^2 for all mortar mixes are presented in Table 3.

Table 3. CWA parameters of mortar with different BF additions on days 1 and 28.

Mortar Code	Curing Period (Days)	S ($kg/m^2 \cdot s^{1/2}$)	D (m^2/s)	R^2
M-0% BF	1	1.15×10^{-2}	8.03×10^{-5}	0.983
	28	8.78×10^{-2}	8.05×10^{-8}	0.995
M-0.5% BF	1	1.18×10^{-1}	9.98×10^{-3}	0.997
	28	8.81×10^{-2}	5.61×10^{-7}	0.997
M-1% BF	1	1.20×10^{-1}	5.60×10^{-2}	0.995
	28	9.31×10^{-2}	6.72×10^{-7}	0.996
M-1.5% BF	1	1.29×10^{-1}	7.81×10^{-2}	0.998
	28	1.25×10^{-1}	2.76×10^{-4}	0.995
M-2% BF	1	1.31×10^{-1}	8.24×10^{-2}	0.998
	28	1.27×10^{-1}	2.07×10^{-3}	0.998

As shown, the sorptivity and diffusion coefficients drop significantly as the curing time increases. The sorptivity coefficient of M-1% BF, for example, decreased from 0.115 $kg/m^2 \cdot s^{1/2}$ at day 1 to 0.0878 $kg/m^2 \cdot s^{1/2}$ at day 28. Also, the diffusion coefficient of M-1% BF dropped from 8×10^{-5} m^2/s at day 1 to 8×10^{-8} m^2/s at day 28. The decrease in sorptivity and diffusion coefficients suggests improved performance, as longer curing durations lead to alterations in the internal structure, making it more challenging for water to penetrate through large pores via capillary actions and finer pores via moisture diffusion [46]. Table 3 displays that as the BF% increased, the sorptivity and the diffusion coefficients also rose. For instance, at 28 days, for mixtures containing 0, 0.5, 1, 1.5, and 2% BF, the sorptivity coefficients are 0.0878, 0.0881, 0.0930, 0.1250, and 0.1270 $kg/m^2 \cdot s^{1/2}$, respectively. Because BF is hydrophobic, it will probably absorb adjoining capillary water. As a consequence, capillary forces in the framework will rise. The rise in the sorptivity coefficient is mainly attributed to the supplemented capillary forces arising from the augmented fiber content in the mixtures. Increasing the percentage of BF introduces additional capillary pathways, resulting in enhanced water uptake and greater sorptivity [48].

Regarding the diffusion coefficients, the values are 8.05×10^{-8}, 5.61×10^{-7}, 6.72×10^{-7}, 2.76×10^{-4}, and 2.07×10^{-3} m^2/s for M-0% BF, M-0.5% BF, M-1% BF, M-1.5% BF, and M-2% BF, respectively. The inclusion of BF in mortar can be ascribed to the fibers' capacity to improve the mortar's permeability by building additional pathways for water flow, which results in an increase in diffusion coefficients. The overall diffusion rate is augmented as the proportion of BF increases, as the fiber-induced channels facilitate greater moisture diffusion into the smaller pores of the mortar [48].

3.6. Relationships between Various Properties

3.6.1. Compressive Strength-UPV Relationship

The correlation between compressive strength and UPV is illustrated in Figure 17. The coefficient of determination R^2 values of 0.94, 0.99, 0.98, 0.98, and 0.99 indicate a strong correlation between compressive strength and the UPV for BF percentages of 0, 0.5, 1, 1.5, and 2%, respectively. These findings indicate that an increase in compressive strength will inevitably lead to an increase in the UPV. This significant relationship between the two variables supports the hypothesis that high UPV values represent high-quality mortar with few voids and, consequently, greater compressive strength [46,49].

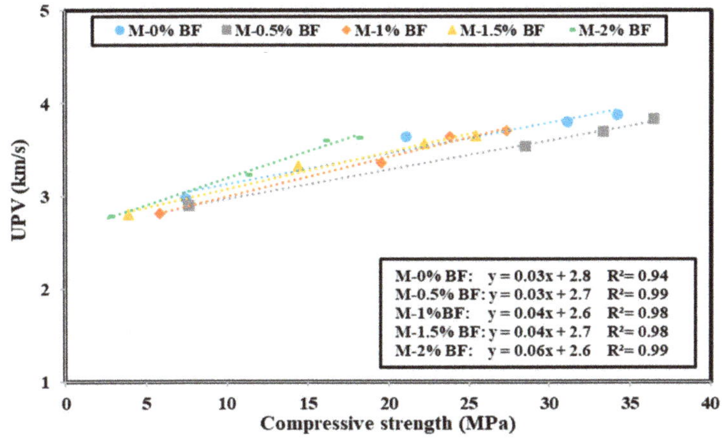

Figure 17. Correlation between compressive strength and UPV.

3.6.2. Compressive Strength–Flexural Strength Relationship

Figure 18 depicts the relationship between compressive strength and flexural strength. For BF percentages of 0, 0.5, 1, 1.5, and 2%, respectively, the coefficient of determination R^2 values of 0.94, 0.99, 0.99, 0.97, and 0.91 demonstrate a strong linear relationship between compressive strength and flexural strength. According to these results, enhancing compressive strength results in increasing flexural strength. As the beams cure, hydration products are formed, which increases the compressive strength of mortar [50]. Similarly, the flexural strength is enhanced due to the beams' enhanced ability to endure stress following peak loading [38].

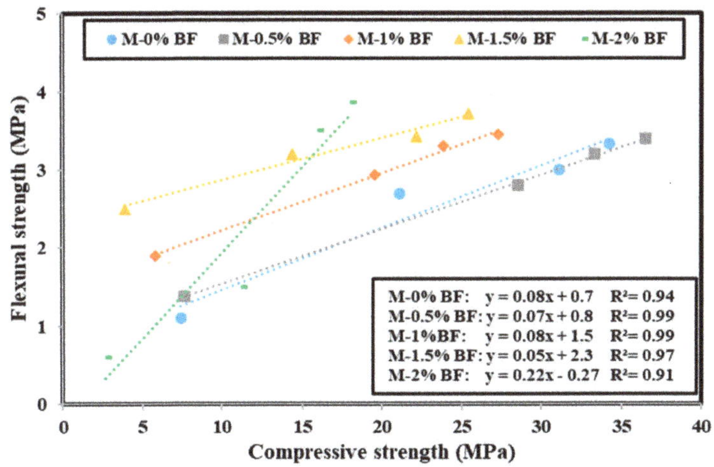

Figure 18. Correlation between compressive strength and flexural strength.

3.6.3. Compressive Strength-(M/A) Relationship

The correlation between compressive strength and M/A is presented in Figure 19. The R^2 values of 0.81, 0.91, 0.84, 0.95, and 0.91 reveal a significant negative relationship between compressive strength and M/A for BF percentages of 0, 0.5, 1, 1.5, and 2%, respectively. These data suggest that an increase in compressive strength will result in a decrease in M/A. As curing days increase, M/A decreases as a result of the augmented density and reduced permeability of mortar [46]. On the other hand, the gradual formation of hydration products enhances the strength of mortar, which increases the compressive strength over time [50].

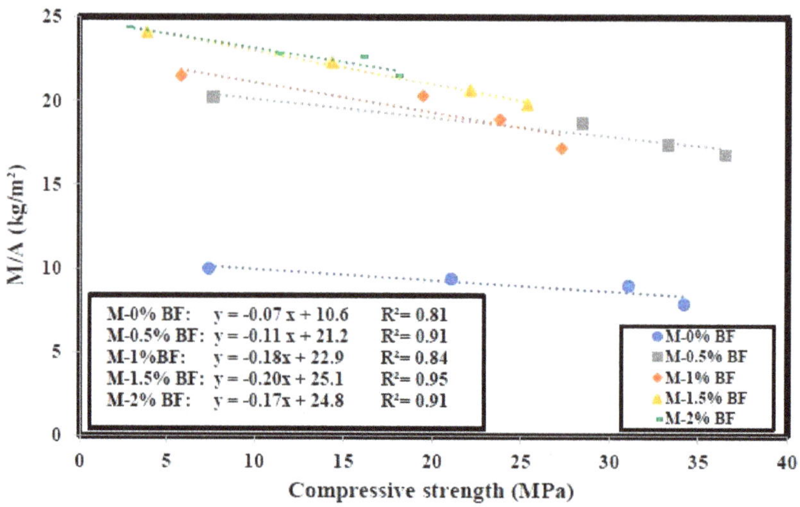

Figure 19. Correlation between compressive strength and M/A.

4. Conclusions

Due to their reduced environmental effect, natural fiber-reinforced cementitious materials offer a practical and long-term substitute for synthetic fiber-reinforced cementitious composites. Banana fiber integration into mortar for pavement applications, including paving blocks, was the focus of the present study. Testing was conducted on the mortar mixtures containing BF to determine their UPV, compressive strength, flexural strength, TWA, sorptivity, and diffusion coefficients. Based on this investigation, the following points can be stated.

- The addition of 0.5% BF resulted in either similar or slightly higher compressive and flexural strength compared to the control. Beyond 0.5%, there is a consistent reduction in mechanical properties.
- A correlation coefficient of $R^2 \sim 0.9$ indicates that the hyperbolic model was effective in predicting the compressive strength over a 90-day period. Both the initial length change and the ultimate length change parameters peaked at 0.5% BF addition and subsequently steadily declined with increasing BF%.
- As the amount of BF added to mortar mixes rises, the TWA increases. With reference to the control, the TWA is increased by 1, 11, 21.8, and 37.1% for mixes containing 0.5%, 1%, 1.5%, and 2% BF, respectively, at 28 days.
- The weight of water absorbed per area increases with the addition of BF to the mortar mixture. M/A is increased by 93, 110, 130, and 152% at 28 days for mixes containing 0.5, 1, 1.5, and 2% BF, respectively, compared to the control mix.
- With a correlation coefficient of $R^2 > 0.99$, the capillary-diffusive model successfully predicted the capillary-diffusive phenomenon as a function of time. The sorptivity and diffusion coefficients exhibited a progressive increase when BF is added and a reduction as the curation period progresses.
- UPV and flexural strength show a strong positive association with compressive strength with high coefficients of correlation ($R^2 \geq 0.9$). On the other hand, M/A is negatively correlated to compressive strength.
- According to this study, BF content of 0.5% in the matrix yields the best mechanical and physical properties, providing appropriate mechanical and durability performance. Future research should consider adding less than 0.5% BF.

Author Contributions: Conceptualization, H.G. and A.E.; Methodology, G.A.-M. and H.G.; Validation, S.E.-Z. and A.E.; Formal analysis, G.A.-M. and H.G.; Writing—original draft, G.A.-M. and H.G.; Writing—review and editing, G.A.-M., J.K., S.E.-Z. and A.E.; Supervision, J.K. and A.E. All authors have read and agreed to the published version of the manuscript.

Funding: This research received no external funding.

Institutional Review Board Statement: Not applicable.

Informed Consent Statement: Not applicable.

Data Availability Statement: Data are contained within the article.

Acknowledgments: The authors express their gratitude for the assistance provided by the staff and technicians at BAU Laboratories.

Conflicts of Interest: The authors declare no conflicts of interest.

References

1. Obi, F.O.; Ugwuishiwu, B.O.; Nwakaire, J.N. Agricultural waste concept, generation, utilization and management. *Niger. J. Technol.* **2016**, *35*, 957–964. [CrossRef]
2. Akinyemi, B.A.; Okonkwo, C.E.; Alhassan, E.A.; Ajiboye, M. Durability and strength properties of particle boards from polystyrene–wood wastes. *J. Mater. Cycles Waste Manag.* **2019**, *21*, 1541–1549. [CrossRef]
3. Tolêdo Filho, R.D.; Ghavami, K.; England, G.L.; Scrivener, K. Development of vegetable fibre–mortar composites of improved durability. *Cem. Concr. Compos.* **2003**, *25*, 185–196. [CrossRef]
4. Al-Massri, G.; Ghanem, H.; Khatib, J.; Kırgız, M.S.; Elkordi, A. Chemical shrinkage, autogenous shrinkage, drying shrinkage, and expansion stability of interfacial transition zone material using alkali-treated banana fiber for concrete. *J. Struct. Integr. Maint.* **2024**, *9*, 2390650. [CrossRef]
5. De Andrade Silva, F.; Mobasher, B.; Toledo Filho, R.D. Cracking mechanisms in durable sisal fiber reinforced cement composites. *Cem. Concr. Compos.* **2009**, *31*, 721–730. [CrossRef]
6. Shah, I.; Li, J.; Yang, S.; Zhang, Y.; Anwar, A. Experimental investigation on the mechanical properties of natural fiber reinforced concrete. *J. Renew. Mater.* **2022**, *10*, 1307. [CrossRef]
7. Khan, M.B.; Shafiq, N.; Waqar, A.; Radu, D.; Cismaș, C.; Imran, M.; Almujibah, H.; Benjeddou, O. Effects of jute fiber on fresh and hardened characteristics of concrete with environmental assessment. *Buildings* **2023**, *13*, 1691. [CrossRef]
8. Prafulla, K.; Nagaraju, A. An experimental study on coir fiber reinforced concrete with ground granulated blast furnace slag and dolomite powder as partial replacement of cement. In *IOP Conference Series: Earth and Environmental Science*; IOP Publishing: Bristol, UK, 2022; Volume 1086, p. 012052. [CrossRef]
9. Machaka, M.; Khatib, J.; Baydoun, S.; Elkordi, A.; Assaad, J.J. The Effect of Adding Phragmites australis Fibers on the Properties of Concrete. *Buildings* **2022**, *12*, 278. [CrossRef]
10. Zouaoui, Y.; Benmahiddine, F.; Yahia, A.; Belarbi, R. Hygrothermal and mechanical behaviors of fiber mortar: Comparative study between palm and hemp fibers. *Energies* **2021**, *14*, 7110. [CrossRef]
11. Li, Q.; Ibrahim, L.; Zhou, W.; Zhang, M.; Yuan, Z. Treatment methods for plant fibers for use as reinforcement in cement-based materials. *Cellulose* **2021**, *28*, 5257–5268. [CrossRef]
12. Mohammed-Ziegler, I.; Oszlánczi, Á.; Somfai, B.; Hórvölgyi, Z.; Pászli, I.; Holmgren, A.; Forsling, W. Surface free energy of natural and surface-modified tropical and European wood species. *J. Adhes. Sci. Technol.* **2004**, *18*, 687–713. [CrossRef]
13. Mohammed-Ziegler, I.; Tánczos, I.; Hórvölgyi, Z.; Agoston, B. Water-repellent acylated and silylated wood samples and their surface analytical characterization. *Colloids Surf. A Physicochem. Eng. Asp.* **2008**, *319*, 204–212. [CrossRef]
14. Arias, P. The World Banana Economy 2003, 1985–2002 (No. 1). Food & Agriculture Org. Available online: https://www.fao.org/3/y5102e/y5102e04.htm (accessed on 15 September 2024).
15. Ali, B.; Azab, M.; Ahmed, H.; Kurda, R.; El Ouni, M.H.; Elhag, A.B. Investigation of physical, strength, and ductility characteristics of concrete reinforced with banana (Musaceae) stem fiber. *J. Build. Eng.* **2022**, *61*, 105024. [CrossRef]
16. Logeshwar, A.B.; Bhuvaneshpandi, M.; Harishkanna, V.; Midhunkumar, R.; Praveenkumar, S. Effect of Banana Fibre on Strength Properties of Bagasse ash & Rice Husk ash blended high performance concrete composite. In Proceedings of the ASPS Conference Proceedings, Rajasthan, India, 19–22 December 2022; Volume 1, pp. 201–206. [CrossRef]
17. Pathan, T.A.; Jhumarwala, R.A. Evaluation of mechanical properties of high strength banana fibre concrete (HSBFC) incorporating fly ash and silica fume. *Int. J. Res. Appl. Sci. Eng. Technol.* **2022**, *10*, 780–785. [CrossRef]
18. Lamichhane, N.; Lamichhane, A.; Gyawali, T.R. Enhancing mechanical properties of mortar with short and thin banana fibers: A sustainable alternative to synthetic fibers. *Heliyon* **2024**, *10*, e30652. [CrossRef]
19. Thanushan, K.; Sathiparan, N. Mechanical performance and durability of banana fibre and coconut coir reinforced cement stabilized soil blocks. *Materialia* **2022**, *21*, 101309. [CrossRef]

20. *ASTM C1437-15 2015*; Standard Test Method for Flow of Hydraulic Cement Mortar. ASTM International: West Conshohocken, PA, USA, 2015.
21. *ASTM C 597*; Standard Test Method for Pulse Velocity through Concrete. American Society for Testing and Materials: West Conshohocken, PA, USA, 2016.
22. Ghanem, H.; Khatib, J.; Elkordi, A. Effect of partial replacement of sand by mswi-ba on the properties of mortar. *BAU J. Sci. Technol.* **2020**, *1*, 4. [CrossRef]
23. *ASTM C109*; Standard Test Method for Compressive Strength of Hydraulic Cement Mortars (Using 2-in. or [50 mm] Cube Specimens. ASTM International: West Conshohocken, PA, USA, 2016.
24. Ghanem, H.; Ramadan, R.; Khatib, J.; Elkordi, A. A Review on Chemical and Autogenous Shrinkage of Cementitious Systems. *Materials* **2024**, *17*, 283. [CrossRef]
25. *ASTM C348*; Standard Test Method for Flexural Strength of Hydraulic-Cement Mortars. American Society for Testing and Materials International: West Conshohocken, PA, USA, 2008.
26. *ASTM C1585*; Standard Test Method for Measurement of Rate of Absorption of Water by Hydraulic-Cement Concretes. ASTM International: West Conshohocken, PA, USA, 2004.
27. Martys, N.S.; Ferraris, C.F. Capillary transport in mortars and concrete. *Cem. Concr. Res.* **1997**, *27*, 747–760. [CrossRef]
28. Villar-Cocina, E.; Valencia-Morales, E.; Vega-Leyva, J.; Munoz, J.A. Kinetics of the water absorption in GGBS-concretes: A capillary-diffusive model. *Comput. Concr. Int. J.* **2005**, *2*, 19–30. [CrossRef]
29. Wongsa, A.; Kunthawatwong, R.; Naenudon, S.; Sata, V.; Chindaprasirt, P. Natural fiber reinforced high calcium fly ash geopolymer mortar. *Constr. Build. Mater.* **2020**, *241*, 118143. [CrossRef]
30. Mathavan, M.; Sakthieswaran, N.; Babu, O.G. Experimental investigation on strength and properties of natural fibre reinforced cement. *Mater. Today Proc.* **2021**, *37*, 1066–1070. [CrossRef]
31. Alatshan, F.; Altlomate, A.M.; Mashiri, F.; Alamin, W. Effect of date palm fibers on the mechanical properties of concrete. *Int. J. Sustain. Build. Technol. Urban Dev.* **2017**, *8*, 68–80. [CrossRef]
32. Khatib, J.M.; Ramadan, R.; Ghanem, H.; Elkordi, A.; Baalbaki, O.; Kırgız, M. Chemical shrinkage of paste and mortar containing limestone fines. *Mater. Today Proc.* **2022**, *61*, 530–536. [CrossRef]
33. De La Grée, G.D.; Yu, Q.L.; Brouwers, H.J.H. Assessing the effect of $CaSO_4$ content on the hydration kinetics, microstructure and mechanical properties of cements containing sugars. *Constr. Build. Mater.* **2017**, *143*, 48–60. [CrossRef]
34. Bilba, K.; Arsène, M.A.; Ouensanga, A. Sugar cane bagasse fibre reinforced cement composites. Part I. Influence of the botanical components of bagasse on the setting of bagasse/cement composite. *Cem. Concr. Compos.* **2003**, *25*, 91–96. [CrossRef]
35. Tolêdo Filho, R.D.; Scrivener, K.; England, G.L.; Ghavami, K. Durability of alkali-sensitive sisal and coconut fibres in cement mortar composites. *Cem. Concr. Compos.* **2000**, *22*, 127–143. [CrossRef]
36. Araya-Letelier, G.; Antico, F.C.; Carrasco, M.; Rojas, P.; García-Herrera, C.M. Effectiveness of new natural fibers on damage-mechanical performance of mortar. *Constr. Build. Mater.* **2017**, *152*, 672–682. [CrossRef]
37. Choi, J.I.; Jang, S.Y.; Kwon, S.J.; Lee, B.Y. Tensile behavior and cracking pattern of an ultra-high-performance mortar reinforced by polyethylene fiber. *Adv. Mater. Sci. Eng.* **2017**, *2017*, 5383982. [CrossRef]
38. Aslon MW, B.; Lakawa, I.; Sulaiman, S.; Hawa, S. Testing the Compressive Strength of Concrete With the Utilization of Rice Husk Ash and Palm Fiber. *Sultra Civ. Eng. J.* **2023**, *4*, 11–19. [CrossRef]
39. Wang, Y.; Cheng, J.; Wang, J. Flexural performance of recycled concrete beam reinforced with modified basalt fiber and nano-silica. *Case Stud. Constr. Mater.* **2023**, *18*, e02022. [CrossRef]
40. Hannawi, K.; Bian, H.; Prince-Agbodjan, W.; Raghavan, B. Effect of different types of fibers on the microstructure and the mechanical behavior of ultra-high performance fiber-reinforced concretes. *Compos. Part B Eng.* **2016**, *86*, 214–220. [CrossRef]
41. Arshad, S.; Sharif, M.B.; Irfan-ul-Hassan, M.; Khan, M.; Zhang, J.L. Efficiency of supplementary cementitious materials and natural fiber on mechanical performance of concrete. *Arab. J. Sci. Eng.* **2020**, *45*, 8577–8589. [CrossRef]
42. Mounanga, P.; Khelidj, A.; Loukili, A.; Baroghel-Bouny, V. Predicting $Ca(OH)_2$ content and chemical shrinkage of hydrating cement pastes using analytical approach. *Cem. Concr. Res.* **2004**, *34*, 255–265. [CrossRef]
43. Kouta, N.; Saliba, J.; Saiyouri, N. Effect of flax fibers on early age shrinkage and cracking of earth concrete. *Constr. Build. Mater.* **2020**, *254*, 119315. [CrossRef]
44. Alzebdeh, K.I.; Nassar, M.M.; Arunachalam, R. Effect of fabrication parameters on strength of natural fiber polypropylene composites: Statistical assessment. *Measurement* **2019**, *146*, 195–207. [CrossRef]
45. Ji, Y.; Ji, W.; Zhang, Z.; Wang, R. Road performance investigation on fiber-reinforced recycled cement base material. *Polymers* **2022**, *14*, 4102. [CrossRef]
46. Ghanem, H.; Machaka, M.; Khatib, J.; Elkordi, A.; Baalbaki, O. Effect of partial replacement of cement by MSWIBA on the properties of mortar. *Acad. J. Civ. Eng.* **2019**, *37*, 82–89. [CrossRef]
47. Khatib, J.; Ramadan, R.; Ghanem, H.; ElKordi, A. Effect of using limestone fines on the chemical shrinkage of pastes and mortars. *Environ. Sci. Pollut. Res.* **2023**, *30*, 25287–25298. [CrossRef]
48. Ren, F.; Zhou, C.; Zeng, Q.; Ding, Z.; Xing, F.; Wang, W. The dependence of capillary sorptivity and gas permeability on initial water content for unsaturated cement mortars. *Cem. Concr. Compos.* **2019**, *104*, 103356. [CrossRef]

49. Khatib, J.; Ramadan, R.; Ghanem, H.; Elkordi, A. Volume stability of cement paste containing limestone fines. *Buildings* **2021**, *11*, 366. [CrossRef]
50. Pipilikaki, P.; Katsioti, M. Study of the hydration process of quaternary blended cements and durability of the produced mortars and concretes. *Constr. Build. Mater.* **2009**, *23*, 2246–2250. [CrossRef]

ceramics

MDPI

Article

Mullite–Silicate Proppants Based on High-Iron Bauxite and Waste from Metallurgical Industry in Kazakhstan

Tolebi Dzhienalyev [1,2,*], Alla Biryukova [1], Bagdaulet Kenzhaliyev [1], Alma Uskenbaeva [1] and Galiya Ruzakhunova [1]

1 Institute of Metallurgy and Ore Beneficiation JSC, Almaty 050010, Kazakhstan; almauskenbaeva@mail.ru (A.U.)
2 Mining and Metallurgical Institute, Satbayev University, Almaty 050013, Kazakhstan
* Correspondence: tolebi992@gmail.com; Tel.: +7-7051239000

Abstract: The suitability of microsilica as a raw material for the production of ceramic mullite–silicate proppants was assessed. The chemical and mineralogical compositions of the initial materials were studied. The mineral composition of bauxite is mainly represented by gibbsite $Al(OH)_3$ and, to a lesser extent, kaolinite $Al_4[Si_4O_{10}](OH)_8$, with impurities of hematite and quartz. It is established that, in order to obtain mullite–silicate proppants, compositions containing 10–20% microsilica are optimal. The sintering of these compositions occurs at 1350–1380 °C. A lightweight ceramic proppant was obtained with a bulk density of 1.21–1.41 g/cm^3, breaking ratio at 51.7 MPa of 19.1–20.3%, and sphericity and roundness of 0.7–0.9, and the optimal roasting temperature was determined as 1370–1380 °C.

Keywords: hydraulic fracturing; bauxite; microsilica; proppants; mullite

1. Introduction

The growth of hard-to-recover hydrocarbon reserves in the modern world has increased the relevance of and demand for the use of hydraulic fracturing in the oil and gas industry. During hydraulic fracturing, proppants (i.e., granular particles of 0.2–2.0 mm in size) are injected under pressure into the well to prevent fracture closure and maintain the permeability of the well after pressure reduction [1–4].

In this regard, the need for proppants is expected to increase and, so, their production technologies must be constantly improved. Commonly used propping agents include frac sand, resin-coated sand, and ceramic proppants. Despite the large selection of propping agents, ceramic proppants with a wide range of compositions and a range of functional properties remain highly popular.

The raw materials for obtaining ceramic proppants include various natural and synthetic alumina- and silica-containing materials. During the firing process, mullite is formed on their basis, which is the main phase in the structure of granules, determining the main functional properties of proppants, such as their strength (breaking ratio), density, chemical resistance, and so on.

One of the most important problems faced by modern manufacturers of functional materials is the creation of competitive products, which are based on traditional raw materials as well as secondary and production waste materials. The advantage of using recycled and production waste materials is that there is no need to extract them. Furthermore, waste disposal will reduce the consumption of primary natural resources, solve the issues of raw material security, reduce production costs, and, at the same time, contribute to improvement of the environmental situation in the region.

An analysis of the available scientific studies in the literature revealed that proppants with aluminosilicate compositions are still the most popular. The works [5–8] intended to produce aluminosilicate proppants through the use of natural clay materials and bauxite with various compositions have been presented. For example, researchers [5] have shown the possibility of obtaining ceramic proppants based on natural aluminosilicate raw materials in Kazakhstan. It was shown that Arkalyk refractory clays and Krasnooktyabrskiy bauxites are basic aluminosilicate raw materials with Al_2O_3 contents of 48.3 and 62.0% (on a calcined substance), respectively. Compositions and technological parameters for the production of lightweight ceramic proppants were also presented, and the efficiency of heat treatment of raw materials at 1000 °C to increase the density and strength of ceramic samples was assessed.

Researchers have come to pay more and more attention to the use of waste materials in the production of proppant materials for the oil industry. In [9], the authors proposed to obtain aluminosilicate proppants based on man-made raw materials (i.e., corundum dust obtained during the crushing of electro-fused corundum and a sintering additive, as well as granodiorite obtained from screenings of crushed stone production). The samples were obtained by pressing a charge containing corundum dust (55 wt%) and granodiorite (45 wt%) at a specific pressing pressure of 20 MPa. Samples of aluminosilicate proppants based on this charge with an apparent density of 2.38–2.76 g/cm^3 and water absorption of 1.4–10.9% were obtained. The strength of the obtained samples increased with an increase in the roasting temperature, ranging from 17 to 32 MPa at a roasting temperature of 1180–1200 °C. The disadvantage of this work was the analysis of proppant quality based on the testing of ceramic samples, rather than the proppants themselves.

In [10], the authors obtained proppants using drill cuttings and modifying additives; in particular, on a wt% basis: aluminum oxide—5.0; cullet glass—20; and NaF (over 100 wt%)—4. The optimal roasting temperature was 1100 °C with a rate of 6 °C/min. However, this composition produced proppants with low sphericity and roundness, which affects their permeability during hydraulic fracturing.

In [11], kaolin from the Borovichsko–Lyubytinskoye deposit was used as the main raw material for the production of ceramic proppants. The authors proposed increasing the pre-calcination temperature of kaolin from 850 to 980 °C. This ensured increased activity in the subsequent sintering of the granulated material, improving the strength properties of proppants after firing both at 1450 °C and 1500 °C. Proppants with a bulk density of up to 1.83–1.90 g/cm^3 were obtained, withstanding destructive pressures up to 70 MPa. However, proppants based on the proposed raw materials require increased energy consumption (firing temperatures up to 1500 °C).

In [12], the authors used kaolinite clay and waste coal to produce aluminosilicate proppants. It was noted that the proppant was obtained at a roasting temperature of 1400 °C. The bulk density was 1.27 g/cm^3, the apparent density was 2.79 g/cm^3, and the breaking ratio was 3.27% at 35 MPa and 8.36% at 52 MPa. The authors suggested that the addition of solid waste could significantly reduce the cost of a ceramic proppant.

A low-density and high-strength ceramic proppant was obtained by the authors of [13], based on high-iron bauxite at 1300–1360 °C. The bulk density of the sample was 1.42 g/cm^3, the apparent density was 2.67 g/cm^3, and the breaking ratio at 52 MPa was 5.1%. The main phase of obtained samples was needle-shaped mullite.

Another study [14] focused on the orthogonal optimization of the design of low-density and high-strength ceramic proppants made of low-grade bauxite and feldspar. The authors studied the effects of the raw material grinding degree and the proppant roasting mode on the properties of the finished product. They found that the optimal parameters for the production of proppants that met the properties of the considered standard were the grinding of bauxite within 4 h, feldspar within 8 h, and a sintering temperature of 1280 °C. Proppants obtained using the optimal combination had a bulk density of 1.48 g/cm^3, an apparent density of 2.7 g/cm^3, a breaking ratio of 4.07% at 52 MPa, and a solubility in acid of 2.15%.

In [15], corundum–mullite ceramic proppants were successfully synthesized using natural bauxite, solid waste coal, and calcium carbonate additives as raw materials. The effects of calcium carbonate supplementation on the phase composition, microstructure, and mechanical characteristics were studied. The results showed that the addition of $CaCO_3$

promotes the formation of a liquid phase at a lower sintering temperature, contributing to the compaction of the samples. Ceramic proppants with an addition rate of 5% sintered at 1350 °C had a breaking ratio of 8.41% at 52 MPa, due to the appropriate amount of liquid phase and grinding of mullite grains.

The authors of [16] noted the advantage of using ultra-lightweight (ULW) proppants with high strength in hydraulic fracturing technologies. Traditional proppants are prone to settling in fracture operations, which can seriously affect the effectiveness of the operation. The article provided a comprehensive review of more than 50 articles published over the past few decades on ultra-lightweight proppants. The purpose of this study was to provide an overview of the current state of development of ultralight proppants concerning the raw materials, manufacturing process, performance characteristics, hydrophobic and lipophilic properties, and practical applications, in order to facilitate research on novel ultralight proppants.

An analysis of the available information reveals that aluminosilicate ceramic proppants can be produced using various alumina- and silicate-containing natural materials, as well as man-made waste. The improvement of technological parameters for the production of aluminosilicate proppants depends on the characteristics of the raw materials used.

Special attention has been paid in recent years to the use of lightweight proppants in the implementation of hydraulic fracturing technology. In this regard, there is an increased interest in the field to create lightweight proppants using various natural types of raw materials, as well as man-made waste.

During the production of metallurgical silicon at KazSilikon LLP (Almaty, Kazakhstan), a huge amount of waste is generated, namely, microsilica, which settles in dust collectors. Using this microsilica in the production of ceramic materials would make it possible to obtain value-added products. The SiO_2 content in this product varies widely, from 80 to 99%, and the average particle size ranges from 0.01 to 1.0 mm.

Microsilica is a microscopic form of amorphous silicon dioxide. Amorphous SiO_2 is essentially not found in nature, and this is even more so the case for its ultrafine form, microsilica. At the same time, the demand for microsilica in the industry is steadily growing [17–21]. It should be noted that, while microsilica is widely used for the production of various building materials, there is almost no information on its application in the field of ceramic proppants.

Therefore, the purpose of this work was to evaluate the utility of microsilica for the production of mullite–silica proppants.

2. Materials and Methods

The raw materials used in the work were high-iron bauxite from the Krasnooktyabrskiy deposit and microsilica (a by-product of KazSilikon LLP, Almaty, Kazakhstan).

The chemical and mineralogical compositions of the starting materials, as well as the structural and phase transformations occurring during the heat treatment of refractory compositions based on these materials, were studied through the use of X-ray fluorescence, X-ray diffraction, thermal, chemical, and microscopic analysis methods.

Samples with particular ceramic compositions were obtained using methods commonly adopted for the production of ceramics and refractories, including the grinding and fine grinding of raw materials, the calculation and composition of charges, the preparation of molding compound, and the pressing of samples from semi-dry masses on a hydraulic press. The granulation of ceramic mixtures for the production of proppant granules was performed using an EL1 Eirich mixer–granulator (Maschinenfabrik Gustav Eirich GmbH & Co KG, Hardheim, Germany).

The determination of the technical properties of ceramics was performed according to the national standards GOST 2409-2014 [22], GOST 4071.1-2021 [23] and GOST 51761-2013 [24].

The structure and phase composition were also studied through electron probe microanalysis (EPMA) using a JEOL JXA-8230 (JEOL, Ltd., Tokyo, Japan) with combined WDS/EDS.

The XRD patterns of the samples were recorded using a Bruker D8 Advance XRD (Bruker Corporation, Billerica, MA, USA) diffractometer (Cu Kα, λ = 1.5406 Å). A thermal analysis of ceramic samples was performed using a simultaneous thermal analyzer (NETZSCH STA 449 F3 Jupiter (NETZSCH group, Selb, Germany).

3. Results and Discussion

3.1. Study of Raw Materials

The chemical composition of bauxite is determined and given in Table 1. Microsilica composition was provided by KazSilikon LLP.

Table 1. Chemical composition of raw materials.

Raw Material	Content of Oxides, Mass. %								
	SiO_2	MgO	CaO	Al_2O_3	Fe_2O_3	K_2O	Na_2O	TiO_2	LOI
Bauxite	7.6	0.1	1.14	47.8	17.3	0.03	0.03	3.5	22.5
Microsilica	95.20	-	-	-	-	-	-	-	4.80

The mineral composition of the bauxite was represented mainly by gibbsite $Al(OH)_3$ and, to a lesser extent, kaolinite $Al_4[Si_4O_{10}](OH)_8$ with characteristic X-ray reflexes (Figure 1). The sample contained a significant amount of iron impurities (in the form of hematite) and, to a lesser extent, quartz.

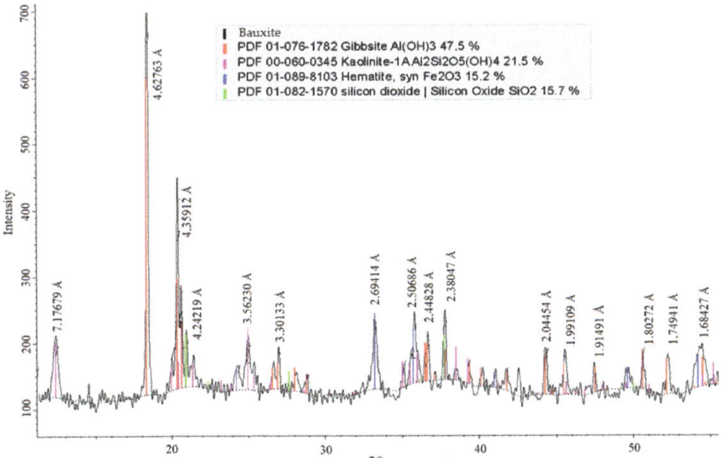

Figure 1. XRD patterns of the bauxite.

The thermal analysis of the bauxite sample established the endothermic effects of varying intensity on the DTA curve (Figure 2), the maximum development of which occurred at 338.40 °C and 553.60 °C. There was also a weak exothermic effect on the DTA curve, with a peak at 961 °C. Additional thermal effects were recorded on the dDTA curve. It is possible to note endothermic effects with extremes at 311.3 °C, 321.5 °C, and 575.4 °C. Two exothermic effects with peaks at 380.90 °C and 570.0 °C appeared.

The endothermic effect (338.4 °C) on the DTA curve may be a manifestation of the dehydration of gibbsite and iron hydroxides. Two endothermic effects (311.3 °C, 321.5 °C) were recorded on the dDTA curve in the area of the development of the above-mentioned effect. This provides evidence for the dehydration processes of gibbsite and iron hydroxides in the temperature range of 150–400 °C.

The presence of an exothermic effect of 380.9 °C on the dDTA curve may indicate the presence of amorphous iron hydroxide in the sample; in particular, this peak indicates formation of the hematite crystal lattice.

The manifestation of kaolinite shows an endothermic effect with maximum development at 553.6 °C and an exothermic effect with a peak at 961 °C on the DTA curve. The exothermic effect recorded on the dDTA curve may be a reflection of the transition of maghemite γ-Fe_2O_3 to the α modification. Endothermic effects indicate the dehydration of aluminum and iron hydroxides, while the exothermic effects indicate the formation of aluminohematite (570 °C dDTA) and the formation of corundum (961 °C DTA). The very weak endothermic effect with an extreme at 575.40 °C on the dDTA curve is a manifestation of quartz impurity inversion.

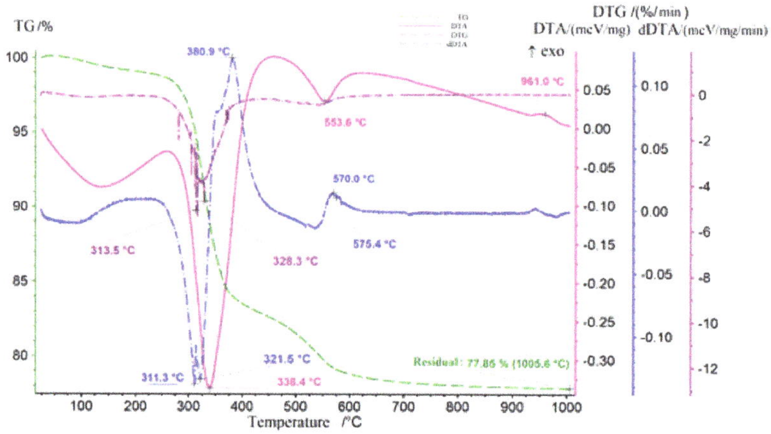

Figure 2. Simultaneous TG/DTA analysis of the bauxite.

The TG curve shows the change in the weight of bauxite during the heat treatment of the sample. In the range up to 300 °C, mechanically bound water is removed, and the mass loss is insignificant. The main change in mass occurs in the temperature range from 300 to 1000 °C. In this temperature range, kaolinite dehydrates with a restructuring of its structure into a metakaolinite structure with the subsequent formation of primary mullite and quartz.

3.2. *Obtaining Ceramic Compositions Based on Bauxite and Microsilica*

To study the effects of the microsilica content on the phase composition and the properties of aluminosilicate proppants based on Krasnooktyabrskiy bauxite, ceramic samples were obtained according to the compositions specified in Table 2.

As shown in the table, the ratio of Al_2O_3/SiO_2 decreases with an increase in the content of microsilica in the composition; that is, the theoretically possible formation of mullite (without taking into account the participation of impurity minerals in the raw material) will also decrease.

Experience from studies of compositions based on Krasnooktyabrskiy bauxite performed earlier [25] indicates that the sintering interval is 1300–1500 °C, depending on the ratio of the initial components. Therefore, molded and dried cylindric samples were fired at 1350 °C. The properties of fired ceramic samples are detailed in Table 3.

Table 2. Chemical contents of compositions.

Composition	Composition of Mixture, Material, Mass. %	Content of Oxides, Mass. %							
		SiO_2	MgO	CaO	Al_2O_3	Fe_2O_3	K_2O	Na_2O	TiO_2
M5	Bauxite, 95 microsilica, 5	14.25	0.14	1.35	58.71	21.2	0.04	0.04	4.27
M10	Bauxite, 90 microsilica, 10	18.62	0.11	1.28	55.69	20.09	0.04	0.04	4.13
M20	Bauxite, 80 microsilica, 20	27.94	0.096	1.12	49.34	17.84	0.03	0.03	3.6
M50	Bauxite, 50 microsilica, 50	54.30	0.09	0.9	31.2	11.2	0.03	0.03	2.25

Table 3. Properties of samples of compositions.

Composition	Open Porosity, %	Apparent Density, g/cm^3	Compressive Strength, MPa
M5	15.30	2.58	32.8
M10	17.45	2.43	30.1
M20	23.05	2.17	26.7
M50	30.66	1.67	10.2

Table 3 shows that the properties of the samples gradually decrease with an increase in the content of microsilica. The structure of the sample becomes more porous with increasing microsilica content, the apparent density and strength under compression decrease, and the appearance of the ceramics suggests under-roasting. This indicates the need to increase the firing temperature with an increase in microsilica in the composition. Obviously, with an increase in the content of microsilica, the formation of low-melting eutectic decreases during the firing process.

As indicated by the X-ray phase analysis results for the samples, the main phases of the compositions were mullite, cristobalite, and spinel phases of the complex composition containing aluminum oxide and impurity oxides of raw materials. The maximum amount of mullite was established in compositions based on bauxite with 10–20% microsilica (Figures 3–6, Table 4).

Table 4. XRD results of compositions.

Composition	Phase	Formula	Quantity, %
M5	α-Al_2O_3, syn	Al_2O_3	54.8
	Mullite, syn	$Al(Al_{0.83}Si_{1.08}O_{4.85})$	23.5
	Hematite, syn	Fe_2O_3	12.8
	Cristobalite, syn	SiO_2	8.9
M10	Mullite, syn	$Al(Al_{0.83}Si_{1.08}O_{4.85})$	69.2%
	Cristobalite, syn	SiO_2	25.6%
	Magnetite, magnesian, syn	$(Mg_{0.22}Fe_{0.78})(Al_{0.40}Mg_{0.78}Fe_{0.82})O_4$	5.3%
M20	Mullite, syn	$Al(Al_{0.83}Si_{1.08}O_{4.85})$	49.9
	Cristobalite, syn	SiO_2	45.7
	Magnetite, magnesian, syn	$(Mg_{0.22}Fe_{0.78})(Al_{0.40}Mg_{0.78}Fe_{0.82})O_4$	4.4

Table 4. *Cont.*

Composition	Phase	Formula	Quantity, %
M50	Mullite, syn	$Al(Al_{0.83}Si_{1.08}O_{4.85})$	23.7
	Cristobalite, syn	SiO_2	69.2%
	Quartz, syn	SiO_2	4.0%
	Magnetite, magnesian, syn	$(Mg_{0.22}Fe_{0.78})(Al_{0.40}Mg_{0.78}Fe_{0.82})O_4$	3.2%

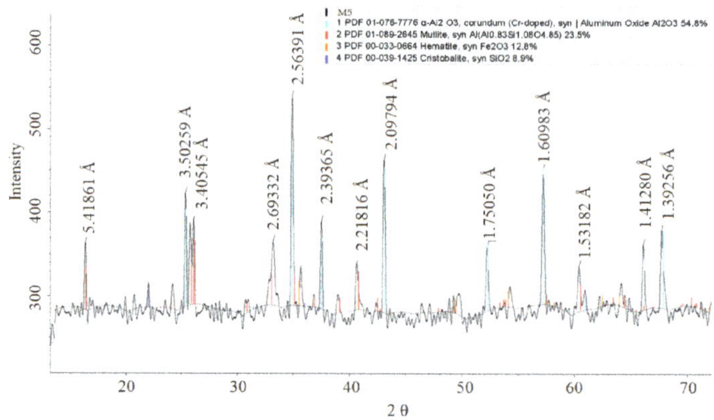

Figure 3. XRD patterns of composition M5 (1350 °C).

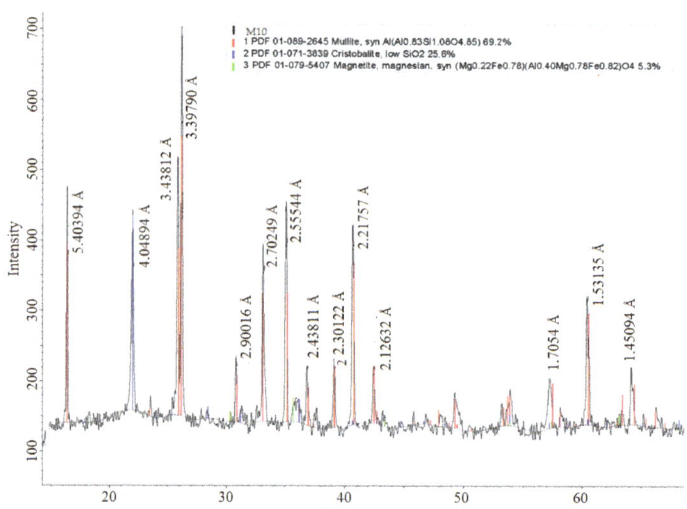

Figure 4. XRD patterns of composition M10 (1350 °C).

An XRD analysis of fired compositions containing 10% and 20% microsilica (Table 4) showed that the main phase is mullite, while the secondary phase is cristobalite and the spinel phase with complex composition (containing elements of aluminum, iron, and other impurities from the raw materials). The predominant phase is cristobalite in the composition containing 50% of microsilica, with mullite present in a smaller amount.

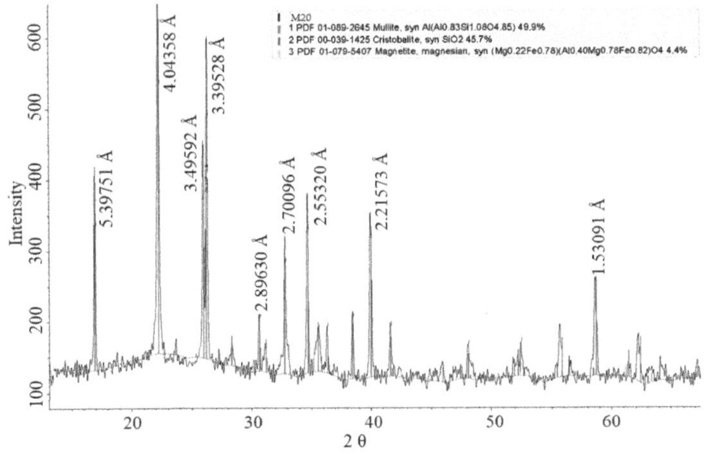

Figure 5. XRD patterns of composition M20 (1350 °C).

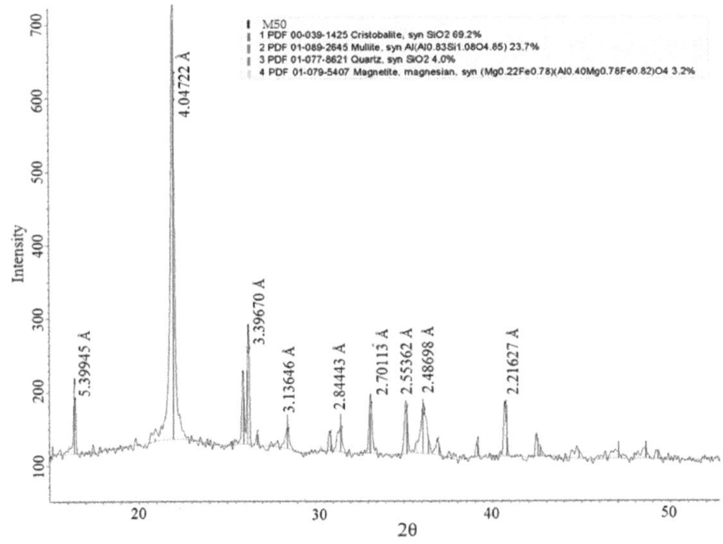

Figure 6. XRD patterns of composition M50 (1350 °C).

The structure of the compositions is porous and multiphasic, consisting of mullite intergrowths between which cristobalite, silicates of complex composition, and spinel phases based on aluminum oxides, iron oxides, and impurities present in the raw materials can be observed (Figure 7).

Our studies showed that with an increase in the content of microsilica up to 50%, the formation of mullite decreases and the amount of cristobalite significantly increases. This results in the deterioration of the strength properties of the samples. The maximum possible formation of mullite was observed in formulations containing 10–20% microsilica.

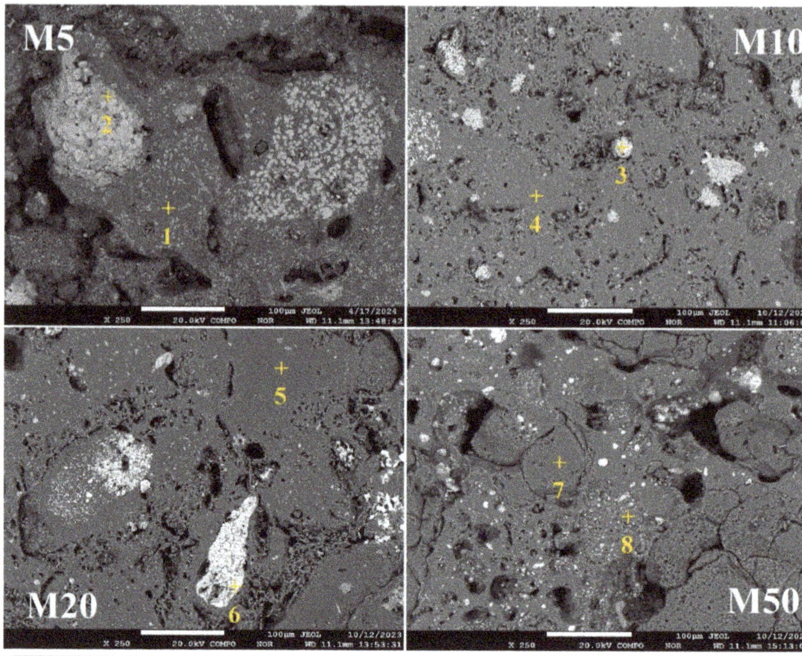

Chemical formula	Point							
	1	2	3	4	5	6	7	8
Al	25.22	17.81	3.97	27.08	21.68	2.20	-	20.81
Si	11.96	-	-	13.75	20.35	-	43.28	21.16
O	50.57	41.35	36.00	52.32	54.05	38.26	56.72	55.12
Fe	10.14	38.39	56.54	6.07	2.72	59.54	-	0.92

Figure 7. SEM images of the compositions of M5, M10, M20, and M50 (1350 °C).

3.3. *Proppants*

As shown in a previous study [22], during the preliminary heat treatment of bauxite, the dehydration of kaolinite and gibbsite occurs with the formation of intermediate compounds. The formation of metastable intermediate compounds with an imperfect structure after the preliminary heat treatment of raw materials contributes to the activation of the sintering process during the subsequent firing of proppants. Bauxite was pre-heat-treated at 1000 °C and milled to a fraction of less than 0.063 mm.

Charges were prepared based on heat-treated bauxite with a microsilica content of 5–20% (M5, M10 and M20). Methylcellulose 1.5% solution in the amount of 40–45 wt% was used as a binder.

The proppants were obtained using a granulator (Eirich Mixer EL1), which consists of a 1 L steel drum and a turboprop swirl equipped with a blade scraper. The swirl speed can be varied from 300 to 6000 rpm. The drum has two rotation modes—clockwise and counterclockwise—and operates at a speed of 85 or 170 rpm.

The granulation process was performed at the drum rotation speed (170 rpm) and swirl rotation speed (1000–2000 rpm).

The obtained granules of mullite–silicate proppants were dried. The proppant granules were then fired in the range of 1330–1400 °C in order to clarify the sintering temperature of the granulated material. The properties of the experimental ceramic proppants are listed in Table 5.

Table 5. Properties of the proppants (1330–1380 °C).

Composition	Open Porosity, %	Apparent Density, g/cm^3	Water Absorption, %
		1330 °C	
M5	51.13	1.80	22.08
M10	52.06	1.72	23.26
M20	50.40	1.65	23.43
		1350 °C	
M5	45.39	1.89	19.34
M10	47.25	1.79	20.87
M20	44.74	1.72	20.71
		1370 °C	
M5	24.09	2.05	10.50
M10	33.42	1.93	14.75
M20	35.08	1.90	15.60
		1380 °C	
M5	18.96	2.12	8.20
M10	23.59	1.97	11.10
M20	22.97	1.98	10.40

According to the results of the experiments (Figure 8), the proppant granules should be roasted in the range of 1370–1380 °C. When the proppant roasting temperature increases above 1380 °C, the overburning and sticking of granules is observed.

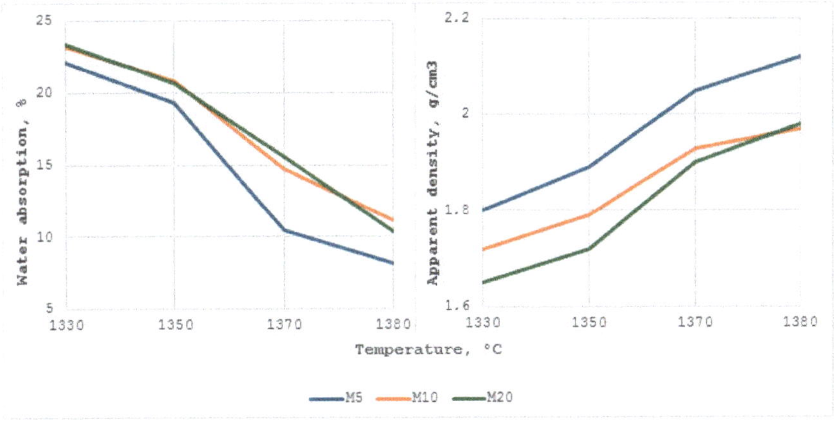

Figure 8. Dependence of water absorption and apparent density of proppants on the roasting temperature.

Experimental ceramic proppants (fraction 0.5–2.0 mm) roasted at 1380 °C with a soaking time of 1 h were tested for compliance with the requirements of the GOST 51761-2013 standard [24] (Table 6).

Table 6. Proppant test results.

Composition	Bulk Density, g/cm^3	Breaking Ratio at 51.7 MPa, %	Solubility in HCl Solution, %	Sphericity	Roundness
M5	1.64	26.7	0.40	0.9	0.9
M10	1.41	20.3	0.40	0.9	0.9
M20	1.21	19.1	0.34	0.7	0.9

4. Conclusions

The by-product of the production of metallurgical silica by KazSilikon LLP, Kazakhstan—namely, microsilica—was evaluated as a raw material for the production of mullite–silicate proppants.

The effects of the microsilica content in compositions based on high-iron bauxite on the phase composition and properties of ceramic samples were studied. It was found that, during the heat treatments of compositions based on bauxite and microsilica, structural and phase changes leading to the formation of mullite and compounds with complex composition (due to the participation of raw material impurities) occurred.

It was established that the introduction of 10–20% microsilica into the composition based on high-iron bauxite allowed for lightweight proppants with a bulk density of 1.21–1.41 g/cm^3, breaking ratio at 51.7 MPa of 19.1–20.3%, and sphericity and roundness of 0.7–0.9 to be obtained, with an optimal roasting temperature of 1370–1380 °C. Furthermore, the obtained proppants met the requirements of the National Standard GOST R 51761-2013.

Author Contributions: Conceptualization, B.K.; methodology, A.B.; validation, T.D. and A.U.; formal analysis, T.D. and G.R.; investigation, T.D.; resources, T.D.; data curation, G.R.; writing—original draft preparation, T.D.; writing—review and editing, A.B.; visualization, A.U.; supervision, B.K.; project administration, B.K.; funding acquisition, B.K. All authors have read and agreed to the published version of the manuscript.

Funding: This research has been funded by the Committee of Science of the Ministry of Science and Higher Education of the Republic of Kazakhstan (grant no. BR21882140 and BR24992757).

Institutional Review Board Statement: Not applicable.

Informed Consent Statement: Not applicable.

Data Availability Statement: The original contributions presented in the study are included in the article, further inquiries can be directed to the corresponding author.

Conflicts of Interest: The authors declare no conflicts of interest.

References

1. Speight, J.G. *Handbook of Hydraulic Fracturing*; John Wiley & Sons, Inc.: Hoboken, NJ, USA, 2016; pp. 195–200.
2. Liang, F.; Sayed, M.; Al-Muntasheri, G.A.; Chang, F.F. Overview of existing proppant technologies and challenges. In Proceedings of the SPE Middle East Oil and Gas Show and Conference, Manama, Bahrain, 8–11 March 2015; p. SPE–172763.
3. Yongpeng, S.; Qihua, W.; Mingzhen, W.; Baojun, B.; Yinfa, M. Experimental study of friction reducer flows in microfracture. *Fuel* **2014**, *131*, 28–35. [CrossRef]
4. Guo, Y.; Zhang, M.; Yang, H.; Wang, D.; Ramos, M.A.; Hu, T.S.; Xu, Q. Friction Challenge in Hydraulic Fracturing. *Lubricants* **2022**, *10*, 14. [CrossRef]
5. Biryukova, A.A.; Dzhienalyev, T.D.; Tikhonova, T.A. Ceramic Proppants Based on Natural Aluminosilicate Raw Materials of Kazakhstan. *New Refract.* **2017**, *6*, 33. [CrossRef]
6. Ma, H.; Bao, C.; Tian, Y.; Li, G. Effects of Feldspar Content on Microstructure and Property for High-Strength Corundum–Mullite Proppants. *Trans. Indian Ceram. Soc.* **2020**, *79*, 18–22. [CrossRef]
7. Liu, Z.; Zhao, J.; Li, Y.; Zeng, Z.; Mao, J.; Peng, Y.; He, Y. Low-temperature sintering of bauxite-based fracturing proppants containing CaO and MnO_2 additives. *Mater. Lett.* **2016**, *171*, 300–303. [CrossRef]
8. Zhao, J.; Liu, Z.; Li, Y. Preparation and characterization of low-density mullite-based ceramic proppant by a dynamic sintering method. *Mater. Lett.* **2015**, *152*, 72–75. [CrossRef]

9. Lodjanaja, V.A.; Shekero, M.A.; Pershina, L.S.; Belonosova, E.K.; Pavlova, I.A. Aluminosilicate Proppants Based on Technogenic Raw Materials. In Proceedings of the International Scientific and Practical Conference of Young Scientists "Energy and Resource Saving. Power Supply. Non–traditional and Renewable Energy Sources. Nuclear Energy. Danilov Readings—2021", Ekaterinburg, Russia, 13–17 December 2021; Available online: http://elar.urfu.ru/handle/10995/128452 (accessed on 8 October 2024). (In Russian)
10. Chumakov, A.A.; Yatsenko, E.A.; Tretiak, A.A. Effect of sodium fluoride and aluminum oxide on the strength characteristics of aluminosilicate proppants obtained based on drill cuttings. Bulletin of the North–Caucasian Region University. *Techn. Sci.* **2022**, *2*, 40–46. (In Russian) [CrossRef]
11. Skripchenko, A.S. Prospects for the use of domestic refractory clay raw materials of the North–West region in the technology of aluminosilicate proppants. In Proceedings of the XVI International Symposium of Students and Young Scientists Named after Academician M.A. Ussov, Tomsk, Russia, 2–7 April 2012. (In Russian)
12. Hao, J.; Hao, H.; Gao, Y.; Li, X.; Qin, M.; Wang, K. Effects of sintering temperature on properties of proppant prepared from coal gangue. *Mater. Sci.* **2020**, *26*, 94–98. [CrossRef]
13. Ding, D.; Fang, Y.; Xiao, G.; Zhu, X.; Fu, P.; Chong, X. Effects of sintering temperature on microstructure and properties of low-grade bauxite-based ceramic proppant. *Int. J. Appl. Ceram. Technol.* **2021**, *18*, 1832–1844. [CrossRef]
14. Li, X.; Hao, J. Optimization design of low–density and high–strength ceramic proppants by orthogonal experiment. *Adv. Compos. Lett.* **2020**, *29*, 2633366X20954875. [CrossRef]
15. Qin, M.; Tian, Y.; Hao, H.; Li, G.; Zhou, Y.; Bai, P. Effects of $CaCO_3$ additive on properties and microstructure of corundum-and mullite-based ceramic proppants. *Int. J. Appl. Ceram. Technol.* **2020**, *17*, 1026–1032. [CrossRef]
16. Liang, F.; Sayed, M.; Al–Muntasheri, G.A.; Chang, F.F.; Li, L. A comprehensive review on proppant technologies. *Petroleum* **2016**, *2*, 26–39. [CrossRef]
17. Perepelicyn, V.A.; Jukseeva, I.V.; Ostrjakov, L.V. Mineral and Raw Material Base of Modern Refractories Production. *Refract. Tech. Ceram.* **2008**, *6*, 53–64. Available online: https://www.elibrary.ru/download/elibrary_15288902_33279702.pdf (accessed on 8 October 2024). (In Russian)
18. Potapov, V.V.; Gorev, D.S. Comparative Results of Concrete Compressive Strength Rising by Addition of Nanosilica and Microsilica. *Mod. High Technol.* **2018**, *9*, 98–102. (In Russian) [CrossRef]
19. Suvorova, O.V.; Motina, A.V.; Manakova, N.K. Thermal insulation materials on the basis of microsilica. *Vestn. MSTU* **2015**, *18*, 149–155. Available online: https://vestnik.mauniver.ru/show.shtml?art=1663 (accessed on 8 October 2024). (In Russian)
20. Baisanov, A.; Vorobkalo, N.; Makhambetov, Y.; Mynzhasar, Y.; Zulhan, Z. Studies of the thermal stability of briquettes based on microsilica. *Kompleksnoe Ispolzovanie Mineralnogo Syra* **2023**, *327*, 57–63. [CrossRef]
21. Sangulova, I.; Selyaev, V.; Kuldeev, E.; Nurlybaev, R.; Orynbekov, Y. Assessment of the influence of the structural characteristics of granular systems of microsilicon on the properties of thermal insulation materials. *Kompleks. Ispolz. Miner. Syra* **2021**, *320*, 5–14. [CrossRef]
22. *GOST 2409-2014*; Refractories. Method for Determination of Bulk density, Apparent and True Porosity, Water Absorption. RU-GOST R: Moscow, Russia, 2014.
23. *GOST 4071.1-2021*; Refractory Products with Less Than 45% True Porosity. Method for Determination of Compressive Strength at Room Temperature. RU-GOST R: Moscow, Russia, 2021.
24. *GOST 51761-2013*; Aluminosilicate Proppants. RU-GOST R: Moscow, Russia, 2013.
25. Dzhienalyev, T.D.; Biryukova, A.A. Ceramic Proppants Based on High–Ferrous Bauxite. *Int. J. Adv. Sci. Technol.* **2019**, *28*, 724–731. Available online: http://sersc.org/journals/index.php/IJAST/article/view/1946 (accessed on 8 October 2024).

ceramics

MDPI

Article

Synthesis of Magnesia–Hercynite-Based Refractories from Mill Scale and Secondary Aluminum Dross: Implication for Recycling Metallurgical Wastes

Praphaphan Wongsawan, Nantiya Boonlom, Muenfahn Vantar and Somyote Kongkarat *

Research Unit in Sustainable Materials and Circular Economy, Faculty of Sciences and Technology, Thammasat University, Khlong Luang, Pathum Thani 12120, Thailand; praphaphan.wong@dome.tu.ac.th (P.W.); nantiya.boo@dome.tu.ac.th (N.B.); muenfahn.van@dome.tu.ac.th (M.V.)
* Correspondence: ksomyote@tu.ac.th

Abstract: This study investigates the synthesis of magnesia–hercynite-based refractories using blends of magnesia powder, aluminum dross (AD), mill scale (MS), and graphite, focusing on the effects of carbon concentration and heating temperature. The results demonstrate successful synthesis at 1550 °C and 1650 °C, with high magnesia content (C80 and D80) leading to the formation of distinct phases, including MgO, $FeAl_2O_4$, $MgFeAlO_4$, $CaMg(SiO_4)$, and $Ca_3Mg(SiO_4)_2$, which influence the ceramic's microstructure and mechanical properties. Increased magnesia content reduces porosity and enhances crushing strength, while heating to 1650 °C significantly improves densification and nearly doubles cold crushing strength, from 43.77–58.97 MPa at 1550 °C to 76.79–95.67 MPa at 1650 °C. These findings suggest that the synthesized refractories exhibit properties comparable to commercial magnesia–hercynite bricks, with potential for the further development for industrial rotary kiln applications.

Keywords: metallurgical wastes; mill scale; aluminum dross; magnesia–hercynite

1. Introduction

High alumina refractories were widely used in the high-temperature zones of cement rotary kilns. As kiln temperatures and capacities increased, magnesia–chrome bricks became the preferred lining because of their durability and ability to retain protective coatings [1]. However, concerns over the environmental hazards of hexavalent chromium (Cr^{6+}) led to the decline in magnesia–chrome bricks, which were replaced by chrome-free alternatives [1–3]. Chromite was initially used for its flexibility, but its environmental impact in the sintering zone prompted a shift to iron–alumina or hercynite spinel as safer options. Hercynite, though rare, is primarily produced through electro-fusion, a process that, while effective, is both time- and energy-consuming [4–7].

Magnesia–hercynite bricks have gained significant attention in the refractory materials industry because of their enhanced mechanical properties, thermal shock resistance, and environmental benefits. The incorporation of hercynite spinel ($FeAl_2O_4$) into magnesia-based refractories improves their thermo-mechanical properties, including density, porosity, and mechanical strength, making them suitable for demanding applications like cement kilns [8]. Studies show that adding spinel structures such as hercynite promotes microcracking from thermal expansion mismatch, enhancing crack propagation resistance [9]. The generated spinel phases, such as $MgFe_2O_4$, contribute to a more compact microstructure, especially with fused hercynite. Research on periclase–hercynite bricks under various atmospheric conditions reveals that iron ion diffusion in oxidizing atmospheres causes ring cracks, while more severe diffusion in CO atmospheres can lead to structural damage [10]. Recent advancements, including plasma fusion synthesis, have made the production of magnesia–hercynite bricks faster, more economical and higher performing [4]. Hercynite spinel is emerging as an environmentally friendly alternative to magnesia–chrome bricks, reducing the environmental risks associated with Cr^{6+} compounds. Overall, these bricks offer superior mechanical properties and thermal stability, with ongoing research expanding their potential applications [4].

Mill scale (MS) is one of the by-products of the iron and steel industries, generated during the hot rolling process of steel slabs, blooms, and billets. It is notable for its high iron content, comprising over 70% metallic iron or more than 90% iron oxides [11,12]. MS has been widely recycled in steelmaking processes, such as being blended with fossil fuels (coal, coke, and graphite) to replace iron ore or scrap iron [13,14]. Research has explored the direct reduction of MS using various reducing agents, including biomass and reducing gases, to produce iron powder and iron-bearing compounds [15–19].

Primary aluminum dross floats on the surface of molten aluminum during the smelting process. It acts as a protective layer that absorbs impurities and minimizes heat loss. It mainly consists of aluminum oxide and some metallic aluminum. Primary aluminum dross is generally re-melted in a rotary kiln to recover the remaining metallic aluminum. This re-melting process produces secondary aluminum dross (AD), primarily composed of aluminum oxide and salts. AD is often utilized in the production of high-value products such as ceramics, refractory materials, construction materials, and calcined alumina [20,21].

Magnesia–hercynite-based refractory is generally produced by mixing magnesia, or periclase (MgO), with hercynite spinel ($FeAl_2O_4$) [22,23]. The spinel can be either naturally occurring or synthesized through various methods, such as electro-fusion and plasma fusion [5–8], using raw materials like alumina powder [4,5], metallic aluminum [7], iron oxide [5,8], and mill scale [7].

Because of the limited availability of natural or virgin material resources, reducing the consumption of these materials in various industrial processes is essential. Rather than relying on virgin materials, this study aims to utilize industrial wastes, specifically mill scale and aluminum dross, as alternative sources of iron oxides and alumina for the production of magnesia–hercynite refractory. This research focuses on synthesizing magnesia–hercynite refractory by blending magnesia powder, mill scale, and aluminum dross in a single-step process, using graphite as a reducing agent in a carbothermic reduction carried out under a normal air atmosphere. The study emphasizes the impact of carbon content (C/O ratios of 1 and 2) and heating temperature on the properties of the resulting ceramics. Morphology, chemical composition, phase analysis, and basic physical and mechanical properties are examined. This study presents a novel approach to industrial waste management by producing magnesia–hercynite ceramics through a single-step blending process followed by firing.

2. Materials and Methods

2.1. Materials Preparation

The magnesia powder used in the present study was supplied by Thai Poly Chemicals Co., Ltd., Samut Sakhon, Thailand. Aluminum dross (AD) in the form of fine powder was collected from Top Five Manufactory Co., Ltd., an aluminum dross melting company, Chachoengsao, Thailand. The AD was generated from the secondary dross melting process, primarily containing alumina as its major component. Mill scale (MS) was provided by UMC Metal Co., Ltd., an electric arc furnace steel mill, Chonburi, Thailand. The MS in the form of small flakes was ground into powder using a ball mill. The raw material was sifted through a sieve to isolate particles with sizes less than 180 μm. An X-ray fluorescence spectrometer (XRF) was employed to determine the chemical compositions of the magnesia, AD, and MS, as shown in Tables 1–3, respectively. The XRF used in the present study was the Rigaku ZSX Primus, Rigaku, Tokyo, Japan. The analysis was conducted using sample powder, scanning the elemental range from boron (B) to uranium (U).

Table 1. Composition of the magnesia powder.

Oxides (wt%)			
MgO	SiO_2	CaO	Fe_2O_3
92.87	3.62	2.56	0.95

Table 2. Composition of the AD powder [24].

Oxides (wt%)												
Al_2O_3	SiO_2	Fe_2O_3	CaO	K_2O	MgO	MnO	Na_2O	SO_3	CuO	TiO_2	ZnO	Others
69.94	5.01	0.54	1.0	0.76	4.91	0.15	10.65	2.46	0.37	0.17	0.25	3.79

Table 3. Composition of the MS powder [24].

Oxides (wt%)								
Fe_2O_3	SiO_2	Al_2O_3	CaO	SO_3	TiO_2	K_2O	P_2O_5	other
93.66	1.42	0.82	0.17	0.08	0.04	0.02	0.04	3.75

The AD and MS were subsequently mixed with graphite powder, following C/O molar ratios of 1 and 2, and then mixed with the magnesia powder at 60–70 wt%, as shown in Table 4. The blends (C60–C80 and D60–D80) were homogeneously mixed in a rolling mill for 30 min. C represents the total moles of carbon in the graphite, while O corresponds to the total moles of oxygen derived from Fe_2O_3 in the MS. The graphite was supplied by Kanto Chemical Co., Tokyo, Japan (Cat. No. 17046-02).

Table 4. Composition of the blends at C/O = 1 (C60–C80) and C/O = 2 (D60–D80).

Blends	AD (wt%)	MS (wt%)	Graphite (wt%)	MgO (wt%)	Total (wt%)
C60	21.00	15.68	3.32	60	100
C70	15.75	11.76	2.49	70	100
C80	10.50	7.84	1.66	80	100
D60	19.40	14.48	6.12	60	100
D70	14.55	10.86	4.59	70	100
D80	9.70	7.24	3.06	80	100

2.2. Material Synthesis

To form cylindrical pellets, approximately 5 g of the raw blends of C60–C80 and D60–D80 were each compressed in a cylindrical mold under a load of 10 kN for 2 min using a hydraulic press. The resulting green pellets had dimensions of approximately 9.76 mm in height and 13.21 mm in diameter, as shown in Figure 1. An overview of the experimental procedure is provided in Figure 2. The produced green pellets were dried in a hot air oven at 90 °C for 48 h to remove moisture. The dried pellets were then placed into alumina refractory crucibles, capable of withstanding temperatures up to 1800 °C, and transferred to a horizontal tube furnace at 1550 °C and 1650 °C for 1 h under normal air atmosphere. To prevent thermal shock, the crucibles were initially held in the cold zone of the furnace for 5 min before being moved to the hot zone. After 1 h, the heated pellets were quenched and prepared for further analysis. The pellets with C/O ratios of 1 (C60–C80) and 2 (D60–D80), after being heated at 1550 °C for 1 h, are presented in Figure 3. Compared with the raw pellets in Figure 1, the resulting ceramic pellets turned brown after the heating process. Based on visual observation, the color changed slightly from dark to light brown as the magnesia content increased from 60 wt% to 80 wt% and became slightly darker with increasing carbon content.

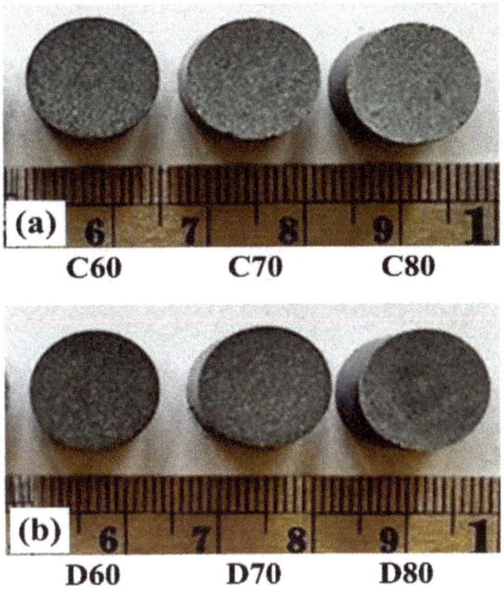

Figure 1. Green pellets of the blends at (**a**) C/O = 1 and (**b**) C/O = 2.

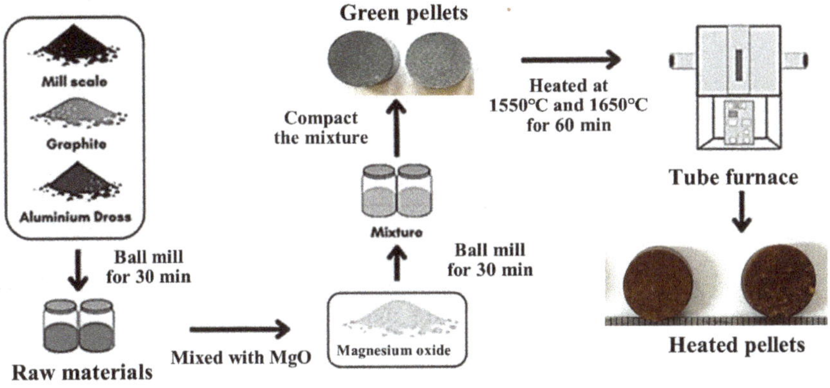

Figure 2. Overview of the experimental procedure.

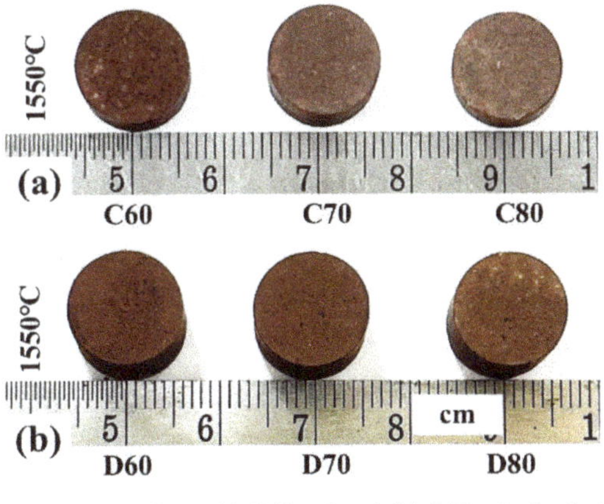

Figure 3. A pellet at (**a**) C/O = 1 and (**b**) C/O = 2 after being heated at 1550 °C for 1 h.

2.3. Formulation of the Reacting Raw Materials

In the magnesia–AD–MS–graphite system, the synthesis of magnesia–hercynite-based material can occur via the formation of hercynite and the reaction with MgO in the magnesia powder. Above 1550 °C, the formation of hercynite can occur through the carbothermic reduction of Fe_2O_3 in the system by solid carbon in graphite, as shown in Equations (1)–(3). The produced CO can further act as a reducing agent for Fe_2O_3, as shown in Equations (4)–(6). The CO_2 produced can react with solid carbon in the pellet to regenerate CO within the system, following the Boudouard reaction, as shown in Equation (7). The FeO produced can directly interact with Al_2O_3 to form hercynite, as represented by Equation (8), with a standard Gibbs free energy of $\Delta G^\circ = -71{,}086 + 11.89 \cdot T\ J \cdot mol^{-1}$ [5,6]. Metallic Fe, derived from Equations (3) and (5), can be oxidized by oxygen to reform FeO. This FeO can then interact with Al_2O_3 to produce hercynite as the final product, as described in Equation (9), which has a standard Gibbs free energy of $\Delta G^\circ = -328{,}348 + 82.004 \cdot T\ J \cdot mol^{-1}$ [5,6]. At 1550 °C, the ΔG° values for Equations (3), (6), (8), and (9) are −124.46, −271.71, −52.67, and −201.24 $kJ \cdot mol^{-1}$, respectively, and they become even more negative at 1650 °C, suggesting that these reactions can occur at this experimental temperature. MgO in the magnesia powder can interact with FeO–Al_2O_3 and other impurity oxides in the system to form a magnesia–hercynite-based material [23], which needs to be investigated.

$$3Fe_2O_3 + C \rightarrow 2Fe_3O_4 + CO \quad (1)$$

$$Fe_3O_4 + C \rightarrow 3FeO + CO \quad (2)$$

$$FeO + C \rightarrow Fe + CO \quad (3)$$

$$3Fe_2O_3 + CO \rightarrow 2Fe_3O_4 + CO_2 \quad (4)$$

$$Fe_3O_4 + CO \rightarrow 3FeO + CO_2 \quad (5)$$

$$FeO + CO \rightarrow Fe + CO_2 \quad (6)$$

$$CO_2 + C \rightarrow 2CO \quad (7)$$

$$FeO\ (l) + Al_2O_3\ (s) \rightarrow FeAl_2O_4\ (s) \quad (8)$$

$$Fe\ (l) + Al_2O_3\ (s) + \frac{1}{2}O_2(g) \rightarrow FeAl_2O_4(s) \quad (9)$$

2.4. Material Analysis

X-ray diffraction (XRD) analysis of the samples was carried out using a Bruker D2 PHASER with Cu Kα (k = 1.54184 Å) radiation, Bruker, Bremen, Germany. XRD scans were carried out over the range (2theta angle) 10° to 80° at a step size of 0.05°/step. Phase identification was conducted using JADE 9.7.0 software and the ICDD 2022 database.

Microstructure examination and elemental analysis were conducted using scanning electron microscopy (SEM) and energy-dispersive X-ray spectroscopy (EDS) with the JSM-7800 F instrument from JEOL, Akishima, Japan. The cold crushing strength of the heated pellets was measured using an H50KS universal testing machine obtained from Tinius Olsen, Philadelphia, PA, USA.

3. Results and Discussion

3.1. Effect of Carbon

In this section, the effect of carbon content in the magnesia–AD-MS–graphite system (C/O = 1 and 2) on the synthesis of magnesia–hercynite-based refractory was investigated.

The heated pellets were cross-sectioned and prepared for microscopic examination. Figures 4 and 5 present the SEM micrographs of the cross-sectioned pellets and their corresponding EDS spectra for pellets with C/O ratios of 1 and 2, respectively. Numerous pores can be observed in the bulk ceramic in both cases. At a magnification of ×100, visual observation reveals that the porosity of the pellets decreases as the magnesia concentration in the blends increases from 60 to 80 wt%. At a magnification of ×1500, the ceramic morphology exhibits a rough surface for the pellets with a C/O ratio of 1 (Figure 4), while a smoother surface is observed as the carbon concentration in the blends increased to a C/O ratio of 2 (Figure 5). EDS analysis showed that the ceramic is primarily composed of Fe-Al-Mg-O-C in all cases. This suggests that the major components of the resulting ceramic are MgO, FeO, and Al_2O_3. An increase in the magnesia concentration in the raw blends from 60 to 80 wt% leads to an increase in the MgO content and a decrease in the FeO and Al_2O_3 contents in the resulting ceramic. However, the MgO content in the resulting ceramic remains lower than the initial MgO content in the raw blends. Increasing the carbon content in the system from a C/O ratio of 1 to 2 resulted in no significant difference in the MgO-FeO-Al_2O_3 contents according to the EDS results. The EDS analysis revealed that the carbon content in the resulting ceramics was lower than the initial carbon input in the raw blends. This reduction is attributed to the consumption of carbon during the carbothermic reduction of Fe_2O_3, as shown in Equations (1)–(3), and material loss during the combustion process.

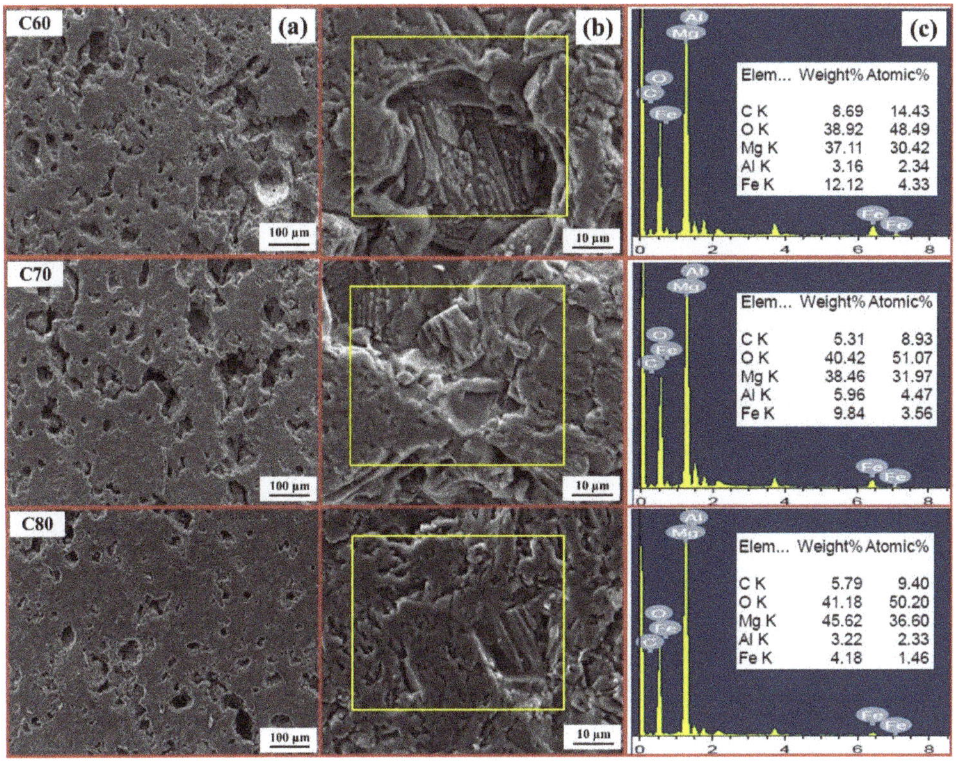

Figure 4. SEM images of cross-sectioned pellets at C/O = 1 after being heated at 1550 °C for 1 h, at the magnifications of (**a**) ×100 and (**b**) ×1500, and (**c**) the corresponding EDS area analysis.

Figure 6 presents the SEM images and EDS elemental distribution of the cross-sectioned pellets for blends C80 and D80 after heating at 1550 °C for 1 h. The images reveal that Mg forms equiaxed grains within the ceramic structure (green), while Al is densely packed in the areas between these grains (red). This occurs across the entire ceramic sample. The distribution of Fe, O, and C atoms is observed across the cross-section of the ceramic. This suggests that the resulting ceramic structure is primarily composed of equiaxed MgO grains (green region), with an FeO-Al_2O_3 phase (hercynite) at the grain boundary (red region). It is also observed that the carbon content in the raw blend, particularly the C/O ratios of 1 and 2, significantly affects the distribution of MgO and Al_2O_3 within the ceramic texture. In the pellet with a C/O ratio of 1, MgO and Al_2O_3 are densely packed in the equiaxed grain and grain boundary regions and also distributed throughout the cross-sectioned area. Conversely, in the pellet with a C/O ratio of 2, MgO and Al_2O_3 are primarily concentrated in the equiaxed grain and grain boundary regions, respectively. This suggests that the formation of MgO-FeO-Al2O4 spinel can occur and be distributed throughout the bulk ceramic. Additional XRF and XRD investigations were carried out for detailed composition analysis and phase identification.

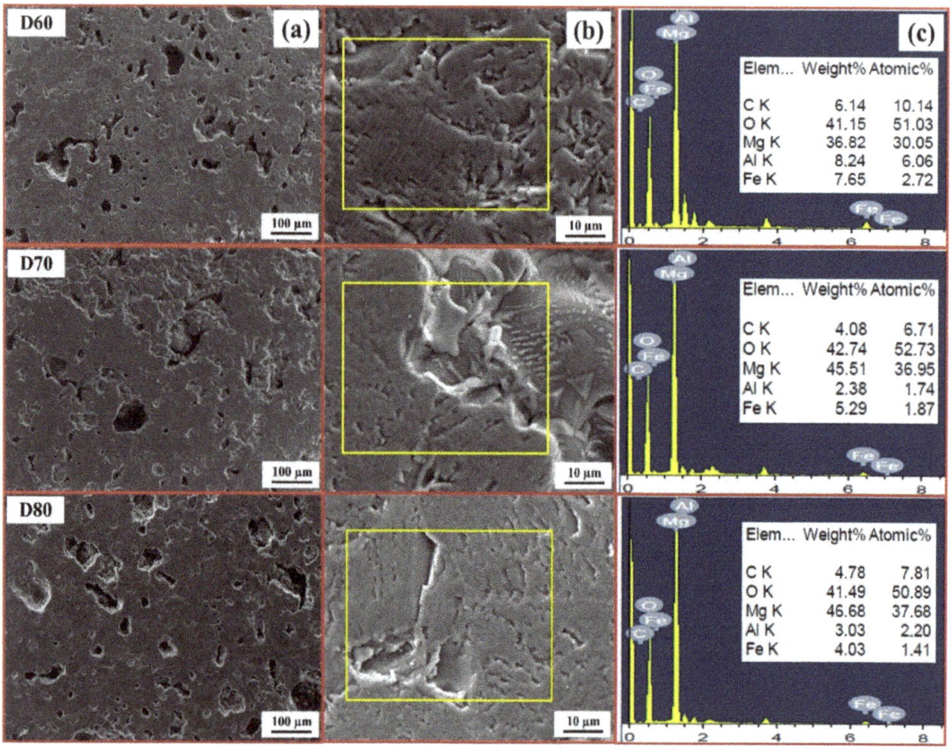

Figure 5. SEM images of cross-sectioned pellets at C/O = 2, after being heated at 1550 °C for 1 h, at the magnifications of (**a**) ×100 and (**b**) ×1500, and (**c**) the corresponding EDS area analysis.

Some of the resulting ceramics were ground into powder for XRF composition analysis, with the results presented in Table 5. In addition to the EDS results, the XRF analysis revealed that the major components of the resulting ceramic are MgO, Fe_2O_3, Al_2O_3, SiO_2, and CaO. The FeO content in the ceramic is represented by the Fe_2O_3 content in the XRF results (FeO = 0.89 × Fe_2O_3). Similar to the EDS results, the MgO content in the ceramic increases with an increase in the magnesia concentration in the raw blends, while the FeO and Al_2O_3 contents decrease. Increasing the carbon content in the system (C60–C80 and D60–D80) results in no significant difference in the chemical composition of the resulting ceramic. Compared with the initial MgO content in the raw blends (60–80 wt%), the MgO content in the resulting ceramics decreases by approximately 20% in samples C60–C80 (51.2–64.6 wt%), and by around 10–30% in samples D60–D80 (43.1–69.4 wt%). This reduction is expected because of the consumption of MgO in the formation of other ceramic compounds and material loss during the combustion process.

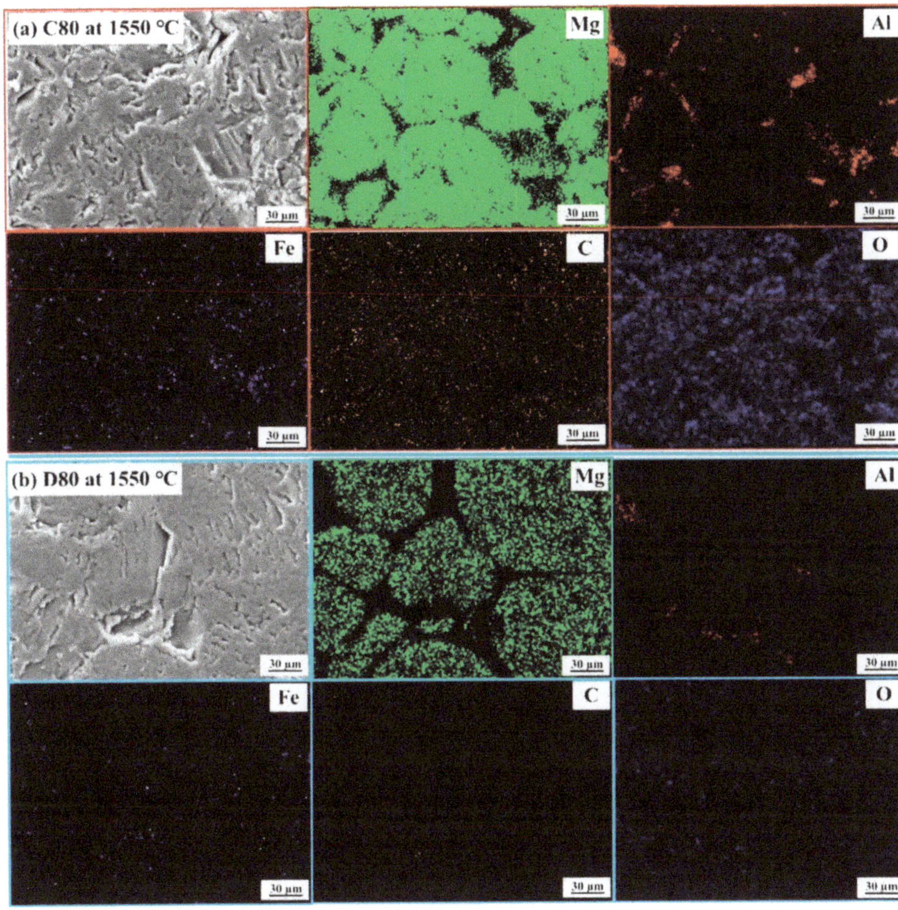

Figure 6. SEM images (×1500) and EDS elemental distribution of cross-sectioned pellets for the (**a**) C80 and (**b**) D80 blends after being heated at 1550 °C for 1 h.

Table 5. XRF analysis of the pellets at C/O = 1 (C60-C80) and C/O = 2 (D60-D80) after being heated at 1550 °C for 1 h.

Pellet	Oxides (wt%)					
	Fe_2O_3	Al_2O3	MgO	SiO_2	CaO	Other
C60	25.52	12.50	51.16	5.93	3.44	1.45
C70	20.88	11.57	55.59	6.14	4.52	1.30
C80	15.29	9.04	64.60	5.37	4.88	0.82
D60	33.28	9.35	43.10	6.58	6.16	1.53
D70	27.81	11.30	49.85	6.15	3.71	1.18
D80	10.04	7.70	69.39	7.32	4.60	0.95

XRD phase identification of the resulting ceramic was conducted, as shown in Figure 7. It was found that MgO, Fe_2O_3, Al_2O_3, SiO_2, and CaO in the ceramic interacted to form various ceramic compounds. The detected phases included periclase (MgO), hercynite ($FeAl_2O_4$), magnesium iron aluminum oxide ($MgFeAlO_4$), calcium magnesium silicate ($CaMg(SiO_4)$), merwinite ($Ca_3Mg(SiO_4)_2$), and carbon. The presence of $FeAl_2O_4$ indicates an interaction between FeO and Al_2O_3, as described in Equations (8) and (9), while the formation of the other compounds suggests interactions between MgO and the other oxides (Fe_2O_3, Al_2O_3, SiO_2, and CaO). The Rietveld refinement method was conducted using JADE 9.7.0 software and the ICDD 2022 database for the quantitative phase analysis of the resulting ceramic in all cases. Figure 8 presents an example of the XRD Rietveld refinement for the pellets of blend C60 (C/O = 1) after heating at 1550 °C for 1 h. The ICDD PDF card numbers of the detected phases are MgO: #98-000-0349, $FeAl_2O_4$: #98-000-0242, $MgFeAlO_4$: #04-018-0090, $CaMg(SiO_4)$: #01-084-1319, $Ca_3Mg(SiO_4)_2$: #98-000-0304, and C: #98-000-0231.

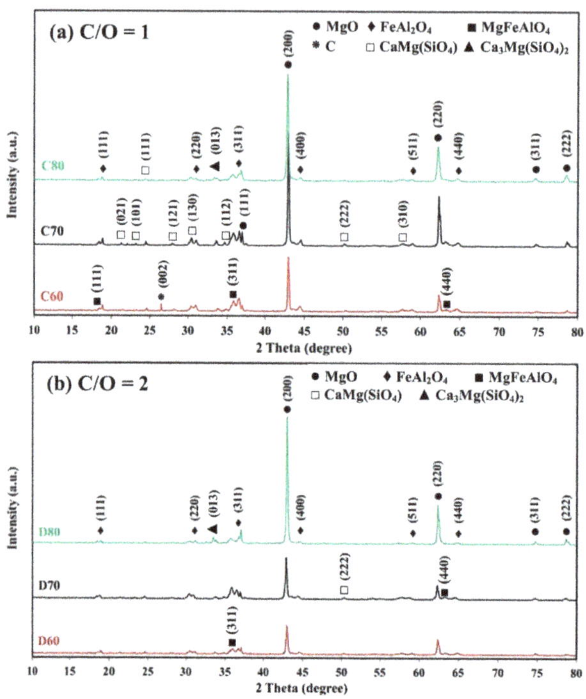

Figure 7. XRD patterns of the pellets at (**a**) C/O = 1 and (**b**) C/O = 2 after being heated at 1550 °C for 1 h.

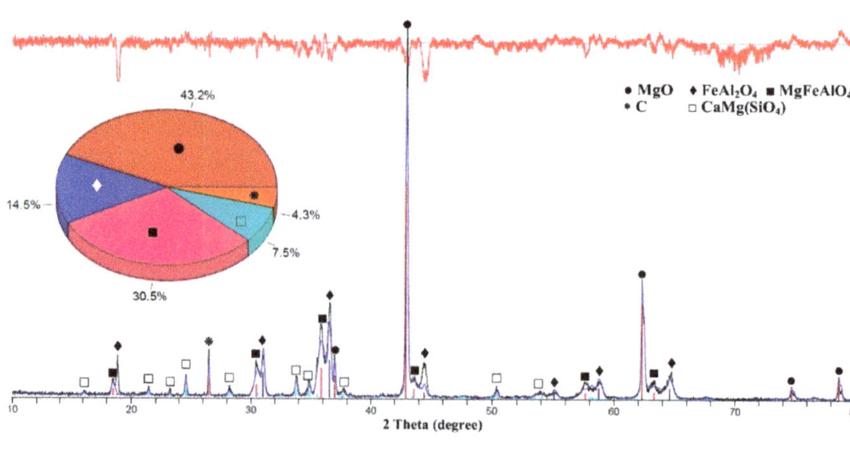

Figure 8. XRD Rietveld refinement of the pellets for blend C60 (C/O = 1) after being heated at 1550 °C for 1 h.

The quantitative phase composition of the pellets with C/O ratios of 1 (C60–C80) and 2 (D60–D80) after heating at 1550 °C for 1 h is shown in Table 6. The major phase in the resulting ceramic is MgO, with contents ranging from 43.2 to 73.3 wt% for the C60–C80 pellets and from 54.4 to 76.2 wt% for the D60–D80 pellets. Increasing the MgO content in the raw blends leads to an increase in the MgO content in the resulting ceramic. As the MgO content increases, the amounts of $FeAl_2O_4$ and $MgFeAlO_4$ tend to decrease. For the pellets with a C/O ratio of 1, C60 contains a moderate amount of MgO, along with significant quantities of $MgFeAlO_4$ and $FeAl_2O_4$. A small quantity of $CaMg(SiO_4)$ is present, and a trace of carbon is detected. In C70, the MgO content is higher than in C60, while the levels of $FeAl_2O_4$ and $MgFeAlO_4$ are reduced. A small amount of $CaMg(SiO_4)$ is also present. C80, which has the highest MgO content among the pellets with a C/O ratio of 1, shows further reductions in $FeAl_2O_4$ and $MgFeAlO_4$. $Ca_3Mg(SiO_4)_2$ is detected, along with a small amount of $CaMg(SiO_4)$. For the pellets with a C/O ratio of 2, D60 exhibits a relatively high MgO content, along with considerable amounts of $MgFeAlO_4$ and $FeAl_2O_4$. It also has the highest $CaMg(SiO_4)$ content among all the pellets, with no $Ca_3Mg(SiO_4)_2$ detected. In D70, the MgO content is lower than in D60, but it has the highest $FeAl_2O_4$ and $MgFeAlO_4$ contents among the pellets with a C/O ratio of 2. The $CaMg(SiO_4)$ content remains similar to that in D60, with no $Ca_3Mg(SiO_4)_2$ detected. D80 has the highest MgO content among all the pellets, with significantly reduced levels of $FeAl_2O_4$ and $MgFeAlO_4$. $Ca_3Mg(SiO_4)_2$ is present at a higher level than in C80, while $CaMg(SiO_4)$ is present in a small amount.

Table 6. Rietveld refinement quantitative phase composition of the pellets at C/O = 1 (C60–C80) and C/O = 2 (D60–D80) after being heated at 1550 °C for 1 h.

Pellet	Oxides (wt%)					
	MgO	$FeAl_2O_4$	$MgFeAlO_4$	$CaMg(SiO_4)$	$Ca_3Mg(SiO_4)_2$	C
C60	43.2	14.5	30.5	7.5	-	4.3
C70	60.7	9.0	24.1	6.2	-	-
C80	73.3	5.7	14.4	2.8	3.8	-
D60	54.4	12.0	25.6	8	-	-
D70	47.3	15.2	28.8	8.2	-	-
D80	76.2	5.2	9.6	2.1	6.9	-

The basic physical properties, including bulk density and apparent porosity, as well as mechanical properties of the resulting ceramic with C/O ratios of 1 and 2 after heating at 1550 °C for 1 h, are shown in Figure 9. It was found that varying the carbon content (C/O ratios of 1 and 2) in the raw blends has an insignificant effect on the bulk density of the resulting ceramic, which ranges between 2.73 and 2.99 g/cm^3. Increasing the amount of magnesia powder in the raw blends also shows no significant impact on the bulk density of the resulting ceramic. The apparent porosity of the resulting ceramic tends to decrease with increasing amounts of magnesia powder in the raw blends. For pellets with a C/O ratio of 1, the apparent porosity decreases from 8.66% to 4.11% vol, while for pellets with a C/O ratio of 2, it decreases from 12.68% to 3.87% vol. For blends containing 60 wt% MgO, the pellet with higher carbon content (D60) exhibits higher apparent porosity (12.68% vol) compared with the pellet with a lower carbon content (C60), which has an apparent porosity of 8.66% vol. However, for blends with higher MgO content (70 and 80 wt%), the trend is reversed, with pellets C70 and C80 showing higher apparent porosity than D70 and D80. This observation is also supported by the visual evidence of pore numbers from the SEM images shown in Figures 4 and 5.

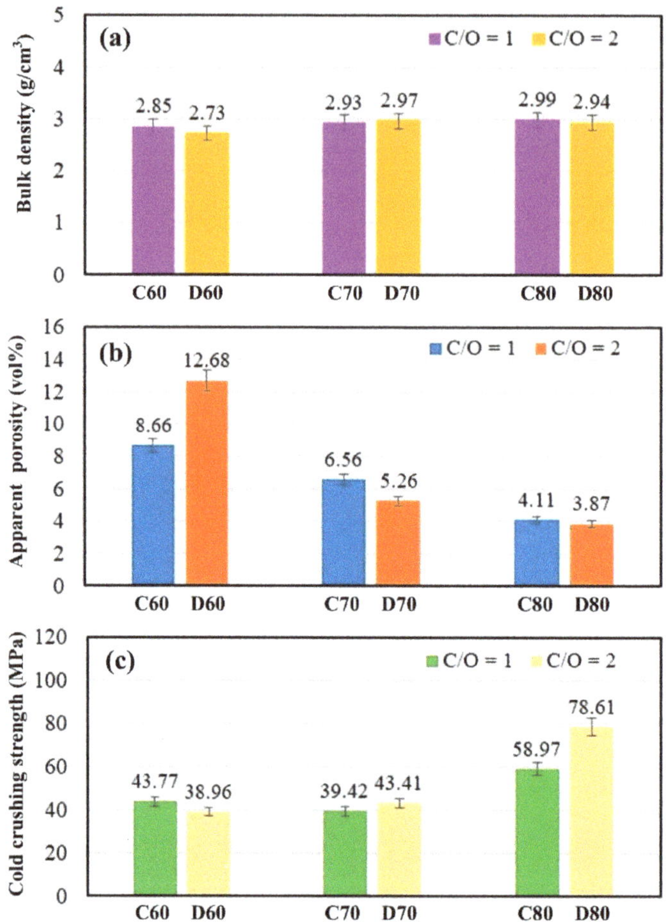

Figure 9. Comparison of the (**a**) bulk density, (**b**) apparent porosity, and (**c**) cold crushing strength of the pellets at C/O = 1 and C/O = 2 after being heated at 1550 °C for 1 h.

In terms of mechanical properties, the cold crushing strength of the resulting ceramic shows a clear trend of increasing with both higher MgO and carbon concentrations in the raw blends. This relationship is evident across different C/O ratios. For the pellets with a C/O ratio of 1, specifically those labeled C60 through C80, the cold crushing strength demonstrates a noticeable increase. The strength ranges from 43.77 MPa for the C60 pellet to 58.97 MPa for the C80 pellet. This trend highlights how the addition of MgO contributes to enhancing the mechanical robustness of the ceramic. In contrast, for the pellets with a C/O ratio of 2, which include D60 through D80, the cold crushing strength is generally higher. These pellets exhibit a range from 38.96 MPa for D60 to an impressive 78.61 MPa for D80. This indicates that the combination of increased MgO content and higher carbon concentrations results in a more substantial increase in crushing strength compared with the pellets with a C/O ratio of 1. The data also reveal that pellets with the highest MgO content, namely, C80 and D80, provide the highest cold crushing strength compared with those with lower MgO content. Specifically, C80, with its high MgO concentration, achieves a strength of 58.97 MPa, whereas D80, which not only has high MgO content but also a higher carbon concentration, reaches a maximum strength of 78.61 MPa.

The increase in carbon content from a C/O ratio of 1 to 2 is observed to enhance the cold crushing strength of the ceramic significantly. This is particularly evident when comparing the strength of 58.97 MPa in C80 (with a C/O ratio of 1) to 78.61 MPa in D80 (with a C/O ratio of 2). This improvement highlights the combined influence of higher MgO levels and increased carbon content on the mechanical properties of the ceramic material. The higher carbon content present in the raw blends provides more carbon for the carbothermic reduction of Fe_2O_3 present in the mill scale, available in both solid form and CO gas. This increased availability of carbon also facilitates other reactions between the oxides in the blends, leading to the formation of various ceramic compounds. As a result, these changes contribute to a reduction in bulk density (2.99 to 2.94 g/cm^3) and apparent porosity (4.11% to 3.87% vol), which, in turn, enhances the cold crushing strength of the ceramic. This relationship suggests that the optimized combination of MgO and carbon not only affects the microstructure but also plays a role in determining the overall mechanical strength of the ceramic.

3.2. *Effect of Temperature*

In this section, the effect of heating temperatures of 1550 °C and 1650 °C on the magnesia–AD-MS–graphite system for synthesizing magnesia–hercynite-based refractory is investigated. The carbon content in the raw blends was fixed at a C/O ratio of 1. Pellets with C/O ratios of 1 (C60–C80), after being heated at 1550 °C and 1650 °C for 1 h, are shown in Figure 10. It can be visually observed that the color of the pellets changes slightly from light brown to a deeper, stronger brown as the heating temperature increases.

Figure 10. A pellet at C/O = 1 after being heated at (**a**) 1550 °C and (**b**) 1650 °C for 1 h.

Figure 11 presents SEM micrographs of the cross-sectioned pellets and their corresponding EDS spectra for pellets with a C/O ratio of 1 after being heated at 1650 °C for 1 h. A number of pores can be observed throughout the bulk ceramic in all samples (C60–C80). However, compared with the results shown in Figure 4, the porosity appears to have decreased visually, and the surface of the ceramic is noticeably smoother. The EDS analysis indicated that the major components in the resulting ceramic are MgO, FeO, and Al_2O_3, which suggests the formation of related compounds within the ceramic, such as $FeAl_2O_4$ and $MgFeAlO_4$. The increase in heating temperature did not significantly affect the EDS analysis results in terms of composition. However, further XRF and XRD analyses are essential for both quantitative and qualitative assessments.

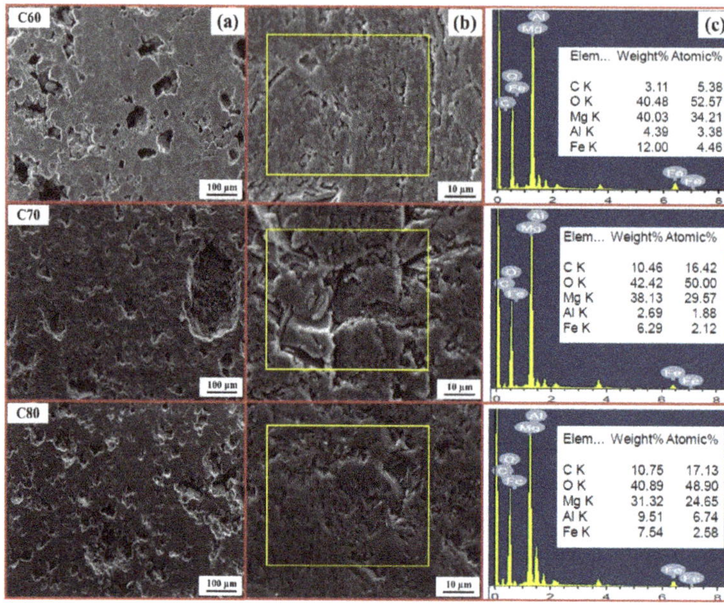

Figure 11. SEM images of cross-sectioned pellets at C/O = 1, after being heated at 1650 °C for 1 h, at the magnifications of (**a**) ×100 and (**b**) ×1500, and (**c**) the corresponding EDS area analysis.

Figure 12 presents the SEM images and EDS elemental distribution of the cross-sectioned C80 pellets after heating at 1650 °C for 1 h. It is observed that Mg (shown in green) forms equiaxed grains within the ceramic structure, while Al (shown in red) is densely packed in areas separate from these equiaxed grains. The distribution of iron (Fe), oxygen (O), and carbon (C) atoms is also observed across the ceramic's cross-section. This suggests that the resulting ceramic structure is primarily composed of equiaxed MgO grains (green regions) and an FeO-Al_2O_3 phase (red regions). Compared with the structure observed at 1550 °C in Figure 6a, the red regions corresponding to the FeO-Al_2O_3 phase increase and are more clearly separated from the green MgO regions. The formation of $MgFeAlO_4$, $CaMg(SiO_4)$, and $Ca_3Mg(SiO_4)_2$ is expected to distribute throughout the bulk ceramic and mainly at the red region or boundary of the MgO grain.

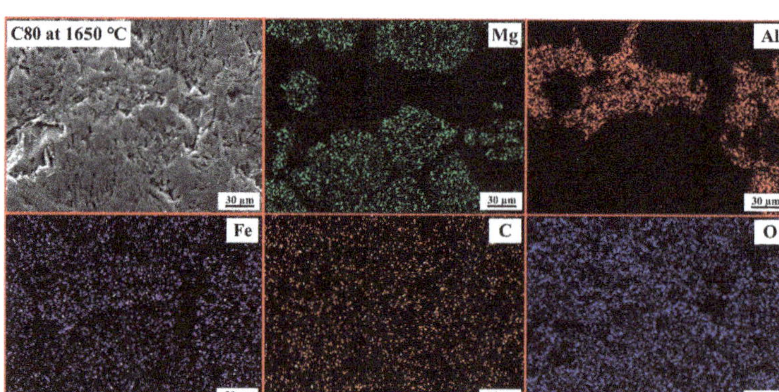

Figure 12. SEM images (×1500) and EDS elemental distribution of the cross-sectioned pellets for the C80 blend after heated at 1650 °C for 1 h.

Table 7 presents the XRF analysis of the pellets with a C/O ratio of 1 (C60–C80) after heating at 1650 °C for 1 h. The resulting ceramic is composed of MgO, Fe_2O_3, Al_2O_3, SiO_2, and CaO, which is not significantly different from the composition observed at 1550 °C, as shown in Table 5. The MgO content in the resulting ceramics of C60 to C80 increases from 46.21 wt% to 60.32 wt%, while the Fe_2O_3 and Al_2O_3 contents decrease from 26.33 wt% to 14.54 wt% and from 13.67 wt% to 10.89 wt%, respectively.

Table 7. XRF analysis of the pellets at C/O = 1 (C60–C80) after being heated at 1650 °C for 1 h.

Pellet	Oxides (wt%)					
	Fe_2O_3	Al_2O_3	MgO	SiO_2	CaO	Other
C60	26.33	13.67	46.21	8.12	4.12	1.55
C70	20.76	12.82	51.06	8.58	5.22	1.56
C80	14.54	10.89	60.32	7.99	5.33	0.93

The XRD analysis in Figure 13 reveals that the phases present in the resulting ceramic include MgO, $FeAl_2O_4$, $MgFeAlO_4$, $CaMg(SiO_4)$, $Ca_3Mg(SiO_4)_2$, and carbon, with variations depending on the magnesia content in the raw blends. The Rietveld refinement results for the quantitative phase composition of the pellets with a C/O ratio of 1 (C60–C80) after heating at 1650 °C for 1 h are shown in Table 8. As the MgO content increases from C60 to C80, there are noticeable changes in the composition of other oxides. In C60, the pellet contains 54.6 wt% MgO, with significant amounts of $MgFeAlO_4$ (29.0 wt%) and $FeAl_2O_4$ (8.9 wt%). The pellet also includes 7.4 wt% $CaMg(SiO_4)$, with no detectable amounts of $Ca_3Mg(SiO_4)_2$ or carbon. In C70, the MgO content rises to 58.1 wt% and $FeAl_2O_4$ increases to 14.7 wt%, while $MgFeAlO_4$ decreases to 19.9 wt%. The $CaMg(SiO_4)$ content remains relatively stable at 7.3 wt%, with no detection of $Ca_3Mg(SiO_4)_2$ or carbon. C80, which has the highest MgO content at 77.9 wt%, shows a significant decrease in $FeAl_2O_4$ and $MgFeAlO_4$ to 7.8 wt% and 12.9 wt%, respectively. The $CaMg(SiO_4)$ content decreases to 3.0 wt%, $Ca_3Mg(SiO_4)_2$ appears at 3.6 wt%, and a small amount of carbon (0.7 wt%) is also detected. Compared with the data in Table 6, increasing the heating temperature from 1550 °C to 1650 °C results in an increase in the amounts of MgO and $MgFeAlO_4$ phases. The quantities of other phases show only slight changes.

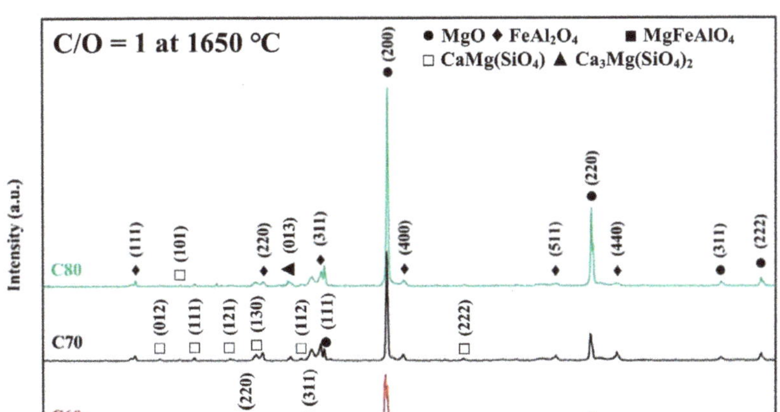

Figure 13. XRD patterns of the pellets at C/O = 1 after being heated at 1650 °C for 1 h.

Table 8. Rietveld refinement quantitative phase composition of the pellets at C/O = 1 (C60–C80) after being heated at 1650 °C for 1 h.

Pellet	Oxides (wt%)					
	MgO	$FeAl_2O_4$	$MgFeAlO_4$	$CaMg(SiO_4)$	$Ca_3Mg(SiO_4)_2$	C
C60	54.6	8.9	29.0	7.4	-	-
C70	58.1	14.7	19.9	7.3	-	-
C80	77.9	7.8	12.9	3.0	3.6	0.7

Figure 14 illustrates the bulk density, apparent porosity, and cold crushing strength of the pellets with a C/O ratio of 1 (C60–C80) after heating at 1550 °C and 1650 °C for 1 h. Increasing the heating temperature from 1550 °C to 1650 °C results in a slight reduction in bulk density. At 1550 °C, the bulk density ranges from 2.85 to 2.99 g/cm^3, while at 1650 °C, it ranges from 2.78 to 2.93 g/cm^3. The apparent porosity shows a significant decrease with increasing temperature. At 1550 °C, the apparent porosity ranges from 8.66% to 4.11% vol, whereas at 1650 °C, it reduces to a range of 4.31% to 1.58% vol. A lower apparent porosity indicates fewer void spaces or pores within the material, as supported by the SEM images in Figure 11. Comparing the pellets heated at 1550 °C and 1650 °C, although the measured densities are nearly the same, the apparent porosity differs. This may be attributed to the types of pores present in the bulk ceramics, specifically open and closed pores. Heating at the higher temperature of 1650 °C may promote the formation of new phases that can fill the pores, thereby reducing porosity. The reduction in apparent porosity of approximately 50% suggests that the ceramic becomes denser and less porous at the higher temperature, reflecting improved material compactness and reduced void content. These observations indicate that heating to 1650 °C enhances the densification of the ceramic material, leading to a more compact structure with lower porosity. As a result, the mechanical properties of the ceramic significantly improve with the increase in heating temperature. The cold crushing strength of the ceramic heated at 1550 °C ranges from 43.77 to 58.97 MPa. However, when the temperature is increased to 1650 °C, the cold crushing strength nearly doubles, reaching values between 76.79 and 95.67 MPa.

3.3. Evaluation of the Magnesia–Hercynite Ceramic Produced from Commercial and Waste Resources

Based on the experimental results, the ceramic pellets of C80 and D80 exhibit the optimal conditions among all tested samples, demonstrating favorable bulk density, low apparent porosity, and high cold crushing strength. These pellets were selected for comparison with commercial magnesia–hercynite bricks, as detailed in Table 9. The following commercial magnesia–hercynite bricks were used for comparison: one from China (RFC#1) and one from India (RFC#2). The MgO source for both RFC#1 and RFC#2 is commercial magnesia. The hercynite in these samples was produced using the electro-fused method with commercial alumina and iron sources. The chemical compositions of RFC#1 and RFC#2 are quite similar. Both bricks have a high MgO content, with RFC#1 containing 85 wt% and RFC#2 containing 87.8 wt%. The major components are followed by Fe_2O_3, which ranges from 3.8 to 4.8 wt%, and Al_2O_3, which ranges from 3.4 to 4.9 wt%. Both samples also include small amounts of CaO and SiO_2. Regarding their basic physical properties, RFC#1 and RFC#2 have nearly identical bulk density and apparent porosity. However, RFC#1 has a cold crushing strength of 70 MPa, whereas RFC#2 has a lower strength of 57.3 MPa. Compared with the commercial magnesia–hercynite bricks, the ceramics produced in the present study have a lower MgO content, ranging from approximately 64 to 70 wt%. In contrast, the Fe_2O_3 and Al_2O_3 components are more than twice as high as those in the commercial ceramics, ranging from approximately 10 to 15 wt% and 8 to 11 wt%, respectively. Because of the waste materials used, the CaO and SiO_2 contents are also significantly higher in the ceramics from the present study. The ceramics produced in the present study exhibit a bulk density similar to that of RFC#1 and RFC#2, ranging from

2.8 to 3 g/cm^3, but with significantly lower apparent porosity. The cold crushing strength is also comparable, ranging from approximately 59 to 95 MPa.

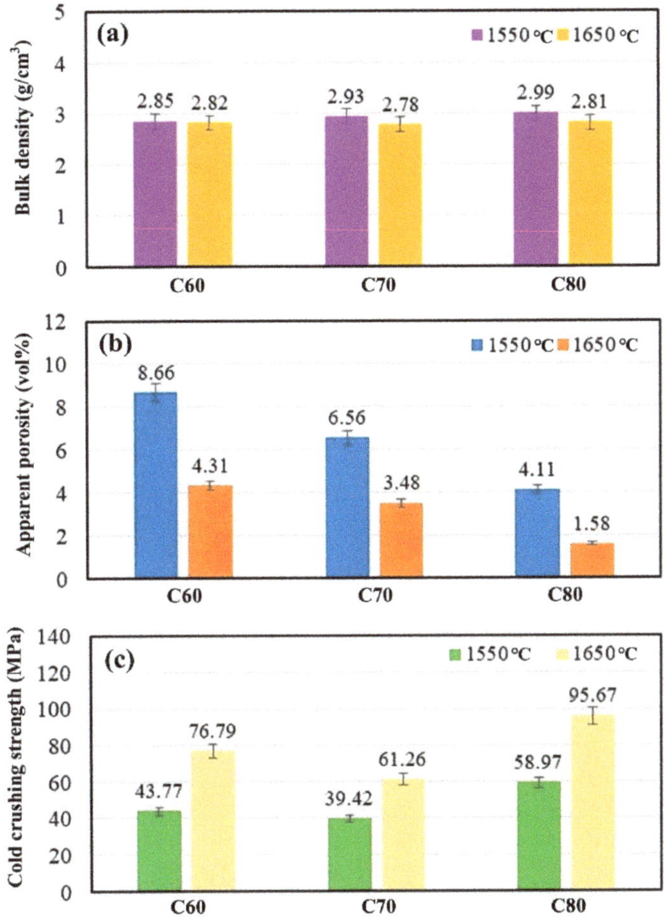

Figure 14. Comparison of the (**a**) bulk density, (**b**) apparent porosity, and (**c**) cold crushing strength of the pellets at C/O = 1 after being heated at 1550 °C and 1650 °C for 1 h.

Ding et al. [22] investigated the effect of hercynite content on the properties of magnesia–spinel composite refractories sintered in different atmospheres. It was reported that the addition of 2–3 wt% FeAl2O4 to magnesia–spinel composite refractories improved sintering and enhanced thermal conductivity. Among the tested specimens, the one with 2 wt% $FeAl_2O_4$ sintered in an argon atmosphere exhibited the highest thermal conductivity and optimal overall properties. The study also indicated that Fe^{2+} was more effective than Fe^{3+} in enhancing the sintering process of $MgO\text{-}MgAl_2O_4$. Jiang et al. [23] studied the morphology characterization of periclase–hercynite refractories produced through reaction sintering. It was concluded that during the reaction sintering of periclase–hercynite bricks at 1600 °C, cation diffusion between periclase and hercynite crystals significantly influenced microstructure development. Fe^{2+}, Fe^{3+}, and Al^{3+} ions from hercynite migrated and reacted with periclase to form spinels with varying Fe/Al ratios. Simultaneously, Mg^{2+} ions from periclase (MgO) migrated into hercynite, leading to spinel formation with a lower Fe/Al ratio. This cation exchange enhanced the sintering process and resulted in a microporous structure in the periclase–hercynite brick.

Table 9. Comparison of magnesia–hercynite ceramic produced from commercial and waste resources.

Refractory Sample	Commercial Resources		Present Study		
	RFC#1 [2]	RFC#2 [4]	C80 at 1550 °C	D80 at 1550 °C	C80 at 1650 °C
			Raw materials		
Magnesia	China	India	Thailand magnesia		
Hercynite	Commercial electro-fused		Mill scale and aluminum dross combustion		
			Chemical composition (wt%)		
MgO	85.0	87.8	64.6	69.93	60.32
Fe_2O_3	3.8	4.8	15.29	10.04	14.54
Al_2O_3	3.4	4.89	9.04	7.7	10.89
CaO	0.7	1.25	4.88	4.6	5.33
SiO_2	0.3	0.6	5.37	7.32	7.99
			Basic physical properties		
Bulk density (g/cm^3)	3.06	2.97	2.99	2.94	2.81
Apparent porosity (vol%)	14.0	16.5	4.11	3.87	1.58
Cold crushing strength (MPa)	70	57.3	58.97	78.61	95.67

In the present study, impurities in the oxides present in the aluminum dross (AD) and mill scale (MS) may account for the variations in chemical composition compared with commercial bricks. These impurities are expected to influence the strength and apparent porosity of the resulting ceramic. In the FeO-Al_2O_3-MgO ternary system, the formation of $FeAl_2O_4$ and $MgFeAlO_4$ phases was observed in the resulting ceramic, with MgO as the primary component. This occurs because of the migration of Mg^{2+} ions from MgO into hercynite, as well as the migration of Fe^{2+}, Fe^{3+}, and Al^{3+} ions from hercynite, which react with MgO [23]. Because of the presence of high amounts of impurity oxides, the combination of MgO, SiO_2, and CaO can form a liquid phase in the ternary SiO_2-CaO-MgO system, which has a melting temperature of 1500 °C [25]. According to the SiO_2-CaO-MgO phase diagram [25], this process results in the formation of $CaMg(SiO_4)$ and $Ca_3Mg(SiO_4)_2$ phases because of the heating temperatures of 1550 °C and 1650 °C used in the present study [23]. The formation of the merwinite phase ($Ca_3Mg(SiO_4)_2$) is observed only in samples with high magnesia content (C80 and D80). The migration of Fe^{2+}, Fe^{3+}, and Al^{3+} ions from hercynite into the liquid phase of the SiO_2-CaO-MgO system transforms the boundary liquid phase from a SiO_2-CaO-MgO system to a SiO_2-CaO-MgO-Fe_2O_3-Al_2O_3 system. This change reduces both the melting temperature and the viscosity of the boundary liquid [26]. These are expected to influence the microstructure of the resulting ceramic, impacting both the cold crushing strength and apparent porosity. As observed from the SEM images and elemental distribution (Figures 6 and 12), the resulting ceramic consists of equiaxed MgO grains (green regions) and an FeO-Al_2O_3 phase (red regions) at the boundaries. The formation of $MgFeAlO_4$, $CaMg(SiO_4)$, and $Ca_3Mg(SiO_4)_2$ is anticipated to occur throughout the bulk ceramic but is primarily concentrated at the red regions or boundaries of the MgO grains.

4. Conclusions

In the present study, the synthesis of magnesia–hercynite-based refractories was conducted using various blends of magnesia powder, aluminum dross (AD), mill scale (MS), and graphite. The effects of carbon concentration and heating temperatures were then investigated. The experimental results can be summarized as follows:

1. Magnesia–hercynite-based refractories can be successfully synthesized from MgO, MS, AD, and graphite via reactions at 1550 °C and 1650 °C for 1 h in a normal air atmosphere. The single-step blending of raw materials was employed for this process.
2. The combination of blend samples with high magnesia content (C80 and D80) and high levels of impurity oxides (CaO and SiO_2) leads to variations in the chemical composition and the formation of distinct phases in the resulting ceramic. The phases present in the resulting ceramic include MgO, $FeAl_2O_4$, $MgFeAlO_4$, $CaMg(SiO_4)$, and $Ca_3Mg(SiO_4)_2$. These phases, in turn, affect the microstructure, as well as the physical and mechanical properties of the produced magnesia–hercynite-based refractory.
3. The produced magnesia–hercynite-based refractories consist of equiaxed MgO grains and an FeO-Al_2O_3 spinel phase at the boundaries. The formation of $MgFeAlO_4$, $CaMg(SiO_4)$, and $Ca_3Mg(SiO_4)_2$ is anticipated to occur throughout the bulk ceramic, but it is primarily concentrated at the red regions or boundaries of the MgO grains.
4. Varying the carbon content (C/O ratios of 1 and 2) has a minimal impact on the bulk density and apparent porosity of the resulting ceramic. Increasing the magnesia powder content (60–80 wt%) does not significantly affect bulk density but tends to reduce apparent porosity. A higher MgO content combined with increased carbon concentration (C/O = 2) leads to a more notable increase in crushing strength compared with a C/O ratio of 1. At 1550 °C, the pellets with the highest MgO content (C80 and D80) exhibit the greatest cold crushing strength.
5. Increasing the heating temperature from 1550 °C to 1650 °C results in a slight reduction in bulk density and a significant decrease in apparent porosity, enhancing the ceramic's densification and compactness. This leads to a notable improvement in mechanical properties, with cold crushing strength nearly doubling from 43.77–58.97 MPa at 1550 °C to 76.79–95.67 MPa at 1650 °C.
6. Blends with high magnesia content (C80 and D80) show optimal conditions for synthesizing magnesia–hercynite-based refractories from the magnesia–dross–scale–graphite system, exhibiting physical and mechanical properties comparable to commercial magnesia–hercynite bricks. However, the quantity and type of phases in the ceramic differ from those in commercial bricks, indicating that further investigation of the ceramic's large-scale and thermal properties is necessary for potential application in industrial rotary kilns.

Author Contributions: Conceptualization, P.W. and S.K.; methodology, P.W. and S.K.; software, S.K.; validation, P.W., N.B., M.V. and S.K.; formal analysis, P.W., N.B., M.V. and S.K.; investigation, P.W., N.B., M.V. and S.K.; resources, S.K.; data curation, P.W., N.B., M.V. and S.K.; writing—original draft preparation, P.W. and S.K.; writing review and editing, S.K.; visualization, P.W. and S.K.; supervision, S.K.; project administration, S.K.; funding acquisition, S.K. All authors have read and agreed to the published version of the manuscript.

Funding: Thailand Science Research and Innovation Fundamental Fund fiscal year 2023, Contract No. TUFF 37/2566.

Data Availability Statement: The raw data supporting the conclusions of this article will be made available by the authors upon request.

Acknowledgments: This work was supported by the Thailand Science Research and Innovation Fundamental Fund fiscal year 2023.

Conflicts of Interest: The authors declare no conflicts of interest.

References

1. Guo, Z.Q. Technical Progress in Basic Refractories for Cement Rotary Kiln. *J. Tech. Assoc. Refract.* **2003**, *23*, 218–225.
2. Guo, Z.; Nievoll, J. An Overview of Magnesia-Hercynite Refractories for Cement Rotary Kilns. *China's Refract.* **2007**, *16*, 65–71.
3. Zhang, X.; Yu, R.; Yu, X. Characteristics of hercynite and its application: In refractories. *China's Refract.* **2012**, *21*, 17–22.
4. Padhi, L.; Sahu, P.; Sahoo, N.; Singh, S.; Tripathy, J. A Novel Process for Synthesis of Iron-Alumina Spinel and Its Application in Refractory for Cement Rotary Kiln. *Trans. Indian Ceram. Soc.* **2017**, *76*, 196–201. [CrossRef]
5. Chen, J.; Su, J.; Yan, M.; Yang, S. Two methods of synthesizing hercynite. *Appl. Mech. Mater.* **2014**, *543–547*, 3830–3833. [CrossRef]

6. Chen, J.; Yu, L.; Sun, J.; Li, Y.; Xue, W. Synthesis of hercynite by reaction sintering. *Eur. Ceram. Soc.* **2011**, *31*, 259–263. [CrossRef]
7. Perdomo-Gonzales, L.; Quintana-Puchol, R.; Alujas-Diaz, A.; Perdomo-Gomez, L.A.; Ruiz-Perez, R.; Cruz-Crespo, A. Aluminothermic synthesis of ceramics from the hercynite-alumina system, using mill scale and aluminium Chips. *DYNA* **2023**, *90*, 106–113. [CrossRef]
8. Rodríguez, E.; Castillo, G.; Contreras, J.; Puente-Ornelas, R.; Aguilar-Martínez, J.; García, L.; Gómez, C. Hercynite and magnesium aluminate spinels acting as a ceramic bonding in an electro fused MgO–$CaZrO_3$ refractory brick for the cement industry. *Ceram. Int.* **2012**, *38*, 6769–6775. [CrossRef]
9. Khlifi, I.; Pop, O.; Dupré, J.; Doumalin, P.; Huger, M. Investigation of microstructure-property relationships of magnesia-hercynite refractory composites by a refined digital image correlation technique. *J. Eur. Ceram. Soc.* **2019**, *39*, 3893–3902. [CrossRef]
10. Liu, M.; Li, Y.; Ma, S.; Xia, W.; Li, Y. Effects of Atmosphere on the Periclase-Hercynite Brick. *Adv. Mater. Res.* **2012**, *476–478*, 1523–1528. [CrossRef]
11. Eissa, M.; Ahmed, A.; El-Fawkhry, M. Conversion of mill scale waste into valuable products via carbothermic reduction. *J. Metall.* **2015**, *2015*, 926028. [CrossRef]
12. Bagatini, M.C.; Zymla, V.; Osorio, E.; Vilela, A.C.F. Characterization and reduction behavior of mill scale. *ISIJ Int.* **2011**, *51*, 1072. [CrossRef]
13. Sen, R.; Dehiya, S.; Pandel, U.; Banerjee, M.K. Utilization of low-grade coal for direct reduction of mill scale to obtain sponge iron: Effect of reduction time and particle size. *Procedia Earth Planet. Sci.* **2015**, *11*, 8–14. [CrossRef]
14. Martin, M.I.; Lopez, F.A.; Torralba, J.M. Production of sponge iron powder by reduction of rolling mill scale. *Ironmak. Steelmak.* **2012**, *39*, 155–162. [CrossRef]
15. Baganiti, M.C.; Kan, T.; Evans, T.J.; Strezov, V. Iron ore reduction by biomass volatiles. *J. Sustain. Metall.* **2021**, *7*, 215–226.
16. Khaerudini, D.S.; Chanif, I.; Insiyanda, D.R.; Destyorini, F.; Alva, S.; Premono, A. Preparation and characterization of mill scale industrial waste reduced by biomass-based carbon. *J. Sustain. Metall.* **2019**, *5*, 510–518. [CrossRef]
17. Sista, K.S.; Dwarapudi, S.; Nerune, V.P. Direct reduction recycling of mill scale through iron powder synthesis. *ISIJ Int.* **2019**, *5*, 787–794. [CrossRef]
18. Shi, J.; Wang, D.R.; He, Y.D.; Qi, H.B.; Wei, G. Reduction of oxide scale on hot-rolled strip steels by carbon monoxide. *Mater. Lett.* **2018**, *62*, 3500–3502.
19. Ye, Q.; Zhu, H.; Zhang, L.; Ma, J.; Zhou, L.; Liu, P.; Chen, J.; Chen, G.; Peng, J. Preparation of reduced iron powder using combined distribution of wood-charcoal by microwave heating. *J. Alloys Compd.* **2014**, *613*, 102–106. [CrossRef]
20. Xie, H.; Guo, Z.; Xu, R.; Zhang, Y. Particle sorting to improve the removal of fluoride and aluminum nitride from secondary aluminum dross by roasting. *Environ. Sci. Pollut. Res.* **2023**, *30*, 54536–54546. [CrossRef]
21. Lin, W.C.; Tsai, C.H.; Zhang, D.N.; Syu, S.S.; Kuo, Y.M. Recycling of aluminum dross for producing calcinated alumina by microwave plasma. *Sustain. Environ. Res.* **2022**, *32*, 50. [CrossRef]
22. Ding, X.; Zhao, H.; Xiang, Z.; Zhang, H.; He, Q.; Li, J. Effect of hercynite content on the properties of magnesia-spinel composite refractories sintered in different atmospheres. *Ceram. Int.* **2016**, *42*, 19058–19062. [CrossRef]
23. Jiang, P.; Chen, J.; Yan, M.; Li, B.; Su, J.; Hou, X. Morphology characterization of periclase- hercynite refractories by reaction sintering. *Int. J. Miner. Metall. Mater.* **2015**, *22*, 1219–1224. [CrossRef]
24. Wongsawan, P.; Srichaisiriwech, W.; Kongkarat, S. Synthesis of Ferroalloys via Mill Scale-Dross-Graphite Interaction: Implication for Industrial Wastes Upcycling. *Metals* **2022**, *12*, 1909. [CrossRef]
25. Huang, W.M.; Hillert, M.; Wang, X. Thermodynamic assessment of the CaO-MgO-SiO_2 System. *Metall. Mater. Trans. A* **1995**, *26*, 2293. [CrossRef]
26. Shankar, A.; Görnerup, M.; Lahiri, A.K.; Seetharaman, S. Experimental Investigation of the Viscosities in CaO-SiO_2-MgO-Al_2O_3 and CaO-SiO_2-MgO-Al_2O_3-TiO_2 Slags. *Metall. Mater. Trans. B* **2007**, *38*, 911. [CrossRef]

ceramics

Article

Anti-Bacterial and Anti-Inflammatory Effects of a Ceramic Bone Filler Containing Polyphenols from Grape Pomace

Elisa Torre, Marco Morra, Clara Cassinelli and Giorgio Iviglia *

Nobil Bio Ricerche Srl, V. Valcastellana 26, 14037 Portacomaro, AT, Italy; etorre@nobilbio.it (E.T.); mmorra@nobilbio.it (M.M.); ccassinelli@nobilbio.it (C.C.)
* Correspondence: giviglia@nobilbio.it; Tel.: +39-0141202547

Abstract: Bone loss is a major burden for society and impacts people's health all over the world. In a changing world looking toward a more conscious use of raw materials, efforts are being made to increasingly consider new promising biomaterials that account for, on one side, the ability to provide specific functional biological activities and, on the other, the feature of being well tolerated. In this regard, the use of phenolic compounds in the field of bone-related bioengineering shows a rising interest in the development of medical solutions aimed at taking advantage of the multiple beneficial properties of these plant molecules. In this work, the anti-bacterial and anti-inflammatory power of a biphasic calcium phosphate synthetic bone filler coated with a mixture of phenolic compounds was investigated by evaluating the minimal inhibitory concentration (MIC) value against *Streptococcus mutans* and *Porphyromonas gingivalis* and the expression of genes involved in inflammation and autophagy by real-time reverse transcription polymerase chain reaction (RT-qPCR) on J774a.1 murine macrophage cells. Results show a MIC of 0.8 μg/mL, a neat anti-inflammatory effect, and induction of autophagy key genes compared to a ceramic bone filler. In conclusion, functionalization with a polyphenol-rich extract confers to a ceramic bone filler anti-bacterial and anti-inflammatory properties.

Keywords: anti-bacterial; anti-inflammatory; ceramic bone filler; calcium phosphates; polyphenols; periodontitis; periimplantitis; antioxidant; *S. mutans*; *P. gingivalis*

1. Introduction

When talking about bone loss, one needs to consider different related diseases and disorders that can be grouped under the term bone diseases. The loss of bone mass during life is a normal phenomenon. In fact, the high prevalence of some of these skeletal diseases, especially osteoporosis and osteopenia, is due primarily to the aging of the population [1]. However, when the mechanisms of homeostasis that maintain the equilibrium between bone formation and bone resorption undergo instability and disruption by several conditions, such as genetic predispositions, the presence of certain diseases, the use of specific drugs, and an unhealthy lifestyle, pathologic bone loss appears [2], representing a chronic burden on both individuals and society [3].

In the oral context, this bone loss mainly occurs in association with periodontal diseases, which are thus considered among the most common medical conditions in the world [4]. In fact, the recently reported prevalence data show a global increase over the years that negatively impacts people's general health and quality of life [5]. Periodontal diseases are defined and characterized by a different degree of the host's inflammatory reactions induced by changes in the oral microbiome, that is, the loss of oral microbial diversity and beneficial microbes in favor of pathogenic bacteria, which can lead to a breakdown of the symbiotic relationship that develops with the host in eubiotic conditions [6]. However, there is a large consensus that dysbiosis and inflammation act according to a reciprocal cause-and-effect relationship, in which the microbiome induces inflammation and inflammation induces the selective growth of specific bacterial species [7].

Normally, the oral cavity is colonized by different types of bacteria, which may accumulate on dental surfaces in the form of biofilm before and plaque after, thus triggering the host's innate immune response to establish an inflammatory state that can result, over time, in chronic inflammation and in the pathogenesis of periodontal diseases, with progressive destruction of the teeth-supporting tissues [8,9].

According to this model of periodontal disease pathogenesis, the Gram-negative anaerobic bacterium *Porphyromonas gingivalis* is considered the keystone pathogen of periodontal disease [10], given that it has a pivotal role in initiating dysbiosis [11] and in progressing the disease [12]. In particular, *P. gingivalis* resides in the sub-gingival plaque and interacts with different bacterial species, such as some of the genus *Streptococcus* spp. [13], from which it benefits in terms of growth and resistance [14]. *Streptococcus mutans* is an early colonizer found in particular in the supra-gingival plaque that determines the formation of an acidic oral environment responsible for the development of dental caries [15]. Although evidence shows that *S. mutans* does not directly interact with *P. gingivalis* [16], the former is found at high levels in the sub-gingival plaque of patients with severe periodontitis, thus representing a high-risk factor for dental root caries development [17–21].

Currently, the first line of therapy for the treatment of periodontitis is the mechanical removal of plaque, the use of antimicrobial agents, and, in severe cases, the use of bone grafts to fill and regenerate the alveolar bone [22]. Grafting materials can be of several types, of natural or synthetic origin, and can be used alone or combined with antimicrobials or biological molecules [23], with the aim of providing structural support for teeth and dental implants and a bioactive action on the surrounding supportive tissues. In this context, plant-derived polyphenols have been shown to positively act on bone regeneration, thanks to their multitude of biological effects, ranging from anti-oxidant to anti-bacterial and anti-inflammatory [24]. Thanks to their ability to interplay with different molecular pathways involved in the promotion of osteogenesis and bone protection [25], their particular properties make them ideal candidates for the treatment of diseases characterized by dysregulation of inflammation and bacterial-induced damage.

In this work, we show the in vitro anti-bacterial and anti-inflammatory action of the Synergoss Red (SR) ceramic bone filler [26], coated with a polyphenol-rich grape pomace extract for which we previously demonstrated the ability to induce osteogenic differentiation in mesenchymal stem cells [25]. In addition to their well-known regenerative properties [27], phenolic compounds from grapes are of interest because they show a high-value potential for their versatile use in circular economy applications. In fact, a consistent amount of these molecules can be recovered from winemaking wastes, which account for about 20% of the starting material [28], thus giving the grape pomace a further valorization chance.

2. Materials and Methods

2.1. Materials

All chemicals were analytical-reagent grade. Ultra-pure (MilliQ) water was used for the preparation of aqueous solutions. All chemicals, namely acetone, acetic acid, Folin–Ciocâlteu reagent, 2,2-Diphenyl-1-picrylhydrazyl, sodium carbonate, beta-tricalcium phosphate gallic acid, quercetin 3-glucuronide, quercetin, tannic acid, catechin, malvidin 3-glucoside, epicatechin, procyanidin B2, myricetin, quercitrin, kaempferol, isorhamnetin, rutin, epigallocatechin gallate (EPGCG), caffeic acid, and trans p-cumaroyl tartaric acid were purchased from Sigma-Aldrich (St. Louis, MI, USA); hydroxyapatite was purchased from Fluidinova (Mala, Portugal); and red grape pomace was purchased from a local wine producer (Croatina grape from ALEMAT, Penango, Italy).

2.2. Synthesis of the Ceramic Bone Filler

Hydroxyapatite (HA) was used for enhancing the mechanical strength of scaffolds, whereas β-tricalcium phosphate (β-TCP) was used for its degradability; they were mixed in a percentage of 70/30 wt.%, respectively, to reach an optimal compromise between the two properties. The ceramic scaffolds were prepared by mixing HA and β-TCP powders (47 wt.%) with the binding agent poly (vinyl alcohol) (3 wt.%) and ultrapure water (50 wt.%) to obtain a ceramic slurry. Dolapix CE 64 was added as a dispersing agent (1 wt.% of the solid load). The polyurethane (PU) sponge impregnation method was used to obtain a macroporous ceramic scaffold [29,30]. A commercial polyurethane (PU) sponge slab (45 ppi) of 200 × 200 × 10 mm^3 was soaked into the ceramic slurry for 90 s, followed by compression along the transverse plane (40 kPa), and left at room temperature for 5 min before repeating the cycle. Impregnation/compression cycles were repeated 3 times. The ceramic-coated PU sponge was left to dry overnight at 37 °C and sintered in a furnace at 1100 °C for 12 h in air (heating rate 5 °C/min) to obtain a porous HA/βTCP slab of 200 × 200 × 10 mm^3 (a volumetric retention of 24% was calculated). The obtained porous slab was ground in a jaw crusher (BB 50, Retsch GmbH, Haan, Germany) and sieved to obtain a 300–2000 μm porous-grained granulate (bone filler control) (BF).

2.3. Preparation of the Polyphenol-Rich Pomace Extracts (PRPE)

Grape pomace was received in dry form from the producer and stored at −20 °C under vacuum until the beginning of the extraction process. In order to make it suitable for the extraction process, the grape pomace was first washed with acidified water, dried in a circulating-air oven (37 °C ± 5 °C), and ground in a bladed mill (GM 200, Retsch). The milled grape pomace (300 g) was extracted in 2000 mL of 50:50 acetone–water (v/v) using an automatic extractor (Micro C TIMATIC, Spello (Pg), Italy). The extraction cycle is fully automatic and alternates a dynamic phase, performed at a programmed pressure and a static phase, in which a forced percolation is generated, which, thanks to the programmable recirculation, ensures a continuous flow of solvent to the interior of the plant matrix, thus avoiding over-saturation [26]. Next, the extracted solution was concentrated under reduced pressure in a rotavap in such a way to eliminate the acetone and to obtain an aqueous extract. The concentrated extract was separated by centrifugation (7000 rpm, 5 min), and the supernatant fraction was used to treat the ceramic granulates. High Performance Liquid Cromatography (HPLC) analysis was conducted on the extract, results are reported in Supplementary Materials.

2.4. Coating of Ceramic Bone Filler with PRPE

Synergoss Red was prepared by soaking the porous ceramic granulate with the polyphenol-rich pomace extract in a ratio of 40:60 wt.%, respectively, and left to evaporate for 24 h at 37 °C to obtain a final ceramic bone filler enriched with polyphenols molecules.

2.5. Morphology Characterization of the Ceramic Granules

The surface morphology of the granules was analyzed through scanning electron microscopy (SEM) analysis. Samples were mounted on aluminum stubs and sputtered with gold at 15 mA for 2 min using Agar Sputter Coater. The morphology of the bone filler was captured using the EVO MA 10 system (Zeiss, 73447 Oberkochen, Germany).

Specific surface areas of the bone filler particles were measured by Kr adsorption–desorption isotherms at 77 K using a Micromeritics ASAP 2020 instrument (Micromeritics, Norcross, GA 30093, USA), and SSA was calculated by the Brunauer–Emmett–Teller (BET) method.

The microstructure of the granulate bone filler was analyzed by using micro-computed tomography (μ-CT), equipped with a desktop μ-CT scanner (SkyScan 1174, Aartselaar, Belgium), set at a voltage of 50 kV and a current of 800 μA, and the sample was scanned at a 9.23 μm pixel resolution. The exposure time per projection was 10,000 ms and a 720 μm aluminum filter was used. A total of 0.05 g of material was used as the sample. The CT-Analyzer software (version 5.30) was used for the imaging analysis.

2.6. Phase Analysis of the Ceramic Granules

The Analytical X'Pert Pro diffractometer (Malvern Panalytical, Malvern, Worcestershire, UK) equipped with an X'Celerator detector powder using Cu Kα radiation generated at 45 kV and 40 mA, was used to characterize the crystal structure of the biphasic calcium phosphate bone filler. The samples were measured in the 2-theta, ranging from 20° to 40° at a rate of 0.2°/min.

2.7. Total Phenolic Content of PRPE

The initial phenolic content of Croatina PRPE was evaluated by using the Folin–Ciocâlteu (FC) method as previously reported [31–33]. The extract was transferred in a 25 mL volumetric flask and diluted 1:50 with distilled water (3 technical replicates). Then, 0.5 g of Folin–Ciocâlteu reagent was added and mixed for 5 min, and 1.5 g of 20% anhydrous sodium carbonate (*w*/*v*) solution was added. After 2 h, absorbance was measured at 765 nm by using water as the compensation liquid and a quartz cell (10 mm path length) in a UV–visible spectrophotometer (PG Instruments Limited, Lutterworth, UK). The absorbance value was used to calculate the concentration of polyphenols by using a calibration curve obtained with gallic acid. The results are expressed as mg/mL of gallic acid equivalents (GAE).

2.8. Calibration Curve

Ten milligrams of gallic acid were diluted in 10 mL of water to obtain 1 mg/mL of stock solution. Aliquots of stock solution were transferred in a 25 mL volumetric flask and diluted in water to the final concentrations of 0.05 mg/mL, 0.025 mg/mL, 0.01 mg/mL, and 0.005 mg/mL. Each standard solution was prepared according to the procedure described above for the PRPE; the absorbance was measured under the same conditions as for PRPE.

2.9. Antioxidant Power of PRPE

The antioxidant power of PRPE (1 mg/mL of GAE) and of Pph released from Synergoss Red was evaluated through the widely used DPPH (2,2-Diphenyl-1-picrylhydrazyl) method [34]. This test measures, through a colorimetric approach, the ability of the test solution to scavenge the DPPH radicals. Briefly, an aliquot of 40 µL of PRPE was added to 1600 µL of water–ethanol 50:50 (*v*/*v*) solution (3 technical replicates). Separately, a DPPH solution (0.1 mg/mL *w*/*v*) in ethanol was prepared and 2 mL of this solution was added to the reaction mixture. Then, the solution was shaken and incubated for 30 min at room temperature in the dark; the absorbance was recorded at 525 nm. The blank solution consisted of a solution of water–ethanol instead of PRPE. The inhibition percentage of the DPPH radicals by the samples was calculated using the following equation:

$$\%\ \text{Reduction} = \left(\frac{A_0 - A_1}{A_0}\right) \times 100 \tag{1}$$

where A_0 is the absorbance of the control sample and A_1 is the absorbance of the test sample.

2.10. Polyphenols Release Study

The release study was performed on 0.2 g of material placed in a 12-well plate, filled with 1 mL of ultrapure water, for 24 h and 48 h at 37 °C. The amount of polyphenols released has been quantified by the Folin–Ciocâlteu method.

2.11. Cell Culture

The cell culture was set up as already described in a previous work [26]. Specifically, the murine macrophage cell line J774a.1, obtained from the Department of Cell Substrates at the Istituto Zooprofilattico Sperimentale of Lombardia and Emilia Romagna "Bruno Ubertini", was used to evaluate the inflammatory response to the bone filler and the SR materials, both incubated in direct contact with cells. Briefly, cells were maintained in Dulbecco's modified eagle medium (Gibco Invitrogen, Cergy-Pontoise, France) supplemented with 10% fetal bovine serum, penicillin (100 U mL^{-1}), streptomycin (100 μg mL^{-1}), and 4 mM l-glutamine and left grow in a 100% humidified incubator at 37 °C with 10% CO_2 and then passaged 2–3 days before use. Two different experiments were conducted with the same bone filler and SR materials: the first was before materials sterilization and the second was after sterilization with gamma rays at 25 kGy; for both experiments, the J774a.1 cells were seeded (1.1×10^6 mL^{-1} for the unsterilized materials and 8.5×10^4 mL^{-1} for the sterilized materials) onto 12-well tissue culture polystyrene plates (Falcon® Corning™, Corning, NY, USA), containing the bone filler and the SR materials both at a concentration of 0.2 g/well.

2.12. *Gene Expression Analysis by Real-Time Quantitative Reverse Transcription PCR (RT-qPCR)*

Gene expression analysis was performed by following the same steps described in a previous article [26]. After 4 h incubation, the RNA from J774.a1 cells were isolated by using the Maxwell® RSC simplyRNA Cells Kit (Promega Corporation, Madison, WI, USA), following the manufacturer's instructions. RNA quantitation was performed using the Quantifluor RNA system kit in the Quantus Fluorometer (both from Promega Corporation), and the obtained total RNA was reverse-transcribed using a HighCapacity cDNA Reverse Transcription Kit in the Thermal Cycler 2720 (both from Thermo Fisher Scientific, Waltham, MA, USA) at the following conditions: 10 min at 25 °C, 120 min at 37 °C, and 5 min at 85 °C, and maintained at 4 °C until further experimentation. RT-qPCR was performed following the fast running protocol of the TaqMan® FastAdvanced Master Mix (Thermo Fisher Scientific, Inc.) in the QuantStudio 5 Real-Time PCR System (Thermo Fisher Scientific, Inc.) using designed murine TaqMan® assays (Thermo Fisher Scientific, Inc.) to quantify gene expression of the following genes: interleukin-1β (IL-1β; ID: Mm01336189_m1), interleukin-6 (IL-6; ID: Mm99999064_m1), tumor necrosis factor alpha (TNF-α; ID: Mm00443258_m1), interleukin 10 (IL-10; ID: Mm99999062_m1), beclin 1 (BECN1; ID: Mm01265461_m1), microtubule-associated proteins 1A/1B light chain 3B (MAP1LC3b; ID: Mm00782868_sH), glyceraldehyde 3-phosphate dehydrogenase (GAPDH; ID: Mm99999915_g1), and tyrosine 3-monooxygenase/tryptophan 5-monooxygenase activation protein ζ (YWHAZ; ID: Mm03950126_s1). The real-time PCR was performed in technical duplicate for all samples and targets in a total volume of 25 μL and the amplification was performed as follows: hold at 50 °C for 2 min, hold at 95 °C for 2 min, 40 cycles at 95 °C for 1 s and 60 °C for 20 s. All transcripts were normalized to the mean of the reference genes YWHAZ and GAPDH. The ΔCT values were calculated as CT(reference gene) − CT(gene of interest). ΔΔCt values represent the mean difference of ΔCt between treated and control samples (polystyrene), thus higher ΔΔCt values reflect induction of gene expression due to the treatment. Results are expressed as Mean ± 95% CI of 3 biological samples (n = 3).

2.13. *Antimicrobial Study to Define the Minimal Inhibitory Concentration (MIC)*

The MIC is the highest dilution (or the lowest concentration) of an antimicrobial agent that inhibits the growth of the test microorganism after 24 h under a standardized set of conditions. On the test items, a study was performed to define the minimal inhibitory concentration (MIC) against A90 *Streptococcus mutans* DSM20523 and B160 *Porphyromonas gingivalis* DSM207009. MIC against *S. mutans* was performed by broth dilution method in tubes and verified by streaking on a plate; MIC against *P. gingivalis* was performed by microbroth dilution method. The test item was a polyphenol-rich extract used to functionalize the Synergoss Red Bone filler, with an initial concentration of 3.2 mg/mL. As reference control, a solution of gallic acid, at the same concentration, was used. The MIC test was performed twice, in three independent replicates. Three series of not less than twelve dilutions of the test item and the reference item, respectively, were prepared and tested.

Dilution Method in Tubes: Tubes containing 1 mL of appropriate dilution to be tested were prepared. Series (1), (2), (4), and (5) were inoculated 1:1 with the test suspension at about 1–2 × 10^6 cfu/mL in order to obtain a final concentration of about 5 × 10^5 cfu/mL. The tubes of series (3) and (6) were not inoculated and were used as microbiological control. Moreover, 3 tubes of uninoculated Muller–Hinton broth calcium adjusted (MHBCA) were used as negative controls and 3 tubes containing MHBA inoculated 1:1 with the test suspension were used as positive controls. All tubes were incubated at 37 °C ± 1 °C for 24 h.

Microbroth Dilution Method (in microplate): One 96-well Microplate containing 0.1 mL/well of appropriate dilution to be tested was prepared. Series (1), (2), (4), and (5) were inoculated 1:1 with the test suspension at about 1–2 × 10^6 cfu/mL in order to obtain a final concentration of about 5 × 10^5 cfu/mL. Wells of series (3) and (6) were not inoculated and were used as microbiological control. Moreover, 3 wells of uninoculated MHCBA were used as negative controls and 3 wells containing MHBCA inoculated 1:1 with the test suspension were used as positive controls. The 96-well microplate was incubated at 37 ± 1 °C for 48 h under anaerobic conditions.

After incubation for optimal growing conditions (temperature/time and tubes/wells) were observed for the growth of the test organism (i.e., obvious button or definite turbidity), and results were recorded.

For the broth dilution method against *Streptococcus mutans* DSM20523, due to the evident turbidity of the test item dilutions in the tube, the MIC was also verified by streaking on TSA plates. For the microbroth dilution method against *Porphyromonas gingivalis* DSM20709, no interfering turbidity of the test item dilutions was observed. Verifying by streak on Columbia Agar plates was not necessary.

The growth was recorded with a (+) when it occurred, whereas a (−) was assigned when no growth was observed. Finally, the MIC value was determined.

The viable microorganism concentration in the inoculum was calculated by applying the following formula:

$$N = \frac{c}{(n_1 + 0.1 n_2) d} \quad (2)$$

where c is the sum of colonies counted on all countable plates; n_1 is the number of counted plates in the lower dilution; n_2 is the number of counted plates in the higher dilution and d is the dilution factor corresponding to the lower dilution. The counting was performed using the number of colonies counted on Petri plates. The highest dilution (or the lowest concentration) of antimicrobial agent that inhibits the growth of the test microorganism after 24 h under a standardized set of conditions is considered the MIC value.

2.14. Statistical Analysis

Statistical analysis was performed using both the Analysis Tool pack in Excel, the PAST software [35] (Release 4.13), and the Real Statistics Resource Pack [36] (Release 5.11) on ΔΔCq data of the different group treatments.

Data are represented as Mean ± 95% CI and include 3 biological replicates per treatment.

Before testing for differences between groups of means, we first assessed the independence of each other of all drawn samples, any deviation from the assumption of normality, by the Shapiro–Wilck statistical test, and any deviation from the assumption of equal variances, by the Levène statistical test.

In the case of all the above assumptions were met, one-way analysis of variance (one-way ANOVA) was performed, followed by the Tukey HSD post hoc test.

For TNF-α data from the experiment before sterilization, the quadratic transformation was performed to meet the assumption of normality.

In the case of unequal variances (as with TNF-α data from the experiment before sterilization and IL-1β data from the experiment after sterilization), the Welch's ANOVA test was performed, followed by the Games–Howell post hoc test.

Transformed data, along with their CI, were backtransformed to their original scale before being reported in the graph.

Results with $p < 0.05$ ($* = p < 0.05$; $** = p < 0.01$; $*** = p < 0.001$ and $**** = p < 0.0001$) are considered statistically significant.

3. Results

3.1. Ceramic Granules Characterization

Surface morphology has a relevant role in the interaction of the biological environment with biomaterial surfaces. The ceramic bone filler was made through a synthesis process that promotes the formation of a rough surface.

Figure 1a shows SEM images of the surface of granules of ceramic bone filler, at different magnifications (50×, 500×, 5000×, and 7500×).

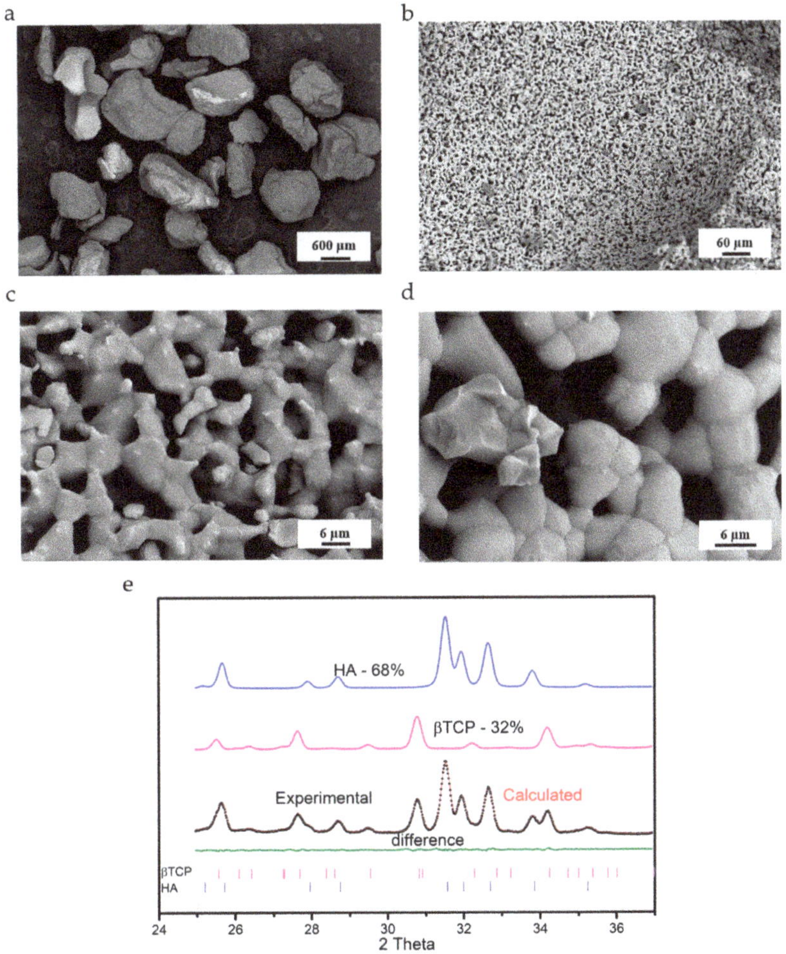

Figure 1. Images of the morphological structure obtained by scanning electron microscopy of the surface of ceramic bone filler at different magnifications (from upper left 50× (**a**), 500× (**b**), 5000× (**c**), and 7500× (**d**)). (**e**) XRD pattern of the sample biphasic bone filler (black dotted line), model patterns of hydroxyapatite (blue line) and βTCP (pink line), experimental data (red line) and the difference between the experimental profile and simulated one (green line). Peak positions for model HA and βTCP are pink and blue tick, respectively.

As it is possible to see in Figure 1a, granules were characterized by a diffused microporosity that influenced the surface area, which was measured using BET analysis and resulted in 1.64 ± 0.06 m^2/g. The surface area also influences the ability of the material to uptake the polyphenol molecules. Different sintering, time, and ratios between HA and βTCP were tested, and the one that showed a higher polyphenols uptake (measured through the release test) and sufficient mechanical strength was selected (data of the optimization process can be provided under request).

Figure 1b represents the XRD pattern of the ceramic bone filler compared with the HA pattern, the βTCP pattern, and the results of a simulation of experimental data. As it is possible to see, the difference between the experimental profile and the simulated one is almost flat, thus meaning that the experimental X-ray pattern revealed the mass phase content of HA and βTCP to be 68% and 32%, respectively, considering the theoretical one of 70–30%.

The porosity, the pore size, and the open porosity were optimized and analyzed using the Micro-CT analysis. The porosity range is between 49.6 and 50.2%, and the pore size is between 100 and 1100 μm, with an open porosity percentage of around 98%. These parameters were obtained by optimizing the production process and the raw material ratio, allowing maximization of the bone filler loading capacity, and keeping the structural resistance and filling ability of the ceramic granulate.

3.2. Polyphenols Release Study from SR Bone Filler

The extract from Croatina grape pomace was analyzed by the Folin–Ciocâlteu method to evaluate the total phenolic content and showed an initial amount of 3.2 mg/mL of gallic acid equivalent (GAE) and a free radical scavenging activity of 80.5%. After functionalization of the ceramic bone filler with the polyphenol-rich extract, an in vitro release study was conducted at 24 h and 48 h. In Table 1 the values of phenolic content, expressed as mg/mL of GAE, and the antioxidant power, expressed as percentage of reduction, are reported.

Table 1. Total phenolic content (GAE mg/mL) and antioxidant power data reported as reduction percentage after 24 h and 48 h of release in ultrapure water.

Time	Total Phenolic Content Released (GAE mg/mL)	Reduction Power of the Released Solution (%)
24 h	0.95	18.3
48 h	1.06	19.6

3.3. Effects of SR on Expression of Genes Involved in Inflammation and Autophagy

The influence of the SR material on the expression of genes involved in inflammation and autophagy was evaluated both prior to and after sterilization. Results from the experiment involving unsterilized materials showed reduced expressions of IL-1β and IL-6 induced by SR, compared to cells in contact with the ceramic bone filler (Figure 2A,B). In contrast, a higher TNF-α expression resulted in cells incubated with SR, compared to both the reference sample polystyrene and the bone filler (Figure 1C). A higher expression of IL-10 was also observed in cells in contact with SR, as compared to polystyrene, whereas the bone filler induced downregulation (Figure 2D).

Expression of BECN1 and MAP1LC3b genes was induced only in cells incubated with SR (Figure 3A,B).

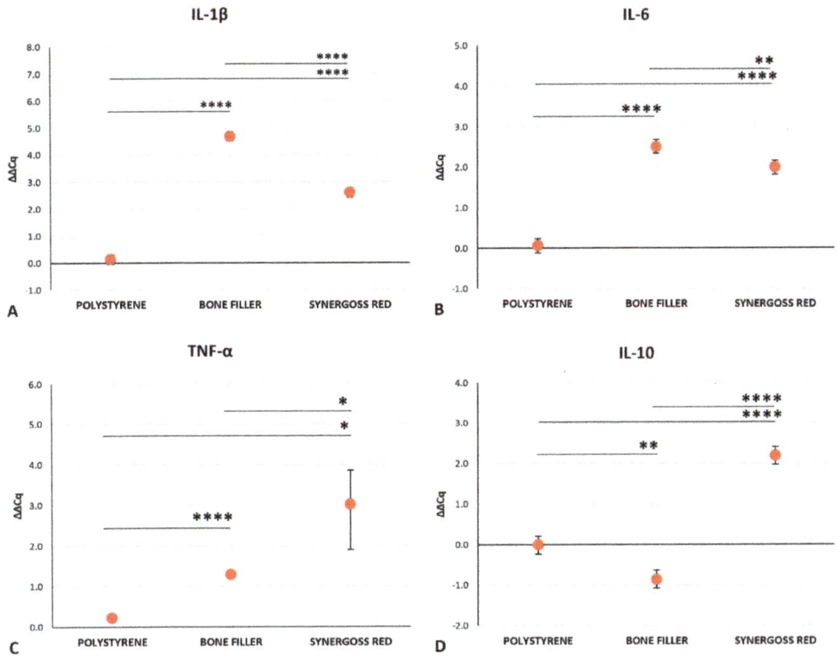

Figure 2. SR shows anti-inflammatory properties before sterilization. Quantitative real-time PCR analysis of mRNA levels of: (**A**) the key-inflammatory mediators IL-1β, IL-6 (**B**), and TNF-α (**C**) and (**D**) the anti-inflammatory cytokine IL-10. * $p < 0.05$, ** $p < 0.01$, **** $p < 0.0001$.

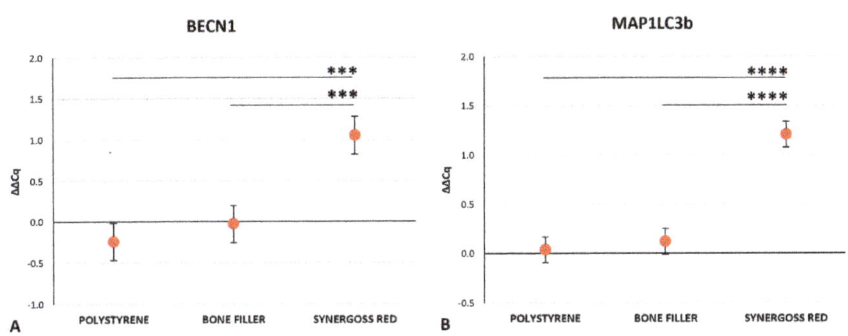

Figure 3. SR induces the expression of autophagy-related genes before sterilization. Quantitative real-time PCR analysis of the key-autophagy genes: (**A**) BECN1 and (**B**) MAP1LC3b. *** $p < 0.001$, **** $p < 0.0001$.

Interestingly, results from the experiment following sterilization (Figure 4A–C) showed reduced expression values for IL-1β, IL-6, and TNF-α for both SR and bone filler, with the lowest levels found in cells in contact with SR.

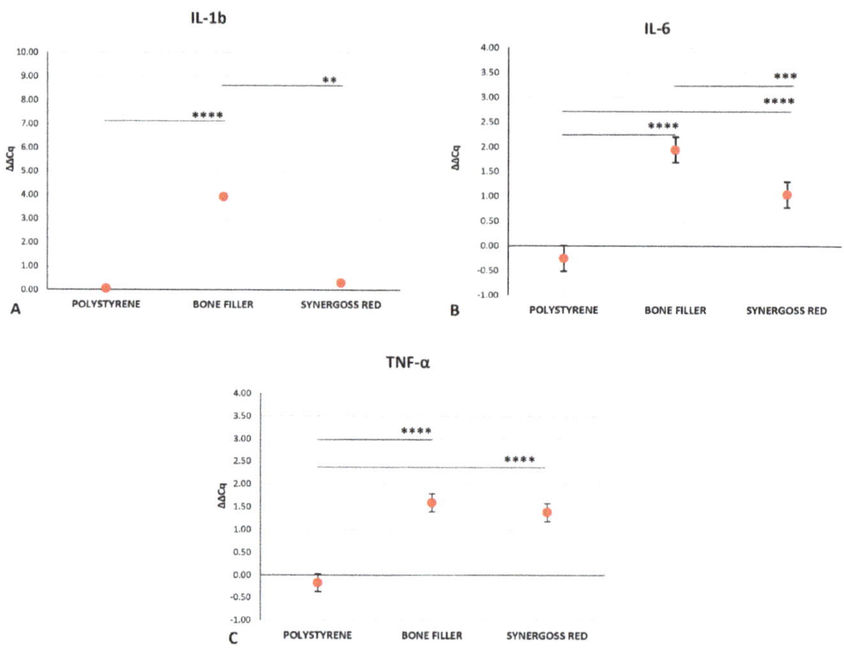

Figure 4. SR maintains its anti-inflammatory properties after sterilization. Quantitative real-time PCR analysis of mRNA levels of the key-inflammatory mediators: (**A**) IL-1β, (**B**) IL-6, and (**C**) TNF-α. ** $p < 0.01$, *** $p < 0.001$, **** $p < 0.0001$.

3.4. Antimicrobial Effect of the Polyphenol-Rich Extract

The results obtained by the antimicrobial study showed an average MIC of 0.8 mg/mL for the polyphenol-rich extract against *S. mutans*, and an average MIC of 0.8 mg/mL against *P. gingivalis*, whereas for the reference gallic acid standard solution, the MIC value cannot be defined against both microorganisms (Table 2).

Table 2. Minimal inhibitory concentration (MIC) of the polyphenol-rich extract and test control against *S. mutans* and *P. gingivalis*.

		MIC (mg/mL)	
Organism	**DSM No.**	**Polyphenol-Rich Extract 3.2 mg/mL**	**Gallic Acid 3.2 mg/mL**
Streptococcus mutans	20523, Type strain	0.8	-
Porphyromonas gingivalis	20709, Type strain	0.8	-

4. Discussion

Replacement of bone loss or restoration of bone defects is still a clinical challenge, especially because of disproportionate inflammatory reactions due to bacterial infections.

The mode of action of most of the commercially available bone fillers is to primarily provide tissues with mechanical support by acting as scaffolds for cell growth. Nowadays, the development of bone filler materials should be devoted to finding active biomolecules coupled with the ceramic material, which should demonstrate firstly biocompatibility (data showed in Supplementary Materials) and then could help in specific cases where the sole mechanical support is insufficient, such as periimplantitis.

Polyphenols represent a wide class of phytochemicals that are extensively studied for different purposes, ranging from medical to cosmetic, nutritional, and engineering applications, thanks to their numerous beneficial properties [37]. The medicinal properties of plant extracts have been known since ancient times, and their benefits also include important anti-bacterial activity against a large number of bacteria [38]. The polyphenol-rich extract used to functionalize our ceramic bone filler, SR, was shown to exhibit anti-bacterial activity against *S. mutans* and *P. gingivalis*. These results are impressive because they show that a natural extract can exert its anti-bacterial effects thanks to the synergistic actions of different phenolic molecules that compose the mixture; in fact, no specific anti-bacterial effects were shown for the reference "gallic acid" phenolic standard. Furthermore, the MIC value of the extract is comparable to the FC results obtained on the solution released by SR (Table 1), thus showing that the bone filler could exert an anti-bacterial effect following implantation. Although interesting and promising, these results need to be deeply investigated in further studies, in order to understand the precise mechanism of the anti-bacterial action of our extract. The ability of polyphenols to regulate inflammation in different biological systems is again well documented in several works [24]. Functionalization of our ceramic bone filler with the polyphenol-rich extract was intended to provide the oral tissues during surgical intervention with a ceramic bone filler with both anti-bacterial and anti-inflammatory actions to control bacterial growth that could drive disproportionate inflammatory reactions. As highlighted by the expression of genes involved in the early inflammatory response, SR determines the downregulation of the pro-inflammatory cytokines IL-1β and IL-6, both prior to and especially after sterilization, as compared to the bone filler. These results demonstrate that the inflammatory cell response is induced by the material itself and that the polyphenol-rich extracts are responsible for the mitigation of this inflammation. These results are in line with those obtained in a previous study [26]. Furthermore, SR was shown to act on early inflammation induced by the material without, however, depleting its related transcripts, a key-factor aspect of the appropriate tissue repair process occurring after a wound or a surgical intervention [39,40]. Looking at TNF-α results both prior and after sterilization, no significant differences between the ceramic bone filler and SR can be appreciated, and this can be translated in a positive aspect; in fact, an early TNF-α signaling is specifically implicated in early fracture repair [41]. During bone repair, inflammation is involved in complex interplays with several biological mechanisms that tailor a sophisticated and controlled process. Among them, autophagy is a highly conserved cellular process implicated in different biological mechanisms, ranging from vesicular transport to energy control and cell death mediation [42]. In our study, expression of the two autophagic key genes BECN1 and MAP1LC3b was enhanced in cells at contact with SR; thus, this action of modulation of the autophagic process can be attributed to the polyphenolic extract itself. In this regard, autophagy has also been recognized as a process involved in immunomodulation; in fact, it directly regulates the production of inflammatory cytokines by inhibiting the inflammasome, with direct consequences on the production of IL-1β, which is proteolytically degraded at the pro-IL-1β stage and is therefore inhibited [43]. The influence of autophagy on cytokine expression also reflects on bone homeostasis; in fact, inhibition of autophagy in periodontal tissues resulted in increased expression of IL-1, IL-6, and TNF-α, leading to a reduced decline in bone density, while restoration of autophagy reduced, in particular, IL-6 expression [44]. It is also proven that autophagy is dysregulated in periodontitis [45] and that bone resorption occurs in cases of abnormal expression of cytokines; for example, overexpression of IL-6 following autophagy inhibition leads to reduced osteoprotegerin gene expression [46]. Furthermore, the ability of autophagy to influence the immune system is also shown by regulation of macrophage polarization. Specifically, induction of autophagy was shown to shift macrophage polarization to the M2 phenotype, the one involved in tissue repair, to attenuate inflammatory reactions [47], thanks to its anti-inflammatory activity and to the expression of the anti-inflammatory cytokine IL-10. Our results show that SR induced a higher expression of IL-10 as compared to the ceramic bone filler and, thus, that the phenolic extract is central to this action, as already shown in a previous work of ours [26]. Phenolic compounds can target different molecular pathways involved in several biological processes, such as energy homeostasis and bone anabolism [24]. In this regard, different polyphenols were shown to regulate the autophagic process, with consequent immunomodulatory effects, among others [48]. In fact, thanks to their antioxidant power, polyphenols can modulate the redox signaling mediated by ROS/RNS, which can give rise to both reversible and irreversible modifications, leading the cell to carry out protective mechanisms, among which autophagy is a pro-survival process. In particular, oxidative stress is responsible for the concomitant induction of autophagy and the antioxidant machinery to decrease oxidative damage and restore cell homeostasis [49]. Autophagy is induced in response to cell damage or other physiological stimuli aimed at restoring and preserving cell homeostasis by promoting the removal of debris and damaged cellular components and, as such, plays a crucial role in the degradation of bacteria and their products. Although it was demonstrated that *P. gingivalis* infection takes advantage of the inactivation of the mTOR pathway in oral epithelial cells [50], activation of autophagy provides an effective defensive response against bacterial infections [51,52] and regulates innate immunity [53]. In this regard, the ability of the ceramic bone filler SR, functionalized with a polyphenol-rich extract, to downregulate inflammation, increase autophagic gene expression, and contrast bacterial growth provides a promising tool for fighting periodontal disease.

5. Conclusions

In the context of periodontal disease, in which the organism is constantly subject to the consequences of a dysbiotic microenvironment, there is a need for innovative applications designed to face the multiple challenges that this pathologic condition poses. The results presented in this work are promising in that they provide the basis for the development of new concepts and strategies to design bone fillers that provide both support by acting as scaffolds for cells and, at the same time, the control of bacterial growth and inflammation thanks to the regenerative properties of wine polyphenols extracted from byproducts.

Further experiments are planned to confirm and clarify the involvement of autophagy in polyphenol action, in addition to evaluating their potential for inhibiting biofilm formation.

Supplementary Materials: The following supporting information can be downloaded at: https://www.mdpi.com/article/10.3390/ceramics7030063/s1, Figure S1: HPLC analysis of grape pomace extract; Table S1: Details of the HPLC gradient method used for chromatographic analysis; Table S2: Polyphenolic quantification through HPLC-DAD analysis; Figure S2: Cell viability assay of L929 fibroblast cells incubated in contact with Synergoss Red bone filler.

Author Contributions: Conceptualization, G.I. and M.M.; methodology, G.I.; validation, M.M., G.I., and C.C.; formal analysis, E.T.; investigation, E.T. and G.I.; resources, M.M. and C.C.; data curation, E.T.; writing—original draft preparation, E.T.; writing—review and editing, E.T. and G.I.; supervision, G.I. and M.M.; project administration, M.M. and G.I.; funding acquisition, M.M. and C.C. All authors have read and agreed to the published version of the manuscript.

Funding: This research received no external funding.

Institutional Review Board Statement: Not applicable.

Informed Consent Statement: Not applicable.

Data Availability Statement: The raw data supporting the data presented in this study are readily available on request from the corresponding author.

Conflicts of Interest: Authors Elisa Torre, Marco Morra, Clara Cassinelli, and Giorgio Iviglia were employed by the company Nobil Bio Ricerche srl. The authors declare that the research was conducted in the absence of any commercial or financial relationships that could be construed as a potential conflict of interest.

References

1. Salari, N.; Darvishi, N.; Bartina, Y.; Larti, M.; Kiaei, A.; Hemmati, M.; Shohaimi, S.; Mohammadi, M. Global prevalence of osteoporosis among the world older adults: A comprehensive systematic review and meta-analysis. *J. Orthop. Surg. Res.* **2021**, *16*, 669. [CrossRef] [PubMed]

2. Barnsley, J.; Buckland, G.; Chan, P.E.; Ong, A.; Ramos, A.S.; Baxter, M.; Laskou, F.; Dennison, E.M.; Cooper, C.; Patel, H.P. Pathophysiology and treatment of osteoporosis: Challenges for clinical practice in older people. *Aging Clin. Exp. Res.* **2021**, *33*, 759–773. [CrossRef]
3. Shen, Y.; Huang, X.; Wu, J.; Lin, X.; Zhou, X.; Zhu, Z.; Pan, X.; Xu, J.; Qiao, J.; Zhang, T.; et al. The Global Burden of Osteoporosis, Low Bone Mass, and Its Related Fracture in 204 Countries and Territories, 1990–2019. *Front. Endocrinol.* **2022**, *13*, 882241. [CrossRef] [PubMed]
4. Yu, B.; Wang, C.Y. Osteoporosis and periodontal diseases—An update on their association and mechanistic links. *Periodontol. 2000* **2022**, *89*, 99–113. [CrossRef] [PubMed]
5. WHO. *Global Oral Health Status Report. Towards Universal Health Coverage for Oral Health by 2030*; WHO: Geneva, Switzerland, 2022.
6. Radaic, A.; Kapila, Y.L. The oralome and its dysbiosis: New insights into oral microbiome-host interactions. *Comput. Struct. Biotechnol. J.* **2021**, *19*, 1335–1360. [CrossRef] [PubMed]
7. Lamont, R.J.; Koo, H.; Hajishengallis, G. The oral microbiota: Dynamic communities and host interactions. *Nat. Rev. Microbiol.* **2018**, *16*, 745–759. [CrossRef]
8. Di Stefano, M.; Polizzi, A.; Santonocito, S.; Romano, A.; Lombardi, T.; Isola, G. Impact of Oral Microbiome in Periodontal Health and Periodontitis: A Critical Review on Prevention and Treatment. *Int. J. Mol. Sci.* **2022**, *23*, 5142. [CrossRef] [PubMed]
9. Hajishengallis, G.; Chavakis, T.; Lambris, J.D. Current understanding of periodontal disease pathogenesis and targets for host-modulation therapy. *Periodontol. 2000* **2020**, *84*, 14–34. [CrossRef] [PubMed]
10. Hajishengallis, G.; Darveau, R.P.; Curtis, M.A. The keystone-pathogen hypothesis. *Nat. Rev. Microbiol.* **2012**, *10*, 717–725. [CrossRef]
11. Hajishengallis, G.; Liang, S.; Payne, M.A.; Hashim, A.; Jotwani, R.; Eskan, M.A.; McIntosh, M.L.; Alsam, A.; Kirkwood, K.L.; Lambris, J.D.; et al. Low-abundance biofilm species orchestrates inflammatory periodontal disease through the commensal microbiota and complement. *Cell Host Microbe* **2011**, *10*, 497–506. [CrossRef]
12. Hajishengallis, G.; Diaz, P.I. *Porphyromonas gingivalis*: Immune Subversion Activities and Role in Periodontal Dysbiosis. *Curr. Oral Health Rep.* **2020**, *7*, 12–21. [CrossRef]
13. Park, Y.; Simionato, M.R.; Sekiya, K.; Murakami, Y.; James, D.; Chen, W.; Hackett, M.; Yoshimura, F.; Demuth, D.R.; Lamont, R.J. Short fimbriae of Porphyromonas gingivalis and their role in coadhesion with Streptococcus gordonii. *Infect. Immun.* **2005**, *73*, 3983–3989. [CrossRef] [PubMed]
14. Hajishengallis, G.; Lamont, R.J. Beyond the red complex and into more complexity: The polymicrobial synergy and dysbiosis (PSD) model of periodontal disease etiology. *Mol. Oral Microbiol.* **2012**, *27*, 409–419. [CrossRef] [PubMed]
15. Lemos, J.A.C.; Abranches, J.; Burne, R.A. Responses of cariogenic streptococci to environmental stresses. *Curr. Issues Mol. Biol.* **2005**, *7*, 95–107. [CrossRef] [PubMed]
16. Demuth, D.R.; Irvine, D.C.; Costerton, J.W.; Cook, G.S.; Lamont, R.J. Discrete protein determinant directs the species-specific adherence of Porphyromonas gingivalis to oral streptococci. *Infect. Immun.* **2001**, *69*, 5736–5741. [CrossRef] [PubMed]
17. Ravald, N.; Birkhedh, D. Factors associated with active and inactive root caries in patients with periodontal disease. *Caries Res.* **1991**, *5*, 13–29. [CrossRef] [PubMed]
18. Contardo, M.S.; Díaz, N.; Lobos, O.; Padilla, C.; Giacaman, R.A. Oral colonization by Streptococcus mutans and its association with the severity of periodontal disease in adults. *Rev. Clin. Periodoncia Implantol. Oral* **2011**, *4*, 9–12. [CrossRef]
19. Estrela, C.R.d.A.; Pimenta, F.C.; de Alencar, A.H.G.; Ruiz, L.F.N.; Estrela, C. Detection of selected bacterial species in intraoral sites of patients with chronic periodontitis using multiplex polymerase chain reaction. *J. Appl. Oral Sci.* **2010**, *18*, 426–431. [CrossRef]
20. Van Der Reijden, W.A.; Dellemijn-Kippuw, N.; Stijne-van Nes, A.M.; De Soet, J.J.; Van Winkelhoff, A.J. Mutans streptococci in subgingival plaque of treated and untreated patients with periodontitis. *J. Clin. Periodontol.* **2001**, *28*, 686–691. [CrossRef]
21. Dani, S.; Prabhu, A.; Chaitra, K.; Desai, N.; Patil, S.; Rajeev, R. Assessment of Streptococcus mutans in healthy versus gingivitis and chronic periodontitis: A clinico-microbiological study. *Contemp. Clin. Dent.* **2016**, *7*, 529–534. [CrossRef]
22. Mancini, L.; Romandini, M.; Fratini, A.; Americo, L.M.; Panda, S.; Marchetti, E. Biomaterials for periodontal and peri-implant regeneration. *Materials* **2021**, *14*, 3319. [CrossRef] [PubMed]
23. Zhao, R.; Yang, R.; Cooper, P.R.; Khurshid, Z.; Shavandi, A.; Ratnayake, J. Bone grafts and substitutes in dentistry: A review of current trends and developments. *Molecules* **2021**, *26*, 3007. [CrossRef]
24. Torre, E. Molecular signaling mechanisms behind polyphenol-induced bone anabolism. *Phytochem. Rev.* **2017**, *16*, 1183–1226. [CrossRef] [PubMed]
25. Torre, E.; Iviglia, G.; Cassinelli, C.; Morra, M.; Russo, N. Polyphenols from grape pomace induce osteogenic differentiation in mesenchymal stem cells. *Int. J. Mol. Med.* **2020**, *45*, 1721–1734. [CrossRef] [PubMed]
26. Iviglia, G.; Torre, E.; Cassinelli, C.; Morra, M. Functionalization with a polyphenol-rich pomace extract empowers a ceramic bone filler with in vitro antioxidant, anti-inflammatory, and pro-osteogenic properties. *J. Funct. Biomater.* **2021**, *12*, 31. [CrossRef] [PubMed]
27. Averilla, J.N.; Oh, J.; Kim, H.J.; Kim, J.S.; Kim, J.S. Potential health benefits of phenolic compounds in grape processing by-products. *Food Sci. Biotechnol.* **2019**, *28*, 1607–1615. [CrossRef] [PubMed]
28. Spanghero, M.; Salem, A.Z.M.; Robinson, P.H. Chemical composition, including secondary metabolites, and rumen fermentability of seeds and pulp of Californian (USA) and Italian grape pomaces. *Anim. Feed Sci. Technol.* **2009**, *152*, 243–255. [CrossRef]

29. Vitale-Brovarone, C.; Baino, F.; Verné, E. High strength bioactive glass-ceramic scaffolds for bone regeneration. *J. Mater. Sci. Mater. Med.* **2009**, *20*, 643–653. [CrossRef] [PubMed]
30. Vitale-Brovarone, C.; Verné, E.; Robiglio, L.; Appendino, P.; Bassi, F.; Martinasso, G.; Muzio, G.; Canuto, R. Development of glass-ceramic scaffolds for bone tissue engineering: Characterisation, proliferation of human osteoblasts and nodule formation. *Acta Biomater.* **2007**, *3*, 199–208. [CrossRef]
31. Russo, N.; Cassinelli, C.; Torre, E.; Morra, M.; Iviglia, G. Improvement of the physical properties of guided bone regeneration membrane from porcine pericardium by polyphenols-rich pomace extract. *Materials* **2019**, *12*, 2564. [CrossRef]
32. Singleton, V.L.; Orthofer, R.; Lamuela-Raventós, R.M. Analysis of total phenols and other oxidation substrates and antioxidants by means of folin-ciocalteu reagent. *Methods Enzymol.* **1999**, *299*, 152–178. [CrossRef]
33. Singleton, V.L.; Rossi, J.A. Colorimetry of Total Phenolics with Phosphomolybdic-Phosphotungstic Acid Reagents. *Am. J. Enol. Vitic.* **1965**, *16*, 144–158. [CrossRef]
34. Brand-Williams, W.; Cuvelier, M.E.; Berset, C. Use of a free radical method to evaluate antioxidant activity. *LWT—Food Sci. Technol.* **1995**, *28*, 25–30. [CrossRef]
35. Hammer, Ø.; Harper, D.; Ryan, P. Past: Paleontological statistics software package for education and data analysis. *Paleontol. Electron.* **2001**, *4*, 9.
36. Zaiontz, C. Real Statistics Using Excel, 2013. (n.d.). Available online: www.real-statistics.com (accessed on 3 December 2023).
37. Rathod, N.B.; Elabed, N.; Punia, S.; Ozogul, F.; Kim, S.K.; Rocha, J.M. Recent Developments in Polyphenol Applications on Human Health: A Review with Current Knowledge. *Plants* **2023**, *12*, 1217. [CrossRef] [PubMed]
38. Manso, T.; Lores, M.; de Miguel, T. Antimicrobial Activity of Polyphenols and Natural Polyphenolic Extracts on Clinical Isolates. *Antibiotics* **2022**, *11*, 46. [CrossRef]
39. Maruyama, M.; Rhee, C.; Utsunomiya, T.; Zhang, N.; Ueno, M.; Yao, Z.; Goodman, S.B. Modulation of the Inflammatory Response and Bone Healing. *Front. Endocrinol.* **2020**, *11*, 386. [CrossRef]
40. Duda, G.N.; Geissler, S.; Checa, S.; Tsitsilonis, S.; Petersen, A.; Schmidt-Bleek, K. The decisive early phase of bone regeneration. *Nat. Rev. Rheumatol.* **2023**, *19*, 78–95. [CrossRef] [PubMed]
41. Mountziaris, P.M.; Mikos, A.G. Modulation of the inflammatory response for enhanced bone tissue regeneration. *Tissue Eng.—Part B Rev.* **2008**, *14*, 179–186. [CrossRef]
42. Ma, Y.; Galluzzi, L.; Zitvogel, L.; Kroemer, G. Autophagy and cellular immune responses. *Immunity* **2013**, *39*, 211–227. [CrossRef]
43. Netea, M.G.; Simon, A.; Van De Veerdonk, F.; Kullberg, B.J.; Van Der Meer, J.W.M.; Joosten, L.A.B. IL-1β processing in host defense: Beyond the inflammasomes. *PLoS Pathog.* **2010**, *6*, e1000661. [CrossRef]
44. Chen, L.; Hua, Y. Autophagy of periodontal ligament inhibits inflammation and reduces the decline of bone density during orthodontic tooth movement of mice. *Arch. Oral Biol.* **2021**, *121*, 104960. [CrossRef] [PubMed]
45. Lorenzo-Pouso, A.I.; Castelo-Baz, P.; Pérez-Sayáns, M.; Lim, J.; Leira, Y. Autophagy in periodontal disease: Evidence from a literature review. *Arch. Oral Biol.* **2019**, *102*, 55–64. [CrossRef] [PubMed]
46. Mayr, A.; Marciniak, J.; Eggers, B.; Blawat, K.; Wildenhof, J.; Bastos Craveiro, R.; Wolf, M.; Deschner, J.; Jäger, A.; Beisel-Memmert, S. Autophagy Induces Expression of IL-6 in Human Periodontal Ligament Fibroblasts Under Mechanical Load and Overload and Effects Osteoclastogenesis in vitro. *Front. Physiol.* **2021**, *12*, 716441. [CrossRef] [PubMed]
47. Ripszky Totan, A.; Imre, M.M.; Parvu, S.; Meghea, D.; Radulescu, R.; Enasescu, D.S.A.; Moisa, M.R.; Pituru, S.M. Autophagy Plays Multiple Roles in the Soft-Tissue Healing and Osseointegration in Dental Implant Surgery—A Narrative Review. *Materials* **2022**, *15*, 6041. [CrossRef] [PubMed]
48. Pallauf, K.; Rimbach, G. Autophagy, polyphenols and healthy ageing. *Ageing Res. Rev.* **2013**, *12*, 237–252. [CrossRef]
49. Gómez-Virgilio, L.; Silva-Lucero, M.D.C.; Flores-Morelos, D.S.; Gallardo-Nieto, J.; Lopez-Toledo, G.; Abarca-Fernandez, A.M.; Zacapala-Gómez, A.E.; Luna-Muñoz, J.; Montiel-Sosa, F.; Soto-Rojas, L.O.; et al. Autophagy: A Key Regulator of Homeostasis and Disease: An Overview of Molecular Mechanisms and Modulators. *Cells* **2022**, *11*, 2262. [CrossRef] [PubMed]
50. Stafford, P.; Higham, J.; Pinnock, A.; Murdoch, C.; Douglas, C.W.I.; Stafford, G.P.; Lambert, D.W. Gingipain-dependent degradation of mammalian target of rapamycin pathway proteins by the periodontal pathogen Porphyromonas gingivalis during invasion. *Mol. Oral Microbiol.* **2013**, *28*, 366–378. [CrossRef] [PubMed]
51. Park, M.H.; Jeong, S.Y.; Na, H.S.; Chung, J. Porphyromonas gingivalis induces autophagy in THP-1-derived macrophages. *Mol. Oral Microbiol.* **2017**, *32*, 48–59. [CrossRef] [PubMed]
52. Bullon, P.; Cordero, M.D.; Quiles, J.L.; Ramirez-Tortosa, M.D.C.; Gonzalez-Alonso, A.; Alfonsi, S.; García-Marín, R.; de Miguel, M.; Battino, M. Autophagy in periodontitis patients and gingival fibroblasts: Unraveling the link between chronic diseases and inflammation. *BMC Med.* **2012**, *10*, 122. [CrossRef]
53. Delgado, M.; Singh, S.; De Haro, S.; Master, S.; Ponpuak, M.; Dinkins, C.; Ornatowski, W.; Vergne, I.; Deretic, V. Autophagy and pattern recognition receptors in innate immunity. *Immunol. Rev.* **2009**, *227*, 189–202. [CrossRef] [PubMed]

ceramics MDPI

Review

Alkali Activation of Glass for Sustainable Upcycling: An Overview

Giulia Tameni and Enrico Bernardo *

Department of Industrial Engineering, Università degli Studi di Padova, 35131 Padova, Italy; giulia.tameni@phd.unipd.it
* Correspondence: enrico.bernardo@unipd.it; Tel.: +39-049-8275-510

Abstract

The recycling of glass presently poses several challenges, predominantly to the heterogeneous chemical compositions of various glass types, along with the waste glass particle size distribution, both of which critically influence the efficiency and feasibility of recycling operations. Numerous studies have elucidated the potential of converting non-recyclable glass waste into valuable materials thanks to the up-cycling strategies, including stoneware, glass wool fibres, glass foams, glass-ceramics, and geopolymers. Among the promising alternatives for improving waste valorisation of glass, alkali-activated materials (AAMs) emerge as a solution. Waste glasses can be employed both as aggregates and as precursors, with a focus on its application as the sole raw material for synthesis. This overview systematically explores the optimisation of precursor selection from a sustainability standpoint, specifically addressing the mild alkali activation process (<3 mol/L) of waste glasses. The molecular mechanisms governing the hardening process associated with this emerging class of materials are elucidated. Formulating sustainable approaches for the valorisation of glass waste is becoming increasingly critical in response to the rising quantities of non-recyclable glass and growing priority on circular economy principles. In addition, the paper highlights the innovative prospects of alkali-activated materials derived from waste glass, emphasising their emerging roles beyond conventional structural applications. Environmentally relevant applications for alkali-activated materials are reported, including the adsorption of dyes and heavy metals, immobilisation of nuclear waste, and an innovative technique for hardening as microwave-assisted processing.

Keywords: glass recycling; up-cycling; mild alkali activation; molecular mechanism

1. Introduction

Glass has been a prevalent material in human history since approximately 3000 BC [1, 2]. Its exceptional versatility can be attributed to several key properties, including optical transparency, chemical inertness, high intrinsic strength, and low permeability [3,4]. These characteristics enable a wide range of applications across various fields, from packaging to construction and technological innovation.

The glass manufacturing is a continuous process and can be classified into four macro-phases: batch preparation, fusion and refining, conditioning and moulding, and finishing [5]. Among these phases, the melting process is a particularly energy-intensive part, as it requires substantial energy input to achieve the elevated temperatures (up to 1500 °C) necessary for turning raw materials into glass. This energy is typically supplied through the combustion of natural gas [6].

The glass manufacturing process, as well as the glass industry, is characterised by high energy consumption, driven by the energy demand per tonne and the large-scale production. Glass container production was recorded at 65 million tonnes [7,8], reflecting a broader trend of growth in global glass production, which doubled until 300 Mt from 2014 [9] to 2021 [10].

Glass manufacturing contributes to greenhouse gas emissions, with approximately 95 million tonnes of CO_2 released annually [11]. These emissions include those directly resulting from combustion of raw materials, the decomposition of carbonates, and the indirect emissions associated with electricity production [12–15]. To meet the emission reduction targets established by the Paris Climate Convention in 2015 [16], it is crucial to consider the entire recycling process.

Currently, the recycling rate of waste glass is not uniformly distributed, showing substantial variation at the international level [17,18]. The estimated worldwide glass recycling rate has been estimated at around 20% [19]. However, Belgium exemplifies best practices with near-total glass recycling, in contrast to Turkey, with a recycling rate of below 10% [20,21].

To comprehend the complexities of glass recycling, it must be realised that, despite its theoretical recyclability, glass cannot be fully recycled in practical scenarios. The implementation of a true 'closed-loop' recycling system faces several limitations. A 'closed-loop' system, in fact, is intended to convert raw materials into equivalent starting products with comparable quality and function [22].

The standard practice for glass recycling is remelting. In terms of energy efficiency, the addition of glass cullet, a granular material produced from recycled glass, results in a reduction of approximately 3% in energy required for melting. This reduction in energy consumption is primarily due to cullet melts at significantly lower temperatures than unprocessed raw materials [23].

Despite its prevalence in industrial practice, 'closed-loop' recycling of glass continues to be limited by several obstacles, which are generally divided into supply-chain barriers and technical limitations. A key limitation in the recycled glass supply chain is the absence of consistent global regulations that guide the glass recycling process. Furthermore, transportation costs associated with moving cullet to treatment and recycling plants, due to the dense nature of glass, may represent up to 33% of total recycling cost [9]. Therefore, despite the prevalence of glass collection worldwide, a significant amount of glass is still disposed of in landfills due to the inadequate facilities capable of carrying out advanced sorting and decontamination processes.

Technically, the integration of glass with diverse chemical compositions into a unified recycling stream presents notable difficulties. In fact, the thermal and processing properties of glass [24] vary considerably depending on chemical formulation. Nevertheless, the remelting of individual glass with the same chemical composition entails certain limitations that prevent total recyclability. Focusing specifically on boro–alumino–silicate glass, trace levels of polymeric and metallic contamination are sufficient to compromise its suitability for pharmaceutical applications. Similarly, the production of cathode ray tube glass (CRT) is discontinued, primarily due to the hazardous emission of lead oxide during the remelting processes. In the context of hazardous emissions, opal glass, used in tableware manufacturing, raises environmental health concerns due to fluorine emissions in high-temperature processing. The main commercial uses and disadvantages of different glass chemical compositions are reported in Table 1.

Table 1. Chemical composition, main commercial use, and limitation in recycling of the most common type of commercialised glass, reprinted from [25].

Glass Type	Composition [wt%]	Commercial Use	Limitations in Recycling
Sodalime (SLG)	73% SiO_2 17% Na_2O 5% CaO 4% MgO 1% Al_2O_3	- Functional or decorative surface coating on plates (screen printing) - Containers	- Permanent fusion of oxides renders coating removal unfeasible - By-products from forming equipment (glass, graphite, carbon, and oil)
Boro–alumino–silicate (BASG)	80% SiO_2 13% B_2O_3 4% Na_2O 2.3% Al_2O_3 0.1% K_2O	Pharmaceutical packaging	- Presence of polymeric and metallic contaminations in the finest fraction
Cathodic Ray Tube (CRT)	63% SiO_2 21% PbO 7.6% Na_2O 6% K_2O 0.6% Al_2O_3 0.3% CaO 0.2% MgO 0.2% B_2O_3	TV screen	- CRTs no longer manufactured - PbO emission occurs in remelting process
Opal	72% SiO_2 12% Na_2O 8% Al_2O_3 5% F_2 2% CaO 2% BaO 1.5% K_2O	Kitchenware	- F emission occurs in remelting process
Liquid-crystal display (LCD)	55% SiO_2 15% Al_2O_3 10% B_2O_3 10% CaO 5% SrO 1% BaO	Electronic device	- Incorporation of functional coatings and optical filters

Overall, an effective incorporation of glass scraps into a 'closed-loop' recycling framework [26] requires strict colour-based sorting and removal of residual impurities.

One significant reason why glass cannot be fully recycled is the issue of glass scrap dimensions (see Figure 1). At the recycling centre, glass is sorted into categories: glass cullet, also known as the coarse glass fraction, and glass fines. Glass cullet is suitable for remelting following the sorting operation. On the other hand, the smallest size fraction, defined as particles smaller than 100 μm, typically generated during the crushing process, contains a considerable level of impurities, which complicates traditional recycling methods [27].

However, even within the finest glass fraction, it is possible to identify two sub-fractions: a purified fraction and one with a higher concentration of impurities. The purified fraction of glass can be suitable for recycling, often through an 'open-loop' recycling process [28]. Conversely, the only viable option for managing the contaminated fraction is the disposal in a landfill [19], as the presence of impurities and contaminants can adversely affect the properties of the newly produced glass [29].

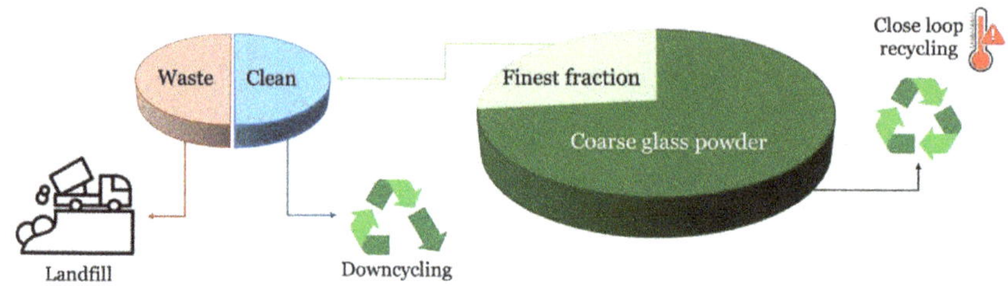

Figure 1. Current condition of glass recycling.

However, innovative recycling strategies have recently demonstrated the feasibility of processing glass materials categories that were considered incompatible with standard recycling practices. For example, to facilitate the development of a circular economy, SCHOTT company has initiated pilot-scale operation to explore the potential for reclaiming used glass-ceramic cooktop panel [Ceran®] and pharmaceutical packaging aimed at generating new products [30]. Ceran® glass is classified as lithium–alumino–silicate (LAS) glass-ceramics [31]. In fact, LAS glass-ceramics require higher processing temperature than typical glass bottles ($T_{melt} \approx 1500$ °C) due to the melting point of neat 1650 °C [32], making them incompatible with the standard glass recycling process. Moreover, the inclusion of LAS glass in the conventional glass recycling process contributes to highly viscous inclusions that compromise the quality of the recycled product [33]. For these reasons, waste LAS glass-ceramics are classified as non-recyclable materials, and typically landfilled [34]. Despite the advancement in recycling, the primary constraints for the material feedstock entering SHOTT pilot plants are the quality of the materials; in fact, the scraps must be free of contaminants and impurities. This requirement creates opportunities for technological innovation in the recycling of more heavily contaminated material fractions.

This work aims to provide an overview of the current state of the art concerning the use of glass fraction that cannot be directly recycled through the remelting process. Particular attention is given to the valorisation of glass waste in the field of alkali activation, initially examining its role as an additive in conventional binders (such as metakaolin) and subsequently exploring its potential as the sole precursor for the synthesis of AAMs. Emphasis is placed on mild alkali activation, molarities below 3 mol/L, and the molecular mechanism that governs this class of materials. Finally, the overview proposes alternative application areas that extend beyond the classical role of replacing conventional construction material, highlighting the versatility of alkali-activated glass-based materials.

Differences Between Waste Glass and Waste-Derived Glass

The discarded glass mentioned in earlier sections can be referred to as 'waste glass'. Vitrification is considered a safe and efficient solution for hazardous waste treatment in the context of alternative waste management strategies [35]. Actually, various industrial waste streams can be vitrified individually or in combination with others through thermal treatment, ultimately yielding glass. During thermal treatment, waste is decomposed into a homogeneous melt, which must be quickly cooled to prevent crystalline phase formation [36].

To achieve a proper glass phase, the process fundamentally requires that initial waste materials include a minimum amount of SiO_2. The final product is termed 'waste-derived' glass. Transforming waste into 'waste-derived' glass is beneficial, primarily due to high chemical resilience, which permits long-term containment and significantly reduces the volume of residual waste [37]. Despite the immobilisation of hazardous inorganic substances and the potential for disposal in landfills, vitrification has disadvantages, as it is associated with financial and energy requirements. The potential release of hazardous gases, including chlorine, resulting from incomplete encapsulation of certain compounds in the glass structures, is a critical issue for this application [38]. An alternative approach to mitigate energy consumption is the conversion of 'waste-derived' glass into useful components that exhibit multifunctional properties. For example, waste-derived glass can serve as a raw material for glass fibres [39] or glass-ceramic tiles [40]. In addition to producing dense materials, researchers have also reported the transformation of 'waste-derived' glass into porous glass-ceramics [38].

The issue of recycling materials derived from the 'waste-derived' waste remains largely unresolved. The lack of a single raw material makes a 'closed-loop' recycling process unfeasible. In this context, alkali activation emerges as an innovative strategy for reusing 'waste-derived' glass precursors to synthesise materials under low-temperature conditions, bypassing conventional high-energy manufacturing steps [41,42].

2. 'Up-Cycling' of Glass: Current Applications

In recent years, there has been growing interest in recycling, accompanied by numerous efforts to enhance the value of scrap materials. Despite these advancements, many technologies and processes lead to materials degradation during recycling, a phenomenon known as 'down-cycling'. While this approach contributes to mitigating waste buildup, it does not maximise the functional capabilities of the product. For glass specifically, the predominant form of downcycling consists of incorporating it as aggregate in road base layers or cementitious construction materials [43].

From a sustainability standpoint, 'up-cycling' emerges as a more attractive process. 'Up-cycling', as defined by Gunter Pauli, is a value-retention strategy whereby post-consumer waste or industrial by-products are upgraded into functionally valuable items [44–46].

A comprehensive summary of the principal products via 'up-cycling' of glasses is provided in the following subparagraphs. Each final product derived from 'up-cycling' exhibits a dual nature, comprising both beneficial aspects, such as the utilisation of materials that would otherwise be sent to a landfill, and negative attributes, primarily concerning energy requirements and potential environmental impacts.

2.1. Stoneware

Traditionally, stoneware is produced using more refractory clays and undergoes firing at high temperatures that can vary from 1100 to 1300 °C, contingent upon flux content [47]. The resulting bodies that are vitrified reach low porosity [48]. Due to its material flexibility, the ceramic industry represents an effective platform for converting waste into secondary resources, such as supplementary additives or auxiliary compounds [49]. By serving as alternative raw materials, glass scraps can be effectively recycled within the production of stoneware tiles, offering a sustainable waste management solution [50]. A wide variety of glass waste materials has been investigated [51]. As a part of a typical porcelain stoneware mixture, feldspar sand and traditional fluxing agents can be substituted by finely ground glass powder [52]. Glass powder, in fact, acts as an effective fluxing agent when incorporated into ceramic mixtures.

The incorporation of glass into ceramic formulations may notably reduce the temperature required for synthesis, reducing the temperature requirement below 1000 °C [50]. A fluxing agent, also known as a softening agent, is an additive used to reduce the sintering temperature in powder-based sintering processes, thereby contributing to energy saving. Moreover, during firing, glass powder enhances the densification process, resulting in a beneficial effect such as reduced open porosity and lower water absorption.

From an industrial point of view, large-scale glass recycling remains challenging: exceeding 10 wt% of waste glass typically results in reduced performance [53]. In fact, the use of glass powder waste may introduce drawbacks, such as increased dimensional shrinkage and elevated closed porosity. The high total porosity (total porosity > 13%) of the samples prevents them from being classified as stoneware [54].

2.2. Glass Wool Fibres

Due to their superior insulation characteristics, flame resistance, low production costs, and simple installation procedures, glass wool and stone wool are among the most widely utilised fibrous materials in building applications [55]. Glass wool consists of thin glass fibres arranged in a random and non-uniform pattern. During the formation process, these fibres enclose numerous voids, which reduces the material's density and enhances the thermal insulation capability by preventing natural convection [56].

The manufacture of glass wools typically utilises a mixture containing up to 85% waste glass (referred to as 'cullet'), while the remainder consists of mineral additives such as sand, limestone, soda, borax, and/or dolomite. However, in certain facilities, glass wool is produced using 100% waste glass, without the incorporation of other virgin raw materials [57,58]. During the melting stage, the raw materials blend must reach temperatures nearing 1400 °C for complete fusion to occur [59]. The glass wool melting process offers improved flexibility over traditional methods by accommodating larger volumes of impure recycled glass, including mixed cullet colours and organic materials, significantly advancing material recycling efforts [60]. The industry is continuously exploring technologies to reduce glass melting temperatures. Notably, the introduction of recycling glass waste helps decrease the energy required for the production of liquid melt [61].

2.3. Glass Foams

Foam glass represents an alternative method for recycling glass waste. Glass foams are cellular glass materials that fulfil strict requirements for fire resistance and structural durability [62]; the materials can be fully derived from glass waste, eliminating the need for virgin raw materials [63,64]. Although glass foams are predominantly utilised for thermal or acoustic insulation [65,66], recent studies highlight their applicability in environmental fields, particularly in filtering dyes and hazardous substances [67].

A gas-releasing additive, commonly referred to as a foaming agent, is frequently derived from finely milled carbon or carbon-based compounds [64]. The thermal treatment of glass powder and foaming agent mixture at temperatures ranging from 850 °C to 1000 °C, approaching the softening threshold of the glass, promotes the viscous flow and bubble expansion [68,69]. During the cooling stage that follows thermal processing, the gas bubbles stabilise and solidify into pores, leading to the development of the foam's characteristic porous architecture [70]. The formation of a uniform structure is governed by a delicate equilibrium between viscous flow and gas release. In the literature, foaming agents are usually classified into two types: redox agents and neutralisation agents [71].

Over 30% of studies related to redox agents focus on carbon-based materials [72]. Optimised foaming performance driven by oxidation of carbon-based agents necessitates sufficient oxygen from both the atmosphere and inter-particle spaces [73]. The available oxygen in the glass is influenced by several factors, including glass composition, presence and concentration of polyvalent ions, and their respective oxidation states. These ions are commonly employed to promote oxidation, with manganese oxide (MnO_2 or Mn_3O_4), iron oxide (Fe_2O_3), or cobalt oxide (Co_3O_4) [74] being among the most effective agents. However, incorporating these additives into glass-based foams increases the cost of production and requires careful control of glass release.

Silicon carbide (SiC) is assessed as an effective foaming agent for foam glass production, capable of producing homogeneous foams with controlled and precise cell sizes [75]. SiC exhibits a higher oxidation temperature than pure carbon, ranging from 950 °C to 1150 °C, which influences the melt viscosity. At elevated temperature, the melt viscosity is low, leading to rapid gas formation that can result in foam collapse [76]. For foams to be stable, it is essential to balance their viscosity with a uniform internal cell structure, ensuring uniform pore size and thin wall [77]. Incorporation of oxidation agents results in a decrease in foaming temperature while promoting greater gas evolution.

Carbonates decompose under thermal treatment, leading to the emission of carbon dioxide. Dolomite ($CaCO_3 \cdot MgCO_3$) and calcite ($CaCO_3$) are commonly employed as foaming additives in industrial processes. The involved reactions promote the decomposition of minerals, resulting in gas release [78]. The primary advantage of using carbonates is their role as fusion agents, reducing the sintering and foaming temperature [62].

In addition to conventional foaming agents, alternative compounds have been explored for their foaming capabilities. One such compound is aluminium nitride (AlN) powder. At elevated temperatures, the glass powder transitions to a viscous liquid state, during which AlN decomposes to produce gas that rapidly expands, resulting in the formation of bubbles within the glass melt. It is crucial for the glass to maintain an appropriate viscosity to facilitate this process [68]. AlN can function on its own or together with other additives, including TiO_2 or Fe_2O_3 [79].

The potential of a range of organic compounds as redox foaming agents has been assessed, showing their ability to create a consistent porous arrangement [80]. One notable advantage of utilising organic compounds is their lower decomposition temperature, which allows for earlier reactions that achieve the desired viscosity. Among these organic materials, glycerol is one of the most used foaming agents; however, other classes of compounds, such as fatty acids, surfactants, and polymers, are also utilised [81].

2.4. Glass-Ceramics

At first, the term 'glass-ceramics' referred to materials with more than 50% crystalline content, but over the years, materials with much lower crystal fraction have also been developed [82]. Materials first formed as glass and later crystallised into glass-ceramics using nucleating agents. Glass-ceramics are high-performance composite materials characterised by their superior mechanical properties, high thermal stability, excellent chemical resistance, and exceptional efficiency in heavy metal immobilisation [83].

The formation of glass-ceramic necessitates a supplementary thermal treatment. In fact, the nucleation stage is performed between 500 and 750 °C, while the crystallisation stage happens between 750 °C and 1000 °C, depending on the chemical composition of the starting glass [84].

To mitigate energy demand, a potential strategy is the utilisation of waste-derived glass. The conversion of waste-derived glass into glass-ceramics has revealed its potential for value-added applications [35,85]. Dense glass-ceramics suitable for building applications such as tiles [86,87] or face bricks [88] can be produced through this method. Additionally, porous glass-ceramics foams have garnered scientific interest for the application in thermal and acoustic insulation [65,89,90], as well as catalyst support [91]. These glass-ceramic foams exhibit excellent physical characteristics, which make them appropriate in the construction field [92]. The transformation of waste-derived glass with correct composition into glass-ceramics can be achieved thanks to the sinter–crystallisation technique [38]. The procedure involves the co-occurrence of glass particles melting, due to the viscous flow sintering, and the development of a crystalline phase.

2.5. Geopolymers

Inorganic polymers, also known as 'geopolymers', have demonstrated their potential in the civil engineering and building sector due to their environmental sustainability [93,94]. In fact, this class of materials acts as substitutes for cements, offering significantly lower CO_2 emissions and energy consumption, reducing consumption by 60–80% [95].

Since the 1970s, Davidovits [96] defined geopolymers as three-dimensional aluminosilicate structures formed through inorganic polycondensation, defined by the formula $M_n[\text{-}(SiO_2)z\text{-}AlO_2]_n \cdot wH_2O$, where M is the alkali metal cation; n indicates the degree of polycondensation; and z can be 1, 2, or 3. However, Kriven et al. [97] revised the definition of geopolymers as inorganic poly silicate–alumino polymers or chemically bonded ceramics.

The two main components of geopolymers are a powdered alumino–silicate-based material and an alkaline activator. The geopolymerisation process consists of converting a significant amount of the amorphous alumino–silicate phase into a dense structure through polycondensation [98].

The properties of geopolymers are greatly influenced by the choice of raw materials. Metakaolin, the anhydrous calcinated form of the clay mineral kaolinite, is commonly used for its high reactivity. Nevertheless, a fundamental conflict exists between the growing need for sustainability in geopolymers and the energy-intensive thermal treatment required to produce them. As a response to this challenge, alternative aluminosilicate precursors, such as ashes and ground granulated blast furnace slag, have been explored. A novel approach to geopolymer production involves using industrial residues and waste materials.

Similar to the conventional concrete production, scrap glass is employed as an aggregate [99]. However, incorporating additional materials, as metakaolin, is essential to fulfil the specific structural needs of geopolymers. Particle size, chemical composition, and temperature play a key role in influencing the reactivity of glass in geopolymeric mixtures [100].

Despite these advancements, the current scientific literature lacks examples of entirely glass-based geopolymers. This indicates that glass recycling is still incomplete, as the presence of glass powder impacts phase development and the following curing times in geopolymer manufacturing. Thus, while waste glass powder offers potential benefits, its integration into geopolymer systems necessitates careful consideration of a complementary position to achieve optimal performance.

The use of glass as an alternative to natural limestone is investigated in the production of geopolymers [101]. However, glass replacement can reduce mechanical properties because of the poor bonding between the matrix and glass. Research has indicated that the optimal inclusion of waste glass should be around 25% [102]. Initially, glass powder acts mainly as a filler, but over time, it is involved in the development of calcium/sodium aluminate silicate hydrate gel (C/N(A)-S-H), which enhances the overall strength of the geopolymer matrix.

Finally, Torres-Carrasco et al. [103] investigated the potential application of urban-derived waste glass as an alkaline activator in geopolymers, specifically in the form of sodium silicate hydrate, or waterglass. Activators are essential in geopolymer production, providing high molarity that results in extreme causticity, high corrosiveness, and notable viscosity. These characteristics require careful handling and safety protocols. Additionally, the significant energy demand for producing these activators increases the cost of geopolymer technology, which may limit its application to small-scale projects [104].

To date, there are no reported examples of geopolymer systems in which the final products are composed solely of glasses, as these formulations consistently require the addition of other synthetic or natural components. The following section, which examines alkali-activated materials, closely related to geopolymers, highlights examples of systems produced entirely from glass, whether specifically formulated or sourced from waste glass by-products.

3. Alkali Activation: Definitions, Mechanisms, and the Role of Waste Glass

Alkali-activated materials (AAMs) are becoming increasingly important in reducing CO_2 emissions, alongside geopolymers. While AAMs and geopolymers share similarities, they are distinguished by their aluminosilicate precursor composition and gelation mechanisms. The terminology surrounding AAMs and geopolymers is still evolving, and several studies have proposed different classification schemes. Alkali-activated materials, alongside inorganic polymers, are considered a broader category within which geopolymers are identified as a distinct subclass. From a chemistry-focused perspective, Van Deventer et al. [105] define alkali-activated materials as the result of a reaction between a calcium-rich aluminosilicate precursor and an alkaline source.

Provis, similarly, categorised alkali-activated material as high-calcium binders [106]. The calcium content influences not only the kinetics of the reaction but also the resulting gel chemistry and microstructure. In calcium-enriched systems, $Ca(OH)_2$ and other calcium species may interact with silicate and aluminate ions in solution. In the early stages, calcium silicate hydrate (C-S-H) forms, and further reaction with Al^{3+}, Si^{4+}, and Ca^{2+} ions with alkalis leads to the development of tobermorite-like calcium–aluminium–silicate hydrate ($CaO{\cdot}Al_2O_3{\cdot}SiO_2{\cdot}nH_2O$, C-A-S-H) gel. In alkali-activated materials, these gel networks are characterised by tetrahedrally coordinated silicate chains with a characteristic pattern, comprising $(3n-1)$ tetrahedra. In composition with moderate calcium concentration, a hybrid gel structure may form, where both C-A-S-H gel and sodium–aluminosilicate hydrate (N-A-S-H) phases [107] have been identified. The variability in gel chemistry highlights the intricate structural nature of AAMs and demonstrates how their properties are strongly influenced by both the composition of the precursor and the conditions under which the activation occurs.

One of the most promising approaches in alkali-activated research involves the incorporation of waste materials or industrial by-products, notably waste glass. Waste glass can function as a supplementary filler or aggregate in alkali-activated materials, often in combination with primary precursors such as fly ash or ground granulated blast furnace slag.

Chemically tailored glass acts as a suitable feedstock for alkali activation and subsequent inorganic polymer formation.

An alternative strategy [108] focuses on converting aluminosilicate sources into reactive binders adapted for alkali activation, offering a further alternative to mere vitrification methods. This process, involving components as SiO_2, Al_2O_3, and Na_2CO_3, leads primarily to the formation of glass thanks to thermal treatment at 1250 °C. The obtained glass is activated thanks to a highly concentrated alkaline solution. The final materials reveal the presence of sodium aluminosilicate hydrate (N-A-S-H) gel, along with the development of secondary zeolitic phases. Despite their technical potential, such a method faces several critical challenges, including the reliance on high-purity reagents, the energy consumption associated with the melting and rapid cooling, and the environmental impact of using highly concentrated alkaline solutions.

To mitigate these limitations on synthetic precursor, the synthesis of compositionally tailored glasses has been refined by integrating naturally sourced materials, such as clay [109]. Nonetheless, the success of these systems remains dependent on careful selection and dosage of pure activators. Poorly chosen combinations may lead to the development of heterogeneous gel structures, where N-A-S-H gel is dispersed in C-A-S-H gel matrix.

Recent studies have proposed using waste glass as the sole precursor in AAM synthesis [110]. To the best of the authors' knowledge, in fact, Cyr and colleagues were the first to report the use of soda–lime waste glass as the sole precursor for the synthesis of geopolymer-like materials activated with diluted alkaline solution. Remarkably, these materials developed sufficient mechanical performance at moderate curing temperatures (40–60 °C) without requiring additional thermal treatment, offering a more energy-efficient and sustainable process. Nevertheless, the activation of soda–lime glass has a significant limitation due to its chemical composition, which hampers the formation of an appropriate zeolitic gel. Therefore, the term 'geopolymer' is also deemed inappropriate for these materials. Despite the variation in chemical composition, further studies conducted by Idir et al. [111] reveal that the optimal concentration of the activating solution is consistently found to be 3 mol/L, giving rise to the concept of weak or mild alkali activation. Despite these advancements, the molecular mechanism underlying hardening in mild alkali-activated materials remains poorly described. For this reason, the next section is intended to explore and address the existing gap in the current understanding and development of the molecular mechanism of mild alkali-activated glass. Briefly, weak alkali activation promotes the development of a gel layer on the particle interface, instead of total dissolution.

In parallel, Rincon et al. [112] proposed a dual approach that combines alkali activation, followed by sintering, to minimise the overall cost of glass-ceramic foams. The hardening of the suspensions is attributed to the formation of calcium silicate hydrate phases. Nonetheless, the produced porous materials exhibit chemical instability, necessitating thermal treatment. While this methodology has been extended to various glass types [113], the gel compounds are often difficult to identify. It is noteworthy that the requirements for thermal treatment, even at low temperatures, constitute a major drawback in terms of environmental sustainability [46]. Moreover, the use of highly concentrated alkaline solutions also represents a limitation in terms of ecological compatibility, in addition to increasing production cost for alkali-activated materials.

These considerations highlight the relevance of mild alkali activation, in which the alkaline solution remains below 3 mol/L, as a more sustainable path.

4. Weak Alkali Activation of Pharmaceutical Glass: Reaction Mechanism

The emergence of COVID-19 has led to a global pandemic. The global healthcare system faced major challenges, but coordinated action among industries and the scientific community enabled accelerated vaccine development and the implementation of an effective distribution framework [114]. The escalation in production has inevitably resulted in a corresponding increase in waste, highlighting the need for careful management of these by-products. Notably, the closed-loop recycling of pharmaceutical boro–alumino–silicate glass (BASG) containers requires exceptionally pure cullet for effective recycling processes.

The potential of employing BASG glass as a raw material for geopolymer synthesis was introduced by Taveri et al. [115]. Their selection is based on the chemical composition of glass residue, which contains critical species, as amorphous silica and alumina, essential for zeolitic network development. Additionally, the presence of boron oxide in the glass chemical composition can partially replace alumina, thereby promoting polycondensation reaction. Boron appears to serve an essential function in the geopolymerisation of BASG activated with a strongly alkaline NaOH solution (13 mol/L). In fact, the ^{11}B NMR spectra of the geopolymers show a notable shift in the relative peaks' intensity: the signal of BO_4, at 1 ppm, persists, whereas the signal for the trigonal boron at 11 ppm has nearly disappeared, suggesting that trigonal boron dissolves and reorganises into a tetrahedral configuration, characteristic of zeolite gels. Unlike soda–lime glass systems, in boro–alumino–silicate glass, the negligible CaO amount prevents the synthesis of calcium silicate hydrate compounds. In contrast to alkali activation of glass containing CaO [112,116], the calcium oxide facilitates the formation of binding products.

Rincon et al. [117] explored the alkali activation of BASG using mild chemical attack with alkali hydroxide activators not exceeding 2.5 mol/L, without incorporating any other silicate or alumino–silicate feedstock. The result was the formation of a complex hydrated gel, predominantly amorphous with a zeolite-like framework, incorporating alkali ions to balance charge.

Further experiments conducted by Metha et al. [118] focused on reducing the firing temperature of boro–alumino–silicate glass to 550–650 °C and exploring complete avoidance of the thermal stage. The incorporation of surfactants significantly affected the nature of the gel phase. Denser pieces from the hardening of suspensions without surfactants remained intact after the boiling test; however, the foams disintegrated. This suggests that alkaline ions in the foams contributed beyond the stabilisation of the boron and aluminium tetrahedral species in the gel matrix, but actively contribute to the development of a highly depolymerised network, constraining the formation of strong Si–O–Si bonds. On the other hand, in compact cold-consolidated materials, the reduced incorporation of alkali ions, caused by hydrated alkali carbonates development, favours a more cross-linked network, strengthening glass particles' bonding and improving the integrity of the final product.

Tameni et al. [119] provided significant advancements in understanding the molecular mechanism of alkali activation of boro–alumino–silicate glass with mild alkaline solution (NaOH/KOH, 2.5 mol/L). The NMR analyses [Figure 2] indicate that, despite the process of alkali activation and potential boiling, there are no significant alterations in the ^{29}Si, ^{27}Al, and ^{11}B MAS NMR spectra. In fact, the alkali activation of BASG using diluted solutions should not be equated with the geopolymerisation process. Boron and aluminium oxides in the glass inhibit the formation of a zeolite-like gel. According to Taveri et al. [115], a stable gel network with a notable increase in BO_4 units can only be achieved under specific activation conditions, such as highly concentrated alkali solution and additives containing a high level of Al_2O_3.

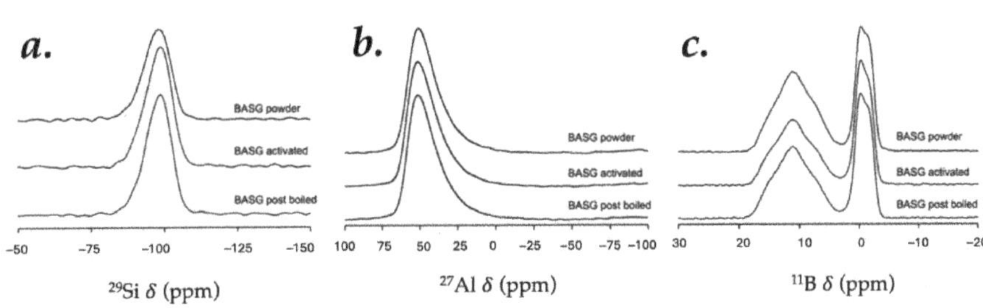

Figure 2. NMR spectra of (**a**) ^{29}Si, (**b**) ^{27}Al, and (**c**) ^{11}B of BASG in as-received, alkali-activated, and boiled conditions. Adapted from [119], Elsevier, 2025.

Structural characterisation using nuclear magnetic resonance, in conjunction with X-ray diffraction (XRD) and Fourier Infrared Spectroscopy (FT-IR) analyses, elucidated that the transition from the 'as-received' glass to alkali-activated glass consists of a cold-consolidated glass network, a gel phase, and alkali carbonates. Notably, the boiling test effectively removed the soluble phases, leading to the reversion to the original phase structure. NMR spectroscopy shows that alkali activation of BASG induces chemical changes only at the particle surface, while the interior of the glass remains unchanged, even after exposure to boiling conditions.

In other words, mild alkali activation of glass suggests a behaviour more closely aligned with corrosion phenomena, a concept that has been extensively documented in the context of nuclear waste [120–122]. The mild alkali activation primarily affects the surface layer of the particles, avoiding substantial dissolution experienced in geopolymers. However, in mild alkaline solutions, glass components released are considered minimal compared to the surface hydration of powdered particles, a phenomenon that leads to the formation of hydroxyl groups through bond cleavage of strong bonds (Si–O–Si, Si–O–Al, and Si–O–B). The condensation of reactive surface groups on adjacent particles facilitates the rebounding. This progressive polycondensation process ends with the formation of a 'skeleton' network of glass particles, with interstitial regions occupied by the secondary phase.

The secondary phase consists of gel, carbonates, and eventually other crystalline phases. The amorphous or semicrystalline gel is generated for partial dismantling of the glass network, comprising borates, silicates, and aluminates. The stabilisation of this gel is contingent upon the activation conditions and the interplay between the chemistry of the glass and the activating agents. Simultaneously, carbonates result from the interaction between alkali ions and CO_2 from the atmosphere. However, only strong bonds forming the primary phase are responsible for the stability of mild alkali-activated glass, as shown in Figure 3.

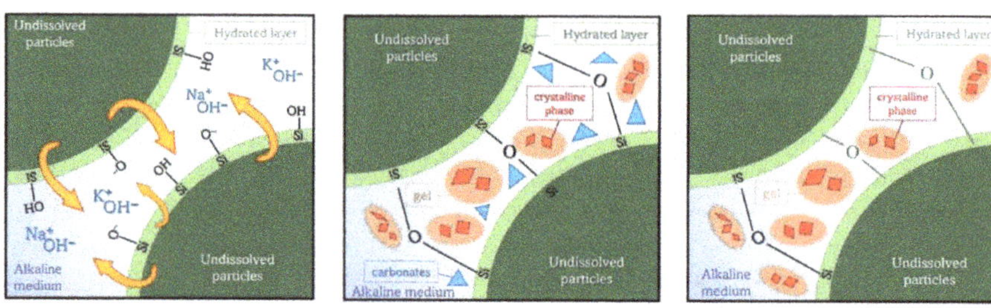

Figure 3. Scheme of alkali activation and cold consolidation.

5. Perspective of Glass Mild Alkali Activation

Beyond presenting the chemical molecular mechanism governing these materials, this section aims to offer new insights into the potential application of mild alkali-activated materials. Progress in using alkali-activated materials (AAMs) to replace traditional cement-based materials in construction and other applications has experienced significant growth in recent years. This section emphasises the innovative applications that are derived from the alkali activation of waste glass.

5.1. Alkali Activation of Glass as Building Materials

Waste glass has been utilised as a partial substitute for other aluminosilicate precursors, such as metakaolin, fly ashes, and ground granulated furnace slag [123–125] in the production of alkali-activated mortars in the building materials context. Nevertheless, an increasing number of studies have focused on the activation of glass as a standalone material.

Redden et al. [126] elucidated the potential for the alkaline activation of soda–lime glass, using an NaOH activating solution (4–8 mol/L), with compressive strength values exceeding 20 MPa regardless of the molarity. Interestingly, prolonged curing durations and high temperatures adversely affect the compressive strength of the final products.

Also, Torres-Carrasco et al. [127] explored the use of soda–lime glass produced by industrial and urban waste treatment plants as a precursor for alkali-activated materials, using various highly concentrated alkaline solutions, including NaOH, NaOH/Na_2CO_3, sodium silicate hydrate, and KOH. Their studies demonstrated that the compressive strength varied significantly, achieving values between 17 MPa and 88 MPa, based on the selected activator and the applied curing methods, indicating the complexity and tunability of glass activation chemistry.

As previously reported, Cyr et al. [110] developed alkali-activated mortars using soda–lime glass powder as the primary binder, employing NaOH and KOH as alkali activators at different concentrations (1, 5, and 10 mol/L). Their findings demonstrated that waterglass was not required for the setting and hardening of the AAMs, and the optimal alkali content is fixed at 5 mol/L. In a subsequent investigation [111], the same authors demonstrated that the maximum compressive strength, which reaches up to 50 MPa, was achieved with alkali-activated materials using KOH at a concentration of 3 mol/L, reinforcing the viability of 'mild' alkali activation.

More recently, boro–alumino–silicate glass is also eligible as a raw material for construction and building alkali-activated materials. Tameni et al. [128] demonstrated that boro–alumino–silicate glass (BASG) can undergo alkali activation (NaOH:KOH, 2.5 mol/L) and cold consolidation by maintaining a temperature of 40 °C for 7 days, yielding compressive strength higher than 22 MPa and a geometrical density of approximately 1.7 g/cm^3. These values place these materials in the range of commercial Plaster of Paris or concrete (structural lightweight).

In an effort to evaluate the robustness of this approach, the same authors also applied it to glass fractions characterised by varying degrees of contamination, including those that have come into contact with oils (BASG_dirty) or grinding machinery (BASG_waste). Through the use of a commercial materials selection software (Ansys Granta EduPack version Ansys 2024 R1, Ansys Granta, Canonsburg, PA, USA), Figure 4 presents a comparative evaluation of new alkali-activated materials and conventional building materials. In the context of construction, Ashby's plot compares the density and the compressive strength to obtain the optimal trade-off [129]. BASG_waste (represented by 'blue item') overlaps with structural lightweight concrete. Moreover, the BASG_clean and BASG_dirty ('green item' and 'orange item', respectively) exhibit lower density and a comparable range of compressive strength of structural lightweight concrete, making them particularly promising solutions as building materials.

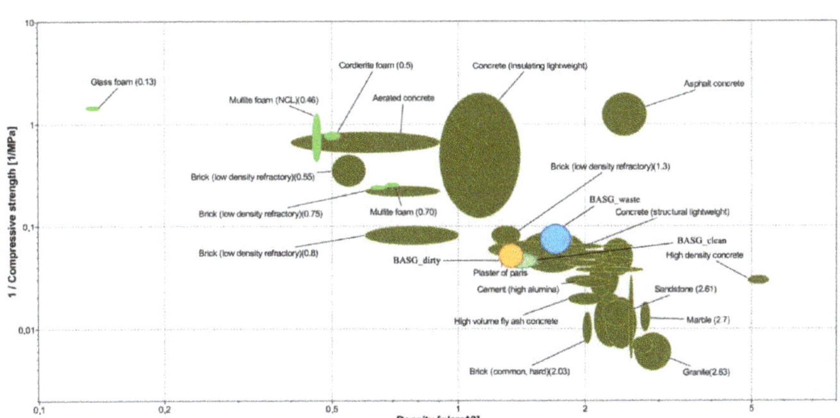

Figure 4. Graph illustrating the trade-off between compressive strength and density of lightweight products (Ansys Granta EduPack version Ansys 2024 R1) in comparison to BASG samples.

5.2. Alkali Activation of Glass as Sorbent for Dyes

Beyond the applications in the construction sector, alkali-activated materials are increasingly being recognised for their potential in water and wastewater treatment. Their intrinsic porosity, characterised by the presence of nano- and micropores, contributes to their effectiveness in adsorbing contaminants and facilitating various purification processes [130]. Thanks to these properties, AAMs are suitable for use as adsorbents, ion-exchangers, membranes, and filtration media [131].

Among the various purification techniques, adsorption constitutes an efficient and economically viable approach for the removal of pollutants in wastewater treatment. This method is characterised by simplicity, cost-effectiveness, and high efficacy, requiring minimal operational complexity [132,133]. In particular, the effectiveness of AAMs in adsorption processes is closely related to their microstructure, especially the surface and the porosity, together with the presence of surface functional groups [134]. These characteristics are particularly critical for dye removal, where both physical and chemical interactions influence adsorption capacity.

It has been demonstrated by Metha et al. [120] that cold-consolidated pharmaceutical glass powder enables the fabrication of innovative glass membranes. Additionally, these materials also function as dye-degrading sorbents when combined with TiO_2.

Moreover, further innovation has been achieved by the integration of additive manufacturing technology. Mahmoud et al. [135] produce masked-stereolithography scaffolds functionalised through alkali activation, resulting in the formation of zeolites and sodium carbonate phases. These innovative filters exhibit a high performance in the filtration and removal of methylene blue, attaining an impressive removal efficiency of 81%. Building upon findings from prior studies, the same research group [136] successfully developed stable glass-based functional scaffolds that do not rely on any organic binders. The reaction rate of new scaffolds achieved 100% degradation of methylene blue in 75 min and retained their performance over five cycles without a significant loss of efficiency. The stability, potential for repeated use, and ease of scaling up mark a significant step toward the filtration technologies based on waste glass AAMs.

5.3. Alkali Activation of Glass as Sorbent for Heavy Metals

Heavy metal contamination in wastewater is a serious risk for both human health and the environment, thereby driving efforts to develop sustainable removal techniques [137]. Wastewater often contains trace metallic elements [138], which are considered persistent in the environment since they do not decompose [139].

Traditional adsorbents, including activated carbon, alumina, and silica, are known for their high efficiency in removing traces of heavy metals. However, their elevated production costs have restricted their widespread use in practical applications [140,141]. In recent years, alkali-activated materials have emerged as an alternative for adsorption and have attracted increasing attention [142] due to low-cost synthesis, variable chemistry, and the capacity to immobilise toxic species.

Boro–alumino–silicate glass activated with a mild alkaline solution (NaOH/KOH, 2.5 mol/L) has been demonstrated to be an excellent matrix for the stabilisation of copper ions [143]. In fact, alkali activation can minimise ion mobility by coupling them with the glass matrix. Specifically, after the optimisation of the pH of $Cu(NO_3)_2$ aqueous solution, adsorption tests indicate an adsorption efficiency of $62 \pm 3\%$, consistent with geopolymeric matrices [144]. The final stabilisation is achieved thanks to the firing process at 700 °C. The efficiency of copper retention was assessed through leaching tests. This suggests that mild alkali activation combined with controlled heat treatment offers an effective strategy for immobilising hazardous ions.

5.4. Alkali Activation of Glass for Nuclear Waste Immobilisation

Waste management necessitates the implementation of effective protocols to diminish the concentrations of radioactive contaminants to levels permissible for the safe discharge of decontaminated liquids into the environment, as well as for the secure disposal of concentrated radionuclides [145,146].

Vitrification is widely regarded as the predominant treatment technology for the immobilisation of high-level radioactive waste, effectively converting hazardous materials into a stable glass matrix that enhances long-term storage and containment characteristics [147]. Boro–silicate glass is a commonly used glass composition for vitrification due to its high chemical and thermal stability [148].

On the other hand, cement solidification represents a well-established and advanced technology that is based on the hydration of cement and its ability to form gels for immobilising radioactive materials. This technology still encounters several technical limitations, [149] such as minimal volume reduction and the relatively high tendency to release contaminants, especially with cesium [150,151]. Indeed, due to the radiation emitted by Cs-137 and Sr-90, ordinary Portland cement-based materials are not appropriate for the immobilisation of these isotopes due to low thermal resistance, high content of free water, and inadequate immobilisation [152,153].

Furthermore, alkali-activated materials offer a viable alternative in encapsulating nuclear waste, due to their cationic binding sites and durability [154–156]. In the context of immobilising simulated nuclear waste through alkali activation, the common matrices are metakaolin [157], fly ash [158], and blast furnace slag [159].

Pioneering studies have been conducted by Lago et al. [160] for the immobilisation of CsOH using alkali-activated boro–alumino–silicate matrices. They successfully achieved an immobilisation of over 20 wt% cesium through a cold consolidation process. A portion of cesium, incorporated into boro-pollucite, is immobilised immediately upon the cold consolidation process. The resulting cold-consolidated blocks resist immersion in boiling water. The sintering process allows further strengthening. Moreover, the leaching tests performed on fired samples revealed minimum cesium release.

To minimise the production of soluble species, an engineered treatment using both CsOH and NH_4OH was employed to maximise the immobilisation of cesium [161]. Despite the absence of increased boro-pollucite formation, the observed enhancement in cesium retention is likely attributable to the formation of a less soluble gel associated with the interaction of Cs^+ ions with an elevated concentration of glass components released into the solution.

5.5. Alkali Activation of Glass and Microwaved-Assisted Process

Microwave heating represents an innovative technological advancement within the cement and concrete industries, attracting significant interest and ongoing enhancements attributable to its numerous advantages [162,163]. First of all, rapid heating rate allow to reduce energy and time: Li et al. [164] outlined that the concrete curing with microwave heating requires one-third of the duration and 10% of energy compared to the traditional process. Moreover, precise and immediate control through electronic systems, the formation of fine microstructure that enhances the properties of the resulting cement, and the absence of secondary waste production are highlighted as benefits [165].

Microwave heating has been applied to alkali-activated materials and geopolymers. Microwave curing has the potential to significantly expedite the hardening phase of these materials [166]. Aschoff et al. [167] reported that the use of microwave treatment leads to an accelerated setting and hardening process in metakaolin and aluminium orthophosphate-based geopolymers. In fact, microwave heating, unlike conventional methods, minimises heat loss and eliminates the need for the heat to be transferred between the heating device and the samples: the energy is directly converted from electromagnetic to thermal energy, resulting in a faster hardening process.

Moreover, microwave treatment leads to a noticeable improvement in the compressive strength of the final product. Hong et al. [168] demonstrate that microwave treatment, applied to coal bottom ash-based geopolymers, after an initial curing phase of 36 h, allows for an increase in compressive strength from 13 to 40 MPa. Carollo et al. [169] performed microwave-assisted cold consolidation on glass waste slurries, revealing that the compressive strength increases from 10 MPa to 18 MPa thanks to 5 min microwave heating at 450 W.

Alkali activation of boro–alumino–silicate glass, combined with microwave curing, provides an opportunity for understanding the molecular mechanisms of 'mild' alkali activation [170]. The application of microwave irradiation facilitates the promotion of carbonate formation associated with potassium (K^+), thereby promoting the formation of less soluble gels, which substantially enhances the overall stability of the material, as shown in Figure 5. Moreover, the microwave treatment has demonstrated a beneficial impact on the mechanical properties of alkali-activated glass, with an enhancement of up to 100% together with an increase in total porosity.

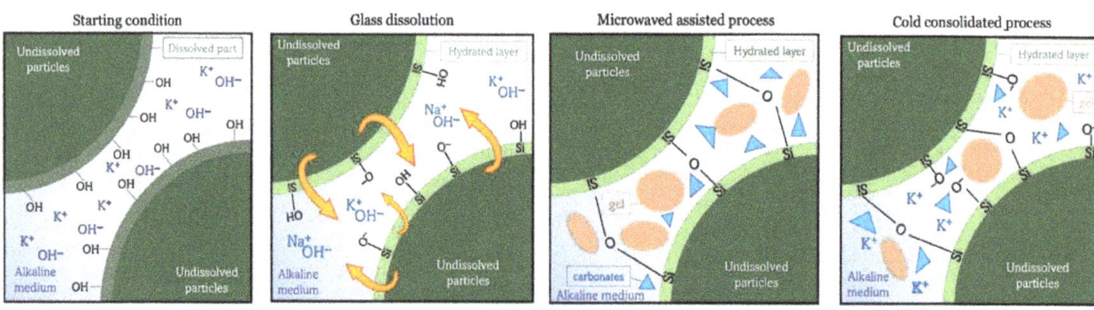

Figure 5. Graphical representation of the molecular mechanism of the microwave-assisted process and cold-consolidated process.

6. Conclusions

Despite glass being widely recognised as one of the most recyclable materials, significant challenges persist in the management of glass waste, particularly concerning the recovery of fine glass fractions. These glass fragments frequently evade efficient recovery through traditional recycling methods and are landfilled. However, emerging trends in recycling and upcycling present promising avenues for addressing this issue. Specifically, the utilisation of glass in the production of geopolymers and alkali-activated materials represents a significant opportunity to convert waste into high-value applications. This comprehensive overview results in the following conclusions:

- Glass has the potential to serve as the sole raw material for alkaline activation, functioning independently of its chemical composition and the molarity of the alkaline solution. This suggests that the molecular mechanism of mild alkali-activated glasses is similar to glass corrosion. Due to the limited molarity, the alkaline solution does not allow for the complete dissolution of the glass used as raw material, but only affects the surface of the glass particles. Moreover, the OH^- ions from the alkaline source led to the cleavage of strong bonds, such as siloxane bonds, resulting in the formation of a hydrated surface layer (–OH). During the hardening phase, there is the rearrangement of Si–O–Si, Si–O–Al, and Si–O–B between the particles. The other constituents in the solution reorganise into a secondary phase that may involve the formation of various compounds, such as carbonates, gel, and potential crystalline phases. The secondary phase does not contribute significantly to the stability of the final product.

- In contrast to their conventional application as substitutes for cements and concrete, alkali materials may be considered for environmental uses. Alkali-activated materials are used as membranes for dye adsorption, such as methylene blue. Moreover, the removal of heavy metals can be performed with an alkali-activated sorbent.
- Alkali-activated glass presents a promising avenue for the immobilisation of simulated nuclear waste, particularly in regard to cesium hydroxide (CsOH). The samples fired at 700 °C exhibited only trace levels of cesium release during leaching tests.
- The microwave-assisted process significantly contributes to reducing the hardening time of materials. Microwave techniques can also provide deeper insights into the molecular mechanism during the hardening process of alkali-activated glass. In fact, the formation of a less depolymerised gel, and consequently a more stable structure, is favoured by a more efficient allocation of potassium ions.

This overview offers a comprehensive overview of the current advancements in alkali activation, with particular focus on mild alkali activation systems. However, several important research gaps remain, most notably the absence of durability studies among the works cited, a key factor in evaluating their long-term viability and application. This is particularly evident in the lack of investigations into resistance to freeze–thaw cycles, chemical attack, and carbonation, which hinders direct comparison with traditional materials such as Portland cement.

Author Contributions: Conceptualisation, E.B. and G.T.; writing—original draft preparation, G.T.; writing—review and editing, E.B. and G.T.; visualisation, E.B. and G.T.; supervision, E.B.; funding acquisition, E.B. All authors have read and agreed to the published version of the manuscript.

Funding: This research was funded by the Italian Ministry of University Research (MUR), through the National project MUR PON R&I 2014–2021 and Next Generation EU funds, PRIN 2022 PNRR project 'GLASS_TREA.S.U.RES' (GLASS-based TREAtments for Sustainable Upcycling of inorganic RESidues, #P2022S4TK2).

Institutional Review Board Statement: Not applicable.

Informed Consent Statement: Not applicable.

Data Availability Statement: No new data were created or analyzed in this study.

Acknowledgments: The authors acknowledge the support of the national projects MUR PON R&I 2014–2021 and GLASS_Trea.S.U.Res (GLASS-based TREAtments for Sustainable Upcycling of inorganic RESidues) (PRIN 2022 PNRR project #P2022S4TK2). They also acknowledge the cooperation with the FunGlass center (D. Galusek) of the Alexander Dubcek University of Trencin (Slovakia) in the framework of the European Union's Horizon 2020 'FunGlass' project and the Centre of Magnetic Resonance (S. E. Ashbrook), University of St Andrews (United Kingdom) in the framework of the Padua-St Andrews Joint Seed Funding Scheme.

Conflicts of Interest: The authors declare no conflicts of interest.

References

1. de Lima, C.J.; Veer, F.; Zhang, H.; Filho, F.F.D.M.; Çopuroglu, O.; Nijsse, R. Thermal, optical and mechanical properties of new glass compositions containing fly ash. *Glass Technol.-Eur. J. Glass Sci. Technol. Part A* **2021**, *62*, 96–108. [CrossRef]
2. Martin, G.; MacFarlane, A. *The Glass Bathyscaphe: How Glass Changed the World*; Profile Books: London, UK, 2011.
3. Jiang, Y.; Ling, T.-C.; Mo, K.-H.; Shi, C. A critical review of waste glass powder—Multiple roles of utilization in cement-based materials and construction products. *J. Environ. Manag.* **2019**, *242*, 440–449. [CrossRef] [PubMed]

4. Lu, J.X.; Zhan, B.J.; Duan, Z.H.; Poon, C.S. Using glass powder to improve the durability of architectural mortar prepared with glass aggregates. *Mater. Des.* **2017**, *135*, 102–111. [CrossRef]
5. Cantini, A.; Leoni, L.; Ferraro, S.; De Carlo, F.; Martini, C.; Martini, F.; Salvio, M. Technological energy efficiency improvement in glass- production industries and their future perspectives in Italy. *Processes* **2022**, *10*, 2653. [CrossRef]
6. Colangelo, S. Reducing the environmental footprint of glass manufacturing. *Int. J. Appl. Glass Sci.* **2024**, *15*, 350–366. [CrossRef]
7. Vinci, G.; D'Ascenzo, F.; Esposito, A.; Musarra, M.; Rapa, M.; Rocchi, A. A sustainable innovation in the Italian glass production: LCA and Eco-Care matrix evaluation. *J. Clean. Prod.* **2019**, *223*, 587–595. [CrossRef]
8. Khan, M.U.A.; Ahmad, S.; Butt, S.I. Environmental impact assessment of the manufacturing of glass packaging solutions: Comparative scenarios in a developing country. *Environ. Impact Assess. Rev.* **2023**, *102*, 107195. [CrossRef]
9. Westbroek, C.D.; Bitting, J.; Craglia, M.; Azevedo, J.M.C.; Cullen, J.M. Global material flow analysis of glass: From raw materials to end of life. *J. Ind. Ecol.* **2021**, *25*, 333–343. [CrossRef]
10. Yuan, X.; Wang, J.; Song, Q.; Xu, Z. Integrated assessment of economic benefits and environmental impact in waste glass closed-loop recycling for promoting glass circularity. *J. Clean. Prod.* **2024**, *444*, 141155. [CrossRef]
11. Wilkinson, C. Glass and a carbon-free United States: What is glass's role in the upcoming green revolution? *J. Am. Ceram. Soc.* **2024**, *107*, 1533–1542. [CrossRef]
12. Atzori, D.; Tiozzo, S.; Vellini, M.; Gambini, M.; Mazzoni, S. Industrial Technologies for CO_2 Reduction Applicable to Glass Furnaces. *Thermo* **2023**, *3*, 682–710. [CrossRef]
13. Hu, P.; Li, Y.; Zhang, X.; Guo, Z.; Zhang, P. CO_2 emission from container glass in China, and emission reduction strategy analysis. *Carbon. Manag.* **2018**, *9*, 303–310. [CrossRef]
14. Caudle, B.; Taniguchi, S.; Nguyen, T.T.H.; Katoaka, S. Integrating carbon capture and utilization into the glass industry: Economic analysis of emissions reduction through CO_2 mineralization. *J. Clean. Prod.* **2023**, *416*, 137846. [CrossRef]
15. Golub, O.V.; Sanzharovskii, A.Y.; Mikhailidi, D.K.; Vartanyan, M.A. Carbon Footprint of the Life Cycle of Glass Containers. *Glass Ceram.* **2022**, *79*, 306–311. [CrossRef]
16. Zier, M.; Stenzel, P.; Kotzur, L.; Stolten, D. A review of decarbonization options for the glass industry. *Energy Convers. Manag. X* **2021**, *10*, 100083. [CrossRef]
17. Harrison, E.; Berenjian, A.; Seifan, M. Recycling of waste glass as aggregate in cement-based materials. *ESE* **2020**, *257*, 100064. [CrossRef]
18. Guo, P.; Meng, W.; Nassif, H.; Guo, H.; Bao, Y. New perspectives on recycling waste glass in manufacturing concrete for sustainable civil infrastructure. *Const. Build. Mater.* **2020**, *257*, 119579. [CrossRef]
19. Ferdous, W.; Manalo, A.; Siddique, R.; Mendis, P.; Zhuge, Y.; Wong, H.S.; Lokuge, W.; Aravinthan, T.; Schubel, P. Recycling of landfill wastes (tyres, plastics and glass) in construction—A review on global waste generation, performance, application and future opportunities. *Resour. Conserv. Recycl.* **2021**, *173*, 105745. [CrossRef]
20. Sena Sağlam, B.B.; Aydın, N. Investigation of waste characteristics and recycling behaviour at educational institutes. *Waste Manag.* **2024**, *180*, 115–124. [CrossRef]
21. Rashad, A.M. Recycled waste glass as fine aggregate replacement in cementitious materials based on Portland cement. *Const. Build. Mater.* **2014**, *72*, 340–357. [CrossRef]
22. Lessard, J.-M.; Habert, G.; Tagnit-Hamou, A.; Amor, B. Tracking the Environmental Consequences of Circular Economy over Space and Time: The Case of Close: The Open-Loop Recovery of Postconsumer Glass. *EST* **2021**, *55*, 11521–11532. [CrossRef]
23. Scalet, B.M.; Garcia Muñoz, M.A.; Sissa, Q.; Roudier, S.; Delgado Sancho, L. *Best Available Techniques (BAT) Reference Document for the Manufacture of Glass*; Joint Research Centre of the European Commission (JRC): Sevilla, Spain, 2013.
24. Bristogianni, T.; Oikonomopoulou, F. Glass up-casting: A review in the current challenges in glass recycling and novel approach for recycling "as-is" glass waste into volumetric glass components. *Glass Struct. Eng.* **2023**, *8*, 255–302. [CrossRef]
25. Tameni, G. Upcycling of Glass Waste in Sustainable Construction Materials. Ph.D. Thesis, Università di Padova, Padova, Italy, 2025.
26. Ingrao, C.; Saja, C.; Primerano, P. Application of Life Cycle Assessment to chemical recycling of post-use glass containers on the laboratory scale towards circular economy implementation. *J. Clean. Prod.* **2021**, *307*, 127319. [CrossRef]
27. Flood, M.; Fennessy, L.; Lockrey, S.; Avendano, A.; Glover, J.; Kandare, E.; Bhat, T. Glass Fines: A review of cleaning and up-cycling possibilities. *J. Clean. Prod.* **2020**, *267*, 121875. [CrossRef]
28. Dilan, R.; Edwin, B.; Sujeeva, S. A new technology of transforming recycled glass waste to construction components. *Const. Build. Mater.* **2021**, *313*, 125539. [CrossRef]
29. Federico, L.M.; Chidiac, S.E. Waste glass as a supplementary cementitious material in concrete—Critical review of treatment methods. *Cem. Concr. Compos.* **2009**, *31*, 606–610. [CrossRef]
30. SCHOTT Launches Pilot Projects on Glass-Ceramics and Specialty Glass for a More Circular Economy. Available online: https://www.schott.com/en-gb/news-and-media/media-releases/2024/schott-launches-pilot-projects-on-glass-ceramics-and-specialty-glass-for-a-more-circular-economy (accessed on 28 November 2024).

31. Kim, Y.; Han, Y.; Kim, S.; Jeon, H.-S. Green extraction of lithium from waste lithium aluminosilicate glass-ceramics using a water leaching process. *PSEP* **2021**, *148*, 765–774. [CrossRef]
32. Musgraves, J.D.; Hu, J.; Calvez, L. *Springer Handbook of Glass*; Springer, Handbooks; Springer International Publishing: Berlin/Heidelberg, Germany, 2019.
33. Lebullenger, R.; Mear, F.O. Glass recycling. In *Springer Handbook of Glass*; Musgraves, J.D., Hu, J., Calvez, L., Eds.; Springer International Publishing: Berlin/Heidelberg, Germany, 2019; pp. 1353–1375. [CrossRef]
34. Daigo, I.; Kiyohara, S.; Okada, T.; Okamoto, D.; Goto, Y. Element-based optimizing of waste ceramics material and glasses recycling. *Resour. Conserv. Recycl.* **2018**, *133*, 375–384. [CrossRef]
35. Bernardo, E.; Scarinci, G.; Colombo, P. Vitrification of Waste and Reuse of Waste-Derived Glass. In *Encyclopedia of Sustainability Science and Technology*; Meyers, R.A., Ed.; Springer: New York, NY, USA, 2012. [CrossRef]
36. Mahmoud, M.; Kraxner, J.; Mehta, A.; Elsayed, H.; Galusek, D.; Bernardo, E. Upcycling waste derived glass into high-performance photocatalytic scaffolds by alkali activation and direct ink writing. *Heliyon* **2024**, *10*, e24737. [CrossRef]
37. Scarinci, G.; Brusatin, G.; Barbieri, L.; Corradi, A.; Lancellotti, I.; Colombo, P.; Hreglich, S.; Dall'Igna, R. Vitrification of industrial and natural wastes with production of glass fibres. *J. Eur. Ceram. Soc.* **2000**, *20*, 2485–2490. [CrossRef]
38. Rincón, A.; Marangoni, M.; Cetin, S.; Bernardo, E. Recycling of inorganic waste in monolithic and cellular glass-based materials for structural and functional applications. *J. Chem. Technol. Biotechnol.* **2016**, *91*, 1946–1961. [CrossRef]
39. Ma, Q.; Ding, L.; Wang, Q.; Yu, Y.; Lida, L.; Hong, L. Preparation and characterization of continuous fly ash derived glass fibers with improved tensile strength. *Mater. Lett.* **2018**, *231*, 119–121. [CrossRef]
40. Ponsot, I.; Detsch, R.; Boccaccini, A.R.; Bernardo, E. Waste derived glass ceramic composites prepared by low temperature sintering/sinter-crystallisation. *Adv. Appl. Ceram.* **2015**, *114*, 17–25. [CrossRef]
41. Hujova, M.; Rabelo Monich, P.; Kankova, H.; Lucas, H.; Xakalashe, B.; Friedrich, B.; Kraxner, J.; Galusek, D.; Bernardo, E. New glass-based binders from engineered mixtures of inorganic waste. *Inter. J. Appl. Glass Sci.* **2021**, *12*, 570–580. [CrossRef]
42. Mehta, A.; Colusso, E.; Kraxner, J.; Galusek, D.; Bernardo, E. Waste-derived glass as a precursor for inorganic polymers: From foams to photocatalytic destructors for dye removal. *Ceram. Int.* **2022**, *48 Pt A*, 27631–27636. [CrossRef]
43. Horodytska, O.; Kiritsis, D.; Fullana, A. Upcycling of printed plastic films: LCA analysis and effects on the circular economy. *J. Clean. Prod.* **2020**, *268*, 122138. [CrossRef]
44. Pauli, G.; Hartkemeyer, J.F. *UpCycling*; Chronik Verlag im Bertelsmann LEXIKON Verlag: Gütersloh, Germany, 1999.
45. Jehanno, C.; Alty, J.W.; Roosen, M.; De Meester, S.; Dove, A.P.; Chen, E.Y.-X.; Leibfarth, F.A.; Sardon, H. Critical advances and future opportunities in upcycling commodity polymers. *Nature* **2022**, *603*, 803–814. [CrossRef]
46. Mazzi, A.; Sciarrone, M.; Bernardo, E. Environmental performance of glass foam as insulation material from waste glass with the alkali activation process. *Heliyon* **2023**, *9*, e19001. [CrossRef]
47. Ribeiro, M.J.; Tulyaganov, D. Traditional Ceramics Manufacturing. In *Ceramics, Glass and Glass-Ceramics*; Baino, F., Tomalino, M., Tulyaganov, D., Eds.; PoliTO Springer Series; Springer: Cham, Switzerland, 2021. [CrossRef]
48. Tite, M.S. Ceramic production, provenance and use—A review. *Archaeometry* **2008**, *50*, 216–231. [CrossRef]
49. Yeşilay, S. Production of stoneware clay bodies by using industrial soda-lime-silica glass waste. *J. Aust. Ceram. Soc.* **2019**, *55*, 747–758. [CrossRef]
50. Andreola, F.; Barbieri, L.; Lancellotti, I.; Leonelli, C.; Manfredini, T. Recycling of industrial wastes in ceramic manufacturing: State of art and glass case studies. *Ceram. Int.* **2016**, *42*, 13333–13338. [CrossRef]
51. Lassinantti Gualteriei, M.; Mugoni, C.; Guandalini, S.; Cattini, A.; Mazzini, D.; Alboni, C.; Siligardi, C. Glass recycling in the production of low-temperature stoneware tiles. *J. Clean. Prod.* **2018**, *197*, 1531–1539. [CrossRef]
52. Bernardo, E.; Esposito, L.; Rambaldi, E.; Tucci, A.; Hreglich, S. Recycle of Waste Glass into "Glass–Ceramic Stoneware". *J. Am. Ceram. Soc.* **2008**, *91*, 2156–2162. [CrossRef]
53. Tucci, A.; Esposito, L.; Rastelli, E.; Palmonari, C.; Rambaldi, E. Use of sodalime scrap-glass as a fluxing agent in a porcelain stoneware tile mix. *J. Eur. Ceram. Soc.* **2004**, *24*, 83–92. [CrossRef]
54. Luz, A.P.; Ribeiro, S. Use of glass waste as a raw material in porcelain stoneware tile mixtures. *Ceram. Int.* **2007**, *33*, 761–765. [CrossRef]
55. Väntsi, O.; Kärki, T. Mineral wool waste in Europe: A review of mineral wool waste quantity, quality, and current recycling methods. *J. Mater. Cycles Waste Manag.* **2013**, *16*, 62–72. [CrossRef]
56. Kapoor, A.; Mudgal, S.; Muruganandam, L. Impact of shock waves on glass wool composition and properties. *Mater. Today Proc.* **2021**, *46*, 7056–7060. [CrossRef]
57. Bennett, T.M.; Allen, J.F.; Garden, J.A.; Shaver, M.P. Low Formaldehyde Binders for Mineral Wool Insulation: A Review. *Glob. Chall.* **2022**, *6*, 2100110. [CrossRef]
58. Yliniemi, J.; Ramaswamy, R.; Luukkonen, T.; Laitinen, O.; de Sousa, A.N.; Huuhtanen, M.; Illikainen, M. Characterization of mineral wool waste chemical composition, organic resin content and fiber dimensions: Aspects for aluminum. *Waste Manag.* **2021**, *131*, 323–330. [CrossRef]

59. Chen, Z.; Liu, T. Development and Application Status of Glass Wool, Rock Wool, and Ceramic Wool. In *Thermal Insulation and Radiation Control Technologies for Buildings. Green Energy and Technology*; Kośny, J., Yarbrough, D.W., Eds.; Springer: Berlin/Heidelberg, Germany, 2022. [CrossRef]
60. Zhang, Y.; Vulfson, Y.; Zheng, Q.; Luo, J.; Kim, S.H.; Yue, Y. Impact of fiberizing method on physical properties of glass wool aluminum. *J. Non Cryst. Solids* **2017**, *476*, 122–127. [CrossRef]
61. Yue, Y. Fiber Forming and Its Impact on Mechanical Properties. In *Fiberglass Science and Technology*; Li, H., Ed.; Springer: Cham, Switzerland, 2021. [CrossRef]
62. Gol'tsman, B.M.; Yatsenkoa, E.A. Modern Methods for Foaming of Glass and Silicate Raw Materials: Review and Analysis. *Theor. Found. Chem. Eng.* **2022**, *56*, 678–686. [CrossRef]
63. Liu, Y.; Xie, J.; Hao, P.; Shi, Y.; Xu, Y.; Ding, X. Study on factors affecting properties of foam glass made from waste glass. *J. Renew. Mater.* **2021**, *9*, 237–253. [CrossRef]
64. Owoeye, S.S.; Matthew, G.O.; Ovienmhanda, F.O.; Tunmilayo, S.O. Preparation and characterization of foam glass from waste container glasses and water glass for application in thermal insulations. *Ceram. Int.* **2020**, *46*, 11770–11775. [CrossRef]
65. Mengguang, Z.; Ru, J.; Zhongmin, L.; Hao, W.; Lili, L.; Zuotai, Z. Preparation of glass ceramic foams for thermal insulation applications from coal fly ash and waste glass. *Const. Build. Mater.* **2016**, *112*, 398–405. [CrossRef]
66. Stochero, N.P.; de Souza Chami, J.O.R.; Souza, M.T.; de Moraes, E.G.; Novaes de Oliveira, A.P. Green Glass Foams from Wastes Designed for Thermal Insulation. *Waste Biomass Valor.* **2021**, *12*, 1609–1620. [CrossRef]
67. Shishkin, A.; Aguedal, H.; Goel, G.; Pecilevica, J.; Newport, D.; Ozolins, J. Influence of waste glass in the foaming process of open cell porous ceramic as filtration media for industrial wastewater. *J. Clean. Prod.* **2021**, *282*, 124546. [CrossRef]
68. El Amir, A.A.M.; Attia, M.A.A.; Fend, T.; Ewais, E.M.M. Production of high-quality glass foam from soda lime glass waste using SiC-AlN foaming agent. *J. Korean Ceram. Soc.* **2022**, *59*, 444–452. [CrossRef]
69. Scheffler, M.; Colombo, P. *Cellular Ceramics: Structure, Manufacturing, Properties and Applications*; John Wiley and Sons: Hoboken, NJ, USA, 2006.
70. Ioana, A.; Paunescu, L.; Constantin, N.; Pollifroni, M.; Deonise, D.; Patcu, F.S. Glass Foam from Flat Glass Waste Produced by the Microwave Irradiation Technique. *Micromachines* **2022**, *13*, 550. [CrossRef]
71. Fernandes, H.R.; Tulyaganov, D.U.; Ferreira, J.M.F. Preparation and characterization of foams from sheet glass and fly ash using carbonates as foaming agent. *Ceram. Inter.* **2009**, *35*, 229–235. [CrossRef]
72. Gol'tsman, B.M.; Yatsenko, E.A. Role of Carbon Phase in the Formation of Foam Glass Porous Structure. *Materials* **2022**, *15*, 7913. [CrossRef]
73. König, J.; Petersen, R.R.; Yue, Y.; Suvorov, D. Gas-releasing reactions in foam-glass formation using carbon and MnxOy as the foaming agents. *Ceram. Int.* **2017**, *43*, 4638–4646. [CrossRef]
74. Petersen, R.R.; König, J.; Yue, Y. Evaluation of foaming aluminum of glass melts by high-temperature microscopy. *Int. J. Appl. Glass Sci.* **2016**, *7*, 524–531. [CrossRef]
75. Abdollahi, S.; Yekta, B.E. Prediction of foaming temperature of glass in the presence of various oxidizers via thermodynamics route. *Ceram. Int.* **2020**, *46 Pt A*, 25626–25632. [CrossRef]
76. Petersen, R.R.; König, J.; Yue, Y. The viscosity window of the silicate glass foam production. *J. Non-Cryst. Solids* **2017**, *456*, 49–54. [CrossRef]
77. Qin, Z.; Li, G.; Tian, Y.; Ma, Y.; Shen, P. Numerical Simulation of Thermal Conductivity of Foam Glass Based on the Steady-State Method. *Materials* **2019**, *12*, 54. [CrossRef] [PubMed]
78. Haesche, M.; Lehmhus, D.; Weise, J.; Wichmann, M.; Mocellin, I.C.M. Carbonates as Foaming Agent in Chip-based Aluminum Foam Precursor. *J. Mater. Sci. Technol.* **2010**, *26*, 845–850. [CrossRef]
79. Lebullenger, R.; Chenu, S.; Rocherullé, J.; Merdrignac-Conanec, O.; Cheviré, F.; Tessier, F.; Bouzaza, A.; Brosillon, S. Glass foams for environmental applications. *J. Non-Cryst. Solids* **2010**, *356*, 2563–2568. [CrossRef]
80. Gol'tsman, B.M.; Yatsenko, E.A. Dynamics of foam glass structure formation using glass waste and liquid foaming mixture. *J. Clean. Prod.* **2010**, *426*, 138994. [CrossRef]
81. Yatsenko, E.A.; Gol'tsman, B.M.; Smolii, V.A.; Gol'tsman, N.S.; Yatsenko, L.A. Study on the Possibility of Applying Organic Compounds as Pore-Forming Agents for the Synthesis of Foam Glass. *Glass Phys. Chem.* **2019**, *45*, 138–142. [CrossRef]
82. Deubener, J.; Allix, M.; Davis, M.J.; Duran, A.; Höche, T.; Honma, T.; Komatsu, T.; Krüger, S.; Mitra, I.; Müller, R.; et al. Updated definition of glass-ceramics. *J. Non-Cryst. Solids.* **2018**, *501*, 3–10. [CrossRef]
83. Zhao, S.Z.; Liu, B.; Ding, Y.J.; Zhang, J.; Wen, Q.; Ekberg, C.; Zhang, S. Study on glass-ceramics made from MSWI fly ash, pickling sludge and waste glass by one-step process. *J. Clean. Prod.* **2020**, *271*, 122674. [CrossRef]
84. Spacil, I.; Li, J. Revealing nucleation kinetics of entrained eutectic Si droplets in high purity melt spun Al-5Si based alloys with additions of Ta, TiB2, Eu and P. *J. Alloys Compd.* **2023**, *968*, 172032. [CrossRef]
85. Zhang, Z.; Li, Z.; Yang, Y.; Shen, B.; Ma, J.; Lina, L. Preparation and characterization of fully waste-based glass-ceramics from incineration fly ash, waste glass and coal fly ash. *Ceram. Int.* **2022**, *48*, 21638–21647. [CrossRef]

86. Tarhan, B.; Tarhan, M.; Aydin, T. Reusing sanitaryware waste products in glazed porcelain tile production. *Ceram. Int.* **2017**, *43*, 3107–3112. [CrossRef]
87. Brusatin, G.; Bernardo, E.; Andreola, F.; Barbieri, L.; Lancellotti, I.; Hreglich, S. Reutilization of waste inert glass from the disposal of polluted dredging spoils by the obtainment of ceramic products for tiles applications. *J. Mater. Sci.* **2005**, *40*, 5259–5264. [CrossRef]
88. Sarkar, T.; Naskar, M.K.; Chakraborty, A.; Roy, P.K.; Chakraborty, S. Preparation of high-strength waste-derived eco-friendly ceramic foam as face brick and its estimation of building energy consumption for thermal insulation. *J. Build. Eng.* **2021**, *88*, 109043. [CrossRef]
89. Kyaw Oo D'Amore, G.; Caniato, M.; Travan, A.; Turco, G.; Marsich, L.; Ferluga, A.; Schmid, C. Innovative thermal and acoustic insulation foam from recycled waste glass powder. *J. Clean. Prod.* **2017**, *165*, 1306–1315. [CrossRef]
90. da Costa, D.L.; Rodrigues da Solva Morais, C.; Rodrigues, A.M. Sustainable glass-ceramic foams manufactured from waste glass bottles and bentonite. *Ceram. Intern.* **2020**, *46A*, 17957–17961. [CrossRef]
91. Zhang, J.; Liu, B.; Zhang, S.A. review of glass ceramic foams prepared from solid wastes: Processing, heavy-metal solidification and volatilization, applications. *Sci. Total Environ.* **2021**, *781*, 146727. [CrossRef]
92. Zhang, J.; Liu, B.; Zhao, S.; Shen, H.; Liu, J.; Zhang, S. Preparation and characterization of glass ceramic foams based on municipal solid waste incineration ashes using secondary aluminum ash as foaming agent. *Const. Build. Mater.* **2020**, *262*, 120781. [CrossRef]
93. Duxson, P.; Fernández-Jiménez, A.; Provis, J.L.; Lukey, G.C.; Palomo, A.; van Deventer, J.S.J. Geopolymer technology: The current state of the art. *J. Mater. Sci.* **2007**, *42*, 2917–2933. [CrossRef]
94. Duxson, P.; Provis, J.L.; Lukey, G.C.; van Deventer, J.S.J. The role of inorganic polymer technology in the development of 'green concrete'. *Cem. Concr. Res.* **2007**, *37*, 1590–1597. [CrossRef]
95. Zareechian, M.; Siad, H.; Lachemi, M.; Sahmaran, M. Advancements in cleaner production of one-part geopolymers: A comprehensive review of mechanical properties, durability, and microstructure. *Const. Build. Mater.* **2023**, *409*, 133876. [CrossRef]
96. Davidovits, J. Geopolymer chemistry and sustainable development. The poly(sialate) terminology: A very useful and simple model for the promotion and understanding of green-chemistry. In Proceedings of the 2005 Geopolymer Conference, Quarrying Australia, Adelaide, SA, Australia, 12–15 October 2005; Volume 1, pp. 9–15.
97. Kriven, W.M.; Leonelli, C.; Provis, J.L.; Boccaccini, A.R.; Attwell, C.; Ducman, V.S.; Ferone, C.; Rossignol, S.; Luukkonene, T.; van Deventer, J.S.J.; et al. Why geopolymers and alkali-activated materials are key components of sustaibnable world: A perspective contribution. *J. Am. Ceram. Soc.* **2024**, *107*, 5159–5177. [CrossRef]
98. Dey, A.; Rumman, R.; Wakjira, T.G.; Jindal, A.; Bediwy, A.G.; Islam, M.S.; Alam, M.S.; Al Martini, S.; Sabouni, R. Towards net-zero emission: A case study investigating sustainability potential of geopolymer concrete with recycled glass powder and gold mine tailings. *J. Build. Eng.* **2015**, *101*, 675–683. [CrossRef]
99. Jani, Y.; Hogland, W. Waste glass in the production of cement and concrete—A review. *J. Environ. Chem. Eng.* **2014**, *2*, 1767–1775. [CrossRef]
100. Varma, D.N.; Singh, S.P. A review on waste glass-based geopolymer composites as a sustainable binder. *Silicon* **2023**, *15*, 7685–7703. [CrossRef]
101. Xiao, R.; Dau, X.; Zhong, J.; Ma, Y.; Jiang, X.; He, J.; Wang, Y.; Huang, B. Toward waste glass upcycling: Preparation and characterization of high-volume waste glass geopolymer composites. *SMT* **2024**, *40*, e00890. [CrossRef]
102. Sarkar, M.; Maiti, M.; Akbar Malik, M.; Xu, S. Waste valorization: Sustainable geopolymer production using recycled glass and fly ash at ambient temperature. *Chem. Eng. J.* **2024**, *494*, 153144. [CrossRef]
103. Torres-Carrasco, M.; Puertas, F. Waste glass in the geopolymer preparation. *Mechanical and microstructural characterisation. J. Clean. Prod.* **2015**, *90*, 397–408. [CrossRef]
104. Namatollahi, B.; Sanjayan, J.; Shaikh, F.U.A. Synthesis of heat and ambient cured one-part geopolymer mixes with different grades of sodium silicate. *Ceram. Int.* **2015**, *41*, 5696–5704. [CrossRef]
105. van Deventer, J.S.J.; Provis, J.L.; Duxon, P.; Brice, D.G. Chemical Research and Climate Change as drivers in the commercial adoption of alkali activated materials. *Waste Biomass Valor.* **2010**, *1*, 145–155. [CrossRef]
106. Provis, J.L. Geopolymers and other alkali activated materials: Why, how, and what? *Mater. Struct.* **2014**, *47*, 11–25. [CrossRef]
107. Myres, R.J.; Bernal, S.A.; San Nicolas, R.; Provis, J.L. Generalized structural description of calcium-sodium aluminosilicate hydrate gels: The cross-linked substituted tobermorite model. *Langmuir* **2013**, *29*, 5294–5306. [CrossRef] [PubMed]
108. Garcia-Lodeiro, I.; Fernández-Jimenez, A.; Pena, P.; Palomo, A. Alkaline activation of synthetic aluminosilicate glass. *Ceram. Int.* **2014**, *40*, 5547–5558. [CrossRef]
109. Ruiz-Santaquiteria, C.; Fernández-Jiménez, A.; Palomo, A. Alternative prime materials for developing new cements: Alkaline activation of alkali aluminosilicate glasses. *Ceram. Int.* **2016**, *42*, 9333–9340. [CrossRef]
110. Cyr, M.; Idir, R.; Poinot, T. Properties of inorganic polymer (geopolymer) mortars made of glass cullet. *J. Mater. Sci.* **2012**, *47*, 2782–2797. [CrossRef]

111. Idir, R.; Cyr, M.; Pavoine, A. Investigations on the durability of alkali-activated recycled glass. *Constr. Build. Mater.* **2020**, *236*, 117477. [CrossRef]
112. Rincon, A.; Giacomello, G.; Pasetto, M.; Bernardo, E. Novel 'inorganic gel casting' process for the manufacturing of glass foams. *J. Eur. Ceram. Soc.* **2017**, *37*, 2227–2234. [CrossRef]
113. Ramteke, D.D.; Hujova, M.; Kraxner, J.; Galusek, D.; Romero, A.R.; Falcone, R.; Bernardo, E. Up-cycling of 8unrecyclable9 glasses in glass-based foams by weak alkali-activation, gel casting and low-temperature sintering. *J. Clean. Prod.* **2021**, *278*, 123985. [CrossRef]
114. Ramakanth, D.; Singh, S.; Maji, P.K.; Lee, Y.S.; Gaikwad, K.K. Advanced packaging for distribution and storage of COVID-19 vaccines: A review. *Environ. Chem. Lett.* **2021**, *19*, 3597–3608. [CrossRef]
115. Taveri, G.; Tousek, J.; Bernardo, E.; Toniolo, N.; Boccaccini, A.R.; Dlouhy, I. Proving the role of boron in the structure of fly-ash/borosilicate glass based geopolymers. *Mater. Lett.* **2017**, *200*, 105–108. [CrossRef]
116. Elsayed, H.; Romero, A.R.; Molino, G.; Vitale Brovarone, C.; Bernardo, E. Bioactive Glass-Ceramic Foam Scaffolds from 'Inorganic Gel Casting' and Sinter-Crystallization. *Materials* **2018**, *11*, 349. [CrossRef]
117. Rincon, A.R.; Tamburini, S.; Taveri, G.; Touaek, J.; Dlouhy, I.; Bernardo, E. Extension of the 'inorganic gel casting' process to the manufacturing of Boro-Alumino-Silicate glass foams. *Materials* **2018**, *11*, 2545. [CrossRef]
118. Metha, A.; Karbouche, K.; Kraxner, J.; Elsayed, H.; Galusek, D.; Bernardo, E. Upcycling of pharmaceutical glass into highly porous ceramics: From foams to membrans. *Materials* **2022**, *15*, 3784. [CrossRef]
119. Tameni, G.; Lago, D.; Kalková, H.; Kalková, L.; Kraxner, J.; Galusek, D.; Dawson, D.M.; Ashbrook, S.E.; Bernardo, E. Alkaline attack of boro-alumino-silicate glass: New insights of the molecular mechanism of cold consolidation and new applications. *Open Ceram.* **2025**, *21*, 100726. [CrossRef]
120. Ma, T.; Jivkov, A.P.; Li, W.; Liang, W.; Wang, Y.; Xu, H.; Han, X. A mechanistic model for long-term nuclear waste glass dissolution integrating chemical affinity and interfacial diffusion barrier. *J. Nucl. Mater.* **2017**, *70–85*, 486. [CrossRef]
121. Rébiscoul, D.; Bruguier, F.; Magnin, V.; Gin, S. Impact of soda-lime borosilicate glass composition on water penetration and water structure at the first time of alteration. *J. Non-Cryst. Solids* **2012**, *358*, 2951–2960. [CrossRef]
122. Gin, S.; Delaye, J.M.; Angeli, F.; Schuller, S. Aqueous alteration of silicate glass: State of knowledge and perspectives. *Mater. Degrad.* **2022**, *5*, 42. [CrossRef]
123. Samarakoon, M.H.; Ranjith, P.G.; Duan, W.H.; Haque, A.; Chen, B.K. Extensive use of waste glass in one-part alkali-activated materials: Towards sustainable construction practices. *Waste Manag.* **2021**, *130*, 1–11. [CrossRef]
124. Su, P.; Eniola, S.; Xie, J.; Zhao, X.; Ugboaja, C.; Li, M.; Si, R.; Dai, Q.; Fei, Y.; Hu, Y.H. Performance evaluation of glass powder as a partial precursor in alkali-activated slag (AAS) binder and recycled glass and steel fibers in AAS mortar. *Constr. Build. Mater.* **2025**, *473*, 140757. [CrossRef]
125. Liang, G.; Li, H.; Zhu, H.; Liu, T.; Chen, Q.; Guo, H. Reuse of waste glass powder in alkali-activated metakaolin/fly ash pastes: Physical properties, reaction kinetics and microstructure. *Resour. Conserv. Recycl.* **2021**, *173*, 105721. [CrossRef]
126. Redden, R.; Naithalath, N. Microstructure, strength, and moisture stability of alkali activated glass powder-based binders. *Cem. Concr. Compos.* **2014**, *45*, 46–56. [CrossRef]
127. Torres-Carrasco, M.; Puertas, F. Waste glass as a precursor in alkaline activation: Chemical process and hydration products. *Constr. Build. Mater.* **2017**, *139*, 342–354. [CrossRef]
128. Tameni, G.; Cammelli, F.; Elsayed, H.; Stangherlin, F.; Bernardo, E. Upcycling of boro-alumino-silicate pharmaceutical glass in sustainable construction materials. *Detritus* **2022**, *20*, 17–21. [CrossRef]
129. Ashby, M.F. Chapter 1, Butterworth-Heinemann. In *Materials Selection in Mechanical Design*; Elsevier: Amsterdam, The Netherlands, 2011; pp. 1–13. [CrossRef]
130. Papa, E.; Medri, V.; Benito, P.; Vaccari, A.; Bugani, S.; Jaroszewicz, J.; Swieszkowski, W.; Landi, E. Synthesis of porous hierarchical geopolymer monoliths by ice-templating. *Microporous Mesoporous Mater.* **2015**, *215*, 206–214. [CrossRef]
131. Luukkonen, T.; Heponiemi, A.; Runtti, H.; Pesonen, J.; Yliniemi, J.; Lassi, U. Application of alkali-activated materials for water and wastewater treatment: A review. *Rev. Environ. Sci. Biotechnol.* **2019**, *18*, 271–297. [CrossRef]
132. Ettahiri, Y.; Bouargane, B.; Fritah, K.; Akhsassi, B.; Pérez-Villarejo, L.; Aziz, A.; Bouna, L.; Benlhachemi, A.; Novais, R.M. A state-of-the-art review of recent advances in porous geopolymer: Applications in adsorption of inorganic and organic contaminants in water. *Constr. Build. Mater.* **2023**, *395*, 132269. [CrossRef]
133. Ettahiri, Y.; Bouddouch, A.; Akhsassi, B.; Khali, A.; Bouna, L.; Benlhachemi, A.; Pérez-Villarejo, L. Optimized porous alkali-activated material for superior dye removal: Synthesis and performance analysis. *Mater. Adv.* **2025**, *6*, 2435–2447. [CrossRef]
134. Xu, H.; Zhu, S.; Lu, K.; Jia, H.; Xia, M.; Wang, F. Preparation of hierarchically floral ZIF-8 derived carbon@polyaniline@Ni/Al layered double hydroxides composite with outstanding removal phenomenon for saccharin. *Chem. Eng. J.* **2022**, *450*, 138127. [CrossRef]
135. Mahmoud, M.; Kraxner, J.; Elsayed, H.; Galusek, D.; Bernardo, E. Advanced dye sorbents from combined stereolithography 3D printing and alkali activation of pharmaceutical glass waste. *Materials* **2022**, *15*, 6823. [CrossRef]

136. Mahmoud, M.; Kraxner, J.; Mehta, A.; Elsayed, H.; Galusek, D.; Bernardo, E. Alkali activation-induced cold consolidation of waste glass: Application on organic-free direct ink writing of photocatalytic dye destructors. *JECS* **2024**, *44*, 5449–5459. [CrossRef]
137. Raji, Z.; Karim, A.; Karam, A.; Khalloufi, S. Adsorption of Heavy Metals: Mechanisms, Kinetics, and Applications of Various Adsorbents in Wastewater Remediation—A Review. *Waste* **2023**, *1*, 775–805. [CrossRef]
138. Qasem, N.A.; Mohammed, R.H.; Lawal, D.U. Removal of heavy metal ions from wastewater: A comprehensive and critical review. *NPJ Clean Water* **2021**, *4*, 36. [CrossRef]
139. Azimi, A.; Azari, A.; Rezakazemi, M.; Ansarpour, M. Removal of heavy metals from industrial wastewaters: A review. *ChemBioEng Rev.* **2017**, *4*, 37–59. [CrossRef]
140. Fu, F.; Wang, Q. Removal of heavy metal ions from wastewaters: A review. *J. Environ. Manag.* **2011**, *92*, 407–418. [CrossRef] [PubMed]
141. Novais, R.M.; Buruberri, L.H.; Seabra, M.-P.; Labrincha, J.A. Novel porous fly-ash containing geopolymer monoliths for lead adsorption from wastewaters. *J. Hazard. Mater.* **2016**, *318*, 631–640. [CrossRef]
142. Cheng, T.W.; Lee, M.L.; Ko, M.S.; Ueng, T.H.; Yang, S.F. The heavy metal adsorption characteristics of metakaolin-based geopolymer. *Appl. Clay Sci.* **2012**, *56*, 90–96. [CrossRef]
143. Lago, D.; Kraxner, J.; Galusek, D.; Bernardo, E. Novel glass-based membranes for Cu adsorption: From alkali activation to sintering. *Helyon* **2023**, *9*, e18221. [CrossRef]
144. Al-Harahsheh, M.S.; Al Zboon, K.; Al-Makhadmeh, L.; Hararah, M.; Mahasneh, M. Fly ash based geopolymer for heavy metal removal: A case study on copper removal. *J. Environ. Chem. Eng.* **2015**, *3*, 1669–1677. [CrossRef]
145. Jain, S.; Onuaguluchi, O.; Banthia, N.; Troczynski, T. Advancements in immobilization of cesium and strontium radionuclides in cementitious wasteforms—A review. *J. Am. Ceram. Soc.* **2024**, *108*, e20131. [CrossRef]
146. El-Kamash, A.M.; El-Naggar, M.R.; El-Dessouky, M.I. Immobilization of cesium and strontium radionuclides in zeolite-cement blends. *J. Hazard. Mater.* **2006**, *136*, 310–316. [CrossRef]
147. Thorpe, C.L.; Neeway, J.J.; Pearce, C.I.; Hand, R.J.; Fisher, A.J.; Walling, S.A.; Hyatt, C.C.; Krugeer, A.A.; Schweiger, M.; Kosson, D.S.; et al. Forty years of durability assessment of nuclear waste glass by standard methods. *npj Mater. Degrad.* **2021**, *5*, 6. [CrossRef]
148. Jahagirdar, P.B.; Warral, P.K. Vitrification of sulphate beraring high level wastes in borosilicate matrix. *Waste Maneg.* **1998**, *18*, 265–273. [CrossRef]
149. Zhu, Y.; Zheng, Z.; Deng, Y.; Shi, C.; Zhang, Z. Advances in immobilization of radionuclide wastes by alkali activated cement and related materials. *Cem. Concr. Compos.* **2022**, *126*, 104377. [CrossRef]
150. Abdel Rahman, R.O.; Zaki, A.A.; El-Kamash, A.M. Modeling the long-term leaching behavior of 137Cs, 60Co, and 152,154Eu radionuclides from cement–clay matrices. *J. Hazard. Mater.* **2007**, *145*, 372–380. [CrossRef] [PubMed]
151. Conner, J.R.; Hoeffner, S.L. A Critical Review of Stabilization/Solidification Technology. *Crit. Rev. Environ. Sci. Technol.* **1998**, *28*, 397–462. [CrossRef]
152. Deng, N.; An, H.; Cui, H.; Pan, Y.; Wang, B.; Mao, L.; Zhai, J. Effects of gamma-ray irradiation on leaching of simulated 133Cs+ radionuclides from geopolymer wasteforms. *J. Nucl. Mater.* **2015**, *459*, 270–275. [CrossRef]
153. Forsberg, C.W. Rethinking high-level waste disposal: Separate disposal of high-heat radionuclides (90Sr and 137Cs). *Nucl. Technol.* **2000**, *131*, 252–268. [CrossRef]
154. Berger, S.; Frizon, F.; Joussot-Dubien, C. Formulation of caesium based and caesium containing geopolymers. *Adv. Appl. Ceram.* **2009**, *108*, 412–417. [CrossRef]
155. Khalil, M.; Merz, E. Immobilization of intermediate-level wastes in geopolymers. *J. Nucl. Mater.* **1994**, *211*, 141–148. [CrossRef]
156. Melar, J.; Renaudin, G.; Leroux, F.; Hardy-Dessources, A.; Nedelec, J.-M.; Taviot-Gueho, C.; Petit, E.; Steins, P.; Poulesquen, A.; Frizon, F. The porous network and its interface inside geopolymers as a function of alkali cation and aging. *J. Phys. Chem. C* **2015**, *119*, 17619–17632. [CrossRef]
157. Mukiza, E.; Phung, Q.T.; Frederickx, L.; Jacques, D.; Seetharam, S.; De Schutter, G. Co-immobilization of cesium and strontium containing waste by metakaolin-based geopolymer: Microstructure, mineralogy and mechanical properties. *J. Nucl. Mater.* **2023**, *585*, 154639. [CrossRef]
158. Fernandez-Jimenez, A.; MacPhee, D.E.; Lachowski, E.E.; Palomo, A. Immobilization of cesium in alkaline activated fly ash matrix. *J. Nucl. Mater.* **2005**, *346*, 185–193. [CrossRef]
159. Komljenović, M.; Tanasijević, G.; Džunuzović, N.; Provis, J.L. Immobilization of cesium with alkali-activated blast furnace slag. *J. Hazard. Mater.* **2020**, *388*, 121765. [CrossRef] [PubMed]
160. Lago, D.; Tameni, G.; Zorzi, F.; Kraxner, J.; Galusek, D.; Bernardo, E. Novel cesium immobilization by alkali activation and cold consolidation of waste pharmaceutical glass. *J. Clean. Prod.* **2024**, *461*, 142673. [CrossRef]
161. Lago, D.; Tameni, G.; Kraxner, J.; Galusek, D.; Bernardo, E. Cesium stabilization by engineered alkaline attack of glass for pharmaceutical containers. *Mater. Lett.* **2024**, *372*, 137097. [CrossRef]

162. Zhu, H.; He, J.; Hong, T.; Yang, Q.; Wu, Y.; Yang, Y.; Huang, K. A rotary radiation structure for microwave heating uniformity improvement. *Appl. Therm. Eng.* **2018**, *141*, 648–658. [CrossRef]
163. Makul, N.; Rattanadecho, P.; Agrawal, D.K. Applications of microwave energy in cement and concrete—A review. *Renew. Sustain. Energy Rev.* **2014**, *37*, 715–733. [CrossRef]
164. Li, S.M.; Gao, X. Acceleration mechanism of nonisothermal microwave heating on strength development of mortar. *Compos. Struc.* **2022**, *279*, 114765. [CrossRef]
165. Leung, C.K.Y.; Pheeraphan, T. Microwave curing of Portland cement concrete: Experimental results and feasibility for practical applications. *Constr. Build. Mater.* **1995**, *9*, 67–73. [CrossRef]
166. Guan, X.; Luo, W.; Liu, S.; Garcia Hernandez, A.; Do, H.; Li, B. Ultra-high eraly strength fly ash-based geopolymer paste cured by microwave radiation. *DIBE* **2023**, *14*, 100139. [CrossRef]
167. Acshoff, J.; Partschefeld, S.; Schneider, J.; Osberg, A. Effect of Microwaves on the Rapid Curing of Metakaolin- and Aluminum Orthophosphate-Based Geopolymers. *Materials* **2024**, *17*, 463. [CrossRef] [PubMed]
168. Hong, S.; Kim, H. Robust synthesis of coal bottom ash-based geopolymers using additional microwave heating and curing for high compressive strength properties. *Korean J. Chem. Eng.* **2019**, *36*, 1164–1171. [CrossRef]
169. Carollo, F.; De Rienzo, E.; D'Angelo, A.; Sgarbossa, P.; Barbieri, L.; Leonelli, C.; Lancellotti, I.; Catauro, M.; Bernardo, E. Cold Consolidation of Waste Glass by Alkali Activation and Curing by Traditional and Microwave Heating. *Materials* **2025**, *18*, 2628. [CrossRef] [PubMed]
170. Tameni, G.; Carollo, F.; Cavazzini, A.M.; Forzan, M.; Bernardo, E. Microwaved assisted cold consolidation of alkali activated suspension of glass waste powders. *Mater. Lett.* **2025**, *389*, 138354. [CrossRef]

ceramics

MDPI

Review

Ceramics—The Forgotten but Essential Ingredients for a Circular Economy on the Moon

Alex Ellery

Department of Mechanical & Aerospace Engineering, Carleton University, Ottawa, ON K1S 5B6, Canada; aellery@mae.carleton.ca; Tel.: +1-613-261-3765

Abstract

Settlement on the Moon will require full exploitation of its resources if such settlements are to be permanent. Such in situ resource utilisation (ISRU) has primarily been focused on accessing water ice at the lunar poles and the use of raw lunar regolith as a compressive building material. Some work has also examined the extraction of metals, but there has been little consideration of the many useful ceramics that can be extracted from the Moon and how they may be fabricated. We introduce a strategy for full lunar industrialisation based on a circular lunar industrial ecology and examine the contribution of ceramics. We review ceramic fabrication methods but focus primarily on 3D printing approaches. The popular direct ink writing method is less suitable for the Moon and other methods require polymers which are scarce on the Moon. This turns out to be crucial, suggesting that full industrialisation of the Moon cannot be completed until the problem of ceramic fabrication is resolved, most likely in conjunction with polymer synthesis from potential carbon sources.

Keywords: ceramics; in situ resource utilisation; lunar industrialisation; lunar industrial ecology; ceramic 3D printing

1. Introduction

It is anticipated that human exploration of the Moon will evolve toward settlement on the Moon and that such settlement will require the exploitation of local lunar resources: in situ resource utilisation (ISRU). Most interest has been focussed on water ice mining in permanently shadowed craters at the South Pole, supplemented by the use of local regolith as a construction material. Regolith is a potential resource that may be 3D printed for the construction of lunar or Martian bases [1]. D-shaping and contour crafting are two of several methods that involve 3D printing regolith with binders or cements for constructing protective shells for such bases [2]. Both D-shaping and contour crafting require bulk volumes of binder imported from Earth. D-shaping utilises $MgCl_2$ binder to react with MgO in lunar regolith (Sorel cement), while contour crafting requires polymer binder. An example of a potentially in situ-resourced cement is sulpho-aluminate cement [3]. These approaches assume minimum processing of the resource to render it useful. A lunar base also requires internal infrastructure that may also be in situ-resourced [4]. Similarly, providing both electrical power and storage may be enabled using in situ resources [5,6]. Given the necessity for supporting human life within lunar bases, lunar water resources will be essential for hydroponic food growth [7]. Although regolith is a 3D printable ceramic in situ resource, we shall not address lunar base construction here, for which there are notably excellent reviews [8].

We are interested in building a lunar infrastructure from lunar resources to minimise the capital cost of launching infrastructure from Earth. This requires significant processing of lunar resources to yield functional materials. We concern ourselves with a specific approach to ISRU that yields a range of lunar mineral-derived materials, including ceramics that have high utility rather than native lunar regolith. Most efforts have been focussed on the extraction of metals from lunar minerals, but we shall consider the neglected ceramic resources on the Moon. It is expected that partial gravity on the Moon will have a negligible effect on processing methods compared with microgravity in orbit [9] or milligravity conditions associated with asteroid processing [10]. Nevertheless, the Moon imposes particular challenges due to its limited diversity of material resources, and this will dominate the feasibility of manufacturing ceramics.

We explore the manufacture and exploitation of ceramics derived from lunar resources. Section 2 briefly describes the Moon's mineralogy and volatiles to provide context to the subsequent sections. Section 3 defines the lunar industrial ecology as the context for exploring ceramics manufacture; Section 4 explores thermal sintering and the problem of brittleness; Section 5 explores ceramic 3D printing methods for the Moon and finds them wanting; Section 6 introduces the importance of polymers; Section 7 explores the use of clays; Section 8 briefly looks at non-clay lunar mineral uses; Section 9 discusses the use of direct thermal heating of ceramics using solar concentrators; and Section 10 concludes that polymers are an essential partner to ceramics and lunar industrialisation will flounder until an alternative to polymers can be found.

2. Lunar Resources

The Moon is covered in a layer of regolith with up to 20% glasses derived from fragmented parental rock with an average grain size of ~60–80 μm to a depth of ~4–5 m in the mare and ~10–15 m in the highlands [11]. Lunar mineralogy is broadly divisible into the anorthositic highlands and the basaltic mare. Anorthosite is dominated by the feldspar anorthite ($CaAl_2Si_2O_8$) while mare basalt may have high-Ti or low-Ti content, the former characterised by 10–20% ilmenite ($FeTiO_3$). The dominant minerals on the Moon are plagioclase feldspar (dominantly anorthite), pyroxene (such as augite ($(Ca,Fe,Mg,Ti,Al)_2Si_2O_6$), olivine (dominated by forsterite Mg_2SiO_4), ilmenite and spinel (dominated by $MgAl_2O_4$) [12], all of which may be exploited as resources [13]. Orthopyroxene, olivine and spinel lithologies are located primarily on the lunar farside and are believed to have originated from large magmatic intrusions (plutons) from the mantle into the lunar crust [14]. Lunar minerals are highly oxidised, representing sources of oxygen through chemical or electrolytic reduction. However, the Moon is depleted in Na, Zn and Cu which imposes the necessity for lunar-constrained industrialisation. Halogens such as Cl and F are highly rarified but are concentrated in rare apatite ($Ca_5(PO_4)_3(OH,F,Cl)$) minerals.

KREEP basalts of the Procellarum KREEP Terrane on the northwest nearside are characterised by more Na-rich plagioclases and relatively elevated concentrations of potassium, rare earth elements (REE), phosphorus, thorium and uranium. The rare earth elements occur primarily in the form of ~μm monazite ($(Ce,La,Nd,Th)PO_4$) inclusions in pyroxene grains. The scarcer yttrobetafite ($(Ca,Y,U,Th,Pb,REE)_2(Ti,Nb)_2O_2$) is a potential source of yttrium, but this requires a complex series of chemical processes and reagents. Granite in (KREEPy) highland rocks is rare on the Moon, but it is a potentially useful resource for [15]: (i) silica minerals (primarily cristobalite rather than quartz); (ii) potassium feldspar $KAlSi_3O_8$; and (iii) zircons $ZrSiO_4$. However, the global abundance of granites is low. Indigenous lunar minerals are supplemented by minor amounts: <0.1% of meteoritic material from asteroid impacts including Fe, Ni, Co and troilite FeS.

Much more rarified is the lunar inventory of volatiles [16]. In particular, C, N and S are scarce within concentrations of ~1–100 ppm in regolith implanted by solar wind, though there are apparent higher concentrations in permanently shadowed craters such as Cabeus: 0.3% NH_3, 0.1% CO_2 and 0.04% CH_4. Similarly, water ice is concentrated in permanently shadowed craters of the lunar poles with concentrations of 5.6 ± 2.9% in the Cabeus crater. The cryogenic temperatures of ~25–40 K in permanently shadowed craters present major challenges for lunar operations. Concentration of ^3He is highly rarified ~20 ppb in mature ilmenite minerals. Although these volatiles may be released by heating ~700 °C, they are too scarce to be a viable resource. Carbon, in particular, averages ~100–120 ppm by weight of regolith so there is ~100 mg carbon per kg of regolith, i.e., extraction of 1 kg carbon would require excavating and heating 10 tonnes of regolith.

Nevertheless, the desirability for in situ resource utilisation is clear—it costs ~$750,000/kg to reach the surface of the Moon (Astrobotic Technology fees), so the use of in situ resources saves the transport cost, but it is essential that production machines' total lifetime mass throughput exceed its own mass. On Earth, this is higher at ~10^5 for excavation of bulk material than for manufacturing machines at ~10^3 which would be expected to be similar for the Moon, i.e., a saving of S750M–$75B per kg of landed lunar manufacturing asset.

The spinel $MgAl_2O_4$ is a resource available on the Moon that may be exploited directly (though contamination with ubiquitous Fe darkens glasses, so it may have to be synthesised from its constituents). Pure $MgAl_2O_4$ can also be synthesised, either from magnesia and alumina powders heated at 1800 °C, or from gibbsite ($Al(OH)_3$) in HCl solution and magnesia powders, heated to 80 °C for 7h, then precipitated when mixed with alkali hydroxide and then calcined at 1000–1400 °C [17], yielding optical properties tailored by the calcination temperature [18]. A common additive is 0.5% CaO which increases the sintering rate and yields maximum densification [19]. Sintering is typically achieved through hot isostatic pressing (HIP) [20]. The chief problem is its brittleness, which renders it difficult to machine, but laser-assisted and ultrasonic vibration-assisted machining of ceramics offer mitigation strategies [20]. $MgAl_2O_4$ has applications as refractory linings of high thermal resistance and low thermal expansion (with a melting point of 2135 °C), electric insulation, transparent optics from UV to IR (0.2–5.5 μm), immunity to chemical corrosion and impact-resistant armour of a very high hardness of 16 GPa. Transparent nanostructured $MgAl_2O_4$ ceramic with grain size of ~40 nm has a hardness of 31.7 GPa and Young's modulus 314 GPa under load that is twice that of large grained ~100–200 μm $MgAl_2O_4$ ceramics, consistent with the Hall–Petch relationship [21]. The addition of Y_2O_3 to $MgAl_2O_4$ grains tailors transparency to the infrared spectrum 0.2–0.6 μm [22]. As well as enhancing optical transparency, the addition of dopants enhances physical properties of strength and toughness by limiting grain boundary growth. The transparent $MgAl_2O_4$ ceramic exhibits enhanced fracture toughness when doped with 500 ppm of large ionic radii Ca and Y, where Y segregates around grain boundaries of Al and Ca segregate around grain boundaries around Mg [23]. High purity α-alumina may be doped with yttrium to segregate at grain boundaries to prevent grain growth and sintered at 1450 °C [24]. Similarly, yttrium-doping migrates to grain boundaries of transparent $ZnAl_2O_4$ ceramic, suppressing grain growth [25]. Ca is readily available from lunar anorthite but Y is more challenging to extract from lunar KREEP minerals requiring a suite of reagents. Rare earth element extraction from rare earth minerals requires dissolution in $(NH_4)_2SO_4$ electrolyte solution to beneficiate from the low-grade ore. Thence, treatment with ammonium bicarbonate ($CO_2 + NH_3 + H_2O \rightarrow (NH_4)HCO_3$) acts as a precipitant which is then burned to yield the rare earth elements. These reagents require N sources which would be imported from

Earth. These then must be separated under the constraint that they have similar physical and chemical properties.

3. Circular Lunar Industrial Ecology

A circular economy for terrestrial industry is premised on several properties that underlie the lunar industrial ecology: (i) maximum efficiency of mining, processing and production; (ii) recycling of waste to reduce pollution; and (iii) refurbishment and repair of products to extend their useful lifetime. Energy efficiency is sacrificed in favour of material closure through recycling or reuse. We have adopted the same philosophy for the lunar industrial ecology [13]. The guide for lunar resources is reflected in the Cheng'E-5 samples, suggesting bulk abundance of SiO_2 (44.9%), Al_2O_3 (17.5%), FeO (12.2%), CaO (12.1%), MgO (8.8%), TiO_2 (2.5%), Na_2O (0.44%), Cr_2O_3 (0.27%) and K_2O (0.11%) [26], though they do not occur in pure oxide form. We limited ourselves to extracting only metals that exceed 1% average concentration, though they exhibit higher concentrations in the commonest four lunar minerals: pyroxene, anorthite, orthoclase and ilmenite. Although there are only a handful of common rock-forming lunar minerals [27], they may be supplemented by iron–nickel asteroid material from impactors [28]. Our lunar industrial architecture emphasised industrial production capacity rather than lunar base infrastructure (Figure 1).

Figure 1. Lunar production architecture showing regolith acquisition by rover fleet, comminution, beneficiation, electrochemical processing and 3D printing facilities, supported by energy charging stations and transportation.

Our lunar industrial ecology was premised on such resources, and we restricted ourselves to building with lunar resources but permitted the import of reagents from Earth that are recycled but not consumed (Figure 2).

We can map the availability of processed lunar resources to the material functional requirements (demandite) of a spacecraft (Figure 3).

Lunar Ilmenite

$Fe^0 + H_2O \rightarrow$ ferrofluidic sealing

$FeTiO_3 + H_2 \rightarrow \mathbf{TiO_2} + H_2O + Fe$

$2H_2O \rightarrow 2H_2 + O_2$

$2Fe + 1.5O_2 \rightarrow Fe_2O_3/Fe_2O_3.CoO$ - ferrite magnets

$3Fe_2O_3 + H_2 \leftrightarrow Fe_3O_4 + H_2O)$ – formation of magnetite at 350-750°C/1-2 kbar

$4Fe_2O_3 + Fe \leftrightarrow 3Fe_3O_4$)

Nickel-Iron Meteorites

W inclusions → Thermionic cathodic material

Mond process:

$W(CO)_6 \leftrightarrow 6CO + W$

$Fe(CO)_5 \leftrightarrow 5CO + Fe$ (175°C/100 bar) → Tool steel

$Ni(CO)_4 \leftrightarrow 4CO + Ni$ (55°C/1 bar) → Electrical steel

$Co_2(CO)_8 \leftrightarrow 8CO + 2Co$ (150°C/35 bar) → Permalloy

S catalyst

Alloy	Ni	Co	Si	C	W
Tool steel				2%	9-18%
Electrical steel			3%		
Permalloy	80%				
Kovar	29%	17%	0.2%	0.01%	

$4FeS + 7O_2 \rightarrow \mathbf{2Fe_2O_3} + 4SO_2$

(Troilite) $SO_2 + H_2S \rightarrow 3S + H_2O$

$FeSe + Na_2CO_3 + 1.5O_2 \rightarrow \mathbf{FeO} + Na_2SeO_3 + CO_2$ at 650°C

KNO_3 catalyst $Na_2SeO_3 + H_2SO_4 \rightarrow Na_2O + H_2SO_4 + Se \rightarrow$ photosensitive Se

↑______|

$Na_2O + H_2O \rightarrow 2NaOH$

$NaOH + HCl \rightarrow NaCl + H_2O$

Lunar Orthoclase

$3KAlSi_3O_8 + 2HCl + 12H_2O \rightarrow KAl_3Si_3O_{10}(OH)_2 + 6H_4SiO_4 + 2KCl$

orthoclase illite silicic acid (soluble silica)

$2KAl_3Si_3O_{10}(OH)_2 + 2HCl + 3H_2O \rightarrow 3Al_2Si_2O_5(OH)_4 + 2KCl$

kaolinite

$[2KAlSi_3O_8 + 2HCl + 2H_2O \rightarrow Al_2Si_2O_5(OH)_4 + 2KCl + \mathbf{SiO_2} + H_2O]$

$KCl + NaNO_3 \rightarrow NaCl + KNO_3$

$2KCl + Na_2SO_4 \rightarrow 2NaCl + K_2SO_4$

Lunar Olivine

$Fe_2SiO_4 + 4\ HCl + 4H_2O \rightarrow 2FeCl_2 + \mathbf{SiO_2} + 2H_2O$

fayalite

$Mg_2SiO_4 + 4HCl + 4H_2O \rightarrow 2MgCl_2 + 2H_4SiO_4$ → Sorel cement

forsterite $MgCl_2 + 2\ NaOH \rightarrow Mg(OH)_2 + 2NaCl$

$Mg(OH)_2 \rightarrow \mathbf{MgO} + H_2O$ → Sorel cement

600-800°C

Lunar Anorthite

$CaAl_2Si_2O_8 + 4C \rightarrow CO + \mathbf{CaO} + \mathbf{Al_2O_3} + 2Si$ at 1650°C → CaO cathode coatings

$CaO + H_2O \rightarrow Ca(OH)_2$

$Ca(OH)_2 + CO_2 \rightarrow CaCO_3 + H_2O$

$CaAl_2SiO_8 + 8HCl + 2H_2O \rightarrow CaCl_2 + 2AlCl_3.6H_2O + \mathbf{SiO_2}$ → fused silica glass + FFC electrolyte

$AlCl_3.6H_2O \rightarrow Al(OH)_3 + 3HCl + H_2O$ at 100°C

↑______|

$2Al(OH)_3 \rightarrow \mathbf{Al_2O_3} + 3H_2O$ at 400°C → $2Al + Fe_2O_3 \rightarrow 2Fe + Al_2O_3$ (thermite)

AlNiCo hard magnets

Al solar sail

Lunar Pyroxene

$Ca(Fe,Al)Si_2O_6 + HCl + H_2O \rightarrow Ca_{0.33}(Al)_2(Si_4O_{10})(OH)_2.nH_2O + H_4SiO_4 + CaCl_2 + Fe(OH)_3$

Augite montmorillonite silicic acid iron hydroxide

Lunar Volatiles

$CO + 0.5\ O_2 \rightarrow CO_2$

$CO_2 + 4H_2 \rightarrow CH_4 + 2H_2O$ at 300°C (Sabatier reaction)

Ni catalyst

850°C 250°C

$CH_4 + H_2O \rightarrow CO + 3H_2 \rightarrow CH_3OH$ 350°C

Ni catalyst Al_2O_3 $CH_3OH + HCl \rightarrow CH_3Cl + H_2O$ 370°C $+nH_2O$

Al_2O_3 $CH_3Cl + Si \rightarrow (CH_3)_2SiCl_2 \rightarrow ((CH_3)_2SiO)_n + 2nHCl$ → silicone plastics/oils

↑______|

$N_2 + 3H_2 \rightarrow 2NH_3$ (Haber-Bosch process)

Fe on $CaO + SiO_2 + Al_2O_3$

$4NH_3 + 5O_2 \rightarrow 4NO + 6H_2O$

Figure 2. *Cont.*

WC on Ni
$3NO + H_2O \rightarrow 2HNO_3 + NO$ (Ostwald process)
↑______________________|
$2SO_2 + O_2 \leftrightarrow 2SO_3$ (low temp)
$SO_3 + H_2O \rightarrow H_2SO_4$

<u>Salt of the Earth</u>
$2NaCl + CaCO_3 \leftrightarrow Na_2CO_3 + CaCl_2$ (Solvay process) → FFC electrolyte
350°C/150 MPa
$Na_2CO_3 + SiO_2(i) \leftrightarrow Na_2SiO_3 + CO_2$ → piezoelectric quartz crystal growth (40-80 days)
1000-1100°C
$CaCO_3 \rightarrow$ **CaO** $+ CO_2$ (calcination)
$NaCl(s) + HNO_3(g) \rightarrow HCl(g) + NaNO_3(s)$
$2NaCl(s) + H_2SO_4(g) \rightarrow 2HCl(g) + Na_2SO_4(s)$

<u>Molten Salt (FFC) Electrolysis Process ($CaCl_2$ electrolyte)</u>
$MO_x + xCa \rightarrow M + xCaO \rightarrow M + xCa + ½xO_2$ where M=Fe, Ti, Al, Mg, Si, etc
$CO + 0.5\ O_2 \rightarrow CO_2$
$CO_2 + 4H_2 \rightarrow CH_4 + 2H_2O$ at 300°C (Sabatier reaction) → $CH_4 \rightarrow C + 2H_2$ at 1400°C → FFC anode regeneration
Ni catalyst

Figure 2. Lunar industrial ecology that converts raw lunar and asteroidal resources into final functional materials—inputs are on the left and applications are on the right (only a few feedback loops are shown).

Functionality (mass fraction)	Lunar-Derived Material
Tensile structures (25%)	Wrought iron Aluminium
Compressive structures (+50%)	Cast iron Regolith + binder
Elastic structures (trace)	Steel springs/flexures Silicone elastomers
Hard structures (3%)	Alumina
Thermal conductor straps (1%)	Fernico (e.g. kovar) Nickel Aluminum
Thermal radiators (3%)	Aluminium
Thermal insulation (3%)	Glass (SiO_2 fibre) Ceramics such as SiO_2
High thermal tolerance (4%)	Tungsten Alumina
Electrical conduction wire (7%)	Aluminium Fernico (e.g. kovar) Nickel
Electrical insulation (1%)	Glass fibre Ceramics (SiO_2, $Al_2O_3$and TiO_2) Silicone plastics Silicon steel for motors
Active electronics devices (vacuum tubes) (12%)	Kovar Nickel Tungsten Fused silica glass
Magnetic materials for actuators (5%)	Ferrite Silicon steel Permalloy
Sensory transducers (5%)	Resistance wire Quartz Selenium
Optical structures (11%)	Polished nickel/aluminium Fused silica glass lenses
Lubricants (trace)	Silicone oils Water
Power system (20%)	Fresnel lens + thermionic conversion Flywheels
Combustible fuels (+250%)	Oxygen Hydrogen

Figure 3. Lunar resources mapped to spacecraft functions—percentages refer to the proportion of a generic dry spacecraft, with excesses indicated for launch propellant to the Gateway and lunar habitat compressive structure.

We focussed primarily on the core of the lunar industrial ecology—the processing of the lunar highlands mineral anorthite. We adopted a geomimetic approach—geomimetics involves inspiration from the synthesis of natural materials through geological processes, such as hydrothermal vent processes at a high temperature and pressure [29]. We have exploited a geomimetic technique inspired by geochemical weathering by weak carbonic acid, such as the dissolution of plagioclase feldspar into kaolinite and calcite [30]:

$$CaAl_2Si_2O_8 + H_2CO_3 + ½O_2 \rightarrow CaCO_3 + Al_2Si_2O_5(OH)_4$$

Kaolinite may be converted into gibbsite via hydrolysis:

$$Al_2Si_2O_5(OH)_4 + 5H_2O \rightarrow 2Al(OH)_3 + 2H_4SiO_4$$

From silicic acid, silica may be precipitated. We adopted HCl acid leaching on anorthite (and other lunar minerals) as our "artificial weathering". The first stage yielded calcium chloride and precipitated silica from silicic acid:

$$CaAl_2SiO_8 + 8HCl + 2H_2O \rightarrow CaCl_2 + 2AlCl_3{\cdot}6H_2O + 2SiO_2$$

Hydrous aluminium chloride decomposed into aluminium hydroxide at 100 °C with HCl partially recycled:

$$2AlCl_3{\cdot}6H_2O \rightarrow 2Al(OH)_3 + 6HCl + 9H_2O$$

Decomposition into alumina occurred at 400 °C:

$$2Al(OH)_3 \rightarrow Al_2O_3 + 3H_2O$$

We have demonstrated this process for the production of calcium chloride, alumina and silica from lunar highland simulant (LHS-1) that is dominantly anorthite [31], i.e., we have utilised 100% of the anorthite mineral. Calcium chloride is feedstock as $CaCl_2$ electrolyte for molten salt electrolysis [32]. Molten salt electrolysis requires much lower temperatures at 800–1000 °C than molten regolith electrolysis at 1650 °C. We have successfully demonstrated the reduction of alumina into 95% pure aluminium metal using molten salt electrolysis [33].

Our concern here is with the ceramics we have produced: alumina and silica. Alumina and silica are applied to situations where high wear resistance, high corrosion resistance, high electrical resistance and high thermal stability are required. We shall illustrate the potential of these two materials in more exotic applications. The float glass technique involves spreading a layer of molten glass onto a bed of molten metal. Fused silica or quartz glass SiO_2, although it has a high melting temperature, is chemically inert. The typical additives of Na_2CO_3 and $CaCO_3$ for sodalime glass, for reduced melting temperature and for hardening, respectively, or B_2O_3 for borosilicate glass of low thermal expansion, are not feasible on the Moon due to the paucity of Na and B but Al_2O_3 additives are feasible for aluminosilicate glass, for high heat tolerance. Glass may be strengthened to reduce its shatterability by heating to a high temperature and cooling it rapidly so the exterior cools faster than the interior. The temperature gradient ensures that the outer surface is in compression while the inner core is in tension (tempering). We do not consider electronics fabrication of Si semiconductors on the Moon to be feasible in the near-term [34]. There are other uses for silica. Silica aerogels may be manufactured from waterglass (sodium silicate) solution, which is acidified into silicic acid followed by gelation using a NH_4OH catalyst and then dried under special conditions [35]. They have multiple uses including high-performance thermal insulation.

It has been suggested that 20–25 m diameter heatshields of mass 40 tonnes for Earth re-entry at temperatures ~3000 °C may be fabricated on the Moon from sintered lunar silicates for the delivery of cargo [36]. Silica extracted from anorthite provides a superior material for refractory heat shield tiles as used on the Space Shuttle. However, their adhesion was substandard. A 5:1 mixture of Al_2O_3:SiO_2 sintered at 1500 °C with a liquid aluminium dihydrogen phosphate ($Al(H_2PO_4)_3 \cdot xH_2O$) binder produces a high shear strength adhesive with a high temperature tolerance [37] that may be suitable. However, this is only feasible if we have access to phosphorus from KREEP (potassium rare earth elements phosphorus) basalt minerals on the Moon which are localised to the Procellarum region. We shall assume that this will be a possible scenario once lunar industrialisation has proceeded to maturity.

HCl "artificial weathering" is a general technique for yielding useful ceramics from lunar minerals. It may be applied to lunar forsterite to yield the ceramic, magnesia:

$$Mg_2SiO_4 + 4HCl + 4H_2O \rightarrow 2MgCl_2 + 2H_4SiO_4$$

$$MgCl_2 + 2\,NaOH \rightarrow Mg(OH)_2 + 2NaCl$$

$$Mg(OH)_2 \rightarrow MgO + H_2O$$

The mare mineral ilmenite may be reduced in the presence of hydrogen at 1000 °C to yield the ceramic rutile: $FeTiO_3 + H_2 \rightarrow Fe + TiO_2 + H_2O$. Typically, the molten Fe metal is tapped at 1600 °C (liquation) and the water is electrolysed to recycle hydrogen and recover oxygen, leaving rutile. HCl "artificial weathering" may also be applied to lunar orthoclase as a source of kaolinite clay ($Al_2Si_2O_5(OH)_4$) via illite clay ($KAl_3Si_3O_{10}(OH)_2$):

$$3KAlSi_3O_8 + 2HCl + 12H_2O \rightarrow KAl_3Si_3O_{10}(OH)_2 + 6H_4SiO_4 + 2KCl$$

$$2KAl_3Si_3O_{10}(OH)_2 + 2HCl + 3H_2O \rightarrow 3Al_2Si_2O_5(OH)_4 + 2KCl$$

Similarly, lunar pyroxene such as augite may be artificially weathered to yield montmorillonite clay:

$$Ca(Fe,Al)Si_2O_6 + HCl + H_2O \rightarrow Ca_{0.33}(Al)_2(Si_4O_{10})(OH)_2 \cdot nH_2O + H_4SiO_4 + CaCl_2 + Fe(OH)_3$$

In fact, acid weathering of pyroxenes at pH~3–4 proceeds more rapidly than that of feldspar, so basalt is enriched in feldspar on Mars [38]. Montmorillonite is an expanding clay, unlike illite and kaolinite, and is the primary ingredient in bentonite: (i) viscous agent in drilling mud; (ii) binder in sand casting—green sand is sand mixed with bentonite clay and water; (iii) absorbent (montmorillonite and kaolinite) in Fuller's earth as a cleaning agent; and (iv) feldspar and kaolinite may be supplemented by bentonite to form pottery. We shall focus on the processing of alumina, silica, magnesia and rutile and exploitation of clays as the ceramic resources available on the Moon. In doing so, we emphasise the multitude of useful properties these ceramics possess that make them so useful on the Moon as a complement to metals. Ceramics are often deployed in conjunction with metals. For example, lunar regolith simulant has been deployed for sand-casting and permanent mould-casting of metal parts such as aluminium [39]. Ceramics also have highly functional applications, e.g., Mg_2Si is a thermoelectric ceramic which can be manufactured from lunar resources. Ball-milled Mg_2Si fine-powder ~2–3 μm blended with an oxygen-free polymer (e.g., polyvinylidene difluoride) and dissolved in a dimethylformadine solution forms a thermoelectric paste which may be 3D printed into any shape and sintered at <400 °C limited by Mg_2Si thermal stability [40]. This requires the use of polymer.

4. Thermal Sintering of Lunar-Derived Ceramics

The most significant problem with the processing of ceramics is their hardness and brittleness at room temperature (an exception being Si_3N_4 which is plastic at room temperature under compressive loading [41]). Ceramics are typically formed through moulding, casting and machining but 3D printing offers an alternative mode of manufacturing. Subtractive processing of ceramics is commonly accomplished through water jets pressurised to 300–600 MPa and laden with an abrasive such as alumina, which can implement cutting, drilling, shock peening and surface structuring. For constructive processing, dry ball milling is the approach for processing single ceramic powders or mixing powders for composites. Thermal sintering involves bonding of particles at ¼–¾ of the melting point. For ceramics, thermal sintering requires sintering temperature from 1000 to 2000 °C range. Reducing such temperatures to 1150 °C requires reducing particle sizes to ~10–100 nm by milling using grinding balls <1 mm in diameter and using additives such as <0.2% CaO and SiO_2 to inhibit grain growth [42]. Selective laser sintering/melting (SLS/M) cannot be employed for direct ceramic melting because of their high melting points, so a polymer, metal or glass binder of lower melting point is required to bind the ceramic particles. One of the chief challenges to direct thermal sintering is the thermal gradients introduced with generate cracks and brittleness. The chief advantage of direct thermal sintering on the Moon is that it requires no reagents and direct solar energy may be concentrated to high temperatures using Fresnel lenses [5].

To obviate high temperature >1000 °C ceramic processing, there are alternative approaches that may be exploited at a lower temperature, e.g., hydrothermal sintering at high pressure for densification, which requires aqueous solution [43]. Hot pressing under high temperature and pressure may be used to sinter particles into a 3D part with homogeneous microstructure. Artificial lithification is a hydrothermal press that compacts loose particles of anhydrous cement at 250 °C and 345 MPa with water into sedimentary rock [44]. Application to crystalline quartz or silica gel with aqueous NaOH solution to improve aqueous dissolution/precipitation reduces the required pressure to 20 MPa [45]. Hydrothermal hot pressing has been demonstrated for alumina, geopolymers, glasses, silica, titania and kaolinite clays. Subjection of nanosized alumina powder with silica gel to 4.5–5.0 GPa at room temperature yielded densified alumina and silica with a hardness of 5.7 GPa and 4.0 GPa, respectively [43]. Cold sintering involves lower applied pressures < 1 GPa and modestly elevated temperatures <350 °C by adding aqueous NaOH solution for lubrication of nanosized ceramic powders as a supersaturated solution, but this has yet to be demonstrated for alumina, silica or magnesia. Hence, hot hydrothermal synthesis under high pressure appears to be a viable approach to the densification of ceramics on the Moon.

One of the chief challenges to machining ceramics is their hardness and brittleness, so a major consideration is the introduction of durability to ceramics. Itacolumite sandstone is a highly flexible rock characterised by angular grain boundary microcracks that may be emulated by sintered aluminium titanate (Al_2TiO_5) powder [46]. Flexibility with flexural strength of 13 MPa is enhanced by a higher sintering temperature and sintering time with optimal values of 1550 °C and 12 h, respectively. Given the availability of TiO_2 on the Moon through hydrogen reduction of lunar ilmenite, this represents an unexplored material option for manufacturing aluminium titanate. Ceramic foams can utilise the superelasticity of alumina Al_2O_3 dispersed with 30% magnesia MgO, which suppresses grain growth of alumina [47]. Although the ductility of MgO-dispersed alumina is inferior to that of 3% yttria-stabilised zirconia, the former can be manufactured from lunar resources and expanded into dense shells. This is unexplored. Ceramic superplasticity occurs in some fine-grained ceramics (ZrO_2, Si_3N_4 and SiC) at a high temperature ~1500 °C through grain boundary sliding of intergranular glass phases [48]. This requires a grain size of <1 μm at a high temperature [49]. The application of an electric field can reduce the viscosity of such glass phases, making the ceramic more plastic [50]. Both these approaches of high temperature and applied electric fields are impractical. Borrowing dislocations in ceramics introduces metals with ceramic through sintering to permit dislocations at metal–ceramic interfaces to enhance ceramic plasticity, e.g., Mo metal applied to La_2O_3 ceramic [51]. Sintered porous Si_3N_4 structures can be infiltrated by molten 6061 Al alloy under high pressure to enhance their fracture toughness [52]. Indeed, the addition of 10% Fe to lunar regolith simulant improved the density of thermally sintering blocks at 1035 °C by reducing cracks, yielding a compressive strength of 150 MPa and tensile strength of 20 MPa [53].

Ceramic matrix composites comprise a ceramic matrix impregnated with glass or metal fibres ~10^{-3} mm to impart non-brittleness. Alumina has high hardness, resistance to corrosion and wear and high thermal conductivity. Alumina particles may be employed as a reinforcement in aluminium metal composites, with optimal stress/strain properties of the composite favoured by small particle size of <10 μm and a higher sintering temperature at 600 °C over a 60 min sintering time [54]. An 80% alumina matrix with a uniform distribution of 20% nickel–iron superalloy steel particles may be fabricated using laser powder bed fusion at 1370 °C [55]. It exhibited porosity even after thermal sintering, due to poor adhesion between steel and alumina and insufficient melting of alumina. Polymer impregnation improved the compressive strength from 56 to 120 MPa. Similarly, digital light processing followed by spark plasma sintering at 1050 °C under 50 MPa pressure to remove a polymer binder has demonstrated the manufacture of fully dense functional grading of metal powder-regolith material [56]. A variation on UV-curable 3D printing involves 25–55% ZrO_2 nanoparticles dispersed in a UV-curable resin as a binder extruded into layers, subsequently cured using a UV laser, heated to 600 °C to remove the resin and sintered at 1450 °C [57]. NanoSiO_2 particle reinforcement may be incorporated into composites to extend their mechanical properties [58]. Bioinspired organic-inorganic materials offer high thermal stability with plastic mouldability, e.g., ceramic $CaCO_3$ with organic thioctic acid (TA) polymerised through S-S bonds, formed by adding $CaCO_3$ to an ethanol solution of TA [59]. Similarly, $CaCO_3$ nanoparticles may be crosslinked with polyacrylic acid to form a hydrogel which is stretchable, shapeable and self-healable [60]. Unfortunately, these materials require organic constituents.

5. Three-Dimensional Printing of Lunar-Derived Ceramics

Traditionally, ceramic parts are manufactured by the casting of slurries in moulds that are then cured and sintered, but such parts cannot assume complex geometries. Three-Dimensional printing of ceramics may be accomplished through fused deposition modelling (FDM), selective laser sintering (SLS), stereolithography (SLA), direct ink writing (DIW) and binder jetting [61]. FDM is an extrusion technique that prints from a ceramic powder-impregnated thermoplastic filament that is melted and extruded. The thermoplastic is typically PLA or ABS, but other higher performance polymers are possible, such as PEEK. PLA is notable because it can be derived from corn starch. Binder jetting is similar, in that powdered ceramic is sprayed with polymer but the binder is removed through polymer debonding at 280–650 °C, followed by thermal sintering at 1400–1500 °C [62]. Al_2O_3 is commonly printed with polymer and sintered in this way. A broad distribution of Al_2O_3 particle sizes yields higher densities and reduced shrinkage on sintering [63]. The use of such additives is common in FDM, but the quality of the parts printed is limited and unsuited to high performance applications such as high temperatures, as the matrix does not permit exploitation of the ceramics. Fibre-reinforced composites comprising high-elasticity fibres within a ceramic matrix such as alumina or silica compensate for brittleness in ceramics whilst retaining their high strength-to-weight ratio and resistance to corrosion and high temperature. Most commonly, the fibre material is SiC or graphite, but silica and alumina may be employed as fibres. Additive manufacturing of such composites favours the adoption of short fibres over continuous fibres, where 3D printing of the pre-formed composite is printed, followed by debinding and sintering [64]. As with ceramic 3D printing, 3D printing may be through SLA, FDM, BJ, DIW or SLS. The adoption of ceramic fibres coated with nanophase iron permits the magnetic alignment of the fibres prior to SLA curing of polymer matrix composites (magnetic 3D printing) [65]. However, there remain challenges with poor bonding at fibre–matrix interfaces compared with traditional composite manufacturing.

Terrestrially, the most promising and versatile mode of 3D printing of ceramics is direct ink writing (DIW) or robocasting due to the wide range of inks [66]. DIW involves the extrusion of a non-Newtonian (shear-thinning) fluid polymer binder loaded with powder at a high density of >50%, such as a colloidal paste, to print 3D parts following sintering. The ink is designed with a yield stress described by the Herschel–Bulkley model: $\tau = \tau_\gamma + K\dot{\gamma}^n$ where τ = applied shear stress, τ_γ = yield stress, $\dot{\gamma}$ = shear rate, K = viscosity, n = degree of shear thinning (n = 1 for Newtonian fluids). During extrusion, shear-thinning is necessary, i.e., a decline in viscosity at higher shear rates to prevent clogging. Beyond τ_γ, shear thinning is described by a power law, $\tau = K\dot{\gamma}^n$. The 10–50 μm powder may be metal, polymer or ceramic but the binder is an organic polymer with additives. Typical polymers are much more complex than FDM polymers and include formaldehyde-phenol-urea, polycarboxylic acid, polyethylene glycol or polyvinyl acetate. For example, the hydrogel Pluronic F-127 comprises a 2:1 mix of polyethylene oxide and polypropylene oxide to provide both hydrophobic and hydrophilic properties. The paste must also include additives such as surfactants and dispersants to disperse the particles and prevent particle aggregation such as stearic acid [67]. After deposition, shape retention requires viscosity to increase as shear decreases. Thermal debinding at 200–600 °C removes the binder and must be slow to prevent crack formation, followed by thermal sintering at 1200–2000 °C, which generally yields 10–20% part shrinkage. Ceramic powder—silica, basalt, etc.—must be highly loaded into the slurry to reduce shrinkage after sintering. Ceramic powder comprises ~40–50% suspensions in the polymer fluid (inks) that are extruded as filaments flowing with decreased viscosity at increased shear [68]. The loading of solid powders imposes the need for larger nozzles than inkjet printing ~400–800 μm; this is because inkjet adopts low ceramic loading (<30%) in suspension for low-viscosity ink compared with high ceramic loading (<60%) for high-viscosity pastes for extrusion or SLA [69]. Inkjet printing is suited to ceramic nanoparticle suspensions in an ink characterised by an inverse Ohnesorge number of 1–10, but only simple structures can be printed in this way. Alumina is an ideal ceramic with high hardness, high compressive strength and resistance to both corrosion and wear that is also thermally and chemically stable at sintering temperatures. Ideal alumina loading is 53–56% with 2% Al metal to minimise shrinkage [70]. Sintering at 1550 °C yields ceramic parts of Al_2O_3 with 98% density. The addition of 3–7% carbon nanotubes imparts electrical conductivity to non-conducting alumina. A loading of 70% regolith simulant powder may be mixed with 30% polylactic-co-glycolic acid and other additives such as dichloromethane to form an ink that may be 3D printed into rubber-like structures with E = 1.8–13 MPa and 250% strain [71]. Although monomers glycolic acid and lactic acid are derivable from urine, the additives are not so readily manufactured from lunar resources. Multi-material 3D printing may be implemented by DIW through the extrusion of different inks with different functional properties [72]. The delivery of multiple composite inks may be through a single paste cartridge or multiple cartridges. Complex parts may be manufactured by DIW, such as gears, electrical components and sensors. Carbon nanotube or graphene in inks can form carbon networks in silicone elastomer matrices whose conductivity responds to pressure and/or temperature.

All these approaches to 3D printing ceramics and/or composites require the use of polymer binders which present major challenges for deployment on the Moon. However, even if carbon sources can be found, the use of DIW is not feasible because it requires careful tailoring of inks which require an Earth-based infrastructure to realise.

6. Problem of Polymers

Carbon is scarce on the Moon, at ~120 ppm embedded in regolith by solar wind impregnation. There are other possible options. It is expected that concentrated asteroidal resources have been deposited into the Moon's surface by impacts [73,74]. A 1 km diameter carbonaceous chondrite striking the Moon at speeds $\leq$ 10 km/s and impact angles $\leq 15^{\circ}$ favour 85% survivability of C-rich (~10^9–10^{10} kg) and N-rich (~10^8–10^9 kg) solids ~20–30 km downrange of the impact site over a 60 km^2 area within <10 m of the surface [75]. Such impacts at <15° and <10 km/s by C-type asteroids are rare (~5 impacts in the last 3 By) but still yield greater C and N concentrations than solar wind impregnation of a typical permanently shadowed crater with ~6 $\times$ 10^5–6 $\times$ 10^8 kg C and N equally. Smaller impactors ~0.1 km would be more common ~500 delivering ~6 $\times$ 10^2–6 $\times$ 10^5 kg C and N equally for each impactor. They may present themselves through their phyllosilicate signature. The identification of graphite in Apollo 12 and Apollo 17 samples provides evidence of the survival of carbonaceous asteroid impacts favoured by porosity, obliquity and low velocity [76]. The Yutu-2 rover of Chang'e 4 has discovered a glassy material in a small 2 m diameter crater with a high 47% concentration of carbonaceous chondrite remnant from an impactor [77]. Even complex organic matter has been associated with pyroclastic beads recovered from Shorty crater by Apollo 17, derived from micrometeoritic kerogen [78]. So asteroid material may exist on the Moon but, if not, asteroids may be manoeuvred into lunar orbit and undergo a controlled descent to the lunar surface [79]. Once the asteroid has soft-landed on the Moon, it may be subject to processing under the Moon's gravity field. The latter scenario of asteroid manoeuvring is not technologically feasible in the near-term. Otherwise, the use of polymers requires its transport from Earth, which imposes a high transport burden. We examine two scenarios: (i) the recovery of diffuse solar-wind-implanted carbon volatiles so small quantities are available; and (ii) the discovery of an asteroid-delivered carbon source presenting copious amounts of carbon, permitting the manufacture of simple polymers.

In case (i), we have a restricted supply of carbon that must be husbanded and consumed frugally—one possibility is to synthesise polymers that use carbon only in its side chains but not its backbone. Inorganic silicone polymers are dominated by siloxanes with Si-O-Si backbones such as polydimethylsiloxane (PDMS), as well as related materials such as polysilazanes with Si–N–Si backbones. However, N is rarer than C on the Moon, ruling out the latter. Siloxanes (silicones) may be manufactured from syngas through the Rochow process via methanol [80]. Silicone can act as a binder for ceramic powders and a source of silica in the sintered ceramic part. Silicones have high temperature tolerance ~300 °C and high UV-resistance compared with C–C bond polymers. Preceramic polymers permit 3D printing of the polymer into complex parts, which is then converted into a ceramic through pyrolysis (ceramisation) at 600–1000 °C in an N_2 atmosphere which decomposes the organic component [81]. Silicone-based pre-ceramic polymer precursors can be 3D printed as extruded polymer and then cured into a SiOC ceramic by vacuum pyrolysis at 1100 °C after deposition. The printed SiOC has a compressive strength of 240 MPa. Pyrolysis in O_2 yields the formation of silicate ceramics by eliminating the carbon required for carbide formation. Silicones reduce the consumption of carbon but nevertheless do consume carbon, requiring recovery of CO/CO_2 during pyrolysis for recycling. Metallosiloxanes such as titanosiloxanes are silicones with metals in the polymer backbone Ti-O-Si, formed from TiO_2 at temperatures >900 °C [82]. They can act as zeolites. Aluminosiloxanes are inorganic polymers based on Al-Si-O backbone but are manufactured from complex precursors [83].

The aluminosiloxane Al(O'Pr),(OSiMe$_3$) may be an organic precursor of $92Al_2O_3$-$8SiO_2$ ceramic [84]. However, complex precursors rule out aluminosiloxanes. Silicones and titanosilicones may be manufacturable in small quantities for specialist applications.

In case (ii), we can assume that a plentiful carbon source has been discovered and that we can synthesise simple plastics. The initial reactions for polymer synthesis from syngas to methanol are the same as for the Rochow process. We focus on synthesising transparent PMMA (polymethyl methacrylate) which has versatility of use, including as a tough acrylic glass, as a resist in electron beam lithography and, most relevant here, as transparent liquid MMA monomers that can act as UV photopolymers in SLA. Although manufacture of both PMMA and its photoinitiator require a source of N, we expect such to be available with C from carbonaceous chondrite deposits. Syngas may be converted to methanol using an alumina catalyst at 200–300 °C and 50–100 bar:

$$CO + 2H_2 \rightarrow CH_3OH$$

Methanol is converted to acetone through dehydrogenation:

$$2CH_3OH \rightarrow CH_3COCH_3 + H_2O + H_2$$

Acetone reacts with HCN to yield acetone cyanohydrin (ACH):

$$CH_3COCH_3 + HCN \rightarrow (CH_3)_2C(OH)CN$$

ACH is converted to transparent liquid MMA using H_2SO_4 at 80–120 °C:

$$(CH_3)_2C(OH)CN + CH_3OH + H_2SO_4 \rightarrow CH_2{=}C(CH_3)COOCH_3 + NH_4HSO_4$$

UV-initiated (benzoin initiator) polymerisation of MMA at 25–60 °C yields PMMA:

$$-[CH_2-C(CH_3)(COOCH_3)-]_n.$$

MMA monomer acts as a UV photopolymer by hardening into PMMA in the presence of the photoinitiator benzoin. Benzoin synthesis begins with conversion of methanol-to-benzene in the presence of a zeolite catalyst at 400–500 °C:

$$6CH_3OH \rightarrow C_6H_6 + 6H_2O$$

Zeolites are tetrahedral MO_4 units where M=Si or Al that act as selective catalysts. Benzene is converted to benzaldehyde through the Gattermann-Koch reaction using an $AlCl_3$ catalyst:

$$C_6H_6 + CO + HCl \rightarrow C_6H_5CHO$$

Benzoin condensation proceeds using an HCN catalyst:

$$2C_6H_5CHO \rightarrow C_6H_5CH(OH)COC_6H_5$$

Hence, we have the UV-photopolymer PMMA with its benzoin photoinitiator to permit the use of SLA. Refractory ceramic powder ~µm size such as silica or alumina may be dispersed at 50–65% by volume in a UV acrylate photopolymer (such as MMA) with photoinitiator (such as benzoin) and subjected to 150–200 µm layered stereolithography by UV laser [85]. The curing depth reachable by the UV laser is $d \propto \frac{1}{\Delta n^2}$ where n = refractive index difference between ceramic and solution derived from the Beer–Lambert law. The photopolymer binder is removed to reveal the ceramic part. Reduced gravity or microgravity imposes criticality to the rheological properties of the ceramic slurry of 50% Al_2O_3 particles in a UV-curable glycol dimethylacrylate resin with a photoinitiator [86]. It behaves as a Bingham fluid in which low shear stress yields high solid-like viscosity, but there is a threshold shear stress that acts as a plastic yield stress. The viscoelasticity of the fluid is increased into a paste by 0.5% cellulose—cellulose is a single additive that can act as dispersant, plasticiser and coagulant that is derivable from vegetation. UV photopolymerisation builds the part layer-by-layer with a layer resolution of 50–100 μm. An 86% dense ceramic results after sintering at 1500 °C.

As we can see, the manufacture of polymers on the Moon would be a boon in opening up 3D printing techniques. Indeed, PMMA in particular exhibits high utility in opening SLA as a 3D printing technique. However, we cannot assume that these resources will be available so we must contend with a paucity of techniques for processing ceramics. The desirability for thermoplastic-type material properties for extrusion-based 3D printing has motivated interest in metallic glasses. Bulk metallic glasses such as $Zr_{44}Ti_{11}Cu_{10}Ni_{10}Be_{25}$ are supercooled liquid metals that soften on heating like thermoplastics suitable for FDM-based extrusion by heating to 460 °C [87]. The printing temperature T determines the incubation time t before recrystallisation according to the Arrhenius equation: $t = t_0e^{-E/RT}$ where t_0 = time constant, E = activation energy, R = universal gas constant, T = annealing temperature. Deposition may proceed at a rate of 5 cm/s until the metallic glass recrystallises into hard metal within 100 s. Bulk metallic glass $La_{60}Ni_{15}Al_{25}$ has been suggested as a recyclable in-space material [88] but this is impractical. Furthermore, bulk metal glasses cannot be readily sourced and alloyed from lunar resources.

The most promising approach to 3D printing ceramics is peculiar to alumina through an inorganic ink for direct ink writing, comprising 58% alumina (Al_2O_3) particles ~80–200 nm within aluminium dihydrogen phosphate ($Al(H_2PO_4)_3$—AP) matrix as a rheological modifier [89]. AP may be synthesised by reacting H_3PO_4 with $Al(OH)_3$. The inorganic AP binder reduces shrinkage during sintering of the 3D printed alumina ceramic at 1250 °C for maximum bending strength. Additives reduce the sintering temperature from the Al_2O_3 sintering temperature of 1700–1800 °C. TheTiO_2 additive dissolves in Al_2O_3 while the CuO additive becomes liquid to fill pores. If the CuO additive can be omitted, this represents a highly suitable ink for use on the Moon that is not reliant on carbon polymers.

7. Lunar-Derived Clays

Here, we adopt clays as a ceramic material with plastic properties to avoid the use of polymers. Clays are hydrated aluminosilicates with ratios of SiO_2/Al_2O_3 of (2.0–5.0):1.0. Kaolinite clay ($2SiO_2 \cdot Al_2O_3 \cdot 2H_2O$) with a water-to-clay ratio of ~0.60 may be 3D printed by DIW through 1.0–1.6 mm diameter nozzles, cured at room temperature for 24 h and sintered at 1100–1200 °C for several hours to yield porcelain with a compressive strength of 20–50 MPa [90]. To prevent clogging of nozzles, clay may be enriched with additives; although sodium polyacrylate is a superior deflocculant to sodium silicate [65], the latter is incorporated into the lunar industrial ecology. The use of embedded temperature and relative humidity sensors for feedback yields strengths of 70 MPa for fired clay and 90 MPa for non-fired clay [91]. Layer-wise slurry deposition uses a slurry feedstock of regolith mixed with 40% hydrated montmorillonite clay, deposited via a blade in 25–100 μm layers [92]. During drying, water from the current layer is drawn into the substrate layer by capillary forces to create a dry high-density powder bed. Either binder jetting of polymer fluid or laser sintering of powder layer may be adopted. The completed 3D part is then sintered at 1130 °C in a furnace. Martian salts $MgSO_4$ binding with SiO_2 to form forsterite Mg_2SiO_4 or enstatite $Mg_2Si_2O_6$ could strengthen the part if Martian resources were available. A lunar version could exploit Sorel cement as a binder at the cost of consuming Cl import from Earth reacting with lunar olivine:

$$3MgO + MgCl_2 + 11H_2O \rightarrow 3Mg(OH)_2{\cdot}MgCl_2{\cdot}8H_2O$$

Bamboo fibre reinforcement may be added [93], a highly versatile material suited to a lunar agricultural facility [7].

Geopolymer is an alkali-activated cement such as the clay-like aluminosilicate metakaolin activated by sodium silicate (waterglass). They form a 3D network of covalent Si-O–Al bonds $M_n\{\text{-}(SiO_2)_z\text{-}AlO_2\}{\cdot}wH_2O$ where M = metal cation such as K^+ and Na^+, n = polymerisation degree, z = 1, 2 or 3, w = bound water. Geopolymer may be constructed from metakaolinite (dehydroxylated kaolinite) comprising 90–95% Al_2O_3 and SiO_2, which dissolves in an activating alkali such as NaOH or KOH. Plasticising fillers can alter the viscoelasticity. These are commonly organics such as sodium carboxymethyl starch which increases viscosity, but magnesium aluminium silicate is a smectite clay derived from feldspar that serves the same purpose. Geopolymer may be mixed with aggregates such as 20% silica sand, which reduces shrinkage. Three-Dimensional printing geopolymers may be via extrusion of geopolymer paste or geopolymer powders which are activated by NaOH solution, which gives a higher compression strength than KOH [94]. The Na_2CO_3 activation reaction proceeds more slowly, which may be desirable but requires a carbon source. Curing at high temperature ~100 °C increases mechanical strength. A lunar regolith simulant-based geopolymer ink, reinforced by carbon fibre/quartz in different patterns, forms a 3D printed composite to reduce the brittleness of the geopolymer; rectangular sandwich patterns yielded the highest compressive strengths of 20 MPa and 34 MPa in y and x directions [95]. Lunar-derived clays offer some of the promise of ceramics, but their true potential requires marriage with polymers.

8. Non-Clay Minerals and Their Uses

There are other uses for non-clay lunar minerals. On the Moon, olivine forsterite (Mg_2SiO_4) may be reacted with quartz (SiO_2) to form the orthopyroxene enstatite ($MgSiO_3$):

$$Mg_2SiO_4 + SiO_2 \rightarrow 2MgSiO_3$$

The hydration of enstatite yields serpentinite ($Mg_3Si_2O_5(OH)_4$) and talc ($Mg_3Si_4O_{10}(OH)_2$):

$$6MgSiO_3 + H_2O \rightarrow Mg_3Si_2O_5(OH)_4 + Mg_3Si_4O_{10}(OH)_2$$

Talc may be employed as a dry lubricant as an alternative to tungsten or molybdenum disulphide [96]. Exploitation of such lunar minerals as ceramic resources has yet to be explored.

9. Direct Solar Sintering/Melting

We consider, finally, direct thermal processing of ceramic materials including regolith. A layer of regolith around 2.5 m thick is required for a radiation shield to reduce exposed radiation to 5 REM. Crucial to the deployment of sintered regolith is compressive strength for load bearing. Implicit is the desirability of manufacturing equipment from lunar resources. Solar concentrators are mirrors or lenses that concentrate light according to the concentrator ratio given by [97]:

$$C = \frac{A_{conc}}{A_{spot}} = \frac{\varepsilon\sigma\left(T_{spot}^4 - T_{amb}^4\right)}{S_0\eta_{conc}}$$

where T_{spot} = spot temperature = 1800 °C to melt regolith, T_{amb} = ambient temperature (270 K) and η_{conc} = concentrator efficiency of 0.85 requiring C~400. Mirrors comprise an aluminium reflector with a cover of fused silica glass (derived from locally extracted SiO_2 or other transparent material such as $MgAl_2O_4$ spinel). Paraboloids may be approximated by multiple appropriately oriented flat mirrors but with some loss in performance. Fresnel lenses offer a higher concentration ratio than mirrors given by the following [97]:

$$C = \left(\frac{n}{n_{amb} sin\theta_{sun}}\right)^2$$

where n = 1.5 for glass, n = 1 for air, θ_{sun} = sun half-angle = 0.275° so C = 97,700 compared to a theoretical C = 45,900 for mirrors. Fresnel lenses comprise fused silica glass with a submicron structured surface constructed by transparent fused silica glass or spinel. Sintering increases regolith-effective thermal conductivity, allowing increased sintering depth by 45% [98]. Solar concentrators generate limited spot size resolution of >10 mm. Large-scale experiments with a 52 m^2 mirror reflecting onto a 147-mirror concentrator concentrated solar 800 W/m^2 onto 4 MW/m^2 [99]. Compressive strength of solar layer-sintered regolith is under 5 MPa which is considered marginal from the minimum required 3 MPa under lunar gravity. Compression strength was enhanced by a higher sintering temperature and shortening the cooling time between layers by reducing thermal stresses. Most experiments on sintered regolith simulants also use simulated solar sources (Xe lamps) to replicate atmosphere-less lunar lighting conditions. The Regolight project deployed solar sintering using Xe lamps concentrated by a parabolic mirror then oriented by a water-cooled flat mirror onto regolith simulant to sinter it layer-by-layer with 100 μm layers into 10 × 20 × 70 mm bricks. They 3D printed interlocking 3D printed tetrahedon bricks and flat dodecahedron tiles using layers ~2 mm thick but feature resolution was limited to >14 mm and compressive strength was poor at 2.5 MPa due to irregular delamination and warping [100–102]. In small-scale experiments using Xe lamps on small regolith samples, compressive strengths of solar sintered regolith can reach ~150 MPa [103]. Successful sintering of regolith simulant bricks with a compression strength of 45 MPa has been achieved through a two-stage process using a polyvinyl alcohol (PVA) binder [104]. The binder was pre-mixed with water, then mixed with regolith simulant and then cast into cubical blocks using a hydraulic press. The brick was then heated in a furnace at 600 °C to remove the binder, then at 1150 °C to sinter the regolith. The conclusion is that solar sintering of regolith is highly restrictive—it has a poor spot size resolution, yet it is limited to small-scale structures. Successful deployment for large-scale application requires the use of polymer binders even for bricks.

Our own preliminary experiments confirm these findings. We used two Fresnel lenses to experiment on the solar melting of LHS-1 highland lunar regolith simulant into glass (Figure 4).

Exolith LHS-1 lunar highland simulant comprises 74.4% anorthosite, 24.7% basalt, 0.4% ilmenite, 0.3% olivine and 0.2% pyroxene. The small/large Fresnel lens numerical aperture was 0.3/0.5 with a focal spot diameter of 8 mm/11 mm. Our experiments were conducted in August in Ottawa. We used the small lens to form melted lines and the large lens to form melted squares. In printing a line, the sun's azimuth movement caused the focus to spread, requiring periodic adjustment to maintain focus on a point for melting (Figure 5a). The first single layer print exhibited a melt depth of 4.37 mm after 3 min of concentrated solar exposure. The three-layer print was printed sequentially by brushing more regolith layers over melted regolith, then melting them, but the final surface was rougher than the underlying layers due to cloud interference (Figure 5b).

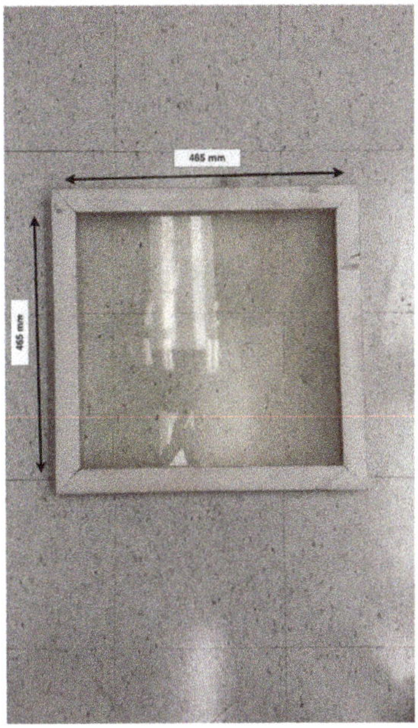

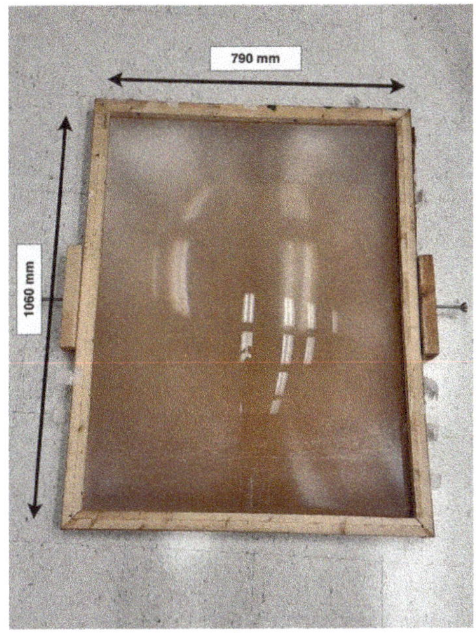

(a) (b)

Figure 4. (**a**) Small Fresnel lens; (**b**) Large Fresnel lens.

(a) (b)

Figure 5. (**a**) line of melted LHS-1; (**b**) square samples 1 to 4 of melted LHS-1.

The final square dimensions were 32.3 mm × 30.7 mm × 10.6 mm with each layer averaging ~3.5 mm (Figure 6).

All samples were compression load tested with squares of approximately the same dimensions on a substrate of unmelted regolith simulant (Figure 7).

Sample 1 was a three-layer sample, but the tests were performed on single-layered samples 2–4 to determine the compressive load limits (Table 1).

Figure 6. Thickness of samples 1 to 4 of melted LHS-1.

(a) (b)

Figure 7. (**a**) Compressive load testing apparatus; (**b**) sample under compression test.

Table 1. Melted sample compressive load limits until cracking.

Sample	Compressive Load Limit (N)
2	239
3	267
4	136 (crack 1)/446 (crack 2)—291 (av)

All the samples cracked into several pieces with a consistent compressive load of ~250 N (including offset tension of 24 N). With a probe cross-section of 110 mm × 29 mm, this gives a critical compressive pressure of 78 kPa, which is insufficient for adequate load bearing on the Moon.

10. Conclusions

The complexity of the ink of DIW and its reliance on polymers renders DIW unsuitable for the Moon even in support of human habitats; organic matter consumed in inks or binders must be either recycled through a closed loop capability or supplied from Earth. Although recycling may be plausible for some simple polymers, DIW inks are too complex to resynthesise. SLA-based methods similarly rely on photopolymers and photoinitiator polymers, but these are more plausibly manufacturable than DIW inks, though they still require a carbon source. More traditional ceramic processing methods appear more suitable for lunar application. Sintering ceramics directly using solar concentrators is not reliant on the manufacture of organic polymers or other complex materials, though there remain issues regarding brittleness. Furthermore, to manufacture such optical devices from lunar resources reintroduces polymers. To exploit the full range and versatility of ceramics derived from lunar resources, carbon sources on the Moon to permit the synthesis of polymers must be searched for at the highest priority. Silicone polymers are preferred for deployment on the Moon for their hardiness to radiation. Although hydrocarbon polymers are susceptible to radiation effects, if used as a binder that is subsequently removed, this problem does not arise. The potential for lunar-derived ceramics is underexplored in comparison with lunar-derived metals. Lunar industrialisation will be critically dependent on the deployment of lunar ceramics, so this requires urgent address.

Funding: This research received no external funding.

Institutional Review Board Statement: Not applicable.

Informed Consent Statement: Not applicable.

Data Availability Statement: No new data were created or analysed in this study.

Acknowledgments: We would like to thank Aidan Chan for performing the solar Fresnel lens tests on the LHS-1 simulant.

Conflicts of Interest: The author declares no conflicts of interest.

References

1. Wang, Y.; Hao, L.; Li, Y.; Sun, Q.; Sun, M.; Huang, Y.; Li, Z.; Tang, D.; Wang, Y.; Xiao, L. In-situ utilisation of regolith resource and future exploration of additive manufacturing for lunar/martian habitats: A review. *Appl. Clay Sci.* **2022**, *229*, 106673. [CrossRef]
2. Mueller, R.; Howe, S.; Kochmann, D.; Ali, H.; Anderson, C.; Burgoyne, H.; Chambers, W.; Clinton, R.; De Kestellier, X.; Ebelt, K.; et al. Automated additive construction (AAC) for Earth and space using in-situ resources. In Proceedings of the Biennial ASCE Conference on Engineering, Science, Construction & Operations in Challenging Environments (Earth & Space 2016), Reston, VA, USA, 11–15 April 2016.
3. Fiske, M.; Ellery, A. *On the Development of an ISRU-Based Calcium Sulphoaluminate Concrete for 3D Printed and Cast Lunar Surface Infrastructure Applications*; AIAA SciTech Forum: Orlando, FL, USA, 2025; p. 1863.
4. Ellery, A. Leveraging in-situ resources for lunar base construction. *Can. J. Civ. Eng.* **2021**, *49*, 657–674. [CrossRef]
5. Ellery, A. Generating and storing power on the Moon using in-situ resources. *Proc. Inst. Mech. Eng. J. Aerosp. Eng.* **2021**, *236*, 1045–1063. [CrossRef]
6. Ellery, A. Can we build nuclear-electric propulsion systems from lunar resources? In *Space Resources Roundtable*; Colorado School of Mines: Golden, CO, USA, 2025; Available online: https://isruinfo.com/public/index.php?page=srr_25 (accessed on 12 August 2025).
7. Ellery, A. Supplementing closed ecological life support systems with in-situ resources on the Moon. *Life* **2021**, *11*, 770. [CrossRef]
8. Azami, M.; Kazemi, Z.; Moazen, S.; Dube, M.; Potvin, M.-J.; Skonieczny, K. Comprehensive review of lunar-based manufacturing and construction. *Prog. Aerosp. Sci.* **2024**, *150*, 101045. [CrossRef]
9. Zocca, A.; Wilbig, J.; Gunster, J.; Widjaja, P.; Neumann, C.; Clozel, M.; Meyer, A.; Ding, J.; Zhou, Z.; Tian, X. Challenges in the technology development for additive manufacturing in space. *Chin. J. Mech. Eng. Addit. Manuf. Front.* **2022**, *1*, 100018. [CrossRef]
10. Ellery, A. Challenges of robotic milli-g operations for asteroid mining. In Proceedings of the Future Technologies Conference 2024, London, UK, 14–15 November 2024; Arai, K., Ed.; Lecture Notes on Networks & Systems. Springer Nature Switzerland AG: Cham, Switzerland, 2024; Volume 1157, pp. 45–64.
11. McKay, D.; Hiken, G.; Basu, A.; Blanford, G.; Simon, S.; Reedy, R.; French, B.; Papike, J. Lunar regolith. In *Lunar Sourcebook*; Heiken, G., Vaniman, D., French, B., Eds.; Cambridge University Press: Cambridge, UK, 1991; pp. 287–356.
12. Klein, C. Lunar materials: Their mineralogy, petrology and chemistry. *Earth Sci. Rev.* **1972**, *8*, 169–204. [CrossRef]
13. Ellery, A. Sustainable in-situ resource utilisation on the Moon. *Planet. Space Sci.* **2020**, *184*, 104870. [CrossRef]
14. Sodha, G.; Dhingra, D. Novel geological framework to understand the origin and diversity of orthopyroxene, olivine, spinel (OOS) lithologies on the Moon. *Sci. Rep.* **2025**, *15*, 2426. [CrossRef]

15. Papike, J.; Taylor, L.; Simon, S. Lunar minerals. In *Lunar Sourcebook*; Heiken, G., Vaniman, D., French, B., Eds.; Cambridge University Press: Cambridge, UK, 1991; pp. 121–181.
16. Crawford, I.; Anand, M.; Barber, S.; Cowley, A.; Crites, S.; Fa, W.; Flahaut, J.; Gaddis, L.; Greenhagen, B.; Haruyama, J.; et al. Lunar resources. *Rev. Mineral. Geochem.* **2023**, *89*, 829–868. [CrossRef]
17. Reveron, H.; Gutierrez-Campos, D.; Rodriguez, R.; Bonassin, J. Chemical synthesis and thermal evolution of MgAl2O4 spinel precursor prepared from industrial gibbsite and magnesia powder. *Mater. Lett.* **2002**, *56*, 97–101. [CrossRef]
18. Ewais, E.; Besisa, D.; El-Amir, A.; El-Sheikh, S.; Rayan, D. Optical properties of nanocrystalline magnesium aluminate spinel synthesised from industrial waste. *J. Alloys Compd.* **2015**, *649*, 159–166. [CrossRef]
19. Lemeshev, D.; Senina, M.; Pedchenko, M.; Boyko, A. Transparent ceramic based on magnesium aluminate spinel for armour. *IOP Conf. Ser. Mater. Sci. Eng.* **2019**, *525*, 012081. [CrossRef]
20. Shi, Z.; Zhao, Q.; Guo, B.; Ji, T.; Wang, H. Review on processing polycrystalline magnesium aluminate spinel ($MgAl_2O_4$): Sintering techniques, material properties and machinability. *Mater. Des.* **2020**, *193*, 108858. [CrossRef]
21. Zhang, J.; Lu, T.; Chang, X.; Wei, N.; Qi, J. Unique mechanical properties of nanostructured transparent $MgAl_2O_4$ ceramics. *Nanoscale Res. Lett.* **2013**, *8*, 261. [CrossRef]
22. Mroz, T.; Goldman, L.; Gledhill, A.; Li, D.; Padture, N. Nanostructured infrared-transparent magnesium aluminate spinel with superior mechanical properties. *Int. J. Appl. Ceram. Technol.* **2012**, *9*, 83–90. [CrossRef]
23. Campos-Quiros, A.; Zughbi, M.; Kundu, A.; Watanabe, M. Correlation between grain boundary segregation behaviours of calcium and yttrium and enhanced fracture toughness in magnesium aluminate spinel. *J. Mater. Sci.* **2025**, *60*, 1826–1852. [CrossRef]
24. Voytovych, R.; MacLaren, I.; Gulgun, M.; Cannon, R.; Ruhle, M. Effect of yttrium on densification and grain growth in α-alumina. *Acta Mater.* **2002**, *50*, 3453–3463. [CrossRef]
25. Xu, P.; Wang, H.; Tu, B.; Gu, H.; Wang, W.; Liu, S.; Ma, C.; Fu, Z. Effect of yttrium-doped grain boundary on sintering behaviour and properties of transparent $ZnAl_2O_4$ ceramics. *J. Eur. Ceram. Soc.* **2024**, *44*, 6597–6606. [CrossRef]
26. Hou, Z.; Gong, Q.; Liu, N.; Jiang, B.; Li, J.; Wu, Y.; Huang, J.; Gu, W. Elemental abundances of Moon samples based on statistical distributions of analytical data. *Appl. Sci.* **2023**, *13*, 360. [CrossRef]
27. Heiken, G.; Vaniman, D.; French, B. *Lunar Sourcebook: A Users Guide to the Moon*; Cambridge University Press: Cambridge, UK, 1991; Available online: https://www.lpi.usra.edu/publications/books/lunar_sourcebook/pdf/LunarSourceBook.pdf (accessed on 12 August 2025).
28. Ellery, A. Mining asteroid versus mining the Moon—Can you have your cake and eat it? In *Space Resources Roundtable*; Colorado School of Mines: Golden, CO, USA, 2025.
29. Unterclass, M. Geomimetics and extreme biomimetics inspired by hydrothermal systems—What can we learn from nature for materials synthesis? *Biomimetics* **2017**, *2*, 8. [CrossRef]
30. Lasaga, A.; Soler, J.; Ganor, J.; Burch, T.; Nagy, K. Chemical weathering rate laws and global geochemical cycles. *Geochim. Cosmochim. Acta* **1994**, *58*, 2361–2386. [CrossRef]
31. Thibodeau, B.; Walls, X.; Ellery, A.; Cousens, B.; Marczenko, K. Extraction of silica and alumina from lunar highland simulant. In Proceedings of the ASCE Earth & Space Conference 2024, Miami, FL, USA, 15–18 April 2024; p. 6962.
32. Ellery, A.; Mellor, I.; Wanjara, P.; Conti, M. Metalysis FFC process as a strategic lunar in-situ resource utilisation technology. *New Space J.* **2022**, *10*, 224–238. [CrossRef]
33. Walls, X.; Ellery, A.; Marczenko, K.; Wanjara, P. Aluminium metal extraction from lunar highland simulant using electrochemistry. In Proceedings of the ASCE Earth & Space Conference 2024, Miami, FL, USA, 15–18 April 2024. Paper No. 7061.
34. Ellery, A. Is electronics fabrication feasible on the Moon? In Proceedings of the ASCE Earth & Space Conference 2022, Denver, CO, USA, 25–28 April 2022; pp. 759–772.
35. Dorcheh, S.; Abbasi, M. Silica aerogel: Synthesis, properties and characterization. *J. Mater. Process. Technol.* **2008**, *199*, 10–26. [CrossRef]
36. Hogue, M.; Mueller, R.; Sibille, L.; Hintze, P.; Rasky, D. Extraterrestrial regolith derived atmospheric entry heat shields. In Proceedings of the 15th Biennial ASCE International Conference on Engineering, Science, Construction, and Operations in Challenging Environments, Orlando, FL, USA, 11–15 April 2016.
37. Chen, Y.; Wang, X.; Yu, C.; Ding, J.; Deng, C.; Zhu, H. Properties of inorganic high temperature adhesive for high temperature furnace connection. *Ceram. Int.* **2019**, *45*, 8684–8689. [CrossRef]
38. McAdam, A.; Zolotiv, M.; Sharp, T.; Leshkin, L. Preferential low pH dissolution of pyroxene in plagioclase-pyroxene mixtures: Implications for martian surface materials. *Icarus* **2008**, *196*, 90–96. [CrossRef]
39. Baasch, J.; Windisch, L.; Koch, F.; Linke, S.; Stoll, E.; Schilde, C. Regolith as substitute mould material for aluminium casting on the Moon. *Acta Astronaut.* **2021**, *182*, 1–12. [CrossRef]
40. Marques, A.; Miglietta, D.; Gaspar, G.; Baptista, A.; Gaspar, A.; Perdigao, P.; Soares, I.; Bianchi, C.; Sousa, D.; Faustino, M.; et al. Synthesis of thermoelectric magnesium-silicide pastes for 3D printing, electrospinning and low-pressure spray. *Mater. Renew. Sustain. Energy* **2019**, *8*, 21. [CrossRef]

41. Frankberg, E. Ceramic that bends instead of shattering. *Science* **2022**, *378*, 359–360. [CrossRef]
42. Kuang, X.; Carotenuto, G.; Nicolais, L. Review of ceramic sintering and suggestions on reducing sintering temperatures. *Adv. Perform. Mater.* **1997**, *4*, 257–274. [CrossRef]
43. Vakifahmetoglu, C.; Karacasulu, L. Cold sintering of ceramics and glasses: A review. *Curr. Opin. Solid. State Mater. Sci.* **2020**, *24*, 100807. [CrossRef]
44. Yamasaki, N.; Yanagisawa, K.; Nishioka, M.; Kanahara, S. Hydrothermal hot-pressing method: Apparatus and application. *J. Mater. Sci. Lett.* **1986**, *5*, 355–356. [CrossRef]
45. Yanagisawa, K.; Nishioka, M.; Ioku, K.; Yamasaki, N. Densification of silica gels by hydrothermal hot-pressing. *J. Mater. Sci. Lett.* **1993**, *12*, 1073–1075. [CrossRef]
46. Chen, W.; Shui, A.; Wang, C.; Li, J.; Ma, J.; Tian, W.; Ota, T.; Xi, X. Preparation of aluminium titanate flexible ceramic by solid phase sintering and its mechanical behaviour. *J. Alloys Compd.* **2019**, *777*, 119–126. [CrossRef]
47. Kishimoto, A.; Obata, M.; Asaoka, H.; Hayashi, H. Fabrication of alumina-based ceramic foams utilising superplasticity. *J. Eur. Ceram. Soc.* **2007**, *27*, 41–45. [CrossRef]
48. Wakai, F.; Kondo, N.; Shinoda, Y. Ceramics superplasticity. *Curr. Opin. Solid. State Mater. Sci.* **1999**, *4*, 461–465. [CrossRef]
49. Pilling, J.; Payne, J. Superplasticity in Al2O3-ZrO2-Al2TiO5 ceramics. *Scr. Metall. Mater.* **1995**, *32*, 1091–1097. [CrossRef]
50. Shen, Z.; Peng, H.; Nygren, M. Formidable increase in the superplasticity of ceramics in the presence of an electric field. *Adv. Mater.* **2003**, *15*, 1006–1009. [CrossRef]
51. Dong, L.; Zhang, J.; Li, Z.; Gao, Y.; Wang, M.; Huang, M.; Wang, J.; Chen, K. Borrowed dislocations for ductility in ceramics. *Science* **2024**, *385*, 422–428. [CrossRef]
52. Peng, L.; Cao, J.; Noda, K.; Han, K. Mechanical properties of ceramic-metal composites by pressure infiltration of metal into porous ceramics. *Mater. Sci. Eng. A* **2004**, *374*, 1–9. [CrossRef]
53. Wan, F.; Zuo, Y.; Pan, Y.; Yu, A.; Qi, J.; Lu, X. Iron-enhanced simulated lunar regolith sintered blocks: Preparation, sintering and mechanical properties. *Constr. Build. Mater.* **2024**, *439*, 137395. [CrossRef]
54. Rahimian, M.; Ehsani, N.; Parvin, N.; Baharvandi, H. Effect of particle size, sintering temperature and sintering time on the properties of Al-Al_2O_3 composites made by powder metallurgy. *J. Mater. Process. Technol.* **2009**, *209*, 5387–5393. [CrossRef]
55. Azami, M.; Siahsarani, A.; Hadian, A.; Kazemi, Z.; Rahmatabadi, D.; Kashani-Bozorg, F.; Abrinia, K. Laser powder bed fusion of alumina/Fe-Ni ceramic matrix particulate composites impregnated with a polymeric resin. *J. Mater. Res. Technol.* **2023**, *24*, 3133–3144. [CrossRef]
56. Laot, M.; Rich, B.; Cheibas, I.; Fu, J.; Zhu, J.-N.; Popovich, V. Additive manufacturing and spark plasma sintering of lunar regolith for functionally graded materials. *Spool* **2021**, *8*, 7–23.
57. Faes, M.; Valkenaers, H.; Vogeler, F.; Vleugels, J.; Ferraris, E. Extrusion-based 3D printing of ceramic components. *Procedia CIRP* **2015**, *28*, 76–81. [CrossRef]
58. Ngo, T.; Kashani, A.; Imbalzano, G.; Nguyen, K.; Hui, D. Additive manufacturing (3D printing): A review of materials, methods, applications and challenges. *Compos. B Eng.* **2018**, *143*, 172–196. [CrossRef]
59. Fang, W.; Mu, Z.; He, Y.; Kong, K.; Jiang, K.; Tang, R.; Liu, Z. Organic-inorganic covalent-ionic molecules for elastic ceramic plastic. *Nature* **2023**, *619*, 293–299. [CrossRef]
60. Sun, S.; Mao, L.-B.; Lei, Z.; Yu, S.-H.; Colfen, H. Hydrogels from amorphous calcium carbonate and polyacrylic acid: Bio-inspired materials for 'mineral plastic'. *Angew. Chem. Int. Ed.* **2016**, *55*, 11765–11769. [CrossRef] [PubMed]
61. Wang, Y.; Wu, T.; Huang, G. State-of-the-art research progress and challenge of the printing techniques, potential applications for advanced ceramic materials 3D printing. *Mater. Today Commun.* **2024**, *40*, 110001. [CrossRef]
62. Clemens, F.; Sarraf, F.; Borzi, A.; Neels, A.; Hadian, A. Material extrusion additive manufacturing of advanced ceramics: Towards the production of large components. *J. Eur. Ceram. Soc.* **2023**, *43*, 2752–2760. [CrossRef]
63. Yeh, T.-S.; Sacks, M. Effect of particle size distribution on the sintering of alumina. *Commun. Am. Ceram. Soc.* **1988**, *71*, C484–C487. [CrossRef]
64. Wang, W.; Zhang, L.; Dong, X.; Wu, J.; Zhou, Q.; Li, S.; Shen, C.; Liu, W.; Wang, G.; He, R. Additive manufacturing of fibre reinforced ceramic matrix composites: Advances, challenges and prospects. *Ceram. Int.* **2021**, *48*, 19542–19556. [CrossRef]
65. Martin, J.; Fiore, B.; Erb, R. Designing bioinspired composite reinforcement architectures via 3D magnetic printing. *Nat. Commun.* **2015**, *6*, 8641. [CrossRef]
66. Romanczuk-Ruszuk, E.; Sztorch, B.; Pakula, D.; Gabriel, E.; Nowak, K.; Przekop, R. 3D printing ceramics—Materials for direct extrusion process. *Ceramics* **2023**, *6*, 364–385. [CrossRef]
67. Hao, L.; Tang, D.; Sun, T.; Xiong, W.; Feng, Z.; Evans, K.; Li, Y. Direct ink writing of mineral materials—A review. *Int. J. Precis. Eng. Manuf. Green. Technol.* **2021**, *8*, 665–685. [CrossRef]
68. Lamnini, S.; Elsaed, H.; Lakhdar, Y.; Baino, F.; Smeacetto, F.; Bernado, E. Robocasting of advanced ceramics: Ink optimization and protocol to predict the printing parameters—A review. *Helyon* **2022**, *8*, e10651. [CrossRef] [PubMed]

69. Chen, Z.; Li, Z.; Li, J.; Liu, C.; Lao, C.; Fu, Y.; Liu, C.; Li, Y.; Wang, P.; He, Y. 3D printing of ceramics: A review. *J. Eur. Ceram. Soc.* **2019**, *39*, 661–687. [CrossRef]
70. Shahzad, A.; Lazoglu, I. Direct ink writing of structural and functional ceramics: Recent achievements and future challenges. *Compos. B* **2021**, *225*, 109249. [CrossRef]
71. Jakus, A.; Koube, K.; Geisendorfer, N.; Shah, R. Robust and elastic lunar and martian structures from 3D printed regolith inks. *Sci. Rep.* **2017**, *7*, 44931. [CrossRef]
72. Rocha, V.; Saiz, E.; Tirichenko, I.; Garcia-Tunon, E. Direct ink writing advances in multi-material structures for a sustainable future. *J. Mater. Chem. A* **2020**, *8*, 15646–15657. [CrossRef]
73. Bland, P.; Cintala, M.; Horz, F.; Cressey, G. Survivability of meteorite projectiles—Results from impact experiments. In Proceedings of the 32nd Lunar & Planetary Science Conference, Houston, TX, USA, 12–16 March 2001; p. 1764.
74. Bland, P.; Artemieva, N.; Collins, G.; Bottke, W.; Bussey, D.; Joy, K. Asteroids on the Moon: Projectile survival during low velocity impacts. In Proceedings of the 39th Lunar & Planetary Science Conference, League City, TX, USA, 10–14 March 2008; p. 2045.
75. Halim, S.; Crawford, I.; Collins, G.; Joy, K.; Davison, T. Assessing the survival of carbonaceous chondrites impacting the lunar surface as a potential resource. *Planet. Space Sci.* **2024**, *246*, 105905. [CrossRef]
76. Joy, K.; Crawford, I.; Curran, N.; Zolensky, M.; Fagan, A.; King, D. Moon: An archive of small body migration in the solar system. *Earth Moon Planets* **2016**, *118*, 133–158. [CrossRef]
77. Yang, Y.; Li, S.; Zhu, M.-H.; Liu, Y.; Wu, B.; Du, J.; Fa, W.; Xu, R.; He, Z.; Wang, C.; et al. Impact remnants rich in carbonaceous chondrites detected on the Moon by the Chang'e 4 rover. *Nat. Astron.* **2022**, *6*, 207–213. [CrossRef]
78. Thomas-Keprta, K.; Clemett, S.; Messenger, S.; Ross, D.; Le, L.; Rahman, Z.; McKay, D.; Gibson, E.; Gonzalez, C.; Peabody, W. Organic matter on the Earth's Moon. *Geochim. Cosmochim. Acta* **2014**, *134*, 1–15. [CrossRef]
79. Ellery, A. Trials and tribulations of asteroid mining. In Proceedings of the ASCE Earth & Space Conference 2024, Miami, FL, USA, 15–18 April 2024; p. 8087.
80. O'Lenick, A. Basic silicone chemistry—A review. Silicone Spect. 2009. Available online: https://www.scientificspectator.com/documents/silicone%20spectator/Silicone_Spectator_January_2009.pdf (accessed on 12 August 2025).
81. Chaudhary, R.; Parameswaran, C.; Idrees, M.; Rasaki, A.; Liu, C.; Chen, Z.; Colombo, P. Additive manufacturing of polymer-derived ceramics: Materials, technologies, properties and potential applications. *Prog. Mater. Sci.* **2022**, *128*, 100969. [CrossRef]
82. Davis, P.; Murugavel, R. Recent developments in the chemistry of molecular titanosiloxanes and titanophosphonates. *Synth. React. Inorg. Met. Org. Nano-Met. Chem.* **2005**, *35*, 591–622. [CrossRef]
83. Unny, R.; Gopinathan, S.; Gopinathan, C. Chelated aluminosiloxanes. *Indian. J. Chem.* **1980**, *19A*, 484–486.
84. Puxviel, J.; Boilot, J.; Poncelet, O.; Hubert-Pfalzgraf, L.; Lecomte, A.; Dauger, A.; Beloeil, J. Aluminosiloxane as a ceramic precursor. *J. Non-Cryst. Solids* **1987**, *93*, 277–286. [CrossRef]
85. Griffith, M.; Halloran, J. Freeform fabrication of ceramics vis stereolithography. *J. Am. Ceram. Soc.* **1996**, *79*, 2601–2608. [CrossRef]
86. Dou, R.; Tang, W.; Hu, K.; Wang, L. Ceramic paste for space stereolithography 3D printing technology in microgravity environments. *J. Eur. Ceram. Soc.* **2022**, *42*, 3968–3975. [CrossRef]
87. Gibson, M.; Mykulowycz, N.; Shim, J.; Fontana, R.; Schmitt, P.; Roberts, A.; Ketkaew, J.; Shao, L.; Chen, W.; Bordeenithikasem, P.; et al. 3D printing metals like thermoplastics: Fused filament fabrication of metallic glasses. *Mater. Today* **2018**, *21*, 697–702. [CrossRef]
88. Liao, C.; Liu MZhang, Q.; Dong, W.; Zhao, R.; Li, B.; Jiao, Z.; Somg, J.; Yao, W.; Zhao, S.; Bai, H.; et al. Low temperature thermoplastic welding of metallic glass ribbons for in-space manufacturing. *Sci. China Mater.* **2020**, *64*, 979–986. [CrossRef]
89. Xu, X.; Zhang, J.; Jiang, P.; Liu, D.; Jia, X.; Wang, X.; Zhou, F. Direct ink writing of aluminium-phosphate-bonded Al_2O_3 ceramic with ultralow dimensional shrinkage. *Ceram. Int.* **2022**, *48*, 864–871. [CrossRef]
90. Revelo, C.; Colorado, H. 3D printing of kaolinite clay ceramics using direct ink writing (DIW) technique. *Ceram. Int.* **2018**, *44*, 5673–5682. [CrossRef]
91. Marquez, C.; Mata, J.; Renteria, A.; Gonzalez, D.; Gomez, G.; Lopez, A.; Baca, A.; Nunez, A.; Hassan, S.; Burke, V.; et al. Direct ink-write printing of ceramic clay with an embedded wireless temperature and relative humidity sensor. *Sensors* **2023**, *23*, 3352. [CrossRef] [PubMed]
92. Karl, D.; Kamutzki, F.; Lima, P.; Gili, A.; Duminy, T.; Zocca, A.; Gunster, J.; Gurlo, A. Sintering of ceramics for clay in situ resource utilisation of Mars. *Open Ceram.* **2020**, *3*, 100008. [CrossRef]
93. Tong, W.; Jaw, W.; Pothunuri, L.; Soh, E.; Le Ferrand, H. Easily applicable protocol to formulate inks for extrusion-based 3D printing. *Ceramica* **2022**, *68*, 152–159. [CrossRef]
94. Lazorenko, G.; Kasprzhitskii, A. Geopolymer additive manufacturing: A review. *Addit. Manuf.* **2022**, *55*, 102782. [CrossRef]
95. Ma, S.; Jiang, Z.; Fu, S.; He, P.; Sun, C.; Duan, X.; Jia, D.; Colombo, P.; Zhou, Y. 3D printed lunar regolith simulant-based geopolymer composites with bio-inspired sandwich architectures. *J. Adv. Ceram.* **2023**, *12*, 510–525. [CrossRef]
96. Abdulkareem, S.; Orkuma, G.; Apasi, A. Talc as a substitute for dry lubricant (an overview). *AIP Proc.* **2011**, *1315*, 1400.

97. *TM-2011-217114/AIAA-2011-704*; Thermal Energy for Lunar In Situ Resource Utilisation: Technical Challenges and Technology Opportunities. NASA: Washington, DC, USA, 2011.
98. Zhang, Y.; Shaw, M.; Brooks, G.; Rhamdhani, A.; Guo, C.; Han, Z.; Jackson, T.; Judkins, G. Investigation of heat transfer processes in multi-sized solar-sintered regolith for lunar ISRY program. *Int. J. Heat. Mass. Transf.* **2023**, *214*, 124387. [CrossRef]
99. Meurisse, A.; Makaya, A.; Willsch, C.; Sperl, M. Solar 3D printing of lunar regolith. *Acta Astronaut.* **2018**, *142*, 800–810. [CrossRef]
100. Imhof, B.; Urbina, D.; Weiss, P.; Sperl, M. Advancing solar sintering for building a base on the Moon. In Proceedings of the 67th International Astronautical Congress, Adelaide, Australia, 25–29 September 2017; IAC-17.C2.9.13.x37414.
101. Imhof, B.; Sperl, M.; Urbina, D.; Weiss, P.; Preisinger, C.; Waclavicek, R.; Hoheneder, W.; Meurisse, A.; Fateri, M.; Gobert, T.; et al. Using solar sintering to build infrastructure on the Moon: Latest advancements in the Regolight project. In Proceedings of the 69th International Astronautical Congress, Bremen, Germany, 1–5 October 2018; IAC-18.E5.1.x47746.
102. Fateri, M.; Meurisse, A.; Sperl, M.; Urbina, D.; Madakashira, K.; Govindaraj, S.; Gancet, J.; Imhof, B.; Hoheneder, W.; Waclavicek, R.; et al. Solar sintering for lunar additive manufacturing. *ASCE J. Aerosp. Eng.* **2019**, *32*, 04019101. [CrossRef]
103. Biswas, D.; Cox, T.; Lapp, J. Sintering behaviour of lunar soil heated by indirect and direct concentrated sunlight. In Proceedings of the ASME 16th International Conference on Energy, Philadelphia, PA, USA, 11–13 July 2022; Paper No. ESS2022-81630.
104. Gupta, N.; Dawara, V.; Kumar, A.; Viswanathan, K. Synthetic space bricks from lunar and martian regolith via sintering. *Adv. Space Res.* **2024**, *74*, 3902–3915. [CrossRef]

ceramics

MDPI

Review

Porous Glass for Thermal Insulation in Buildings with a Focus on Sustainable Materials and Technologies: Overview and Challenges

Francesco Baino [1,*] and Pardeep Kumar Gianchandani [1,2]

1 Institute of Materials Physics and Engineering, Department of Applied Science and Technology, Politecnico di Torino, 10129 Torino, Italy; pardeep.gianchandani@polito.it
2 Department of Textile Engineering, Mehran University of Engineering & Technology, Jamshoro 76062, Sindh, Pakistan
* Correspondence: francesco.baino@polito.it

Abstract: In response to environmental challenges and primary resource scarcity, sustainable approaches that rely on recycling and reusing waste materials are becoming valuable and highly appealing options in modern society. This paper deals with the usage of porous glass and glass-ceramic products derived from waste in the field of thermal insulation in buildings. After providing an overview of the current state of the art with a focus on existing commercial products and related manufacturing methods (foaming strategies), this review discusses the emerging trends toward greener approaches, including the use of by-products or waste substances as foaming agents (e.g., eggshells or mining residues), the use of vitrified bottom or fly ashes from municipal solid waste incinerators as starting materials, the application of surface treatment to reduce post-processing temperatures, and the promise of additive manufacturing technologies in this field. The increased use and spread of sustainable practices are expected to significantly contribute to glass recycling, to minimize landfilling, and to generally reduce energy consumption as well as greenhouse emissions.

Keywords: glass; glass-ceramic; waste; porosity; foaming; additive manufacturing

1. Introduction

The use of properly designed thermal insulating materials in residential and commercial buildings not only contributes to create a more comfortable living and working environment, but it has also been widely recognized as a conceptually easy strategy to reduce energy losses. This material-based, "inherent" (or "passive") type of thermal insulation is useful for potentially reducing energy losses over the whole year while maximizing the efficiency of cooling and heating "active" systems (plants) [1]. Furthermore, the use of proper thermal insulating materials in construction carries two additional advantages, i.e., (i) the decrease in running cost of electricity consumed by cooling and heating systems, and (ii) the reduction of the fixed costs related to the initial installation of cooling and heating plants [2]. It is worth underling that, besides in carbon emissions, with a positive impact on the environment [3]. In this regard, electricity generation stands as the top contributor to global CO_2 emissions. According to the 2023 report published by the Energy Information Administration [4], electricity generation accounts for about one-third of global carbon emissions, and residential consumption of purchased electricity is forecast to increase between about 14 and 22% from 2022 to 2050 across all cases—including the heating and cooling domestic systems reaching between 5.9 and 6.3 quadrillion British thermal units (quads).

Therefore, the development of thermal insulating materials able to tackle the above-mentioned challenges will be of utmost importance over the next few decades.

Many types of thermal insulating materials for buildings are available on the market and/or have been experimentally proposed; according to their nature, they are typically classified in six major groups (Figure 1) [5]:

(1) Organic materials;
(2) Inorganic materials;
(3) Metallic or metallized reflective membranes;
(4) Aerogels;
(5) Thermal insulators from waste;
(6) Composite materials.

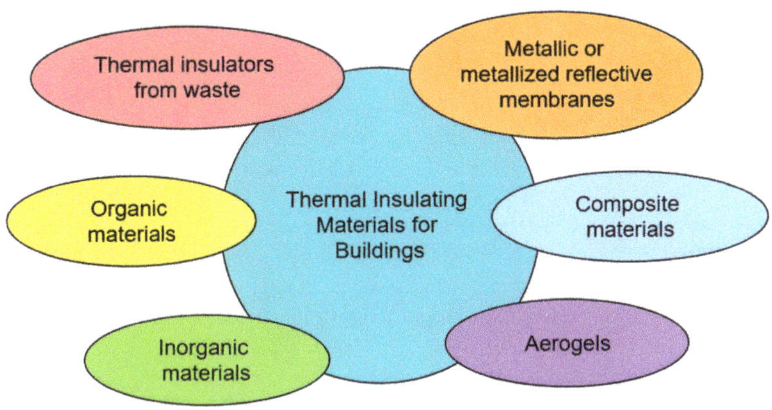

Figure 1. Classification of thermal insulating materials for building.

It should be underlined that such a classification, although being commonly accepted, suffers from some overlaps among the groups. For example, just to mention a few, materials in class (4) formally belong to class (2) (because aerogels are actually silicate materials); class (5) materials can encompass materials from classes (1) (e.g., natural waste polymers, like vegetal fibers from agri-food industry) and (2) (e.g., waste glass or vitrified bottom/fly ashes from municipal solid waste incinerators (MSWIs)).

If we want to adopt a simplified categorization, most of the available thermal insulating materials for building can substantially fall under the following basic types: inorganic materials (e.g., glass/rock wool, ceramic and glass products), organic materials (natural polymers such as cellulose, cotton, wood, cane, etc., and synthetic polymers such as polystyrene, polyethylene, polyurethane, etc.), and composites. Polymers are generally known to be good thermal insulating materials due to their relatively stable physical and chemical properties as well as versatility of processing and shaping. However, mechanical properties are low and make these materials unsuitable for concurrent load-bearing applications, which require improvement by the addition of inorganic fillers to enhance strength or the design of composite products [6].

There are also some advanced thermal insulating materials/systems, including the vacuum insulation panels (VIPs), aerogels, and nanomaterials like nanocellulose, which show great promise for use in building applications. These materials, however, still need to be optimized as they suffer from various drawbacks such as low mechanical properties, high cost, and potential toxicity. Major limitations of all VIPs include high cost, lack of flexibility, and possible changes in thermal conductivity under severe boundary conditions. In fact, air and moisture diffusion through the barrier envelope and into the core of the panel can lead to an irreversible increase in thermal conductivity over time, which is otherwise very low (0.004 W/(m·K)) [7]. VIPs have a long service life in perfect conditions (>50 years) but also exhibit a fragile nature and require protection against perforation of the foil, which carries additional costs.

Aerogels are produced using the sol–gel method and exhibit a unique microstructure of nano-sized particles with mesopores having diameters ranging from 2 to 50 nm and ultra-high porosity (>99 vol.%). Aerogels can also be used as a core for VIPs [8]. The thermal conductivity of aerogels was found to be quite low, ranging between 0.017 and 0.04 W/(m·K) [9]. Despite these attractive properties, the widespread commercial use of aerogels is limited due to the high production cost. The expensiveness is mainly due to the cost of raw materials (high-purity chemicals) and supercritical drying processes, the latter also carries a healthy risk for operators. On the other hand, advancements in aerogel production technologies as well as the push toward mass production suggest that aerogel thermal insulating products will be fully competitive with conventional thermal insulating materials in the coming years.

Some studies have also been performed to design thermal insulating materials starting from agri-food and industrial waste. This approach has some important added value. First, cost reductions of the thermal insulating materials can be inherently achieved by using natural materials and/or wastes as a part of the main matrix. Second, this also implies an obvious reduction of CO_2 emission. Finally, a possible solution to the waste management problem is found by reusing the waste materials as an alternative to combustion or landfilling [10]. Different wastes like agri-food residues, textile waste, scrap rubber, and waste paper can be blended together and structured into cheap products for thermal and even acoustic insulation, with performances comparable to those of commercial options. Specifically, the use of nanomaterials, like nanocelluloses prepared from wood biomass, is another increasing trend in the design of thermal insulating materials [11]. A key challenge to tackle in the case of biopolymer-based thermal insulating materials is their poor resistance to fire and their moisture sensitivity. Furthermore, mechanical properties are often unsatisfactory if these materials are used alone.

Although the above-mentioned advanced solutions for thermal insulation in buildings are indeed fascinating and promising, there is still room for improvement in the field of more conventional materials, too. Within the class of inorganic materials for thermal insulation, glass and glass-ceramic foams provide a unique combination of attractive properties, including low density and low thermal conductivity that are comparable to those of insulating polymers, accompanied by higher durability and mechanical properties (especially strength) [12]. Using recycled or, more generally, glass derived from waste is highly appealing in the circular economy approach. It helps to reduce the consumption of energy and resources as well as the greenhouse emissions associated with the production of "primary" glass by melting. In fact, it was estimated that 16.9 MJ of waste heat and 0.57 kg of CO_2 are produced for the processing of 1 kg of clear glass [13].

This review article aims at providing a comprehensive picture of the state of the art and the challenges ahead related to "sustainable" porous glass products for thermal insulation in construction. Porous glass-ceramics obtained, for example, through sinter-crystallization of the waste-derived parent material are also considered. A comparison with the properties of ceramic materials used for the same purpose will also be reported for the sake of completeness.

2. Factors Affecting the Functional Properties

2.1. Porosity and Porous Microstructure

The relationship between the pore structures and the key properties of porous glass is a complex interplay that significantly influences thermal conductivity, mechanical strength, and fire resistance. Porous glass, characterized by its own pore size, porosity, pore shape, and distribution, exhibits varied properties based on these structural parameters. Understanding these relationships is crucial for optimizing porous glass for specific applications.

With regard to thermal properties, higher porosity generally leads to lower thermal conductivity due to increased air content, which acts as an insulator. However, the size of the pores also plays a critical role. Smaller pores can reduce thermal conductivity more effectively than larger ones, as they disrupt heat flow more efficiently [14,15].

The shape and distribution of pores affect the path of heat flow. Spherical and evenly distributed pores tend to lower thermal conductivity more than irregularly shaped or unevenly distributed pores [16,17].

In general, porous ceramics may exhibit a wide range of thermal conductivity values, influenced by factors such as porosity, pore size, and material composition (see also Section 2.2). Some quantifications are presented in Table 1. Porous silicon carbide (SiC) ceramics, for instance, have thermal conductivities ranging from 37.9 to 5.8 W/(m·K)) as porosity increases from 30% to 63% [18]. It should be noted that SiC has a significantly higher conductivity compared to the majority of ceramics, which makes it an exception in this class of materials. On the contrary, the thermal conductivity of porous anorthite ceramics is notably low, between 0.018 and 0.13 W/(m·K), with porosities ranging from 69% to 91% [19]. Similarly, porous mullite ceramics demonstrate thermal conductivities as low as 0.09 W/(m·K) at high porosities of 73 to 86 vol.% [20]. Porous alumina ceramics, prepared using starch as a pore-forming agent, show a thermal conductivity that decreases with increasing porosity, fitting a modified exponential relation [21]. Porous yttria-stabilized zirconia (YSZ) ceramics achieve ultra-low thermal conductivities as low as 0.06 W/(m·K), with porosities between 52 and 76 vol.% [22]. The thermal conductivity of fibrous ceramics decreases from 0.18 to 0.06 W/(m·K) as porosity increases from 73 to 90 vol.% [23]. These variations highlight the critical role of porosity and microstructural characteristics in determining the thermal properties of porous ceramics, with higher porosity generally leading to lower thermal conductivity across different ceramic types [24–26].

Focusing on the applications in construction, foamed glasses and cellular ceramics are widely used not only for thermal but also for acoustic insulation in buildings due to their high porosity, which significantly reduces thermal conductivity and enhances sound absorption. The porosity of these ceramic- or glass-based construction materials typically ranges from about 60 to over 96 vol.%, depending on the specific type and preparation method. For instance, high closed porosity foamed ceramics prepared from coal gangue waste exhibit a porosity of 83.95 vol.%, which is beneficial for reducing thermal conductivity to 0.11 W/(m·K) [27]. Similarly, Al_2O_3 hollow glass sphere foam ceramics achieve a porosity above 94%, resulting in a thermal conductivity as low as 0.0244 W/(m·K), classifying them as superinsulating materials [28]. Glass-ceramic foams produced from zeolite-poor rock and eggshells show a density range from 0.54 to 1 g/cm^3, with thermal conductivities between 0.07 and 0.4 W/(m·K) [29]. High-entropy ceramic foams exhibit porosities between 90.13 and 96.13 vol.%, with ultralow thermal conductivities ranging from 0.0343 to 0.0592 W/(m·K) [30]. For sound absorption, cellular ceramic foams with porosities of about 60–75 vol.% are optimized for frequencies between 200 and 4000 Hz, demonstrating that higher porosity generally improves sound absorption performance [31]. Lightweight ceramic foams with porosities from 74.1 to 83.7 vol.% also show enhanced sound absorption, with larger pores providing superior performance [32]. These examples illustrate the versatility and effectiveness of foam glasses and ceramic foams in building insulation applications, with porosity being a critical factor in their thermal and acoustic properties.

Furthermore, it should be pointed out that the fire resistance of porous glass is closely linked to its thermal insulation properties. Materials with lower thermal conductivity due to high porosity and small pore sizes can better resist heat penetration, enhancing fire resistance [14,33].

With regard to mechanical properties, increased porosity typically results in decreased mechanical strength, as the solid matrix supporting the structure is reduced. However, the presence of smaller, well-distributed pores can mitigate this effect by providing a more uniform stress distribution [34,35]. Smaller pores can enhance mechanical strength by preventing the coalescence of cracks, while larger pores may act as stress concentrators, leading to failure under load [33,34]. Highly interconnected pore networks can compromise mechanical integrity, whereas isolated pores may help maintain strength [36,37].

Table 1. Thermal conductivity and porosity values of selected porous glass-based and ceramic materials.

Material	Thermal Conductivity (W/(m·K))	Porosity (vol.%)	Reference
Porous silicon carbide (SiC) ceramics	37.9–5.8	30–63	[18]
Porous anorthite ceramics	0.018–0.13	69–91	[19]
Porous mullite ceramics	0.09	73–86	[20]
Porous yttria-stabilized zirconia (YSZ) ceramics	0.06	52–76	[22]
Highly porous fibrous ceramics	0.18–0.06	73–90	[23]
Glass wool	0.03	-	[38]
Vitrified bottom ash-based porous granules	0.13	-	[39]
40 wt.% coal fly ash and 60 wt.% waste glass with 30 wt.% borax and 0.5 wt.% calcium carbonate	0.36	-	[40]
Aerogel	0.017–0.04	-	[9]
Al_2O_3 hollow glass sphere foam	0.0244	94	[28]
Glass-ceramic foams produced from zeolite-poor rock and eggshells	0.07–0.4	Only the density is reported: 0.54–1 g/cm^3	[29]
High-entropy ceramic foams	0.0343–0.0592	90.13–96.13	[30]

2.2. Material Composition

The presence of certain impurities or additives can influence the fire resistance by altering the thermal stability of the glass matrix [14]. Various studies have demonstrated that the incorporation of specific additives can enhance thermal performance by modifying the microstructure and pore characteristics of the materials.

The addition of nano-sized TiO_2 powders in silica insulation media resulted in a 46% reduction in thermal conductivity at high temperatures due to improved radiation scattering efficiency [41]. This effect deserves to be mentioned for the sake of completeness, although it is appreciable at a significantly higher thermal range (around 800 °C) than the common temperatures reached in buildings. Therefore, it is of interest only in very special cases, such as fire propagation.

Incorporating chlorides into the charge mixture of porous glass ceramics led to a finely porous structure, achieving a minimum thermal conductivity of 0.065 W/(m·K) [42].

The incorporation or inherent presence of Fe_2O_3 in CaO-MgO-Al_2O_3-SiO_2 porous glass-ceramics facilitated the formation of a closed pore structure, which significantly reduced thermal conductivity and enhanced mechanical strength [43]. This effect will be discussed in more detail in the Section 3 of the present review.

The introduction of organic additives in granular glass-ceramic materials improved the insulation properties by optimizing the pore formation process during high-temperature firing [44]. The generation of pores through the burn-off of sacrificial organic phases will be discussed in more detail in Sections 3 and 4.2 of the present review.

While these findings highlight the benefits of specific additives in enhancing thermal insulation, it is essential to consider that excessive impurities may adversely affect the structural integrity and thermal performance of the materials (in other words, higher porosity yields a better thermal insulating effect and worse mechanical properties). Therefore, balancing composition is crucial for optimizing the properties of porous glass structures.

While the above points highlight the direct relationships between pore structures and material properties, it is important to consider the trade-offs involved. For instance, enhancing mechanical strength by reducing porosity often leads to increased thermal conductivity, which may not be desirable for thermal insulation applications. Conversely, optimizing for low thermal conductivity might compromise mechanical strength. Therefore, the design of porous glass must balance these competing requirements to achieve the desired performance for specific applications.

3. Porous Glasses for Thermal Insulation in Buildings: Overview and Market Survey

Porous glasses and glass-ceramics can be used in a plethora of high-tech applications such as filtering and adsorption in agriculture and industry, catalysis, and construction [45]. For use as thermal insulators, porous ceramics and glasses are required to exhibit a thermal conductivity of less than 0.2 W/(m·K) [20,40,46]. Multiple studies indicate that less than 0.2 W/(m·K) is a common requirement for porous ceramics used in thermal insulation. For example, the allowable thermal conductivity range for porous ceramics like Y_2SiO_5 and Al_2O_3 has been reported as 0.081–0.283 W/(m·K) [46], and a conductivity below 0.2 W/(m·K) is typically necessary for insulation applications [20]. In order to obtain such low thermal conductivity, it is essential for the porous ceramic or glass material to contain a large volume of void space (i.e., high porosity).

Perhaps the most commonly used glass-based thermal insulating product is glass wool, in which the porosity is attributed to the void spaces among the fibers. Glass wool is a sustainable material for construction because it can be produced with up to 80% of recycled glass. Glass wool is typically produced from borosilicate glass at a temperature around 1400 °C, where the heated mass is pulled through rotating nozzles, thus creating fibers [47]. According to current European regulations [48], glass wool must be biosoluble in order to avoid the risk of toxicity to humans upon inhalation (carcinogenicity). While glass wool products are relatively well-established and available on the market for decades owing to bio-safety (e.g., [49]) and significantly low thermal conductivity (around 0.03 W/(m·K) [38]), other types of porous glasses are still under investigation and development. In this regard, cellular glass, also known as foamed glass, represents a wide class of products characterized by low density, dictated by their porous or even highly porous structure (60–85 vol.%), which is particularly suitable for thermal and acoustic insulation in buildings. Load-bearing capabilities can be allowed or not depending on the material and pore characteristics. Other remarkable properties include excellent resistance to fire; thermal, chemical, and dimensional stability; and resistance to water and moisture. The higher production cost of foamed glass compared to some polymeric options still represents a limiting factor for the widespread diffusion of these materials, which can however be mitigated by the use of waste materials (e.g., recycled glass).

The alveolar (also called "cellular") structure of foamed glass consists of a dispersion of macropores with dimensions typically varying between 10 μm and 5 mm, divided by interpore glass walls. The type and morphology of the pores, combined with the microstructural characteristics that can be obtained in these materials, are key to determine the potential applications of the cellular glass. Specifically, the pores (or cells) can be either completely isolated from each other, yielding a product with a totally closed porosity, or interconnected, leading to a predominantly open porosity.

Two main processes are typically used to fabricate glass foams: the first relies on the direct introduction of gas into the molten glass, while the second involves the sintering of fine glass powders using viscous flow accompanied by the foaming of the glass mass due to the decomposition or oxidation of specific additives acting as "foaming agents" (e.g., silicon carbide, polymethylmethacrylate, calcium carbonate, etc.). In general, carbonates and sulfates give rise to decomposition reactions, while oxidation takes place between carbon-containing substances (e.g., C, SiC) and oxygen in the atmosphere of the oven. Both reactions cause the emission of gas that forms bubbles inside the viscous mass, thus allowing the foaming process to occur. Concurrently with or after the foaming process, thermal consolidation takes place via sintering. Especially in an industrial scenario, large amounts of starting glasses from various compositions/sources (including waste) are processed. The foaming of glass from waste can be complex due to the tendency of these materials to devitrify during heating (sinter-crystallization). The presence of a porous structure corresponds to a high specific surface area, which promotes surface nucleation of crystal nuclei. As a result, crystallization and sintering/foaming processes may take place simultaneously. In this regard, a comprehensive study was reported by Tulyaganov et al. [50] who produced highly porous (>87 vol.%) glass-ceramic foams using cullet from sheet glass (i.e., broken pieces of clear glass) as starting material and 0.25–1 wt.% SiC as a pore-forming agent (Figure 2). Moderate amounts of alkali earth aluminosilicate glass powders (1–10 wt.%) were also added as they tended to crystallize into diopside and anorthite, thus increasing the compressive strength of the foams up to 2.6 MPa.

Composition	C	SiC	AD
1	99.75	0.25	–
2	99.5	0.5	–
3	99.0	1.0	–
4	98.0	1.0	1.0
5	96.0	1.0	3.0
6	94.0	1.0	5.0
7	89.0	1.0	10.0

Figure 2. Rectified cuboids of porous glass obtained by foaming mixtures of cullet (C) and alkali earth aluminosilicate glass powders (AD) with different amounts of SiC. Seven formulations are considered, as illustrated in the table on the left. Reproduced from [50].

Another study on the use of SiC as a pore-forming agent was reported by Bernardo et al. [51] who produced glass-ceramic foams from waste glass fibers. MnO_2 particles were added to the mixture to promote the oxidation of SiC and, thus, the development of gaseous products at temperatures below 1000 °C.

Problems of microstructural inhomogeneity in the foam can be overcome by adding soda–lime glass to the batch. These additions, which may come from clear recycled glass, allow decreasing viscosity of the melt upon heating and the foaming to occur without being hindered by crystallization.

Cellular glass can also be obtained by foaming processes mediated by the presence of appropriate oxides, with lower costs of production. This is the typical case of glass derived from waste containing a high amount of Fe_2O_3, in which iron has a valence state +3. Ferric oxide plays an important role in foaming processes: in fact, upon being reduced to ferrous oxide (FeO) and passing to a valence state of +2, it releases oxygen that generates large pores in the fired body, thus yielding a density decrease [52]. Iron oxide then acts as an oxidation promoter for carbon-based foaming agents and, depending on its amount in the mixture, can also influence the magnetic and electrical properties of the final product. In this regard, a well-known example is represented by the production process of the "Foamglas®" cellular glass developed by the Pittsburgh Corning Company (now Owens Corning, Toledo, OH, USA), in which a mixture consisting of recycled glass and carbon black as a foaming agent is sintered; an iron-rich glass fraction is also added to promote carbon oxidation [52].

Porosity in cellular glass can be open and interconnected or closed with isolated pores; in the former case, glass products are typically permeated by fluids (liquid or gas) and mostly used for filtration, separation processes, etc. On the other hand, glass foams characterized by a predominantly closed porosity are more commonly used for thermal and acoustic insulation. They resist compressive loads and do not allow water or other liquids to penetrate inside, making them also useful against the proliferation of fungi, mold, and other micro-organisms. Foamglas® belongs to the latter category and can be processed to make sheets, panels, and other special shapes to perform particular functions. It has to be noted, however, that the effective thermal conductivity of porous structures with open pores is smaller than that of porous structure with closed pores due to the larger thermal resistance of the solid matrix when the porosity and the thermal conductivity of the solid matrix are the same [53]. In general, glass foams are incombustible, which is an advantage over polymeric insulators.

In the Foamglas® production process, the mixture made up of recycled glass, and "new" glass is pulverized in a ball mill together with the foaming agent (carbon black). Then, the mix is cast into molds to fill them partially and, subsequently, thermally treated inside an oven at a temperature of 850 °C to allow foaming and sintering processes to occur. Upon completion of this operation, the glass foam blocks are passed through a cooling oven where they undergo a long annealing treatment in order to eliminate internal thermal stress and stabilize the product; the blocks are then cut to obtain the final shape/size and eventually packed and stored for selling and commercialization.

In summary, the main features of commercial Foamglas®—which should also be ideally exhibited by any other thermal insulating porous glass-derived product—include the following:

- Impermeability to water (it does not absorb humidity and does not swell, remaining dimensionally/geometrically stable over time);
- Incombustibility (in case of fire, it does not spread flames, burn-off, or develop harmful gases);
- High thermal insulation (or low thermal conductivity);
- Resistance to parasites, rodents, microorganisms, and bacteria (this is due to its inorganic nature and can be an advantage over polymers);
- Constant compressive strength, even over a long period, and possible use also in load-bearing conditions (structural applications);
- Dimensional stability to variations in temperature (low coefficient of thermal expansion) and humidity, so that no cracks or shrinkage are generated;
- Resistance to acids and organic solvents;
- Ease of processing, cutting, and shaping;

- Recyclability (glass foams keep their properties unaltered for long periods of time, so they can be reused even after disposal as a filler or insulating granulate);
- Very low density (between 100 and 170 kg/m^3).

In addition to the Foamglas® products described above, there are other materials under commercial development belonging to the porous glass family and obtained from waste, and the fields of application for these materials are of considerable interest for the building and industrial sectors.

Specifically, Misapor® products (Zizers, Switzerland) are not manufactured in panels but in the form of cellular glass gravel, with piece sizes ranging from 5–10 mm to 50–75 mm. The foamed glass gravel is obtained through a continuous process in which the starting recycled glass is crushed to reach particle dimensions between 75 and 150 μm that are subsequently mixed with the foaming agent. The prepared mixture is then passed through an oven at a temperature of about 950 °C on a conveyor belt to be finally crushed in the form of porous gravel. An interesting field of application of Misapor® products is the introduction of foamed gravel in concrete mixes to obtain lightweight building materials with both structural and thermal insulating properties.

The company Liaver (Ilmenau, Germany) also developed the Reapor® product, which is a panel obtained by sintering glass foam granules (or pellets). It is produced by inserting the starting mixture into a pelletizer that feeds a rotary kiln where the foaming of the granules takes place at temperatures between 750 and 900 °C; the obtained glass foamed granules can be used in the production of insulating panels by sintering the glass foamed granules together. Due to its acoustic insulation properties, Reapor® also finds various applications in the field of industrial plants and machines as well as a motorway barrier.

The major fields of application of commercial porous materials deriving from recycled glass are summarized in Table 2. An overview of these commercial products is also displayed in Figure 3.

Table 2. Cellular glass products and applications.

Product	Applications
Foamglas® (without mechanical loads)	- Facades - Internal insulation
Foamglas® (with mechanical loads)	- Flat and sloping roofs - Facades - Insulation of floors and perimeters - Metal roofs and special roofs - Internal insulation for walls and ceilings
Misapor®	- Coupling with concrete - Perimeter insulation and foundation slabs - Roof insulation - Vertical insulation of walls in contact with the ground
Reapor®	- Fire insulation - Insulation of railway tunnels and ventilation shafts - Insulation of industrial plants and machines (thermal and acoustic) - Indoor insulation (thermal and acoustic) - Highway barriers (thermal and acoustic)

Figure 3. Cellular glasses for building: Foamglas® panels and products (**top left**); Reapor® products in the form of single porous granule (**top right**), pellets (**bottom left**), and sintered panel (**bottom right**).

The major advantages of cellular glasses with closed pores emerge when compared to traditional insulation materials, such as polymeric foams (polystyrene and polyurethanes), which are flammable and exhibit poor structural strength. Furthermore, cellular glass allows minimizing some criticalities, such as the formation of thermal bridges in the insulation of facades or foundations. In addition, being impermeable and having high resistance to water vapor diffusion allows insulating internal environments without the risk of condensation and formation of mold, thereby avoiding damage from humidity and infiltrations. Finally, the low density of cellular glass also carries additional advantages such as design flexibility, lower construction times and costs, reduced manual processing, and lower transportation costs as compared to other massive building materials like stones or concrete.

4. Toward the Future: New Sources of Waste Materials, Technologies, and Approaches

4.1. The Potential of Vitrified Ashes from MSWIs

It cannot be ignored that, at present, the environmental impact associated with the fabrication of "conventional" ceramic or glass insulating foams may still be quite high due to the overall energy consumption and the inherent cost of raw materials to be used, thus leaving room for further improvement [54]. In the particular case of Foamglas®, for example, the advantage of reusing recycled glass is counterbalanced by (i) the addition of a secondary glass fraction derived from an expensive melting process in order to incorporate ferric oxide acting as an oxidizer and (ii) the need for a thermal treatment of foaming and sintering. The additional component (i) is useful in providing oxygen by being converted into ferrous oxide, in turn reacting with carbon black that is used as a foaming agent. Oxidizing compounds may be used as additive in mixtures of glass and foaming agent, but they still imply additional costs and a strict control of the process [55].

In order to maximize the environmental benefits related to using glass-based foams for thermal insulation, focus can be directed to the starting material to be foamed. Raw materials can encompass recycled glass from different sources or even vitrified bottom and/or fly ashes. The composition of the ashes—and, hence, of the vitrified product—may vary within a large range depending on the source of waste that is incinerated; there is also a significant variability depending on the geographic location (i.e., from country to country). Significant amounts of bottom ashes are continuously produced by MSWIs and then often confined in controlled landfills [12]. An interesting alternative to landfilling involves bottom ash vitrification, which allows reducing the volume of these residues significantly (up to 80%), thus extending the life of landfills, and makes the melted glass slag chemically stable and non-toxic [56]. Bottom ashes are essentially formed by metal oxides and, therefore, can be considered as suitable raw materials for the preparation of glass. In the field of construction materials, vitrified bottom ashes have been reused to produce glass-ceramic tiles for decorative applications [57], glass fibers [58], and low-cost granules for lightweight concrete [59], as well as to partially replace ordinary Portland cement and/or sand in mortars [60]. The last strategy is noteworthy as the production of ordinary Portland cement, which is the most commonly used binder worldwide in concrete and mortar design, accounts for 5 to 7% of global CO_2 emissions. Therefore, policies contributing to the reduction or replacement of cement consumption would have a significant impact on reducing the greenhouse gases [61]; comprehensive overviews of this topic were recently reported in [62,63]. Vitrified bottom ash-based porous granules with mechanical strength higher than that of Foamglas® (2 vs. 1 MPa) and thermal conductivity around 0.13 W/(m·K) have been recently produced using a foaming process, in which vitrified Cinderlite® particles (Iris Ambiente Srl, Conselve, Italy) were mixed with borax and calcium carbonate acting as fluxing and foaming agent, respectively [39]. Uniaxially pressed cylindrical compacts of powders (diameter 10 mm) were thermally treated at 820 °C to remove the pore-forming agents and sinter-crystallize the glass particles, thus obtaining porous glass-ceramics. Bloating of iron-containing vitrified bottom ashes from MSWIs was also reported, using an approach similar to that adopted for producing commercial Foamglas® [64].

The use of fly ashes, although being more problematic compared to bottom ashes due to the higher content of leachable heavy metals (e.g., Zn, Pb, Cd, Cr, etc.), has also been considered as an additional fraction in the manufacturing of ceramic or glass-ceramic foams mainly consisting of recycled glass; a good overview of this topic was recently provided in [65]. Fernandes et al. [66] fabricated highly porous (82–84 vol.%) and relatively strong (compressive strength 2.40–2.80 MPa) glass-ceramic foams using a mixture containing around 80 wt.% of sheet glass and 20 wt.% of fly ash with low added amounts of carbonates (1–2 wt.%) as foaming agents. The sintering temperature was set in the range of 750 to 950 °C, which was above the softening point of glass, thereby ensuring the necessary low viscosity for effective foaming, and was also consistent with the decomposition temperatures of carbonates. However, thermal functional properties were not assessed in that work.

In another study, Zhu et al. [67] explored the simultaneous use of borax and calcium carbonate as fluxing and foaming agents, respectively, in a mixture of coal fly ash and waste glass. In an effort to study the influence of processing parameters on the properties of the final porous product, it was observed that the bulk density and compressive strength of porous ceramics increased with the increment of sintering temperature, while the apparent porosity decreased, as expected (Figure 4). The best results, in term of mechanical properties (compressive strength > 5 MPa with bulk density of 0.46 g/cm^3) and low thermal conductivity (0.36 W/(m·K)) of the glass-ceramic foams, were achieved using 40 wt.% coal fly ash and 60 wt.% waste glass in the mixture, along with 30 wt.% borax and 0.5 wt.% calcium carbonate, with sintering at 800 °C for 45 min. However, thermal conductivity was above the threshold (0.2 W/(m·K) [40]) recommended for good thermal insulating materials.

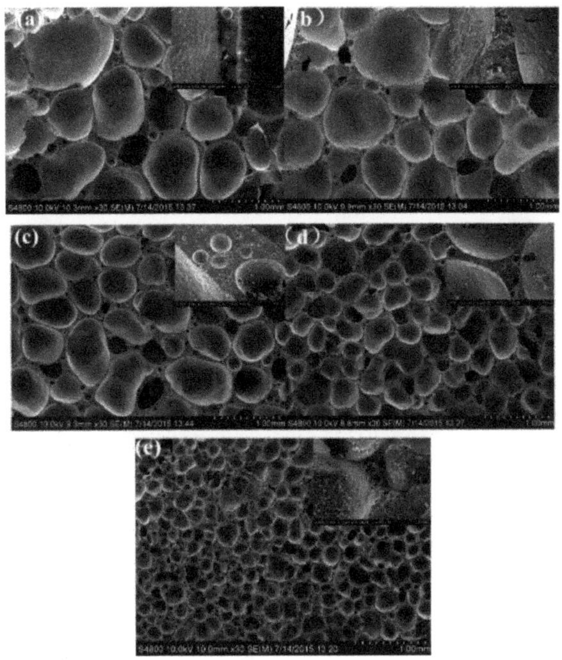

Figure 4. SEM images of waste glass-based foams sintered at 800 °C with different coal fly ash content: (**a**) 35 wt.%, (**b**) 40 wt.%, (**c**) 45 wt.%, (**d**) 50 wt.%, and (**e**) 55 wt.%. Reproduced from [67].

Another key factor to be considered is the particle size of ashes, which can significantly affect the sinterability and final properties of sintered glass-ceramics. It was shown that the sintering of smaller fly ash particles was easier compared to that for larger ones as the activation energy of the former was lower [68]. However, regardless of particle size, the flexural strength increased while the porosity and water absorption of sintered bodies decreased with increasing sintering temperature. It was also reported that when the fly ash size decreased, the linear shrinkage, bulk density, compressive strength, acid resistance, and thermal conductivity of lightweight insulation materials increased, but the apparent porosity decreased [69].

Therefore, using vitrified waste materials can be of interest for the production of thermal insulating glass or glass-ceramic components for construction applications. In order to skip the vitrification process, the use of ashes as-such was also proposed. For example, Li et al. [70] produced porous insulating ceramics by gel-cast foaming using fly ash as the raw material followed by thermal treatment at 1300 °C for 2 h. Compressive strength and thermal conductivity were 3.27 MPa and 0.13 W/(m·K), respectively, with the latter being acceptable for thermal insulating applications. However, a high-temperature treatment was still necessary—with relevant energy consumption. In addition, the risk of heavy metal release from the non-vitrified ashes could be a concern. Optimizing the vitrification process by mixing the fly ashes with cullet as a source of silica in an attempt to decrease the treatment temperature is an alternative approach that also allows for obtaining safer products from a toxicological viewpoint [71].

The main features of waste sources and vitrified products from the studies discussed above are provided in Table 3.

Table 3. Selection of studies about MSWI ash vitrification: main features of wastes and vitrified products.

<table>
<tr><th>Type of Waste</th><th>Composition of Waste</th><th>Resulting Vitrified Product</th><th>Composition of the Vitrified Product and Crystalline Phases (If Glass-Ceramic)</th><th>Reference</th></tr>
<tr><td>Glass powder from obsolete cathode ray tubes</td><td>
<table>
<tr><th>Oxide</th><th>Amount (wt.%)</th></tr>
<tr><td>SiO_2</td><td>61.04</td></tr>
<tr><td>Al_2O_3</td><td>2.37</td></tr>
<tr><td>Na_2O</td><td>7.74</td></tr>
<tr><td>K_2O</td><td>7.11</td></tr>
<tr><td>MgO</td><td>0.32</td></tr>
<tr><td>CaO</td><td>0.84</td></tr>
<tr><td>SrO</td><td>7.82</td></tr>
<tr><td>BaO</td><td>9.66</td></tr>
<tr><td>ZrO_2</td><td>1.52</td></tr>
<tr><td>TiO_2</td><td>0.34</td></tr>
<tr><td>Fe_2O_3</td><td>0.18</td></tr>
<tr><td>ZnO</td><td>0.20</td></tr>
<tr><td>PbO</td><td>0.23</td></tr>
<tr><td>Sb_2O_3</td><td>0.43</td></tr>
</table>
</td><td>Glass</td><td>-</td><td>[55]</td></tr>
<tr><td>Different dry bottom ashes (BA) from MSWIs and wastes from an aluminum alloy industry (AW)</td><td>
<table>
<tr><th>Element</th><th>AW (wt.%)</th><th>BA from Bergamo (wt.%)</th><th>BA from Reggio Emilia (wt.%)</th></tr>
<tr><td>Si</td><td>6.2</td><td>19.9</td><td>21.1</td></tr>
<tr><td>Al</td><td>54.6</td><td>9.3</td><td>5.1</td></tr>
<tr><td>Fe</td><td>0.4</td><td>4</td><td>2.7</td></tr>
<tr><td>Cu</td><td>1.4</td><td>0.2</td><td>-</td></tr>
<tr><td>Mn</td><td>0.2</td><td>0.1</td><td>-</td></tr>
<tr><td>Pb</td><td>0.1</td><td>0.16</td><td>0.01</td></tr>
<tr><td>Cr</td><td>0.02</td><td>0.03</td><td>0.03</td></tr>
<tr><td>Ni</td><td>0.05</td><td>0.02</td><td>-</td></tr>
<tr><td>Zn</td><td>0.4</td><td>0.9</td><td>0.26</td></tr>
<tr><td>Ca</td><td>1.3</td><td>11.7</td><td>13.4</td></tr>
<tr><td>Mg</td><td>5.1</td><td>1.4</td><td>1.3</td></tr>
<tr><td>Na</td><td>3.2</td><td>9.6</td><td>3.4</td></tr>
<tr><td>K</td><td>1.1</td><td>1.5</td><td>1.1</td></tr>
<tr><td>Ti</td><td>0.1</td><td>1.5</td><td>0.56</td></tr>
<tr><td>Other</td><td colspan="3">Balance</td></tr>
</table>
</td><td>Glass, glass-matrix composites</td><td>
<table>
<tr><th>Oxide</th><th>BA-derived Glass (wt.%)</th><th>S10 glass (wt.%)</th></tr>
<tr><td>SiO_2</td><td>42.2</td><td>56.6</td></tr>
<tr><td>Al_2O_3</td><td>17.8</td><td>12.4</td></tr>
<tr><td>Fe_2O_3</td><td>3.5</td><td>4.4</td></tr>
<tr><td>CuO</td><td>0.1</td><td>0.2</td></tr>
<tr><td>MnO</td><td>0.1</td><td>0.2</td></tr>
<tr><td>PbO</td><td>0.1</td><td>0.2</td></tr>
<tr><td>Cr_2O_3</td><td>0.1</td><td>0.2</td></tr>
<tr><td>NiO</td><td>2.10^{-1}</td><td>-</td></tr>
<tr><td>ZnO</td><td>0.8</td><td>0.4</td></tr>
<tr><td>CaO</td><td>16.3</td><td>19.9</td></tr>
<tr><td>MgO</td><td>2.0</td><td>2.7</td></tr>
<tr><td>Na_2O</td><td>11.6</td><td>4.2</td></tr>
<tr><td>K_2O</td><td>1.4</td><td>1</td></tr>
<tr><td>TiO_2</td><td>2.0</td><td>1.0</td></tr>
<tr><td>Other</td><td colspan="2">Balance</td></tr>
</table>
</td><td>[57]</td></tr>
<tr><td>Bottom ash (BA) from MSWIs, sludge, cullet glass</td><td>
<table>
<tr><th>Oxide</th><th>BA (wt.%)</th><th>Venice sludge (wt.%)</th><th>Glass cullet (wt.%)</th></tr>
<tr><td>SiO_2</td><td>45.13</td><td>42.3</td><td>70.3</td></tr>
<tr><td>Al_2O_3</td><td>9.37</td><td>13.2</td><td>2.18</td></tr>
<tr><td>CaO</td><td>18.78</td><td>22.5</td><td>9.30</td></tr>
<tr><td>MgO</td><td>2.20</td><td>7.00</td><td>2.13</td></tr>
<tr><td>Na_2O</td><td>4.64</td><td>2.85</td><td>13.32</td></tr>
<tr><td>K2O</td><td>1.29</td><td>0.50</td><td>1.19</td></tr>
<tr><td>BaO</td><td>0.17</td><td>410 ppm</td><td>0.15</td></tr>
<tr><td>B_2O_3</td><td>-</td><td></td><td>0.14</td></tr>
<tr><td>ZrO_2</td><td>0.29</td><td>285 ppm</td><td>0.06</td></tr>
<tr><td>PbO</td><td>0.01</td><td>500 ppm</td><td>0.074</td></tr>
<tr><td>TiO_2</td><td>0.93</td><td>0.72</td><td>0.068</td></tr>
<tr><td>Fe_2O_3</td><td>3.84</td><td>7.50</td><td>0.293</td></tr>
<tr><td>Cr_2O</td><td>0.04</td><td>115 ppm</td><td>0.071</td></tr>
<tr><td>ZnO</td><td>0.32</td><td>0.40</td><td>0.12</td></tr>
<tr><td>SO_3</td><td>3.07</td><td>0.70</td><td>-</td></tr>
<tr><td>SrO</td><td>-</td><td>300 ppm</td><td>-</td></tr>
<tr><td>P_2O_5</td><td>1.25</td><td>0.90</td><td>-</td></tr>
<tr><td>MnO</td><td>-</td><td>0.11</td><td>-</td></tr>
</table>
</td><td>Glass, glass fibers</td><td>-</td><td>[58]</td></tr>
</table>

Table 3. *Cont.*

<table>
<tr><th>Type of Waste</th><th>Composition of Waste</th><th>Resulting Vitrified Product</th><th>Composition of the Vitrified Product and Crystalline Phases (If Glass-Ceramic)</th><th>Reference</th></tr>
<tr><td>Waste glass
(clear and colored)</td><td>
<table>
<tr><th>Element</th><th>Clear Glass (wt.%)</th><th>Colored glass (wt.%)</th></tr>
<tr><td>SiO_2</td><td>68.1</td><td>68.7</td></tr>
<tr><td>Al_2O_3</td><td>0.9</td><td>1.0</td></tr>
<tr><td>Fe_2O_3</td><td>0.6</td><td>0.9</td></tr>
<tr><td>CaO</td><td>14.5</td><td>12.0</td></tr>
<tr><td>MgO</td><td>1.8</td><td>1.8</td></tr>
<tr><td>K_2O</td><td>0.8</td><td>1.0</td></tr>
<tr><td>Na_2O</td><td>12.2</td><td>13.3</td></tr>
<tr><td>SO_3</td><td>0.4</td><td>0.1</td></tr>
</table>
</td><td>Glass, also combined with mortar</td><td>-</td><td>[59]</td></tr>
<tr><td>Bottom ashes from MSWIs
W1: vitrified bottom ashes collected from a Japanese MSWI equipped with a direct melting system (DMS)
W2: bottom ashes collected from Lomello MSWI (Alessandria, Italy) and placed outdoors for 3 months under ambient humidity and CO_2
(this step (weathering) is recommended by a large body of research on the stabilization of heavy metals in bottom ashes)</td><td>-</td><td>Glass, glass-ceramics</td><td>
<table>
<tr><th>Oxide/element</th><th>W1 (wt.%)</th><th>W2 (wt.%)</th></tr>
<tr><td>SiO2</td><td>38.35</td><td>36.72</td></tr>
<tr><td>Al_2O_3</td><td>14.23</td><td>14.82</td></tr>
<tr><td>CaO</td><td>36.99</td><td>20.08</td></tr>
<tr><td>FeO</td><td>-</td><td>4.9</td></tr>
<tr><td>Fe_2O_3</td><td>2.66</td><td>13.8</td></tr>
<tr><td>Na_2O</td><td>2.26</td><td>-</td></tr>
<tr><td>K_2O</td><td>0.51</td><td>-</td></tr>
<tr><td>MgO</td><td>1.94</td><td>2.53</td></tr>
<tr><td>TiO_2</td><td>1.03</td><td>-</td></tr>
<tr><td>Cl^-</td><td>0.09</td><td>0.118</td></tr>
<tr><td>S</td><td>0.3</td><td>-</td></tr>
<tr><td>Al</td><td>-</td><td>4.0</td></tr>
<tr><td>BaO</td><td>0.17</td><td>-</td></tr>
<tr><td>Cr_2O_3</td><td>0.08</td><td>-</td></tr>
<tr><td>CuO</td><td>0.04</td><td>-</td></tr>
<tr><td>MnO</td><td>0.25</td><td>-</td></tr>
<tr><td>NiO</td><td>0.04</td><td>-</td></tr>
<tr><td>P_2O_5</td><td>0.55</td><td>-</td></tr>
<tr><td>SrO</td><td>0.08</td><td>-</td></tr>
<tr><td>ZnO</td><td>0.05</td><td>-</td></tr>
<tr><td>ZrO_2</td><td>0.36</td><td>-</td></tr>
<tr><td>Other</td><td>0.02</td><td>3.03</td></tr>
</table>
W1 contains gehlenite
W2 contains quartz, gehlenite, anorthite, hematite, calcite</td><td>[60]</td></tr>
<tr><td>Dry bottom ash (BA) from MSWIs of Bergamo and Vercelli (Italy) were vitrified</td><td>
<table>
<tr><th>Oxide</th><th>Bergamo BA (wt.%)</th><th>Vercelli BA (wt.%)</th></tr>
<tr><td>SiO_2</td><td>42.5</td><td>40.85</td></tr>
<tr><td>Al_2O_3</td><td>17.6</td><td>13.6</td></tr>
<tr><td>Fe_2O_3</td><td>5.7</td><td>5.7</td></tr>
<tr><td>CuO</td><td>0.2</td><td>0.2</td></tr>
<tr><td>MnO</td><td>0.1</td><td>0.1</td></tr>
<tr><td>PbO</td><td>0.2</td><td>0.3</td></tr>
<tr><td>Cr_2O_3</td><td>4×10^{-2}</td><td>7×10^{-2}</td></tr>
<tr><td>NiO</td><td>2×10^{-2}</td><td>6×10^{-2}</td></tr>
<tr><td>ZnO</td><td>1.1</td><td>0.4</td></tr>
<tr><td>CaO</td><td>16.4</td><td>12.2</td></tr>
<tr><td>MgO</td><td>2.3</td><td>3.1</td></tr>
<tr><td>Na_2O</td><td>12.9</td><td>10.9</td></tr>
<tr><td>K_2O</td><td>1.8</td><td>1.8</td></tr>
<tr><td>TiO_2</td><td>2.5</td><td>1.2</td></tr>
</table>
</td><td>Glass-ceramics</td><td>
<table>
<tr><th>Oxide</th><th>Bergamo BA glass</th><th>Vercelli BA glass</th></tr>
<tr><td>SiO_2</td><td>44.2</td><td>41.1</td></tr>
<tr><td>Al_2O_3</td><td>17.8</td><td>14.0</td></tr>
<tr><td>Fe_2O_3</td><td>3.5</td><td>4.1</td></tr>
<tr><td>CuO</td><td>0.1</td><td>0.4</td></tr>
<tr><td>MnO</td><td>0.1</td><td>0.2</td></tr>
<tr><td>PbO</td><td>0.1</td><td>4×10^{-2}</td></tr>
<tr><td>Cr_2O_3</td><td>0.1</td><td>0.1</td></tr>
<tr><td>NiO</td><td>2×10^{-2}</td><td>1×10^{-2}</td></tr>
<tr><td>ZnO</td><td>0.8</td><td>0.3</td></tr>
<tr><td>CaO</td><td>16.3</td><td>13.9</td></tr>
<tr><td>MgO</td><td>2.0</td><td>4.6</td></tr>
<tr><td>Na_2O</td><td>11.6</td><td>10.3</td></tr>
<tr><td>K_2O</td><td>1.4</td><td>3.2</td></tr>
<tr><td>TiO_2</td><td>2.0</td><td>2.9</td></tr>
</table>
</td><td>[64]</td></tr>
</table>

Table 3. *Cont.*

<table>
<tr><th>Type of Waste</th><th>Composition of Waste</th><th>Resulting Vitrified Product</th><th>Composition of the Vitrified Product and Crystalline Phases (If Glass-Ceramic)</th><th>Reference</th></tr>
<tr>
<td>Fly ash (FA) from thermal power plant
+
cullet of commercially produced sodium-calcium-silicate sheet glass
Commercial dolomite and sludge from a marble cutting-polishing plant consisting of calcite were used as foaming gents</td>
<td><table>
<tr><th>Oxide</th><th>FA (wt.%)</th><th>Glass cullet (wt.%)</th></tr>
<tr><td>SiO_2</td><td>49.6</td><td>70.64</td></tr>
<tr><td>Al_2O_3</td><td>20.9</td><td>0.68</td></tr>
<tr><td>Fe_2O_3</td><td>6.6</td><td>0.18</td></tr>
<tr><td>CaO</td><td>0.5</td><td>9.93</td></tr>
<tr><td>MgO</td><td>1.8</td><td>3.55</td></tr>
<tr><td>Na_2O</td><td>0.4</td><td>13.66</td></tr>
<tr><td>K_2O</td><td>2.7</td><td>0.29</td></tr>
<tr><td>TiO_2</td><td>0.9</td><td>-</td></tr>
<tr><td>MnO</td><td>0.3</td><td>-</td></tr>
<tr><td>P_2O_5</td><td>0.5</td><td>-</td></tr>
<tr><td>LOI</td><td>15.8</td><td>-</td></tr>
<tr><td>SO_3</td><td>-</td><td>0.21</td></tr>
</table></td>
<td>Glass-ceramic</td>
<td>Crystalline phases: quartz, tridymite, pargasite, and augite</td>
<td>[66]</td>
</tr>
<tr>
<td>Coal fly ash (FA) from a thermal power plant
+
waste glass
Borax added as fluxing agent, calcium carbonate chosen as foaming agent</td>
<td><table>
<tr><th>Oxide</th><th>FA (wt.%)</th><th>Waste glass (wt.%)</th></tr>
<tr><td>SiO_2</td><td>35.42</td><td>66.55</td></tr>
<tr><td>Al_2O_3</td><td>39.40</td><td>1.55</td></tr>
<tr><td>Fe_2O_3</td><td>2.36</td><td>0.26</td></tr>
<tr><td>CaO</td><td>10.04</td><td>8.71</td></tr>
<tr><td>MgO</td><td>1.85</td><td>4.55</td></tr>
<tr><td>Na_2O</td><td>0.14</td><td>17.26</td></tr>
<tr><td>K_2O</td><td>0.40</td><td>0.55</td></tr>
<tr><td>TiO_2</td><td>1.19</td><td>0.027</td></tr>
<tr><td>SO_3</td><td>4.26</td><td>0.281</td></tr>
<tr><td>P_2O_5</td><td>0.217</td><td>0.022</td></tr>
<tr><td>LOI</td><td>4.09</td><td>0.24</td></tr>
</table></td>
<td>Glass-ceramics</td>
<td><table>
<tr><th>Oxide</th><th>B35 (wt.%)</th><th>B40 (wt.%)</th><th>B45 (wt.%)</th><th>B50 (wt.%)</th><th>B55 (wt.%)</th></tr>
<tr><td>SiO_2</td><td>42.8</td><td>41.6</td><td>40.4</td><td>39.2</td><td>38.0</td></tr>
<tr><td>Al_2O_3</td><td>11.4</td><td>12.8</td><td>14.3</td><td>15.8</td><td>17.2</td></tr>
<tr><td>Fe_2O_3</td><td>0.8</td><td>0.9</td><td>1.0</td><td>1.1</td><td>1.2</td></tr>
<tr><td>CaO</td><td>7.1</td><td>7.1</td><td>7.2</td><td>7.2</td><td>7.3</td></tr>
<tr><td>MgO</td><td>2.8</td><td>2.7</td><td>2.6</td><td>2.5</td><td>2.4</td></tr>
<tr><td>Na_2O</td><td>8.7</td><td>8.1</td><td>7.4</td><td>6.8</td><td>6.1</td></tr>
<tr><td>K_2O</td><td>0.4</td><td>0.4</td><td>0.4</td><td>0.4</td><td>0.4</td></tr>
<tr><td>TiO_2</td><td>0.3</td><td>0.4</td><td>0.4</td><td>0.5</td><td>0.5</td></tr>
<tr><td>SO_3</td><td>1.4</td><td>1.6</td><td>1.7</td><td>1.9</td><td>2.1</td></tr>
<tr><td>P_2O_5</td><td>0.1</td><td>0.1</td><td>0.1</td><td>0.1</td><td>0.1</td></tr>
<tr><td>B_2O_3</td><td>0.2</td><td>0.2</td><td>0.2</td><td>0.2</td><td>0.2</td></tr>
<tr><td>LOI</td><td>1.2</td><td>1.4</td><td>1.5</td><td>1.7</td><td>1.8</td></tr>
</table>
Crystalline phases: diopside, quartz</td>
<td>[67]</td>
</tr>
<tr>
<td>Coal fly ash (FA) from a thermal power plant in three size fractions:
CFA = 12.6 μm
20CFA = 7.5 μm
40CFA = 4.9 μm</td>
<td><table>
<tr><th>Oxide</th><th>CFA (wt.%)</th><th>20CFA (wt.%)</th><th>40CFA (wt.%)</th></tr>
<tr><td>SiO_2</td><td>53.31</td><td>53.28</td><td>53.34</td></tr>
<tr><td>Al_2O_3</td><td>26.55</td><td>26.58</td><td>26.52</td></tr>
<tr><td>CaO</td><td>5.88</td><td>5.86</td><td>5.85</td></tr>
<tr><td>Fe_2O_3</td><td>4.41</td><td>4.44</td><td>4.40</td></tr>
<tr><td>K_2O</td><td>2.42</td><td>2.40</td><td>2.45</td></tr>
<tr><td>SO_3</td><td>1.42</td><td>1.40</td><td>1.43</td></tr>
<tr><td>TiO_2</td><td>1.06</td><td>1.09</td><td>1.05</td></tr>
<tr><td>Other</td><td>4.95</td><td>4.95</td><td>4.96</td></tr>
<tr><td>LOI</td><td>3.02</td><td>3.03</td><td>3.03</td></tr>
</table></td>
<td>Glass-ceramics</td>
<td><table>
<tr><th>Phase/Element</th><th>Granular crystallite (at %)</th><th>Spherical particle (at %)</th><th>Glassy matrix (at %)</th></tr>
<tr><td>O</td><td>47.98</td><td>48.98</td><td>17.84</td></tr>
<tr><td>Al</td><td>18.56</td><td>25.38</td><td>9.90</td></tr>
<tr><td>Si</td><td>27.37</td><td>25.64</td><td>54.73</td></tr>
<tr><td>Ca</td><td>6.08</td><td>-</td><td>6.49</td></tr>
<tr><td>Fe</td><td>-</td><td>-</td><td>11.04</td></tr>
</table>
Crystalline phases: quartz, mullite and hematite, with amorphous metastable glassy phase forms</td>
<td>[68]</td>
</tr>
<tr>
<td>Fly ash, flint clay, kyanite, clay, saw dust
+
organic binder polyvinyl alcohol (2 wt.% solution)</td>
<td><table>
<tr><th>Comp wt %</th><th>Fly ash</th><th>Flint clay</th><th>Kyanite</th><th>Clay</th></tr>
<tr><td>SiO_2</td><td>31.87</td><td>47.26</td><td>36.34</td><td>28.85</td></tr>
<tr><td>Al_2O_3</td><td>13.98</td><td>47.76</td><td>58.90</td><td>51.39</td></tr>
<tr><td>Fe_2O_3</td><td>1.60</td><td>0.95</td><td>0.91</td><td>1.80</td></tr>
<tr><td>CaO</td><td>0.87</td><td>0.36</td><td>0.02</td><td>0.21</td></tr>
<tr><td>TiO_2</td><td>0.57</td><td>0.82</td><td>0.02</td><td>0.57</td></tr>
<tr><td>IL</td><td>50.0</td><td>0.06</td><td>0.23</td><td>16.12</td></tr>
</table></td>
<td>Glass-ceramics</td>
<td>Crystalline phase: mullite</td>
<td>[69]</td>
</tr>
<tr>
<td>Fly ash floating beads with polycarboxylate (dispersant), dextrin (binder), sodium carboxymethyl cellulose (thickener), and sodium alkyl sulfate (foaming agent)</td>
<td><table>
<tr><th>Oxide</th><th>Fly ash (wt.%)</th></tr>
<tr><td>SiO_2</td><td>58.52</td></tr>
<tr><td>Al_2O_3</td><td>35.02</td></tr>
<tr><td>K_2O</td><td>1.73</td></tr>
<tr><td>Fe_2O_3</td><td>1.34</td></tr>
<tr><td>TiO_2</td><td>0.975</td></tr>
<tr><td>Na_2O</td><td>0.646</td></tr>
<tr><td>MgO</td><td>0.606</td></tr>
<tr><td>SO_3</td><td>0.175</td></tr>
<tr><td>P_2O_5</td><td>0.114</td></tr>
</table></td>
<td>Glass-ceramics</td>
<td>Amount of crystalline phases of porous glass-ceramics prepared at different temperatures
<table>
<tr><th>Temperature (°C)</th><th>Mullite (wt.%)</th><th>Cristobalite (wt.%)</th></tr>
<tr><td>1100</td><td>66</td><td>34</td></tr>
<tr><td>1200</td><td>67</td><td>33</td></tr>
<tr><td>1300</td><td>70</td><td>30</td></tr>
<tr><td>1400</td><td colspan="2" rowspan="2">Amorphous phases</td></tr>
<tr><td>1500</td></tr>
</table></td>
<td>[70]</td>
</tr>
</table>

Table 3. *Cont.*

Type of Waste	Composition of Waste	Resulting Vitrified Product	Composition of the Vitrified Product and Crystalline Phases (If Glass-Ceramic)	Reference
Fly ashes from a waste-to-energy plant from two different sources (FA1, FA2) mixed in 50 wt.% combination with glass cullet and silica sand	See composition of waste below	Glass-ceramics	See composition of vitrified product below	[71]

Composition of Waste:

Oxide/element	FA1 (wt.%)	FA2 (wt.%)	Glass cullet (wt.%)	Silica sand (wt.%)
Na_2O	5.83	5.55	11.38	0
MgO	6.53	2.29	0.93	0
Al_2O_3	22.73	18.19	3.38	0.20
SiO_2	14.02	14.35	64.46	99.75
P_2O_5	0.82	1.23	0	0
Cl^-	14.82	13.72	0	0
K_2O	3.08	3.22	1.97	0
CaO	20.22	32.49	9.73	0.05
TiO_2	0.76	0.76	0.78	0
Cr_2O_3	0.08	0.02	0.01	0
Fe_2O_3	0.56	0.62	0.15	0
CoO	0	0	0	0
NiO	0	0	0	0
CuO	0	0.58	0	0
ZnO	1.30	0.44	0.06	0
Br	0.04	0.10	0	0
SrO	0.02	0.06	0	0
ZrO_2	0	0	6.69	0
CdO	0	0	0	0
SnO2	0.08	0	0	0
Sb_2O_3	0.08	0.04	0	0
HgO	2.40	1.99	0	0
PbO	6.61	4.37	0.45	
Total	100	100	100	100

Composition of the Vitrified Product:

Oxide/element	G1 (wt.%)	G2 (wt.%)	G3 (wt.%)	G4 (wt.%)
Na_2O	2.26	1	3.99	8.82
MgO	5.41	1.36	8.63	2.23
Al_2O_3	13.47	12.9	9.45	23.58
SiO_2	54.04	53.69	43.07	43.2
P_2O_5	1.09	1.59	1.46	1.76
Cl^-	ND	ND	ND	0.01
K_2O	1.21	0.93	0.28	1.01
CaO	15.3	23.18	26.06	12.1
TiO_2	0.94	0.57	1.35	0.67
Cr_2O_3	0	0	0	0.11
Fe_2O_3	0.86	0.75	1.8	0.65
CoO	0.15	ND	0.3	ND
NiO	0.13	0.04	0.32	ND
CuO	0.09	ND	ND	0.52
ZnO	0.06	0.24	0.44	0.13
Br	2.33	2.67	2.1	3.3
SrO	ND	ND	ND	ND
ZrO_2	0.82	0.74	0.52	1.04
CdO	0.28	ND	ND	0.24
SnO2	ND	ND	ND	ND
Sb_2O_3	ND	ND	ND	ND
HgO	0.47	0.32	0.21	0.56
PbO	1.09	0.03	ND	0.09
Total	100	100	100	100

Crystalline phases: wollastonite (traces)

4.2. The Search for "Sustainable" Pore-Forming Agents

The "sustainability" of porous foams from waste glass can be further increased by paying special attention when selecting the foaming additives. At present, one of the most-commonly used is SiC, which, however, may be quite expensive and have a negative impact on the overall cost/benefit balance. On the other hand, SiC could be sourced from the waste originated by the polishing procedures (e.g., waste grit paper) of many other "hard" materials, such as traditional ceramics, glass, and stones [66]. By-products of common industrial processes, such as carbonates and carbon-containing waste, can also be used as foaming agents to be added in glass-based mixes [72,73]. Other inorganic "waste foaming agents" include dolomite (by-product of mining activities) [66] and eggshells [74], which are entirely constituted by $CaCO_3$. In all these cases, foaming should take place in a pyroplastic mass, the viscosity of which is dictated by the softening of glass powders [75] (Figure 5). Therefore, the optimal thermal treatment conditions indeed vary depending on the glass used, and the preferable viscosity range for foaming remains an arguable issue. In fact, different authors described different foaming "windows" depending on the specific material investigated. For example, Fernandes et al. [74] studied how to develop glass foams using pulverized cathode ray tube waste glass and eggshells in the thermal range of 600 to 800 °C and reported that the best condition was when foaming occurred while the melt viscosity was within 10^7–10^8 Pa·s (Figure 6).

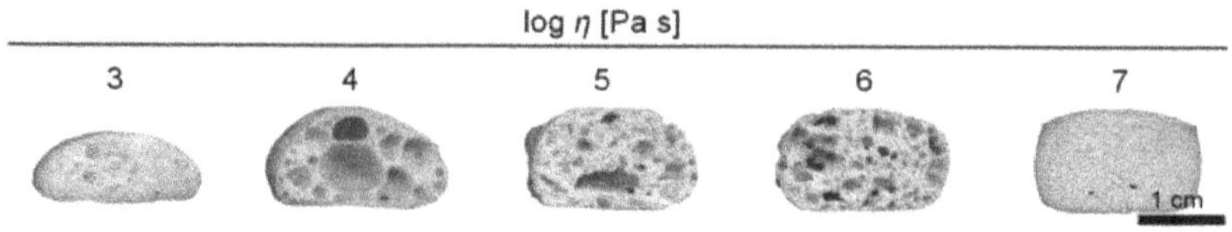

Figure 5. Glass foams prepared from soda–lime–silica glass and 2 wt% $CaCO_3$ at different viscosities (values of log η = 3, 4. . . 7 are reported on the top) Reproduced from [75].

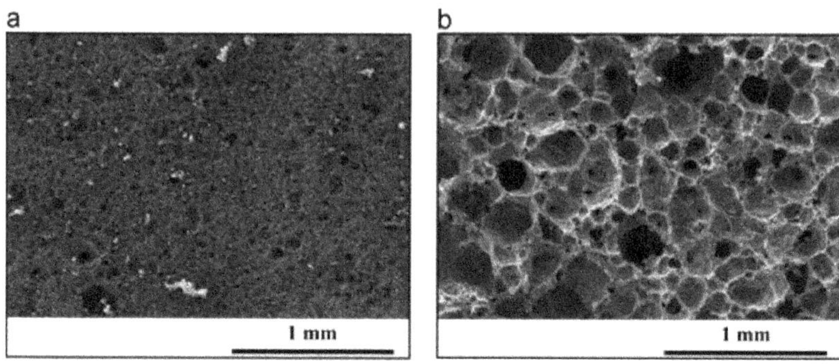

Figure 6. Microstructures of glass foams produced using eggshells as pore-forming agent under thermal treatment at (**a**) 600 °C (small homogeneously distributed pores) and (**b**) 700 °C (larger pores, foam-like appearance). Reproduced from [74].

Using industrial waste, such as ash and slag from power stations, to produce porous glass is recognized a highly interesting approach. The process involves preparing a batch mixture with broken glass, ash, slag, and additional crystallization initiators like chromium oxide and zirconium dioxide. This method results in foam glass with a uniform porous structure and varying crystalline phases, contributing to its thermal insulation properties [76].

Foamed glass with thermal and acoustic insulation properties can be produced from a blend of 97% glass waste and 3% expanding agents, such as carbon and manganese oxides from spent alkaline batteries. This process involves heating the mixture in a ventilated furnace, resulting in glass foam with moderate thermal and acoustic insulation properties (Figure 7) [77].

The use of agro-industrial residues, such as rice husk ash, in combination with waste glass powder, has been shown to produce closed-pore glass foams with excellent thermal insulation properties. The addition of rice husk ash reduces the viscosity of the glass, facilitating pore formation and enhancing the insulating capabilities of the material (Figure 8) [78].

Porous glass-ceramics can be produced from granite dust and maize cob using one-step sintering technology. This method results in materials with high porosity and low thermal conductivity, making them viable for thermal insulation of residential buildings [79].

In addition to inorganic agents, organic substances are also used in foaming procedures. The most effective organic foaming agents used in the production of porous glass for thermal insulation applications include glycerol, eggshell, coffee ground waste, and yerba mate, among others. Glycerol is frequently utilized due to its ability to form a carbon phase during thermal decomposition, which aids in the foaming process by reacting with sulfate ions in the glass to release gases, thus enhancing porosity [80,81]. Additionally, glycerol derived from biodiesel production, which contains impurities such as methanol and fatty acids, has been shown to produce foam glass with favorable thermal properties when combined with sodium carbonate [82]. Natural waste materials like eggshells and coffee grounds also serve as effective foaming agents. Eggshells, when used with soda–lime–silica glass, result in excellent thermal insulation properties, while coffee grounds, although less effective, still contribute to the foaming process [83]. Yerba mate waste, when combined with pulverized glass from green bottles, has been shown to produce glass foams with high porosity and low thermal conductivity, making it a viable option for thermal insulation [84]. Other organic compounds such as acetic acid, gelatin, starch, saccharose, and paraffin have also been explored for their potential as pore-forming agents, with varying degrees of success depending on the specific composition and processing conditions [85].

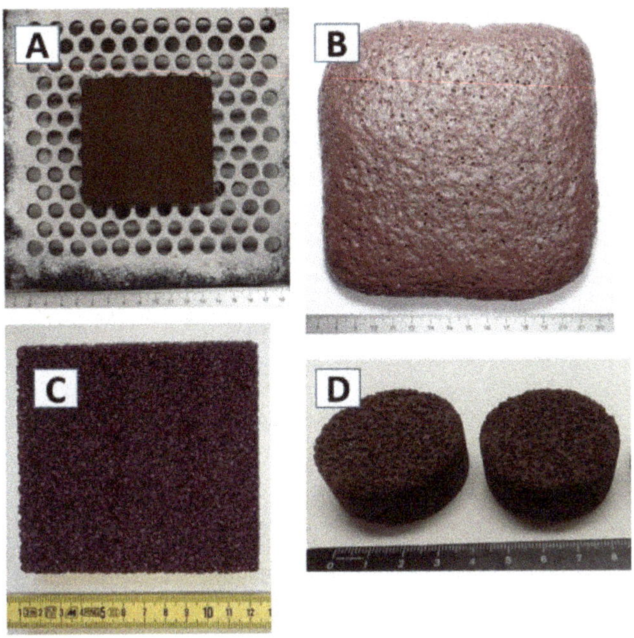

Figure 7. Foaming of waste glass using a mixture of manganese oxides and graphite recovered from the cathodes of exhausted alkaline batteries: green sample before heat treatment (**A**); expanded sample after heat treatment (**B**); square-based cut sample (**C**), and cut cylindrical samples (**D**). Reproduced from [77].

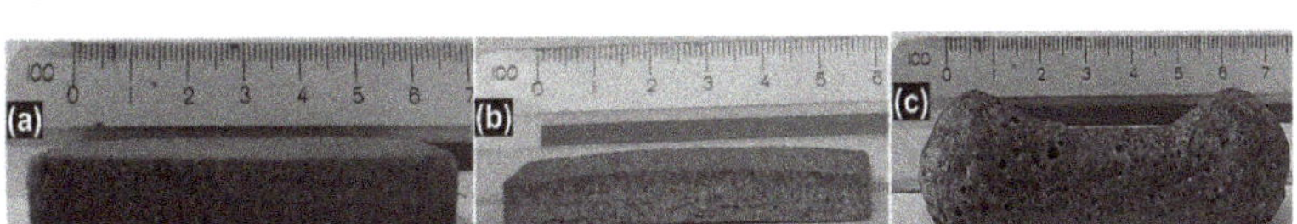

Figure 8. Foamed glass sintered at (**a**) 750 °C, (**b**) 800 °C, and (**c**) 850 °C using rice husk as a pore-forming agent. Reproduced from [78].

Pine scales have also been used as a pore-forming agent in the production of glass foams from waste glass bottles. This method results in foams with high porosity and low thermal conductivity, suitable for thermal insulation applications [86] (Figure 9).

The use of these organic foaming agents not only enhances the thermal insulation properties of foam glass but also supports sustainable practices by utilizing waste materials, thereby contributing to a circular economy [78]. Overall, the choice of foaming agent significantly influences the thermal, mechanical, and structural properties of the resulting foam glass, making it crucial to select the appropriate agent based on the desired application and performance criteria.

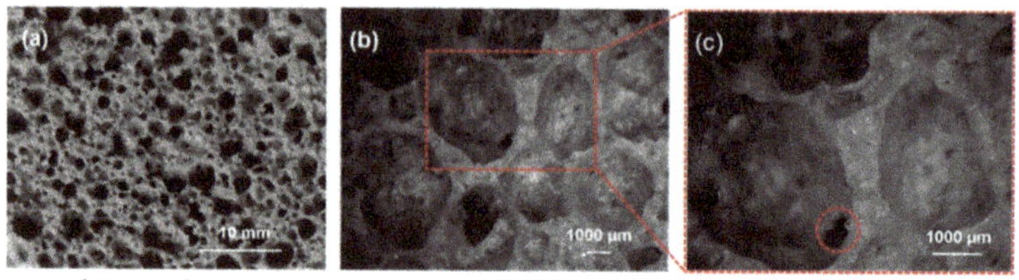

Figure 9. Morphology of a glass foam, at different magnifications, produced using recycled glass and pine scale powder (foaming agent) with 50 vol.% of pine scale. Reproduced from [86].

Looking at the fate of the foaming agents during and after the thermal treatment, it cannot be ignored that all the substances mentioned above will produce CO and CO_2 gases, thus yielding an increased greenhouse effect. In order to avoid this drawback, the use of NaOH as a pore-forming agent in sintered waste glass or glass-ceramic products was proposed as a possible alternative [87]; however, the overall cost behind the production of NaOH should be also considered to definitely assess whether this approach is less impactful on the environment. Mixing secondary aluminum ash with a mixture of bottom/fly ashes and pickling sludge was also found beneficial to gas generation. Bloating was attributed to the oxidation of AlN in the aluminum ash, yielding the development of N_2, NO, and NH_3 when the temperature reached 700 °C, thereby allowing the production of porous glass-ceramics [88].

4.3. The Potential of Alkali-Activation Treatment

The fact that the foaming and consolidation (sintering) processes are concurrent phenomena may give rise to some criticalities, such as the difficult control of the final pore shape/size; hence, the functional properties (e.g., mechanical behavior, permeability, thermal conductivity) are dictated by the pore characteristics. Furthermore, if vitrified ashes are used—for the production of which a high-energy vitrification process was previously applied—the method of transformation into porous glass or glass-ceramic products should be particularly simple and affordable to comply with the circular economy paradigm.

In this regard, a new foaming approach was proposed relying on the alkali activation of glass or glass-containing mixtures. For example, porous glass microspheres can be fabricated from alkali-activated fiber glass waste. The activation process involves using highly concentrated alkali solutions, which promote pore formation during heat treatment, resulting in microspheres with potential applications in thermal insulation [89]. The alkali activation of glass powder aqueous suspensions [90,91] or glass/slag mixtures [92,93] was also reported. This method comprised two phases: (i) first, a controlled alkaline attack (e.g., using KOH) was applied to the glass particles, and, thus, the suspensions underwent progressive hardening through the partial dissolution of glass and the formation of calcium-silicate-hydrated (C-S-H) compounds as the binding phase; and (ii) then, before reaching complete setting, the mixtures were foamed by vigorous mechanical stirring with the help of a surfactant. These "green" foams eventually underwent sintering or sinter-crystallization (if the glass was prone to surface nucleation). Shrinkage, geometrical alterations, and pore reshaping may take place during sintering. However, if crystallization is important and concurrent to densification, the increase in the viscosity of the softened glass caused by crystal inclusions can "freeze" the microstructural evolution; thus, the open-cell 3D architecture of the porous greens is preserved in the final sintered samples. The pore size and shape can be properly designed by varying the solid/liquid ratio in the suspension, by setting the parameters of mechanical stirring at low temperature (foaming process), and/or by applying a "pre-curing" treatment.

In summary, sinter-crystallization is decoupled from the foaming process and can occur at relatively moderate temperatures (800–900 °C). This strategy was successfully applied to vitrified bottom ashes from MSWIs (Figure 10) [94], blends of waste soda–lime glass, porcelain stoneware residue, plasma-processed asbestos-containing waste [95], and vitrified mixtures of fly ashes/bottom ash slag [96]. In all these cases, the compressive strength (2–6 MPa) was suitable for applications in construction, and the leaching tests revealed a very low release of potentially toxic heavy metals, thus fulfilling the international recommendations for the safe use of these materials [97]. However, thermal properties were not reported and should be investigated in the future.

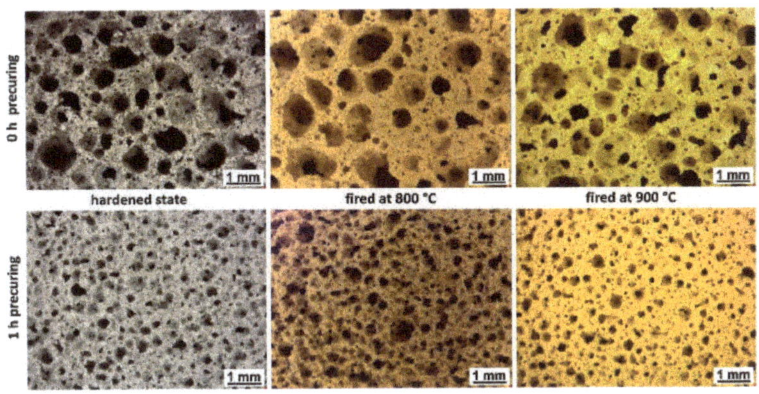

Figure 10. Microstructural morphology of foams from vitrified bottom ashes in the hardened state and after firing treatments (the images report the actual colors of the samples). Reproduced from [94].

4.4. *The Potential of Additive Manufacturing*

As previously underlined, the total porosity, pore size distribution, and pore interconnectivity play a key role in dictating the final properties of porous ceramics and glasses. Thus, careful design and control of all these aspects should be performed. Direct foaming allows tailoring, to some extent, the porous structure (e.g., pore size and distribution), but this process cannot produce porous glasses or glass-ceramics with complex and well-defined shapes. Furthermore, geometrical variations (volumetric expansion) due to foaming may also be not fully predictable and controllable, unless a closed mold is used (like in the production of Foamglas®). The foams produced typically require additional post-sintering steps (e.g., cutting, polishing) to fit the desired geometry [98]. In this regard, additive manufacturing has emerged over the last years as a valuable group of technologies for producing three-dimensional (3D) components with complex shapes, customizable microstructures, and precise control of their internal architecture as well as overall shape and size. The final geometry of 3D-printed products is regular, eliminating the need for rectification/cutting and ultimately reducing the volume of waste generated. In a typical approach, the component is first designed in CAD (computer-aided design) and then given as an input virtual model to the printing system (CAM—computer-aided manufacturing). A wide variety of 3D printing techniques are available to ceramics/glass manufacturers, including selective laser sintering, robocasting, and vat photopolymerization [99]. Indeed, additive manufacturing can enter the circular economy approach through the use of recycled materials, including ceramics, glasses, and composites thereof, as well as sustainable approaches. A comprehensive overview of this specific topic was recently provided by Villa et al. [100].

While some special, high-added-value glasses (e.g., bioactive glasses [101]) have been processed using additive manufacturing and investigated in both the literature and industry, attempts to print recycled or, in general, waste glass are relatively scarce. Most of relevant studies use waste glass as a replacement for natural aggregates [102] or cement paste [103] in 3D-printed concrete. Both approaches are valuable as the former allows saving natural resources (sand, crushed stones) and the latter is useful to reduce the energy consumption and CO_2 emissions associated with cement production. Pulverized cullet was also used as a precursor material in the 3D printing of geopolymer concrete based on fly ashes and slag because of its beneficial influence on the geo-polymerization rate and extrudability [104]. Other more sophisticated examples include the extrusion of suspensions of ground waste glass, gelatin, and agar [105] as well as borosilicate glass and Pluronic F127 [106] to fabricate porous glass panels and rings, respectively, using robocasting.

With regard to porous glass, Guzi de Moraes et al. [107] reported the production of foam-like glass structures with robocasting, using bentonite (6.5–9.5 wt.%) to overcome the problem of low plasticity of recycled glass powders (median particle size <4 μm). After being fired at 700 °C for 1 h, the materials exhibited total porosity of up to 53 vol.%, pore sizes in the range of 1090 to 1870 μm, and a compressive strength of 18 MPa (Figure 11); however, thermal conductivity was not assessed. Therefore, additive manufacturing deserves to be further investigated in the field of thermal insulating materials for building as a valuable strategy to produce ceramic, glass, glass-ceramic products, or even waste glass-containing (polymer-matrix or polymer-bonded) composites.

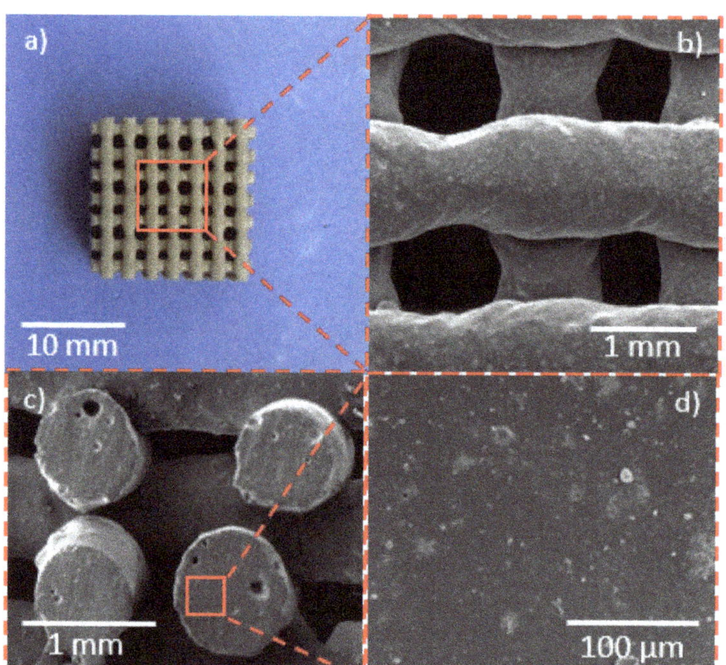

Figure 11. Morphology of robocast recycled glass-based porous structures containing 8.5 wt.% of bentonite: (**a**) overview of the 3D-printed architecture, (**b**,**c**) SEM micrographs showing the outer surface and internal cross-sections of filaments, and (**d**) internal microstructure of the filaments. Reproduced from [107].

While these innovative fabrication techniques offer promising solutions for producing thermal insulation materials from waste, challenges remain in optimizing the mechanical properties and scalability of these porous materials for widespread use. Additionally, the environmental impact of the production processes and the long-term performance of the materials in real-world applications require further investigation. Nonetheless, the use of waste materials in manufacturing porous glass for thermal insulation aligns with sustainable development goals and offers a pathway toward more environmentally friendly construction practices.

5. Conclusions and Challenges

Glass is a highly versatile material used in our lives, the applications of which range from daily to high-tech applications. Glass is fully recyclable as it can be reprocessed indefinitely without undergoing any significant loss of performance. The transformation of cullet or even vitrified bottom/fly ashes from MSWIs into porous products for thermal insulation is convenient provided that the additional processing is relatively simple and cheap. Furthermore, if we look at the circular economy approach, some specific considerations are necessary.

Innovative techniques for manufacturing porous glass from waste materials for thermal insulation in buildings have been explored extensively in recent research. These techniques focus on utilizing various waste materials, such as industrial by-products, glass waste, and agro-industrial residues, to produce foam glass and other porous materials with desirable thermal insulation properties. The use of waste materials not only reduces production costs but also addresses environmental concerns by promoting recycling and reducing landfill waste.

The common methods used to produce porous glass or glass-ceramics for thermal insulation—even at an industrial level—belong to the family of foaming strategies, which always require a high-temperature consolidation step (sintering of glass particles). This additional energy consumption can be justified by the high added value of the final product; moreover, the temperatures used for sintering are typically lower than those required for direct re-melting of recycled glass. Recent studies have proposed the use of vitrified bottom or fly ashes as starting materials for the foaming process; in this case, the large energy consumption associated with the vitrification process is somehow balanced by the advantages of reducing landfilling and obtaining a toxicologically safe final product, from which the release of heavy metals is minimized. In general, the use of bottom or fly ashes to produce foamed glass/glass-ceramics not only provides a valuable path to recycle industrial waste but also offers economic and environmental benefits. This approach reduces the need for natural raw materials that are commonly used in primary glass production (e.g., quartz or silica sand, limestone, and other carbonates), thereby preserving natural resources while mitigating the environmental impact associated with hazardous ash disposal, such as landfilling in dedicated sites preceded by special treatment to make these materials as chemically inert as possible.

Before being used in foaming procedures, recycled glass or vitrified material should be carefully screened from a compositional viewpoint for wise selection; furthermore, the particle size as well as the type and amount of pore-forming agent(s) have to be taken into account to obtain an optimal result. In fact, the characteristics of glass or glass-ceramic foams depend on a delicate balance between viscous flow sintering and gas evolution that is dictated by some factors like the composition and particle size of raw materials, the heat-treatment temperature, the heating rate, and the gas atmosphere in the furnace. Crystallization may occur concurrently to sintering, thus affecting the final physical, mechanical, and thermal properties of the foam (e.g., density, compressive strength, and thermal conductivity). In order to avoid excessive crystallization while maintaining an acceptable control on both foaming and crystallization, waste glasses with a limited tendency toward crystallization should be selected as starting materials, or mixtures combining waste glass from different sources (e.g., vitrified bottom/fly ashes and cullet) can be prepared. Using waste foaming agents, e.g., by-products from mining industry (dolomite, limestone) or cooking (eggshells), can further enhance the "green" fingerprint of the process.

Low-temperature strategies to promote foaming, such as alkali activation followed by mechanical stirring, are potentially applicable to any kind of waste glass and contribute to reducing the energy consumption of the additional pore-forming process.

Big hopes are recently rising from the application of additive manufacturing technologies in the circular economy approach. At present, however, there is a paucity of reports specifically focusing on 3D-printed waste glass. In principle, additive manufacturing allows saving raw materials as well as achieving a better control on pore shape and pore size distribution. These methods are also scalable, allowing mass production with high reproducibility of the goods. On the other hand, if obtaining cellular glass for thermal insulation is the specific goal, it is likely that foaming approaches are still more affordable than 3D printing methods, which may require—especially when implemented on large scales—high investment costs in terms of equipment. Furthermore, there are some important sustainability-related issues that need to be fully considered, such as the formulation of the printable ink. In fact, the use of polymer-based slurries or inks, for example in robocasting and vat photopolymerization, can lead to additional CO_2 emissions generated during post-printing treatments, like high-temperature debinding that aims at removing all the organics. Other aspects deserving consideration within the framework of the circular economy philosophy include the appropriate selection of polymers for designing the ink—bio-based formulations could be an option—and the minimization of waste solvents, for example by using aqueous slurries.

However, these open issues reveal a great opportunity for future research and development toward greener strategies for the fabrication of porous glasses and glass-ceramics for a more sustainable world. Indeed, there are specific challenges in the manufacturing of porous glass that must be addressed. The first challenge is related to material consistency and quality. The variability in waste material composition can lead to inconsistencies in the final product properties. For example, the use of different waste sources like granite dust and maize cob can result in varying porosity and mechanical strength [79]. The second challenges concern energy consumption. The high temperatures required for sintering and foaming processes can lead to significant energy consumption, which may offset some of the environmental benefits of using waste materials [108]. The third challenge is about the mechanical properties. In fact, while high porosity is desirable for insulation, it often results in reduced compressive strength, as seen in materials with compressive strengths ranging from 0.7 to 9.7 MPa [79].

To overcome these challenges, several solutions are being explored. Optimizing the process parameters by adjusting temperature, time, and materials ratios can help optimize the balance between porosity and strength. For instance, varying the processing temperature and the amount of pore-forming agents can control the properties of anorthite ceramics [109]. Furthermore, innovative techniques like colloidal suspension foaming allows for precise control over pore size and structure, enhancing the mechanical properties of the final product [110]. Lastly, incorporating special additives can improve the mechanical properties and reduce the energy required for processing [111].

Therefore, while the use of waste materials for producing porous glass insulation presents several challenges, it also offers significant environmental benefits. The development of innovative manufacturing techniques and optimization of existing processes can help overcome these challenges, making porous glass a viable and sustainable option for building insulation. The integration of waste materials not only reduces the environmental impact but also provides a cost-effective solution for thermal insulation needs.

Author Contributions: Conceptualization, F.B.; methodology, F.B. and P.K.G.; investigation, F.B. and P.K.G.; writing—original draft preparation, F.B.; writing—review and editing, P.K.G.; supervision, F.B.; project administration, F.B.; funding acquisition, F.B. All authors have read and agreed to the published version of the manuscript.

Funding: This research work was performed within the framework of a project funded under the National Recovery and Resilience Plan (NRRP), Mission 4 Component 2 Investment 1.3—Call for tender No. 1561 of 11.10.2022 of Ministero dell'Università e della Ricerca (MUR); the project was funded by the European Union—NextGenerationEU. Project details: project code PE0000021, Concession Decree No. 1561 of 11.10.2022 adopted by Ministero dell'Università e della Ricerca (MUR), CUP E13C22001890001, Project title "Network 4 Energy Sustainable Transition—NEST".

Institutional Review Board Statement: Not applicable.

Informed Consent Statement: Not applicable.

Data Availability Statement: No new data were created or analyzed in this study.

Conflicts of Interest: The authors declare no conflicts of interest.

References

1. Cao, L.; Li, T.; Wang, F. Heat insulation and thermal insulation method of passive low energy consumption residential building exterior envelope structure based on BIM. *Results Eng.* **2024**, *23*, 102734. [CrossRef]
2. Ali, A.; Issa, A.; Elshaer, A. A Comprehensive Review and Recent Trends in Thermal Insulation Materials for Energy Conservation in Buildings. *Sustainability* **2024**, *16*, 8782. [CrossRef]
3. Fragkos, P.; Tasios, N.; Paroussos, L.; Capros, P.; Tsani, S. Energy system impacts and policy implications of the European Intended Nationally Determined Contribution and low-carbon pathway to 2050. *Energy Policy* **2017**, *100*, 216–226. [CrossRef]
4. Kearney, D. EIA's Outlook Through 2035. In *Annual Energy Outlook 2010*; Energy Information Administration: Washington, DC, USA, 2010; pp. 1–15.
5. Al-Homoud, M.S. Performance characteristics and practical applications of common building thermal insulation materials. *Build. Environ.* **2005**, *40*, 353–366. [CrossRef]
6. Liu, Q.; Zhang, Y.; Xu, H. Properties of vulcanized rubber nanocomposites filled with nanokaolin and precipitated silica. *Appl. Clay Sci.* **2008**, *42*, 232–237. [CrossRef]
7. Villasmil, W.; Fischer, L.J.; Worlitschek, J. A review and evaluation of thermal insulation materials and methods for thermal energy storage systems. *Renew. Sustain. Energy Rev.* **2019**, *103*, 71–84. [CrossRef]
8. Zhuang, J.; Ghaffar, S.H.; Fan, M.; Corker, J. Restructure of expanded cork with fumed silica as novel core materials for vacuum insulation panels. *Compos. Part B Eng.* **2017**, *127*, 215–221. [CrossRef]
9. Hoseini, A.; McCague, C.; Andisheh-Tadbir, M.; Bahrami, M. Aerogel blankets: From mathematical modeling to material characterization and experimental analysis. *Int. J. Heat Mass Transf.* **2016**, *93*, 1124–1131. [CrossRef]
10. Nadir, N.; Bouguettaia, H.; Boughali, S.; Bechki, D. Use of a new agricultural product as thermal insulation for solar collector. *Renew. Energy* **2019**, *134*, 569–578. [CrossRef]
11. Apostolopoulou-Kalkavoura, V.; Munier, P.; Bergström, L. Thermally insulating nanocellulose-based materials. *Adv. Mater.* **2021**, *33*, 2001839. [CrossRef]
12. Chao, C.-W.; Liao, C.-J. Approaches to eliminate waste and reduce cost for recycling glass. *Waste Manag.* **2011**, *31*, 2414–2421. [CrossRef] [PubMed]
13. Nodehi, M.; Mohamad Taghvaee, V. Sustainable concrete for circular economy: A review on use of waste glass. *Glass Struct. Eng.* **2022**, *7*, 3–22. [CrossRef]
14. Pavlenko, A.; Cheilytko, A.; Ilin, S.I.; Koshlak, H. Porous structures and their effect on thermophysical properties of thermal protection elements. *Solid State Phenom.* **2019**, *291*, 20–27. [CrossRef]
15. Guo, T.; Liu, Z.; Yu, C.; Ding, J.; Yu, P.; Deng, C. Effect of pore structure evolution on mechanical properties and thermal conductivity of porous SiC-Mullite ceramics. *Ceram. Int.* **2023**, *49*, 33618–33627. [CrossRef]

16. Neumann, M.; Gräfensteiner, P.; Santos de Oliveira, C.; Martins de Souza e Silva, J.; Koppka, S.; Enke, D.; Huber, P.; Schmidt, V. Morphology of nanoporous glass: Stochastic 3D modeling, stereology and the influence of pore width. *Phys. Rev. Mater.* **2024**, *8*, 045605. [CrossRef]
17. Zhao, Z.; Zhou, X.-P. Investigating effects of pore-scale variables and pore geometries on the thermal behaviors of porous concretes. *J. Build. Eng.* **2023**, *75*, 106895. [CrossRef]
18. Kultayeva, S.; Ha, J.-H.; Malik, R.; Kim, Y.-W.; Kim, K.J. Effects of porosity on electrical and thermal conductivities of porous SiC ceramics. *J. Eur. Ceram. Soc.* **2020**, *40*, 996–1004. [CrossRef]
19. Han, Y.; Li, C.; Bian, C.; Li, S.; Wang, C.-A. Porous anorthite ceramics with ultra-low thermal conductivity. *J. Eur. Ceram. Soc.* **2013**, *33*, 2573–2578. [CrossRef]
20. Gong, L.; Wang, Y.; Cheng, X.; Zhang, R.; Zhang, H. Porous mullite ceramics with low thermal conductivity prepared by foaming and starch consolidation. *J. Porous Mater.* **2014**, *21*, 15–21. [CrossRef]
21. Živcová, Z.; Gregorová, E.; Pabst, W.; Smith, D.S.; Michot, A.; Poulier, C. Thermal conductivity of porous alumina ceramics prepared using starch as a pore-forming agent. *J. Eur. Ceram. Soc.* **2009**, *29*, 347–353. [CrossRef]
22. Hu, L.; Wang, C.-A.; Huang, Y. Porous yttria-stabilized zirconia ceramics with ultra-low thermal conductivity. *J. Mater. Sci.* **2010**, *45*, 3242–3246. [CrossRef]
23. Sun, J.; Hu, Z.; Zhuo, J.; Wang, X.; Sun, C. Thermal properties of highly porous fibrous ceramics. In Proceedings of the 5th International Conference on Porous Media and Their Applications in Science, Engineering and Industry, Kona, HI, USA, 22–27 June 2014.
24. Құлтаева, Ш.М.; Құрманбекова, Э.Б. Tuning the Thermal Conductivity of Porous Silicon Carbide Ceramics: A Review. *Вестн. КзГ АСА* **2024**, *3*, 160–171.
25. Smith, D.S.; Alzina, A.; Bourret, J.; Nait-Ali, B.; Pennec, F.; Tessier-Doyen, N.; Otsu, K.; Matsubara, H.; Elser, P.; Gonzenbach, U.T. Thermal conductivity of porous materials. *J. Mater. Res.* **2013**, *28*, 2260–2272. [CrossRef]
26. Gregorová, E.; Pabst, W.; Sofer, Z.; Jankovský, O.; Matějíček, J. Porous alumina and zirconia ceramics with tailored thermal conductivity. *J. Phys. Conf. Ser.* **2012**, *395*, 012022. [CrossRef]
27. Li, X.; Pan, M.; Tao, M.; Liu, W.; Gao, Z.; Ma, C. Preparation of high closed porosity foamed ceramics from coal gangue waste for thermal insulation applications. *Ceram. Int.* **2022**, *48*, 37055–37063. [CrossRef]
28. Wang, C.; Rong, Y.; Zhang, B.; Yang, J. Facile Method for Preparing Hierarchical Al_2O_3–Glass Foam Ceramics with Superior Thermal Insulating Property. *Langmuir* **2022**, *38*, 1141–1150. [CrossRef]
29. Ibrahim, J.E.F.; Gömze, L.A.; Koncz-Horvath, D.; Filep, Á.; Kocserha, I. Preparation, characterization, and physicomechanical properties of glass-ceramic foams based on alkali-activation and sintering of zeolite-poor rock and eggshell. *Ceram. Int.* **2022**, *48*, 25905–25917. [CrossRef]
30. Yang, R.; Liang, Y.; Xu, J.; Meng, X.; Zhu, J.; Cao, S.; Wei, M.; Zhang, R.; Yang, J.; Gao, F. Rare-earth-niobate high-entropy ceramic foams with enhanced thermal insulation performance. *J. Mater. Sci. Technol.* **2022**, *116*, 94–102. [CrossRef]
31. Sun, J.; Liu, P. Optimization of structural parameters for the sound absorption performance of a cellular ceramic foam. *Multidiscip. Model. Mater. Struct.* **2021**, *17*, 1108–1118. [CrossRef]
32. Chen, J.; Liu, P.; Sun, J. Sound absorption performance of a lightweight ceramic foam. *Ceram. Int.* **2020**, *46*, 22699–22708. [CrossRef]
33. Lucio, M.D.S.; Kultayeva, S.; Kim, Y.-W. Improved mechanical strength and thermal resistance of porous SiC ceramics with gradient pore sizes. *J. Eur. Ceram. Soc.* **2022**, *42*, 6785–6794. [CrossRef]
34. Niyogi, S.; Gupta, B.S. Mechanical properties and pore size distribution in athermal porous glasses. *Soft Matter* **2021**, *17*, 9716–9724. [CrossRef] [PubMed]
35. Miyazaki, N.; Omoto, Y.; Hoshide, T. Numerical simulation on effects of pore characteristics on strength in porous ceramics. *Mech. Eng. J.* **2019**, *6*, 19-00234. [CrossRef]
36. Pereira, C.; Rice, R.; Skalny, J. Pore structure and its relationship to properties of materials. *MRS Online Proc. Libr.* **1988**, *137*, 3. [CrossRef]
37. Jung, D.; Kwon, Y. Investigation of Structure-Property Relations in Porous Metals Using Finite Element Simulation. *Korean J. Met. Mater.* **2019**, *57*, 747–754. [CrossRef]
38. Jeon, C.-K.; Lee, J.-S.; Chung, H.; Kim, J.-H.; Park, J.-P. A study on insulation characteristics of glass wool and mineral wool coated with a polysiloxane agent. *Adv. Mater. Sci. Eng.* **2017**, *2017*, 3938965. [CrossRef]
39. Baino, F.; Ferraris, M. Production and characterization of ceramic foams derived from vitrified bottom ashes. *Mater. Lett.* **2019**, *236*, 281–284. [CrossRef]
40. Dele-Afolabi, T.; Hanim, M.A.; Norkhairunnisa, M.; Sobri, S.; Calin, R. Research trend in the development of macroporous ceramic components by pore forming additives from natural organic matters: A short review. *Ceram. Int.* **2017**, *43*, 1633–1649. [CrossRef]

41. Park, S.; Kwon, Y.-P.; Kwon, H.-C.; Lee, H.-W.; Lee, J.C. Effect of composition on thermal conductivity of silica insulation media. *J. Nanosci. Nanotechnol.* **2008**, *8*, 5052–5056. [CrossRef]
42. Rodin, A.; Ermakov, A.; Erofeeva, I.; Erofeev, V. Effect of Chlorides Content on the Structure and Properties of Porous Glass Ceramics Obtained from Siliceous Rock. *Materials* **2022**, *15*, 3268. [CrossRef]
43. Jiang, K.; Xia, M.; Tang, Y.; Xu, Y.; Deng, T.; Li, B.; Chen, W. Formation of closed pore structure in CaO-MgO-Al2O3-SiO2 (CMAS) porous glass-ceramics via Fe2O3 modified foaming for thermal insulation. *J. Eur. Ceram. Soc.* **2023**, *43*, 1689–1697. [CrossRef]
44. Apkaryan, A.; Kulkov, S. Formation of structure and closed porosity under high-temperature firing of granules of porous glass-ceramic material. *Inorg. Mater. Appl. Res.* **2018**, *9*, 286–290. [CrossRef]
45. Kim, H.D.; Baek, C.R.; Jang, Y.C. Advancing glass recycling and environmental applications with porous glass: A mini-review. *J. Mater. Cycles Waste Manag.* **2024**, *26*, 2620–2633. [CrossRef]
46. He, R.; Qu, Z.; Cheng, X. Effects of starch addition amount on microstructure, mechanical properties and room temperature thermal conductivity of porous Y_2SiO_5 ceramics. *Ceram. Int.* **2016**, *42*, 2257–2262. [CrossRef]
47. Müller, A.; Leydolph, B.; Stanelle, K. Recycling mineral wool waste: Technologies for the conversion of the fiber structure, Part 1. *Interceram* **2009**, *58*, 378–381.
48. Jagadeeswaran, I.; Sriram, H. EU 1272/2008—Classification, Labelling and Packaging of Substances and Mixtures. In *Medical Device Guidelines and Regulations Handbook*; Springer: Berlin/Heidelberg, Germany, 2022; pp. 261–295.
49. Mesa Sanchez, M.; Giro Guasch, E.; Pasalaigua Huguet, J. Biosoluble Composition of Glass Fibres for the Production of Glass Wools and Similar Patent ES2212962T3, 16 August 2004.
50. Tulyaganov, D.; Fernandes, H.; Agathopoulos, S.; Ferreira, J. Preparation and characterization of high compressive strength foams from sheet glass. *J. Porous Mater.* **2006**, *13*, 133–139. [CrossRef]
51. Bernardo, E.; Scarinci, G.; Bertuzzi, P.; Ercole, P.; Ramon, L. Recycling of waste glasses into partially crystallized glass foams. *J. Porous Mater.* **2010**, *17*, 359–365. [CrossRef]
52. Owens Corning FOAMGLAS® Thermal Insulation Made of Cellular Glass. Available online: https://www.foamglas.com/it-ch (accessed on 10 December 2024).
53. Liu, H.; Zhao, X. Thermal conductivity analysis of high porosity structures with open and closed pores. *Int. J. Heat Mass Transf.* **2022**, *183*, 122089. [CrossRef]
54. Ye, L.; Hong, J.; Ma, X.; Qi, C.; Yang, D. Life cycle environmental and economic assessment of ceramic tile production: A case study in China. *J. Clean. Prod.* **2018**, *189*, 432–441. [CrossRef]
55. König, J.; Petersen, R.R.; Yue, Y.; Suvorov, D. Gas-releasing reactions in foam-glass formation using carbon and MnxOy as the foaming agents. *Ceram. Int.* **2017**, *43*, 4638–4646. [CrossRef]
56. Sakai, S.-i.; Hiraoka, M. Municipal solid waste incinerator residue recycling by thermal processes. *Waste Manag.* **2000**, *20*, 249–258. [CrossRef]
57. Ferraris, M.; Salvo, M.; Smeacetto, F.; Augier, L.; Barbieri, L.; Corradi, A.; Lancellotti, I. Glass matrix composites from solid waste materials. *J. Eur. Ceram. Soc.* **2001**, *21*, 453–460. [CrossRef]
58. Scarinci, G.; Brusatin, G.; Barbieri, L.; Corradi, A.; Lancellotti, I.; Colombo, P.; Hreglich, S.; Dall'Igna, R. Vitrification of industrial and natural wastes with production of glass fibres. *J. Eur. Ceram. Soc.* **2000**, *20*, 2485–2490. [CrossRef]
59. Islam, G.S.; Rahman, M.; Kazi, N. Waste glass powder as partial replacement of cement for sustainable concrete practice. *Int. J. Sustain. Built Environ.* **2017**, *6*, 37–44. [CrossRef]
60. Sharifikolouei, E.; Canonico, F.; Salvo, M.; Baino, F.; Ferraris, M. Vitrified and nonvitrified municipal solid wastes as ordinary Portland cement (OPC) and sand substitution in mortars. *Int. J. Appl. Ceram. Technol.* **2020**, *17*, 573–583. [CrossRef]
61. Maddalena, R.; Roberts, J.J.; Hamilton, A. Can Portland cement be replaced by low-carbon alternative materials? A study on the thermal properties and carbon emissions of innovative cements. *J. Clean. Prod.* **2018**, *186*, 933–942. [CrossRef]
62. Sabbrojjaman, M.; Liu, Y.; Tafsirojjaman, T. A comparative review on the utilisation of recycled waste glass, ceramic and rubber as fine aggregate on high performance concrete: Mechanical and durability properties. *Dev. Built Environ.* **2024**, *17*, 100371. [CrossRef]
63. Hamada, H.; Alattar, A.; Tayeh, B.; Yahaya, F.; Thomas, B. Effect of recycled waste glass on the properties of high-performance concrete: A critical review. *Case Stud. Constr. Mater.* **2022**, *17*, e01149. [CrossRef]
64. Appendino, P.; Ferraris, M.; Matekovits, I.; Salvo, M. Production of glass–ceramic bodies from the bottom ashes of municipal solid waste incinerators. *J. Eur. Ceram. Soc.* **2004**, *24*, 803–810. [CrossRef]
65. Fernandes, H.R.; Gunduz, O.; Tulyaganov, D.U. Synthesis of Bulk-Nucleated Glass–Ceramics and Porous Glass–Ceramic Composites through Utilization of Fly Ashes. *Ceramics* **2024**, *7*, 1014–1029. [CrossRef]
66. Fernandes, H.; Tulyaganov, D.; Ferreira, J. Preparation and characterization of foams from sheet glass and fly ash using carbonates as foaming agents. *Ceram. Int.* **2009**, *35*, 229–235. [CrossRef]
67. Zhu, M.; Ji, R.; Li, Z.; Wang, H.; Liu, L.; Zhang, Z. Preparation of glass ceramic foams for thermal insulation applications from coal fly ash and waste glass. *Constr. Build. Mater.* **2016**, *112*, 398–405. [CrossRef]

68. Zeng, L.; Sun, H.-j.; Peng, T.-j.; Zheng, W.-m. The sintering kinetics and properties of sintered glass-ceramics from coal fly ash of different particle size. *Results Phys.* **2019**, *15*, 102774. [CrossRef]
69. Chen, R.; Li, Y.; Xiang, R.; Li, S. Effect of particle size of fly ash on the properties of lightweight insulation materials. *Constr. Build. Mater.* **2016**, *123*, 120–126. [CrossRef]
70. Li, M.; Gao, Q.; He, Q.; Li, S.; Quan, Y.; Wu, M.; Li, C. Low-cost porous thermal insulation materials with tunable pore structures derived from fly ash by foam-gelcasting. *Int. J. Appl. Ceram. Technol.* **2024**, *21*, 267–277. [CrossRef]
71. Sharifikolouei, E.; Baino, F.; Salvo, M.; Tommasi, T.; Pirone, R.; Fino, D.; Ferraris, M. Vitrification of municipal solid waste incineration fly ash: An approach to find the successful batch compositions. *Ceram. Int.* **2021**, *47*, 7738–7744. [CrossRef]
72. Ding, L.; Ning, W.; Wang, Q.; Shi, D.; Luo, L. Preparation and characterization of glass–ceramic foams from blast furnace slag and waste glass. *Mater. Lett.* **2015**, *141*, 327–329. [CrossRef]
73. Rincón, A.; Marangoni, M.; Cetin, S.; Bernardo, E. Recycling of inorganic waste in monolithic and cellular glass-based materials for structural and functional applications. *J. Chem. Technol. Biotechnol.* **2016**, *91*, 1946–1961. [CrossRef]
74. Fernandes, H.R.; Andreola, F.; Barbieri, L.; Lancellotti, I.; Pascual, M.J.; Ferreira, J.M. The use of egg shells to produce Cathode Ray Tube (CRT) glass foams. *Ceram. Int.* **2013**, *39*, 9071–9078. [CrossRef]
75. Petersen, R.R.; König, J.; Yue, Y. The viscosity window of the silicate glass foam production. *J. Non-Cryst. Solids* **2017**, *456*, 49–54. [CrossRef]
76. Fedosov, S.; Bakanov, M.; Grushko, I. Using anthropogenic raw materials in the process of synthesizing foam glass with heterogeneous microstructure. *Vestn. MGSU* **2024**, *19*, 258–269. [CrossRef]
77. Cozzarini, L.; De Lorenzi, L.; Fortuna, L.; Bevilacqua, P. Recycling of glass waste and spent alkaline batteries cathodes into insulation materials. *Sustain. Mater. Technol.* **2023**, *38*, e00767. [CrossRef]
78. Fernandes, F.A.d.S.; Costa, D.d.S.d.O.; Martin, C.A.G.; Rossignolo, J.A. Vitreous foam with thermal insulating property produced with the addition of waste glass powder and rice husk ash. *Sustainability* **2023**, *15*, 796. [CrossRef]
79. Odewole, O.P.; Kashim, I.B.; Akinbogun, T.L. Investigation into the viability of the properties of porous glass-ceramics produced from granite dust and maize cob for use in thermal insulation of external walls of residential buildings. *J. Mech. Eng. Sci.* **2022**, *16*, 8943–8952. [CrossRef]
80. Goltsman, B.M.; Yatsenko, E.A. Role of carbon phase in the formation of foam glass porous structure. *Materials* **2022**, *15*, 7913. [CrossRef]
81. Paunescu, L.; Axinte, S.M.; Dragoescu, M.F.; Cosmulescu, F.; Paunescu, B.V. Simultaneous Use of Liquid and Solid Foaming Agents by a Nonconventional Technique to Obtain a High-Strength Glass Foam with Fine Porosity. *Nonconv. Technol. Rev.* **2021**, *25*, 3–9.
82. Sooksaen, P.; Sudyod, N.; Thongtha, N.; Simsomboonphol, R. Fabrication of lightweight foam glasses for thermal insulation applications. *Mater. Today Proc.* **2019**, *17*, 1823–1830. [CrossRef]
83. Chaima, S.; Andrea, S.; Peter, B. Investigation of foamed glass using natural waste materials as foaming agent. *Multidiszcip. Tudományok* **2022**, *12*, 280–291.
84. Da Silva, L.L.; Ribeiro, L.C.N.; Santacruz, G.; Arcaro, S.; Alves, A.K.; Bergmann, C.P. Glass Foams Produced from Glass and Yerba Mate (*Ilex paraguarinensis*) Waste. *FME Trans.* **2018**, *46*, 70–79.
85. Yatsenko, E.; Gol'tsman, B.; Smolii, V.; Gol'tsman, N.; Yatsenko, L. Study on the possibility of applying organic compounds as pore-forming agents for the synthesis of foam glass. *Glass Phys. Chem.* **2019**, *45*, 138–142. [CrossRef]
86. Stochero, N.; de Souza Chami, J.; Souza, M.; de Moraes, E.; de Oliveira, A.N. Green glass foams from wastes designed for thermal insulation. *Waste Biomass Valorization* **2021**, *12*, 1609–1620. [CrossRef]
87. Da Silva, R.C.; Kubaski, E.T.; Tenório-Neto, E.T.; Lima-Tenório, M.K.; Tebcherani, S.M. Foam glass using sodium hydroxide as foaming agent: Study on the reaction mechanism in soda-lime glass matrix. *J. Non-Cryst. Solids* **2019**, *511*, 177–182. [CrossRef]
88. Zhang, J.; Liu, B.; Zhao, S.; Shen, H.; Liu, J.; Zhang, S. Preparation and characterization of glass ceramic foams based on municipal solid waste incineration ashes using secondary aluminum ash as foaming agent. *Constr. Build. Mater.* **2020**, *262*, 120781. [CrossRef]
89. Mahmoud, M.; Kraxner, J.; Kaňková, H.; Hujová, M.; Chen, S.; Galusek, D.; Bernardo, E. Porous glass microspheres from alkali-activated fiber glass waste. *Materials* **2022**, *15*, 1043. [CrossRef]
90. Rincón, A.; Giacomello, G.; Pasetto, M.; Bernardo, E. Novel 'inorganic gel casting'process for the manufacturing of glass foams. *J. Eur. Ceram. Soc.* **2017**, *37*, 2227–2234. [CrossRef]
91. Elsayed, H.; Rincón Romero, A.; Ferroni, L.; Gardin, C.; Zavan, B.; Bernardo, E. Bioactive glass-ceramic scaffolds from novel 'inorganic gel casting' and sinter-crystallization. *Materials* **2017**, *10*, 171. [CrossRef]
92. Rincón, A.; Desideri, D.; Bernardo, E. Functional glass-ceramic foams from 'inorganic gel casting'and sintering of glass/slag mixtures. *J. Clean. Prod.* **2018**, *187*, 250–256. [CrossRef]
93. Monich, P.R.; Romero, A.R.; Höllen, D.; Bernardo, E. Porous glass-ceramics from alkali activation and sinter-crystallization of mixtures of waste glass and residues from plasma processing of municipal solid waste. *J. Clean. Prod.* **2018**, *188*, 871–878. [CrossRef]

94. Romero, A.R.; Salvo, M.; Bernardo, E. Up-cycling of vitrified bottom ash from MSWI into glass-ceramic foams by means of 'inorganic gel casting'and sinter-crystallization. *Constr. Build. Mater.* **2018**, *192*, 133–140. [CrossRef]
95. Monich, P.R.; Romero, A.R.; Rambaldi, E.; Bernardo, E. Case studies of up-cycling of partially crystallized ceramic waste in highly porous glass-ceramics. *Constr. Build. Mater.* **2020**, *261*, 119971. [CrossRef]
96. Zhang, J.; Liu, B.; Zhang, X.; Shen, H.; Liu, J.; Zhang, S. Co-vitrification of municipal solid waste incinerator fly ash and bottom slag: Glass detoxifying characteristics and porous reformation. *Ecotoxicol. Environ. Saf.* **2022**, *243*, 113995. [CrossRef] [PubMed]
97. *UNI EN 12457-2:2004*; Caratterizzazione dei Rifiuti–Lisciviazione–Prova di Conformità per la Lisciviazione di Rifiuti Granulari e di Fanghi. In Parte 2: Prova a Singolo Stadio, con un Rapporto Liquido/Solido di 10 L/kg, per Materiali con Particelle di Dimensioni Minori di 4 mm (con o Senza Riduzione Delle Dimensioni). UNI: Rome, Italy, 2004.
98. Souza, M.T.; Onghero, L.; Passos, A.B.; Simão, L.; Piva, R.H.; Repette, W.L.; de Oliveira, A.P.N. Sustainable glass foams produced with stone waste as a pore-forming agent: Assessing the role of heating rate in foamability and glass foams recyclability. *J. Clean. Prod.* **2022**, *338*, 130596. [CrossRef]
99. Chen, Z.; Li, Z.; Li, J.; Liu, C.; Lao, C.; Fu, Y.; Liu, C.; Li, Y.; Wang, P.; He, Y. 3D printing of ceramics: A review. *J. Eur. Ceram. Soc.* **2019**, *39*, 661–687. [CrossRef]
100. Villa, A.; Gianchandani, P.K.; Baino, F. Sustainable Approaches for the Additive Manufacturing of Ceramic Materials. *Ceramics* **2024**, *7*, 291–309. [CrossRef]
101. Simorgh, S.; Alasvand, N.; Khodadadi, M.; Ghobadi, F.; Kebria, M.M.; Milan, P.B.; Kargozar, S.; Baino, F.; Mobasheri, A.; Mozafari, M. Additive manufacturing of bioactive glass biomaterials. *Methods* **2022**, *208*, 75–91. [CrossRef]
102. Cuevas, K.; Chougan, M.; Martin, F.; Ghaffar, S.H.; Stephan, D.; Sikora, P. 3D printable lightweight cementitious composites with incorporated waste glass aggregates and expanded microspheres–Rheological, thermal and mechanical properties. *J. Build. Eng.* **2021**, *44*, 102718. [CrossRef]
103. Deng, Q.; Zou, S.; Xi, Y.; Singh, A. Development and Characteristic of 3D-Printable Mortar with Waste Glass Powder. *Buildings* **2023**, *13*, 1476. [CrossRef]
104. Sheng, Z.; Zhu, B.; Cai, J.; Han, J.; Zhang, Y.; Pan, J. Influence of waste glass powder on printability and mechanical properties of 3D printing geopolymer concrete. *Dev. Built Environ.* **2024**, *20*, 100541. [CrossRef]
105. Derevianko, O.; Derevianko, O.; Zakiev, V.; Zgalat-Lozynskyy, O. 3D printing of porous glass products using the robocasting technique. *Powder Met. Metal Ceram.* **2022**, *60*, 546–555. [CrossRef]
106. Nan, B.; Gołębiewski, P.; Buczyński, R.; Galindo-Rosales, F.J.; Ferreira, J.M. Direct ink writing glass: A preliminary step for optical application. *Materials* **2020**, *13*, 1636. [CrossRef]
107. De Moraes, E.G.; Ferreira, I.M.; Teixeira, L.B.; Cartapati, L.H.; Souza, M.T.; de Oliveira, A.P.N. Additive manufacturing of cellular structures from recycled soda-lime glass printing inks by robocasting. *Ceram. Int.* **2023**, *49*, 6554–6562. [CrossRef]
108. Smolii, V.; Kosarev, A.; Yatsenko, E. Cellular Heat Insulation Building Glass Materials Based on Wastes from Thermal Power Plants and Ferrous Metallurgy. *Glass Ceram.* **2017**, *74*, 52–54. [CrossRef]
109. Omerašević, M.; Pavkov, V.; Rosić, M.; Egerić, M.; Nenadović, S.; Bučevac, D.; Potkonjak, N. Fabrication of Porous Anorthite Ceramic Insulation Using Solid Wastes. *Materials* **2024**, *17*, 1478. [CrossRef] [PubMed]
110. Huo, W.; Yan, S.; Wu, J.M.; Liu, J.; Chen, Y.; Qu, Y.; Tang, X.; Yang, J. A novel fabrication method for glass foams with small pore size and controllable pore structure. *J. Am. Ceram. Soc.* **2017**, *100*, 5502–5511. [CrossRef]
111. Mann, C.J.; Hee, K.D.; Gyu, K.S. Porous Ultra Light Ceramic Insulator for Building External Insulation System Using Waste Glass and Coal Ash and Manufacturing Method Thereof Patent KR20160137716A, 1 December 2016.

ceramics

MDPI

Review

Synthesis of Bulk-Nucleated Glass–Ceramics and Porous Glass–Ceramic Composites through Utilization of Fly Ashes

Hugo R. Fernandes [1,*], Oguzhan Gunduz [2] and Dilshat U. Tulyaganov [3,*]

1 Department of Materials and Ceramic Engineering, CICECO-Aveiro Institute of Materials, University of Aveiro, 3810-193 Aveiro, Portugal
2 Centre for Nanotechnology & Biomaterials Research, Marmara University, Istanbul 34722, Turkey; oguzhan@marmara.edu.tr
3 Department of Natural-Mathematical Sciences, Turin Polytechnic University in Tashkent, Tashkent 100095, Uzbekistan
* Correspondence: h.r.fernandes@ua.pt (H.R.F.); tulyaganovdilshat@gmail.com (D.U.T.)

Abstract: Coal combustion in power plants for electric power generation produces millions of tons of residues that are generally disposed of in landfills or ponds occupying vast land, resulting in serious environmental pollution. Fly ash (FA) is one of the main solid wastes generated in coal-based thermal power plants, representing the largest fraction of coal combustion residues (65–95%). Unfortunately, the enormous amount of FA residue is utilized only partly, mainly in the cement industry and building materials field. An alternative approach to using FA is its incorporation into ceramic, glass and glass–ceramic production, aligning with circular economy principles and reducing the environmental footprint of both the energy and ceramic sectors. In this review article, the topics of the composition, properties, classification, and utilization of fly ashes from thermal power plants are discussed. The main objective of this work is a critical analysis of the experimental trials directed to the involvement of FA as a raw material in the fabrication of glass–ceramics and porous ceramic composites.

Keywords: glass-ceramics; fly ashes; wastes; crystallization

1. Introduction

Coal, being available in abundance, is a cheap fuel compared to other sources and a driving force for the energy generation needed to meet the global energy demand. The coal combustion in power plants for electric power generation produces millions of tons of residues. Residues produced by coal burning comprise boiler slag, bottom ash, fly ash, fluidized bed combustion ash, etc. [1]. Fly ash (FA) is the dust collected by dust extraction equipment, such as electrostatic precipitators or filter bags, from flue gas after coal combustion. FA is one of the main solid wastes generated in coal-based thermal power plants, representing the largest fraction of coal combustion residues (65–95%). For instance, a 1000 MW power plant uses around 12,000 tonnes of coal and generates around 2000 tonnes of fly ashes daily [2]. The disposal of these enormous amounts of FA is a major matter of concern because it represents a significant negative environmental impact including the leaching of potentially toxic substances into soils and groundwater, changes in the elemental composition of vegetation growing close to the ashes, and the accumulation of toxic elements in the food chain [3–5].

It is very difficult to handle FA in the dry state (mainly due to its fineness) because the flying fine particles of ash may corrode structural surfaces and negatively affect agriculture and the environment. It is also dangerous to the general population since long exposure to and inhalation of FA causes severe health problems like silicosis, fibrosis of the lungs, bronchitis, and pneumonitis [6].

Over the years, there has been an increasing interest in the use of FA for a variety of applications such as as a raw material or as an additive in the cement and concrete industry, a raw materials substitute in the ceramic industry, soil stabilization for roads and embankments, landfill liners or covers, in mine filling, wastewater treatment, synthesis of glass–ceramics, zeolites, agriculture, etc. [7–9].

The main objective of this brief review is a critical analysis of the experimental trials directed to the involvement of FA as a raw material in the fabrication of glass–ceramics and foamed glass products. The discussion of the results is strictly confined to the physical, mechanical, and environmental properties of the bulk-nucleated glass–ceramics and porous glass–ceramic composites based on coal FA and their practical suitability.

2. Fly Ashes from Thermal Power Plants: Properties, Composition, and Classification

In the global scenario, more than 750 million tons of FA are produced annually worldwide [10]. The production and utilization of FA in the main world producers are depicted in Figure 1.

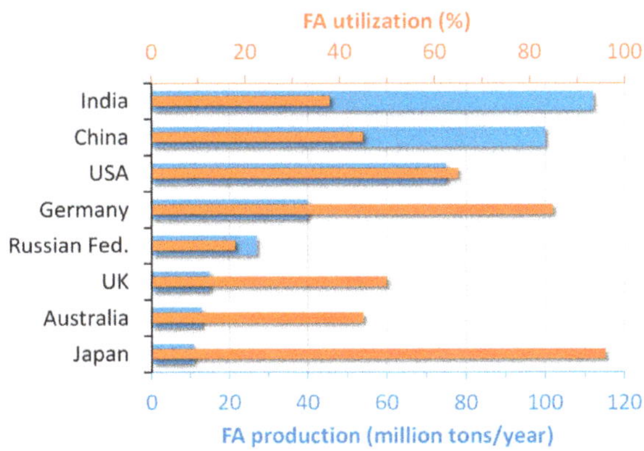

Figure 1. Production and utilization of FA in some countries (adapted with permission from [11], Elsevier 2019).

The physical, chemical, and mineralogical properties of FA are closely dependent on the nature of the used coal, the combustion conditions, the type of emission control device, and the storage and handling methods [12]. For instance, FA generated by the combustion of anthracite, bituminous, and lignite coal could present significant, different compositions and properties. The physical properties of fly ash vary widely depending on the coal type, boiler type, ash content in coal, combustion method, and collector setup [13].

FA is a solid powder at room temperature, occurring as fine particles with low to medium bulk density, high surface area, and light texture [12]. The appearance of FA is similar to cement, presenting a color ranging from light grey to grey black. The color of FA can reflect the amount of carbon content and, to a certain extent, the fineness of the FA: the darker the color, the finer the particle size of the FA and the higher the carbon content. For instance, the FA powders generated from anthracite and bituminous coal contain higher carbon residues, resulting in a dark brown color. On the other hand, FA powder from lignite and sub-bituminous coal contains less carbon residue and traces of calcium or lime show a grey color [14].

Regarding morphology, FA is a heterogeneous material containing powders of different shapes and sizes. Generally, it includes particles mostly spherical in size, along with rounded particles, although particles with angular and irregular shapes may also be present [14]. FA powder particles may be hollow (cenospheres) or filled with amorphous powder (plerospheres) [15]. The sizes of the spherical particles of FA may vary between 0.5 and 200 μm, while the other particles (with irregular and angular shapes) may be bigger in size [16]. For example, particle sizes of FA from bituminous and anthracite coals are exceptionally similar to that of silt (<75 μm), whereas sub-bituminous and lignite FA particles are larger compared to other types (>75 μm) [11].

The bulk density of fly ash generated from different sources without any classification is quite different, most of which are between 1.9 and 2.9 $g \cdot cm^{-3}$. However, the bulk density among each particle may also vary due to their different compositions and structures, e.g., there are floating beads with a bulk density less than 1 $g \cdot cm^{-3}$ and iron beads with a bulk density greater than 3 $g \cdot cm^{-3}$ [17].

The composition of FA significantly depends on the coal type and technological processes used in thermal power plants. FA is usually produced at relatively high temperatures in the range of 1200–1700 °C and generated from various inorganic and organic constituents present in the used coals [18]. The most common elements presented in FA are O, Si, Al, Ca, Fe, C, K, Mg, H, Na, Ti, N, P, and Ba. Less common elements such as Mn, Sr, F, and Cl can also be present in FA [19].

FA is generally constituted by three different parts: (1) an inorganic, (2) organic, and (3) fluid constituent. The inorganic constituent is mainly composed of non-crystalline (amorphous, glassy particles) and crystalline (mineral) matter, such as crystals, grains, and aggregates of various minerals. The organic part comprises char materials and organic minerals, while the fluid constituent comprises liquid, gas and gas–liquid inclusions associated with both inorganic and organic matter [20]. Besides the glassy (amorphous) phase, the common phases present in FA include (in the order of decreasing amounts) mullite, quartz, char, hematite–magnetite, anhydrite–gypsum, feldspars, lime–portlandite, clay and mica minerals, cristobalite–tridymite, calcite–ankerite, corundum, jarosite, and some Ca and Ca–Mg silicates [19,21]. The chemical compositions of some types of FA produced in different counties are presented in Table 1.

Table 1. Summary of typical chemical composition (in wt.%) ranges of FA from some countries [16,22–24].

Country	SiO_2	Al_2O_3	Fe_2O_3	CaO	K_2O	MgO	SO_3	TiO_2	Na_2O	P_2O_5	LOI
Australia	31–69	17–33	1–27	0–5	0–3	0–2	0–0.6	1.2–3.7	0–1.5	0–4	na
China	36–57	19–55	2–19	1–7	0.8–0.9	0.7–5	1–3	0.2–0.7	0.6–1.3	1–1.5	nd
France	47–51	26–34	7–9	2–3	na	1.5–2	0.1–0.6	na	2–6	na	0.5–5
Germany	20–80	1–19	1–22	2–52	0–2	0.5–11	1–15	0.1–1	0–2	na	0–5
India	50–60	14–32	3–17	0.6–9	0.2–5	0.1–2	na	0.3–2.7	0.2–1.2	na	0.5–7
Japan	54–63	18–26	4–6	2–8	0.6–3	1–2	0.3–1.4	0.8–1.2	1.1–2.1	na	0.5–2
Russia	41–49	23–26	na	7–13	2–3	3–4	na	0.5–0.6	1.2–1.5	0.3–0.4	na
US	35–59	19–29	3–26	0.7–22	1–3	0.5–5	0.1–2	1–1.6	0.2–1.8	0.1–1.3	0.2–21

Note: na = not available; nd = not determined.

FA is generally classified into two chemical types for industrial application (mostly in concrete and cement production): Class F and Class C. The first type of FA comprises ≥70% of $SiO_2 + Al_2O_3 + Fe_2O_3$, whereas the second type has a sum of these oxides in the range of 50–70% [25]. The Class F type of fly ash is generated by the combustion of anthracite or bituminous coal, whereas Class C is generated by burning sub-bituminous coal or lignite. Class C fly ash usually exhibits cementitious properties along with siliceous and aluminous characteristics due to the presence of lime, whereas type F is rarely cementitious. Class C fly ash usually comes from coals which may generate ash with higher lime content generally, more than 15%, which in turn exhibits self-hardening characteristics due to the presence of more CaO. In other words, the availability of Class F is on a larger scale, and it generally has less lime, usually containing less than 15%, with a greater content of alumina, silica, and iron [26]. Other chemical requirements such as the ASTM classification of classes C and F include SO_3 content (which should be ≤5%), moisture (should be ≤3%), Na_2O (≤1.5%), particle size (34% ≤ 5% retained on 45 mm), and loss on ignition (LOI) (6–12% for Class F, considering performance).

However, FA can be classified according to different standards which are not universal, in terms of chemical composition (oxide content), chemical groups according to some chemical and physical characteristics, loss of ignition (LOI), fineness, pH (acid and basic FA type), etc. [20]. Different methods are used by the European Union, Russia, and Canada to classify FA, and there is currently no international grouping system [27]. Table 2 presents a summary of some common classifications of FA.

Table 2. Summary of some common classifications of FA [20,22,25,28].

Criteria	Classification	Indicators
Chemical types	Class F Class C	≥70% $SiO_2 + Al_2O_3 + Fe_2O_3$ 50–70% $SiO_2 + Al_2O_3 + Fe_2O_3$
Chemical groups	Group I—silico-aluminate Group II—alumino-silicate Group III—lime sulphate Group IV—basic FA	$SiO_2/Al_2O_3 \geq 2$ and CaO < 15% $SiO_2/Al_2O_3 < 2$, CaO < 15%, SO_3 < 3% CaO > 15% and SO_3 > 3% CaO > 15% and SO_3 < 3%
Particle size and specific surface area (SSA)	fine-grained medium-grained coarse-grained	<25% are >75 μm, SSA > 3.0 $m^2 \cdot g^{-1}$ 0–75% are <75 μm, SSA 1.5–3.0 $m^2 \cdot g^{-1}$ <40% are <75 μm, SSA < 1.5 $m^2 \cdot g^{-1}$
Activity (content of free CaO)	inactive or very little active slightly active active very active	<3.5% CaO 3.5–7% CaO 7.0–14.0% CaO >14% CaO
pH value	acidic low alkaline highly alkaline	11–138–91.1–7.2 11–138–91.1–7.2 11–138–91.1–7.2

3. Bulk-Nucleated Glass–Ceramic from Fly Ashes

FA is basically composed of silica, alumina, iron oxides, alkali and alkaline earth metal oxides plus several heavy metals and transition metal oxides, which are components commonly presented in the raw materials used in the ceramic and glass industry. By incorporating FA into ceramic and glass production, manufacturers can divert significant amounts of waste from landfills, aligning with circular economy principles and reducing the environmental footprint of both the energy and ceramic sectors. FA has also found application in the processing of glass–ceramic materials. Although the sintering process is steadily gaining more interest in the fabrication of glass ceramics from FA, in this article, the main attention is focused on glass–ceramic that is conventionally produced through devitrification of bulk glass by a single- or two-stage heat treatment [29–31].

Thus, the FA from Çayırhan Thermal Power Plant in Turkey was used as the principal precursor in glass fabrication with no additives and nucleating agents. As received, the FA sample had a chemical composition (in wt.%) of 42.82 SiO_2, 16.38 CaO, 5.85 MgO, 7.01 Fe_2O_3, 13.36 Al_2O_3, 5.06 Na_2O, 1.83 K_2O, and 6.47 SO_3 and comprised crystalline phases, namely quartz, mullite, enstatite, anorthite, and hematite. The fly ash sample was melted in Pt-crucibles at 1400 °C for 4 h, and then the melt was poured into the water; glass frit, after drying, was crushed, pulverized, and remelted at the same temperature for 10 h, followed by casting in a preheated, graphite mold. To fabricate the as-cast glass–ceramic, the glass was heat-treated at 680 °C for 5, 10, and 15 h for nucleation and then at 924 °C for 20 min for crystal growth. The diopside-based glass–ceramic sample nucleated at 680 °C for 5 h demonstrated the highest microhardness value of 907 $kg.mm^{-2}$ (8.89 GPa) [32].

Vitrification of a glass batch composed of merely coal FA was found to be difficult, and crystallization of the molten glass was not possible without modifying the composition of the FA through the addition of selected oxides [33]. Ark shell (as a source of calcium) and TiO_2 (as a nucleating agent) were added to the glass batch in order to decrease the melting temperature and promote internal crystallization. The iron and titanium oxides present in the ash components and added materials have a combined effect on the nucleation. The glass devitrified through a single-stage heat treatment, during a non-isothermal process:

melting proceeded in Pt-Rh crucibles at 1450 °C for 3 h, and then the melt was poured into a graphite mold preheated to 750 °C. Crystallization via nucleation and crystal growth is processed as a simple cycle heating process at 950–1050 °C for 10–240 min. The glass–ceramics produced demonstrated a density of 2.77–2.81 $g \cdot cm^{-3}$, a bending strength of 98–176 MPa, a fracture toughness of 1.12–3.10 $MPa \cdot m^{1/2}$, an elastic modulus of 91–100 GPa, and a CTE of (5.5–6.5) $\times 10^{-6} K^{-1}$. Generally, with a temperature and holding time rise, the aspect ratio of the crystal increased, leading to a higher mechanical strength of resulting glass–ceramics. Importantly, the effect of temperature on the crystal growth was more pronounced than that of holding time. It was found that at 1050 °C, only a short time was needed to obtain glass–ceramics with a high strength, while at a temperature above 1050 °C the material deformed after 30 min due to approaching a low-viscosity region.

The experimental results demonstrated that an appropriate ratio between the glass network former and modifiers is crucial to obtain suitable glass and glass–ceramic products. Consequently, glasses and glass–ceramics were produced by incorporating up to 50 wt.% of Italian or Spanish coal fly ash with glass cullet and float dolomite. The batch was melted at approximately 1500 °C (with thermal cycles of 5 h), after which the glass was poured into a steel mold, annealed at 550 °C for 2 h, and heat-treated in the interval 900–1200 °C for 0.5–8.0 h [34]. In all the investigated fly ash glass compositions, no induction time of nucleation and crystallization was detected since the crystal growth rate was very fast, occurring from very short thermal treatment times. The mechanical properties of the glass–ceramic products were as follows: fracture toughness 0.70–3.00 $MPa \cdot m^{1/2}$ (KIC), elastic modulus 42–92 GPa (E), and microhardness 4.6–7.3 GPa (HV). The experimental glass 4T had a composition in wt.% of 50.07 SiO_2, 17.00 CaO, 8.7 MgO, 3.62 Fe_2O_3, 14.71 Al_2O_3, 4.39 Na_2O, 0.69 K_2O, and 0.82 TiO_2 after heat treatment at 1200 °C, devitrified to feldspar, pyroxene, and spinel phases, and demonstrated high KIC, E, and HV values.

The possibility of the effective utilization of FA originating from a coal-burning thermal power plant was investigated as one of the starting materials to synthesize wollastonite-based glass ceramics [35]. FA supplied from a Rayong (Thailand) coal-burning thermal power plant had the following chemical composition, in wt.%: 41.62 SiO_2, 23.11 Al_2O_3, 14.92 CaO, 1.59 MgO, 6.03 Fe_2O_3, 1.15 Na_2O, and 1.59 K_2O. The raw FA comprised some quantity of glassy phase and crystalline phases, α-Quartz (SiO_2), anorthite ($CaAl_2Si_2O_8$), gehlenite ($Ca_2Al_2SiO_7$), and the percent crystallinity was evaluated to be around 2%. CaF_2 was added as a nucleating agent and spodumene [$LiAl(SiO_3)_2$] was also added for a reduction in the thermal expansion coefficient. Glass was prepared by mixing and melting the batch containing 35% of fly ash in a Pt/Rh10 crucible using an electric furnace at 1450 °C for 2 h. The melt was poured onto an iron plate, and then crushed and remelted under the same conditions. This refined melt was again poured onto an iron plate. Bubble-free and dark green glass was obtained. The samples were then annealed at just above the T_g for 30 min and cooled to room temperature in the furnace. They were then heat-treated for nucleation and crystallization under various conditions. Crystallization exothermic peaks appeared at 970 °C for G0 (zero CaF_2) glass and 900 °C for G3 (3% CaF_2) and G5 (3% CaF_2 and 20% spodumene) glasses. It was clear that the addition of CaF_2 and spodumene reduced the crystallization temperature. Experimental glasses were heat-treated at 750 °C for 10 h and further at 950 °C for 5 h for crystal growth. Surface crystallization was revealed for G0 glass while CaF_2-containing glasses G3 and G5 exhibited bulk crystallization resulting in the highest values of bending strength of 206 and 230 MPa, respectively [35].

A preliminary study of the co-utilization of wastes, namely ferronickel slag FNS (65–80 wt.%) and coal fly ash FA (20–35 wt.%), for preparing glass–ceramics revealed that the suitable glass contained 75 wt.% FNS and 25 wt.% FA and its chemical composition was as follows, in wt.%: 49.55 SiO_2, 21.36 CaO, 3.76 MgO, 9.39 FeO, and 10.15 Al_2O_3. By controlling the crystallization of the melted mixture at 867 °C for 2 h, an enstatite–spinel-based glass–ceramic was obtained featuring a density of 3.11 $g \cdot cm^{-3}$, a bending strength of 116 MPa, an acid resistance of 99.97%, and an alkali resistance of 99.70% [36].

As a continuation of the previous work [36], the effect of Cr_2O_3 content (varied from 0 to 2.0 wt.%) on the crystallization of glass containing 75 wt.% FNS and 25 wt.% FA was studied [37]. The flow chart of the preparation of glass–ceramics from coal fly ash (CFA) and ferronickel slag (FNS) that are typical for preparation of bulk glass–ceramic is shown in Figure 2. The raw materials were mixed and then melted at 1550 °C for 1 h to obtain homogeneous glass. The as-cast glass was annealed at 650 °C for 2 h followed by heat treatment in non-isothermal conditions with a heating rate of 10 K·min^{-1} to produce glass–ceramics.

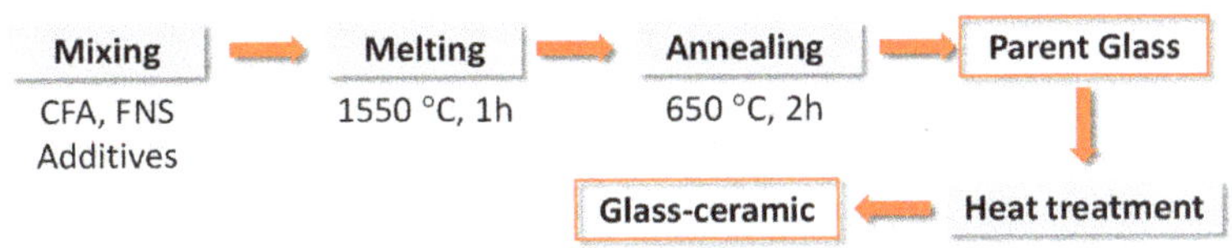

Figure 2. Flow chart of preparation of glass–ceramics from coal fly ash (CFA) and ferronickel slag (NS) (adapted with permission from [37], Elsevier 2023).

Crystallization kinetics studies demonstrated that with the rise in Cr_2O_3 content from 0 (Cr0) to 1.6 (Cr1.6) wt.%, the crystallization mode of the parent glass changed from surface crystallization to bulk crystallization whilst the activation energy decreased from 448.23 to 232.65 kJ. mol^{-1}. With a further Cr_2O_3 content increment to 2.0 (Cr2.0) wt.%, the activation energy for crystallization increased to 304.21 kJ. mol^{-1} and the crystallization mode went back to surface crystallization. The properties of synthesized bulk glass–ceramics are demonstrated in Figures 3 and 4. An optimal content of Cr_2O_3 was established to be 1.6 wt.%, which caused the formation of a dense glass–ceramic having a main crystalline phase of enstatite and a secondary crystalline phase of spinel. This glass–ceramic material featured a density of 3.17 g·cm^{-3}, a bending strength of 175 MPa, a Vickers microhardness of 9.51 GPa, an acid resistance of 99.99%, and an alkali resistance of 99.97%.

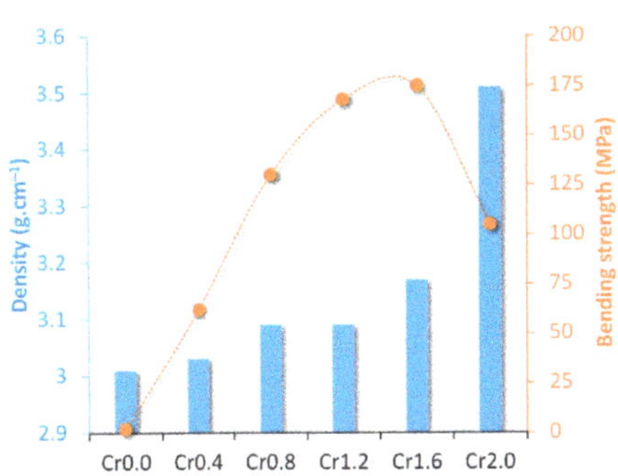

Figure 3. Densities and bending strengths of Cr0-Cr2.0 bulk-nucleated glass–ceramics (adapted with permission from [37], Elsevier 2023).

It was well documented elsewhere [38–40] that Cr_2O_3 significantly increases the crystallization rate only for Fe-containing compositions as the corresponding mechanism includes the formation of spinels, which, in turn, actively catalyze the formation of pyroxene phases. The first step of the heat treatment, at a lower temperature, is connected with the formation of heterogeneous nuclei, i.e., small crystallites promoting the growth of the major crystal phase. The higher the number of nuclei formed, the finer the structure of the glass–ceramics and the better the properties of the material.

Cr_2O_3, as a nucleating agent, was used to produce mono-mineral augite-type GCs from low-silica and high-calcium fly ash [41–44]. This FA derives from the burning of lignite in the thermal power stations of Ptolemaida (Northern, Greece). Lignite plays an important role in Greece's energy sector as it satisfies over 67% of the country's needs in electric power. Lignite deposits are located in three regions of Greece, namely Ptolemais–Amynteo, Florina, and Megalopolis, where the annual production of lignite is around 60 million tons, while each year, approximately 10 million tons of lignite fly ash is produced in Greece [45]. FA from the thermal power stations of Ptolemaida was used as the principal raw material in glass–ceramic fabrication. Its average chemical composition in wt.% was 30.03 SiO_2, 13.67 Al_2O_3, 38.87 CaO, 5.32 Fe_2O_3, 4.42 MO, 0.63 Na_2O, 0.59 K_2O, 6.17 SO_3, and 0.30 TiO_2; the XRD spectra and an SEM image of FA are shown in Figure 5.

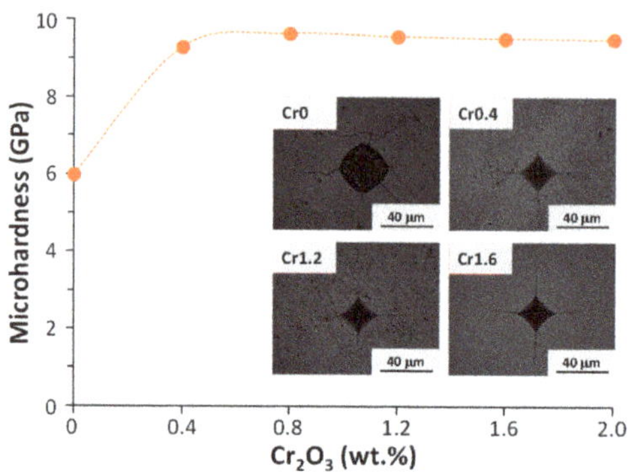

Figure 4. Vickers microhardness of Cr0-Cr2.0 bulk-nucleated glass–ceramics (adapted with permission from [37], Elsevier 2023).

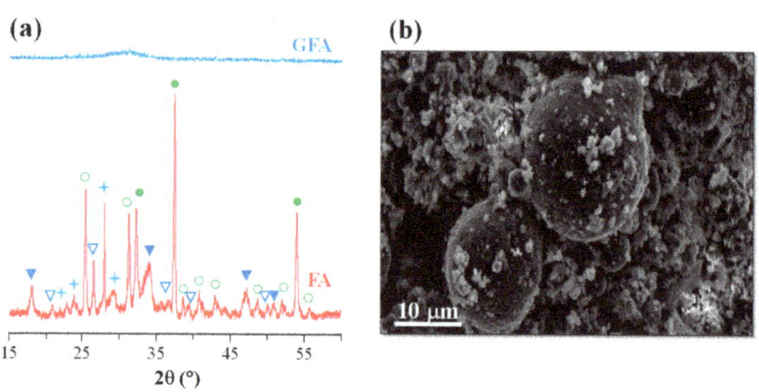

Figure 5. (**a**) XRD spectra of as-received FA (• calcium oxide (CaO); ○ anhydrite ($CaSO_4$); ▽ quartz SiO_2; ▼ portlandite $Ca(OH)_2$; ✦ calcium aluminum silicate) and (**b**) SEM image of as-received FA (adapted from [42], SAGE Publications, 2009).

A parent glass (C1) with nominal composition intended to be close to $CaMg_{0.75}Al_{0.4}Fe_{0.1}Si_{1.75}O_6$ was produced with silica, alumina, and magnesium carbonate added to the batch; Cr_2O_3 was introduced in the amounts of 0.50 wt.%, 0.75 wt.%, and 1.0 wt.% (over 100% of nominal composition) which resulted in the compositions C2, C3, and C4, respectively. Homogeneous mixtures of the batches were melted in alumina crucibles at 1500–1550 °C for 1 h and glasses in bulk form were produced by casting of the molten glass on preheated (500 °C) bronze molds. The T_g of the glasses ranged between 665 °C and 690 °C and the T_s between 761 °C and 778 °C. All investigated compositions featured a single exothermic crystallization peak (T_p) between 900 °C and 1000 °C (Figure 6a)—this peak is shallow and occurs at a high temperature in the case of C1 but it is much sharper and occurs at lower temperatures in C2 and more pronouncedly in C3 and C4. The effect of the Cr_2O_3 on nucleation (promoted by heat treatment close to the T_g, ~700 °C, for 1 h) and crystallization (promoted by heat treatment in the range of the T_p for 2 h) was evaluated by the difference in density between the parent glasses and the GCs (Table 3), which reflects the degree of conversion and the formation of crystalline phases (e.g., for the same temperature of heat treatment, C3 and C4 feature the highest degree of crystallization). Basically, the density of Cr_2O_3-nucleated C2, C3 and C4 compositions was not significantly influenced by heat treatment temperature rise (Table 3).

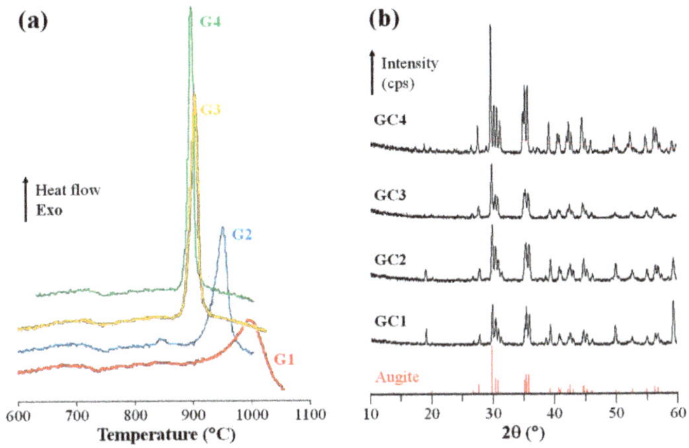

Figure 6. (**a**) DTA plots of fine powders (glass frit) of the four investigated glasses and (**b**) X-ray diffractograms the GCs crystallized at 900 °C for 2 h (adapted from [42], SAGE Publications, 2009).

Table 3. Density of the parent glasses and glass–ceramics crystallized at different temperatures for 2 h (standard deviations < 0.01 g·cm^{-3}) [37].

Content of Cr_2O_3 (wt.%)	Density (g·cm^{-3})			
	Initial Glass	2 h at 900 °C	2 h at 950 °C	2 h at 1000 °C
0	2.91	2.93	2.93	2.96
0.5	2.91	3.05	3.04	3.04
0.75	2.92	3.11	3.08	3.08
1.00	2.93	3.10	3.11	3.10

The heat treatment at 900 °C predominantly results in a single crystalline phase of augite (Ca, Mg, Fe, Na)(Mg, Fe, Al, Ti)(Al, Si)$_2$O$_6$, which belongs to the family of pyroxenes (Figure 6b). Increasing the Cr_2O_3 content causes an improvement in crystallinity, which is evidenced by the increasing intensity and sharpening of the diffraction peaks.

Figure 7 illustrates the impact of Cr_2O_3 on the microstructure of the glass–ceramics. The Cr_2O_3-free GC C1, crystallized at 1000 °C (Figure 7a), presents a structure with coarse crystal layers embedded in a highly chemical-resistant glassy phase. Observations at the sample edges confirmed that crystallization initiates from the surface.

The presence of Cr_2O_3 led to fine-grained bulk-crystallized GCs after heat treatment at a relative low temperature (900 °C), which is consistent with DTA results. The microstructure of GC C3 (0.75% Cr_2O_3) crystallized at 900 °C (Figure 7b) shows submicron crystals perfectly bordered one to the other. Observations of the bulk and surface of GCs C2 and C3 revealed no surface crystallization, which confirms that the Cr_2O_3-free C1 glass tends to surface crystallization, whereas the presence of Cr_2O_3 promotes bulk crystallization, as seen in GCs C2, C3, and C4.

GC C3 crystallized at 900 °C, demonstrating a high flexural strength that was ~200 MPa. The CTE values of the GCs C2, C3, and C4 crystallized at 900 °C, determined by the slope of dilatometry curves, were 8.81×10^{-6} K^{-1}, 9.55×10^{-6} K^{-1}, and 9.36×10^{-6} K^{-1}, respectively. The calculated electrical resistivity for GC 3 was 1.40×10^6 Ω·cm and for GC4 it was

5.78×10^5 Ω·cm (at 1000 °C), which are comparable with the data obtained from conventional electrical insulators made from alumina (7×10^6 Ω·cm) [46] and cordierite–anorthite composites (4.6×10^5 Ω·cm) [47] at 1000 °C. Bulk and porous GCs were produced through bulk nucleation and controlled crystal growth of parent glass manufactured from the fly ash of St. Dimitrios Power Plant, situated in Western Macedonia, Greece [48]. Ash samples, following dehumidification at 170 °C for 1 h, were stratified by sieving to obtain grain sizes below 65 μm. The glass batch comprising fly ash (30 wt%) and glass cullet (70 wt%) was melted in alumina crucibles at 1400–1500 °C for 0.5 h, but was first heat-treated at 1200 °C, ensuring the embodiment of the fly ash particles within the melt and aiming for proper homogenization. The bulk glasses were produced by casting the molten glass in preheated (500 °C) bronze molds and annealing them at 700 °C for 1h in order to release residual stresses. Nucleation and crystallization were conducted at a rate of 1.5 K·min^{-1} in the interval of 900–1050 °C (with 50 °C intervals) for 1.5 h. Mechanical testing revealed the influence of the heating strategy on the products' strength characteristics. Thus, glass samples subjected to heat treatment at 1050 °C exhibited a 21% increment in bending strength (~205 MPa) compared to those treated at 900 °C. This was attributed to the formation of a finer microstructure, confirmed by SEM investigation. The differences in strength observed for the samples produced at the two different melting temperatures of 1400 °C and 1500 °C were found to be statistically insignificant.

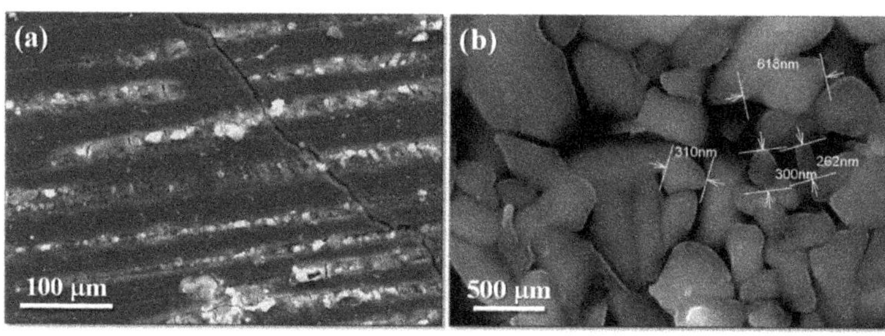

Figure 7. Microstructures of (**a**) the Cr_2O_3-free glass–ceramic C1 crystallized at 1000 °C, and (**b**) the 0.75% Cr_2O_3-containing glass–ceramic C3 crystallized at 900 °C (adapted from [42], SAGE Publications, 2009).

Elsewhere [49], a systematic evaluation of the sustainability of recycling fly ash, from thermal power plants, into novel glass–ceramics (GCs) was provided and concluded that using fly ash as substitute of conventional materials during the production of high-added-value products has a significant environmental potential, next to the obvious benefit of raw material saving. Table 4 provides a comparison of the properties of some bulk-nucleated GCs produced from different fly ashes [50,51].

Table 4. Properties of the presented bulk-nucleated GCs produced from different fly ashes.

Ref.	Nucleating Agent (%)	Micro-Hardness (GPa)	Flexural Strength (MPa)	CTE ($\times 10^{-6}$ K^{-1})	Density (g·cm^{-3})	Fracture Toughness (MPa·m$^{1/2}$)
[32]	Not added	8.89	n/r	n/r	n/r	n/r
[33]	2–11 TiO_2	6.20–7.60	98–176	5.5–6.5	2.77–2.81	1.12–3.10
[34]	Not added	4.6–7.3	n/r	n/r	n/r	0.70–3.00
[35]	3 CaF_2	n/r	135–230	8.5–9.0	2.66–2.72	n/r
[36]	Not added	n/r	116	n/r	3.11	n/r
[37]	0.4–2.0 Cr_2O_3	9.28–9.63	62–175	n/r	3.03–3.51	n/r
[41,42]	0.5–1.0 Cr_2O_3	n/r	200	8.8–9.5	3.04–3.11	n/r

"n/r" stands for "not reported".

4. Porous Glass–Ceramic Composites from Fly Ashes

FA is a cheap and abundant product, rich in minerals containing silicon and aluminum, making it suitable as a starting material for foamed ceramics. Foamed ceramics have the advantages of high porosity, large specific surface area, high strength, and high temperature resistance. They are widely used in industrial sewage treatment, automobile exhaust equipment, heat and sound insulation material, and the filtration of molten metal extending to the field of aviation, electronics, and biochemicals [52–56].

The utilization of FA to produce foam glass and glass–ceramic composites not only provides a valuable way to recycle industrial waste but also offers economic and environmental benefits. It reduces the need for natural raw materials like kaolin or bauxite, quartz, [52,53] and silica micro powder, which are typically used in foamed-ceramic production, thus conserving natural resources while mitigating the environmental impact associated with fly ash disposal, such as landfilling.

Foam glass and glass–ceramic foam are porous, heat-insulating, and soundproof materials, with a true porosity up to 90 vol.%. Generally, they are produced by the powder method when a mixture of glass powders and special additives facilitating the formation of a gaseous phase is sintered upon heating. These additives introduced into a glass foam batch in small quantities are called pore-forming or gas-forming agents. Under thermal treatment of such a mixture, when the temperature exceeds the softening temperature glass particles start sintering and form a continuous sintered body. Particles of the pore-forming agent become insulated by the softened glass. After a certain temperature is reached, they start emitting gases, frothing the glass melt. Due to gas emission, pores emerge in all parts of the sintered body where the particles of the pore-forming agent have been blocked. The shape of pores and the properties of foam glass obtained largely depend on the concentration and type of the foaming agent used. In its physical aspect, foam glass is a heterogeneous system consisting of gaseous and the solid phases. The solid phase is amorphous or devitrified glass that forms thin walls of single cells several micrometers thick. The cells are filled with the gaseous phase.

Nowadays, extensive research has been conducted to produce glass–ceramic foams using metallurgical slag, municipal solid waste, waste glass and fly ash, polishing porcelain stoneware tile residue, etc. [2,53–64]. Likewise, new technologies, such as replica, sacrificial template, and direct foaming methods, have been developed. Direct foaming is the simplest and most common method, involving the sintering of raw materials with a small amount of additives [53,56,60]. These additives, known as pore-foaming agents, create porosity by releasing gas that froths the glass ceramic melt at relatively high temperatures. The concentration and type of pore-foaming agent can significantly affect the pore size and structure of the resulting glass ceramic foams.

Porous glass–ceramic composites based on waste silicomanganese slag (SMS) and fly ash (FA) as raw materials and silicon carbide (SiC) as a foaming agent were systematically investigated [56]. Figure 8 presents a schematic illustration of the foamed-glass–ceramic preparation process. The mixture of FA and SMS was thoroughly blended with a specific quantity of water, pressed, dried, and then sintered in a furnace at 1050–1170 °C for 45 min at a heating rate of 5 $K \cdot min^{-1}$. Figure 9 presents the cross-sectional morphology of the specimens sintered at various temperatures, evidencing the increasing of the size and number of pores with the rising of the temperature [53]. The best overall performance was achieved for compositions comprising 20 wt.% SMS and 80 wt.% FA as raw materials, with a SiC addition of 1.0 wt.% and a sintering temperature of 1100 °C (Figure 9), compressive strength, bulk density, and total porosity of 8.09 MPa, 0.57 $g \cdot cm^{-3}$, and 71.04%, respectively. These properties make it suitable for application as an insulating material or as a decorative material for building partitions [53].

Figure 8. Schematic illustration of the foamed glass–ceramic preparation process (adapted from [53], MDPI, 2023).

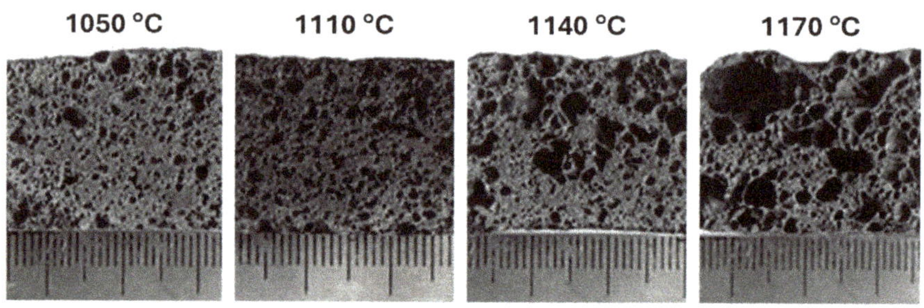

Figure 9. The cross-section macrostructure of the samples at different temperatures (adapted from [53], MDPI, 2023).

Glass–ceramic composites were prepared by the direct foaming method using coal fly ash (FA) and waste glass as primary materials, with borax and calcium carbonate serving as fluxing and foaming agents, respectively [60]. The study systematically investigated the effects of coal FA additions, foaming time, heating rate, and sintering temperature on bulk density, porosity, mechanical properties, and thermal conductivity. The bulk density and compressive strength of porous ceramics increase with the increasing of sintering temperature, while the apparent porosity and water absorption decrease. The experimental results indicated that the optimal parameters for fabricating glass–ceramic foams were achieved after sintering at 800 °C for 45 min, using 40 wt.% coal FA, 60 wt.% waste glass, 30 wt.% borax, and 0.5 wt.% calcium carbonate. The obtained foams exhibited a low bulk density of 0.46 $g{\cdot}cm^{-3}$, a compressive strength exceeding 5 MPa, and a low thermal conductivity ~0.36 $W{\cdot}m^{-1}K^{-1}$ [60].

Glass–ceramic composites were synthesized utilizing cullet from sheet glass and FA derived from thermal power stations, incorporating carbonates such as commercial dolomite and calcite-based sludges to act as foaming agents [2]. The study assessed how the type and content of foaming agents, along with the sintering temperature, affected the resulting foam glass's apparent density, compressive strength, microstructure, and crystalline phase composition. To achieve the objective of integrating a suitable quantity of FA, five distinct carbonate-free mixtures were formulated consisting solely of glass and FA in varying proportions of glass cullet (G) and fly ash (A), i.e., G/A: G90A10, G80A20, G70A30, G60A40, and G50A50. The preparation process involved dry mixing the components in a cylindrical rotary mixer for 30 min. Cylindrical pellets measuring 20 mm in diameter and 3 mm in thickness were produced through uniaxial pressing at 80 MPa and sintered at temperatures of 800 and 900 °C for 20 min. Subsequent evaluations included measuring the apparent densities and shrinkage and visual ex-amination of the samples to achieve the highest feasible ash integration into the mixture and the capability to create porous structures; a composition comprising 80 wt.% glass and 20 wt.% FA was selected to produce the foams, to which varying quantities of carbonates were incorporated [2].

The processing temperature for compositions containing carbonate was set between 750 and 950 °C. This range is well above the dilatometric softening point of glass (590 °C), ensuring the necessary low viscosity for effective foaming. Additionally, this temperature interval aligns with the decomposition temperatures of the selected foaming agents. It was found that the density decreased with an increasing sintering temperature, reaching minimum values within the temperature range of around 850–900 °C; however, further increasing the heat treatment temperature to 950 °C caused an increase in the apparent density. Figure 10 illustrates the standard appearance of the glass foams post-rectifying.

Figure 10. The typical aspect of the glass–ceramic composites after rectifying (adapted with permission from [2], Elsevier, 2009).

The study revealed that it is possible to achieve homogenous microstructures characterized by large pores by incorporating 1–2 wt.% carbonates and employing a low sintering temperature (850 °C). This process results in the production of foams with an apparent density and compressive strength in the ranges of approximately 0.36–0.41 $g \cdot cm^{-3}$ and 2.40–2.80 MPa, respectively. A strong correlation was found between the compressive strength, apparent density, and microstructural features such as pore size, strut thickness, and internal porosity. Furthermore, the crystalline phases within the glass foams evolved with increasing temperature; for example, compositions GAC1 and GAD1 exhibited the presence of quartz (ICDD card 46-1045) and tridymite (ICDD card 42-1401) at 750 °C, whereas the pyroxene phase augite, with the composition $Ca(Mg,Fe,Al)(Si,Al)_2O_6$ (ICDD card 00-41-1483), was observed to crystallize at 800 °C.

In general, glass foams depend on a similarly delicate balance between viscous flow sintering and gas evolution that is governed by many factors such as the nature and particle size of raw materials, firing temperatures and heating rate, gas atmosphere in a furnace, etc. Crystallization may occur as well, affecting the final physical, mechanical, and thermal properties of foam, namely density, compressive strength, thermal conductivity, and diffusivity [30]. To avoid excessive crystallization and to control both foaming and crystallization, the waste glasses with a limited tendency towards crystallization are used as the most suited starting material. In turn, the crystallization may be intentionally stimulated by using glass cullet mixed with foaming agents as well as with substances more prone to devitrification. Although substantial crystallization is negative for the development of highly porous foams, its advantage is the formation of an open-cell morphology of a three-dimensional trabecular structure.

Special attention must be given to choosing foaming additives. Although SiC may be quite expensive and have a negative impact on the overall cost/benefit balance, SiC could derive from the waste originated by the polishing of glass or traditional ceramics. Likewise, carbonaceous residues, as foaming agents, may derive from common industrial wastes [2,30,63]. Importantly, the foaming reaction must take place in a pyroplastic mass, determined by the softening of glass powders, with a specified viscosity (in the order of 10^3–10^5 Pa·s [64]).

Another important issue to be considered is the particle size of FA. Thus, the influence of particle size on the sinterability and properties of sintered glass–ceramics has been discussed elsewhere [65]. Coal fly ash was separated into three size fractions, namely CFA, 20CFA, and 40CFA, with the median particle size (d50) of the powder being 12.6 μm, 7.5 μm, and 4.9 μm for CFA, 20CFA, and 40CFA, respectively. The sintering of the smaller particles was easier than for the larger ones, since the sintering activation energy of CFA was higher than 20CFA and 40CFA. However, regardless of particle size, the flexural strength of sintered glass–ceramics increased with a rise in temperature. The porosity and water absorption of the sintered samples gradually decreased when increasing the sintering temperature. Noticeably, with the finer the coal fly ash particles, a better sintering effect and better performance of the sintered glass–ceramics was achieved [65]. Likewise, it was demonstrated [66] that when the particle size of FA decreased, the linear shrinkage, bulk density, compressive strength, acid resistance, and thermal conductivity of lightweight insulation materials increased, but the apparent porosity decreased.

5. Conclusions

In the global scenario, the disposal of enormous amounts of FA is a major matter of concern because it represents a significant negative environmental impact, including the leaching of potentially toxic substances into soils and groundwater. Over the years, there has been an increasing interest in the use of FA for a variety of applications, since the reutilization of waste materials into novel high-added-value products is widely recognized as a challenge of high technologic interest. Likewise, encouraging results have been demonstrated in the preparation of both monolithic and cellular ceramics utilizing fly ash and other waste materials. In particular, the results discussed in this brief review demonstrate that FA can be successfully recycled into both bulk-nucleated glass–ceramics and porous glass–ceramic composites through the simultaneous control of both the formulations and manufacturing processes.

The composition of FA significantly depends on the coal type and technological processes used in thermal power plants. Vitrification is undoubtedly easier for wastes rich in glass-forming oxides, considering the fact that suitable glass-based products can be obtained only when an appropriate ratio between the glass network former and modifiers is attained.

Vitrification of a glass batch composed of merely coal FA was found to be difficult, or almost impossible. To produce glass–ceramic, the glass composition containing FA has to be modified through the addition of other ingredients in the form of natural raw materials, selected oxides, industrial wastes, etc. A key feature of FA-containing glasses is the availability of nucleating agents from the waste stream such as Fe_2O_3, SO_3, TiO_2, and P_2O_5. However, nucleating agents such as CaF_2, TiO_2, and Cr_2O_3 are intentionally added to the formulation of certain base glasses containing FA to promote controlled internal crystallization with the formation of desired crystalline phases resulting in dense glass–ceramics with high mechanical properties. Indeed, bulk-nucleated glass–ceramics demonstrate superior mechanical properties compared to sinter-crystallized glass—ceramics of the same compositions or to glass–ceramics without a preliminary vitrification step, i.e., direct sintering of wastes, especially if combined with recycled glasses.

Porous glass–ceramic composites are mainly produced by the powder method when a mixture of glass powders and pore-forming agents facilitating the formation of a gaseous phase is sintered upon heating. Thus, glass-based foams represent a further variant of glass sintering and depend on a delicate balance between viscous flow sintering and gas evolution. New technologies have been developed such as replica, sacrificial template, and direct foaming methods and the most straightforward and widely used way is direct foaming, namely, sintering a mixture of raw materials and a small number of additives. The type and content of foaming agents, along with the sintering temperature, affect the resulting foam glass's apparent density, compressive strength, microstructure, and crystalline phase composition. The driving force in the sintering process is the surface energy of the powder, which is closely related to the particle size of the powder. Therefore, the introduction of different particle size of fly ash has a great influence on the properties of porous ceramic materials.

There are still many problems associated with high-temperature processing of bulk-nucleated glass–ceramics that should be taken into account upon designing new formulations containing FA.

Another open issue is the necessity to provide samples' toxicity characteristics, considering the presence of potentially toxic substances in FA not accommodated in the crystalline structures and remaining in glassy phase.

New technologies including additive manufacturing will be potentially useful in the transition from results obtained from laboratory trials to the industrial scale when dealing with the utilization of FA in glass-based materials.

Author Contributions: Conceptualization, H.R.F. and D.U.T.; methodology, H.R.F. and D.U.T.; validation, H.R.F., O.G. and D.U.T.; data curation, H.R.F., O.G. and D.U.T.; writing—original draft preparation, H.R.F., O.G. and D.U.T.; writing—review and editing, H.R.F., O.G. and D.U.T.; supervision, H.R.F. and D.U.T. All authors have read and agreed to the published version of the manuscript.

Funding: H.R.F. thanks the CICECO-Aveiro Institute of Materials, UIDB/50011/2020, UIDP/50011/2020, and LA/P/0006/2020 (through the FCT/MCTES (PIDDAC)).

Conflicts of Interest: The authors declare no conflicts of interest.

References

1. Nadaf, M.B.; Dutta, S.; Mandal, J.N. Fly Ash as Backfill Material in Slopes Using Waste PET Bottles as Reinforcement. In *Waste Management and Resource Efficiency*; Ghosh, S., Ed.; Springer: Singapore, 2019; pp. 905–913. [CrossRef]
2. Fernandes, H.R.; Tulyaganov, D.U.; Ferreira, J.M.F. Preparation and characterization of foams from sheet glass and fly ash using carbonates as foaming agents. *Ceram. Int.* **2009**, *35*, 229–235. [CrossRef]
3. Panias, D.; Giannopoulou, I.P.; Perraki, T. Effect of synthesis parameters on the mechanical properties of fly ash-based geopolymers. *Colloids Surf. A Physicochem. Eng. Asp.* **2007**, *301*, 246–254. [CrossRef]
4. Das, D.; Rout, P.K. Synthesis, Characterization and Properties of Fly Ash Based Geopolymer Materials. *J. Mater. Eng. Perform.* **2021**, *30*, 3213–3231. [CrossRef]
5. Roy, R.; Das, D.; Rout, P.K. Mullite Ceramics Derived from Fly Ash Powder by Using Albumin as an Organic Gelling Agent. *Biointerface Res. Appl. Chem.* **2023**, *13*, 339. [CrossRef]
6. Whiteside, M.; Herndon, J. Aerosolized Coal Fly Ash: Risk Factor for COPD and Respiratory Disease. *J. Adv. Med. Med. Res.* **2018**, *26*, 1–13. [CrossRef]
7. Yu, X.; Cui, Y.; Chen, Y.; Chang, I.S.; Wu, J. The drivers of collaborative innovation of the comprehensive utilization technologies of coal fly ash in China: A network analysis. *Environ. Sci. Pollut. Res.* **2022**, *29*, 56291–56308. [CrossRef]
8. Abbass, W.; Abbas, S.; Aslam, F.; Ahmed, A.; Ahmed, T.; Hashir, A.; Mamdouh, A. Manufacturing of Sustainable Untreated Coal Ash Masonry Units for Structural Applications. *Materials* **2022**, *15*, 4003. [CrossRef]
9. Daniel, C.W.; Tsang, L.W. (Eds.) *Low Carbon Stabilization and Solidification of Hazardous Wastes*, 1st ed.; Elsevier: Amsterdam, The Netherlands, 2021.
10. Koshlak, H. Synthesis of Zeolites from Coal Fly Ash Using Alkaline Fusion and Its Applications in Removing Heavy Metals. *Materials* **2023**, *16*, 4837. [CrossRef]
11. Gollakota, A.R.K.; Volli, V.; Shu, C.M. Progressive utilisation prospects of coal fly ash: A review. *Sci. Total Environ.* **2019**, *672*, 951–989. [CrossRef]
12. Jala, S.; Goyal, D. Fly ash as a soil ameliorant for improving crop production—A review. *Bioresour. Technol.* **2006**, *97*, 1136–1147. [CrossRef]
13. Ukwattage, N.L.; Ranjith, P.G.; Bouazza, M. The use of coal combustion fly ash as a soil amendment in agricultural lands (with comments on its potential to improve food security and sequester carbon). *Fuel* **2013**, *109*, 400–408. [CrossRef]
14. Das, D.; Rout, P.K. A Review of Coal Fly Ash Utilization to Save the Environment. *Water Air Soil Pollut.* **2023**, *234*, 128. [CrossRef]
15. Niraj, N.; Pandey, K.M.; Dey, A. Tribological behaviour of Magnesium Metal Matrix Composites reinforced with fly ash cenosphere. *Mater. Today Proc.* **2018**, *5*, 20138–20144. [CrossRef]
16. Wesche, K. *Fly Ash in Concrete—Properties and Performance*; CRC Press: London, UK, 1991.
17. Feng, S.; Li, Y. Study on coal fly ash classified by bulk density. *J. Phys. Conf. Ser.* **2021**, *1732*, 012127. [CrossRef]
18. Radenović, A. Inorganic constituents in coal. *Kem. Ind. Chem. Chem. Eng.* **2006**, *55*, 65–71.

19. Vassilev, S.V.; Vassileva, C.G. Methods for Characterization of Composition of Fly Ashes from Coal-Fired Power Stations: A Critical Overview. *Energy Fuels* **2005**, *19*, 1084–1098. [CrossRef]
20. Vassilev, S.V.; Vassileva, C.G. A new approach for the classification of coal fly ashes based on their origin, composition, properties, and behaviour. *Fuel* **2007**, *86*, 1490–1512. [CrossRef]
21. Vassilev, S.V.; Vassileva, C.G. Mineralogy of combustion wastes from coal-fired power stations. *Fuel Process. Technol.* **1996**, *47*, 261–280. [CrossRef]
22. Blissett, R.S.; Rowson, N.A. A review of the multi-component utilisation of coal fly ash. *Fuel* **2012**, *97*, 1–23. [CrossRef]
23. Vichaphund, S.; Aht-Ong, D.; Sricharoenchaikul, V.; Atong, D. Characteristic of fly ash derived-zeolite and its catalytic performance for fast pyrolysis of Jatropha waste. *Environ. Technol. UK* **2014**, *35*, 2254–2261. [CrossRef]
24. Franus, W.; Wiatros-Motyka, M.M.; Wdowin, M. Coal fly ash as a resource for rare earth elements. *Environ. Sci. Pollut. Res.* **2015**, *22*, 9464–9474. [CrossRef] [PubMed]
25. *ASTM C618-12*; Standard Specification for Coal Fly Ash and Raw or Calcined Natural Pozzolan for Use in Concrete. American Society for Testing and Materials International: West Conshohocken, PA, USA, 2019. [CrossRef]
26. Hardjito, D.; Wallah, S.E.; Sumajouw, D.M.J.; Rangan, B.V. On the development of fly ash-based geopolymer concrete. *ACI Mater. J.* **2004**, *101*, 467–472. [CrossRef]
27. Xu, G.; Shi, X. Characteristics and applications of fly ash as a sustainable construction material: A state-of-the-art review. *Resour. Conserv. Recycl.* **2018**, *136*, 95–109. [CrossRef]
28. Ciećko, Z.; Żołnowski, A.C.; Madej, M.; Wasiak, G.; Lisowski, J. Long-Term effects of hard coal fly ash on selected soil properties. *Pol. J. Environ. Stud.* **2015**, *24*, 1949–1957. [CrossRef] [PubMed]
29. Rawlings, R.D.; Wu, J.P.; Boccaccini, A.R. Glass-ceramics: Their production from wastes—A Review. *J. Mater. Sci.* **2006**, *41*, 733–761. [CrossRef]
30. Rincón, A.; Marangoni, M.; Cetin, S.; Bernardo, E. Recycling of inorganic waste in monolithic and cellular glass-based materials for structural and functional applications. *J. Chem. Technol. Biotechnol.* **2016**, *91*, 1946–1961. [CrossRef] [PubMed]
31. Vu, D.H.; Wang, K.S.; Chen, J.H.; Nam, B.X.; Bac, B.H. Glass-ceramic from mixtures of bottom ash and fly ash. *Waste Manag.* **2012**, *32*, 2306–2314. [CrossRef] [PubMed]
32. Erol, M.; Genç, A.; Öveçoğlu, M.L.; Yücelen, E.; Küçükbayrak, S.; Taptik, Y. Characterization of a glass-ceramic produced from thermal power plant fly ashes. *J. Eur. Ceram. Soc.* **2000**, *20*, 2209–2214. [CrossRef]
33. Kim, J.M.; Kim, H.S. Processing and properties of a glass-ceramic from coal fly ash from a thermal power plant through an economic process. *J. Eur. Ceram. Soc.* **2004**, *24*, 2825–2833. [CrossRef]
34. Barbieri, L.; Lancellotti, I.; Manfredini, T.; Queralt, I.; Rincon, J.M.; Romero, M. Design, obtainment and properties of glasses and glass-ceramics from coal fly ash. *Fuel* **1999**, *78*, 271–276. [CrossRef]
35. Waraporn, E.; Shigeki, M. Preparation of Glass-Ceramics Using Fly Ash as a Raw Material. *J. Sci. Technol.* **2005**, *13*, 137–142.
36. Shang, W.; Peng, Z.; Xu, F.; Tang, H.; Rao, M.; Li, G.; Jiang, T. Preparation of enstatite-spinel based glass-ceramics by co-utilization of ferronickel slag and coal fly ash. *Ceram. Int.* **2021**, *47*, 29400–29409. [CrossRef]
37. Yu, J.; Peng, Z.; Shang, W.; Chen, Q.; Zhu, G.; Tang, H.; Rao, M.; Li, G. Dual roles of Cr_2O_3 in preparation of glass-ceramics from ferronickel slag. *Ceram. Int.* **2023**, *49*, 15947–15958. [CrossRef]
38. Karamanov, A.; Pelino, M. Evaluation of the degree of crystallisation in glass-ceramics by density measurements. *J. Eur. Ceram. Soc.* **1999**, *19*, 649–654. [CrossRef]
39. Alizadeh, P.; Marghussian, V.K. The effect of compositional changes on the crystallization behaviour and mechanical properties of diopside-wollastonite glass-ceramics in the SiO_2-CaO-MgO (Na_2O) system. *J. Eur. Ceram. Soc.* **2000**, *20*, 765–773. [CrossRef]
40. Karamanov, A.; Taglieri, G.; Pelino, M. Iron-rich sintered glass-ceramics from industrial wastes. *J. Am. Ceram. Soc.* **1999**, *82*, 3012–3016. [CrossRef]
41. Vasilopoulos, K.C.; Tulyaganov, D.U.; Agathopoulos, S.; Karakassides, M.A.; Ferreira, J.M.F.; Tsipas, D. Bulk nucleated fine grained mono-mineral glass-ceramics from low-silica fly ash. *Ceram. Int.* **2009**, *35*, 555–558. [CrossRef]
42. Vasilopoulos, K.C.; Tulyaganov, D.U.; Agathopoulos, S.; Karakassides, M.A.; Ribeiro, M.; Ferreira, J.M.F.; Tsipas, D. Vitrification of low silica fly ash: Suitability of resulting glass ceramics for architectural or electrical insulator applications. *Adv. Appl. Ceram.* **2009**, *108*, 27–32. [CrossRef]
43. Tulyaganov, D.U.; Dimitriadis, K.; Agathopoulos, S.; Baino, F.; Fernandes, H.R. Wollastonite-containing glass-ceramics from the CaO–Al_2O_3–SiO_2 and CaO–MgO–SiO_2 ternary systems. *Open Ceram.* **2024**, *17*, 100507. [CrossRef]
44. Tulyaganov, D.U.; Dimitriadis, K.; Agathopoulos, S.; Fernandes, H.R. Glasses and glass-ceramics in the CaO–MgO–SiO_2 system: Diopside containing compositions—A brief review. *J. Non. Cryst. Solids* **2023**, *612*, 122351. [CrossRef]
45. Koukouzas, N.K.; Zeng, R.; Perdikatsis, V.; Xu, W.; Kakaras, E.K. Mineralogy and geochemistry of Greek and Chinese coal fly ash. *Fuel* **2006**, *85*, 2301–2309. [CrossRef]
46. Gitzen, W.H. *Alumina as a Ceramic Material*; American Ceramic Society: Columbus, OH, USA, 2006; ISBN 978-0-916094-46-1.
47. Tulyaganov, D.U.; Tukhtaev, M.E.; Escalante, J.I.; Ribeiro, M.J.; Labrincha, J.A. Processing of cordierite based ceramics from alkaline-earth-aluminosilicate glass, kaolin, alumina and magnesite. *J. Eur. Ceram. Soc.* **2002**, *22*, 1775–1782. [CrossRef]
48. Ganatsios, S.; Maropoulos, S.; Papanikolaou, S.; Gheorghe, G.; Dontu, O.; Seritan, G.; Besnea, D. Novel bulk and porous glass ceramics produced by waste materials. *Rom. Rev. Precis. Mech. Opt. Mechatron.* **2015**, *48*, 186–189.

49. Ganatsios, S.; Maropoulos, S.; Papanikolaou, S.; Dontu, O.; Besnea, D.; Spânu, A.; Gheorghe, G.I. Sustainability of waste material reutilization into novel glass ceramics: A life cycle analysis. *Rom. Rev. Precis. Mech. Opt. Mechatron.* **2016**, *50*, 53–56.
50. Dimitriadis, K.; Moschovas, D.; Tulyaganov, D.U.; Agathopoulos, S. Development of novel bioactive glass-ceramics in the Na_2O/K_2O-CaO-MgO-SiO_2-P_2O_5-CaF_2 system. *J. Non-Cryst. Solids* **2020**, *533*, 119936. [CrossRef]
51. Kaur, G.; Kumar, V.; Baino, F.; Mauro, J.C.; Pickrell, G.; Evans, I.; Bretcanu, O. Mechanical properties of bioactive glasses, ceramics, glass-ceramics and composites: State-of-the-art review and future challenges. *Mater. Sci. Eng. C* **2019**, *104*, 109895. [CrossRef] [PubMed]
52. Wang, M.; Xu, S. Preparation and Applications of Foam Ceramics. *IOP Conf. Ser. Earth Environ. Sci.* **2018**, *186*, 012066. [CrossRef]
53. Yu, G.; Gao, W.; Yao, Y.; Zhang, T.; Fu, Y.; Kong, X. Recycling of Silicomanganese Slag and Fly Ash for Preparation of Environment-Friendly Foamed Ceramics. *Materials* **2023**, *16*, 6724. [CrossRef] [PubMed]
54. Cao, W.; Cheng, X.; Gong, L.; Li, Y.; Zhang, R.; Zhang, H. Thermal conductivity of highly porous ceramic foams with different agar concentrations. *Mater. Lett.* **2015**, *139*, 66–69. [CrossRef]
55. Zhu, L.; Li, S.; Li, Y.; Xu, N. Novel applications of waste ceramics on the fabrication of foamed materials for exterior building walls insulation. *Constr. Build. Mater.* **2018**, *180*, 291–297. [CrossRef]
56. Zhou, L.Z.; Wang, C.A.; Huang, Y. Effects of the Starting Powders on Microstructure and Properties of Porous Silica Ceramics. *Key Eng. Mater.* **2010**, *434–435*, 674–677.
57. Chen, T.; Peng, T.; Sun, H.; Li, X.; Xiao, S.; You, D.; Chu, L.; Wu, M.; Zhang, T.; Tang, Y.; et al. Diopside-like based ceramics prepared via one-step sintering of extracted titanium slag and coal fly ash. *Ceram. Int.* **2024**, *50*, 14040–14053. [CrossRef]
58. Wang, S.; Li, X.; Zou, B.; Ma, Z.; Yu, W. Design and performance of lightweight high–temperature resistant microwave penetrating foamed ceramics. *Ceram. Int.* **2024**, *50*, 13974–13981. [CrossRef]
59. Dong, X.; Han, F.; Yu, N.; Li, M. Preparation and characterization of novel spontaneous foam ceramics based on all-solid waste. *J. Alloys Compd.* **2024**, *976*, 173135. [CrossRef]
60. Zhu, M.; Ji, R.; Li, Z.; Wang, H.; Liu, L.L.; Zhang, Z. Preparation of glass ceramic foams for thermal insulation applications from coal fly ash and waste glass. *Constr. Build. Mater.* **2016**, *112*, 398–405. [CrossRef]
61. Bai, J.; Yang, X.; Xu, S.; Jing, W.; Yang, J. Preparation of foam glass from waste glass and fly ash. *Mater. Lett.* **2014**, *136*, 52–54. [CrossRef]
62. Zhang, R.; Feng, J.; Cheng, X.; Gong, L.; Li, Y.; Zhang, H. Porous thermal insulation materials derived from fly ash using a foaming and slip casting method. *Energy Build.* **2014**, *81*, 262–267. [CrossRef]
63. Ding, L.; Ning, W.; Wang, Q.; Shi, D.; Luo, L. Preparation and characterization of glass-ceramic foams from blast furnace slag and waste glass. *Mater. Lett.* **2015**, *141*, 327–329. [CrossRef]
64. Sarkar, T.; Naskar, M.K.; Chakraborty, A.; Roy, P.K.; Chakraborty, S. Preparation of high-strength waste-derived eco-friendly ceramic foam as face brick and its estimation of building energy consumption for thermal insulation. *J. Build. Eng.* **2024**, *88*, 109043. [CrossRef]
65. Zeng, L.; Sun, H.-J.; Peng, T.-J.; Zheng, W.-M. The sintering kinetics and properties of sintered glass-ceramics from coal fly ash of different particle size. *Results Phys.* **2019**, *15*, 102774. [CrossRef]
66. Chen, R.; Li, Y.; Xiang, R.; Li, S. Effect of particle size of fly ash on the properties of lightweight insulation materials. *Constr. Build. Mater.* **2016**, *123*, 120–126. [CrossRef]

MDPI AG
Grosspeteranlage 5
4052 Basel
Switzerland
Tel.: +41 61 683 77 34

Ceramics Editorial Office
E-mail: ceramics@mdpi.com
www.mdpi.com/journal/ceramics

www.ingramcontent.com/pod-product-compliance
Lightning Source LLC
Chambersburg PA
CBHW041015270926
51850CB00024B/199
* 9 7 8 3 7 2 5 8 7 1 8 8 9 *